2013 International Conference on Advanced Optoelectronics and Lasers

(CAOL 2013)

Sudak, Ukraine
9 – 13 September 2013

IEEE Catalog Number: CFP13814-POD
ISBN: 978-1-4799-0019-0

Copyright © 2013 by the Institute of Electrical and Electronic Engineers, Inc
All Rights Reserved

Copyright and Reprint Permissions: Abstracting is permitted with credit to the source. Libraries are permitted to photocopy beyond the limit of U.S. copyright law for private use of patrons those articles in this volume that carry a code at the bottom of the first page, provided the per-copy fee indicated in the code is paid through Copyright Clearance Center, 222 Rosewood Drive, Danvers, MA 01923.

For other copying, reprint or republication permission, write to IEEE Copyrights Manager, IEEE Service Center, 445 Hoes Lane, Piscataway, NJ 08854. All rights reserved.

***This publication is a representation of what appears in the IEEE Digital Libraries. Some format issues inherent in the e-media version may also appear in this print version.**

IEEE Catalog Number: CFP13814-POD
ISBN 13:   978-1-4799-0019-0
ISSN: 2160-1518

Additional Copies of This Publication Are Available From:

Curran Associates, Inc
57 Morehouse Lane
Red Hook, NY 12571 USA
Phone: (845) 758-0400
Fax: (845) 758-2633
E-mail: curran@proceedings.com
Web: www.proceedings.com

CAOL*2013 program committee

Igor A. Sukhoivanov	*Conference PC Co-Chair*, University of Guanajuato, Mexico
Marian Marciniak	*Conference PC Co-Chair*, National Institute of Telecommunications, Warsaw, Poland
Vasiliy A. Svich	*Conference PC Co-Chair*, V. N. Karazin National University, Kharkiv, Ukraine
E. Alvarado Mendez	University of Guanajuato, Mexico
J. A. Andrade Lucio	University of Guanajuato, Mexico
G. Belenky	State University, New York, USA
T. Benson	University of Nottingham, Nottingham, UK
I. V. Blonskiy	Institute of Physics, National Academy of Science of Ukraine, Kyiv Ukraine
P. Chamorro-Posada	Universidad de Valladolid, Valladolid, Spain
R. De La Rue	University of Glasgow, UK
A. Desyatnikov	Nonlinear Physics Centre, Australian National University, Canberra, Australia
S. Donati	University of Pavia, Italy
S. F. Dyubko	V. N. Karazin National University, Kharkov, Ukraine
I. V. Dzedolik	Taurida National University, Simferopol, Crimea, Ukraine
N. N. Elkin	State R&D Center TRINITI, Troitsk, Russia
W. Freude	University Karlsruhe, Germany
A. F. Glova	State Research Center of the Russian Federation Troitsk Institute for Innovation and Fusion Research, Troitsk, Russian Federation
H. Guo	Laboratory of Light Transmission Optics, South China Normal University, Guangzhou, PR China and School of Electronics Engineering & Computer Science, Peking University, Beijing, PR China
C. Jagadish	Australian National University, Canberra, Australia
H. Kawaguchi	Yamagata Univ., Japan
V. K. Kononenko	Institute of Physics BAS, Minsk, Belarus
M. Koshiba	Hokkaido University, Japan
A. V. Kudryashov	Moscow State Open University, Moscow, Russia
V. V. Lysak	Chonbuk National University, Jeonju, South Korea
V. A. Makarov	International Teaching and Educational Laser Centre of the Moscow State University
V. A. Maslov	V. N. Karazin National University, Kharkov, Ukraine
V. K. Miloslavsky	V. N. Karazin National University, Kharkov, Ukraine
A. M. Negriyko	Institute of Physics, National Academy of Sciences of Ukraine, Kyiv, Ukraine
V. G. Niziev	Institute of Laser and Information Technologies, Russian Academy of Sciences, Shatura, Russia
A. I. Nosich	Institute Radiophys and Electronics of National Academy of Sciences of Ukraine, Kharkiv, Ukraine
M. F. Pereira	Sheffield Hallam University, UK
R. Rojas-Laguna	University of Guanajuato, Mexico
B. Sahraoui	Laboratoire POMA CNRS UMR 6136, Universiteé d'Angers, France
R. Schatz	KTH, Stockholm, Sweden
O. V. Shulika	University of Guanajuato, Mexico
S. V. Svechnikov	Institute of Semiconductor Physics of National Academy of Sciences of Ukraine, Kiev, Ukraine
R. Vlokh	Institute of Physical Optics, Ukraine

CAOL*2013 organizing committee

I. A. Sukhoivanov	*Organization Chair*, University of Guanajuato, Mexico
V. A. Maslov	*Co-chair, Local Management*, V. N. Karazin National University, Kharkiv, Ukraine
I. V. Dzedolik	*Local Chair*, Taurida National V. I. Vernadsky University, Simferopol, Crimea, Ukraine
O. V. Shulika	*Coordinator, Publication Chair, Web-maintenance*, University of Guanajuato, Mexico
A. V. Kublik	*Secretary*
A. V. Degtyarev	*Local Organization*, V. N. Karazin National University, Kharkiv, Ukraine
O. V. Gurin	*Local Organization*, V. N. Karazin National University, Kharkiv, Ukraine
V. I. Fesenko	*Local Organization*, Institute of Radio Astronomy of National Academy of Sciences of Ukraine, Kharkiv, Ukraine
T. B. Gryschenko	*Social Program*, National University of Radio Electronics, Kharkiv, Ukraine
S. G. Gryschenko	*Local Organization*, National University of Radio Electronics, Kharkiv, Ukraine
I. V. Guryev	*Web- maintenance*, University of Guanajuato, Mexico
S. O. Iakushev	*Local Organization*, National University of Radio Electronics, Kharkiv, Ukraine
M. V. Klymenko	*Local Organization*, University of Liege, Liege, Belgium
A. Levchenko	*Local Organization*, V. N. Karazin National University, Kharkiv arkov, Ukraine
T. F. Ruban	*Local Organization*, V. N. Karazin National University, Kharkiv, Ukraine
A. N. Topkov	*Local Organization*, V. N. Karazin National University, Kharkiv, Ukraine
A. V. Vasyanovich	*Local Organization*, National University of Radio Electronics, Kharkiv, Ukraine
V. Boiko	*Local Organization*, Taurida National V.I.Vernadsky University, Simferopol, Ukraine
V. Gryaznova	*Local Organization*, Taurida National V.I.Vernadsky University, Simferopol, Ukraine
I. Khomenko	*Local Organization*, Taurida National V.I.Vernadsky University, Simferopol, Ukraine
M. Pikula	*Local Organization*, Taurida National V.I.Vernadsky University, Simferopol, Ukraine
A. Shevchenko	*Local Organization*, Taurida National V.I.Vernadsky University, Simferopol, Ukraine
C. Shimko	*Local Organization*, Taurida National V.I.Vernadsky University, Simferopol, Ukraine

Contents

Advanced Optoelectronics

(Invited) From Order to Chaos and Back: Recent Advances in Optical Cryptography of Transmitted Data
S. Donati, V. Annovazzi-Lodi 1

(Invited) Advances in optoelectronic approaches for wideband and programmable processing of ultrafast signals
R. A. Minasian, E. H. W. Chan, X. Yi. 7

(Invited) Grating Resonances on Periodic Arrays of Sub-Wavelength Wires and Strips: Historical Narrative and Possible Applications
A. I. Nosich, V. O. Byelobrov, O. V. Shapoval, D. M. Natarov, T. L. Zinenko, M. Marciniak, J. Ctyroky 10

(Invited) Dynamic singular vector speckle fields by the Hurst exponent time analysis.
M. S. Soskin, V. I. Vasilev 14

(Invited) Frequency shifted feedback lasers: from basic physics to applications
L. P. Yatsenko 17

(Invited) Applications of whispering gallery modes resonances of silica rods and microcapillaries
E. Rivera-Perez, A. Diez, J. L. Cruz, A. Rodriguez-Cobos, and M. V. Andres 19

(Invited) Surface acoustic wave modulation of quantum cascade lasers
P. Harrison 22

(Invited) Recent progress in polarization bistable VCSELs and their applications for all-optical signal processing
H. Kawaguchi, T. Katayama 25

(Invited) Superradiant lasing and collective dynamics of active centers with polarization lifetime exceeding photon lifetime
Vl. V. Kocharovsky, E. R. Kocharovskaya, V. V. Kocharovsky 28

(Invited) Characterization and Fusion Splicing of Single-Mode Photonic Crystal Fiber
K. Borzycki, K. Schuster 31

(Invited) Application of multipass modes for high quality generation in stable open resonators
V. G. Niziev 35

Photonics of (Quasi)Periodic Media

(Invited) Antiguided resonant multicore fiber with improved intermodal discrimination
N. N. Elkin, A. P. Napartovich, D.V. Vysotsky 38

(Invited) Dye-doped polymer optical fiber pulse illuminator for microscopes
D. Kiesewetter, P.G. Gabdullin, A.Y. Savina, A.I. Bodrov, N.O. Stelmakova, V.M. Levin, G.G. Baskakov 41

Scattering of the plane wave from a periodically perforated dielectric slab
V. Byelobrov, T. M. Benson 43

Design of 3D optical media with periodically distributed emitting centers
S. O. Klimonsky, T. Bakhia, N.S. Borodinov, A.V. Knotko, N. Yu. Vereschagina 46

The PDLC photonic structures diffraction characteristics managing by the spatially non-uniform electric field
A. O. Semkin, S. N. Sharangovich 48

Defect Modes of a Two-Dimensional Photonic Crystal in the Presence of Acoustic Wave
Yu. V. Pilgun 50

Simulation study on the mode cutoff frequency in large mode area photonic crystal fibers
Xiaojian Shu, Xianfeng Bao, Xiaojun Wang 53

Ternary Comb-like Silicon Photonic Crystal: Oxidation, Intrinsic Modes, Reflection Windows and Contrastivity
E. Ya. Glushko 56

Investigation of the 2-D Photonic Crystal Filter
A. I. Filipenko, A. N. Donskov 58

Tapered double-clad optical fibers as gain medium for high power lasers and amplifiers
V. E. Ustimchik, Yu. K. Chamorovskii, V. N. Filippov, J. Kerttula, A. E. Ulanov, S. A. Nikitov 60

II

Investigation of modal content of radiation in multilayer cylindrical W- fibers
A. E. Ulanov, S. A. Nikitov, V. E. Ustimchik, Yu. K. Chamorovskii 63

Localized optical modes in photonic liquid crystals and low threshold DFB lasing
V. A. Belyakov 65

The researches of optical fibers speckle at the presence of optical vortices
D. Kiesewetter, N.V. Ilin 68

Spectral dependence of all-solid photonic bandgap fiber transmittance
A. Plastun, A. Konyukhov, E. Romanova T. Benson,
G. Athanasiou, J. Lousteau, G. Scarpignato, E. Mura, N. Boetti, D. Milanese 71

Non-reciprocal beam behavior in two-dimensional photonic crystal slab
A.P. Ostroukh, R.A. Lymarenko, V.A. Kolyadenko, V.B. Taranenko 74

Bragg resonance in multilayer structure with compound nonlinear layer
D. Sidorov, V. Borulko 77

Transmission and reflection spectra of photonic crystal with plasmonic defect layer
S. G. Moiseev, V. A. Ostatochnikov, D. I. Sementsov 80

Magneto-optical spectra of microcavity one-dimensional magnetophotonic crystals with double layer
V. N. Berzhansky, A. N. Shaposhnikov, T. V. Mikhailova, A. R. Prokopov, A. V. Karavainikov,
Yu. M. Kharchenko, I. M. Lukienko, O. V. Miloslavskaya, M. F. Kharchenko 82

Research of misalignments and cross-sectional structure influence on optical loss in photonic crystal fibers connections
A. I. Filipenko, O. V. Sychova 85

Advanced Lasers and Applications

(Invited) Influence of water medium and flame on the processes of laser materials treatment
A. F. Glova, S. V. Gvozdev, V. Yu. Dubrovskii, S. T. Durmanov,
A. G. Krasyukov, A. Yu. Lysikov, G. V. Smirnov 88

(Invited) Adaptive system for high power (more than 100 kW) CW CO_2 lasers
A. Kudryashov, A. Alexandrov, V. Samarkin, A. Rukosuev 91

(Invited) The control of energy, temporal and spatial characteristics a microchip laser with active output mirror
V.V. Kiyko, S.V. Gagarsky, V.I. Kislov, V.A. Kondratyev, E.N. Ofitserov, A.N. Sergeev 93

(Invited) Study of Stable Laser Cavity With Hole-coupling Output Mirror
Z. S. Tian, Y. C. Zhang, Z. H. Sun, S. Y. Fu, Q. Wang 95

The technique of measuring the velocity of melt removal in gas-laser cutting technology using multi-channel pyrometer
Y. N. Zavalov, A. V. Dubrov, V. D. Dubrov, N. G. Dubrovin, E. S. Makarova A. N. Antonov 98

Intracavity singly-resonant optical parametric oscillator pumped by a semiconductor disk laser
Yu. A. Morozov, M. Yu. Morozov 100

Infrared laser emission in a compact CW and quasi-CW diode pumped Nd3+: GdLuCOB laser
C. A. Brandus, L. Gheorghe, T. Dascalu 102

Large aperture bimorph deformable mirror for extremely high power laser system
J. Sheldakova, V.Samarkin, A. Kudryashov, A.Rukosuev 104

Segments of Oversized Waveguides in Open Resonant Systems
I. K. Kuzmichev, A.Yu. Popkov 106

Degree of paraxiality for monocromatic beams
A. B. Katrich 109

Narrow-band DFB laser based on a dye-doped volume Bragg grating
T. N. Smirnova, O. V Sakhno, V. M. Fitio, J. Stumpe 111

Broadband superluminescent diodes of NIR range with quasi-Gaussian spectra
E. V. Andreeva, S. N. Ilchenko, Yu. O. Kostin, M. A. Ladugin, P. I. Lapin, A. A. Marmalyuk, S. D. Yakubovich 114

Numerical analysis of the performance of AlGaAs/GaAs multi-quantum well superluminescent diodes
A. Asgari, P. Navaeipour 117

III

Modeling of XeCl excilamps with barrier discharge in frequency regime of work
S. S. Anufrik, A. P. Volodenkov, K. F. Znosko 120

Broadband semiconductor optical amplifiers of NIR range based on nanoheterostructures
E. V. Andreeva, S. N. Ilchenko, M. A. Ladugin, A. A. Lobintsov, A. A. Marmalyuk, M. V. Shramenko, S. D. Yakubovich 123

Dependence of efficiency of generation of the dye laser from a wave length of microsecond pumping
S. S. Anufrik, V. Yu. Kurstak, V. V. Tarkovsky 126

Spectral-luminescent and lasing properties of coumarin dyes
N. Kh. Ibrayev, E. V. Seliverstova, T. N. Kopylova, R. M. Gadirov, V. I. Alekseeva, L. E. Marinina, L. P. Savvina 129

Super/Subradiant Frequency Doubling by Quantum Wells Coupled to External Resonator
G.A. Koganov, R. Shuker 131

Manifestations of optical transient nutation in frequency- modulated CW laser beams
I. I. Plastun, A. G. Misurin 134

Propagation of frequency-modulated CW laser beams in coherent population trapping conditions
A. N. Bokarev, I. L. Plastun 137

The new laser media of dipyrromethene complexes with boron fluoride
R. T. Kuznetsova, Yu. V. Aksenova, T. A. Solodova, T. N. Kopylova,
E. N. Telminov, G. V. Mayer, M. B. Berezin, A. S. Semeikin 139

Photophysical properties of laser active elements based on dyes in new aliphatic polyurethane matrix
P. V. Bezrodna, L. F. Kosyanchuk, A. M. Negryiko, M. S. Stratilat, T. T. Todosiichuk 142

Influence of the electric field on laser material processing
A. Yu. Ivanov, S. V.Vasiliev 145

Modeling of XeCl excilamps with glow discharge in frequency regime of work
S. S. Anufrik, A. P. Volodenkov, K. F. Znosko 148

Spectral features of some red and NIR laser dyes in silica matrices
I. M. Pritula, O. N. Bezkrovnaya, V. M. Puzikov, V. V. Maslov, A. G. Plaksiy, A. V. Lopin, Yu. A. Gurkalenko 151

Faraday rotator for the optical switch of a two-wave light flux in fiber-optic networks
G. D. Basiladze, V. N. Berzhansky, A. I. Dolgov 155

Realization of two-electron mechanism of two charged ions creation upon multiphoton ionization of barium atoms by infrared.
I. I. Bondar, V. V. Suran 158

Formation of a quasi-uniform output beam in the waveguide CO_2 laser
O. V. Gurin, A. V. Degtyarev, V. A. Maslov, V. A. Svich, A. N. Topkov, T. F. Ruban 160

Conical 90 degree mirrors in a terahertz gas-discharge laser
V.P. Radionov, V.K. Kiseliov 164

High efficient vertical LED with pattern surface texture
M. H. Mustary, V. V. Lysak 166

End diode pumped solid state mini-lasers passive Q-switched by colored polymeric matrix
A. O. Yaskovets 169

THz Photonics, Plasmonics, Nanophotonics

(Invited) Hyperbolic Airy beams
V. Kotlyar, A. Kovalev 170

(Invited) Dielectric properties of some practical-use materials in the low-frequency part of the terahertz band
V. V. Meriakri, E. E. Chigryai, I. P. Nikitin 173

(Invited) Gyrotropic Metamaterials and Polarization Experiment in the Millimeter Waveband
S.I. Tarapov, S.Yu. Polevoy 176

(Invited) Electromagnetic wave diffraction by periodic structures with nonlinear inclusions
V. V. Khardikov, P. L. Mladyonov, S. L. Prosvirnin, V. R. Tuz 179

Frequency and wave-vector dispersion of the microwave mobility of drifting electron gas in GaN
V. V. Korotyeyev, G. I. Syngayivska and V. A. Kochelap 186

Zero reflection phenomenon in terahertz crystalline spectra
S. G. Felinskyi, G. S. Felinskyi ...189

Amplification of terahertz radiation by plasmons in graphene with a planar Bragg grating
O. V. Polischuk, V. V. Popov, S. A. Nikitov, V. Ryzhii, T. Otsuji, M. S. Shur ...192

Modeling of functional optical coatings based on plasmonic nanocomposites.
S.G. Moiseev ...195

Left-Handed Photonic Crystal Waveguide Sensors
D. El-Amassi, M. M. Shabat ...197

Fluid pumping cell of photonic - plasmonic microcavity sensor for biomedical application
V. A. Saetchnikov, E. A. Tcherniavskaia, A. V. Saetchnikov, G. Schweiger, A. Ostendorf ...199

Interaction wave at diffraction on slit
R. A. Lymarenko ...202

Interaction of optical pulses in nonlinear plasmonic systems
D. O. Ignatyeva, A. P. Sukhorukov ...204

Laser irradiation effect on ZnO nanoparticles
W. A. Farooq, W. Tawfik, A. Fatehmulla, S. M. Ali, M. Aslam ...207

Signal forming of chromophores secondary emission near noble metals plasmon films
N. D. Strekal, V. F.Askirka, A. E.German, S. A. Maskevich ...209

Effect of excitation intensity and surface morphology on the photoluminescence of ZnO films under the influence of surface plasmon resonance
S. I. Rumyantsev, V. M. Markushev, M. V. Ryzhkov, A. P. Tarasov,
Ch. M. Briskina, A. A. Lotin, O. A. Novodvorsky, V. L. Lyaskovskii ...211

Finite Comb-Like Silver Nanostrip Grating in the Optical Range: Interplay of Resonances
O. V.Shapoval, J. Ctyroky ...214

Photonics of new metal alkanoate composites contained semiconductor nanoparticles
A. Lyashchova, D. Fedorenko, G. Klimusheva, I. Dmitruk, S. Bugaychuk, T. Mirnaya ...217

Selection rules for angular-resolved photoionization of single P donor atom in silicon
M. V. Klymenko, F. Remacle ...220

Resonant enhancement of electromagnetic wave in the structure with refractive index gradient
A. A. Abramov ...223

Nanosized structures with the field localization: new opportunities for precise control of photonic and electronic device
A. N. Yakunin, G. G. Akchurin, N. P. Abanshin, B. I. Gorfinkel ...225

Custom types of waves spatially periodic structure of microwires
V. A. Boiko, V. I. Ponomarenko ...227

Hybrid plasmon resonances in the scattering and absorption of light by a circular silver nanotube
E. A. Velichko, A. I. Nosich ...229

The diffraction of the laser radiation on the two-band axial microelement
D. A. Savelyev, S. N. Khonina ...231

Subwavelength elliptical focal spot generated by a binary zone plat
M. Kotlyar, S. Stafeev ...234

Tight light localization in a hyperbolic secant planar slit lens
A. G. Nalimov, V. V. Kotlyar ...237

A solution of the wave equation for planar gradient waveguides in a frequency domain
V. M. Fitio, V. V. Romakh, Y. V. Bobitski ...240

Plasmon resonance, periodical structures and absorption spectra induced by laser beam in composite waveguide AgCl-Ag films
L. A. Ageev, V. K. Miloslavsky, V. M. Reznikova, E. D. Makovetsky ...243

Shift of surface plasmon-polaritons resonances via acoustic waves in hybrid metal-semiconductor structures
N. E. Khokhlov, V. I. Belotelov, B. A. Glavin ...245

V

Effective coupling between THz electromagnetic radiation and 2D plasmons
V. V. Korotyeyev, Yu. M. Lyaschuk 248

Theory of near field management via magnetoplasmon tunneling
A. N. Kalish, V. I. Belotelov, S. N. Andreev, V. P. Tarakanov, A. K. Zvezdin 251

Gaussian beam in linear gradient-index medium
A. A. Kovalev, V. V. Kotlyar 254

Nature of extraordinary transmission through a metal grating at frequencies closed to the Rayleigh-Wood anomaly
M. I. Panov, A. A. Shmat'ko 257

Gaussian beam tunneling through a gyrotropic -nihility finely stratified structure
V. R. Tuz, V. I. Fesenko 260

Nonlinear & Ultrafast Optics

(Invited) Spiral Beams: New Results and Application
V. Volostnikov, S. Kishkin, S. Kotova 263

(Invited) Broadband similariton: applications to ultrafast optics and photonics
L. Kh. Mouradian, A. S. Zeytunyan, G. L. Yesayan, F. Louradour, A. Barthélémy, R. Zadoyan 265

(Invited) Transient processes and steady state regimes in dynamic WGM microcavities with time dependent material parameter
N. K. Sakhnenko 268

(Invited) Interpretation of the ultrafast optical measurements of time delay in the ionization of Coulomb systems
V. L. Derbov, V. V. Serov, T.A. Sergeeva 271

Impact of electric field on dissipative micro-patterns in cholesteric-nematic mixtures (COC-5CB) doped by multiwalled carbon nanotubes
M. S. Soskin, V. V. Ponevchinsky, L. N. Lisetski, O. Deriabina A. I. Goncharuk, N. I. Lebovka 275

Optimal spectral decomposition of Raman gain profile using time response test
M. Y. Dyriv, G. S. Felinskyi, P. A. Korotkov 278

Submicron structures for nonlinear photonics in the mid-infrared spectral range
E. A. Romanova, A. I. Konyukhov, E. V. Borisov, D. S. Zhivotkov 281

Pulse shape transformation upon reflection from Bragg resonators with asymmetric feedback
V. Borulko, O. Drobakhin, D. Sidorov 284

Optical properties of layered organic-inorganic perovskite $(CH_3NH_3)_2PbBr_4$
S. Ahmadi-Kandjani, H. Ghanbari, M. S. Zakerhamidi 287

Femtosecond Parabolic Pulse Formation in All-Normal Dispersion Photonic Crystal Fiber
I. A. Sukhoivanov, S. O. Iakushev, J. A. Andrade Lucio, A. García Pérez, O. Ibarra Manzano 290

Similariton pulse compression using hybrid grating-prism compressor
A. Zeytunyan, G.Yesayan, L.Mouradian 293

2D wave structure induced by femtosecond laser pulse in semiconductor
V. A. Trofimov, M. M. Loginova, V.A.Egorenkov 296

Nonlinear interference effects in different types of three-level quantum systems
A. A. Orudzhev, I. L. Plastun, V.L. Derbov 299

A dynamically optical fiber loop memory using a nonlinear regeneration element
I. A. Malevich, A. V. Polyakov, S. I. Chubarov 301

Acoustic Modes on $As_2S_3/Y^{+128}X$ $LiNbO_3$ Crystal
R. M. Taziev 303

Dispersion oscillating fibers for the fusion of optical solitons
M. A. Dorokhova, A. I. Konyukhov 305

Spatial anisotropy of electro- and nonlinear optical effects in the $LiNbO_3$ and KTP crystals: calculation and experiment
A. S. Andrushchak, O. V. Yurkevych, I. M. Solskii, A. Rusek 307

Light beams interaction in the cell with thermal optical nonlinearity at the presence of a feedback system
G. A. Knyazev, D. A. Davtyan, A. P. Sukhorukov 309

Vibration spectra evaluation of dyes from their stimulated Raman scattering in multiple scattering media
V. P. Yashchuk, A. A. Sukhariev 312

Problem of coherence in modern optoelectronics and relaxed optics
P. P. Trokhimchuck, I. P. Dmytruk 314

Optical Characterization & Instrumentation

(Invited) About fluorescence excitation spectrums
N. Kh. Gomidze, Z. Kh. Shashikadze, K. A. Makharadze, M. R. Khajishvili, O. M. Nakashidze 317

(Invited) Precise laser measurings of a material index refraction on Brewster angle
E. A. Tikhonov, A. K. Lyamets 320

A nondestructive validation of reverse impact experiment based on shape measurement using high speed photograph
D. Khodadad, T. Sjöberg 322

Features of Optical Image Jitter in a Random Medium with a Finite Outer Scale
L. A. Bolbasova, P. G. Kovadlo, V. P. Lukin, V. V. Nosov, A. V. Torgaev 325

Using digital camera as metering device in geometrical, spectral and intensity measurements
A. V. Kraisky, T. V. Mironova, T. Sultanov 327

Scanning devices for the beam profile measurement of laser irradiation
N. G. Kokodij, B. V. Safronov, I. A. Priz, V. P. Balashin, M. P. Perepechaj 330

Measurement of microparticle sizes by digital processing of light scattering patterns
N. G. Kokodij, M. V. Kaydash, V. O. Timaniuk, Yu. O. Derecha 332

Absolute Calibration of Profile Thin-Wire Bolometric Gauge of Laser Pulse Energy
S. Pogorelov 334

Features of the wavefront sensor based on the Talbot effect
D. V. Podanchuk, A. A. Goloborodko, M. M. Kotov 337

Modern approach for estimating uncertainty of a precision optoelectronic phase noise measurement
P. Salzenstein, E. Pavlyuchenko 340

Pulsed laser/ion beam treatment of Ge/Si and Ge/Al$_2$O$_3$ thin film structures
R. I. Batalov, R. M. Bayazitov, H. A. Novikov, V. A. Shustov,
I. A. Faizrakhmanov, N. M. Lyadov, K. N. Galkin, P. I. Gaiduk, G. D. Ivlev 342

Some features of estimation of the diffusion length of minority carriers in cathodoluminescence microscopy
Yu. E. Gagarin, N. N. Miheev, N. A. Nikiforova, M. A. Stepovich 344

The possibilities of the colorimetric method for measuring of wavelength's
distribution of light radiation in standard and the RAW
A. V. Kraiski, T. V. Mironova, T. T. Sultanov, V. A. Postnikov 346

Peculiarities of interferometric studies of surfaces with phase fluctuations of various statistical characteristics
O. V. Gnatovskyi, L. A. Derzhypolska, A. M. Negriiko 349

Agitated reactor with in situ nanoparticle size control by light scattering photon correlation spectroscopy
A. G. Lazarenko, A. N. Andreev, M. Ben Amar, K. Chhor, A. V. Kanaev 352

Development of algorithms for image forward using finite element method with florescent molecular tomography
Sima Saleh 355

Laser-microwave spectroscopy of singlet Mg I atoms in S,P,D,F,G Rydberg state
S. F. Dyubko, V. A. Efremov, A. S. Kutsenko, N. L. Pogrebnyak, K. B. MacAdam 358

Optical properties of composite thick films based on silver nanoparticles in the acrilic polymer matrices
N. M. Ushakov, I. D. Kosobudsky, P. A. Muzalev, V. Ya. Podvigalkin 361

Dynamic holographic grating in liquid crystalline polymer
E. O. Berezhniy, M. M. Burykin, S. G. Ilchenko, A. P. Ostroukh, R. A. Lymarenko 363

Differential method of doppler laser anemometry of the objects with retroreflecting sheet
G. N. Dolya, A.M.Kryukov,V.G.Mudrik 365

VII

Microstructure and chemical composition of heterogeneous crystal GaSe:AgGaS$_2$
V. V. Atuchin, Yu. M. Andreev, N. F. Beisel, A. R. Tsygankova, T. A. Gavrilova, L. D. Pokrovsky, A. I. Saprykin 368

Pyroelectric effect in X-cut LiNbO3 optical modulators
R. S. Ponomarev, A. B. Volyntsev, I. S. Azanova, E. D. Voblikov 371

Fabrication of controllable holographic gratings to manage the energy transfer
S. Bugaychuk, L. Pryadko, I. Pryadko, O. Kolesnyk,2, R. Conte, V. Gnatovskiy, A. Negriyko 373

Some optical properties of Zn$_{1-x}$Mn$_x$Te semimagnetic films
O. Klymov, D. Kurbatov, O. Levchenko 376

Angular Spectra of Phase Diffraction Gratings Illuminated by Interference Field
V. O. Gnatovskyy, S. A. Bugaychuk, A. M. Negriyko, I. I. Pryadko, A. V. Sidorenko 378

Crystal Growth Sector Effect on Dielectrical Properties of Carbamide Doped KDP Crystal
A. N. Levchenko, I. M. Pritula, V. B. Tyutyunnik, A. O. Penkina, A. V. Kosinova, M. I. Kolybayeva 381

Research of the time of the transverse relaxation using the incoherent NMR-spectrometer
M. O. Pikula, K. V. Shimko 384

Ge-Ga-S/Se glasses studied with PALS technique in application to chalcogenide photonics
A. Ingram , H. Klym, O. Shpotyuk 386

LIDAR hyperspectral system for detecting chemical and biological agents
V. V. Gnidenko, V. O. Yatsenko 388

Opto-cryogenic sensitive element with ultrasensitive laser interferometer and microprocessor controller
M. V. Nalyvaychuk, V. O. Yatsenko 391

Modern polarization-independent trap detectors
K. I. Muntean 393

Thermal processes in the bolometric measurer of laser radiation characteristics
A. O. Pak, N. G. Kokodiy 396

Non-steady-state photoelectromotive force in AlN crystal
M. Bryushinin, V. Kulikov, E. Mokhov, S. Nagalyuk, I. Sokolov 399

Spin controlled optical radiation pressure
G.Tkachenko, E. Brasselet 402

Bio- & Chemi-photonics

(Invited) Biomedical effect the phenomenon of in vivo blood oxyhemoglobin photodissociation
M. M. Asimov, R. M. Asimov, D. B. Vladimirov, A. N. Rubinov 405

Method of generating a pulsed X-ray
M. Samoylovich 408

Optical sensor network for detecting chemical and biological agents
V.Yatsenko 411

Diagnostics of biomedical agents by whispering gallery mode optical resonance based sensor
V. A. Saetchnikov, E. A. Tcherniavskaia, A. V. Saetchnikov, G. Schweiger, A. Ostendorf 412

Photo-Induced Birefringence Investigation of Azo Polymer with Cyano Substituted group
E. Bagherzadeh Khajeh Marjan, S. Ahmadi-Kandjani, M. S. Zakerhamidi 415

The change response of holographic glucose sensors when changing pH of the solution
V. A. Postnikov, A. V. Kraiski, M. A. Shevchenko 418

Application of the digital holographic interference and electron microscopy methods for study of blood erythrocytes 3D
T. V. Tishko, D. N. Tishko, V. P. Titar, O. O. Prikhodko, V. I. Bumeister 420

Nonlinear holographic model of physiological optics
V. P. Titar 423

(Invited) Picosecond 175 ~ 210 nm tunable deep-ultraviolet laser
Shen-Jin Zhang, Yong Bo, Feng-Feng Zhang, Feng Yang, Zhi Xu1, Feng-Liang Xu, Zhi-Min Wang,
Qin-Jun Peng, Jing-Yuan Zhang, Xiao-Yang Wang, Chuang-Tian Chen, Wen-Qiang Lei, and Zu-Yan Xu 426

From Order to Chaos and Back: Recent Advances in Optical Cryptography of Transmitted Data

(Invited Paper)

Silvano Donati[1,2], *Life Fellow, IEEE,* and Valerio Annovazzi-Lodi[1], *Senior Member, IEEE,*

[1] Department of Industrial and Information Engineering, University of Pavia, Pavia, Italy

[2] Graduate Institute of Precision Engineering, National Chung Hsing University, Taichung, Taiwan

Abstract: We outline the fundamentals of optical cryptography with coupled semiconductor lasers, and the associated tools of optical chaos generation and synchronization, as well as chaos coding of transmitted information, with special reference to CM (Chaos masking) and CSK (Chaos Shift Keying). We present different versions of such schemes which have been numerically and/or experimentally tested both with discrete component crypto-systems and with a PIC (Photonic Integrated Circuit) realization of the chaotic source. Finally, we outline some recent developments on the chaos secure transmission technique.

1 Introduction

Chaos is a quite interdisciplinary subject of research, with founding contributions coming from mathematics and physics [1], and with applications being developed almost everywhere in applied science. Indeed, chaos is the novel phenomenon we first encounter when the level of system complexity [2,3] - the number of individual entities or number of differential equations and of nonlinearities in the system - departs markedly from the ground level of the Lagrangian system of the nineteenth century. Because of the added complexity classical reductionism fails, and the pseudo-random - or chaos - evolution of the system unveils new conceptual phenomena unpredictable with the lower level description, as well as it opens a new ground of applications for development.

In optics, Haken [4,5] has been the first to point out that the Maxwell-Block equations describing the laser [5,6] can be brought to coincide with the Lorentz equations describing convective flow in the atmosphere, a well known paradigm for chaos. Since then, the rate equations in three variables for the laser are known as Lorenz-Haken (L-H) equations, and they were soon shown by Arecchi [6] to be the base to identify classes of stable and unstable lasers.

2 Unveiling Chaos

Actually, semiconductor lasers belong to class B [6], and are stable unless a new term is added into the L-H equations, such as external injection from another laser or the feedback from a remore mirror. With this term, the L-H become the well known Lang and Kobayashi equations (or, Lamb's equation when excitation is decoupled from electric field) and system may experience chaos when driven heavily enough into feedback - a favourable condition because it allows easy control of the desired working regime.

So, the scheme we have considered to study the high-level dynamical regimes has been that of Fig.1, where we can have either a double-source arrangement (either in the mutual, symmetrical or asymmetrical, or in the unidirectional case) or

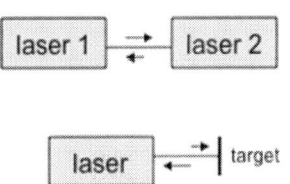

Fig. 1. Coupled-lasers: top, mutual (can be symmetrical or asymmetrical) bottom, self-coupling

a single-source aided by the retroreflection from an external target (the so called self-coupling or self-mixing case). Self-mixing is interesting because it has two distinct applications in two fields far away from each other: (i) at weak injection level (when the return is barely 10^{-8} to 10^{-3} of emitted power) perturbations of the laser field are frequency and amplitude (AM and FM) modulations of the cavity field, depending on external amplitude and phase of the return, and the system can be used in instrumentation for measurements of the external pathlength (the self-mixing interferometer [7]) or as an optical radar detecting weak echoes [8,9], also known as a coherent injection detector [9]; while (ii) at moderate/large coupling (10^{-3} to a few10^{-2}) the system enters a high-level dynamical regime and starts generating periodic and multiperiodic oscillations and chaos, opening new avenues in communications and information technology.

An exhaustive literature is available on optical chaos and its applications, and recently ample reviews have appeared on the physics principles [10] and on the application repertoire [11] of chaos and related phenomena, to which we direct the interested reader.

About the analytical approach to study the new chaos phenomena, the Lang and Kobayashi equations [12,13] provide a powerful tool, well confirmed by experimetal evidence. Using the L-K equation to study the double-source unidirectional coupling case (Fig.1, top), the route to high dynamics behavior and chaos is readily unveiled, as seen in Fig.2. The parameter used to depict the system evolution was here the beating amplitude $S=E_1E_2^*$ of oscillations in laser 1 and 2 (master and slave arrangement), vs. the coupling strength K (fraction of field injected from laser 1 into laser 2) [14]. The regimes are identified based on three indicators: (i) the time series [or amplitude S(t)], (ii) the frequency spectrum S(f), and the state diagram, S vs dS/dt [5,11].

In unperturbed conditions, the time series is a sinusoid, the spectrum is a single peak and the state diagram is a circle.

978-1-4799-0019-0/13 $31.00 © 2013 IEEE

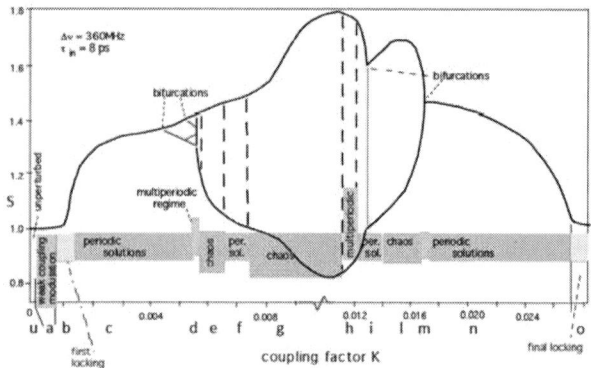

Fig. 2. Beating amplitude of master and slave oscillations of the coupled system, versus the strength of coupling K, starting from the unperturbed regime (region u) and unveiling regimes of periodic solutions (regions c,f,i,n), multiperiodicity (d,h,m) and chaos (e,g,l), passing through first locking (b) to final locking (region o).

Then, we encounter a narrow weak-coupling modulation region, followed by a locking state, and signal S disappears.

But, at increased K, S reappears and now the time series has a distortion every other period, the spectrum carries subharmonics and the state diagram is a double loop figure, all indicative of the so-called period-1 oscillation regime. At even larger K, the subharmonics increase in number and amplitude, and the state diagram has multiple loops: it is the multiperiodic regime. Last, when harmonics accompany the sub-harmonics and broaden the spectrum widely, the time series is random-like looking, and the state diagram spread all over the available space of coordinates, we have chaos [11,14].

An example of the three diagrams for multiperiodic solutions and chaos is reported in Fig.3.

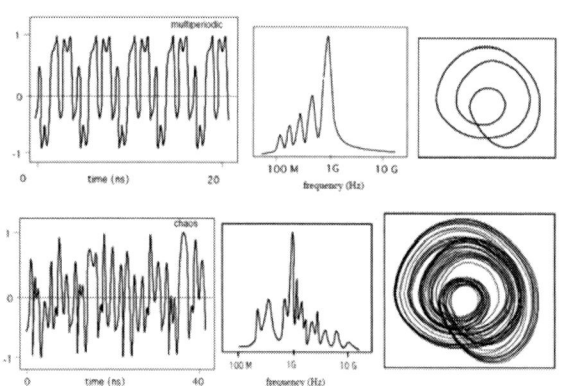

Fig.3 Time series (left), frequency spectrum (center) and state diagram (right) of the beating signal S for the multiperiodic regime (regions d,h,m in Fig.2) and chaos (egl) (From Ref.[11])

2 Taming Chaos

The new waveforms can be thought as a sort of free-response or eigenfunctions of the complex system, similar to sinusoidal oscillations being the free response of a second-order system or oscillator.

On this analogy, we may also think of injecting from the external a chaos waveform into the system, in the conjecture that the system will adjust itself to follow the dynamical evolution of the injected signal, or become *synchronized*. And, like when attempting to lock a linear second-order oscillator we need a frequency close to the free oscillation frequency, thus we need to stay close to the system free response to be able to synchronize the chaos generator.

In Ref.[15] we have considered the synchronized scheme of Fig. 4, with two identical coupled-lasers systems, system 1 with LD1 and LD2, and system 2 with LD3 and LD4. The output of system 1 (out 1) is connected to the sum node in system 2, at the input of LD4; in addition, the LD4 output is sent in subtraction to the node in system 2. By this arrangement, when the LD2 and LD4 outputs differ, system 1 applies a correction to system 2 and brings it closer, and when equality is reached system 2 is synchronized and can evolve freely.

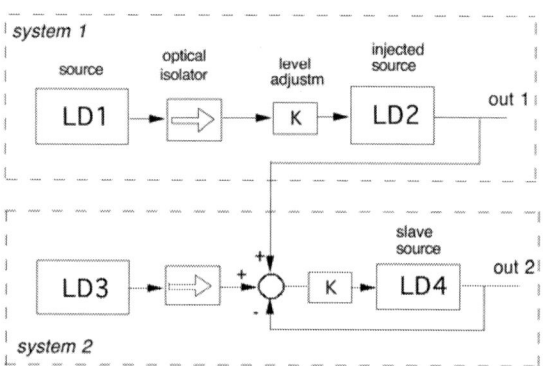

Fig. 4. Scheme of synchronization: two identical coupled-laser systems (LD1/LD2 and LD3/LD4) are used. Out 1 is sent to the sum node at the input of slave laser LD4 of system 2, and the output of LD4 is sent back in subtraction, so that when out 2 is equal to out 1 (is synchronized) no further correction is applied. (From Ref.[15])

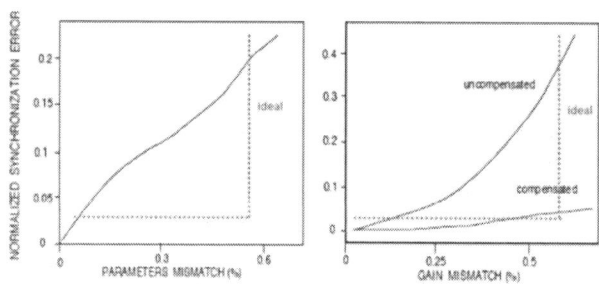

Fig.5 Synchronization error versus relative mismatch of system parameters (left) and vs gain mismatch, in uncompensated and amplitude-normalized conditions (right) (From Ref.[15])

By using L-K equations to model the synchronization scheme (Fig.4), we find that output of system 2 reaches the steady-state solution irrespective of the arbitrary starting point, with an amplitude error $(E_2-E_1)/E_0$ damping quickly to zero after a few cycles of oscillation with a small residual error, for all the different chaos waveforms freely generated.

One critical issue is *sensitivity* to system parameters: we wish that small deviations from nominal values of the parameters are tolerated, whereas beyond a certain threshold (for example, 0.5%) synchronization isn't achieved any longer. Red dotted lines indicate this ideal condition (Fig.5). For a real system, the error $(E_2-E_1)/E_0$ is a function of the rms mismatch allowed for the parameters (Fig.5, left); removing the effect of gain mismatch (of E_2 and E_1) we obtain the graph of Fig.5, right. Interesting to note, upon changing the laser parameters over a wide range of reasonable values doesn't affect the general trend of synchronization. And, for the error to be small we need a parameter mismatch less than ~0.2%, while error becomes large and synchronization is lost for a mismatch larger than ~0.8%, typically. The sensitivity of chaos synchronization to system parameters is the key of secure transmission, as discussed in the next paragraph.

3 Using Chaos to Encrypt Messages

A first scheme of cryptography, CM (Chaos Masking) readily follows from synchronization: summing (incoherently) to the generated chaos a small signal carrying the desired message will not impair synchronization of the slave because the (small) deterministic signal will be ignored, and upon subtracting the synchronized output (chaos) from the received signal (chaos+ message) we will free out the message [15].

However, the CM scheme has some drawbacks. First, if the sum of signal to chaos is made at a slightly different λ, clever filtering could reveal the message. Second, the small amplitude of the message (e.g., 5% of the transmitted power) makes the transmitted power mostly used by chaos rather than message, and thus SNR is poor. So, the quest is on a method using message as large as the chaos.

The answer is chaos shift keying (CSK) [15]. In CSK, we code with a different chaos waveform the "0"s and "1"s bits of the digital message, by acting on one of the several parameters governing the dynamical evolution of the system, e.g., the drive

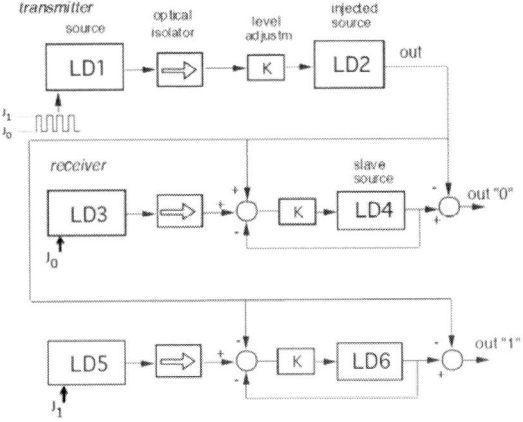

Fig.6 CSK cryptography: the binary message modulates levels J_0 and J_1 of the bias current, and lasers LD1/LD2 generate a sequence of chaos waveforms, for the "0" and "1" of the message. At the receiver end two twin systems are biased at J_0 (LD3) and J_1(LD5) and injection of transmitted waveform makes them synchronize on its designated bit, "0" for system LD3/LD4 and "1" for LD5/LD6. (From Ref.[15])

current of the laser [16].

Thus, each bit uses the entire chaos waveform and associated power, and we fully exploit the available photons (and SNR).

As shown in Fig.6, at the transmitter the drive current of laser LD1 is switched from J_0 to J_1, to code the bit "0" and bit "1" of the message, and we get a sequence of piece-wise chaos waveforms for the coded message. At the receiver end, two twin systems are set at bias J_0 (LD3/LD4) and J_1 (LD5/LD6). Injection of the received waveform synchronizes the designated bit, "0" for system LD3/LD4 and "1" for system LD5/LD6.

The CSK and CM cryptography schemes, however, are difficult to implement with the two-laser structure of the basic coupled-system cell, which is not minimum part-count. To go in a practical system, we need a simpler scheme, and this is indeed possible with the self-mixing (also called DOF - Delayed Optical Feedback) chaos cell (Fig.7).

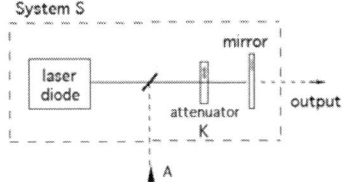

Fig.7 DOF (delayed optical feedback) chaos generator: the laser is subject to a self-injection-coupling regime from the mirror. The beamsplitter injects an external signal for synchronization.

In a typical all-fiber DOF setup (Fig.8 top), the laser diode is conjugated through a lens to a single-mode fiber, whose end-face is angled (8-12 deg. typ.) to avoid back-reflection.

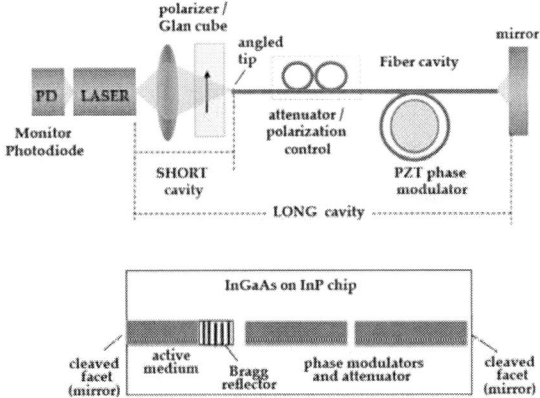

Fig.8 Technologies for implementing the DOF scheme: all-fiber (top) and integrated optics (bottom).

The fiber ends on a mirror, partially reflecting back the outgoing signal. By varying the distance of fiber tip to the mirror we can adjust the level K of feedback, while a PZT phase modulator serves to add the phase-coded message. In this case a so-called long-cavity system is realized.

We can also implement the DOF cell by an integrated optics technology (Fig.8 bottom), incorporating in a single photonic integrated circuit (PIC) the active source (a DFB laser), an active waveguide for amplitude and/or phase modulation, and the retroreflector (the mirror-like cleaved facet of the chip). This solution offers a more stable short-cavity system.

A picture of the chip, fabricated in InP [17] under an FET European research Contract (Picasso) is shown in Fig.9.

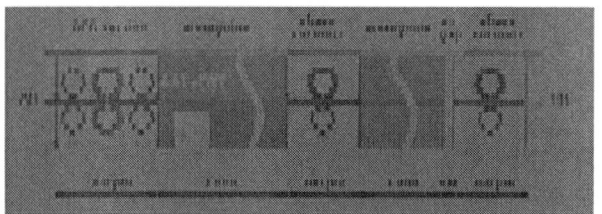

Fig.9 A DOF short-cavity PIC-chip, fabricated by InGaAs waveguides on a InP substrate, and incorporating a DFB laser, two 5-mm waveguides and two phase modulators, to realize a phase-coded scheme of transmission (see below) (From Ref.[17])

Practical solutions of CM and CSK schemes are available in the literature. An example of a CSK-PM system is reported in Fig.10 [18], using a short-cavity DOF generator and a LiTaO$_3$ phase modulator to impress the message as a phase $\Delta\psi_{in}$ added to the optical pathlength 2ks. As the short-cavity DOF is sensitive to the external cavity phase, the chaos waveform is somehow coded by phase $\Delta\psi_{in}$. At the receiver end, an identical DOF generator has the phase modulator set at zero voltage. Thus, the receiver is synchronized only for zero input message or phase $\Delta\psi_{in}$.

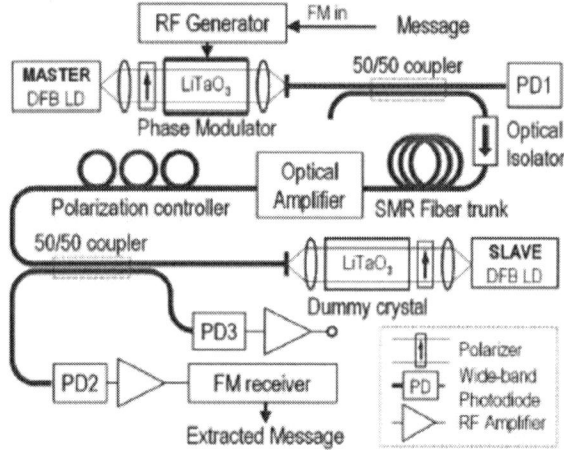

Fig.8 A short-cavity DOF-CSK system, uses a LiTaO$_3$ phase modulator to impress the message as a phase variation in the cavity and hence in the chaos waveform generated by the DOF. At the receiver, a dummy modulator synchronizes only when phase is zero. Photodetector and an FM receiver act as phase-to-amplitude converter, thus extracting the message. (From Ref.[18]).

Correlation of receiver and transmitter chaos waveforms progressively decreases at the increase of the phase difference, and this process is a sort of phase-to-amplitude conversion. With photodetection followed by a PM conversion we get a signal proportional to $\Delta\psi_{in}$ and hence to input signal.

In all the proposed solutions for secure transmission using chaos, security is based on synchronization sensitivity to laser parameters. The two users must share a 'twin' laser pair, i.e.,

two lasers of very similar parameters (selected from the same wafer). Thus, for an eavesdropper, it is very difficult to find a laser compatibile with the twin pair, to synchronize chaos to decode the message.

With the standard two-laser system, chaotic transmission of digital signals in the GHz range has been demonstrated on the metropolitan network of Athens [19]. Also, analog transmission of RF signals has been experimentally tested (Fig.9) [20].

Several basic building blocks for future long-distance transmission have been already proposed, such as a chaotic wavelength converter for WDM transmission [21] and a chaotic repeater [22]. Moreover, coding solutions for a more efficient signal protection have been studied [23].

Alternative solutions based on the electro-optical feedback have been also proposed [13].

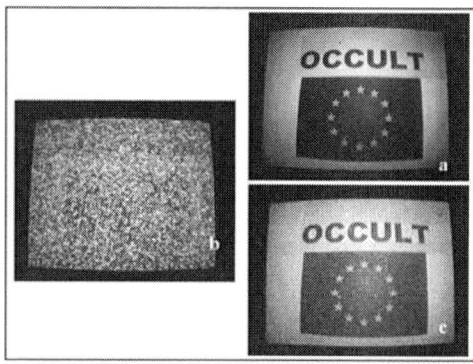

Fig.9 Experiment of transmission of a video signal: a) plaintext message; b) message hidden in chaos; c) recovered message (from Ref. [20])

4 Recent Developments

In addition to the basic two-laser scheme, more advanced configurations have been developed, that offers a better synchronization quality to the authorized subscribers, who share the matched laser pair, as well as a better security to an eavesdropper attack. Both results are based on such schemes being symmetrical (differently from the basic two-laser solution), since both Tx and Rx lasers are injected by a third (possibly unmatched) chaotic laser (the driver Drv).

A CM implementation setup of such three-laser scheme is shown in Fig.10. In this case, the Drv laser is routed to chaos by delayed optical feedback, while Rx and Tx may be open-loop, or may experience a weak local optical feedback by a mirror. Numerical results for RF power spectra of Drv, Tx and Rx, and their difference, are shown in Fig. 11. In Fig.12, transmission of a 5Gb/s message is simulated, showing the RF spectra of the plaintext message, of the same hidden in chaos, and of the recovered message after chaos subtraction. The eye-diagram of the recovered message is also shown.

In experiments, such a high cancellation level cannot be achieved. However, the advantage of the three-laser over the two-laser scheme has been confirmed. The requirement of using a supplementary laser is not really a drawback, since in a network the same Drv can be used to assist several interconnections between couples of users.

978-1-4799-0019-0/13 $31.00 © 2013 IEEE

The three-laser scheme can be easily modified for use in free space propagation. The recent interest in FSOLs (Free Space Optical Link), especially for countries rapidly expanding their communication infrastructure, also requires a careful design for security, since the open optical beams can be easily intercepted by an eavesdropper.

A suitable three-laser scheme for such scenario, where attenuation is usually high, is found in Fig. 13, where the optical emission of Drv is photodetected and amplified before injection into Rx and Tx. The same is done for Tx to Rx injection. Electrical amplification is a convenient solution, with respect to optical amplification, and results in a lower cost, easier to align system.

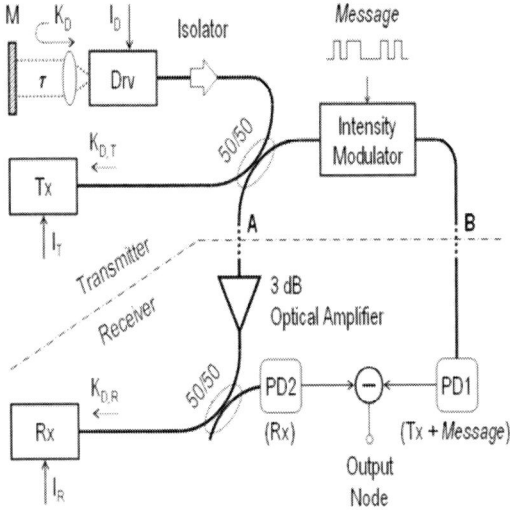

Fig.10 A three-laser CM crypto-system uses a common source driver (Drv) to route into chaos and synchronise a pair of twin lasers (Tx and Rx). The message is recovered as in the basic scheme by chaos cancellation at the receiver (from Ref. [24])

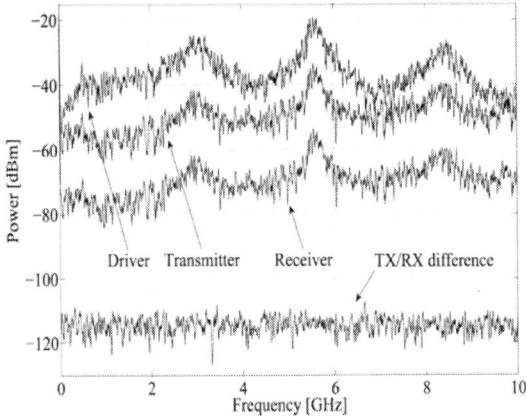

Fig. 11 Numerical RF power spectra for Drv, Tx and Rx in the scheme of Fig.10. For better visualization, the traces of Drv and Rx have been shifted upwards by 20 db and downwards by 20 dB, respectively. The difference signal is also shown (from Ref.[24]).

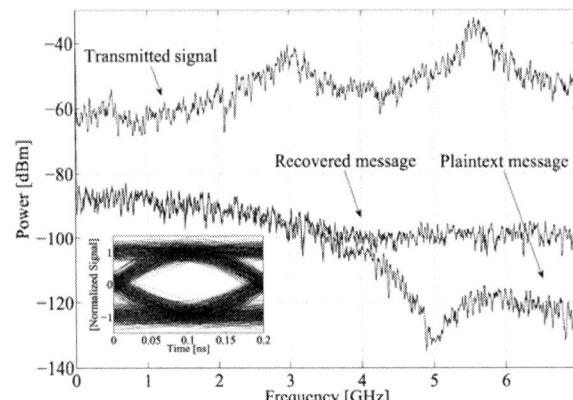

Fig.12 Numerical RF spectra for CM transmission of a 5Gb/s digital message with the scheme of Fig.10. The eye diagram of the recovered message is shown in the inset (from Ref.[24]).

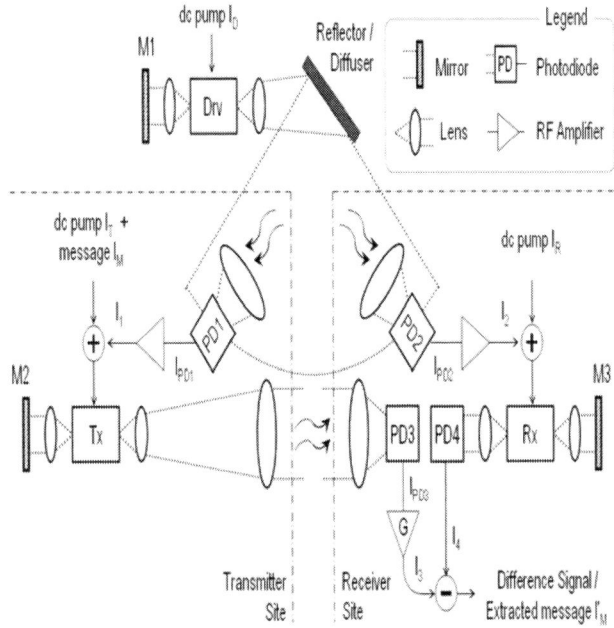

Fig. 13 Configuration for secure data transmission in free-space. Due to high channel attenuation, the optical emission of the Drv is photodetected and amplified before injection into Rx and Tx. The same is done for Tx to Rx injection (from Ref.[25])

The same scheme offers a viable solution for indoor secure transmission, in a room or on a train or airplane, where many users share the same diffused channel, as well as the Drv, which can be installed, e.g., on the ceiling.

Another evolution of the standard setup has been in the direction of providing a real multi-user network protected by chaos.

All the schemes considered so far are practically restricted to transmission between two specific users, sharing a twin laser pair. However, in most cases, it is required to organize a network of several users that can freely transmit data to one-another in a secure way. In practice, it is difficult to find more than 3 or 4 matched devices on a wafer, which can synchronize efficiently, and thus multi-user transmission would require that each user holds a laser (of a twin pair) for each potential partner. This is clearly unpractical. Alternatively, one could increase the message amplitude, so that a lower synchronization level can be tolerated, but in this case the security level would be poor.

A possible solution has been recently proposed [26], deriving from the well-known *public-key* cryptography. With this new approach, it is possible to exchange data between any couple of subscribers in a network. This is possible by introducing a provider, who is responsible of the network security.

For each subscriber a couple of twin chaotic lasers is required, one being held by the provider and the other by the subscriber himself. The basic two-laser (or the three-laser) scheme of chaotic transmission is then used for secure data exchange, both between a subscriber and the provider, and between two subscribers. In the second case, the provider routes the suitable chaotic carrier to Tx, where data are added and sent to Rx. Several arrangements can be considered, based on this architecture, which is studied in detail in [26].

References

[1] see for example International Journal of Bifurcation and Chaos, vol.23, 2013. WSP (Singapore) ISSN: 0218-1274

[2] see for example: Journal of Complexity, vol.29, 2013. Elsevier B.V.(Amsterdam) ISSN: 0885-064X

[3] H. Tsoukas: "Chaos, Complexity and Organization Theory" Organization, vol. 5, 1998 pp. 291-313

[4] H. Haken: "Analogy between higher instabilities in fluids and lasers" Physics Letters 53A: (1975) pp.77–78

[5] K. Ohtsubo: "Semiconductor Lasers: Stability, Instability and Chaos" 2nd edition, Springer Series Optical Sciences vol.111, Springer-Verlag, New York, 2009.

[6] F.T. Arecchi, G.L. Puccioni, J.R. Tredicce: "Deterministic Chaos in Laser with Injected Signal", Optics Commun. vol.51 (1984), pp.308–314.

[7] S. Donati: "Developing Self-Mixing Interferometry for Instrumentation and Measurements" Laser and Photonics Review, vol.6 (2012), pp. 393–417 (DOI) 10.1002/lpor.201100002.

[8] S. Donati: "Responsivity and Noise of Self-Mixing Photodetection Schemes", IEEE Journ. Quantum El., vol.47, 2011, pp.1428-1433.

[9] S. Donati: "Photodetectors", Prentice Hall, Upper Saddle River, N.J., 2000, see Sect.8.4.

[10] M.C.Soriano, J.Garcia-Ojalvo, C.Mirasso, I.Fischer: "Complex Photonics: Dynamics and Applications of Delay-Coupled Semiconductors Lasers", *Review Modern Physics*, vol. 85, pp.421–470 (2013).

[11] S. Donati, S.-K. Hwang: "Chaos and High-Level Dynamics in Coupled Lasers and their Applications", *Progress in Quantum Electronics* (2012), vol.36, Issues 2–3, March–May 2012, pp. 293–341.

[12] R. Lang and K. Kobayashi, "External Optical Feedback Effects on Semiconductor Injection Laser Properties", *IEEE J. Quant. Electr.*, vol.QE-16 (1980), pp.347-355.

[13] S. Donati, C. Mirasso (Editors): "Optical Chaotic Cryptography", Feature Issue of: *IEEE Journal of Quantum Electronics*, vol. QE-38 (Sept 2002), pp.1138-1184.

[14] V. Annovazzi-Lodi, S. Donati, M. Manna: "Chaos and Locking in a Semiconductor Laser due to External Injection", *IEEE Journal of Quantum Electronics*, vol. QE-30 (1994), pp.1537-1541.

[15] V. Annovazzi-Lodi, S. Donati, A. Scirè: "Synchronization of Chaotic Injected-Laser Systems and its Application to Optical Cryptography", *IEEE J. of Quant. Electr.*, vol. QE-32 (1996), pp.953-959, see also: S. Donati, V.Annovazzi Lodi: "Crittografia Ottica Caotica", *Proc. Conf. Fotonica'95*, Sorrento, May 1-4,1995, pp.463-466.

[16] V. Annovazzi-Lodi, S. Donati, A. Scirè: "Synchronization of Chaotic Lasers by Optical Feedback for Cryptographic Applications", IEEE Journal of Quant. Electr., vol. QE-33 (1997), pp.1449-1454.

[17] D. Syvridis, A. Argiris, A. Bogris, M. Hamacher, I. Giles, " Integrated Devices for Optical Chaos Generation and Communications Applications", IEEE J. of Quantum Electron. Vol. 45,no.11, pp.1421-1428, Nov. 2009.

[18] V. Annovazzi-Lodi, M. Benedetti, S. Merlo, T. Perez, P. Colet, C. Mirasso: "Message Encryption by Phase Modulation of a Chaotic Optical Carrier" *IEEE Phot. Techn. Lett.*, vol.19 (2007), pp.76-78

[19] A. Argyris, D. Syvridis, L. Larger, V. Annovazzi, P. Colet, I. Fischer, J. Garcia-Ojalvo, C. Mirasso, L. Pasquera, K.A. Shore: "Chaos-based Communication Link at high Bit Rate Using Commercial Fiberoptic Link", *Nature Letters*, 2005 pp.343-346.

[20] V. Annovazzi-Lodi, M. Benedetti, S. Merlo, M. Norgia, B. Provinzano: "Optical Chaos Masking of Video Signals" *IEEE Photonic Technology Letters*, vol. 17, (2005), pp.1995-1997.

[21] V.Annovazzi-Lodi, G. Aromataris, M. Benedetti, I. Cristiani, S. Merlo, P. Minzioni: "All-Optical Wavelength Conversion of a Chaos Masked Signal" *IEEE Phot. Techn. Lett.*, vol.19 (2007), pp.1783-1785.

[22] M. W. Lee, K. A. Shore, "Demonstration of a Chaotic Optical Message Relay Using DFB Laser Diode," *IEEE Phot. Tech. Lett.*, vol. 18, pp. 169-171, Jan.2006.

[23] L. Ursini, M. Santagiustina, V. Annovazzi-Lodi: "Enhancing Chaotic Communication Performances by Manchester Coding", *IEEE Photonics Technology Letters,* vol.20, 2008, pp. 401-403.

[24] V. Annovazzi-Lodi, G. Aromataris, M. Benedetti, S. Merlo, "Private Message Transmission by Common Driving of Two Chaotic Lasers", *IEEE Journal of Quantum Electronics*, vol.46, 2010, pp.258-264.

[25] V. Annovazzi-Lodi, G. Aromataris, M. Benedetti, S. Merlo, "Secure Optical Transmission on a Free Space Optics Data Link", *IEEE Journal of Quantum Electronics*, vol.44, 2008, pp.1089-1095.

[26] V. Annovazzi-Lodi, G. Aromataris, M. Benedetti, "Multi-User Private Transmissionwith Chaotic Lasers", *IEEE Journal of Quantum Electronics*, vol.48, 2012, pp.1095-1101.

Advances in optoelectronic approaches for wideband and programmable processing of ultrafast signals

(Invited Paper)

R. A Minasian, *Fellow, IEEE,* E. H. W. Chan, and X. Yi

School of Electrical and Information Engineering, Institute of Photonics and Optical Science,
University of Sydney, NSW, Sydney, 2006, Australia

Abstract: Recent photonic signal processing techniques that realize ultra-wideband phase shifters for phased array beamforming, single passband microwave photonic filters, and programmable switchable microwave photonic tunable filters, are presented. These processors provide new capabilities for the realisation of high-performance and high-resolution signal processing.

Photonic signal processing, using photonic approaches to condition microwave and radio frequency (RF) signals, is attractive due to the inherent advantages of high time-bandwidth product and immunity to electromagnetic interference (EMI) [1-3]. Photonic signal processing also enables dynamic reconfiguration of the processor characteristics over multi-GHz bandwidths, which may not be possible with conventional electrical approaches. In this paper, we present new structures that enable the realization of ultra-wideband phase shifters for phased array beamforming, single passband microwave photonic filters, and programmable switchable microwave photonic tunable filters.

WIDEBAND PHASE SHIFTERS FOR OPTICALLY CONTROLLED PHASED ARRAY BEAMFORMING

Optically controlled beamforming techniques for phased array antennas are of significant interest because they can offer wide operating bandwidth, remote antenna feeding, and immunity EMI. Programmable phase shifters are required for beamforming, and a structure for a microwave photonic phase shifter having full 360° phase shift capability with very little RF signal amplitude variation, with operation over a very wide frequency range is shown in Fig. 1. It is based on controlling the amplitude and phase of the optical carrier and the two RF modulation sidebands via the DC bias voltages to a dual-parallel Mach-Zehnder modulator (DPMZM) [4]. The two RF modulation sidebands, which have different amplitudes and phases, beat with the optical carrier at the photodetector to generate the output phase shifted RF signal. The amplitude and phase of this RF output depends on the amplitude and phase of the optical carrier and the differing amplitudes and phases of the two RF modulation sidebands which are controlled by the DPMZM bias voltages. The structure is simple, and has the advantage of having a very convenient control of the RF phase shift which is set by adjusting DC voltages.

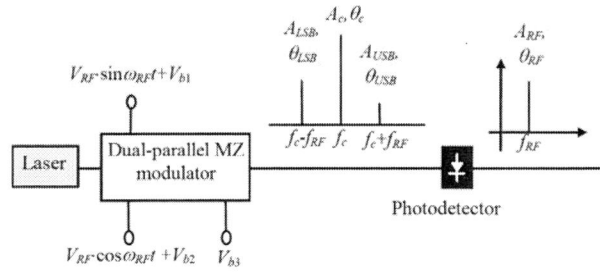

Fig. 1. Microwave photonic RF phase shifter showing the optical and RF spectrums.

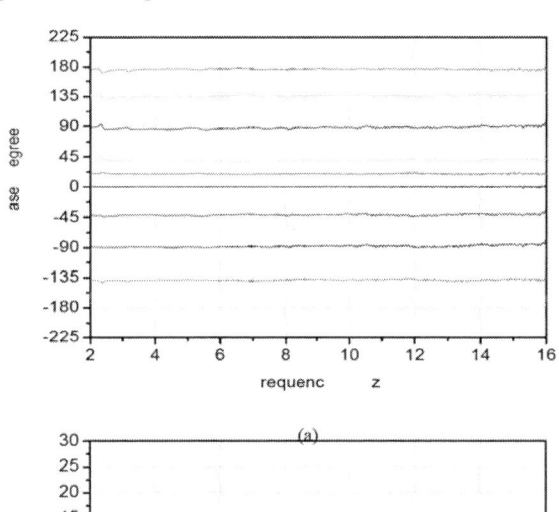

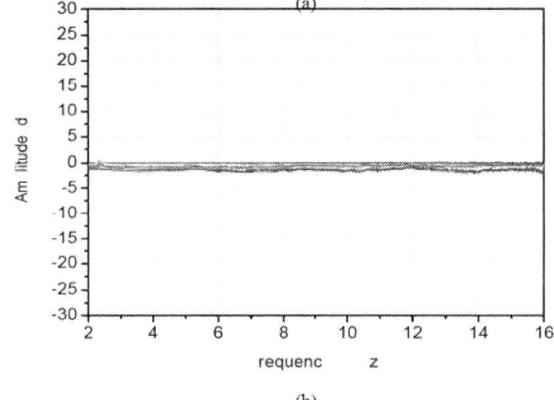

Fig. 2. Measured (a) phase, (b) amplitude responses of the photonic RF phase shifter for different modulator bias voltages.

The measured phase and amplitude response of the phase shifter over a wide frequency range from 2 GHz to 16 GHz for different modulator bias voltages, shown in Fig. 2, demonstrate a flat phase response performance over a wideband 8:1 frequency range. The phase ripple standard deviation is < 2° and there are < 3 dB changes in the RF signal amplitude over the entire broadband frequency range during the 0° to 360° phase shift operation, which is one of the flattest responses reported for a wideband RF phase shifter.

SINGLE PASSBAND MICROWAVE PHOTONIC FILTERS

The ability to realize programmable and reconfigurable signal processors is important for adaptive applications. Moreover, it is essential to have a single passband response. However, discrete time signal processors are intrinsically limited by the presence of multiple harmonic passbands in their frequency response. A technique to overcome this problem in spectrum sliced filters is to use both the dispersion induced carrier suppression effect and the RF decay effect arising from the spectrum slice width, to introduce a low pass filter characteristic whose notches coincide with the expected microwave photonic filter harmonic response frequencies [5].

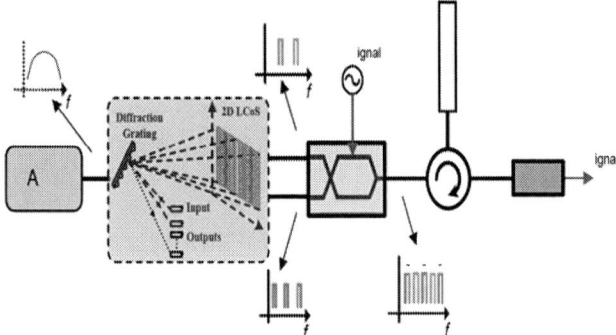

Fig. 3. Structure of single passband microwave photonic filter.

The structure of a true single passband microwave photonic filter is illustrated in Fig. 3. Amplified spontaneous emission (ASE) from an erbium doped fibre amplifier is spectrally sliced using a two-dimensional array of liquid crystal on silicon (LCoS) pixels, which can be programmed to provide optical filters with arbitrary centre wavelengths, bandwidths and attenuations. As each wavelength is manipulated by a specific vertical column through advanced phase modulation techniques of the phase front, arbitrary attenuations of different optical wavelengths are obtained by accordingly steering a portion of light to a discard location instead of one of the two output ports. The LCoS acts as optical slicing filters and also as an optical wavelength selective switch that routes wavelengths to one of the two output ports. Bipolar-coefficients are generated due to the modulation scheme which causes 180° phase difference between the optical fields that come from the two different input ports of the modulator, and the dispersive medium i.e. chirped fibre Bragg grating CFBG gives wavelength dependent group delays.

The optimum design chooses a spectral slice width that simultaneously minimizes the attenuation of the first (desired) passband and which maximizes the suppression of the higher order passbands suppression relative to the first passband. The measured frequency response for a 53-tap microwave photonic filter designed using these principles and using a Gaussian weighted bipolar tap filter coefficient profile shown in Fig. 4(a), is displayed in Fig. 4(b).

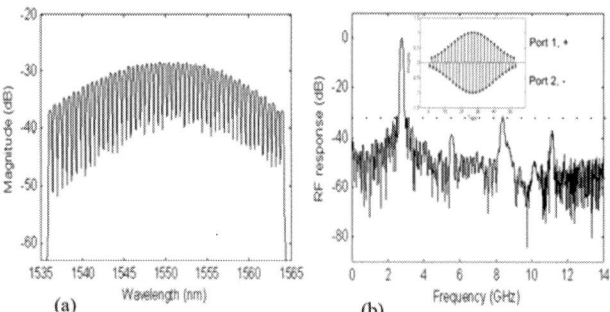

Fig. 4. Microwave photonic filters with (a) Gaussian weighted bipolar tap profile; (b) corresponding measured RF response.

SWITCHABLE MICROWAVE PHOTONIC FILTERS

Multi-function applications require the ability to switch the filtering function between a bandpass filter and a notch filter to provide both channel selection and channel rejection. A structure that enables the realization of a switchable microwave photonic filter that can be switched between a bandpass filter and a stopband notch filter response, using simple and rapid control is shown in Fig. 5 [6]. It is based on a stimulated Brillouin scattering (SBS) technique in conjunction with a dual drive Mach-Zehnder modulator (DDMZM) that processes the sidebands of the RF modulated signal. Switching of the filter function is simply and conveniently obtained by changing the DC bias to the DDMZM. In addition, the centre frequency of the switchable filter can be tuned over a wide frequency range.

The pump light is modulated at RF frequency f_p by a low-biased intensity modulator (IM) in the lower branch which generates a double-sideband suppressed-carrier signal, and which is used to tune the filter frequency by changing the positions of the SBS gain and loss. The filter switching function is achieved by changing the DC bias voltage V_{DC} of the DDMZM to control the relative phase differences between the carrier and the sidebands. The SBS process occurs in the fiber between the RF modulated signal and the pump signal. Each sideband of the pump signal introduces both SBS gain and SBS loss spectra at frequency f_B away from the sideband, where f_B is the Brillouin frequency shift.

In order to realize bandpass filtering, the bias is selected so that the DDMZM operates as a phase modulator. Only the RF signal with the frequency of $f_{RF} = f_p - f_B$ undergoes the SBS effect which consequently breaks the amplitude equality between the sidebands since the lower sideband of the RF modulated signal is significantly attenuated while the upper sideband is amplified, as shown in the right-hand side of Fig. 5(b), and this creates a bandpass response. In order to realize notch filtering, we select the bias so that the phase difference between the upper sideband and the carrier is $\pi/4$ while the phase difference between the lower sideband and carrier is $3\pi/4$, and in addition there is a large amplitude imbalance between the sidebands, as shown in Fig. 5(b). Hence, an RF signal is obtained at the photodetector output at all frequencies except when the input RF frequency is $f_{RF} = f_p - f_B$, and the amplitudes of the two sidebands are equalized by properly choosing the SBS pump power, as shown in the right-hand side of Fig. 5(b), when the beatings between the carrier and the two sidebands are out-of-phase and fully cancel, thus creating a notch. Consequently, a switchable microwave photonic filter at the centre frequency of $f_p - f_B$ is realized, in which the switching function is easily accomplished by changing the DC bias of the DDMZM. The frequency where the switched filtering occurs can also be tuned by changing the drive frequency f_p.

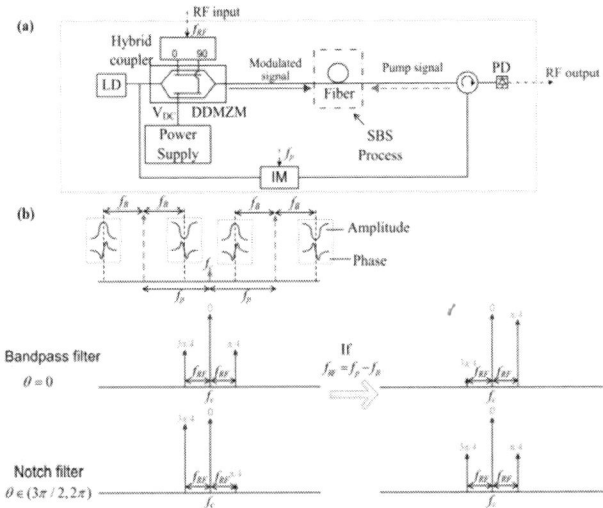

Fig. 5. Operational principle of the switchable filter (a) structure; (b) spectra.

Experimental results are shown in Fig. 6 which demonstrate the ability of this structure to switch between a high-resolution bandpass filter and a high-resolution notch filter at each frequency, with Q values around 400 to 500, and the ability to operate over a wide frequency range from 2-20 GHz. Tuning of the filter was obtained by changing the pump frequency f_p. Switching between high–resolution bandpass and notch filtering with 3-dB widths of 30-40 MHz, at all frequencies from 2 GHz to 20 GHz with shape-invariant response is demonstrated.

CONCLUSION

Recent new methods in photonic signal processing that enable the realization of ultra-wideband phase shifters for phased array beamforming, single passband microwave photonic filters, and programmable switchable microwave photonic tunable filters, have been presented. These processors provide new capabilities for the realisation of high-performance and high-resolution signal processing.

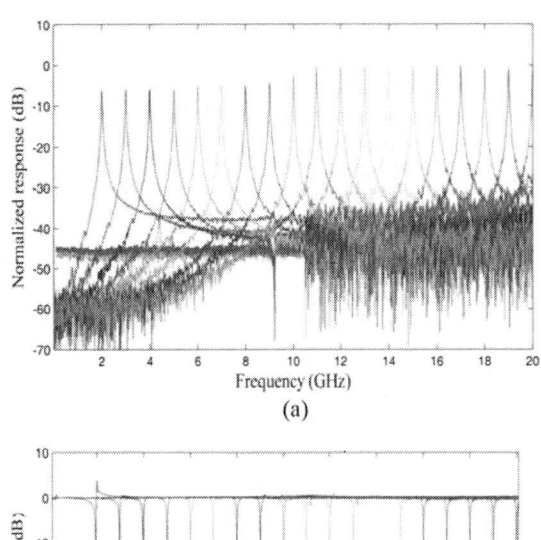

(a)

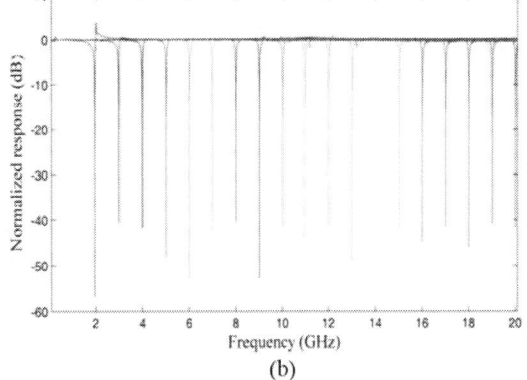

(b)

Fig. 6. Measured normalized frequency response of the switched tunable filter between (a) bandpass response; and (b) notch response.

ACKNOWLEDGMENT

This work was supported by the Australian Research Council. Thanks are extended to T.X. Huang and W. Zhang for their valuable contributions to this work.

REFERENCES

[1] R. A. Minasian, "Photonic signal processing of microwave signals", IEEE Trans. Microwave Theory Tech., vol. 54, pp. 832-846, 2006.

[2] J. Capmany, B. Ortega, and D. Pastor, "A tutorial on microwave photonic filters," J. Lightwave Technol. vol. 24, pp. 201-229, 2006.

[3] J. P. Yao, "Microwave photonics," J. Lightwave Technol., vol. 27, pp. 314-335, 2009.

[4] E. H W Chan, W. Zhang and R. A. Minasian, "Photonic RF phase shifter based on optical carrier and RF modulation sidebands amplitude and phase control", IEEE Journal of Lightwave Technology, Vol. 30, No. 23, pp. 3672-3678, 2012.

[5] T. X. Huang, X. Yi, and R. A. Minasian, "Single passband microwave photonic filter using continuous-time impulse response", Optics Express, vol. 19, No. 7, pp. 6231-6242, 2011.

[6] W. Zhang and R. A. Minasian, "Switchable and tunable microwave photonic Brillouin-based filter", IEEE Photonics Journal, Vol. 4, No. 5, pp. 1443-1455, 2012.

Grating Resonances on Periodic Arrays of Sub-Wavelength Wires and Strips: Historical Narrative and Possible Applications

(Invited Paper)

Alexander I. Nosich[1], *Fellow, IEEE*, Volodymyr O. Byelobrov[1], *Student Member, IEEE*,
Olga V. Shapoval[1], *Student Member, IEEE*, Denys M. Natarov[1], *Student Member, IEEE*,
Tatiana L. Zinenko[1], *Member, IEEE*, Marian Marciniak[2], *Senior Member, IEEE*
and Jiří Čtyroký[3], *Senior Member, IEEE*

[1]Laboratory of Micro and Nano Optics, Institute of Radio-Physics and Electronics NASU, Kharkiv, Ukraine
[2]National Institute of Telecommunications, Szachowa 1, 00-894 Warsaw, Poland
[3]Institute of Photonics and Electronics AS CR, v.v.i., Chaberská 57, 18251 Prague 8, Czech Republic
Tel: 380(57) 720-3782, Fax: 380(57) 315-2105, e-mail: anosich@yahoo.com

Abstract: This paper reviews the history of discovery and the study of the nature of the high-quality natural modes existing on periodic arrays of sub-wavelength scatterers as specific periodically structured open resonators. Here, the arrays can be finite and infinite, and their elements can be dielectric and metallic. These grating modes (G-modes), like any other natural modes, are the "parents" of corresponding resonances in the electromagnetic-wave scattering and absorption. In the scattering cross-sections, they are usually observed as Fano-shape (double-extremum) resonances, while in the absorption they always display conventional Lorentz-shape peaks. Thanks to high tunability, the G-resonances can potentially supplement or even replace the better known surface-plasmon resonances in the design of nanosensors, nanoantennas, and nanosubstrates for surface-enhanced Raman scattering.

Noble-metal nanowires and nanostrips are known to display intensive surface-plasmon (SP) resonances in the visible range if illuminated with the H-polarized light (i.e. polarized orthogonally to the scatterer generatrix). The SP resonance wavelengths depend primarily on the shape of the scatterer cross-section. For instance, a sub-wavelength circular metal wire of dielectric permittivity ε_{met} located in the infinite host medium with $\varepsilon_h > 0$, has a single broad peak in the scattering and absorption cross-sections near the wavelength λ^P where $\mathrm{Re}\,\varepsilon_{met}(\lambda^P) = -\varepsilon_h$. The plane-wave scattering by such a wire can be studied analytically using the separation of variables. The resulting expressions can be further simplified using the small-argument asymptotics of cylindrical functions. This study shows that the wire possesses infinite number of closely spaced double-degenerate SP eigenmodes of azimuth orders $n \neq 1$, appearing as complex poles of the field as a function of the wavelength. However the corresponding resonance peaks overlap because the noble metals are lossy in the visible range, although the largest contribution comes from the dipole terms with $n = \pm 1$. Non-circular wire scattering analysis needs more elaborated techniques such as volume or boundary integral equations. They also reveal shape dependent SP-modes of different types and symmetries.

Thus the wavelengths of SP-resonances are specific for every metal and host medium that makes possible the "sensing" of the host medium refractive index by means of measuring of the SP wavelength. Still the Q-factors of SP-resonances are low, of the order of $\mathrm{Re}\,\varepsilon_{met} / \mathrm{Im}\,\varepsilon_{met} \approx 10$ in the visible range.

Although the pairs or small clusters of coupled metal wires or strips have been well documented, the optical properties of periodic ensembles of them, i.e. chains, arrays and gratings, remain less studied and their interpretation is still controversial. Here is a brief historical narrative of related publications.

The scattering of plane waves by free-standing infinite periodic gratings of *circular cylinders (wires)* made of metals and dielectrics has been extensively studied as a canonical scattering problem since the late 1890s [1-7].

It was in 1979 when K. Ohtaka and H. Numata reported, apparently for the first time, that the scattering of light by infinite one-period grating of thin dielectric cylinders showed narrow total-reflection resonances near specific wavelengths $\lambda_m^R = (d / m)(1 \pm \cos \beta)$, $m = \pm 1, \pm 2, \dots$ depending on the period d and angle of incidence β [8]. However that effect did not attract any specific attention of research community. This is a good example of discovery that was done ahead of time and remained unclaimed for the next 25 years.

Although the G-resonances in the cases of both E- and H-polarization can be noticed in the figures of papers published in the 1980s-2000s (for instance, Figs. 2 and 3 of [7], they became an object of specific investigation only in 2006 – see papers [9,10]. In these papers, the authors used the dipole approximation to show that total reflection resonances appeared just above the Rayleigh anomalies or "passing-off wavelengths," for a grating of thin dielectric wires. By that moment the manufacturing and measuring techniques have reached maturity, and soon the experimental verification of this effect was published in [11].

As known, the scattering resonances of various types are caused by the presence of the "parent" complex-valued poles of the field as a function of the wavelength while the Rayleigh

978-1-4799-0019-0/13 $31.00 © 2013 IEEE

anomalies are associated with the branch points and exist only for the infinite gratings. Therefore one can guess that the reason of overlooking the G-resonances in the most of studies before 2006 was their extreme proximity to the branch-point Rayleigh wavelengths λ_m^R, especially for thin-wire gratings.

Full-wave analysis of both wave-scattering and eigenvalue problems for the dielectric-wire gratings was presented in [12,13] and fully supported earlier findings of [8-10]. Effects of both G-resonances and SP-resonances on infinite gratings of silver wires (in the H-polarization case) have been studied numerically in [13,14]. In [14], new asymptotic expression for the complex-valued frequencies of G-modes has been derived; it has shown that if the wire radius or its dielectric contrast goes to zero then their natural frequencies go to the Rayleigh anomalies and the associated Q-factors rise to infinity. What is also new, in [12,13] it has been discovered that if the grating is made of quantum wires (i.e. can be pumped to display gain) then the G-modes demonstrate ultra-law thresholds of lasing that can be much lower than the threshold of the SP-mode.

It is interesting to check how these optical effects manifest themselves on finite gratings; such a study has been published in [21,22] for finite silver nanowire gratings. It has shown that the G-type resonances become visible in the reflectance and transmittance (see [21] for the definition of these quantities for finite gratings) provided that the number of wires is at least around $N = 10$. If it gets larger, the mode Q-factors tend to their limit values observed for infinite gratings. Important finding related to the case of high-quality G-resonance on a grating of many dozens or hundreds of wires tuned exactly to the wavelength of much lower-quality SP-resonance. In this situation, the presence of the G-mode induces a narrow band of optical transparency cutting through the much wider band of intensive reflection associated with the SP mode.

Flat gratings made of thin *noble-metal strips* have been always attractive in optics as easily manufactured components able to provide wavelength and polarization discrimination. The scattering by strip gratings had been initially studied (see [1,16-18]) assuming their infinite extension, zero thickness, perfect electric conductivity (PEC), and free-space location. Under these rude assumptions, the strip gratings show only the Rayleigh anomalies. In contrast, a gold-strip grating lying on a dielectric substrate displays both SP and G-resonances [19] provided that the substrate is sufficiently thick; and even a PEC-strip grating on a dielectric substrate has no SP-resonances however has strong G-resonances, as found in [20].

The G-resonances on the free-standing *infinite* non-PEC strip gratings were found at first for thin dielectric strips in 1998 [21] and later for silver nanostrips [22] (see Figs. 1,2). In [22], it has been shown analytically that the wavelengths of G-modes tend to λ_m^R if the strip width or thickness gets smaller. Numerical study of both SP and G-resonances on finite gratings of silver strips has been published in [23].

It should be added that the G-resonances have been also studied theoretically and experimentally on the chains and various gratings of 3-D particles – see, for instance, [24-34].

The controversy around the G-resonances on various gratings of metal scatterers consists in the fact that, in the early studies, they were frequently mixed up with more conventional SP resonances. The failure to recognize their specific nature can be seen in the use of plasmon-related terminology such as "radiatively non-decaying plasmons," "supernarrow plasmon resonances," and "plasmon resonances based on diffraction coupling of localized plasmons." This started changing recently when the terms like "collective resonance" of [31-33] and "photonic resonance" of [34] appeared. Today it is known that the G-resonances exist in the scattering by the gratings of both metallic and dielectric elements and in the both of two principal polarizations. Hence it is clear they are caused solely by the periodicity and have nothing common with plasmons.

To highlight the inter-relation between the conventional SP-resonances and G-resonances in the visible-light scattering by periodic noble-metal scatterers, we present some numerical data computed using the convergent algorithm, based on the analytical regularization [21], for an infinite grating of thin silver strips illuminated by a normally incident (i.e. along the *x*-axis in Fig. 1(d)) H-polarized plane wave of the unit amplitude.

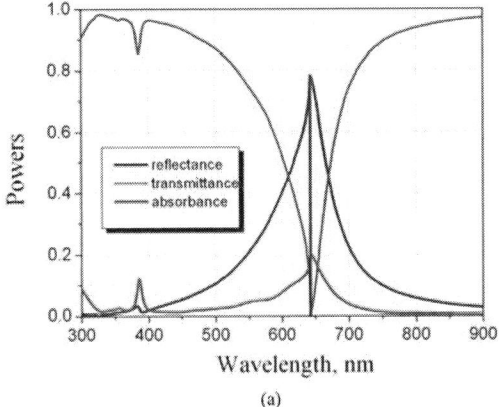

(a)

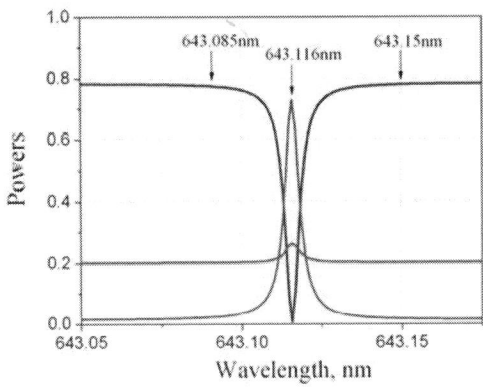

(b)

Fig. 1. Reflectance, transmittance, and absorbance as a function of the wavelength for the scattering of the H-polarized plane wave from the grating of silver nanostrips. The angle of incidence is $\varphi = 0^0$, the strip width is $2w = 150$nm, the strip thickness is $h = 10$nm, and the grating period is $d = 643$nm.

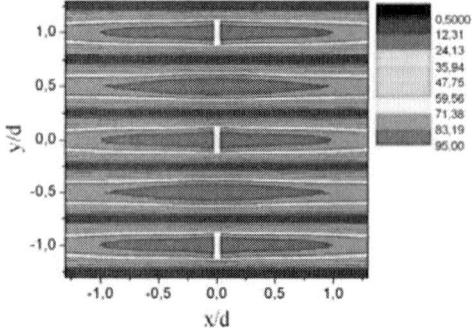

Fig. 2. The near-field pattern on three periods for the scattering of the H-wave from the grating of thin silver nanostrips in the combined SP/G-resonance (λ=643.116nm). Other parameters are the same as for Fig. 1.

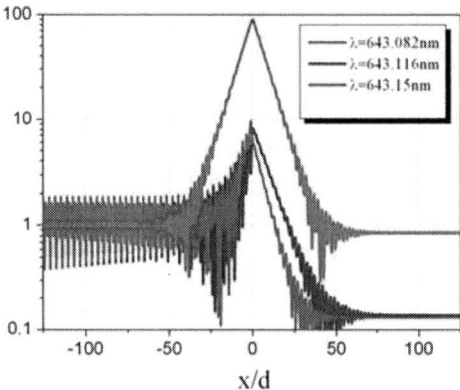

Fig. 3. The profile of the near field magnitude along the line y=0 for the scattering of the H-wave around combined SP/G-resonance at λ^P=643.116nm, λ=643.082nm, and λ=643.15nm. Other parameters are the same as for Fig. 1.

The dispersion of the complex dielectric permittivity of silver has been taken into account using the measured data for the both parts from the classical paper of Johnson and Christy.

The plots of reflectance, transmittance and absorbance as a function of the wavelength are presented in Figs. 1(a) and 1(b). They demonstrate two broad SP-resonances of enhanced reflection and absorption associated with the first and third-order standing-wave modes built on the short-range surface plasmon wave bouncing between the edges of each strip. Besides of them, one can see much sharper G-resonance at the wavelength slightly larger than the period; this resonance has the shape of reflectance minimum, i.e. leads to the optically induced transparency.

According to [22], in the normal-incidence case the normalized frequencies $\kappa = d / \lambda$ of the G-modes on material strip gratings are $\kappa_m^{GH} = m - m^3 (2\pi wh)^2 d^{-4} + O(|\varepsilon| m^4 h^4 d^{-4})$, where $2w$ and h are the strip width and thickness, respectively. If the incident wave length approaches the real part of the m–th natural G-mode wavelength, then the m–th Floquet harmonic amplitude a_m takes a large value not restricted by the power conservation law because it still exponentially decays in the

normal direction as $\operatorname{Re}\kappa_m < m$. In resonance, under the normal incidence, the optical field near the grating is dominated by the intensive standing wave built of two identical Floquet harmonics with numbers $\pm m$. For the plots in Fig. 1, $m = 1$ and $H \approx 2a_1 e^{(k\alpha_1|x|)} \cos(k\beta_1 y) \approx Q_{G1} \exp(-|x/d| Q_{G1}^{-1}) \cos(2\pi y/d)$.

This is fully consistent with the near field patterns observed in Figs. 2 and 3. Note that in the G-resonance very large values of the near field stretch to the distance of some 50 periods on the both sides of the silver-strip grating and has the peak value of 95. This is ~25 times larger than in the SP-resonance whose near-field bright spots are small and stick to the strips [22].

In the case of finite silver-strip gratings, far field scattering patterns demonstrate intensive sidelobes in the plane of grating, explained by the mentioned Floquet modes excitation [23].

We have demonstrated that the grating or lattice resonances on the long periodic chains of finite number of wires or strips may have the Q-factors that are much higher than those usually associated with the plasmon resonances. These Q-factors grow up if the number of the grating periods gets larger. Therefore the grating resonances may serve as a superior alternative to localized surface plasmons for various applications in chemical and biological sensing and SERS. Their interplay depends on the angle of incidence, period of the grating, and the width and thickness of each strip. Choosing these parameters in optimal manner may help design periodic sensors, absorbers, and SERS substrates with improved characteristics.

This work was supported, in part, by the National Academy of Sciences of Ukraine via the State Target Program "Nanotechnologies and Nanomaterials," the European Science Foundation via the travel grants of the Research Networking Programs "Plasmon-Bionanosense" and "Newfocus" to A.I.N., V.O.B., O.V.S. and D.M.N., the International Visegrad Fund via the Ph.D. Scholarships to V.O.B., O.V.S., and D.M.N., and the Rennes-Metropole mobility grant to D.M.N.

REFERENCES

[1] H. Lamb, "On the reflection and transmission of electric waves by a metallic grating," *Proc. London Math. Soc.*, vol. 29, pp. 523-544, 1898.

[2] Lord Rayleigh, "On the dynamical theory of gratings," *Proc. Royal Soc. London*, vol. A-79, pp. 399-416, 1907.

[3] V. Twersky, "On a multiple scattering theory of the finite grating and the Wood anomalies," *J. Appl. Phys.*, vol. 23, pp. 1099-1118, 1952.

[4] A W. K. Pursley, "The transmission of electromagnetic radiation through wire gratings," Tech. Report Project 2351, Engineering Res. Inst., Univ. Michigan Ann Arbor, 1956.

[5] A. Z. Elsherbeni and A. A. Kishk, "Modeling of cylindrical objects by circular dielectric and conducting cylinders," *IEEE Trans. Antennas Propagat.*, vol. 40, pp. 96-99, 1992.

[6] D. Felbacq, G. Tayeb, and D. Maystre, "Scattering by a random set of parallel cylinders," *J. Opt. Soc. Am. A*, vol. 11, pp. 2526–2538, 1994.

[7] K. Yasumoto, H. Toyama, and T. Kushta, "Accurate analysis of 2-D electromagnetic scattering from multilayered periodic arrays of circular cylinders using lattice sums technique," *IEEE Trans. Antennas Propagat.*, vol. 52, pp. 2603-2611, 2004.

[8] K. Ohtaka and H. Numata, "Multiple scattering effects in photon diffraction for an array of cylindrical dielectrics," *Phys. Lett.*, vol. 73-A, no 5-6, pp. 411-413, 1979.

[9] R. Gomez-Medina, M. Laroche, J.J. Saenz, "Extraordinary optical reflection from sub-wavelength cylinder arrays," *Opt. Exp.*, vol. 14, no 9, pp. 3730-3737, 2006.

[10] M. Laroche, S. Albaladejo, R. Gomez-Medina, and J. J. Saenz, "Tuning the optical response of nanocylinder arrays: an analytical study," *Phys. Rev. B*, vol. 74, no. 9, pp. 245422/10, 2006.

[11] P. Ghenuche, G. Vincent, M. Laroche, N. Bardou, R. Haidar, J.-L. Pelouard, and S. Collin, "Optical extinction in a single layer of nanorods," *Phys. Rev. Lett.*, vol. 109, pp. 143903/5, 2012.

[12] V. O. Byelobrov, J. Ctyroky, T.M. Benson, R. Sauleau, A. Altintas, and A.I. Nosich, "Low-threshold lasing modes of infinite periodic chain of quantum wires," *Optics Lett.*, vol. 35, no 21, pp. 3634-3636, 2010.

[13] V.O. Byelobrov, T.M. Benson, and A.I. Nosich, "Binary grating of sub-wavelength silver and quantum wires as a photonic-plasmonic lasing platform with nanoscale elements," *IEEE J. Selected Topics Quant. Electron.*, vol. 18, no 6, pp. 1839-1846, 2012.

[14] D.M. Natarov, V.O. Byelobrov, R. Sauleau, T.M. Benson, and A.I. Nosich, "Periodicity-induced effects in the scattering and absorption of light by infinite and finite gratings of circular silver nanowires," *Opt. Exp.*, vol. 19, pp. 22176-22190, 2011.

[15] D. M. Natarov, R. Sauleau, and A.I. Nosich, "Periodicity-enhanced plasmon resonances in the scattering of light by sparse finite gratings of circular silver nanowires," *IEEE Photonics Techn. Lett.*, vol. 24, no 1, pp. 43-45, 2012.

[16] Z. S. Agranovich, V. A. Marchenko, and V. P. Shestopalov, "Diffraction of a plane electromagnetic wave from plane metallic lattices," *Sov. Phys. Tech. Phys.*, vol. 7, pp. 277–286, 1962.

[17] T. Uchida, T. Noda, and T. Matsunaga, "Spectral domain analysis of electromagnetic wave scattering by an infinite plane metallic grating," *IEEE Trans. Antennas Propagat.*, vol. 35, no 1, pp. 46–52, 1987.

[18] A. Matsushima and T. Itakura, "Singular integral equation approach to plane wave diffraction by an infinite strip grating at oblique incidence," *J. Electromagn. Waves Applicat.*, vol. 4, no 6, pp. 505–519, 1990.

[19] A. Christ, T. Zentgraf, J. Kuhl, S.G. Tikhodeev, *et al.*, "Optical properties of planar metallic photonic crystal structures: experiment and theory," *Phys. Rev. B*, vol. 70, no 12, pp. 125113/15, 2004.

[20] R. Rodriguez-Berral, F. Medina, F. Mesa, and M. García-Vigueras, "Quasi-analytical modeling of transmission/reflection in strip/slit gratings loaded with dielectric slabs,"

IEEE Trans. Microwave Theory Tech., vol. 60, no 3, pp. 405-418, 2012.

[21] T. L. Zinenko, A. I. Nosich, and Y. Okuno, "Plane wave scattering and absorption by resistive-strip and dielectric-strip periodic gratings," *IEEE Trans. Antennas Propag.*, vol. 46, pp. 1498-1505, 1998. [Note: in the H-case, narrow grating resonances were missed because of too coarse grid of computation points.]

[22] T.L. Zinenko, M. Marciniak, and A.I. Nosich, "Accurate analysis of light scattering and absorption by an infinite flat grating of thin silver nanostrips in free space using the method of analytical regularization," *IEEE J. Sel. Topics Quant. Electron.*, vol. 19, no 3, pp. 9000108/8, 2013.

[23] O.V. Shapoval and A.I. Nosich, "Finite gratings of many thin silver nanostrips: optical resonances and role of periodicity," *AIP Advances*, vol. 3, pp. 042120/13, 2013.

[24] K. T. Carron, W. Fluhr, M. Meier, A. Wokaun, and H.W. Lehmann, "Resonances of two-dimensional particle gratings in surface-enhanced Raman scattering," *J. Opt. Soc. Am. B*, vol. 3, no 3, pp. 430-440, 1986.

[25] S. Zou, N. Janel, and G. C. Schatz, "Silver nanoparticle array structures that produce remarkably narrow plasmon lineshapes," *J. Chem. Phys.*, vol. 120, pp. 10871/5, 2004.

[26] E. M. Hicks, *et al.*, "Controlling plasmon line shapes through diffractive coupling in linear arrays of cylindrical nanoparticles fabricated by electron beam lithography," *Nano Lett.*, vol. 5, pp. 1065-1070, 2005.

[27] N. Felidj, *et al.*, "Grating-induced plasmon mode in gold nanoparticle arrays," *J. Chem. Phys.*, vol. 123, pp. 221103/5, 2005.

[28] F. J. G. Garcia de Abajo, "Colloquium: Light scattering by particle and hole arrays," *Rev. Mod. Phys.*, vol. 79, no. 4, pp. 1267-1289, 2007.

[29] Y. Chu, E. Schonbrun, T. Yang, and K.B. Crozier, "Experimental observation of narrow surface plasmon resonances in gold nanoparticle arrays," *Appl. Phys. Lett.*, vol. 93, no 18, pp. 181108/3, 2008.

[30] V. G. Kravets, *et al.*, "Extremely narrow plasmon resonances based on diffraction coupling of localized plasmons in arrays of metallic nanoparticles," *Phys. Rev. Lett.*, vol. 101, pp. 087403/4, 2008.

[31] B. Auguie and W. L. Barnes, "Collective resonances in gold nanoparticle arrays," *Phys. Rev. Lett.*, vol. 101, pp 143902/4, 2008.

[32] V. Giannini, G. Vecchi, and J. Gomez Rivas, "Lighting up multipolar surface plasmon polaritons by collective resonances in arrays of nanoantennas," *Phys. Rev. Lett.*, vol. 105, pp. 266801/4, 2010.

[33] S.R.K. Rodriguez, M.C. Schaafsma, A. Berrier, and J. G. Rivas, "Collective resonances in plasmonic crystals: size matters," *Phys. B*, vol. 407, pp. 4081-4085, 2012.

[34] T.V. Teperik and A. Degiron, "Design strategies to tailor the narrow plasmon-photonic resonances in arrays of metallic nanoparticles," *Phys. Rev. B*, vol. 86, pp. 245425/5, 2012.

Dynamic singular vector speckle fields by the Hurst exponent time analysis

(Invited Paper)

M.S. Soskin, V.I. Vasil'ev,

Institute of Physics, National Academy of Sciences of Ukraine, Kiev, Ukraine

Abstract: The "optical-damage" effect, or random changes of local refractive index in photorefractive media, is unique phenomenon for creation of the generic **dynamic singular vector** speckle fields. They were realized in the photorefractive LiNbO3: Fe crystal under illumination by He-Ne laser beam. This paper is devoted to investigation of **topological dynamics** of the developing singular vector fields and its thorough analysis by the Hurst exponent firstly used in dynamic singular optics.

The "optical-damage" effect, or random changes of local refractive index in photorefractive media [1], is unique phenomenon for creation of generic dynamic vector fields. They were realized in the photorefractive LiNbO3: Fe crystal under illumination of propagating He-Ne laser beam [2]. This paper is devoted to investigation of topological dynamics of developing vector fields and its thorough analysis by the Hurst exponent technique [3] firstly used in singular optics.

Most important kind of scattered speckle fields are the *vector* speckle fields, which consist from the elliptically polarized areas. The anisotropic $LiNbO_3$:Fe crystal is very suitable for their creation (Fig. 1).

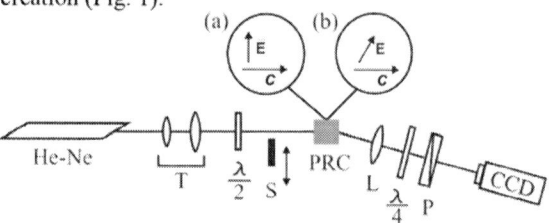

Fig. 1. The set-up for creation and real-time stokes polarimetry of the developing vector speckle fields. T – telescope, $\lambda/2$ - half-wave plate, S – the movable screen, PRC –photorefractive crystal $LiNbO_3$:Fe, L – collimating lens, $\lambda/4$ – quarter-wave plate , P – polarizer, CCD – camera with input screen 5.7x4.6 mm^2 and 576x720 pixels for recording speckle field structure.

The developing vector speckle fields were created sequentially at two shown orientations of the polarization vector in the incident laser beam. This allowed realize full gamut of diffraction gratings and the polarization ellipses accordingly. The **movable screen S** blocks the incident laser beam during optical elements readjustments needed for measurement of the Stokes parameters. This eliminates fully possible distortions of measured dynamics for the developing speckle fields.

Each polarization ellipse can be decomposed on the **right** and **left circular** components with different amplitude and phase values. The stronger component of an ellipse defines its **handedness**. When smaller component becomes zero in the center of OV, ellipse transforms automatically to the circularly polarized **singular C** point. During circumference around some point the phase of polarization ellipses changes on modulo π because it rotates *twice* [4]. Each C point orients the long axes of surrounding ellipses in three possible morphological configurations: **stars**, **monstars** and **lemons.** During circumference around C points ellipses rotate *clockwise* for stars, and *counterclockwise* for monstars and lemons. Therefore, their indices equal $-\frac{1}{2}$ for stars and $+\frac{1}{2}$ for monstars and lemons. It's important that morphological forms of C points don't depend on their handedness. Helicoidal wave front of OVs under C points contains **circularly** polarized points only contrary to well known OVs in scalar wave fields which contain *linearly* polarized points [4].

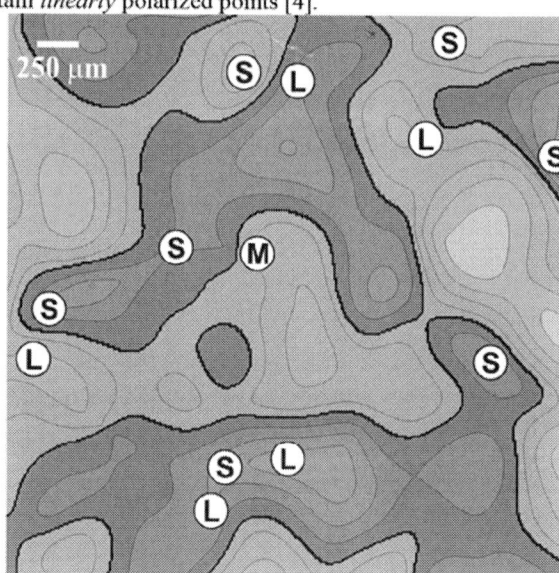

Fig. 2. Fragment of the dynamic singular vector speckle field presented as amplitude b(x,y) of ellipses small axis and field handedness. Left-hand and right-hand areas are marked by dark and light gray areas. C points morphology Stars, Monstars and Lemons are marked by circles with corresponded letters. Bold black lines are the linearly polarized L lines [4]. Thin black lines mark contours of the ellipses b-axis with different values.

978-1-4799-0019-0/13 $31.00 © 2013 IEEE

The typical fragment of the dynamic vector speckle field (Fig. 2) consists from speckled left- and right-polarized areas delimited by linearly polarized L lines [3] and multitude pairs of singular C points with the *same* sign.

Contrary, pairs of underlined optical vortices possess the opposite topological charges ±1. This is possible due to indefinite phase of their zero-amplitude centers (1):

$$\Psi_l(\vec{r}) = u(r, z)\, e^{-ikz}\, e^{-il\theta} \qquad (1)$$

Simultaneously, handedness of wavefronts around zero-amplitude centers in underlying OVs impose direction of rotation of polarization ellipses during circumference around upper C points of arbitrary handedness that is their morphologies: stars for clockwise and monstars/lemons for counterclockwise wavefronts in underlying OVs. This resolves marked above the "**sign paradox**". Direct experiment has confirmed this conclusion. It shows that star and monstar/lemon morphological forms of C points are realized both for *left-* and *right*-polarized C points (Fig. 3).

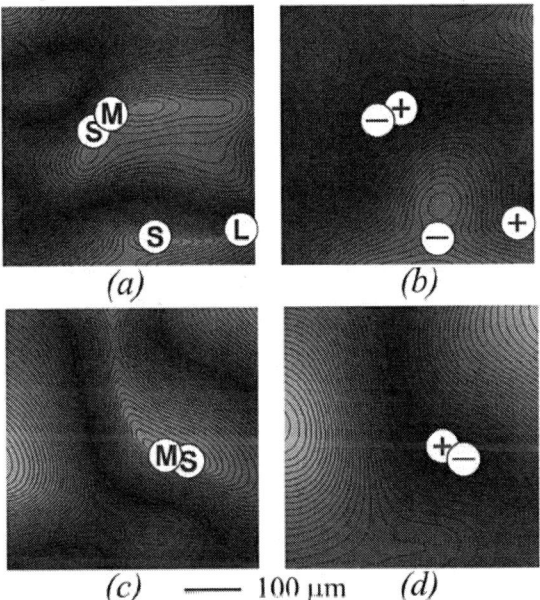

(c) ——— 100 μm *(d)*

Fig. 3. The sign of a C point is defined by handedness of correspondent speckle. *(a)*, *(b)* present left-hand polarized (LHP) area (1620 s), *(c)*, *(d)* present right-hand polarized (RHP) area, (2280 s). C point morphologies are marked as in Fig 2. OVs sign is marked by circle with + (charge +1) and − (charge − 1). Contour maps on *(a)* and *(c)* mark the values of ellipses b-axis. Contour maps on *(b)* and *(d)* mark the distribution of amplitude of OVs underlining corresponding C points.

Contrary, its morphology is defined by sign of the helical wavefront of the underlying OV (LHP for the upper and RHP for the lower C point pairs).

Most interesting is the topological dynamics of the developing vector speckle fields. It was shown that they evolve trough loop and chain trajectories of C point pairs with equal signs and opposite charged underlying OVs. The chain trajectories are in principle not limited in time.

But what is the time correlation during real-time development of chain reactions? English scientist Hurst has discovered new statistical method of normalized scope (R/S) method [3]. Mathematics of the Hurst exponent **H** is next:

$$Z_t = \sum_{i=1}^{t}\left(X_i - \frac{1}{N}\sum_{k=1}^{N} X_k\right)$$

$$S = \sqrt{\frac{1}{N}\sum_{i=1}^{N}\left(X_i - \frac{1}{N}\sum_{i=1}^{N} X_i\right)}$$

$$R = \max(Z_1, Z_2 ... Z_N) - \min(Z_1, Z_2 ... Z_N),$$

where X_i are coordinates of C point in *i*-th instant of time. Dependence of measured values of **Ln(R/S)** as the function of **Ln(N)** is built then. Finally, the Hurst component equals H = tan(*fi*) of experimental direct line (Fig. 5). $1 > H > 0.5$ corresponds to the *correlated* series. Contrary, $0.5 > H > 0$ corresponds to the *uncorrelated* series with long-term switching between high and low values lasting a long time into future. It follows that the angle of measured line to Ln(N) axis can't reach pi/4.

Fig. 4 shows the measured chain reaction with found H values for its links. The found direct lines of Ln(R/S) = *f*(LnN) are presented in Fig. 5.

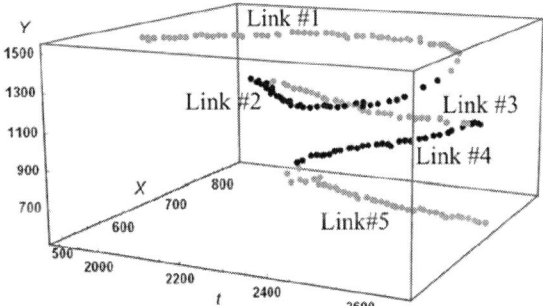

Fig. 4. Example of the chain reaction consisting from 5 links.

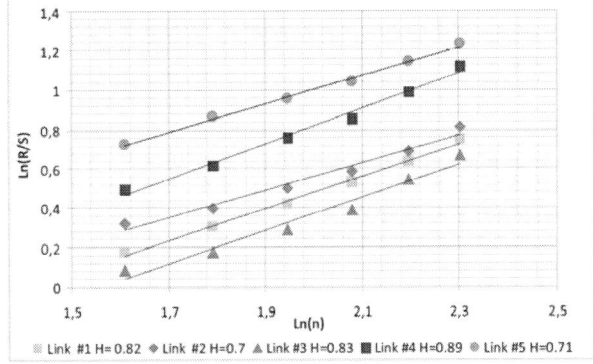

Fig. 5. Example of H values definition from links of measured chain reaction.

978-1-4799-0019-0/13 $31.00 © 2013 IEEE

These very high values are realized due to the high correlated character of optical-damage effect followed by the *random* changes of speckles parameters and distribution of speckle structure in the output of propagating laser beam. To check the connection of evolution of actual speckles structure with movement, transformations and annihilation of nested C points we have measured both of them at the characteristic moments of chain reaction development (Fig. 6).

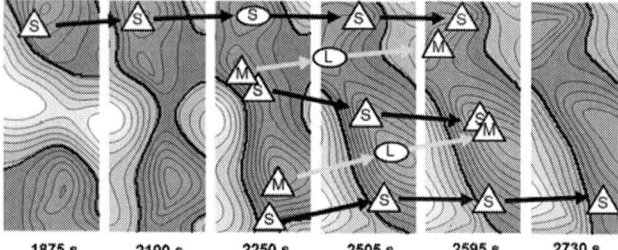

1875 s   2100 s   2250 s   2505 s   2595 s   2730 s

Fig. 6 Correlation of actual speckles shape evolution with trajectories of nested C points. This explains completely their nucleation and annihilation during a chain reaction development. Thin black lines mark contours of the ellipses b-axis with different values. Two elliptic and 16 hyperbolic structures of C points are marked by ovals and triangles correspondingly.

As it is seen, all actual speckles possess *equal* LH handedness as has to be [5]. This explains fully the high values of the Hurst exponents. Due to our knowledge, it is the first case when observed high values of the Hurst exponent where not only fixed but *explained physically!*

It's well known that both elliptic and hyperbolic optical diabolos are connected with detailed shape of speckles, which produce them [5]. Presented trajectories are in full agreement with this topological low. Indeed, hyperbolic C point exists on the slope of above speckle local maximum at 1875 s. Both LH speckles interflow at 2100 s with formation of the additional local maximum. Two pairs of C points nucleate around mean and lower speckles maxima at 2250 s. Meanwhile, upper C point occupies upper maximum and transforms to an elliptic. To 2505 s all five C points move across speckle middle local maximum disappears and nested their two C points both joint upper and lower local maxima. At 2595 s, the upper and lower C points p[air attract. They annihilate finally to 2730 s and remained lowest C point prolongs its movement to next link of the considered chain reaction.

Finally, it has to be fixed that all chain reactions are happen in the speckles with equal handedness in the frames of closed L contours because each ellipse transforms to linear polarization on L contour and any topological connections between LHP and RHP speckles and C points are impossible.

Summary

1. The generic developing vector speckle fields were realized first by the "**optical-damage**" effect in photorefractive crystal LiNbO3: Fe. Their investigation by advanced methods of dynamic singular optics and technique of the **Hurst exponent** allowed establish topological regularities of their development and their time correlation.

2. All **C** point pairs inside **linearly** polarized **L contours** possess **equal** handedness contrary to +1/-1 handedness of underlined OVs pairs due to the **phase uncertainty** for zero-amplitude centers of circularly polarized OVs. **Morphological** form of ellipses arrangement around C point of **any sign** is defined **completely** by **helicity** of **underlying** circularly polarized **OVs**. Namely this "**sign paradox**" guaranties conservation of zero total topological charge during development of vector speckle fields.

3. Measured high values of the **Hurst** exponent **H ≈ 0.7 ÷ 0.89** for the **C** points **chain reactions** witness realization of the <u>**long-term positive autocorrelation**</u> processes in producing them **speckles** during **generic** development of singular **vector speckle fields.**

4. Nucleation, trajectories, topological evolution and annihilation of polarization singularities (C points) are determined **completely** by evolution of generating those speckles.

REFERENCES

[1] S. G. Odoulov, B. I. Sturman, "Photorefraction with the Photovoltaic Charge Transport" in: "Progress in Photorefractive Nonlinear Optics" (Taylor & Frances, London, New York (2002).

[2] Marat S. Soskin and Vasyl I. Vasil'ev, "Space–time design of the tangled C-points and optical vortex chain and loop reactions in paraxial dynamic elliptic speckle fields", J. Opt. A, **15**, 044022 (2013).

[3] H E Hurst, R P Black, Y M Simeika, "Long-term Storage: An Experimental Study" (London, Comlake 1965).

[4] J. F. Nye, "Natural Focusing and Fine Structure of Light" (Bristol: Institute of Physics Publishing, 1999).

[5] Roman I. Egorov and Marat S. Soskin, Isaac Freund, "Experimental optical diabolos", Opt. Lett., **31**, 2048 (2006).

Frequency shifted feedback lasers: from basic physics to applications

(Invited Paper)

L.P. Yatsenko

Institute of Physics, National Academy of Sciences of Ukraine Prospect, Nauki 46, Kiev-39, 03650, Ukraine

Abstract

We discuss basic concepts of the frequency-shifted feedback (FSF) laser and present our approach to modelling the physics of this device. Such apparatus offers potential for length measuring with micrometer accuracy over tens meter distances. We present experimental results, obtained using a Yb^{3+} and Er^{3+} fiber ring laser, that demonstrate the usefulness of such a device.

The remarkable properties of FSF lasers have drawn attention as broad-band light sources and, more recently, as tools for metrology (see [1] and references therein). In essence, an FSF laser comprises a closed optical path, in which light undergoes not only the usual gain of energy from excited states but also, during each cavity round trip, a discrete frequency shift Δ (see Fig. 1).

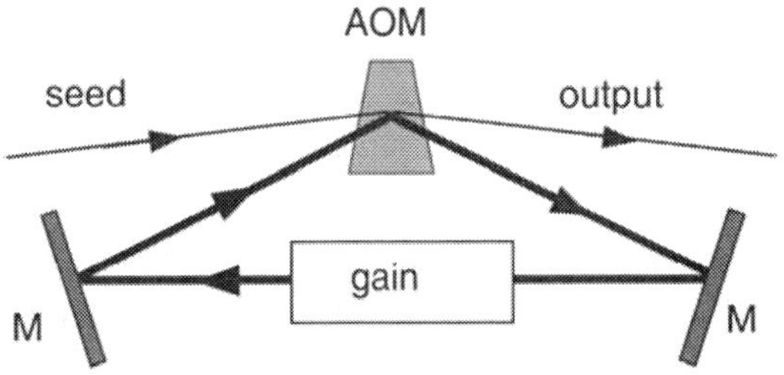

This frequency shift provides FSF lasers with several unusual properties not found with ordinary lasers. Most notably, the spectrum of a FSF laser has no well-defined longitudinal modes because the frequency shifter destroys the constructive interference inside the cavity. This makes a FSF laser a potential source of wideband continuous radiation, a property that has already found many applications.

Another important feature of a FSF laser is that, because of periodic frequency shifting on each round trip, the output radiation can be

Fig. 1. Representative ring layout for frequency-shifted feedback laser, showing acousto-optic modulator (AOM), cavity mirrors (M) and gain medium.

regarded as chirped. After passing through a Michelson interferometer any chirped laser will produce an amplitude modulation at a frequency proportional to the difference in length of the two interferometer arms. This makes the FSF laser a promising tool for optical ranging of distances up to several kilometers [2,3].

In this talk we present an approach to the description of the output from a FSF laser seeded by a continuous-wave (CW) laser [1]. We show the equivalencemof two common viewpoints of the FSF laser output as either a moving combmof equidistant frequencies or as a fixed set of discrete frequencies. We also present theoretical analysis, using correlation functions, of the coherence properties of the utput from a FSF laser seeded simultaneously by an external seed laser and by spontaneous emission (SE). We show that the output of a FSF laser is a cyclostationary process, for which the second order correlation function is not stationary, but periodic. From the fourth order correlation function of the output of a Michelson interferometer we obtain the essential characteristics of the radio-frequency (RF) spectrum, needed or describing the use of the FSF laser for optical-ranging metrology. We show that, ven for a FSF-laser seeded by SE, the RF spectrum comprises a sequence of doublets, hose separation gives directly a measure of the length difference between the nterferometer arms. This doublet structure is a result of the correlation of interference terms of individual components of the cyclostationary stochastic process.

The potential advantages of chirped pulses for very precise measurement of distance, through frequency-domain ranging, has prompted consideration of FSF lasers as sources of interferometer light. We provide a first-principles prediction for the optical ranging signals obtained when using a FSF laser system, seeded by a phase-modulated laser. Such a system has many useful advantages over other alternative laser techniques. The use of a phase-modulated seed to the FSF laser dramatically improves the signal-to-noise ratio, enabling distance measurements with the accuracy expected of optical interferometry. We present an intuitively accessible description of the physics that underlies this dramatic enhancement of optical ranging signals. Unlike a free-running FSF laser, each one of the many equidistant frequency components of the seeded FSF laser spectrum (typically $>10^4$) has a definite amplitude, and a phase which varies with component number and modulation frequency Ω of the seed radiation. Suitable adjustment of Ω gives all components a common phase; the resulting constructive interference enhances the signal by orders of magnitude.

We present experimental characteristics of an Yb^{3+}-doped fiber ring laser operating with FSF through an acousto-optic modulator (AOM) and seeded by both a stationary CW laser and SE. We show the spectrum and output characteristics for operations with several effective gain bandwidths, as established by Fabry-Perot etalons inside the cavity. Observation using a high finesse Fabry-Perot interferometer shows that, as expected from earlier work, although the spectrum of the FSF laser without seeding is continuous, when seeded by a CW-laser the spectrum consists of a comb of discrete modes, each offset from the seed by an integer number of AOM frequency shifts. The experimental results are in excellent quantitative agreement with the theory.

We experimentally demonstrate that a FSF laser, when seeded by a phase modulated CW laser, is a powerful tool for distance measurements to accuracy better than 10 μm and resolution better than 100 μm, for distances of a few meters. In such measurements the unknown distance forms one arm of a Michelson interferometer, in which the intensity of the output signal is modulated at the phase-modulation frequency of the seed. The amplitude of the output-signal modulation exhibits a resonance for every distinct signal delay, i.e. for each distinct distance within the laser spot on the target. The results are in excellent agreement with theoretical predictions for the resolution limit and high signal-to-noise ratio for this new technique. We describe the operation of an all-fiber FSF laser that uses an Er^{3+}-doped active iber as the gain medium. We demonstrate that the resulting system is capable of ast and precise measurements. With the bandwidth limitations of our system we achieved an accuracy better than 0.1 mm.

[1] L. Yatsenko et al., Opt. Comm. **236**, 183 (2004).
[2] K. Nakamura et al., IEEE J. Quantum Electronics **36**, 305 (2000).
[3] L. Yatsenko et al., Opt. Commun., **242**, 581 (2004).

Applications of whispering gallery modes resonances of silica rods and microcapillaries

(Invited Paper)

E. Rivera-Pérez[1,2], A. Díez[1], J. L. Cruz[1], A. Rodríguez-Cobos[2] and M. V. Andrés[1], *Member, IEEE*

[1]Dept. Física Aplicada – ICMUV, Universidad de Valencia, Dr. Moliner 50, 46100 Burjassot, Spain
[2]Instituto de Investigación en Comunicación Óptica, UASLP, Av. Karakorum 1470, 78210 SLP, Mexico

Abstract: Cylindrical microcavities, as solid rods and microcapillaries, have several specific features compared to microspheres, microrings and microtoroids that make them particularly well suited for some applications. Making the optical resonances tunable is essential for many applications, and the cylindrical topology enables the implementation of several tuning techniques. In addition to thermal and strain tuning, slightly tapered rods, Erbium doped microresonators and liquid filled microcapillaries can be used to implement different tuning techniques.

Optical fine tuning of whispering gallery modes (WGM) resonances is demonstrated using Erbium doped microresonators and an auxiliary 980 nm pump diode. Using a slightly tapered silica rod, the reflection produced by high Q WGM resonances can be exploited to make a tunable single-frequency fiber laser. WGM resonances of thin microcapillaries provide a unique solution for refractive index sensor applications, enabling the use of liquids with refractive higher than that of silica and offering an optimum compatibility between optical microcavities and microfluidics.

I. INTRODUCTION

Optical microresonators (MR) based on whispering gallery modes (WGM) resonances exhibit an interesting combination of high quality factor, small mode volume, and good mechanical stability [1,2]. The applications include narrowband filtering [3], nonlinear frequency conversion [4,5], microlasers [6-7], and sensing [8]. Although the optical MR topologies most commonly used are microspheres, microdisks, and microtoroids, some studies using optical MR with cylindrical geometry, as microcylinders and thin-walled microcapillaries, have been reported more recently [9-11].

The capability of tuning optical microcavities is an essential feature for many applications. Thermo-electrical tuning has been demonstrated on a planar microresonator device [12]. The resonances of microcapillaries can be tuned by filling the capillary with liquids of different refractive index [9]. WGM resonances of solid microcylinders can be tuned by using slightly tapered devices [13]. Strain tuning of bottle microresonators [14] has been also demonstrated by applying stress from the extremities.

II. TUNABLE ERBIUM-DOPED SILICA MICRORESONATOR

Here we present tuning of the resonances by using a silica MR doped with erbium and an auxiliary 980 nm optical pump that enables optical control of the resonances.

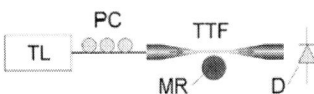

Fig. 1. Diagram of the experimental arrangement: (TL) tunable laser, (PC) polarization controller, (TTF) thin tapered fiber, (MR) microresonator, (D) optical detector.

Figure 1 gives a diagram of the experimental arrangement. The MR is a section of uncoated erbium-doped fiber (Fibercore M12/980/125, 11.6 dB/m absorption at 980 nm), that could be tapered previously down to a certain waist diameter. One pigtail of the MR was fusion spliced to a fibered 980 nm diode laser. The excitation of the WGM resonances is performed using a thin tapered fiber (TTF) [15]. Measuring the transmittance spectrum through the TTF, the resonances will be detected as a narrow dip. A fine adjustment of the separation between the MR and the TTF permits to control the visibility of the resonances. A polarization controller permits to adjust the input polarization for a selective excitation of the TE^z and TM^z WGM of the MR.

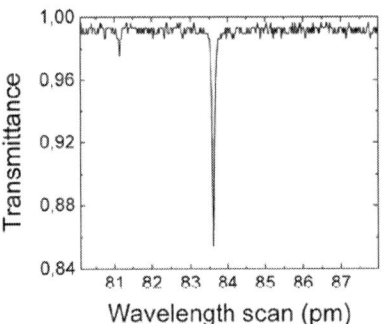

Fig. 2. Transmittance spectrum.

Figure 2 gives an example of the transmittance spectra. The WGM resonances of uniform cylindrical MR made of silica can exhibit Q factors above 10^6, i.e., a linewidth below 1 pm. In Fig. 2, the resonance in centered at 1530 nm and its linewidth is 0.068 pm (Q = 2.2×10^7). When the erbium-doped MR is pumped with 980 nm, the fiber will be heated up and the thermal effects will produce a shift of the resonances. Such a shift will be determined by the material thermal expansion, α_T, and the thermal coefficient of the refractive index. In the case of silica MR, the wavelength shift is estimated to be 10 pm/°C, as in the case on fiber Bragg gratings.

Figure 3 gives the wavelength shift versus the pump power provided by the 980 nm diode laser. A wavelength range of about 50 pm can be scanned with 350 mW of pump power. The tuning is stable and can be repeated without observing hysteresis or short term drift. The linewidth of the resonance is preserved along the 50 pm tuning range.

Moreover, if the wavelength shift is properly calibrated by measuring $\delta\lambda_R/\delta T$, then a precise measurement of the temperature effects in doped fiber or at any critical fiber point of a high power system could be performed. If we assume that the spectral resolution of our measurements is half the linewidth of the resonances, then we estimate a temperature resolution of 3×10^{-3} °C.

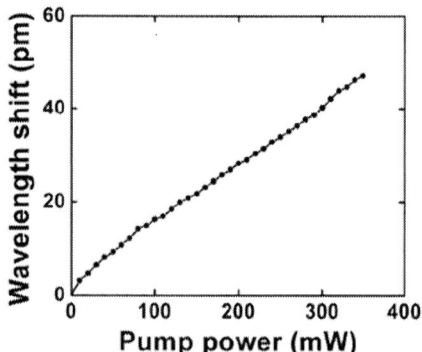

Fig. 3. Wavelength shift versus 980 nm pump power.

III. TUNABLE SINGLE FREQUENCY FIBER LASER

Surface roughness and inhomogeneities cause mode coupling between propagating and counterpropagating WGM. Thus, in an experimental arrangement, as that depicted in Fig. 1, the excitation of a WGM resonance produces a dip in the transmittance spectrum and, simultaneously, a peak in the spectrum of backward-propagating light. The amount of light reflected by a WGM resonance can be large enough for a number of applications. Figure 4 (a) gives the transmission and reflection spectra of a resonance in which the separation between the TTF and the MR was adjusted to have an optimum visibility. A single-frequency erbium–doped fiber laser has been demonstrated using the reflection from a WGM resonance of a microsphere to feedback the cavity [16]. In order to make this type of single-frequency fiber laser tunable, we have investigated the use of a slightly tapered cylindrical MR as the laser cavity feedback element.

Figure 4 (b) gives a diagram of the laser arrangement. A relatively large ring fiber laser cavity was combined with a TTF-MR system. The reflection from the MR was used as the frequency selective element and tuning was achieved by sliding the MR onto the TTF and taking advantage of the radius variation along the slightly tapered MR. Fine tuning of the pump power and the TTF-MR separation is required to achieve single frequency emission. In our experiments, we were able to preserve single frequency emission along a tuning range of

1.16 nm, using the MR of 60 μm diameter, which corresponds to a diameter change of 46 nm. The laser linewidth was below 35 kHz (i.e., 2.8 fm).

(a)

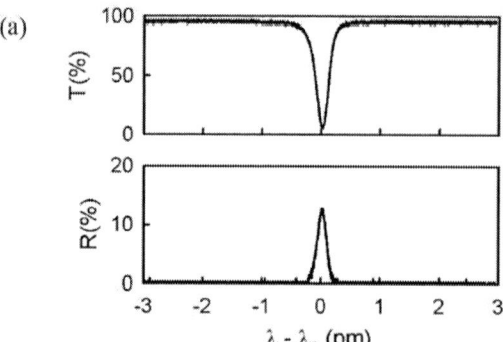

(b)

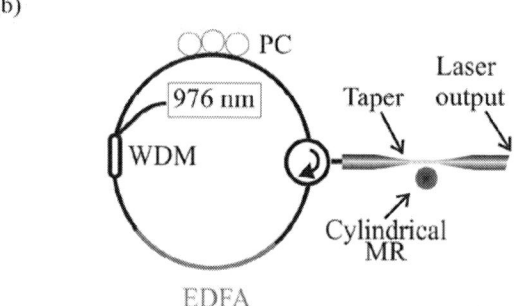

Fig. 4. (a) Transmittance and reflectivity spectra of a given WGM resonance. (b) Diagram of the laser arrangement: (PC) polarization controller, (EDFA) erbium-doped fiber, (WDM) wavelength division multiplexer. The laser output was angle cleaved to prevent unwanted reflections.

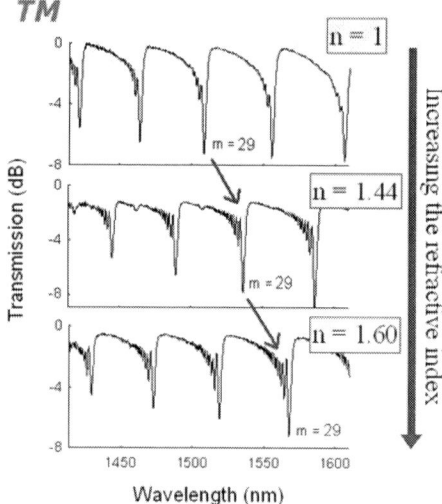

Fig. 5. TM^2 WGM resonances of a capillary with 11 μm diameter and 0.8 μm wall thickness (the resonance with azimuthal order 29 is labeled. The refractive index of the liquid that fills the capillary is indicated in each spectrum.

IV. SENSOR APPLICATIONS OF MICROCAPILLARIES

The development of two different techniques for the preparation of thin silica microcapillaries [18, 9] permits the implementation of WGM refractive index sensors applications with direct compatibility with microfluidics. WGM resonances of microcapillaries with submicrometric wall thickness exhibit high sensitivity to small changes of the refractive index of liquids filling the capillary. Since the external medium is air, the total internal reflection is preserved even when liquids with refractive index higher than that of silica are used (see Fig. 5). Sensitivities as high as 850 nm per refractive index unit have been achieved [10].

V. CONCLUSIONS

WGM resonances of cylindrical microresonators can be tuned more easily than other type of microcavities as microspheres, microrings or microtoroids. Using erbium-doped MR, WGM resonances can be tuned by pumping the MR with an auxiliary diode laser, enabling an all-optical fine tuning of the resonances. We think that the same idea can be exploited the other way round, in order to develop a high resolution temperature sensor for the characterization of thermal effects generated in doped fibers when being used in fiber amplifiers and lasers. Using a slightly tapered microresonator, we have demonstrated also a tunable single frequency fiber laser. Finally, microcapillaries are directly compatible with microfluidics and exhibit high sensitivity as refractive index sensor. Liquids with a refractive index higher than that of silica can be used.

ACKNOWLEDGMENTS

This work was supported by the Ministerio de Economía y Competitividad and the Generalitat Valenciana of Spain (projects TEC 2008-05490 and PROMETEO/2009/077, respectively). E. Rivera-Pérez acknowledges financial support from the Consejo Nacional de Ciencia y Tecnología of Mexico (grant 290618). A. Rodríguez-Cobos acknowledges partial support from PIFI 2012-24MSU0011E-06-02.

REFERENCES

[1] M. L. Gorodetsky, A. A. Savchenkov, and V. S. Ilchenko, "Ultimate Q of optical microsphere resonators," *Opt. Lett.*, vol. 21, pp. 453-455, Apr. 1996.

[2] D. K. Armani, T. J. Kippenberg, S. M. Spillane, and K. J. Vahala, "Ultrahigh-Q toroid microcavity on a chip," *Nature*, vol. 421, pp. 925–928, Feb. 2003.

[3] A. A. Savchenkov, W. Liang, A. B. Matsko, V. S. Ilchenko, D. Seidel, and L. Maleki, "Narrowband tunable photonic notch filter," *Opt. Lett.*, vol. 34, pp. 1318-1320, May 2009.

[4] T. Carmon, and K. J: Vahala, "Visible continuous emission from a silica microphotonic device by the third harmonic generation," *Nature Phys.*, vol. 3, 430–435, Jun. 2007.

[5] P. Del'Haye, A. Schliesser, O. Arcizet, T. Wilken, R. Holzwarth, and T. J. Kippenberg, "Optical frequency comb generation from a monolithic microresonator," *Nature*, vol. 450, pp. 1214-1217, Dec. 2007.

[6] M. Cai, O. Painter, K. J. Vahala, and P. C. Sercel, "Fiber-coupled microsphere laser," *Opt. Lett.*, vol. 25, pp. 1430-1432, Oct. 2000.

[7] M. Bumki, T. J. Kippenberg, K. J. Vahala, "Compact, fiber-compatible, cascaded Raman laser," *Opt. Lett.*, vol. 28, pp. 1507-1509, Sep. 2003.

[8] N. M. Hanumegowda, C. J. Stica, B. C. Patel, I. White and X. Fan, "Refractometric sensors based on microsphere resonators," *Appl. Phys. Lett.*, vol. 87, art. 201107, 2005.

[9] V. Zamora, A. Díez, M. V. Andrés, and B. Gimeno, "Refractometric sensor based on whispering-gallery modes of thin capillaries," *Opt. Express*, vol. 15, pp. 12011-12016, Sep. 2007.

[10] V. Zamora, A. Díez, M. V. Andrés, and B. Gimeno, "Cylindrical optical microcavities: Basic properties and sensor applications," *Photon. Nanostruct. Fundam. Appl.*, vol. 9, pp. 149–158, 2011.

[11] C. P. K. Manchee, V. Zamora, J. W. Silverstone, J. G. C. Veinot, and A. Meldrum, "Refractometric sensing with fluorescent-core microcapillaries," *Opt. Express*, vol. 19, pp. 21540-21551, Oct. 2011.

[12] D. Armani, B. Min, A. Martin, and K. J. Vahala, "Electrical thermo-optic tuning of ultrahigh-Q microtoroid resonators," *Appl. Phys. Lett.*, vol. 85, pp. 5439-5441, Nov. 2004.

[13] T. A. Birks, J. C. Knight and T. E. Dimmick, "High-resolution measurement of the fiber diameter variations using whispering gallery modes and no optical alignment," *IEEE Photon. Technol. Lett.*, vol. 12, pp. 182-184, Feb. 2000.

[14] M. Pöllinger, D. O'Shea, F. Warken, and A. Rauschenbeutel, "Ultrahigh-Q tunable whispering-gallery-mode microresonator," *Phys. Rev. Lett.*, vol. 103, art. 053901, Jul. 2009.

[15] J. C. Knight, G. Cheung, F. Jacques, and T. A. Birks, "Phase-matched excitation of whispering-gallery-mode resonances by a fiber taper," *Opt. Lett.*, vol. 22, pp. 1129-1131, Aug. 1997.

[16] K. Kieu and M. Mansuripur, "Fiber laser using a microsphere resonator as a feedback element," *Opt. Lett.*, vol. 32, pp. 244-246, Aug. 2007.

[17] P. Horak and Wei H. Loh, "On the delayed self-heterodyne interferometric technique for determining the linewidth of fiber lasers," *Opt. Express*, vol. 14, pp. 3923-3928, May 2006.

[18] I. M. White, H. Oveys and X. Fan, "Liquid-core optical ring-resonator sensors," *Opt. Lett.*, vol. 31, No. 9, pp. 1319-1321, May 2006.

Surface acoustic wave modulation of quantum cascade lasers

J. D. Cooper, Z. Ikonić, J. E. Cunningham, P. Harrison, M. Salih, A. G. Davies and E. H. Linfield
School of Electronic & Electrical Engineering,
University of Leeds, LS2 9JT, U.K.

(Invited Paper)

Abstract—In this work, a description is given of a simulation technique employed to model the interaction between surface acoustic waves and ridge-waveguide quantum cascade lasers (QCLs). Firstly, a finite-difference time-domain (FDTD) scheme for modelling acoustic wave propagation in arbitrary semiconductor structures is outlined, and verified by comparison with experimental measurements of the frequency response of surface acoustic wave transmission between interdigitated transmitters and receivers on a bulk crystal. The model is developed further to represent the ridge-waveguide as a prominence above the surface and the active region of the laser is accounted for by a free-charge region buried within the structure. The modulation of this free charge, or carrier concentration by the propagating surface acoustic wave, is then used as an input to a rate equation model of a QCL to show how the gain will be affected. It is this control of the gain through the amplitude of the surface acoustic wave which will allow for modulation of the mid-infrared or terahertz output of the laser and hence its incorporation in many new applications.

I. INTRODUCTION

QUANTUM cascade lasers are n-type unipolar semiconductor heterostructure lasers fabricated from many repeats of an active region unit cell that is itself comprised of several quantum wells. The electron energy levels and lifetimes within an active region are engineered to create a population inversion between two levels which when coupled with a resonant cavity or waveguide can lead to gain (amplification). GaAs-based devices give quantum wells that are one or two hundred meV deep, with a spacing between electron energy levels of a few tens of meV and hence transitions between these states are typically in the mid- or far-infrared (terahertz) regions of the spectrum. These wavelengths have already been shown to be useful for chemical and biological sensing[1]. It is then of interest to achieve precise and continuous dynamical tuning of the laser wavelength, certainly within the limits set by the active transition linewidth. One possibility for this is to employ distributed feedback (DFB) lasers, rather than the conventional end-mirror resonator lasers, where the distributed feedback is provided by gain and refractive index modulation caused by a surface acoustic wave (SAW). The latter are generated by applying alternating voltages to interdigitated metallic fingers deposited on a surface which then form a transducer, producing a mechanical wave through the piezoelectric effect. This wave in turn will modulate the electron density within the active region of the laser, and hence the gain and the refractive index, providing an optical feedback. The operating frequency is thus tuned by changing the SAW frequency, i.e.

the DFB grating period. Modulation of the laser intensity is also possible via the acoustic wave modulation depth, i.e. its power.

The schematic structure of the SAW-modulated QCL is shown in Fig. 1, where the QCL is placed on top of the SAW substrate in between two interdigitated transducers (IDTs), which can either generate or detect a SAW. Physically, the reason for having a second IDT is to check that a working SAW device has been fabricated. This is shown in this schematic for completeness and need not be included within the simulation when modelling the SAW propagation through the QCL ridge. Since the carrier concentration within the QCL active region will be modulated by the electric field of the SAW, generated via the piezoelectric effect, a portion of the SAW energy must move up into the QCL from the substrate for any modulation to be seen. The purpose of the simulation of SAW propagation is to determine how much of the SAW energy moves from the substrate, where it is generated by the transmitting IDT (TxIDT), into the QCL and modulates the active region. Because the structure does not vary in the direction parallel to the SAW wavefront (x_2), it is assumed that there is no variation in this direction therefore making the problem two-dimensional and greatly reducing the computational expense of finding a solution.

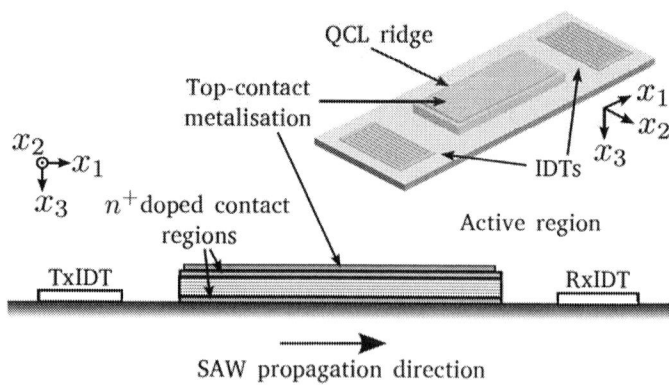

Fig. 1. Schematic diagram of the simulated device showing the transmitting and receiving IDTs (TxIDT/RxIDT) with the QCL ridge in the center. (Inset) Three-dimensional schematic of the device.

II. Theory

The equations of motion which govern acoustic wave propagation within a piezoelectric crystal are:

$$\rho \frac{\partial u_i}{\partial t} = \frac{\partial v_{ij}}{\partial x_j} \qquad \text{for } i, j = 1, 2, 3, \tag{1}$$

where ρ is the density of the material, u_i is displacement inside the material along the three orthogonal axes x_i and v is an auxiliary field whose time differential takes the form:

$$\frac{\partial v_i}{\partial t} = \sigma_i = C_{ij}\epsilon_j + e_{ik}^T \frac{\partial \phi}{\partial x_k}$$

$$\text{for } i, j = 1, \ldots, 6, \quad k = 1, 2, 3, \tag{2}$$

where this auxiliary field is expressed in matrix notation rather than the tensor notation in equation (1), as described in [2]. σ is the stress, C is the elastic constant, ϵ is the strain, and ϕ is the potential. The strain, ϵ may be described in terms of displacement:

$$\begin{aligned}
\epsilon_i &= \frac{\partial u_i}{\partial x_i} \qquad \text{for } i = 1, 2, 3, \\
\epsilon_4 &= \frac{1}{2}\left(\frac{\partial u_3}{\partial x_2} + \frac{\partial u_2}{\partial x_3}\right), \\
\epsilon_5 &= \frac{1}{2}\left(\frac{\partial u_3}{\partial x_1} + \frac{\partial u_1}{\partial x_3}\right), \\
\epsilon_6 &= \frac{1}{2}\left(\frac{\partial u_2}{\partial x_1} + \frac{\partial u_1}{\partial x_2}\right),
\end{aligned} \tag{3}$$

to complete the set of equations. These are solved by discretising onto an interlaced mesh and using an FDTD forward-stepping algorithm, as described in [3], including perfectly-matched-layer boundary conditions to stop artificial reflections from the edges of the simulation domain. The potential within the piezoelectric crystal is assumed to be adiabatic and found by solving Poisson's equation for the strain-induced charge displacement, which is given by,

$$\rho = \frac{\partial}{\partial x_i} e_{ij}\epsilon_j \qquad \text{for } i = 1, 2, 3, \quad j = 1, \ldots, 6. \tag{4}$$

Because surface-bound propagation modes are to be simulated, i.e. SAWs, a surface boundary condition need to be imposed within the simulation domain. This requires that the components of stress which act across the surface boundary therefore vanish, defining the surface as perpendicular to the x_3 axis and at the point $x_3 = 0$,

$$\sigma_3 = \sigma_4 = \sigma_5 = 0 \qquad \text{at } x_3 = 0. \tag{5}$$

In order to excite the SAW, we mimic the potential profile generated by the IDT. Since the simulation domain is assumed invariant in the direction parallel to the wavefront, the potential is assumed to be invariant along the length of each IDT finger. The potential across each finger is also assumed to be constant as the movement of charge is very fast compared to the SAW propagation and is therefore considered adiabatic, as already stated. Mimicking the potential profile around the IDT therefore entails fixing the potential within the solution to Poisson's equation at the surface boundary condition imposed within the simulation domain, such that there are spatially alternating regions of positive and negative potential which oscillate from positive to negative in the time domain. This method of excitation is very general as it not only allows for any IDT structure (which is invariant along the wavefront) to be modelled by changing the positions and dimensions of the areas of fixed potential, but also the frequency and size of the applied field may be altered.

As the acoustic wave equations of motion will support all modes of acoustic wave propagation, the validity of using this method to excite SAWs must be checked. Initially, a simple visual check may be used to ensure that the majority of the acoustic wave energy exists near the surface, therefore ensuring a surface bound mode has been excited.

III. Comparison with Experiment: SAW Propagation in Bulk

The simulated frequency response of a SAW device, consisting just of transmitting and receiving IDTs to generate and detect the SAW, similar to Fig. 1 but *without* the QCL ridge, was compared with experimental results. The results of these simulations are shown in the inset of Fig. 2 for the particular case of IDTs each with 40 finger-pairs. It can be seen that the form of the frequency response from the simulations matches that of the experimental measurements very well, though there is a consistent difference in the losses which may be accounted for by the lack of impedance matching between the co-axial lines and the transducers in the experiment. The full-width at half-maximum (FWHM) of the central peak, as evident in the inset, is a well-known characteristic of the response of IDTs and varies with the number of transducer finger pairs. With this in mind, several SAW devices were fabricated with varying numbers of finger pairs in the transmitting and receiving IDTs, and their measured frequency responses are compared to simulations of the same structures in the main part of Fig. 2. Again there is an excellent agreement between experiment and simulation which serves to validate the approach to simulating SAW propagation described here.

IV. Simulations of SAW Propagation in a QCL

In order to simulate SAW propagation through a QCL ridge, the surface boundary condition must be altered to accommodate ridge structures. This may be done by using a similar surface boundary for vertical surfaces, such that the stresses across the vertical surface vanish, and then combining the two boundary conditions at the corner points by considering which elements of the stress tensor are zero. To simulate the screening effect of the charge within the QCL active region, as well as to determine the magnitude of the modulation of free carriers within the active region, a layer of free electrons is placed within the ridge structure along with a positive charge density to account for the ionised dopant atoms which donate the free charge. Since the quantum wells within the active region run parallel to the SAW propagation direction, and the in-plane mobility of electrons is much higher than in the direction through the quantum well structure, it is assumed that the free charge can only move in the direction parallel to the SAW propagation. It is also assumed that movement of the

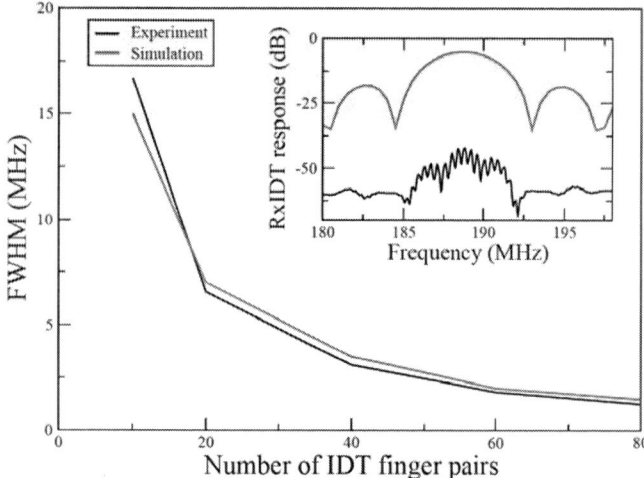

Fig. 2. Comparison of experimental and simulated FWHM of the response of transmitting and receiving IDT pairs with varying numbers of finger pairs. (Inset) Experimental and simulated frequency response for IDTs with 40 finger-pairs.

free charge is very fast, and therefore adiabatic compared to the SAW propagation. The electron density within this layer is then given as:

$$\rho_e = \rho_{dop.} + \rho_{SAW} \quad (6)$$

where $\rho_{dop.}$ is the charge density from the ionised dopant atoms and ρ_{SAW} is the charge density induced from the SAW, as in [4]. The bottom inset of figure 3 shows the resulting modulation in carrier concentration for a small section of a 5 μm-high ridge, showing that the SAW electric field completely depletes some regions of free charge. A self-consistent one-dimensional scattering rate model[5] was used to determine the gain characteristics of the QCL active region from Ref. [6] for a range of free-carrier concentrations, the results of which are shown in figure 3.

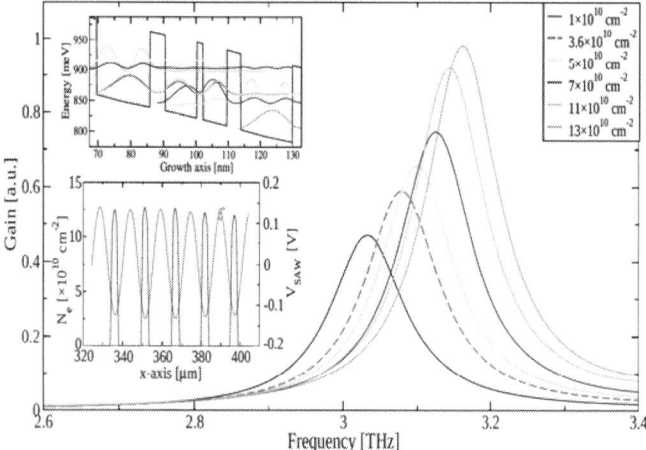

Fig. 3. Variation in the gain of the three-well resonant phonon QCL active region from [6] (top inset) for a range of free-carrier concentrations. Bottom insert: Modulation of free carrier concentration from a SAW propagating through a 5 μm high QCL ridge.

It can be seen from the main graph of Fig. 3 that changes in the carrier density within the active region, of the scale that can be produced by SAWs, are sufficient to modulate the peak gain

of the QCL by $\pm20\%$ around the central non-modulated value, and by even more when considering a particular terahertz frequency, e.g consider the change in gain at 3.15 THz. This is quite sufficient to strongly modulate the output of the QCL in order to transmit digital data or to use in applications with coherent detection to improve signal-to-noise ratios.

V. SUMMARY

In summary, we have presented a method for determining the strength of the modulation of free-carrier concentration within a QCL active region when a SAW is passed through the QCL ridge. The results show that the SAW contains enough energy to fully deplete areas of the QCL active region of carriers, indicating that SAW modulation of QCLs is feasible. The presented method may be exploited further to gain a better insight into the device dimensions required (e.g. QCL ridge height) to give good levels of modulation and may therefore be used as a design tool. Furthermore, the results from this model, translated into the gain and the refractive index modulation, may be used in the coupled wave description of the distributed feedback laser, in order to determine the lasing threshold conditions and the magnitude of frequency modulation achievable.

ACKNOWLEDGMENTS

The authors would like to acknowledge funding from EPSRC (U.K.) and the European Research Council grants 'NOTES' and 'TOSCA' and valuable discussions with D. Indjin.

REFERENCES

[1] W. Withayachumnankul, G. M. Png, X. X. Yin, S. Atakaramians, I. Jones, H. Y. Lin, B. S. Y. Ung, J. Balakrishnan, B. W. H. Ng, B. Ferguson, S. P. Mickan, B. M. Fisher and D. Abbot. T-ray sensing and imaging. *Proceedings of the IEEE*, 95(8):1528–1558, 2007

[2] J. F. Nye and R. B. Lindsay. *Physical Properties of Crystals: Their Representation by Tensors and Matrices.* Clarendon Press, Oxford, 1957.

[3] J.D. Cooper, A. Valavanis, Z. Ikonic, P. Harrison, and J. E. Cunningham. Stable perfectly-matched-layer boundary conditions for finite-difference time-domain simulation of acoustic waves in piezoelectric crystals. *Journal of Computational Physics*, Pending Publication.

[4] Robert L Miller. *Acoustic Charge Transport: Device Technology and Applications.* Artech House, 1992.

[5] D. Indjin, P. Harrison, R. W. Kelsall and Z. Ikonić. Self-consistent scattering theory of transport and output characteristics of quantum cascade lasers. *Journal of Applied Physics*, 91(11):9019–9026, 2002.

[6] H. Luo, S.R. Laframboise, Z.R. Wasilewski, G.C. Aers, H.C. Liu, and J.C. Cao. Terahertz quantum-cascade lasers based on a three-well active module. *Applied Physics Letters*, 90(4):041112–041112–3, 2007.

Recent progress in polarization bistable VCSELs and their applications for all-optical signal processing

H. Kawaguchi, *Member, IEEE*, and T. Katayama. *Member, IEEE*
Nara Institute of Science and Technology, Nara, Japan

(Invited Paper)

Abstract: We summarize recent results of research on polarization bistable vertical-cavity surface-emitting lasers (VCSELs) and their applications for all-optical signal processing. 980-nm polarization bistable VCSELs with an oxide confinement structure were fabricated for low power consumption operation. The threshold current of the VCSEL was reduced to 0.22 mA and all-optical flip-flop operation was demonstrated at a record low bias current of 0.85 mA. We proposed an optical header processing system using a 1.55-μm polarization bistable VCSEL as an all-optical AND gate and a holding element. The payloads of 40-Gb/s NRZ format were send to designated ports depending on the state of a selected bit in the 4-bit header with a 500-Mb/s RZ format. In these operations, the input optical power for the VCSEL was much smaller than the output power. Therefore, the polarization bistable VCSEL is a promising candidate as an all-optical bistable device to realize an all-optical packet router.

I. INTRODUCTION

Optical packet switching is emerging as a potentially important technology in optical networks. The increasing speed in optical telecommunications has focused attention on high-speed optical header processing. A vertical-cavity surface-emitting laser (VCSEL) is a promising candidate as a suitable bistable device to realize small and low power consumption systems due to its easy construction of an array and small operation current. We previously investigated polarization bistability in VCSELs [1] and their application to all-optical signal processing [2]. An ultrafast polarization switching time of 7 ps has been achieved [1]. 4-bit all-optical buffer memory [3] and 20-Gb/s RZ and 40-Gb/s NRZ bit memory [4] have also been demonstrated.

In this presentation, we report polarization bistable characteristics of a 980-nm VCSEL with an oxide confinement structure [5]. We also present a demonstration of all-optical flip-flop operation by injection of optical pulses at a bias current of 0.85 mA. Moreover, we propose and demonstrate a novel optical header recognition system using a 1.55-μm polarization bistable VCSEL [6, 7]. The VCSEL operates as an all-optical AND gate and a holding element. Depending on the state of a selected bit in the 4-bit header, the VCSEL held one of two orthogonal polarization states. The light of the VCSEL output through a polarizer switched and held the output state of an optical switch. This enabled the payload to be sent to the destination determined by the header.

II. ALL-OPTICAL FLIP-FLOP OPERATION WITH LOW POWER CONSUMPTION USING AN OXIDE CURRENT CONFINEMENT STRUCTURE

Oxide current confinement structures have been widely used to achieve lasing of VCSELs at a low threshold current. If an oxide current confinement structure is introduced into the polarization bistable VCSELs, the bias current is expected to drastically reduce. A schematic structure of the 980-nm VCSEL is shown in Fig. 1(a). The VCSEL consists of a three-quantum-well InGaAs active layer, a bottom 27-pair and a top 38-pair $Al_{0.16}GaAs/Al_{0.9}GaAs$ DBR mirrors on an n-type (001) GaAs substrate. The size of a rectangular mesa of the p-DBR is 5.5×6.0 μm, and the sides are aligned to the $[\bar{1}10]$ and $[110]$ crystal orientations. A 50-μm square mesa consists of a current confinement layer, an active layer, and an n-DBR. A 30-nm thick current confinement layer was made of $Al_{0.98}Ga_{0.02}As$, which was oxidized from the sides of the 50-μm square mesa in high-temperature steam at 430°C. When the oxidation was interrupted, the front edge of oxidation was able to be observed with an IR-CCD camera from the top of the device in the chamber. In our oxidation process, the depth of oxidation increased in proportion to the time exposed to high-temperature steam. After observation, oxidation was resumed at the same

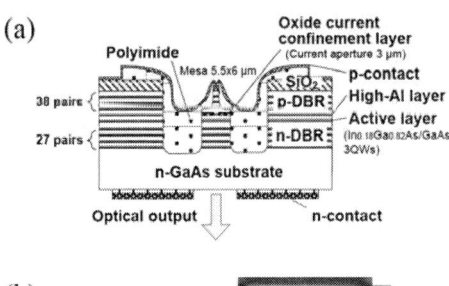

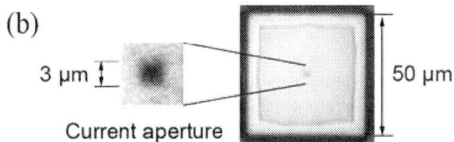

Fig. 1: (a) Schematic cross-section of polarization bistable VCSEL with oxide current confinement structure.
(b) Infrared transparent image of oxide current confinement layer.

rate. We fabricated a current aperture of about 3 μm square at the center of the mesa, as shown in Fig. 1(b).

Voltage and polarization-resolved light output versus current (*V-I* and *L-I*) curves measured at 10°C are shown in Fig. 2. The threshold current was 0.22 mA, which is the lowest value for polarization bistable VCSELs to the best of our knowledge. The average threshold current of the VCSELs in one piece of the VCSEL array was 0.94 mA. The threshold current was dramatically reduced from 5.7 mA to 0.94 mA by introducing the oxide current confinement structure. The orthogonal polarization suppression ratio (OPSR) was 21 dB at 1.15 mA.

Measured polarization-resolved spectra showed that the polarization bistable VCSEL lased in a single longitudinal mode for both polarization modes. The polarization-resolved near field patterns showed that the VCSEL oscillated in the lowest order transverse mode in both polarizations. These results clearly demonstrate that the oxide current confinement mesa structure VCSEL exhibits polarization bistability under a single frequency and the lowest transverse mode operation.

The lasing polarization of the VCSEL with the oxide confinement structure switched to the orthogonal polarization when the orthogonally polarized optical pulse was injected. As shown in Fig. 3, we performed all-optical flip-flop operation using the polarization bistable VCSEL. The VCSEL was operated at 0.85 mA corresponding to a power consumption of 1.7 mW, and 10°C, and its output power was 258 μW. Under the initial condition, the VCSEL oscillated with 90° polarization. When a 0° polarization trigger pulse with a 200-ps duration was injected, the polarization of the VCSEL switched from 90° to 0°. The injected pulse had power of 3.6 μW. The input power of the VCSEL was much smaller than the output power. This means that the polarization switching of the VCSEL could operate with a high optical gain of 22 dB. The lasing polarization remained at 0° for a holding time of 7.3 ns, which was determined by the interval between the set (0°) and

the reset (90°) pulses. When a 90° polarization trigger pulse was injected, the polarization of the VCSEL switched from 0 to 90°. Thus, all-optical flip-flop operation using the polarization bistable VCSEL with the oxide confinement structure was achieved between the two orthogonally polarized modes.

III. ALL-OPTICAL HEADER RECOGNITION AND PACKET SWITCHING SYSTEM

The overall concept of optical header recognition using a polarization bistable VCSEL is shown in Fig. 4. The data signal and the set pulse are 0° polarization and the reset pulse is 90° polarization. All three of these signals are injected into the polarization bistable VCSEL. The injection power of both the data signal and the set pulse are set to less than the polarization switching threshold of the VCSEL. When both the data signal and the set pulse are injected simultaneously, the injection power exceeds the polarization switching threshold and the lasing polarization of the VCSEL is switched from 90° to 0°. This is an optical AND gate operation. Until the reset pulse is injected, the lasing polarization state of the VCSEL is held at 0°. The VCSEL output light through the polarizer with its polarization axis oriented at 0° is input into the control port of an all-optical switch to switch the output port from the original one to another. Due to limitations imposed by our experimental instruments, we used a photodiode and a LiNbO₃ switch (LNSW) instead of an all-optical switch. Thus, depending on the state of a selected bit in the header, 1 or 0, the output of the payload is switched between output ports 0 and 1.

The experimental results are shown in Fig. 5. The data signals consisted of a header and a payload. The format of the header was RZ 4-bit length of 500-Mb/s. The payloads were NRZ pseudorandom-bit-sequence (PRBS) data signals as long as a 2^{11}-1-bit of 40-Gb/s. To recognize the 2nd bit of the headers, the timing of the set pulses was adjusted to the 2nd bit of the headers. In this system, the headers were not extracted from the data signals. Thus the headers were also output from one of the output ports. The peak power of the data pulses, the

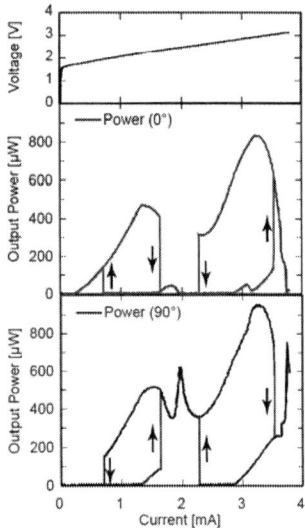

Fig. 2: *V-I* and polarization-resolved *L-I* curves of VCSEL at 10°C under CW operation.

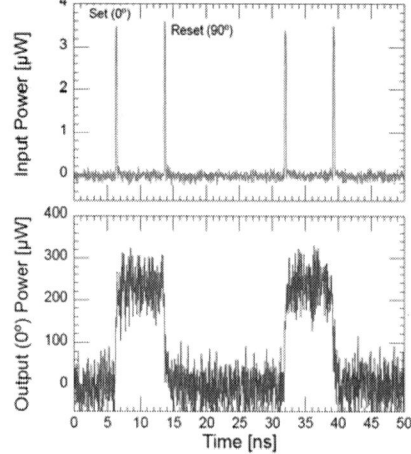

Fig. 3: Experimental demonstration of all-optical flip-flop operation at 0.85 mA bias current and 10°C.

set pulses, and the reset pulses at the VCSEL input were 0.4 µW (10 aJ), 0.6 µW (0.6 fJ), and 2.0 µW (2.0 fJ), respectively, much smaller than the VCSEL output power of 658 µW. The optimum frequency detuning of the data and set pulses was –2.0 GHz from the lasing frequency of the VCSEL and that of the reset pulses was –0.3 GHz. When the 2nd bit of the header was "1," the VCSEL output light through the 0° polarizer (VCSEL output (0°)) was observed. Therefore, when the headers were "1101" and "1111," the payloads were output from port 1 of the LNSW. In contrast, when the headers were "1011" and "1001", the payloads were output from port 0 of the LNSW. The recognition results were correctly reflected to the output of the LNSW in this operation condition. We also successfully demonstrated the packet switching using a 500 Mb/s payload with almost the same operating conditions. These results show that this system can operate for diffent speed and length of payloads with slight change in operating conditions.

IV. CONCLUSION

We introduced an oxide current confinement structure to 980-nm polarization bistable VCSELs. A threshold current as small as 0.22 mA was achieved. We demonstrated all-optical flip-flop operation at a bias current of 0.85 mA, which is a record low to the best of our knowledge. We demonstrated a novel optical header processing system that switches 40-Gb/s NRZ payloads to a designated port depending on the state of a selected bit in the 4-bit header with 500-Mb/s RZ format using a 1.55-µm polarization bistable VCSEL. Toward the realization of an all-optical router, we successfully demonstrated the optical packet switching system using the polarization bistable VCSEL.

ACKNOWLEDGEMENT

This work was supported in part of "Industrial Technology Research Grant Program in 2011" from NEDO, "R&D Promotion Scheme Funding International Joint Research" promoted by NICT, and JSPS KAKENHI Grant Number 24226011 and 23560396.

REFERENCES

[1] H. Kawaguchi, "Bistable Laser Diodes and Their Applications: State of the Art," *IEEE J. Sel. Topics Quantum Electron.*, vol. 3, no. 5, pp. 1254-1270, Oct. 1997.

[2] T. Mori, Y. Sato, and H. Kawaguchi, "Timing Jitter Reduction by All-Optical Signal Regeneration Using a Polarization Bistable VCSEL," *J. Lightw. Technol.*, vol. 26, no. 16, pp. 2946-2953, Aug. 2008.

[3] T. Katayama, T. Ooi, and H. Kawaguchi, "Experimental Demonstration of Multi-Bit Optical Buffer Memory Using 1.55-µm Polarization Bistable Vertical-Cavity Surface-Emitting Lasers," *IEEE Journal of Quantum Electronics*, vol. 45, no. 11, pp. 1495-1504, Nov. 2009

[4] J. Sakaguchi, T. Katayama, and H. Kawaguchi, "All-optical memory operation of 980-nm polarization bistable VCSEL for 20-Gb/s PRBS RZ and 40-Gb/s NRZ data signals," *Opt. Exp.*, vol. 18, no. 12, pp. 12362-12370, May 2010.

[5] T. Katayama, A. Yanai, K. Yukawa, S. Hattori, K. Ikeda, S. Koh, and H. Kawaguchi, "All-Optical Flip-Flop Operation at 1mA Bias Current in Polarization Bistable Vertical-Cavity Surface-Emitting Lasers With an Oxide Confinement Structure," *IEEE Photon. Technol. Lett.*, vol. 23, no. 23, pp. 1811-1813, Dec. 2011.

[6] T. Katayama, T. Okamoto, and H. Kawaguchi, "All-Optical Header Recognition and Packet Switching Using Polarization Bistable VCSEL," *IEEE Photon. Technol. Lett.*, vol. 25, no. 9, pp. 802-805, May 2013.

[7] T. Katayama, T. Okamoto, and H. Kawaguchi, "Optical Packet Switching by All-Optical Header Recognition Using 1.55-µm Polarization Bistable VCSEL," *CLEO/EUROPE 2013*, CI-5.1, Munich, Germany, May 2013.

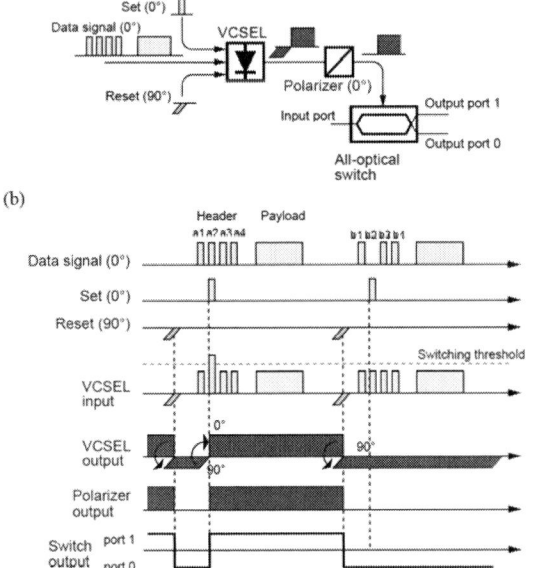

Fig. 4: Operation principle of optical header recognition using 1.55-µm polarization bistable VCSEL. (a) implementation, (b) timing chart

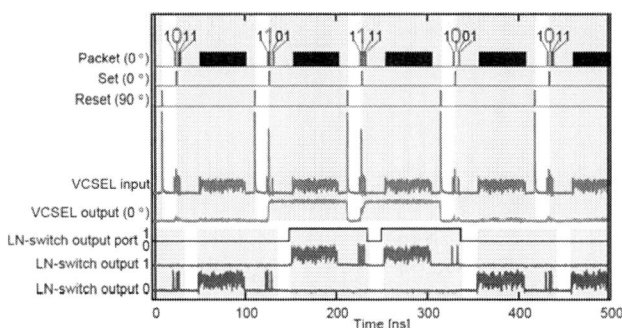

Fig. 5: Experimental results of 500 Mb/s RZ header recognition and 40 Gb/s PRBS NRZ payload packet switching.

Superradiant lasing and collective dynamics of active centers with polarization lifetime exceeding photon lifetime

(Invited Paper)

Vl. V. Kocharovsky[1], E. R. Kocharovskaya[1], V. V. Kocharovsky[1,2]

[1]Institute of Applied Physics, Russian Academy of Science, Nizhny Novgorod, Russia
[2]Department of Physics and Astronomy, Texas A&M University, College Station, TX, USA

Abstract: Analytic theory and qualitative analysis of the threshold conditions, nonlinear formation and spectral features of the pulsed superradiant emission and cooperative radiative behavior of many-particle systems in a low-Q Fabry-Perot cavity with distributed feedback of counter-propagating electromagnetic waves are carried out. Novel superradiant lasers are shown to be promising in the information optics and condensed matter physics.

We summarize the main results of the recent experiments on pulsed superfluorescence and discuss prospects of CW superradiant lasing in various active media as well as outline the expected non-stationary regimes of that lasing, which requires a low-Q cavity. Under typical conditions of superradiance, the latter means that a photon lifetime in a cavity should be much less than a polarization lifetime of an active center. That situation corresponds to the so-called class D lasers [1, 2], which differ in many respects from the standard lasers of A, B, and C classes introduced by Arecchi and Harrison [3].

The cardinal difference is related to the collective dynamics of the active centers, that changes completely the dynamical spectra of their population inversion and cavity field as compared to the standard lasers. As a result, one could easily manage the time profile, the correlation features, and the spectral properties of the radiation generated by the class D lasers. At the same time, that laser radiation imprints collective behavior of the active centers and, thus, may be used as a diagnostic tool for many-particle phenomena in the dense ensembles, for example, for the phase transitions involving a radiative interaction of particles.

We present a general theory of mode superradiant lasing for a low-Q Fabry-Perot cavity combined with a distributed feedback along the cavity. First, we consider an active medium in a general case, when an inhomogeneous broadening is about a homogeneous one, and describe spectra of various modes, laser thresholds (both 1st and 2nd), and necessary conditions of mode superradiance leading to intriguing cooperative dynamics of particles. Then, we focus on the limiting cases of the almost homogeneous broadening and strong inhomogeneous broadening when the lasing modes are the so-called polariton modes with negative energy and electromagnetic modes with positive energy, respectively (see Fig. 1).

In both cases we analyze in detail physical mechanisms responsible for various superradiance regimes and compare possible dynamical spectra of laser radiation and inversion of an active medium. Special attention is paid to the number of lasing modes and their correlated behavior or independent generation, which may be used for managing the time profile of the outgoing radiation.

As an example related to an active medium with almost homogeneous broadening, we consider the exciton recombination lasing in heterostructure traps for Bose–Einstein condensation of dipolar excitons and show that it is possible to achieve lasing on the polariton modes in nowadays experiments of that type [4-6]. If observed, that lasing could raise a series of fundamental problems in laser physics and lead to a breakthrough in studying condensed-matter physics of exciton/polariton ensembles.

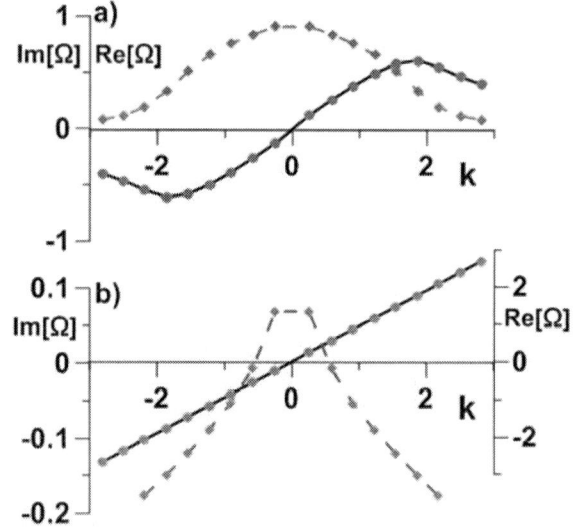

Fig. 1. Typical dimensionless growth rates, $\mathrm{Im}[\Omega]$, (dashed line) and frequency shifts, $\mathrm{Re}[\Omega]$, (solid line) of modes in a 1D distributed feedback laser vs. a wave number shift, k, for a) unstable polariton modes in the case of an active medium with homogeneous broadening and b) unstable electromagnetic modes in the case of an active medium with strong inhomogeneous broadening. The values $\Omega = 0$, $k = 0$ correspond to the Bragg resonance of the distributed feedback turned on a central frequency of the two-level transition in an active medium.

As an example related to an active medium with strong in-homogeneous broadening, we consider the semiconductor lasers based on multilayered heterostructures of sub-monolayer quantum dots with distributed feedback and show that mode superradiant emission under CW pumping is possible in a wide range of laser parameters [7, 8]. Examples of the single-mode and multimode superradiant lasing are given in Figs. 2 and 3. Also to this point, pay attention to the talk by *Kocharovskaya E.R.* et al. *"Class D lasers vs. class B lasers: Dynamical spectra analysis"* at the parallel conference LFNM 2013.

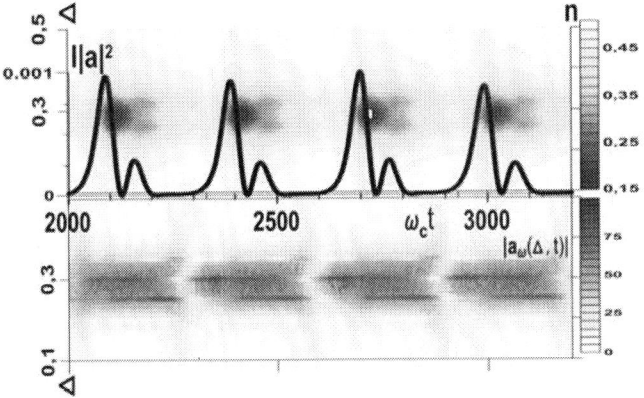

Fig. 2. Single-mode superradiant lasing in a combined distributed feedback Fabry-Perot cavity in the case of an active medium with strong inhomogeneous broadening. Holes burned in the dynamical spectrum of inversion, n, (blue) correspond to peaks in the dynamical spectrum of the field, $|a_\omega(\Delta,t)|$, (green) and peaks in the oscillogram of intensity of the output field, $I|a|^2$, (black thick line). All values are dimensionless, including the frequencies, Δ, of the field harmonics and active centers.

A remarkable regime of such lasing appears if only one or a few modes are superradiant and most of other modes are almost in the steady-state regime. Namely, in addition to the bunches of the superradiance pulses, there exists a periodic series of the coherent pulses with the period being equal to one or one half of the cavity round-trip time (see Fig. 4). That self-mode-locking effect supported by the superradiant lasing does not require any mode-locking technique (passive or active). Another quite general feature of the field dynamical spectra is a specific order of the mode switching on and switching off which is responsible for a quite regular frequency shift in the consecutive mode superradiant pulses.

We outline a systematic description of the above-mentioned and other new nonlinear phenomena inherent to rich dynamics of the class D lasers and discuss their relation to the specific problems of cooperative radiative behavior of many-particle systems in the condensed matter physics.

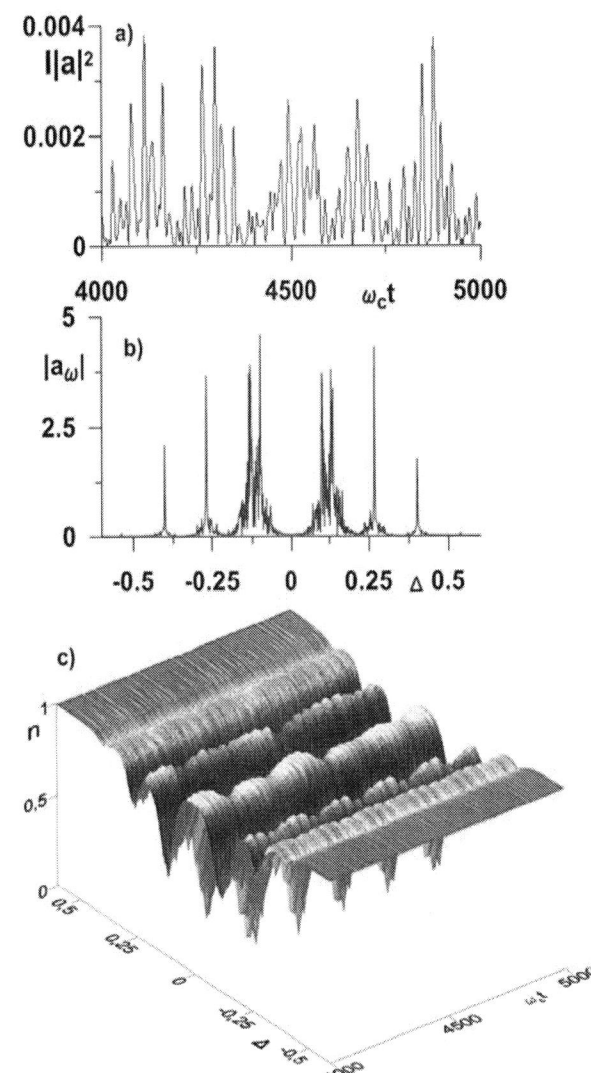

Fig. 3. A multimode superradiant lasing in a combined distributed feedback Fabry-Perot cavity in the case of an active medium with strong inhomogeneous broadening. a) An intensity oscillogram, $I|a|^2$, shows a quasi-periodic sequence of the bunches of the superradiant pulses. b) A spectrum of the output field contains two superradiant modes on the edges of photonic band gap and four almost steady-state modes. The former make much deeper holes in a dynamical spectrum of the inversion, n, (c) than the latter. Again, all values are dimensionless, including the frequencies, Δ, of the field harmonics and active centers.

978-1-4799-0019-0/13 $31.00 © 2013 IEEE   29

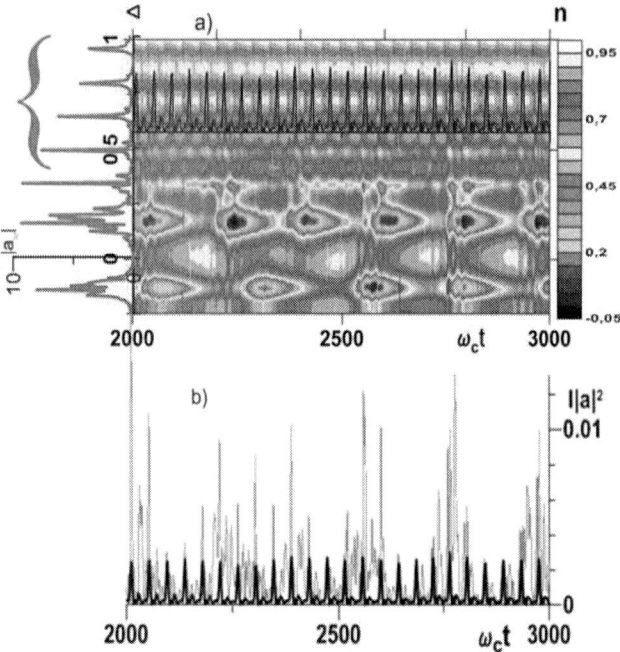

Fig. 4. Spontaneous self-mode-locking in a superradiant laser with a strong inhomogeneous broadening of an active medium in a combined distributed feedback Fabry-Perot cavity. The deep holes of the inversion (blue drops in the bottom part of an inversion dynamical spectrum (a)), caused by the superradiant pulses of the main two modes, play part of a saturable absorber and ensure synchronization of the other four steady-state modes (marked by a brace at the upper left corner). The latter produce an output field (thick black lines on the plots a) and b)) which is periodic and responsible for about 30% of the laser output power, according to an oscillogram (b). Again, all values are dimensionless, including the frequencies, Δ, of the field harmonics and active centers.

ACKNOWLEDGEMENTS

The research was supported by the Russian Foundation for Basic Research (project no. 12-02-00855), the program of Fundamental Research of the Presidium of the RAS no. 21, and the programs of Fundamental Research of the Department of Physical Sciences of RAS nos. III.7 and IV.12.

REFERENCES

[1] Ya .I. Khanin, *Fundamentals of Laser Dynamics*, Cambridge International Science Publishing, 2006, p.10.

[2] A. A. Belyanin, V. V. Kocharovsky, and Vl. V. Kocharovsky, *Quantum Semiclass. Opt. J. Eur. Opt. Soc.*, 1997, vol. B9, pp. 1–44.

[3] F. T. Arecchi and R. G. Harrison, *Instabilities and Chaos in Quantum Optics*, N.Y. etc., Springer Verlag. 1987, p. 47.

[4] P. A. Kalinin, V. V. Kocharovsky, Vl. V. Kocharovsky, *Laser Phys.*, 2010, vol. 20, pp. 2011–2014.

[5] P. A. Kalinin, V. V. Kocharovsky, Vl. V. Kocharovsky, *Radiophys. Quantum Electron.*, 2011, vol. 54, pp. 316–333.

[6] P. A. Kalinin, V. V. Kocharovsky, and Vl. V. Kocharovsky, *Solid State Communications*, 2012, vol. 152 (12), pp. 1008-1011.

[7] Vl. V. Kocharovsky et al., *Proc. of II symposium on coherent optical radiation of semiconductor compounds and structures*, – LPI RAS, Moscow, 2010, pp. 68-77, in Russian.

[8] E. R. Kocharovskaya et al., *Proc. of III symposium on coherent optical radiation of semiconductor compounds and* structures, – LPI RAS, Moscow, 2012, pp. 71-81, in Russian.

Characterization and Fusion Splicing of Single-Mode Photonic Crystal Fibers

(Invited Paper)

K. Borzycki[1], K. Schuster[2],

[1]National Institute of Telecommunications (NIT), Warsaw, Poland
[2]Institute of Photonic Technology (IPHT), Jena, Germany

Abstract: Characterization of two single mode silica photonic crystal fibers strongly doped with GeO_2 included also arc fusion splicing to standard single mode fibers for connections to test instruments, made successfully with machine designed for telecom fibers. Tested with OTDR, both fibers exhibited very strong backscattering and fairly high attenuation: 45-70 dB/km at 1310 nm and 1550 nm. Despite very similar design, their polarization mode dispersion (PMD) differed significantly (1127 ps/km and 118 ps/km) and was apparently produced by different mechanisms, as suggested by dependence on temperature and fiber twist.

INTRODUCTION

This paper presents work on characterization of two silica photonic crystal fibers (PCFs). Splicing and measurements were done with equipment and tools for telecom single mode fibers. PCFs were spliced to pigtails with Corning SMF-28 or similar telecom single mode fiber (SMF) for connections to test instruments. While splicing procedure had to be tailored to each PCF, it was relatively easy in comparison to splicing of un-doped silica PCFs, in particular of suspended-core type.

FIBERS

PCFs were made of fused silica and had small, segmented core doped with variable amount of GeO_2 up to 36% mol, in order to ensure single mode operation and high optical nonlinearity, surrounded by 5 layers of holes. Fibers were fabricated at IPHT Jena, employing MCVD process to make preform for core area, and stack-and-draw method for photonic structure and cladding [1]. Single layer acrylate coating was about 40 μm thick. Fiber cross-sections are shown in Fig. 1, while dimensions are listed in Table 1.

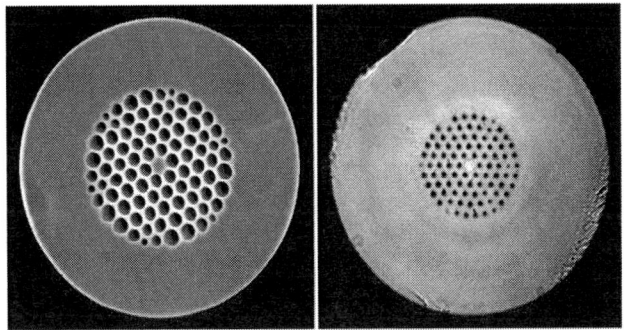

IPHT252b5   IPHT282b4

Fig. 1. Cross-sections of fibers tested.

TABLE I
FIBER DIMENSIONS

Parameter	Unit	IPHT 252b5	IPHT 282b4
Cladding diameter	μm	82.7	124.4
Hole diameter (d)	μm	3.6	0.7
Hole spacing (Λ)	μm	4.2	4.2
Diameter of holey structure	μm	42.8	43.0
Diameter of doped core	μm	0.5/2.0/4.1	1.2/3.9/7.3

Such PCFs exhibit strong nonlinearity and birefringence, useful for supercontinuum generation [1], wavelength conversion by four wave mixing or generation of polarization mode dispersion (PMD).

FUSION SPLICING

As majority of test equipment available was equipped with SMF interfaces, having FC/PC connectors, PCF samples were spliced to SMF pigtails with identical connectors at both ends, 2-3 m long, as shown in Fig. 2.

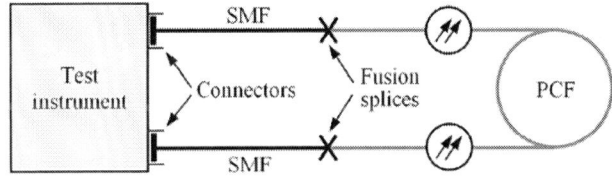

Fig. 2. Connections between PCF sample and instruments.

For measurements with optical time domain reflectometer (OTDR), much longer sections of SMF were connected to both ends of the sample to eliminate dead zone and compare intensity of backscattered radiation from PCF and SMF.

During splicing, insertion loss of sample was monitored with 1558 nm laser source and optical power meter (HP8153A), and procedure optimized to minimize loss at this wavelength. More details and experiments with other fibers are presented in paper [2]. Publication [3] includes additional information on PCF preparations for splicing, factors influencing splice loss, etc. Detailed analysis of arc fusion splicing in general can be found in book [4].

A. Fusion splicing of 125 μm PCF

IPHT 282b4 was a reasonable match to Corning SMF-28 fiber in the pigtails, which had 8.2 μm core with 4% mol. GeO_2

doping and 125 μm cladding [5]. Therefore, splicing procedure was simple, but fusion duration had to be short to minimize collapse of holes. Current and duration for 3 steps of fusion process (pre-fusion, fusion, annealing) were: 9 mA – 3 s, 17 mA – 0.5 s, and 9 mA – 3 s, respectively. Electrode gap was 1 mm. Typical fusion duration for splicing SMF to SMF on the same machine is 1-1.2 s. Fiber stripping and cleaving was performed with standard tools, but use of any solvent for cleaning of PCF was strictly avoided. Otherwise, solvent deeply infiltered holes, producing high and variable loss.

Fibers before and after fusion are shown in Fig. 3. Splice strength was excellent, and typical loss was 1.25 dB at 1558 nm. PCF holes collapsed over length of 300 μm, but light guiding was still provided by doped core. Collapse of holes could be largely avoided by shortening fusion to approx. 0.25 s, but splices made this way were fragile and frequently broke during handling. This compromise between splice strength and damage to photonic structure applies to all PCFs.

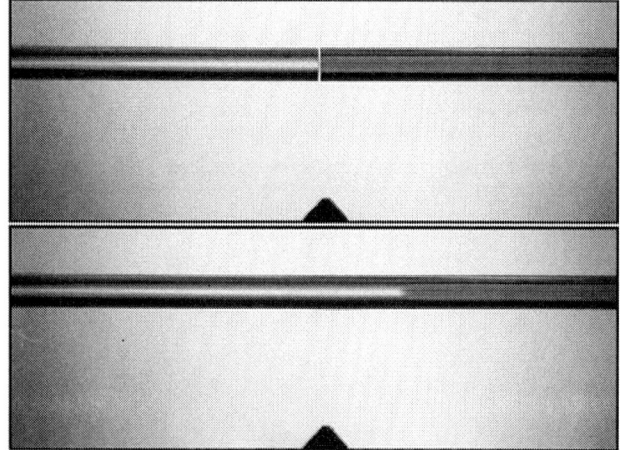

Fig. 3. SMF (left) and IPHT 282b4 (right) photographed before (top) and after arc fusion (bottom). Electrode tip is visible at the bottom of each image.

Finished splice was protected with standard heat-shrinkable sleeve, 60 mm long; no loss change was observed after this operation. Such sleeves later served as holders to apply tensile and torsion forces to PCF samples during mechanical tests.

B. Fusion splicing of 80 μm PCF

Splicing of IPHT 252b5 to SMF was difficult due to considerable mismatch of diameters. Procedure described above failed due to uneven temperature of fibers and forming of gas bubble at the PCF-SMF border. Better method described in [6] included melting fiber tips into ball lenses before fusion and extra post-fusion heating. In our experiments, 1 dB splice loss was achieved with careful control of fiber movements, but splicing was time-consuming. Fiber temperatures were equalized by introducing 200 μm offset of contact point vs. axis of electrodes. Again, splice strength was excellent and no noticeable excitation of higher order modes was observed.

Photos taken during splicing are shown in Fig. 4. More details and images can be found in [2] and [3].

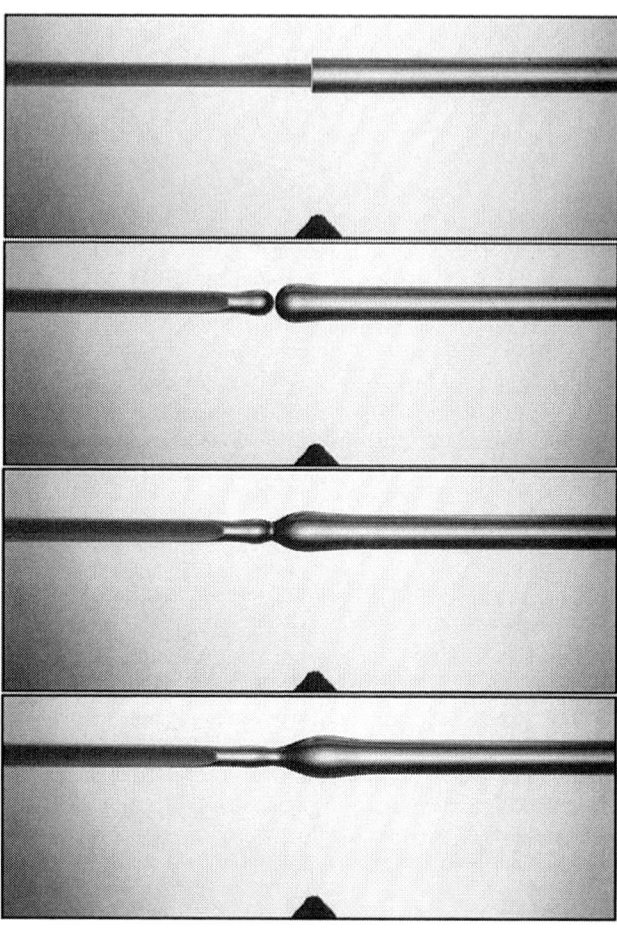

Fig. 4. Splicing of IPHT 252b5 (left) to SMF. Top to down: fibers cleaved, fibers with balls positioned for fusion, fibers after fusion and after final heating.

OPTICAL MEASUREMENTS

A. Measurements with OTDR

OTDR allowed accurate measurements of PCF attenuation even in short (10-15 m) samples, free of errors introduced by variable connector and coupling losses. Performance of OTDR (Tektronix TFP2) was improved by very strong backscattering produced in both PCFs (and other of similar design), ca. 100x stronger than in SMF [7,8], producing considerable upwards shift in trace of PCF with respect to SMF spliced before or after. Example of OTDR trace is shown in Fig. 5.

IPHT 282b4 fiber, a longer sample of which was available, had uniformity comparable to telecom fibers (±0.1 dB). Attenuation of both PCFs was high – see Table 2.

TABLE II
OTDR TEST RESULTS

Parameter	Unit	IPHT 252b5		IPHT 282b4	
Wavelength	nm	1310	1550	1310	1550
PCF attenuation	dB/km	44.5	58.3	69.3	60.2
Trace shift vs. SMF	dB	11.3	10.95	8.7	9.5

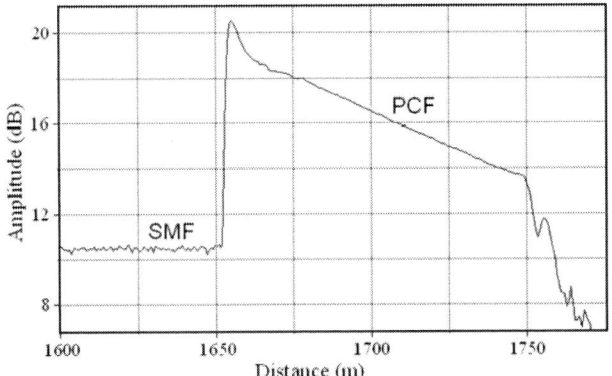

Fig. 5. OTDR trace of IPHT 282b4 fusion spliced to SMF. PCF length: 104 m. Wavelength: 1550 nm.

B. Spectral Loss

PCF samples spliced to SMF pigtails were measured, with loss of two splices and connector added to loss of fiber under test. Test setup included a supercontinuum source (Koheras SuperK Compact) and optical spectrum analyzer (Yokogawa AQ-6315B). Results are presented as Fig. 6 and 7.

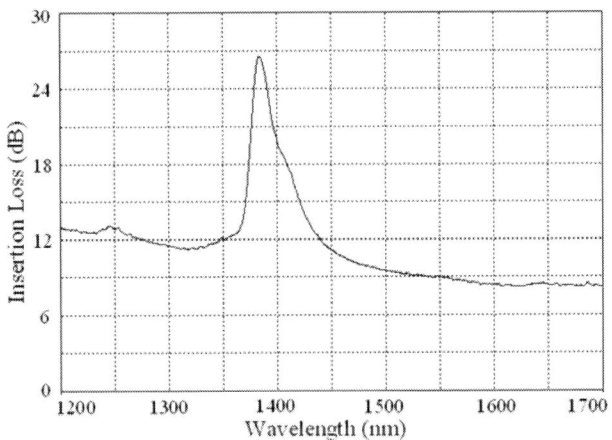

Fig. 6. Loss spectrum: IPHT 282b4, 104 m long.

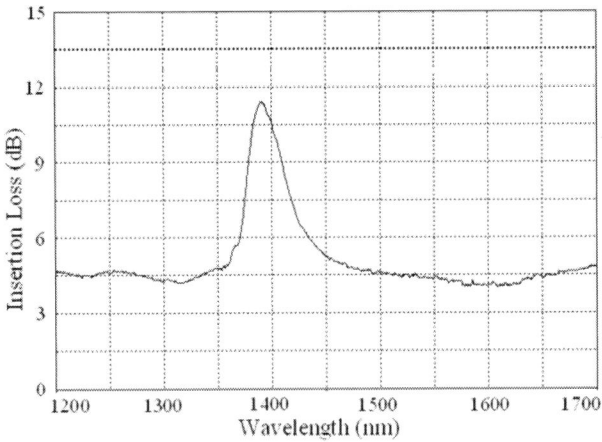

Fig. 7. Loss spectrum: IPHT 252b5, 16.08 m long.

Added attenuation at water peak around 1385 nm was 445 dB/km for IPHT 252b5 and 146 dB/km for IPHT 282b4, corresponding to OH^- content of 10.1 ppm and 3.3 ppm, respectively. These values reflect difficulty with removing traces of water from fiber preforms made using stack-and-draw method.

C. Polarization Mode Dispersion

This parameter was tested using Jones Matrix Eigenanalysis (JME) method, with Adaptif Photonics A2000 PMD analyzer. Spectra of Differential Group Delay (DGD) for both fibers are shown in Fig. 8 and 9.

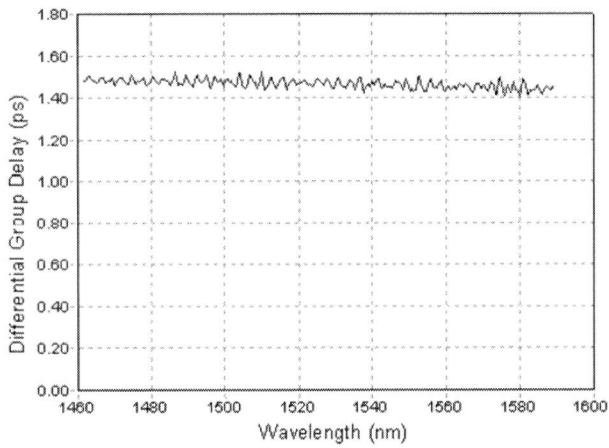

Fig. 8. DGD spectrum of IPHT 282b4. Length: 12.40 m.

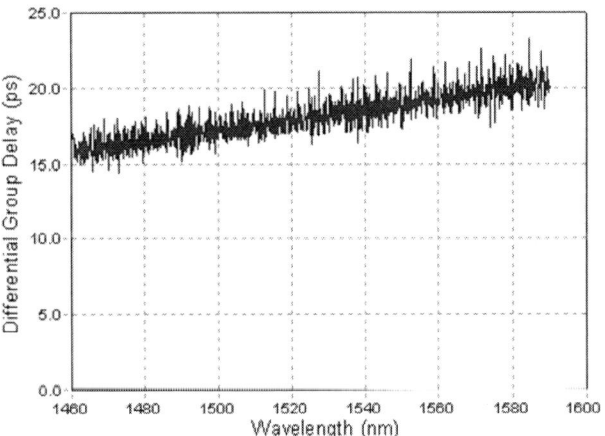

Fig. 9. DGD spectrum of IPHT 252b5. Length: 16.08 m.

PMD coefficients for 1440-1590 nm band were 117.9 ps/km for IPHT 282b4 and 1127 ps/km for IPHT 252b5. Opposite directions of slope of DGD with wavelength suggest that PMD is produced by different mechanism in each fiber tested.

MECHANICAL AND THERMAL TESTS

PMD changed with both twist (Fig. 10) and temperature (Fig. 11 and 12). The 80 μm fiber was less sensitive to twist, but withstood considerably higher twist rate before breaking.

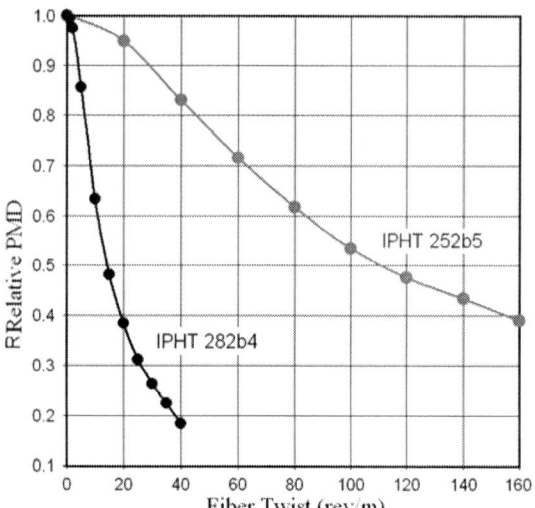

Fig. 10. Variations of PMD with fiber twist.

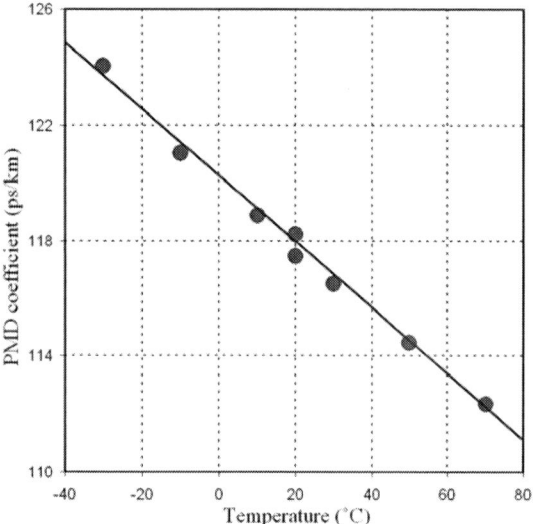

Fig. 11. PMD in IPHT 282b4 as function of temperature.

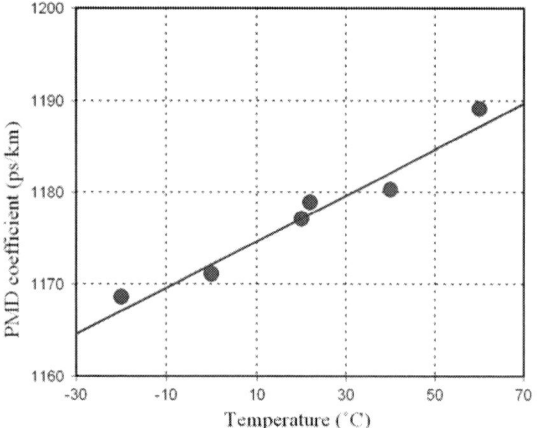

Fig. 12. PMD in IPHT 252b5 as function of temperature.

Temperature coefficients of PMD were $2,13*10^{-4}$/K for IPHT 252b5 fiber (PMD measured in 1540-1560 nm band), and $-9,7*10^{-4}$/K for IPHT 282b4 (1490-1590 nm). PMD in the latter PCF is produced by strain resulting from differential shrinking of its parts after drawing.

CONCLUSIONS

Results of attenuation and PMD measurements agree with those published previously [1] for the 80 μm fiber. Despite similar design of core and photonic structure, as well as identical fabrication method, two PCFs tested exhibited notably different properties, especially in case of polarization dispersion and its sensitivity to external factors. PCF manufacturing, despite all progress made, is still in need of improvement with respect to consistency of fibers.

ACKNOWLEDGMENT

Research at NIT presented in this paper was carried out within COST Action 299 "FIDES" and financially supported by Polish Ministry of Science and Higher Education as special research project COST/39/2007.

REFERENCES

[1] K. Schuster, et al., "Microstructured fibers with highly nonlinear materials". *Opt. Quant. Electron.*, vol. 39, pp. 1057-1069, 2007.

[2] K. Borzycki, J. Kobelke, K. Schuster, and J. Wójcik, "Arc fusion splicing of photonic crystal fibers to standard single mode fibers", *Proc. SPIE* 7714-38, 2010.

[3] K. Borzycki, and K. Schuster, "Arc fusion splicing of photonic crystal fibres", *Photonic Crystal Fibres – Book 1*, ISBN 978-953-308-135-9, pp. 175-200, Intech Publishing, Rijeka, 2012.

[4] A. Yablon, *Optical Fiber Fusion Splicing*, Springer, ISBN 3-540-23104-8, Berlin - Heidelberg - New York, 2005.

[5] Corning Inc., *Corning SMF-28e Optical Fiber Product Information*, PI1344, 2008.

[6] Y. Wang, et al., "Splicing Ge-doped photonic crystal fibers using commercial fusion splicer with default discharge parameters". Optics Express, vol. 16, no. 10, pp. 7258-7263, 2008.

[7] K. Borzycki, J. Kobelke, P. Mergo, and K. Schuster, "Characterization of photonic crystal fibers with OTDR", *Proc. ICTON-2011*, paper We.B4.5, Stockholm, Sweden, June 26-30, 2011.

[8] K. Borzycki, J. Kobelke, K. Schuster, and J. Wójcik, "Optical, thermal and mechanical characterization of photonic crystal fibers: results and comparisons", *Proc. SPIE* 7714-31, 2010.

Application of multipass modes for high quality generation in stable open resonators

(Invited Paper)

V.G. Niziev

Institute on Laser and Information Technologies, RAS, Shatura, Moscow Region, 140700, Russia
Phone: +7 49645 25995, Fax: +7 49645 22532, E-mail: niziev@yahoo.com

Absrtact: Newly developed numerical simulation of axially symmetrical resonators was employed in study of transverse mode formation. Single pass wave and multipass ray modes were obtained. Possibilities to obtain quality radiation from power lasers are discussed.

The resonator numerical model was based on the Fox and Li iteration approach [1], and applied for axially symmetric resonator. The analytical solution of the diffraction problem for a narrow ring slit of radius r0 was used, see (1) in square brackets [2]. Reflection of an incident wave with given amplitude-phase distribution E0(r0) from a mirror is regarded as a Green problem [3]:

$$E(L,\theta) \approx \int_0^{r_m} E_0(r_0) \left[2\pi r_0 ik \frac{e^{ikL}}{L} e^{ik\frac{r_0^2}{2L}} J_0(kr_0\theta) \right] \exp\left(-ik\frac{r_0^2}{R} \right) dr_0 \tag{1}$$

Here θ is the polar angle, k is the wave vector, J_0 is the Bessel function for homogeneous polarization, r_m is mirror size and R is its curvature radius. For azimuthal (or radial) polarization J_1 must be used instead of J_0. The consequent iteration step uses the incident field specified on the resonator mirror and calculated at the previous step. The calculations start from any initial wave front and finish after (quasi) stationary state is established. In our study we considered only stable resonators.

In studying field amplitude oscillations at the mirror for an empty resonator we faced with a highly principal problem of relationship between the wave and geometric optics in the description of open resonators. Using wave approach of resonator description we've observed as the single-pass principal Laguerre-Gaussian mode TEM00 formed after multiple reflections, so multipass modes of geometric paraxial resonances [4].

The parameters of resonators for resonance oscillations are in conformity with the condition of paraxial resonance derived in the context of resonator describing by the methods of geometric optics [8]:

$$g_1 \cdot g_2 = \frac{1+\cos\theta}{2}; \quad \theta = 2\pi \frac{K}{N}; \quad 0 \le K \le N/2 \tag{2}$$

N is the number of resonator round-trips required to form a closed ray trajectory. $g_i = 1 - L/R_i$; $i = 1,2$ are the parameters of stability diagram of open resonators.

In all numerical experiments of stable resonators with Fresnel number more than one the quasi stationary state is established with different kind of oscillations depending on resonator parameters. The reason for this behaviour of the field is that the single-pass higher-order Laguerre-Gaussian modes cannot meet competition with the "multipass modes" of paraxial resonances with traces situated in the radial plane. With the similar typical size of the field along the mirror radius single-pass Laguerre-Gaussian modes exhibit higher diffraction losses at the mirror edges than "multipass modes". The exception is the principal TEM00 mode, in the scope of which a consideration in terms of the beam is not applicable.

Fig.1 illustrates what happens if we increase the mirror radius keeping other parameters the same. The time of the principal mode establishment depends on the tube radius (a,b). Finally, in Fig.1c, we obtain a quasi-stationary situation of "multimode generation". In this case it is the regular oscillations with a period of three bounces, because the chosen resonator parameters correspond to the conditions of three-pass paraxial resonance at $g_1 = g_2 = 0.5$ [4]. If the resonator parameters are out of paraxial resonance the oscillations of "quasi stationary generation" are irregular.

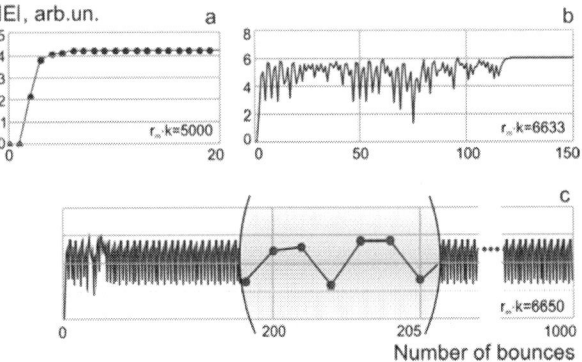

Fig.1. Evolution of the field amplitude at the mirror center in the laser for different mirror radii $r_m k$=5000, 6633, 6650 for a, b, c correspondingly. Parameters: $L\cdot k$=2543000, (k is wave vector), $R_1 = R_2 = 2L$.

So for Fresnel number being considerably in excess of unity the picture is drastically changed. With all the examined parameters of the stable resonators $0 < g_1, g_2 < 1$ the field does not display stationary distribution. The steady-state mode reveals the quasi-stationary oscillations of the field (Fig.2). In the stable resonator, at F>1 the multipass pattern of beam propagation has a competitive advantage over the classical single-pass modes. With the similar typical size of the field

978-1-4799-0019-0/13 $31.00 © 2013 IEEE

along the mirror radius, the single-pass Laguerre-Gaussian modes show higher diffraction loss at the mirror edges than the multipass modes. The loss of single-pass modes at the mirror edges is the same for each pass. The field radius of a multipass mode is cyclically changed. So the averaged loss of the multipass mode is lower.

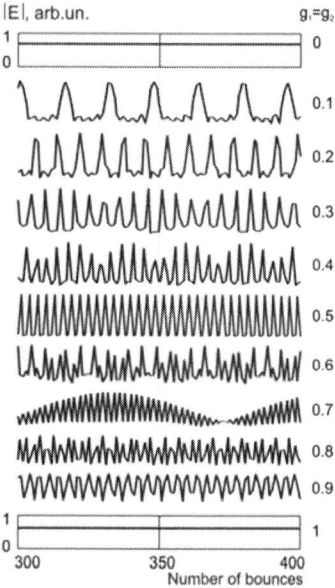

Fig.2. Field amplitude on the mirror under steady-state stationary (or quasi-stationary) generation in the stable resonators at different stability parameters $g_1=g_2$.

A quite different pattern of quasi-stationary oscillations of the field in Fig.2 has the following explanation. With the parameters corresponding to paraxial resonance, the oscillations are regular and related to generation of a multipass mode with the closed ray traces. Nearby the resonances beats are observed. Away from the resonances the mode of oscillations is nearly chaotic. As Fig. 2 presents the results for $g_1=g_2$ varying from 0 to 1 with a constant increment of 0.1, the coincidence with the parameters of paraxial resonance occurred only at $g_1=g_2=0.5$, formula (2). In other cases the calculation points did not fall within the resonances. At the resonator parameters corresponding to the stability boundary in the steady-state regime of generation no oscillations were observed. The results obtained do not contradict to the classical work of Fox and Li [1] as they have only made the calculations for the resonators being at the stability boundary. It is evident that in the resonator with plane mirrors the closed trajectory of the multipass modes cannot be realized; this is the reason for the establishment of stationary distribution.

Other paraxial resonances, Fig.3, can be also obtained in exact accordance with formula (2) obtained in [4].

In principle multipass modes can be applied for generation of high quality radiation in wide aperture lasers (Fresnel number is more than one). At first the parameters of resonator corresponding to some paraxial resonance should be chosen. Let's take, for example N=3. In this case the generation can develop along three main trajectories: azimuthal angle multipass mode, ecliptic and non-ecliptic modes, Fig.4. For high quality radiation the generation along one of them should be provided. Others must be depressed by specially inserted resonator losses.

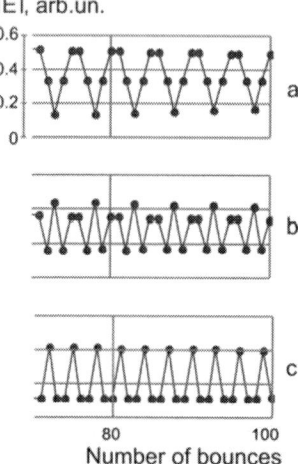

Fig.3. Examples of paraxial resonance realization in a symmetric resonator. The field amplitude is as a function of the number of bounces. The points on the curve correspond to consequent reflections from the mirror. (a) – seven-pass resonance $g_1=g_2= 0.222$; (b) – five-pass resonance $g_1=g_2= 0.309$; (c) – five-pass resonance $g_1=g_2= 0.809$.

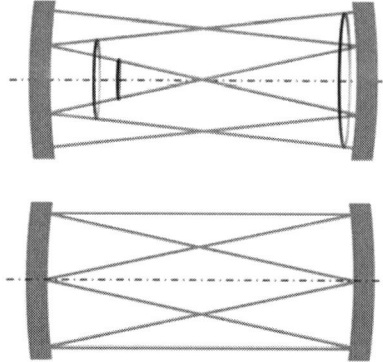

Fig.4. Rays trajectories at three pass paraxial resonance $g_1=g_2=0.5$. Ecliptic (above). Closed non-ecliptic trajectory (below).

The most convenient way to depress azimuthal angle multipass modes is to use azimuthally (radially) polarized radiation. Such radiation has ring form of field distribution. Typically they use diffraction mirror with ring grooves and high local polarization selectivity as one resonator mirror to obtain such radiation. The resonator design to suppress ecliptic trajectory and output radiation from non-ecliptic trajectory is shown in Fig.4.

The results of calculations of the field amplitude and phase distributions on the right mirror (Fig.5), are presented in Fig.6.

The amplitude-phase distribution is similar to Laguerre Gaussian mode TEM_{11*} with azimuthal (radial) polarization

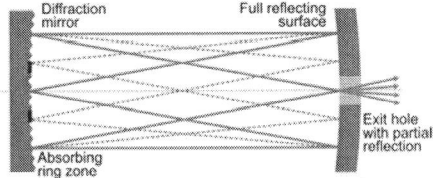

Fig.5. The resonator design with depressing ecliptic trajectory of generation (dash lines). The output of radiation of non-ecliptic trajectory (solid lines) is through the exit zone with partial reflection. Three pass resonance $g_1=g_2=0.5$.

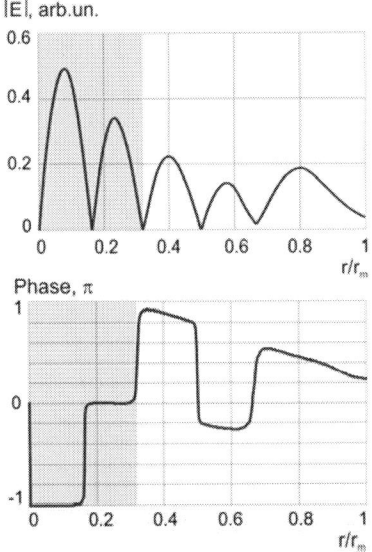

Рис.6. The radial distribution of field amplitude (left) and phase (right) on the surface of the right mirror (Fig.5). Calculation parameters: $g_1=g_2=0.5$, N=3, azimuthal polarization. The zone filled by grey is the exit of radiation from the resonator. Partial reflectivity is 50%. The Fresnel number F=4.5.

Another example of high quality radiation for F=10.1 using multipass pass mode with N=8 is presented in Fig.7,8. In the contrary upper case here ecliptic trajectory is used for generation.

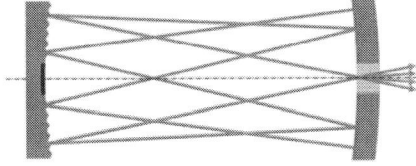

Fig.6. The resonator design for depressing non-ecliptic trajectory of generation and the outputing radiation of ecliptic trajectory through the exit zone with partial reflection. Eight pass resonance $g_1=g_2=0.383$.

As for active medium, it influence essentially to the transverse structure of output radiation. It has been shown in particular in [5]. Knowing the properties of active media of the concrete laser we are able to calculate predicted field distribution of laser beam.

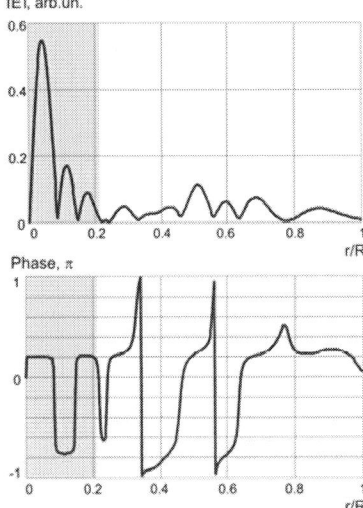

Рис.7. The radial distribution of field amplitude (left) and phase (right) on the surface of the right mirror (Fig.6). Calculation parameters: $g_1=g_2=0.383$, N=8, azimuthal polarization. The zone filled by grey is the exit of radiation. Partial reflectivity is 50%, F=10.1.

The results are presented of numerical experiments on amplitude-phase distribution of output radiation for the three- four- and eight-pass paraxial modes in the selected trajectories, which support the effectiveness of the proposed solutions for F=4.5-10. The quality of output radiation is high in all the ecliptic and non-ecliptic trajectories at even and odd multiplicities of paraxial resonance despite the particular qualitative features of beam propagation in these cases.

[1] A. G. Fox and T. Li, "Resonant modes in a maser interferometer," Bell Sys. Tech. J. 40, 453-458 (1961).
[2] A.V. Nesterov and V.G. Niziev, Vector solution of the diffraction task using the Hertz vector, Physical Review E **71**, 4, 046608, 2005.
[3] V. G. Niziev and R. V. Grishaev, "Dynamics of Mode Formation in an Open Resonator" Appl. Opt. 49, 6582-6590 (2010).
[4] I.A. Ramsay and J.J. Degnan, A Ray Analysis of Optical Resonators Formed by Two Spherical Mirrors Appl. Opt. **9**, 385-398 (1970).
[5] V. G. Niziev and D. Toebaert, "Formation of transverse mode in axially symmetric lasers," Appl. Opt. 51, 954-962 (2012)

Antiguided resonant multicore fiber with improved intermodal discrimination
(Invited Paper)

N. N. Elkin[1,2], A. P. Napartovich[1,2], *Member, IEEE*, D.V. Vysotsky[1]

[1]SRC RF Troitsk Institute of innovation and fusion research, Troitsk, Moscow, Russian Federation

[2]Moscow Institute for Physics and Technology, Dolgoprudny, Moscow region, Russian Federation

Abstract: Novel fiber laser design strategy is proposed based on in-phase mode selection in a circular antiguided resonant structure. A complete set of guided modes of the structure was calculated using 2D mode solver. Parameters of 7-core array are found providing the maximal gain overlap factor for the in-phase mode.

Rare-earth-doped fiber lasers have established themselves as a power scalable laser concept for a variety of operation regimes. However, its maximum single mode output power is limited by nonlinear effects, which increase with the light intensity and the fiber length [1]. Multi-core fiber approach promises to result in a design of a comparatively short device with high-power output beam of high quality [2] provided the single transverse mode operation is achieved. Authors of [3] have introduced a new design for the ribbon fiber laser utilizing an effect of congruence between spatial profiles of the gain and of the in-phase mode wave field intensity. Cross-sectional view of a simplified ribbon structure is shown in Fig. 1. More general design of a ribbon structure with intentionally done index profile was analyzed in ref. [4]. It was predicted that a proper choice of ribbon fiber parameters allows one to phase-lock a large number of active cores. In this paper the next step in development of the ribbon-like design is done. A new idea is to wrap a composite ribbon with alternating active and passive regions around the cylindrical pump beam core, as shown in Figs. 4 and 7. Gain overlap factors were calculated numerically for a manifold of optical modes at various construction parameters to search conditions for the best in-phase mode discrimination against all higher-order modes.

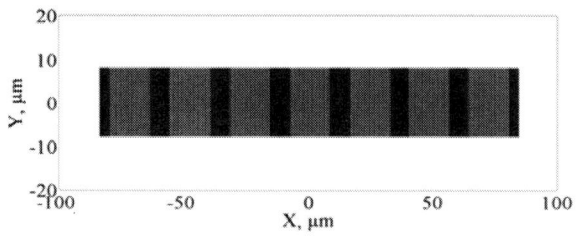

Fig. 1. Cross-sectional view of a ribbon fiber with doped low-index regions (dark) alternating with high-index passive regions (black). The outer region (white) is lower-index cladding.

The cross-sectional view of the constructions under consideration is shown in Figs. 1, 4 and 7. The refractive index

in active segments, n_a, is of intermediate magnitude between that of the central core (outer region), n_c, and of the passive segments, n_g:

$$n_c < n_a < n_g.$$

As known from theory of diode laser arrays [5], in a resonant array of active elements separated by higher refractive index inter-element regions the gain overlap factor is maximal one for the in-phase mode. It can be shown that the out-of-phase mode at an odd number of elements does not compete with the in-phase mode, because of much smaller gain overlap factor. In contrast, at an even number of elements the out-of-phase mode is expected to be dominant. Therefore, we may limit our studies considering constructions with only an odd number of elements.

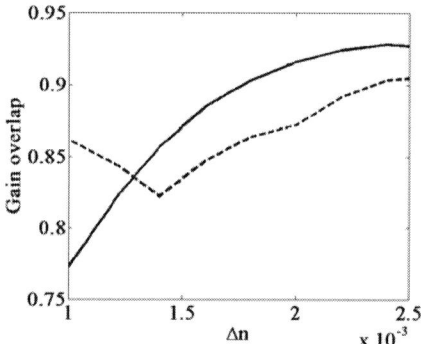

Fig. 2. Gain overlap (Γ) of the in-phase mode (solid line) and of the next highest overlap mode (dashed line) for the structure shown in Fig. 1. $\Delta n = n_g - n_a$.

A 2D mode solver has been developed [6] for the calculation of fiber modes in the passive structure. Scalar approximation was implemented, and sufficiently small refractive index variations were assumed. A complete set of guided modes of the structure was calculated. As a result, the modal propagation constants and gain overlap factors

$$\Gamma = \left(\int |E(x,y)|^2 g(x,y) dx dy \right) \left(\int |E(x,y)|^2 dx dy \right)^{-1}$$

were found. Here $E(x,y)$ is the wave electric field, $g(x,y)$ is a function with value 1 in those portions of the fiber that are gain loaded (gray-tone regions in Figs. 1, 4 and 7) and has value 0 in other regions.

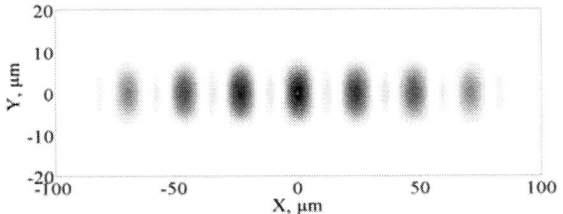

Fig. 3. The calculated in-phase mode intensity profile for the structure shown in Fig. 1 at Δn=0.0018.

Calculation were performed for the wavelength $\lambda = 1.1\,\mu$m. The refractive index in the central core and outer cladding was fixed in all the structures, $n_c = 1.456$. The refractive index in the passive segments was fixed also, $n_g = 1.462$. The refractive index in active segments was varied to find the best in-phase mode discrimination against all higher-order modes. The results are presented as a function of $\Delta n = n_g - n_a$.

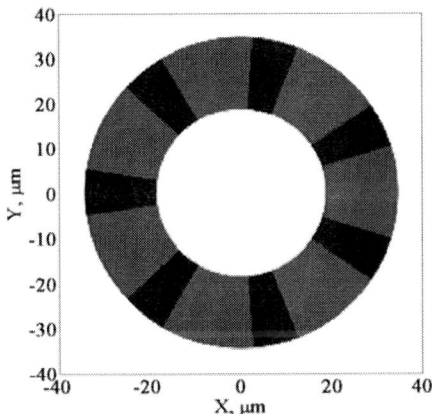

Fig. 4. Cross-sectional view of a fiber with doped low-index regions (dark) alternating with high-index passive regions (black). Central region (white) is lower-index pump core. The outer region (white) is lower-index cladding.

The first series of numerical calculations was performed for the ribbon structure shown in Fig. 1. The waveguide cross section area is equal to $16\times168\,\mu$m^2. The doped antiguided elements have the width in the x direction $d = 16\,\mu$m, while high-index regions are $s = 8\,\mu$m. The edging high-index regions have the width $s/2 = 4\,\mu$m. We use the lateral resonance condition [4] applicable to 1D ribbon structure:

$$\frac{1}{s^2} - \frac{1}{d^2} = \frac{8n_g\Delta n}{\lambda^2},$$

as a first estimate of design parameters. The 1D resonance took place at $\Delta n = 0.001215$. Fig. 2 demonstrates that at this point the in-phase mode has not highest gain overlap. Actually, a rather wide interval $0.0015 < \Delta n < 0.0025$ exists where the in-phase mode possesses the highest gain. The in-phase mode profile at Δn=0.0018 is shown in Fig. 3.

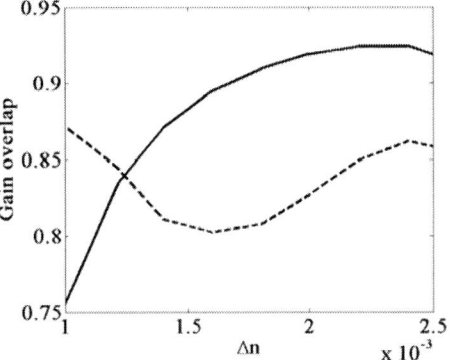

Fig. 5. Gain overlap (Γ) of the in-phase mode (solid line) and of the next highest overlap mode (dashed line) for the structure shown in Fig. 4. $\Delta n = n_g - n_a$.

The next series of calculations was performed for the structure shown in Fig. 4. The radius of the circular median is equal to 26.5 μm. This structure can be viewed as the straight ribbon structure wrapped around the cylinder of radius $18.5\,\mu$m. Fig. 5 demonstrates that at the index step higher than 0.0013 the in-phase mode has the highest gain overlap factor. The in-phase mode profile is shown in Fig. 6. Apparently, absence of borders in the ring structure leads to improvement of gain discrimination.

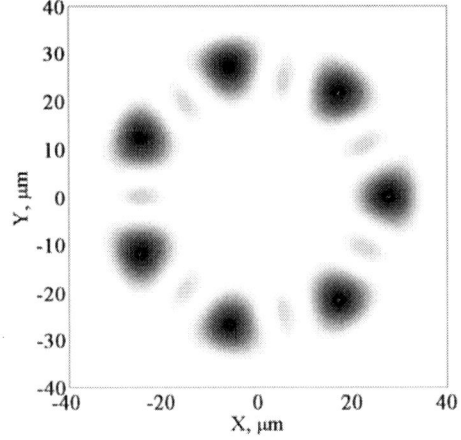

Fig. 6. The calculated in-phase mode intensity profile for the structure shown in Fig. 4. Δn=0.0018.

Fabrication of fiber structure in a form of angular segments, as shown in Fig. 4, may meet technological problems. One more option of fiber design with circle elements shown in Fig. 7 could be more practical. We have performed calculations for this structure. The circular median was taken of the same radius as one in Fig. 4. Fig. 8 demonstrates that the range of index step values with reasonably good mode discrimination has changed. The in-phase mode profile (Fig. 9) is similar to that observed for the structure with angular segments.

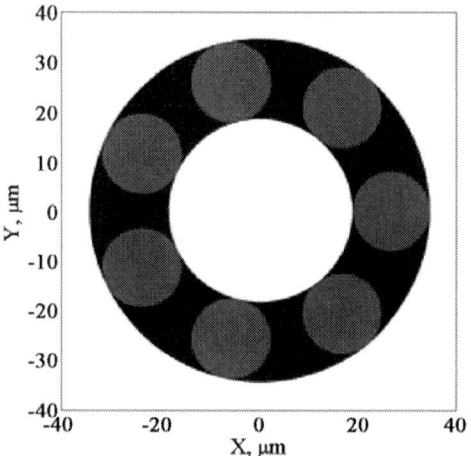

Fig. 7. Cross-sectional view of a fiber with doped low-index regions (dark) alternating with high-index passive regions (black). Central region (white) is lower-index pump core. The outer region (white) is lower-index cladding.

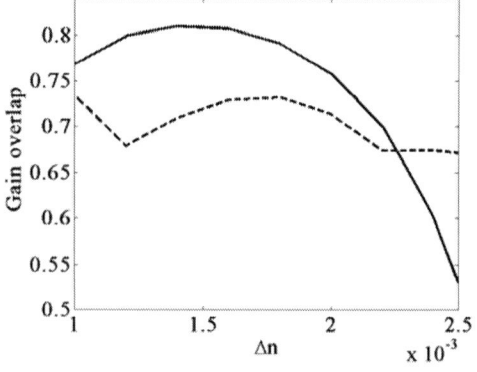

Fig. 8. Gain overlap (Γ) of the in-phase mode (solid line) and of the next highest overlap mode (dashed line) for the structure shown in Fig. 7. $\Delta n = n_g - n_a$.

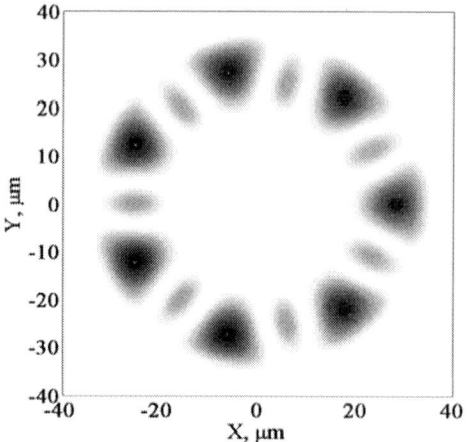

Fig. 9. The calculated in-phase mode intensity profile for the structure shown in Fig. 7. $\Delta n = 0.0012$.

Further progress in increasing the total emitting aperture area can be done by increasing of the ring area at a deliberate choice of such parameters as ring thickness, the set of refractive indices and the number of active cores.

REFERENCES

[1] D. J. Richardson, J. Nilsson, W. A. Clarkson "High power fiber lasers: current status and future perspectives", JOSA B, vol. 27, pp. B63-B92, 2010.

[2] P. Glas, M. Naumann, A. Schirrmacher, Th. Pertsch,"The multicore fiber — a novel design for a diode pumped fiber laser", Opt. Commun., vol. 151, pp. 187–195 1998.

[3] R. J. Beach, M. D. Feit, R. H. Page, et al., "Scalable antiguided ribbon laser", JOSA B, vol. 19, issue 7, pp. 1521-1534, 2002.

[4] D. V. Vysotsky, A. P. Napartovich, A. G. Trapeznikov, "Optical mode selection in ribbon fiber with gain modulation", Quantum Electron., vol. 33, no. 12, pp. 1089–1095, 2003.

[5] Dan Botez "Monolithic phase-locked semiconductors laser arrays", in Diode Laser Arrays, D. Botez and D. R. Scifres, ed. Cambridge Univ. Press, Cambridge, UK, 1994.

[6] N. N. Elkin, A. P. Napartovich, V. N. Troshchieva, D. V. Vysotsky, "Diffraction modeling of the multicore fiber amplifier", J. Lightwave Technol., vol. 25, no. 10, pp. 3072–3077, 2007.

Dye-doped polymer optical fiber pulse illuminator for microscopes

(Invited Paper)

D.V. Kiesewetter[1], P.G. Gabdullin[1], A.Y. Savina[1], A.I. Bodrov[2], N.O. Stelmakova[2], V.M. Levin[3], G.G. Baskakov[3]

[1]Saint Petersburg State Polytechnical University, Saint Petersburg, Russia
[2]Turbotekt - Saint-Petersburg, Co Ltd, Saint Petersburg, Russia
[3]Technological center of the polymer optical fibers, Tver, Russia

Abstract: Application of dye-doped polymer optical fiber used for conversion of the pulse coherent in the incoherent radiation for the creation of the illuminator to the microscope is described. The technical requirements to the fiber are determined. The manufacturing technology of the fiber and the experimental setup are describes. Experimentally measured the contrast speckles. The images of microscopic moving particles are present.

The devices for obtaining the images of moving microscopic particles needed for many technical applications. In this case, the main difficulty in using optical microscope with a television camera is a necessity of application of pulsed light sources with high intensity and short duration. Such parameters have only lasers. However, laser's radiation is coherent. With the passage of the coherent radiation of the optical inhomogeneous medium - optical windows, studied objects appears random interference pattern, which is a randomly located dark and bright spots called «speckles». This pattern reduces the quality of the image and seriously complicates the determination of the size and number of particles.

As a source of incoherent radiation with pulse duration τ of 10 ns was used YAG: Nd^{3+} laser with a segment of the dye-doped polymer optical fiber (DD POF). This allowed obtaining high-quality images of particles with sizes from 0.5 mkm or more, moving with a speed up to 100 m/s. The scheme created experimental setup is shown in Fig. 1. DD POF pumped with second harmonic of radiation of the YAG laser (λ_{2h}=532 nm). Since that rhodamine dyes absorb radiation mainly in the green range of wavelengths, the first harmonic of laser was cut off with the special optical filter. Intermediate optical fiber (OF) with quartz core (QC) applied to prevent the destruction of the end of the POF by laser's radiation. Optical cell through which pumped flow of technical oils with contaminants was located in the focal plane of the lens of the microscope.

The principle of the device is the following. Pulse coherent laser's radiation, after passing the filter 11, which cuts off infrared light, introduced into the quartz optical fiber 9 through the input end. After the QC OF 9, through the connector 8, radiation gets in the DD polymer optical fiber 2. In the DD POF 2 is absorbed coherent pump radiation at the wavelength of 532 nm, which is converted into incoherent radiation of fluorescence. Output light beam 6 is converted into the parallel beam of light by the collimator 3. In the focal plane of the microscope 5 set the cell, through which the pumped liquid or gas with the test microscopic particles. The beam of light,

having the cell, enters the micro lens of the microscope. Application of pulsed laser radiation allows obtaining images of moving particles at high speed.

The fibers were made of polymethyl methacrylate (PMMA). As a core material of the DD polymer optical fibers used methyl methacrylate copolymer with methyl acrylate in the ratio of 95%/5% with the addition of rhodamine (6G and C) respectively, 1, 2, 5, 10, 25, 50 mg/kg, and as reflecting cladding used poly fluoroacrylates. Dye doped optical fibers were obtained by extrusion at temperature of 195^0 - 200^0 C. Parameters of core: refractive index (RI) – 1.4895, diameter - 950 microns; parameters reflecting cladding: RI – 1.4201, thickness 25 microns. The samples of DD POF length up to 100 meters and the polymeric fibers, without doped dye were manufactured.

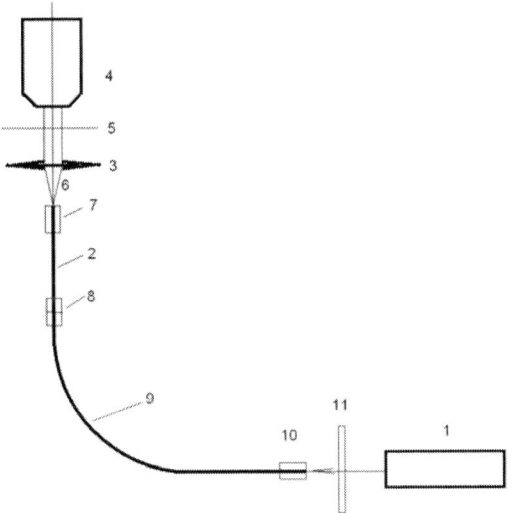

Fig. 1. Experimental setup: 1 – YAG: Nd^{3+} laser, 2 – dye-doped polymer optical fiber, 3 – collimator, 4 – micro lens of the microscope, 5 – focal plane of the lens, 6 – beam of radiation coming out of the DD POF, 7 – output connector of the DD POF, 8 – input connector of the quartz core optical fiber, 9 – QC optical fiber, 10 – input connector of the QC optical fiber, 11 – filter.

978-1-4799-0019-0/13 $31.00 © 2013 IEEE

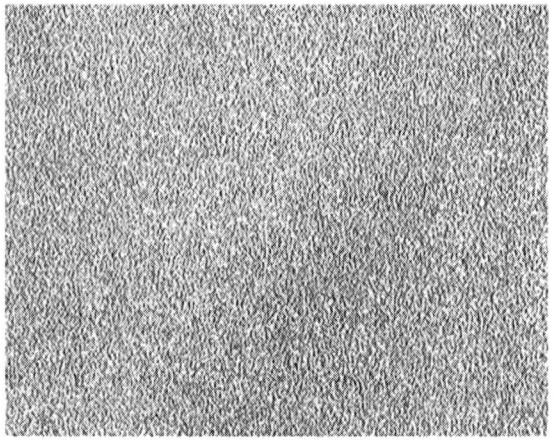

Fig. 2. The image obtained in the coherent light.

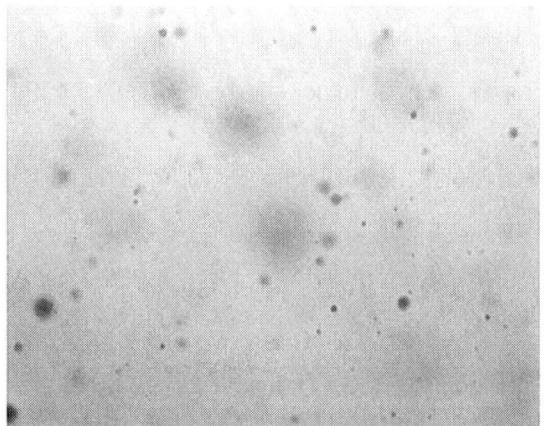

Fig. 3. Image of contaminants particles in the flow of technical oil obtained using the described above setup.

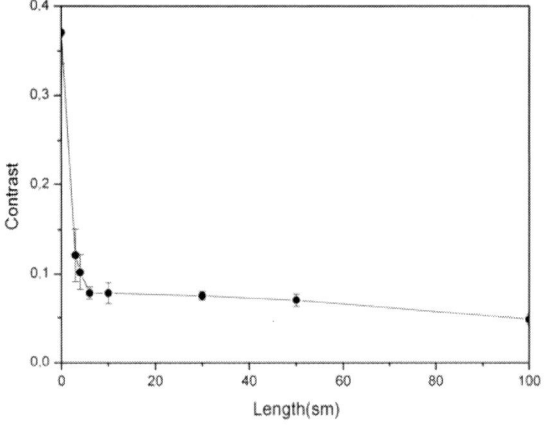

Fig. 4. Contrast of speckles as a function of the length of the DD polymer optical fiber; concentration of rhodamine 6G in the DD POF is 5 mg/kg.

For the efficient operation of the device, necessary to note the following: coherent radiation of the laser must be fully absorbed in the DD POF, the impulse duration of incoherent radiation at the output of the DD POF should not exceed the given value. The first condition is achieved by selecting the length DD fiber at a given concentration of dye: $L_{DDPOF} \gg 1/\alpha(\lambda_{2h})$, where $\alpha(\lambda_{2h})$ is the coefficient of attenuation of the radiation at the wavelength λ_{2h}. To meet the second condition when using rhodamine dyes enough to choose the length of L_{DDPOF} so that inter-mode dispersion does not exceed the duration of the impulse τ. The necessary parameters of the DD optical fibers were determined experimentally [2].

Quality of images when using the pulse illuminator compared with using of light of emission diodes or light of incandescent lamps is actually determined by the contrast of speckles, which depends on the ratio of powers of incoherent and coherent radiation at the output of DD POF. As an example, in Fig. 2 shows the image obtained in the coherent light without using DD optical fiber. In the Fig. 3 shows the image of microscopic particles in the flow of technical oils, moving with the velocity 1 m/s when using the pulse light of DD POF. The difference between the images is obvious. In the Fig. 3 particle image cannot be detected at the background of speckles. In the Fig 3 the particle image can be easily identified and can be defined the geometric parameters of the particles.

For the impartial evaluation of the quality of images calculation of contrast of speckles in the absence of contaminants was carried out: $C = d/I_a$, where d is the standard deviation of the intensity of the pixels of the image of the average I_a by the image area. Experimentally measured dependence $C(L_{DDPOF})$ is shown in Fig. 4. The notable difference of the contrast from zero at the length of L_{DDPOF} 0.2 m and more can be explained by the presence of radiation of the first harmonic of laser at the output of the DD POF and by the own noise of CCD television camera. Contrast of speckle can be reduced through the use of the filter 11 greater thicknesses (Fig. 1) or the application DD optical fiber with the smaller concentration of dye, but greater length.

The developed device allows obtaining images of microscopic particles moving at speeds up to 100 m/s and more. Laboratory model of the device is successfully used in the systems of cleaning of industrial oils to determine the concentration, size and type of contaminants.

REFERENCES

[1] A.I. Bodrov, N.O. Stelmakova, P.G. Gabdulling, D.V. Kiesewetter, A.Y. Savina, V.M. Levin, G.G. Baskakov, N.V. Ilin, N.V. Bankul, "Fiber-optic illuminator", Patent № 122187 RU, *Bul. Izobr.*, 32, 20.11.2012.

[2] Kiesewetter D.V., Savina A.Yu, Levin V.M., Baskakov G.G. "The measurement of attenuation in polymer optical fibers doped with fluorescent dye", *Nauch. Tech. Vedomosti SPbSTU*, №2(146), pp. 119-124, 2012.

Scattering of the Plane Wave from a Periodically Perforated Dielectric Slab

Volodymyr O. Byelobrov[1] *Student Member, IEEE* and Trevor M. Benson[2], *Senior Member, IEEE*

[1]Laboratory of Micro and Nano Optics, Institute of Radio-Physics and Electronics NASU, Kharkiv, Ukraine

[2] George Green Institute for Electromagnetics Research, University of Nottingham, Nottingham, UK

Tel: 380(57) 720-3782, Fax: 380(57) 315-2105, e-mail: volodia.byelobrov@gmail.com

Abstract: An infinite grating of identical dielectric or metal cylinders or strips illuminated by a plane wave has high-quality optical resonances near the Rayleigh anomalies [1-16]. These resonances are present for any periodic 2-D structures and have striking properties as their Q-factors increase if the grating period becomes larger. Besides, in such a resonance the domain of high-intensity scattered field occupies a very wide domain along the grating. In this paper we investigate a configuration where we have a dielectric slab and an infinite grating of air holes inside it. This structure is closer to practical applications in nano and micro-optics and our study shows the presence of the grating resonances having complicated behaviour.

In this work we investigate a dielectric slab (Fig. 1) of the refractive index ν_s with an embedded into it dielectric grating of circular cylinders that is placed at the distances d_1 and d_2 from the upper and lower boundaries, respectively. The grating period is p and its cylinders have radius a and refractive index $\nu = 1$ or in fact are the air holes. In computations, we have taken the slab refractive index to be $\nu_s = 1.4142$, i.e. small to eliminate excessive impact of the slab modes.

In the preceding studies we have investigated the scattering of a plane wave normally incident on a passive periodic structure of lossless dielectric circular cylinders in free space, parallel to the z-axis and periodic along the x-axis. The reflectance of the plane wave from such a structure has shown the presence of specific grating resonances near the Rayleigh anomalies [7,13]. These resonances possess interesting feature as they become sharper if the grating becomes sparser. A sample relief of the reflectance of the plane wave normally incident on such a grating is shown in Fig. 2. Here the grating resonances are seen as bright ridges that become sharper with the growth of the period and eventually merge with the Rayleigh anomalies, i.e. the lines $\sigma = 1, 2, ...,$ where $\sigma = p / \lambda$ is the dimensionless frequency and λ stands for the wavelength in free space. Additionally the scattered field at a grating-resonance frequency forms a standing wave along the grating that stretches far away from it in the normal direction

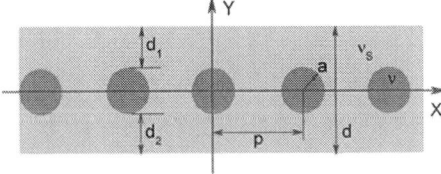

Fig. 1 Cross-sectional geometry of a periodically perforated dielectric slab

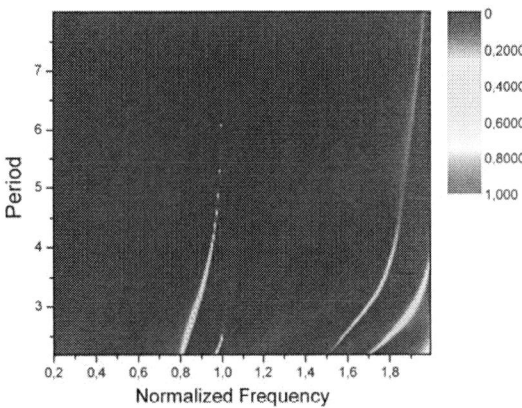

Fig.2 The reliefs of reflectance of the H-polarized normally incident plane wave from the grating of circular dielectric cylinders in free space $\nu_s = 1.0$ and $\nu = 1.4142$, as a function of the normalized frequency and the period-to-radius ratio.

[8]. Although there is infinite number of natural modes near each Rayleigh anomaly, only few first ones can be observed for realistic material and geometrical parameters of the resonator. If the grating elements are not magnetic, then the Q-factors of the grating resonances for the H-polarization are higher than for the E-polarization.

We solve the perforated-slab scattering problem using the method of separation of variables and transfer matrices. Firstly,

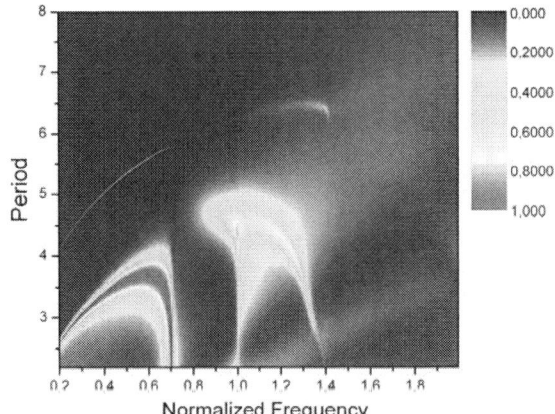

Fig.3. The same as in Fig. 2 for a lossless dielectric slab with a grating of circular air holes placed in the middle with $d_1 = d_2 = 0.05a$; slab refractive index is $\nu_s = 1.4142$

we construct the scattering matrices of the grating; thereto we represent the filed inside a hole and the incident one using the Fourier expansions with Bessel functions in coefficients, while the scattered field series involve the Hankel functions of the first order. The order is chosen according to the time dependence and the radiation condition. In analytical derivations, among several mathematical operations infinite summation of the Hankel functions appears.

Direct calculation of such sums, in view of their extremely slow convergence, is not efficient. For acceleration we use the lattice-sums technique [17]. Exclusion of the coefficients of the internal field brings us to an infinite matrix equation for the

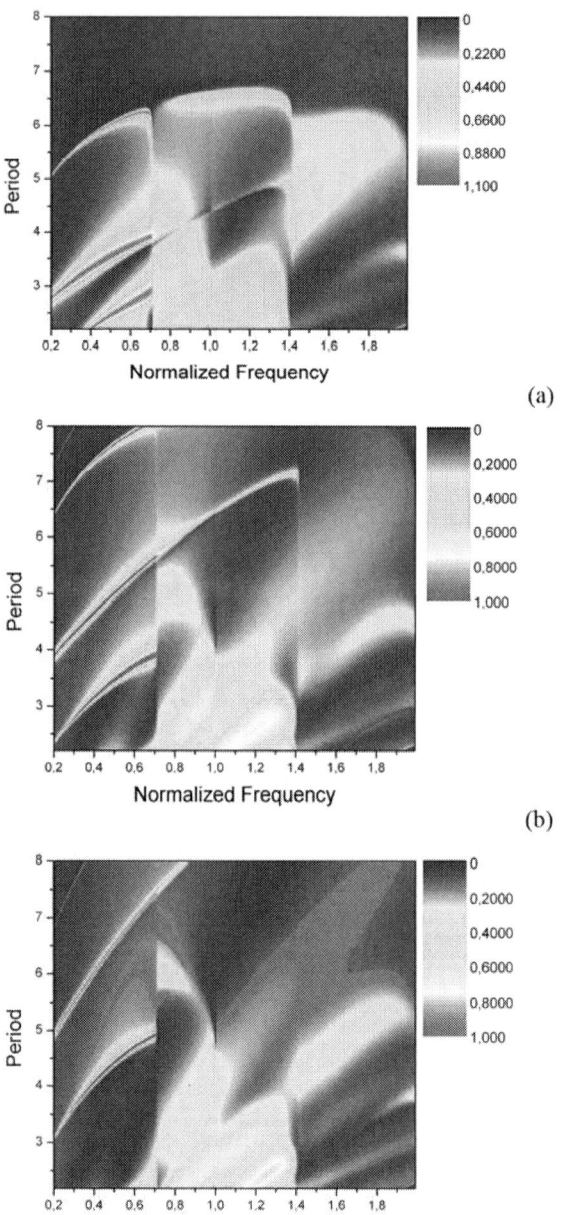

(a)

(b)

(c)

Fig.4. The reliefs of reflectance of the H-polarized normal incident plane wave on the dielectric slab ($\nu_s = 1.4142$) with periodic air holes ($\nu = 1.0$), where $d_1 = 0.05a$ and $d_2 = 0.45a$ for (a), $d_2 = 0.95a$ for (b), and $d_2 = 1.45a$ for (c).

Fourier coefficients of the external field. This matrix equation is the Fredholm equation of the second kind. As a result the algorithm built on solving the truncated equation has good stability and provides high accuracy for rather small truncation order of the matrix. The Fourier coefficients of the scattered field have to be converted to the amplitudes of the Floquet harmonics, which are the elements of the grating scattering matrix. The planar boundaries of the slab are represented using the diagonal Fresnel matrixes. Compiling both kinds of matrices we build the scattering matrix of the whole structure. The overall scheme of the solution coincides with one explained in [13,18].

In Fig 3, we present the relief of the reflectance of a perforated dielectric slab depending on the perforation period and the normalized frequency. Here one can see that besides of the free-space Rayleigh anomalies, new set of anomalies appears at the frequencies $\sigma = n / \nu_s$, $n = 1, 2...$. Besides, unlike the grating of cylinders in free space, here the most intensive resonance in the strip $\sigma < 0.7$ does not exist for large enough values of period; instead it seems to vanish in the Rayleigh anomaly $\sigma = 1/\nu_s \approx 0.7$ (branch point) at a finite value of period-to-radius ratio. Similar dynamics is demonstrated by the higher order resonance visible as a thin curve at larger values of period. The resonances in the strip between the first branch point at $\sigma \approx 0.7$ and the second one at $\sigma \approx 1.41$ show similar behaviour, however they are weaker.

We have also investigated the impact of the slab width on the resonances associated with periodic perforations. We looked at the three different geometries, where the distance between the grating of air holes and upper boundary, d_1, stays the same but the slab width varies. In Fig. 4, three reliefs of the reflectance depending on the perforation period and the normalized frequency are shown for identical material parameters however different geometries: for (a) $d_1 = 0.05a$ and $d_2 = 0.45a$ therefore the width of slab is $d = 1.5p$; for (b) $d_2 = 0,95a$ and $d = 2p$; and for (c) $d_2 = 1.45a$ and $d = 2.5p$. Comparing Figs. 3 and 4, we can see that now the number of bright resonances in the strip $\sigma < 0.7$ is at least three, and each of them is accompanied by the almost total reflection in a narrow range. One should note that for wider slabs the grating resonances are more intensive and have larger Q-factors.

In conclusion we may say that a perforated slab possesses high-quality resonances whose nature is strongly connected to the periodicity of perforation. They may appear for any value of the normalized frequency between the branch points and are also in complicated interplay with the resonances of the homogeneous slab perturbed by the perforations.

This work was supported, in part, by the National Academy of Sciences of Ukraine via the State Target Program "Nanotechnologies and Nanomaterials" and the European Science Foundation via the travel grants of the Research Networking Program "Newfocus."

REFERENCES

[1] K. Ohtaka, H. Numata, "Multiple scattering effects in photon diffraction for an array of cylindrical dielectrics," *Phys. Lett.*, vol. 73-A, no 5-6, pp. 411-413, 1979.

[2] T.L. Zinenko, A.I. Nosich, Y. Okuno, "Plane wave scattering and absorption by resistive-strip and dielectric-strip periodic gratings," *IEEE Trans. Antennas Propagat.*, vol. 46, no 10, pp. 1498-1505, 1998.

[3] R. Gomez-Medina, M. Laroche, J.J. Saenz, "Extraordinary optical reflection from sub-wavelength cylinder arrays," *Opt. Exp.*, vol. 14, no 9, pp. 3730-3737, 2006.

[4] M. Laroche, S. Albaladejo, R. Gomez-Medina, J. J. Saenz, "Tuning the optical response of nanocylinder arrays: an analytical study," *Phys. Rev. B*, vol. 74, no. 9, pp. 245422/10, 2006.

[5] A. Christ, T. Zentgraf, J. Kuhl, S. G. Tikhodeev, *et al.*, "Optical properties of planar metallic photonic crystal structures: experiment and theory," *Phys. Rev. B*, vol. 70, no 12, pp. 125113/15, 2004.

[6] F. J. G. Garcia de Abajo, "Colloquium: Light scattering by particle and hole arrays," *Rev. Mod. Phys.*, vol. 79, no. 4, pp. 1267-1289, 2007.

[7] V.O. Byelobrov, J. Ctyroky, T.M. Benson, R. Sauleau, A. Altintas, A.I. Nosich, "Low-threshold lasing eigenmodes of an infinite periodic chain of quantum wires," *Optics Lett.*, vol. 35, no 21, pp. 3634-3636, 2010.

[8] V.O. Byelobrov, T.M. Benson, A.I. Nosich, "Near and far fields of high-quality resonances of an infinite grating of sub-wavelength wires," *Proc. European Conf. Microwaves (EuMC-11)*, Manchester, 2011, pp. 858-861

[9] S. R. K. Rodriguez, M. C. Schaafsma, A. Berrier, J. G. Rivas, "Collective resonances in plasmonic crystals: size matters," *Phys. B*, vol. 407, pp. 4081-4085, 2012.

[10] T.V. Teperik, A. Degiron, "Design strategies to tailor the narrow plasmon-photonic resonances in arrays of metallic nanoparticles," *Phys. Rev. B*, vol. 86, pp. 245425/5, 2012.

[11] P. Ghenuche, G. Vincent, M. Laroche, N. Bardou, R. Haidar, J.-L. Pelouard, S. Collin, "Optical extinction in a single layer of nanorods," *Phys. Rev. Lett.*, vol. 109, pp. 143903/5, 2012.

[12] R D.M. Natarov, R. Sauleau, A.I. Nosich, "Periodicity-enhanced plasmon resonances in the scattering of light by sparse finite gratings of circular silver nanowires," *IEEE Photonics Technology Lett.*, vol. 24, no 1, pp. 43-45, 2012.

[13] V.O. Byelobrov, T.M. Benson, A.I. Nosich, "Binary grating of sub-wavelength silver and quantum wires as a photonic-plasmonic lasing platform with nanoscale elements," *IEEE J. Sel. Topics Quant. Electron.*, vol. 18, no 6, pp. 1839-1846, 2012.

[14] R. Rodríguez-Berral, F. Medina, F. Mesa, M. García-Vigueras, "Quasi-analytical modeling of transmission/ reflection in strip/slit gratings loaded with dielectric slabs," *IEEE Trans. Microwave Theory Tech.*, vol. 60, no 3, pp. 405-418, 2012.

[15] T.L. Zinenko, M. Marciniak, A.I. Nosich, "Accurate analysis of light scattering and absorption by an infinite flat grating of thin silver nanostrips in free space using the method of analytical regularization," *IEEE J. Sel. Topics Quant. Electron.*, vol. 19, no 3, pp. 9000108/8, 2013.

[16] O.V. Shapoval, A.I. Nosich, "Finite gratings of many thin silver nanostrips: optical resonances and role of periodicity," *AIP Advances*, vol. 3, no 4, pp. 042120/13, 2013.

[17] K. Yasumoto, H. Toyama, T. Kushta, "Accurate analysis of 2-D electromagnetic scattering from multilayered periodic arrays of circular cylinders using lattice sums technique," *IEEE Trans. Antennas Propagat.*, vol. 52, pp. 2603-2611, 2004.

[18] H. Toyama, K. Yasumoto, "Electromagnetic scattering from periodic arrays of composite circular cylinder with internal cylindrical scatterers," *Progress in Electromagn. Research*, vol. 52, no 10, pp. 321-333, 2005.

978-1-4799-0019-0/13 $31.00 © 2013 IEEE

DESIGN OF 3D OPTICAL MEDIA WITH PERIODICALLY DISTRIBUTED EMITTING CENTERS

S. O. Klimonsky [1], T. Bakhia [1], N.S. Borodinov [1], A.V. Knotko [1], N. Yu. Vereschagina [2]

[1] Department of Materials Science, M.V. Lomonosov Moscow State University, 119991 Moscow, Russia
[2] D.I. Mendeleev University of Chemical Technology of Russia, 125047 Moscow, Russia

Abstract: Media with periodically distributed absorbing or emitting centers open new possibilities in a design of microoptical devices. However three dimensional (3D) structures of such type have not been obtained up to now. This question is considered in the present work. A multistage approach based on the colloid crystal technique has been suggested and structures consisted of CdSe quantum dots periodically distributed in ETPTA photoresist matrix have been prepared.

Structures with periodically distributed absorbing or emitting centers (PDAC and PDEC structures, correspondingly) open new possibilities in design of microoptical devices. Particularly absorbing structures permit to create photonic crystals with complex dielectric index contrast. Especially interesting case is realized when the photonic resonance connected with the refractive index contrast overlaps with the absorption feature, hence creating a "resonantly absorbing photonic crystal". A one dimensional (1D) multilayer Bragg stack referred to as resonantly absorbing bragg reflector was designed from an absorbing semiconductor and demonstrated efficient optical switching [1]. Then a 2D absorbing structure with periodically distributed dye was created and investigated in [2, 3]. This structure behaved not only as an imaginary refractive index photonic crystal, but also as a microscopic waveguide array with a controlled dispersion opening the way to the efficient color separation of light [3, 4]. In the present study we analyze possible ways to create 3D PDAC and PDEC structures based on the colloid crystals technique.

The worked out approach consists of the following stages:

1. Synthesis of SiO_2 microspheres with a narrow dispersion by the multistep modification of the Stöber method [5].
2. Growing of opal-type thin film photonic crystals by the vertical deposition technique [6] (Fig. 1a).
3. Filling of the structure with ETPTA photocurable resin followed by a photopolymerization and a creation of the opal-polymer composite.
4. Inversion of the structure by dissolving of SiO_2 microspheres in HF (Fig. 1b).
5. Incorporation of CdSe quantum dots into the colloid crystal films and into the inverse structures.
6. Filling of the inverse structure cavities with the photocurable resin followed by the photopolymerization and creation of the 3D PDEC structure.

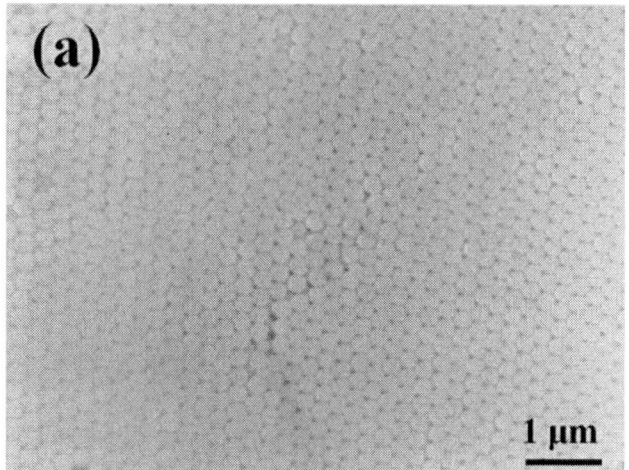

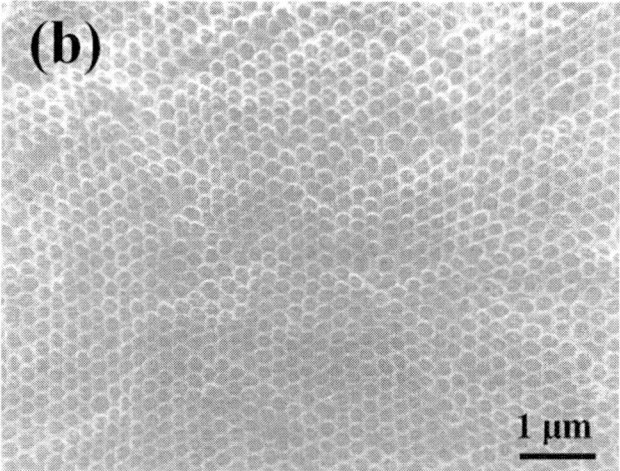

Fig. 1. Scanning electron microscopy images of SiO_2 opal-type film (a) and inverse ETPTA film (b).

Some of transmittance spectra obtained for our samples are shown in Fig. 2. The modification of curve 4 in comparison with curve 3 is accounted for CdSe quantum dots absorption (see curve 1). The photonic band gap is responsible for deep drops in curves 3, 4 near 560-570 nm. The band gap almost disappears in SiO_2-ETPTA composite (curve 5) due to close values of SiO_2 and ETPTA refractive indices (n = 1.45 and 1.47, correspondingly). The transmission spectrum of SiO_2-ETPTA composite with the periodically distributed quantum dots (curve 6) demonstrates a drop near 617 nm also, which is presumably connected with the photonic band gap induced by the imaginary refractive index contrast.

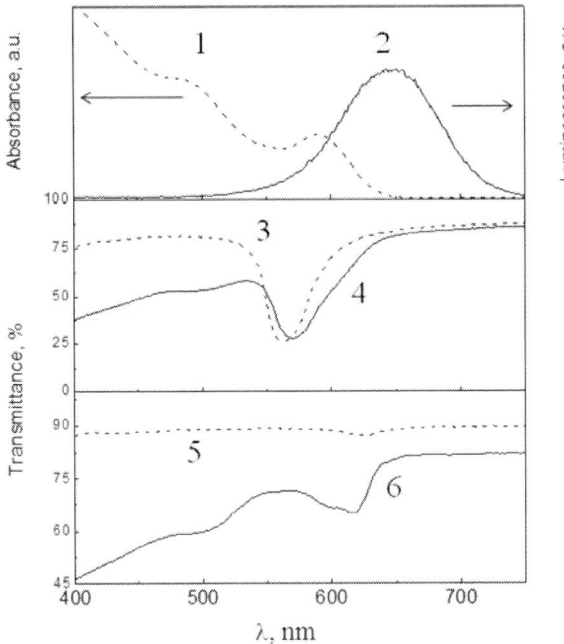

Fig. 2. 1 – absorbance spectrum of CdSe quantum dots with the diameter of 5.1 nm; 2 – photoluminescence spectrum of the same quantum dots (300 nm excitation); 3 – transmittance spectrum of the opal-type film shown in Fig. 1a; 4 – transmittance spectrum of the same film after the quantum dots incorporation; 5 - transmittance spectrum of a similar film filled by the photoresist (SiO_2-ETPTA composite without quantum dots); 6 - transmittance spectrum of a similar film with quantum dots incorporated into the opal matrix before filling with the photoresist (SiO_2-ETPTA composite containing periodically distributed quantum dots).

The work is supported by the Russian Foundation for Basic Research (grant number 13-03-91151).

REFERENCES

[1] J. P. Prineas, J. Y. Zhou, J. Kuhl, H. M. Gibbs, G. Khitrova, S. W. Koch, A. Knorr, "Ultrafast ac Stark effect switching of the active photonic band gap from Bragg-periodic semiconductor quantum wells," *Appl. Phys. Lett.*, vol. 81, pp. 4332-4334, 2002.

[2] J.T. Li, B. Liang, Y.K. Liu, P.Q. Zhang, J.Y. Zhou, S.O. Klimonsky, A.S. Slesarev, Yu.D. Tretyakov, L. O'Faolain, T.F. Krauss, "Photonic Crystal Formed by the Imaginary Part of the Refractive Index," *Adv. Mater.*, vol. 22, pp. 2676-2679, 2010.

[3] M. Feng, Y. Liu, Y. Li, X. Xie, J. Zhou. "Light propagation in a resonantly absorbing waveguide array," *Optics Express*, vol. 19, pp. 7222-7229, 2011.

[4] Y.K. Liu, S.C. Wang, Y.Y. Li, L.Y. Song, X.S. Xie, M.N. Feng, Z.M. Xiao, S.Z. Deng, J.Y. Zhou, J.T. Li, K.S. Wong, T.F. Krauss. "Efficient color routing with a dispersion-controlled waveguide array," *Light: Science & Applications*, vol. 2, p. e52, 2013.

[5] V.M. Masalov, N.S. Sukhinina, E.A. Kudrenko and G.A. Emelchenko. "Mechanism of formation and nanostructure of Stober silica particles," *Nanotechnology*, vol. 22, p. 275718, 2011.

[6] P. Jiang, J.F. Bertone, K.S. Hwang, V. Colvin, "Single-crystal colloidal multilayers of controlled thickness," *Chem. Mater.*, vol. 11. pp. 2132-2140, 1999.

The PDLC photonic structures diffraction characteristics managing by the spatially non-uniform electric field

A.O Semkin, S.N. Sharangovich

Tomsk State University of Control Systems and Radioelectornics (TUSUR)

Abstract: We develop an analytical model of light beams diffraction on the holographic photonic PDLC structures under the influence of a spatially non-uniform electric field. We propose an additional way to control the diffraction characteristics of these structures.

The ability to control the flow of optical radiation without any electro-optical conversions in the modern communication systems are currently of great interest. Such devices can be constructed based on a lot of physical phenomena, and one of the most promising of them is the light beams diffraction on manageable holographic photonic structures (GPhS) formed in the composite photopolymer materials with liquid crystal component (PDLC). Some methods of the GPhS diffraction characteristics control for PDLCs with different liquid crystal (LC) concentrations are shown in [1-3]. Besides, in [2,3] complicated amplitude-phase diffraction characteristics dependence on the value of a spatially homogeneous control field for materials with a high concentration of LC (> 90% volume) is analytically proved.

The aim of this study is to investigate the influence of a spatially non-uniform electric field on the diffraction characteristics of photonic structures formed in the PDLC with high concentrations of LC.

The optical properties heterogeneity of GPhS in the PDLC has a periodic and smooth character. In this case, smooth electro-optical inhomogeneity caused by the orientation effects in the LC and appears only in the external electric field. Changes in the optical properties of the PDLC samples with high (> 90% by volume) concentration of the liquid crystal molecules is described by the Fredericks model [2-4].

Let us consider a two-dimensional Bragg diffraction of extraordinary waves on the photonic structure in a sample of the PDLC, excluding light beams self-action. An alternating control electric field is formed by applying the heterogeneous electrode structure topology on the surface of the sample (Fig. 1).

In figure 1 the following notations are made: $E_p(r)$ - the incident beam, for simplicity we divide it into two, diffracted on the impact areas of the different polarity electric fields, the corresponding $E_{00}(r)$ and $E_{01}(r)$ - for transmitted and $E_{10}(r)$ and $E_{11}(r)$ - for the first order diffracted beams; $T_d(\Delta, E)$ - the resulting transfer function for the first-order diffraction beam,

which will be defined below; $+E$ and $-E$ indicate the parameters of the field of opposite polarity; r - the radius vector.

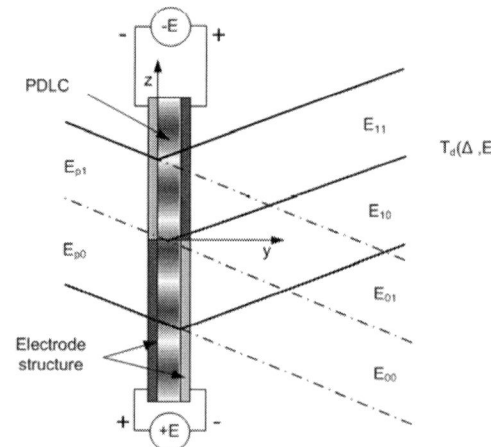

Fig. 1. Geometry of the light beams diffraction on the PDLC photonic structure.

Changing the polarity of the control voltage leads to a rotation of the liquid crystal director in the opposite side. Local vector diagrams of the light beams diffraction is shown in Fig. 2.

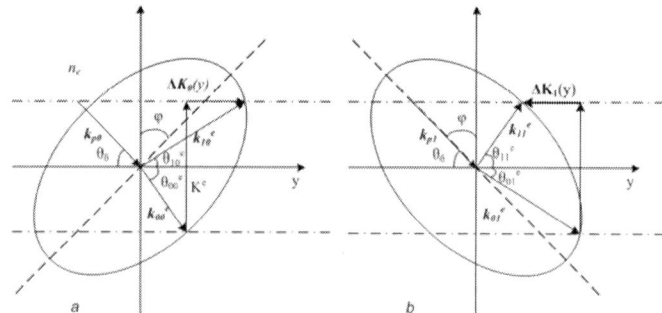

Fig. 2. Local vector diagrams of the light beams diffraction under the influence of different polarities control fields.

In the figure 2 $k_{p0,p1}$ - wave vectors of the incident beams; $k^e_{i,j}(r,E) = (\omega/c)n^e_{i,j}N^e_{i,j}$ - wave vectors of the diffracted beams; $i = 0,1$ - the diffraction order; j - number of the beam as shown in Fig. 1; $n^e_{i,j}$ refraction indexes; $N^e_{i,j}$ - the refractive-wave normal; e corresponds to the extraordinary wave; κ - grating vector; $\Delta K_j(r,E)$ - vectors of the local phase mismatch, φ - angle of rotation of the LC director.

The amplitude profiles $E_{i,j}(r)$ will be found from solution of coupled-wave equations (CWE) [2]:

$$N^e_{r0,j} \cdot \nabla E^e_{0,j}(r) = -iC^e_{1,j}(r)E^e_{1,j}(r)n^1(r)\exp(+i\Theta_j(r)), \quad (1)$$

$$N^e_{r1,j} \cdot \nabla E^e_{1,j}(r) = -iC^e_{0,j}(r)E^e_{0,j}(r)n^1(r)\exp(-i\Theta_j(r)), \quad (2)$$

where $C^e_{0,j}(\bar{r}) = \dfrac{1}{4}\dfrac{\omega}{c_c n^e_{0,j}}e^e_{1,j}(r)\cdot\Delta\hat\varepsilon\cdot e^e_{0,j}(r)$ - the local amplitude coupling coefficients; $e^e_{i,j}(r)$ - the polarization vectors; $\Delta\hat\varepsilon$ - perturbation of the dielectric permittivity of the sample due to the formation of a photopolymer and the liquid crystal gratings; $n^1(r)$ - normalized amplitude profile of the grating first harmonics; $N^e_{r0,1}$ - group normals; c_c - the speed of light; $\Theta_j(r)$ - parameter of an integral phase mismatch:

$$\Theta_j(y,E) = \int_0^y \Delta K_j(y',E)dy', \quad (3)$$

where $\Delta K_j(y,E)$ - the modulus of the phase mismatch, which characterizes the diffraction geometry changes (Fig. 2a,2b) under the influence of an electric field:

$$\Delta K_i(r,E) = k[n^e_{0i}(r,E)(N^e_{0i}\cdot y_0) - n^e_{1i}(r,E)(N^e_{1i}\cdot y_0)] + \kappa\cdot y_0. \quad (4)$$

Appearing in (4) functions $N^e_{i,j}(r,E)$ can be found by solving the eikonal equation. The refractive indexes are defined by the following expressions:

$$n^e_{i,0} = n_o n_e[n^2_e\cdot\sin^2(\varphi(r,E)\pm\theta^e_{i,0}) + n^2_o\cdot\cos^2(\varphi(r,E)\pm\theta^e_{i,0})]^{-1/2}, \quad (5)$$

$$n^e_{i,1} = n_o n_e[n^2_e\cdot\sin^2(\pi-\varphi(r,E)\pm\theta^e_{i,1}) + n^2_o\cdot\cos^2(\pi-\varphi(r,E)\pm\theta^e_{i,1})]^{-1/2}, \quad (6)$$

where $\theta^e_{i,j}$ - the angles of incidence and diffraction of extraordinary waves (Fig. 1), and the dependence $\varphi(r,E)$ is determined from Fredericks equation:

$$\frac{1}{\xi_E(E)}\left(\frac{d}{2}+y\right) = \int_0^\varphi\left(\sin^2\varphi_m(r,E) - \sin^2\varphi\right)^{-1/2}d\varphi, \quad (7)$$

where $\xi_E(E)$ - optical coherence length, d - thickness of the sample; φ_m - the maximum angle of rotation of the LC director.

Using the solution of CWE (1), (2) in an optically inhomogeneous medium in the approximation of a given field [2], we represent the amplitude distribution of the diffraction light field from the angular spectrum (AS) plane waves:

$$E_d(\theta,E) = E_{p0}(\theta)T_{d,0}(\Delta,E) + E_{p1}(\theta)T_{d,1}(\Delta,E), \quad (8)$$

where we have introduced the partial transfer functions (TF) of GPhS:

$$T_{d,j}(\Delta,E) = \frac{1}{d}\cdot\int_0^d C(y,E)\cdot n^1(y)\cdot\exp\left[i\cdot\left(\Delta\frac{y}{d}+\Theta_j(y,E)\right)\right]dy. \quad (9)$$

where $E_{pj}(\theta)$ - AS of the incident light field; $C(y,E)$ - amplitude coupling coefficient [2]; $\Delta(\vartheta)$ - relative phase mismatch; $\vartheta = \theta - \theta_B$ - the deviation from the Bragg angle θ_B, the angle θ here characterizes the direction of plane-wave AS components $E_{d,j}(\theta,E)$ relatively to the light beams wave normals $N_{d,j}$.

Assuming $E_{p0}(\theta) - E_{p1}(\theta)$ we introduce the resulting PF:

$$T_d(\Delta,E) = T_{d0}(\Delta,E) + T_{d1}(\Delta,E). \quad (10)$$

The results of numerical simulation of TF modules based on expressions (9) - (10) are shown in Fig. 3.

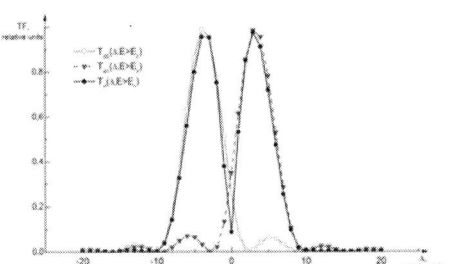

Fig. 3. Results of numerical simulations.

As can be seen from Fig. 3, under the influence of an alternating electric field on the structure, intensity of the slightly diverging light beam which falls on the change of control field polarity area (area of zero relative mismatch Fig. 3) is lower than when the homogeneous control field is there. This phenomenon is caused by taking into account the phase of the TFs in (10) and suggests a more efficient management of GPhS in PDLC diffraction characteristics by reducing the required strength of the electric field.

Thus, the resulting mathematical model describes the light beams diffraction on the holographic photonic PDLC structures when subjected to a spatially non-uniform control field. Shows an additional method of the GPhS in PDLC diffraction characteristics control if slightly divergent light beams fall to the area of the control field polarity change.

REFERENCES

[1] B.F. Nozdrevatykh, S.V. Ustyuzhanin and S.N. Sharangovich *TUSUR papers*, no. 1(21), p. 109, 2010.

[2] B.F. Nozdrevatykh, S.V. Ustyuzhanin and S.N. Sharangovich *TUSUR papers*, no. 2(16), p. 192, 2007.

[3] B.F. Nozdrevatykh, S.V. Ustyuzhanin and others *Electronic devices and control systems*, part 1, pp. 241-244, 2007.

[4] A.S. Sonin "Introduction to the physics of liquid crystals", M.: Nauka, 1983.

Defect Modes of a Two-Dimensional Photonic Crystal in the Presence of Acoustic Wave

Yu. V. Pilgun

Laboratory of Acousto-Optics
Department of Quantum Radiophysics
Taras Shevchenko National University of Kyiv
64/13 Volodymyrska St, Kyiv 01601, Ukraine

Abstract-**Defect modes of a photonic crystal modulated by propagating acoustic wave are investigated using supercell plane-wave expansion method. It is show that as a result of dielectric permittivity modulation resonant frequencies of defect modes change harmonically in time with the same frequency as the acoustic wave. Modulation level is larger for low-frequency acoustic wave and decrease with increasing frequency of sound due to spatial averaging of dielectric permittivity changes in defect region.**

Photonic crystals (PhCs) have been attracting much attention because of their promising applications in integrated optics. One particularly interesting aspect of these systems is the possibility of creating crystal defects that confine light in localized modes. Better understanding of these defect modes has stimulated the design of photonic-crystal waveguides, resonant cavities, filters, and other optical components. Recently, there have been many efforts to enable dynamic control of device's parameters and create active components based on PhC. The modulation can be achieved in numerous ways such as changing the temperature, external electric field, mechanical strain, or carrier density [1]. Also modulation caused by an acoustic wave is used [2-4].

In present paper we study possibility of controlling defect mode parameters with an acoustic wave. Spatial modulation of refractive index caused by the acoustic wave leads to change of resonant frequency of defect modes in PhC.

There are two possible topologies of a two-dimensional photonic crystal composed of dielectric material with air spacing. The first one is dielectric rods placed in air. The second one is inverse structure: dielectric base with holes filled with air. Obviously, only second type could be used for modulation with the acoustic wave because ultrasound propagates efficiently in bulk material but not in the air. In principle holes could be filled with another dielectric of different permittivity, but we consider air-filled structure in hope it will be easier to manufacture such a structure experimentally. Most simple and most studied structures are based on square and hexagonal lattices. Square lattice of circular holes in dielectric has significant drawback for our application. It's hard to create sufficiently large photonic band gap in such a structure if we use small radius of holes to keep structure mechanically rigid [5, 6]. Thus we limit our attention to hexagonal lattice of air holes in dielectric base, where large optical band gaps are possible for reasonable radius of holes.

The optical response of a photonic band gap crystal can be described by its dispersion relation. We use plane-wave expansion (PWE) method [5] to investigate mode frequencies and field distribution for defect modes in two-dimensional PhC. Supercell approach is applied to introduce a defect to periodic structure of the crystal [6]. Configuration of the structure is show in fig. 1. Dielectric permittivity of base material is assumed to be $\varepsilon_b = 11.56$, which corresponds to GaAs with refractive index $n = 3.4$ at light wavelength $\lambda = 1.55$ μm. Respective period of the structure $a = 375$ nm. Dielectric permittivity for air columns $\varepsilon_a = 1$. Radius of holes $r/a = 0.29$. We consider only in-plane x-y propagation of light and sound waves for two-dimensional structure with infinite size along z-axis.

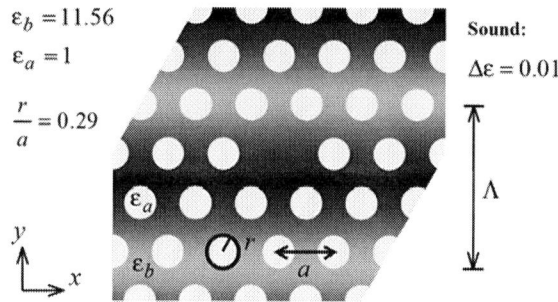

Fig. 1. Photonic crystal with low-frequency acoustic wave propagating along y-axis. Modulation of dielectric permittivity caused by acoustic wave is shown as shades of gray. Changes of geometrical sizes are not considered.

Photonic crystal we analyzed has optical band gap for TE polarization in normalized frequency range $0.209 \div 0.269$. The defect created by one missing hole introduces two very close modes with normalized frequencies 0.24099 and 0.24110 which have different field configuration. Distribution of z-component of magnetic field for defect modes is shown in fig. 2.

The acoustic wave is modeled as sinusoidal modulation of dielectric permittivity of base material. As the speed of light is five orders of magnitude higher than the speed of sound, quasistatic approximation is used. Dielectric constant of holes assumed to be unchanged by sound and equals to 1. Change of lattice geometry due to acoustic wave is not included in the model. Displacement caused by acoustic wave is too small compared to period of the structure and resolution of PWE method is not enough to take it into account. Perturbations of sound field near walls of holes are also neglected, which is

978-1-4799-0019-0/13 $31.00 © 2013 IEEE

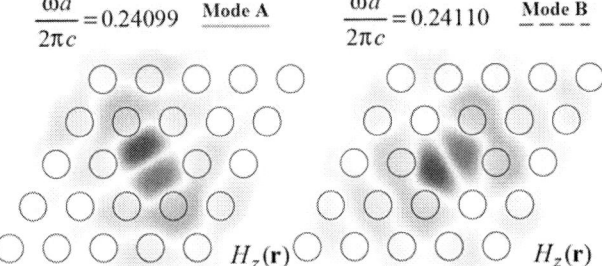

$$\frac{\omega a}{2\pi c} = 0.24099 \quad \underline{\text{Mode A}} \qquad \frac{\omega a}{2\pi c} = 0.24110 \quad \underline{\text{Mode B}}$$

Fig. 2. Optical field distribution of two defect modes with no sound applied. Neighboring regions of high amplitude have opposite sign.

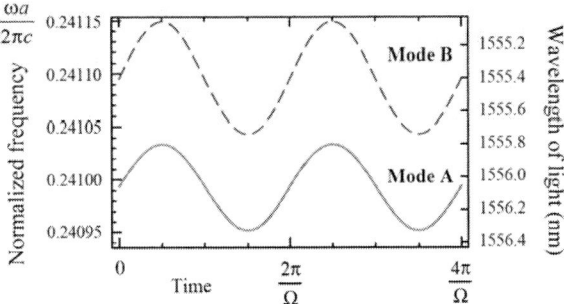

Fig. 3. Modulation of defect modes frequencies with time due to acoustic wave movement. Wavelength of sound $\Lambda = 3a$.

adequate approximation for low-frequency case when wavelength of sound is much bigger than period of the structure. The case when wavelength of sound is comparable with the structure's period should be treated separately, because the structure becomes a phononic crystal in addition to being the photonic crystal and manifests much more complex distribution of sound field.

Ultrasound of frequency higher than several gigahertzes is rarely used in practice because of difficulties of excitation and high material losses. Assuming density of base material is $\rho = 5300$ kg/m^3 and speed of longitudinal sound wave is $v_L = 4700$ m/s, wavelength of sound corresponding to frequency 1 GHz equals $\Lambda = 4.7$ μm, which is still one order of magnitude larger than period of the structure. This allows us not to take into account phononic properties of periodic structure and assume that acoustic wave propagates like in bulk material. Usually modulation of refractive index caused by acoustic wave is relatively small, so changes of optical band gap boundaries would be hard to detect experimentally. Changes in optical properties of defect modes should be easier to detect because of their resonant nature. We estimate that modulation of dielectric permittivity needed to produce noticeable effect on defect mode frequencies should be of order $\Delta\varepsilon = 0.01$. This corresponds to quite large energy flux of acoustic wave $\Phi = 60$ W/mm^2, assuming effective elastooptic constant of the material is $p_{eff} = -0.16$. It remains the question if it is possible to excite bulk acoustic wave of such a power in small volume of the photonic crystal, so actual realization will probably use surface acoustic wave for modulation like in [3]. Amplitude of particle displacement for such sound intensity and frequency is estimated as $A = 0.35$ nm. Displacement will be bigger at lower frequencies, but nevertheless a value is small enough not to consider geometrical changes of the structure.

To simulate acoustic wave propagation, we repeated calculation of mode frequencies for different spatial position of superimposed dielectric permittivity modulation. With change of time variable, frequency of defect modes shifts in response to modulation. As can be seen in fig. 3, change of frequency is harmonic in time and replicate time dependence of permittivity changes induced by acoustic wave. In current work we intentionally avoid conditions where classical acousto-optical interaction could arise, so present effect is solely due to time-dependent variation of dielectric parameters of resonant cavity formed by defect. Moreover,

only small region around defect is responsible for introduced modulation. Defect modes are highly localized in space, so they "feel" changes of dielectric permittivity only in immediate vicinity. We confirmed this conclusion with additional computations where only central region is modulated by acoustic wave. Masking out modulation outside circle of radius $3a$ around defect is enough to reproduce effectively the same result as with full modulation of the whole crystal. This gives important insight how frequency of defect modes could be varied. When acoustic wave has large wavelength, change of dielectric permittivity is almost constant over the defect, so biggest modulation level should be expected. If acoustic wave has small wavelength, several periods of wave would fill the defect. Net result of such modulation will be lower, because changes in dielectric permittivity will be averaged over the area. To illustrate this

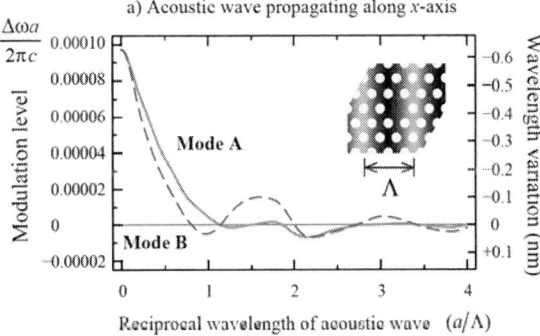

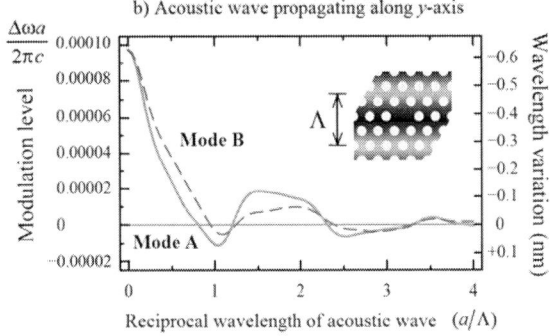

Fig. 4. Optical frequency modulation as a function of reciprocal wavelength of acoustic wave. Negative values correspond to antiphase modulation. Direction of acoustic wave propagation is shown on inset.

978-1-4799-0019-0/13 $31.00 © 2013 IEEE

idea we calculated modulation level for varied wavelength of sound. Plots for different direction of acoustic wave propagation are shown in fig. 4. It is seen than modulation level depends not only on wavelength of acoustic wave, but also on orientation of the wave. Defect modes have complex optical field distribution (fig. 2), which is altered by the presence of acoustic wave. Obviously, differently oriented acoustic waves interfere differently with optical field. In some cases one mode is affected by sound wave while other is not.

We should warn the reader that plots for modulation level at fig. 4 are not strictly valid for wavelength of sound less than structure periodicity size ($a/\Lambda > 1$). When wavelength of sound is comparable with size of the structure periodicity, acoustic field distribution inside crystal becomes much more complex and can't be described with the simple model we used here. In fact according to Bloch theorem envelope of acoustic field inside phononic crystal is a superposition of periodic functions with different spatial frequencies. So plots obtained here are still useful, since they give hints how different spectral components of resulting sound field distribution can influence optical modes.

CONCLUSIONS

It is shown than modulation of resonant frequency of defect mode in photonic crystal caused by an acoustic wave is possible without direct photon-phonon interaction. The effect appears due to time-harmonic modulation of dielectric permittivity in the defect region. Such situation is typical when wavelength of sound is larger than period of the structure. The effect is quite small, so high intensity of the acoustic wave is required to obtain noticeable modulation. The effect becomes even smaller when wavelength of sound is shorter than size of the defect region, because changes of dielectric permittivity get averaged over an area.

There are several possibilities to increase level of modulation caused by the acoustic wave. One is to use surface acoustic waves instead of bulk waves to obtain higher acoustic power density. Also surface acoustic waves have lower speed of wave propagation, which further increase changes of dielectric permittivity. Another approach is to take advantage of special acousto-optical materials such as tellurium dioxide (TeO_2), having particularly low speed of sound propagation in selected directions. Yet another possibility is to regard the periodic structure as a phononic crystal. In such case slowdown of sound wave or acoustical mode localization in periodic structure can be used to enhance changes of dielectric permittivity. To make it possible wavelength of sound should be comparable with the period of the structure, thus higher frequency of sound or a material with lower speed of sound needs to be used. All these approaches are still to be studied.

REFERENCES

[1]   H. Ruda, N. Matsuura, "Nano-Engineered Tunable Photonic Crystals in the Near-IR and Visible Electromagnetic Spectrum" in *Springer Handbook of Electronic and Photonic Materials*, Edited by S. Kasap, P. Capper, Springer, 2006, pp. 1011–1014.
[2]   S. Krishnamurthy and P. V. Santos, "Optical modulation in photonic band gap structures by surface acoustic waves," *J. Appl. Phys.*, Vol. 96, No. 4. pp. 1803–1810, August 2004.
[3]   N. Courjal, S. Benchabane, J. Dahdah, G. Ulliac, Y. Gruson, and V. Laude, "Acousto-optically tunable lithium niobate photonic crystal," *Appl. Phys. Lett.*, Vol. 96, 131103, March 2010.
[4]   I. E. Psarobas, N. Papanikolaou, N. Stefanou, B. Djafari-Rouhani, B. Bonello, and V. Laude, "Enhanced acousto-optic interactions in a one-dimensional phoxonic cavity," *Phys. Rev. B*, Vol. 82, 174303, 2010.
[5]   R.D. Meade, K.D. Brommer, A.M. Rappe, and J. D. Joannopoulos, "Existence of a photonic band gap in two dimensions," *Appl. Phys. Lett.*, Vol. 61, No. 4, pp. 495–497, July 1992.
[6]   J. D. Joannopoulos, S. G. Johnson, J. N. Winn, R. D. Meade, *Photonic Crystals: Molding the Flow of Light*, 2nd ed., Princeton and Oxford: Princeton University Press, 2008.

Simulation study on the mode cutoff frequency in large mode area photonic crystal fibers

Xiaojian Shu, Xianfeng Bao, Xiaojun Wang

Institute of Applied Physics and Computational Mathematics, Beijing 100088, China

Abstract: The number of transverse modes and the field distribution in large-mode area photonic crystal fibers (PCFs) are simulated. Through adjusting the PCF structure parameters, the mode properties can be controlled to meet the actual needs. The single mode cutoff in the triangular or square lattice large mode area PCFs is analyzed. The single mode cutoff frequency is obtained, and the V parameter and the normalized cutoff frequency V* are redefined. The result shows that, by increasing the number of the missing central air holes, V* is more and more close to 2.405 for triangular lattice PCF, and π/2 for square lattice PCF.

The properties of transverse modes in a photonic crystal fiber (PCF) [1] are often characterized by the so-called V parameter [2]. N.A. Mortenson et al [3] introduced a definition for V parameter, which using the natural size of PCFs core Λ as the effective core radius and replacing the refractive index of the silica core with the effective index of the fundamental mode. With their definition it can be shown that the condition for the higher-order mode cutoff can be formulated as the normalized cutoff frequency V*=π. However, their analysis only focused on the one-rod core triangular PCFs. Most large-mode area (LMA) photonic crystal fibers are usually obtained by removing more central surrounding air holes in the fiber transverse section. As the natural size of LMA PCFs core is not equal to Λ any more, their V definition and the V* need to be corrected for different structures.

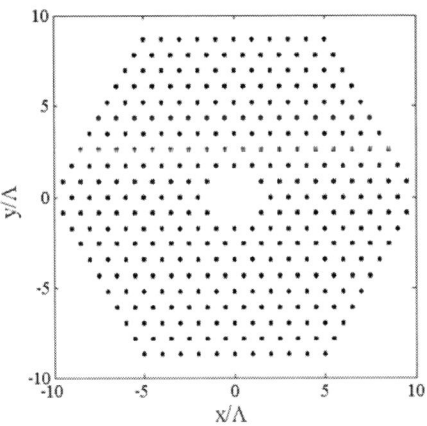

Fig. 1. Cross section of a 9-ring 7-rod core triangular lattice PCF.

We firstly redefine the V parameter and V* for 7-rod, 19-rod core triangular-lattice PCFs and 9-rod, 25-rod core square-lattice PCFs. Then, the analysis will focus on the single-mode

cutoff frequency kΛ* of these LMA PCFs. Finally, the information about the single-mode regime has been used to evaluate the cutoff value V*. The result shows that, by increasing the number of the missing central air holes, V* is more and more close to 2.405 for triangular lattice PCF, and π/2 for square lattice PCF.

N.A. Mortenson et al [3] introduced the following V parameter for one-rod core triangular PCF

$$V_{PCF} = k\Lambda \cdot (n_{FM}^2 - n_{FSM}^2)^{1/2} \qquad (1)$$

where k is the free-space wave number, Λ is the pitch, n_{FM} and n_{FSM} are the effective indices of the fundamental mode (FM) and the fundamental space-filling mode (FSM), respectively.

According to the different natural size of LMA PCF, specifically, for 7-rod core triangular PCFs and 9-rod core square PCFs, We correct Eq.1 to

$$V_{PCF} = k(2\Lambda) \cdot (n_{FM}^2 - n_{FSM}^2)^{1/2} \qquad (2)$$

and for 19-rod core triangular PCFs and 25-rod core square PCFs,

$$V_{PCF} = k(3\Lambda) \cdot (n_{FM}^2 - n_{FSM}^2)^{1/2} \qquad (3)$$

The number of transverse modes and the field distribution are simulated with two dimensional finite-difference time-domain (FDTD) method [4, 5]. The eigenvector represents the field distribution of an mode and the eigenvalue the eigen-frequencies (kΛ) at a fixed value of the propagation constant (β₀Λ), and the effective index of the mode can be obtained by n_{eff}=β₀/k.

The regime of the single mode can be determined by comparing the effective index of the first higher-order mode (n_{HOM1}) and that of the fundamental mode (n_{FSM}) of space filling PCF for a fixed d/Λ value [1, 6]. The first higher-order mode at a certain frequency (kΛ) is no longer guided if its n_{HOM1} is lower than the n_{FSM} at the same kΛ. So the cutoff frequency kΛ* is obtained with the condition n_{HOM1}=n_{FSM}.

The effective index of the fundamental mode, the first higher-order mode and the fundamental mode of space filling PCF as a function of the frequency kΛ is shown in Fig. 2. The cutoff frequency kΛ* is obtained with the condition n_{HOM1}=n_{FSM}, so kΛ* is 10, 5.5, 8, and 6.2 for the case of (a), (b), (c) and (d), respectively. It is noteworthy that the cutoff frequency kΛ* of 9-ring 7-rod core triangular PCFs with d/Λ=0.2 is in perfect agreement with the result of Ref. [7].

As shown in Fig. 3, VPCF and V* have been calculated. V* =V(kΛ*) is 2.6, 2.405, 2.0, and1.8 for the case of (a), (b), (c) and (d), respectively. It is noteworthy that V* for these PCFs is not equal to π, as that in one-rod core triangular PCFs. By

978-1-4799-0019-0/13 $31.00 © 2013 IEEE

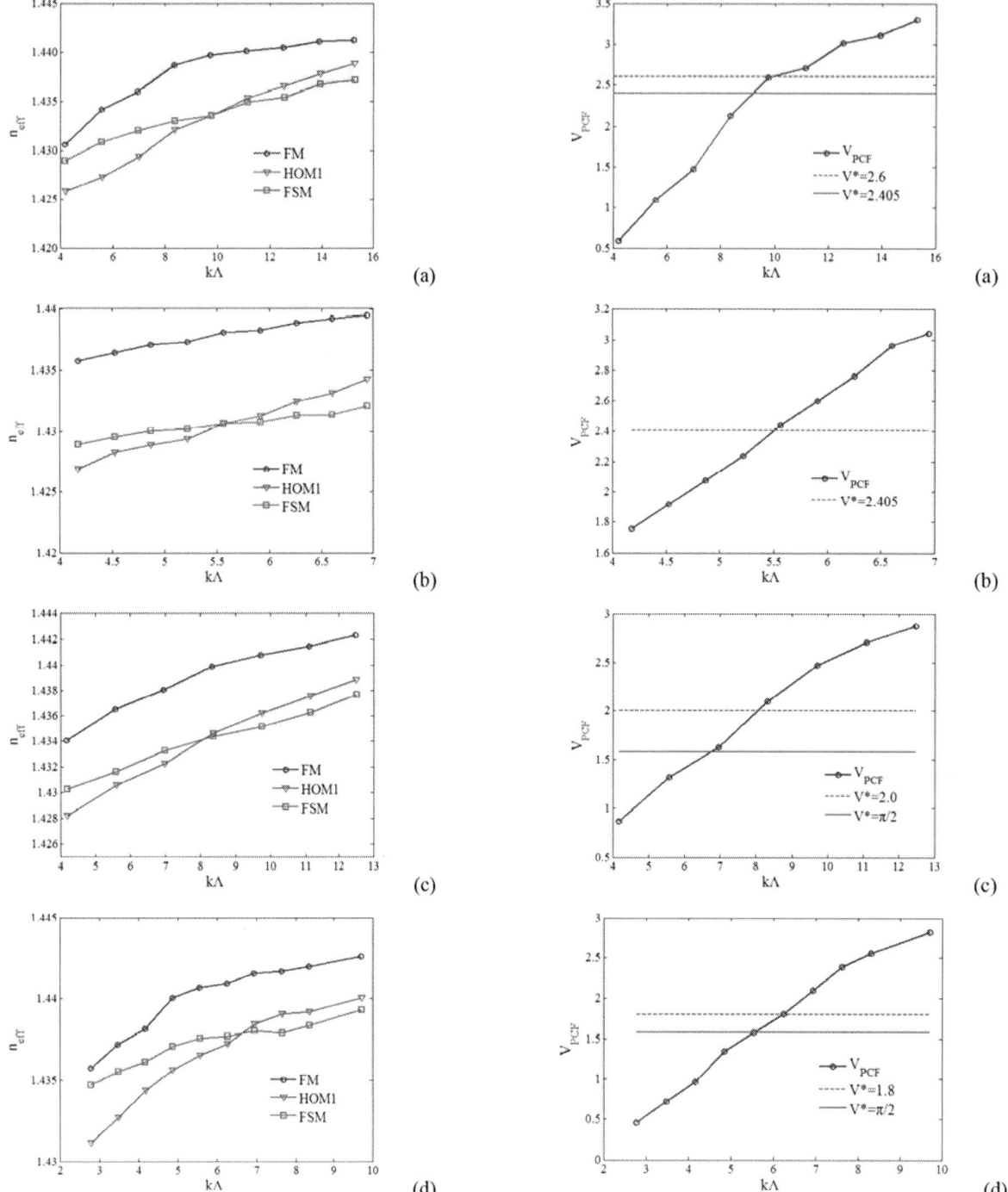

Fig. 2. Effective index of the FM, HOM1, FSM versus the frequency kΛ, for (a) 9-ring 7-rod core triangular-lattice PCFs with d/Λ=0.2, (b) 8-ring 19-rod core triangular-lattice PCFs with d/Λ=0.2, (c) 9-ring 9-rod core square-lattice PCFs with d/Λ=0.2, (d) 8-ring 25-rod core square-lattice PCFs with d/Λ=0.15.

Fig. 3. V_{PCF} as a function of the frequency kΛ for (a) 9-ring 7-rod core triangular-lattice PCFs, (b) 8-ring 19-rod core triangular-lattice PCFs, (c) 9-ring 9-rod core square-lattice PCFs, (d) 8-ring 25-rod core square-lattice PCFs. The dashed line indicates $V^*=V(kΛ^*)$.

978-1-4799-0019-0/13 $31.00 © 2013 IEEE   54

increasing the number of the missing central air holes, V* is more and more close to 2.405 for triangular-lattice PCFs, and $\pi/2$ for square-lattice PCFs.

In one-rod core triangular PCFs, the effective wavelength of the second-order mode is $\lambda_\perp^* \approx 2\Lambda$, which is just the size of the defect region where the mode fits in, and the normalized cutoff frequency becomes $V^* = \frac{2\pi}{\lambda_\perp^*}\Lambda \approx \pi$. In the same way, for a 7-rod core triangular-lattice PCFs which is obtained by removing the central air-hole and the first six surrounding ones, that effective wavelength becomes to $\lambda_\perp^* \approx 4\Lambda$, and $V^* = \frac{2\pi}{\lambda_\perp^*}2\Lambda \approx \pi$. But it is important to point that the 7-rod core triangular-lattice PCFs has much smaller air holes and larger core radius than that of one-rod core PCFs, leading to the weaker constraint of the electromagnetic field. So the second order mode effective transverse wavelength should be higher than 4Λ in 7-rod core triangular PCFs and the V* will be lower than π, which is 2.6 from our numerical simulations as shown in Fig. 3 (a). One cannot simply extend the derivation of V* for one-rod core triangular PCFs to the cases of LMA PCFs.

Furthermore, by removing more central surrounding air holes in the fiber transverse section, the shape of core in LMA triangular PCFs is close to a circle, so which should have the same mechanism of guidance as a cylinder waveguide, and it leads to V* in LMA triangular PCFs tend to the solution of the first zero of the Bessel function, 2.405. For a square PCF, which is similar to a square waveguide, therefore V* tends to $\pi/2$. So, by increasing the number of the missing central air holes, V* is more and more close to 2.405 for triangular lattice PCF, and $\pi/2$ for square lattice PCF.

REFERENCES

[1] N. A. Mortensen, M. D. Nielsen, J. R. Folkenberg, et al, "Improved large-mode-area endlessly single-mode photonic crystal fibers," *Optics letters*, vol. 28, no. 6, pp.393-395, 2003.

[2] T. A. Birks, J. C. Knight and P. S. J. Russell, "Endlessly single-mode photonic crystal fiber," *Optics letters*, vol. 22, no. 13, pp. 961-963, 1997.

[3] N. A. Mortensen, J. R. Folkenberg, M. D. Nielsen, et al, "Modal cutoff and the V parameter in photonic crystal fibers," *Optics letters*, vol. 28, no. 20, pp. 1879-1881, 2003.

[4] A. Cucinotta, S. Selleri, L. Vicetti and M. Zoboli, "Holey fiber analysis through the finite-element metho," *Photonics Technology Letters*, vol. 14, no. 11, pp. 1530-1532, 2002.

[5] Min Qiu, "Analysis of guided modes in photonic crystal fibers using the finite-difference timedomain Method," *Microwave and Optical Technology Letters* vol. 30, no. 5, pp. 327-330, 2001

[6] M. Nielsen, N. A. Mortensen, J. R. Folkenberg, A. Petersson, and A. Bjarklev, "Improved all-silica endlessly single-mode photonic crystal fiber," *Proc.* *Optical Fiber Communications Conference OFC 2003*, Atlanta, Georgia, USA, Mar. 23-28, 2003.

[7] M. Foroni, F. Poli, L. Rosa, et al. "Cutoff properties of large-mode-area photonic crystal fibers," *Proc. of 2005 IEEE/LEOS Workshop on. IEEE*, pp. 41-46, 2005.

Ternary Comb-like Silicon Photonic Crystal: Oxidation, Intrinsic Modes, Reflection Windows and Contrastivity

E. Ya. Glushko

Institute of Semiconductor Physics of NAS of Ukraine, Kiev Ukraine

Abstract: A ternary photonic crystal - A/B/A/C periodic structure is investigated analytically and numerically in the framework of transfer matrix formalism. The influence of oxidation to photonic gaps and positions of reflection windows for $(SiO_2/Si/SiO_2/Air)_N$ structure is calculated. It was shown that intrinsic optical contrastivity has a non-monotone behavior during the process of oxidation of silicon. The found results will allow to determine the optimal regimes of oxidation to obtain needed photonic device properties.

Introduction. The ternary comb-like photonic structures attract attention of investigators due to their more extensive list of useful properties in comparison with binary analogues. To characterize the general ability of a complicated photonic structure to create well expressed gaps in spectrum we will proceed from general definition of contrastivity of a multi

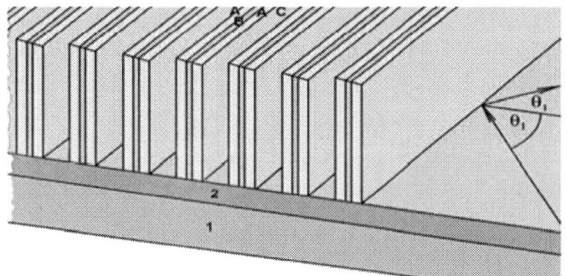

Fig. 1. Three component $(A/B/A/C)^{N-1}(A/B/A)$ photonic crystal – a part of a logic gate. 1, substrate; 2, protective anti-oxidizing layer, C, air voids, A, oxide layer, B, matrix material layer; θ_l, angle of incidence of external plane wave.

component system given in [1, 2]. A simple expression like $C_{ext}=(n_1-n_2)/(n_1+n_2)$ describing external optical contrastivity imply two contacting semiinfinite media with a plain boundary. For more complicated systems with well expressed intrinsic structure, like photonic crystals have, the definition of C_{ext} used above should be modified. Taking into account a circumstance that "the more is the internal contrastivity between constituents of an optical structure the more the relative gap is in photonic spectrum" following the definition introduced in [1, 2] we have for internal optical contrastivity C_{int}:

$$C_{int} = \frac{gaps}{gaps + bands}. \qquad (1)$$

In Fig. 1, a comb-like layered structure grown on a substrate 1 and consisting of N periods of alternating from left to right materials A, B, A and C with refractive indices, n_a, n_b, n_c. A

protective layer 2 provides an opportunity to control the A-layer thickness a due to, for example, the oxidation process.

Bandgap structure and reflection windows. In Fig. 2, calculated united bandgap and reflection window diagrams for a ternary layered 10-period $(SiO_2/Si/SiO_2/Air)_{10}$ structure are shown. At the upper θ_b panel, the photon energy dependence of field band and local modes inside the total reflection region of $(SiO_2/Si/SiO_2/Air)$ structure in the angular interval $(17°{-}37°)$

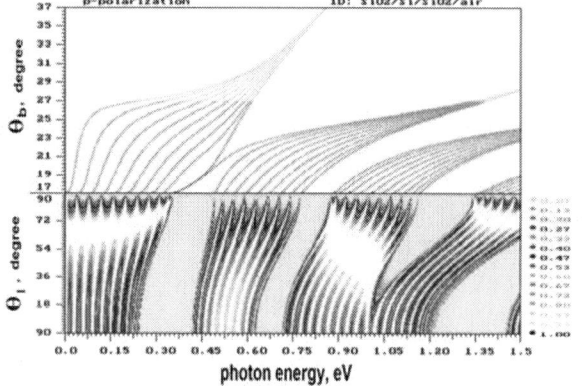

Fig. 2. Bandgap and reflection window diagram of a ternary layered structure. θ_b panel: bandgap frequency dependence of $(SiO_2/Si/SiO_2/Air)_{10}$ structure; θ_b, plane wave incidence angle inside the silicon b-layer, shown are $17°<\theta_b<37°$; $b=0.201~\mu m$, Si-layer thickness; $a=0.339~\mu m$, air voids thickness $c=0.120~\mu m$; $\varepsilon_a=2.40$; $\varepsilon_b=11.56$. θ_l panel (below): angular frequency color diagram for the reflection of external incident light; θ_l, plane wave external incidence angle $0°<\theta_l<90°$.

is plotted. Here, the interval of energy was chosen a little wider than the silicon transparency region $(0{-}1.0)~eV$. The thickness of Si-layer is taken $b=0.201~\mu m$, air voids, $c=0.120~\mu m$ and oxidized layer, $a=0.339~\mu m$, the total internal reflection angle for silicon is $\varphi_{TIR}=17.105°$. The high optical contrast is manifested in relatively big photonic gaps and quickly narrowing bands in the depth of the TIR region. At the lower panel, plotted is the color angle-energy diagram of reflection. In general, the reflection diagram consists of alternating reflection windows $R=1$ and windows of transmission with well expressed modal structure. In our case of a ternary system, the total transmission band observed for silicon at the Brewster angle of incidence near $\theta_l=72°$ is suppressed here by existing

978-1-4799-0019-0/13 $31.00 © 2013 IEEE

A-layers of silicon dioxide. Both diagrams, upper and lower, are matched in area of whispering incidence at the angle of incidence $\theta_f \sim 90°$ where sharp peaks of transmission transforms above into the resonator photonic modes of the TIR region.

Local states of two types – external and intrinsic relatively the layers A, may arise in a ternary comb structure which is placed in air - at that material C also is air. In Fig. 2, the weakly detached A-external local states arise at given geometry in the first gap at $21.4 > \theta_b > 17.1$ degree and in interval of photon energy between 0.338 and 0.488 eV. First signs of a local state exhibit itself when the thickness of A-layer begins to exceed 56.8 nm. At $a=90$ nm local state arises also in the second gap, then vanishes there at approximately 250 nm, and arises in the third gap at approximately 900 nm width of silicon dioxide. The local states of external relatively the A-layer kind may exist only in a filled comb structure.

Oxidation. The initial non-oxidized structure has the every silicon layer thickness b_0, and air void thickness c_0. Due to the protective anti-oxidation layer 2 (Fig.1) the system period $d_0 = b_0 + c_0$ remains constant during the process of oxidation. We assume in accordance with [3] that for every unit thickness of silicon consumed, 2.27 unit thicknesses of oxide will appear during the oxidation process. In Fig. 3, the relative gap dependence on photon energy is shown for $(SiO_2/Si/SiO_2/Air)_{10}$ structure with initial parameters $b_0 = 1$ μm, $c_0 = 1$ μm. The positive inclined parts of curves correspond to gap contribution into the bandgap ratio (1) and negative inclination - to bands.

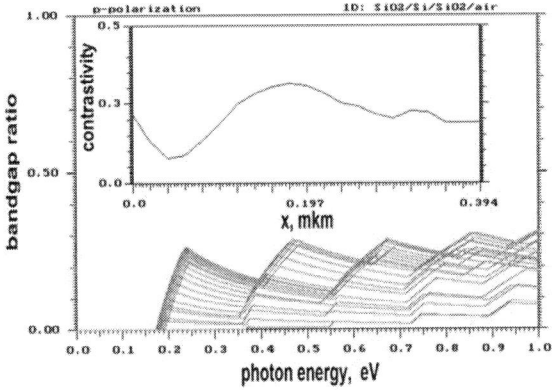

Fig. 3. SiO$_2$/Si/SiO$_2$/Air periodic structure. Relative gap dependence on photon energy. **Inset**: contrastivity C_{int} vs oxidation parameter x.

Both bands and gaps were taken into account inside the energy interval of transparency of silicon $(0-1.0)$ eV. The behavior of contrastivity in dependence on the oxidation parameter x is shown on the inset, where initial part of the curve corresponds to non-oxidized silicon whereas the right part of the inset, $x=0.394$, corresponds to completely oxidized silicon in the considered system. During the process of oxidation the following correlations between layer thicknesses are valid [3]:

$$\begin{cases} a = 2.27x \\ b = b_0 - 2x \\ c = c_0 - 2.54x \\ 2a + b + c = b_0 + c_0 \end{cases} \quad (2)$$

Depending on the etching temperature and the type of oxidation, wet or dry, time of the process may be as long as several hours [3] for the chosen above parameters. The contrastivity non-monotony observed in Fig. 3 is explained by complicate character of bandgap structure behavior during the process of oxidation: bands and gaps leave region of transparence with different velocity.

Contrastivity. The intrinsic contrastivity determines the ability of an optical material to form gaps and reflection windows in spectrum. Therefore, a non-contrastive medium should have continuous spectrum of electromagnetic waves with absent gaps. We have calculated the optical contrast x-c_0 surface for

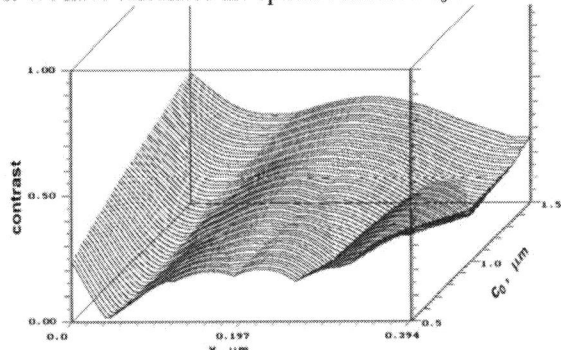

Fig. 4. Contrastivity C_{int} dependence of parameter x on initial thickness of air voids c_0 for SiO$_2$/Si/SiO$_2$/Air periodic structure. Optical transparency region $(0, 1$ eV$)$. $\varphi_{TIR}=17.105°$, C_{int} varies from 0.017 at $a=0.093$ μm , $c_0=0.5$ μm to 0.521 at $a=0$, $c_0=1.5$ μm.

the oxidized silicon periodic matrix SiO$_2$/Si/SiO$_2$/Air with starting thickness of silicon layer $b_0=1.0$ μm. The result is shown in Fig. 4, the gap-band ratio was calculated in region of optical transparency of Si for 60 different initial widths of air voids c_0 which were changed in interval from 0.5 μm to 1.5 μm. The surface has several ridges. Absolute minimum of $C_{int} = 0.017$ is found at $a=0.093$ μm, $c_0=0.5$ μm, absolute maximum 0.521 is found at $a=0$, $c_0=1.5$ μm. An intermediate maximum 0.420 is placed at $x=0.219$ μm, $a=0.497$ μm, $c_0=1.5$ μm.

Summary. Proceeding from the obtained x-c_0 contrastivity map of ternary SiO$_2$/Si/SiO$_2$/Air oxidized photonic crystal needed bandgap structure may be found. Controlled oxidation of prepared comb-like silicon photonic crystals is a means to create optical media with predetermined properties.

REFERENCES

[1] E.Ya. Glushko, "Optical contrastivity and sensitivity of multicomponent ordered structures," Opt. Commun. Vol. 285, no. 13-14, pp. 3133–3136, 2012.

[2] E.Ya. Glushko, "Optical contrastivity of complicated materials: finite layered structures," European Physical Journal D Vol. 66 no. 1, Article 23 –doi: 10.1140/epjd/e2011-20301-3, 2012.

[3] R. C. Jaeger, "Thermal Oxidation of Silicon," *Introduction to Microelectronic Fabrication*. Upper Saddle River: Prentice Hall. 2001.

Investigation of the 2-D Photonic Crystal Filter

A. I. Filipenko, A. N. Donskov

Kharkov National University of Radio Electronics, Kharkov, Ukraine

Abstract: We investigated the impact of the pillars diameter on the frequency band in the case of 2D photonic crystal guide. The obtained results show that the bandwidth (and its borders positions) changes with changing of the pillars diameter. We also have shown that by changing the diameter of some pillars we can design filters with various characteristics.

The photonic crystal is a periodic structure of alternating layers of materials with different refractive indices. A photonic crystal waveguide can be created by removing some pillars in a photonic crystal structure. Waveguides that are confined inside of a photonic crystal can have very sharp low-loss bends, which may enable an increase in integration density of several orders of magnitude. The distance between the pillars prevents light of certain wavelengths to propagate into the crystal structure. Depending on the distance between the pillars and on the diameter of pillars, waves within a specific frequency range are reflected instead of propagating through the crystal. This frequency range is called the photonic bandgap [1, 2]. By removing some of the GaAs pillars in the crystal structure it's possible to create a guide for the frequencies within the bandgap. Light can then propagate along the outlined guide geometry. In this work we consider the wave propagation in a photonic crystal that consists of GaAs pillars.

To investigate photonic crystal waveguide we use finite element method (FEM) [3].

The main research tasks are listed as follows:

- study TE waves propagation through the crystal;
- investigation of the photonic bandgap variations at a variation of GaAs pillars diameters;
- study the possibility of creating a photonic bandgap filter.

The geometry of an investigated photonic crystal is a triangular lattice of GaAs pillars surrounded by an air. The pillars diameters of an investigated photonic crystal is d = 125 nm and the distance between pillars centers (pitch) Λ = 380 nm.

To study TE waves propagating through the crystal we used a scalar equation for the transverse electric field component E_Z,

$$-\nabla \cdot \nabla E_Z - n^2 k_0^2 = 0$$

where n is the refractive index and k_0 is the free-space wave number.

The geometry of the investigated photonic crystal guide and the propagation of the wave through the guide are shown in Fig. 1.

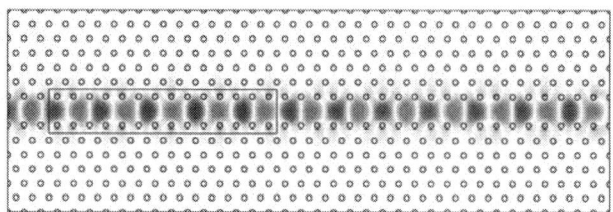

Fig. 1. The geometry of the investigated photonic crystal guide and the propagation of the wave through the guide.

To study TE waves propagation through the investigated photonic crystal waveguide and to find photonic bandgap we calculated total transverse electric field in a range from 0.75 μm up to 1.07 μm (200 values). The results that we obtained are shown in Fig. 2

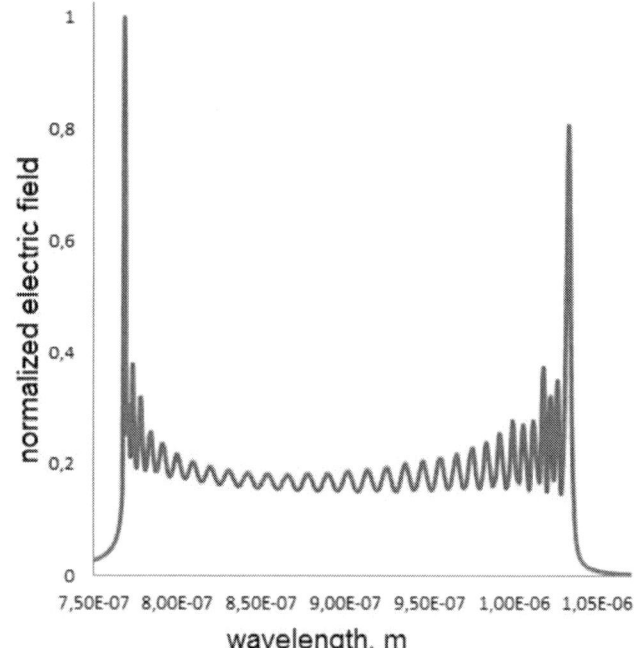

Fig. 2. The bandwidth of the investigated fiber.

Note that Fig.2 shows that total electric field of incoming wave bounces at the borders of bandgap.

To investigate the photonic bandgap variations at a variation of GaAs pillars diameters we changed pillars diameters in the range from 65 nm up to 195 nm. Fig.3 (a) shows the photonic bandgaps of the guides with different pillars diameters.

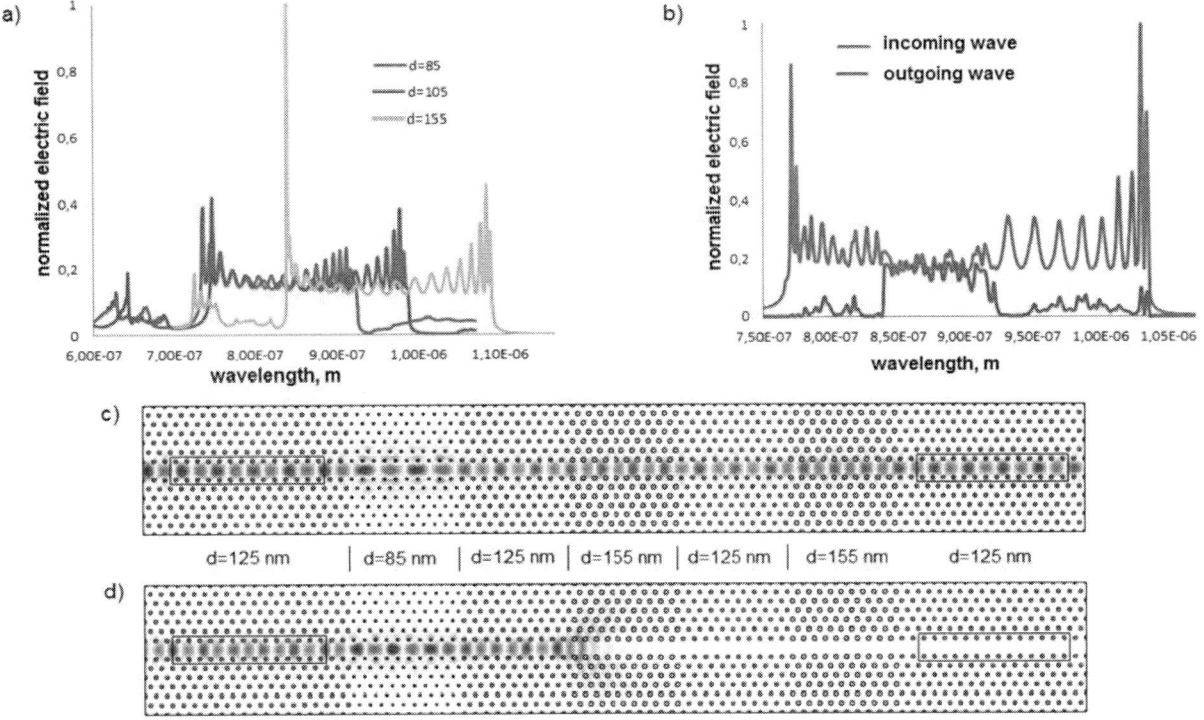

Fig. 3. (a) The photonic bandgaps of the guides with the different pillars diameters; (b) the photonic bandgaps of the considered filter; (c) the geometry of an investigated photonic crystal filter and the propagation of the wave through the filter; (d) wave doesn't propagate through the outlined geometry.

The result presented on Fig. 3 (a) show that the photonic bandgap of the considered type of waveguide shifts to the higher wavelength with the increasing of pillars diameters. Results also show the possibility of creating a photonic bandgap filter by creating the structure in which the incoming light wave will propagate through parts of photonic waveguide with different diameter of circular pillars of GaAs.

As you can see from Fig.3 (a) we can shift right border of bandgap to the lower wavelength if we implement to the waveguide structure the region with the lower diameters of the pillars and, properly, we can shift left border to the higher wavelength if we implement to the waveguide structure the region with the higher diameters of the pillars.

In this work we consider photonic bandgap filter with the structure in which the incoming light waves propagate through the waveguide with the pillars diameters in the first and last part d=125 nm (Fig. 3 (c)). In this filter we implement regions with an array of circular pillars with d=85 nm and d=155 nm. Note that to decrease peaks at lover wavelengths we implemented two regions with diameters of pillars 2 times. Fig.3 (b) shows bandwidth of the incoming and the outgoing waves. The propagation of the wave through the considered photonic crystal filter has shown on Fig.3 (c).

Fig. 3 (d) also shows that if the wavelength of the incoming wave is less or higher than the cutoff frequency of the waveguide, the wave does not propagate through the outlined guide geometry.

Presented results show that that the photonic bandgap of the considered types of photonic crystal waveguide shifts to the higher wavelength with the increasing of pillars diameters We also have shown the photonic bandgap filter in which incoming light wave will propagate through parts of photonic waveguide with different diameters of circular pillars of GaAs.

We believe that all presented results will be helpful for creating ultra-compact optical components and also believe that next research will add functionality to the photonic crystal waveguides.

REFERENCES

[1] J.D. Joannopoulus, R.D. Meade, and J.N. Winn, "Photonic Crystals", Princeton University Press, 1995.

[2] M. Skorobogatiy, J. Yang, "Fundamentals of Photonic Crystal Guiding", Cambridge University Press, 2009. – 263p.

[3] L. Oyhenart, V. Vigneras, "Photonic Crystals – Introduction, Applications and Theory", Published by InTech, pp. 267–290, March 2012.

Tapered double-clad optical fibers as gain medium for high power lasers and amplifiers

V.E. Ustimchik [1,2,*], Yu.K. Chamorovskii[1], V.N. Filippov[3], J. Kerttula[3], A.E. Ulanov[1,2] S.A. Nikitov[1,2]

[1]Institute of Radio-engineering and Electronics of the Russian Academy of Sciences,
Mokhovaya st. 11, bld.7, 125009 Moscow, Russia
[2]Moscow Institute of Physics and Technology (State University),
Institutskiy per. 9, 141700, Dolgoprudniy, Moscow region, Russia
[3]Optoelectronics Research Centre, Tampere University of Technology, 33101 Tampere, Finland
Tel: +79265739183; E-mail: ustimchikv@gmail.com

Abstract: We report advantages of tapered active double-clad fibers as gain medium for high power fiber lasers and amplifiers with diffraction limited output beam quality and extremely large output beam cross-section up to 100 μm in diameter

Fiber lasers possess a great number of advantages over other laser types; that is why they are used as reliable, efficient, compact, inexpensive, and high power sources in many scientific applications. Most applications require high output power and good quality of the laser beam. For this purpose, it is necessary to obtain a high power fundamental mode radiation with large beam cross-section, using pumping sources with a relatively large aperture and low brightness, and therefore, relatively inexpensive.

This work demonstrates advantages of the concept of high power fiber lasers and amplifiers, with nearly diffraction-limited output beam quality, based on a tapered optical double-clad fiber (T-DCF) as an active medium [1–3]. In contrast to standard multimode cylindrical fiber which is traditionally used in fiber lasers, tapered fibers allow to gain fundamental mode more effectively. Moreover having large input cladding diameters, these tapered fibers can be pumped by high-power, low-brightness laser diode bars. Also due to specific longitudinal geometric profile, tapered fibers have inherent mechanism of high mode filtration.

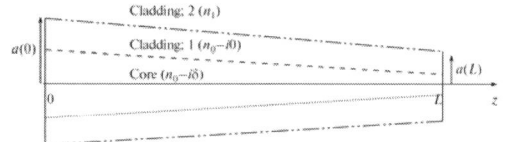

Fig.1 Schematic of double-clad tapered fiber.

The general structure of a tapered fiber is shown in Fig. 1. It consists of a core doped with rare earth ions (in particular in our case it is Yb^{3+}), optical guiding cladding 1 in which the pump radiation propagates, and outer cladding 2. The taper diameter gradually decreases

from the wide end a(0) to the narrow one a(L), which makes it possible to obtain singlemode radiation at the output end with using multimode pump coupled into wide end. The active tapered double-clad fiber is an axially nonuniform double-clad active optical fiber with the diameter varied along its length. It is practical to use the wide end of the T-DCF for launching the pump radiation from low-brightness sources.

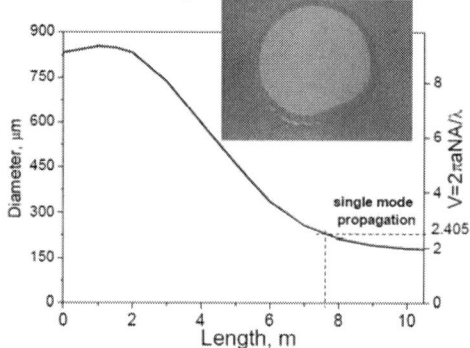

Fig.2 T-DCF clad diameter and normalized frequency as function of fiber length

Technology of T-DCF manufacturing has been developed, which allows to manufacture fiber tapers with different parameters such as concentration of active ions, tapering ratio, ratio of core/clad diameter and etc. Preforms were made by either SPCVD or MCVD technology. Ytterbium doped T-DCF samples were made by drawing on a tower with special pulling regime. The total length of samples varied from 3 m to over 15 m and cladding diameter was dependent on the length and example of one sample is shown in Fig. 2. For samples core/clad diameter ratios were typically from 1/5 to 1/30 and numerical apertures of core and cladding were about 0.114 and 0.22, respectively. The cross section of fiber is shown in the Fig. 2 as an inset. The core and cladding diameters of the wide part of the tapered fiber (as shown in fig.2) were 27 and 834 μm, respectively; the core and cladding diameters of the narrow part were 5.8 and 177 µm for this sample of fiber. The length-dependent outer diameter and corresponding normalized frequency V with a maximum value up to V = 38 was achieved. For other samples of T-DCF core diameters could reach over 100 µm in the wide end and had the same value of 5.8 µm at the narrow one, corresponded to singlemode operation regime with V = 2.4. Thereby tapering ratio could become the value over 17.

High operational parameters of T-DCF were experimentally and theoretically confirmed.

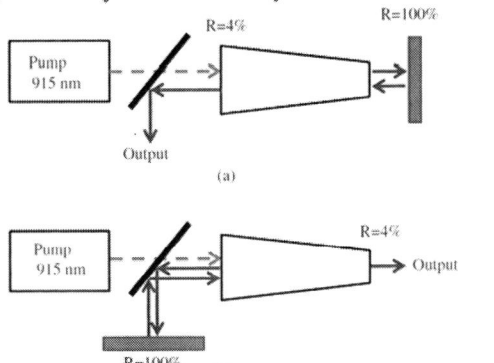

Fig.3 (a) Counter-propagation and (b) co-propagation laser schemes.

We demonstrated high power fiber lasers and amplifiers operated on both pulse and CW regimes.

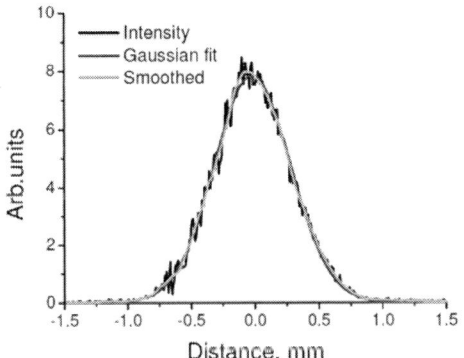

Fig.4 Output beam profile with $M^2 = 1.08$

For CW laser with output power about 600 W T-DCF showed $M^2 = 1.08$ and highly Gaussian distribution of output beam intensity (Fig.4) [13].

Comparison of the laser performance based on tapered and regular fibers shows that the slope efficiency of a tapered fiber laser is higher both for co-propagation and counter-propagation geometries fig.3 [5,6]. Additionally, the slope efficiency for the counter-propagation scheme is superior with both fiber types as shown both numerically and experimentally [5,6]. Furthermore, the counter-propagating design offers a substantially higher self-pulsing threshold compared with co propagating tapered lasers and lasers based on regular fibers. Although mode coupling exists in a coiled T-DCF, and the counter-propagating scheme uses a multimode output port, the output emission retains good beam quality. It was also found from modeling and measurements that pump absorption in a T-DCF is higher than that in a regular fiber with an equivalent

volume of active material. This feature originates from the mode mixing introduced by the taper shape, the larger active volume of a brightness-equivalent T-DCF, and the nonuniform distribution of active volume along the T-DCF, i.e., most of the gain material is located in its wide section at pump input end. Accordingly, pump absorption per unit length in a T-DCF is higher than in the equivalent regular fiber. This allows for shorter length of active fiber with the same density of the dopants, which in turn decreases the intracavity losses and improves the slope efficiency.

The results of simulations and measurements allow us to conclude that tapered fiber lasers offer significant advantages especially in terms of low-brightness pump launching, self-pulsing threshold, and pump conversion efficiency compared to lasers built from regular fibers [6]. The distinctive feature of tapered laser design is that enhanced power characteristics can be combined with good beam quality despite the mode coupling occurring in coiled tapers. Therefore, we can conclude that the T-DCF concept offers a practical alternative for bright, high-power fiber oscillators pumped by cost-effective low-brightness sources.

The mode composition of multimode fiber output is determined mainly by three factors: initial mode excitation, local (e.g. due to a splice) or distributed (e.g. due to microbending) mode coupling, and differential modal attenuation or amplification. Thus, high-quality initial excitation and absence of significant mode coupling sources are essential for robust singlemode propagation. Adiabatic, splice-free fiber tapering is an excellent method for fundamental mode excitation in a large-core, highly-MM fiber. In an ideal taper, significant mode coupling can be caused either by changes of the core diameter [8, 9] or by local bending [10, 11]. Mode coupling in the tapered fiber due to varying core radius has been considered theoretically in [8, 9], and found negligible in long adiabatic tapers with small tapering angle. Furthermore, mode coupling in tapered fibers caused by local bends was studied in [10, 11], and the effect was shown to be strong only for local bends with radii of a few millimeters. This allows assuming that mode coupling in a long adiabatically tapered fiber is insignificant for practical coiling radii of tens of centimeters, given that the taper is free from imperfections such as abrupt diameter changes.

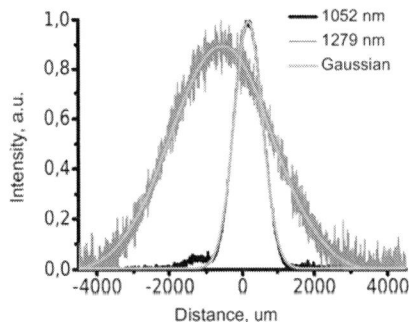

Fig. 5 Output beam profile using SLD 1052 nm and 1279 nm in comparison with Gaussian distribution

In [12], robust propagation of the solitary fundamental mode in long, step-index, highly-MM fiber tapers has been demonstrated for the first time with V up to 38 and fiber lengths of several meters. This became possible by the record tapering ratio (up to 17), and the absence of any significant structural flaws and splicing points, although some variation of core diameter due to small oscillation of the pull speed regulation system remains possible. The solitary fundamental mode propagation has been confirmed by several experiments (Fig.5). Parameter M^2 was less then 1.6 with effective output beam diameter 90 μm.

To conclude we would like to mention, that different schemes of fiber lasers and amplifiers were made based on T-DCF. We demonstrated high power lasers up to 600 W in robust singlemode regime [13]. Also we demonstrated a high-gain, 110W single stage narrowband amplifier based on tapered ytterbium-doped double-clad fiber. Maximum achievable gain was found to be 25.4 dB with a narrowband input signal, and 38.9 dB with a broadband input signal [7]. Furthermore it was investigated, that with the same intracavity optical power tapered fibers have much less optical power density in comparison with regular double-clad fibers [5]. That is why the nonlinear threshold is higher and maximum achievable output power in singlemode regime is higher too. Based on this we can expect further progress in application of active double-clad optical fibers for high average power, peak power lasers and amplifiers, improvement of beam quality.

REFERENCES

[1] V. Filippov, Y. Chamorovskii, J. Kerttula, K. Golant, M. Pessa, and O. G. Okhotnikov, "Double clad tapered fiber for high power applications," Opt. Express 16, 1929–1944 (2008).

[2] V. Filippov, Y. Chamorovskii, J. Kerttula, A. Kholodkov, and O. G. Okhotnikov, "600 W power scalable single transverse mode tapered double-clad fiber laser," Opt. Express 17, 1203–1214 (2009).

[3] V. Filippov, J. Kerttula, Y. Chamorovskii, K. Golant, and O. G. Okhotnikov, "Highly efficient 750 W tapered double-clad ytterbium fiber laser," Opt. Express 18, 12499–12512 (2010).

[4] D. Marcuse, Light Transmission Optics, (Van Nostrand Reinhold Company, New York, 1972), Chap. 9.

[5] V.E. Ustimchik, S.A. Nikitov, Yu.K. Chamorovskii, Simulation of Radiation Generation in an Active Double-Cladding Conical Fiber, Journal of Communications Technology and Electronics, 2011, Vol. 56, No. 10, pp. 1249, 1255

[6] Juho Kerttula, Valery Filippov, Yuri Chamorovskii, Vasily Ustimchik, Konstantin Golant, and Oleg G. Okhotnikov, Principles and Performance of Tapered Fiber Lasers: from Uniform to Flared Geometry, Applied Optics, Vol. 51 Issue 29, pp.7025-7038 (2012)

[7] Vasily Ustimchik, J. Kerttula, V. Filippov, Y. Chamorovskii, K. Golant, and O. G. Okhotnikov, Tapered fiber amplifier with high gain and output power, Laser Physics, Vol. 22, No. 11, pp. 1734-1738 (2012)

[8] A. W. Snyder, "Coupling of modes on a tapered dielectric cylinder," IEEE Trans. Microw. Theory Tech. 18(7), 383–392 (1970).

[9] D. Marcuse, "Mode conversion in optical fibers with monotonically increasing core radius," J. Lightwave Technol. 5(1), 125–133 (1987).

[10] P. M. Shankar, L. C. Bobb, and H. D. Krumboltz, "Coupling of modes in bent biconically tapered single-mode fibers," J. Lightwave Technol. 9(7), 832–837 (1991).

[11] L. C. Bobb, P. M. Shankar, and H. D. Krumboltz, "Bending effects in biconically tapered single-mode fibers," J. Lightwave Technol. 8(7), 1084–1090 (1990).

[12] Juho Kerttula, Valery Filippov, Yuri Chamorovskiy, Vasily Ustimchik, Oleg G Okhotnikov, Mode evolution in long tapered fibers with high tapering ratio, Optics Express, Vol. 20 Issue 23, pp.25461-25470 (2012)

[13] V. Filippov, Y. Chamorovskii, J. Kerttula, A. Kholodkov, and O. G. Okhotnikov, "600 W power scalable single transverse mode tapered double-clad fiber laser," Opt. Express 17(3), 1203–1214 (2009).

Investigation of modal content of radiation in multilayer cylindrical W- fibers

A. E. Ulanov[1,2], S. A. Nikitov[1,2], V. E. Ustimchik[1,2], Yu. K. Chamorovskii[2]

[1] Moscow Institute of Physics and Technology (State University), Dolgoprudny, Russia
[2] Institute of Radio Engineering and Electronics of the Russian Academy of Sciences, Moscow, Russia

Abstract: Computer simulation and experimental investigation of properties propagating and leaky modes in multilayer cylindrical W-fibers were presented in this paper. It was shown that leaky modes might significantly influence on properties of radiation propagation into cylindrical W-fibers depending on structure and they should be taken into account in the study and design of fibers.

Fiber technologies are developed rapidly in nowadays. For applications in telecommunications, industry, medicine it is necessary to use large mode area fibers. Important properties of these fibers are maintaining of single-mode propagation of radiation, low losses and low sensitivity to bending. The problem such fiber designing is solved in different ways. Involving bending into the fiber [1], chirally-coupled core fibers [2], microstructured fibers[3]. Another way to resolve this problem is usage of W-fibers. W-fibers have simpler structure than analogs, and therefore a simpler method of producing than above mentioned methods. In addition a big variety of parameters such as number and radii of claddings, the refractive index profile can change properties of W-fiber in a very wide range.

The aim of this paper is to study the mode content of radiation propagating in W-fibers. W-fibers with two and three claddings were studied in this paper. Both propagating and leaky modes were investigated. Leaky modes have complex constant of propagation β and effective refractive index of leaky modes is lower than refractive index of outer cladding. Leaky modes can significantly influence on properties of radiation propagation in the fiber. The research based on the computer model of radiation propagation in cylindrical W-fibers, earlier similar analysis had been done only in planar waveguides [4, 5]. Transmission matrix approach was used in numerical analysis of W-fibers [6]. In research were calculated propagation constants and leakage losses for propagating and leaky modes. Special attention was paid to the properties of leaky modes such as dependence of leakage losses on the fiber parameters, such as refractive index and radius of claddings, and the radiation wavelength. Figure 1 shows the dependence of the leakage losses on the radii of the claddings, where c is radius of second cladding and δ is difference between core radius and radius of first cladding. Figure 2 shows the dependence of the leakage losses on the differences of the refractive indices of the core and claddings, where n_1 is refractive index of core, n_2 is refractive index of first cladding and n_3 is refractive index of second cladding.

It can be seen in this figure that ranges of fiber parameters where loss of first mode is small exist while losses of higher modes are sufficient. With certain fiber parameters loss of first leaky mode in order of value may even equals to grey losses of propagating modes. It means that leaky mode will spread to significant distance without considerable losses. Consequently, in such structures wave front is determined by propagating and leaky modes. Therefore, propagating modes, leaky modes and their interaction should be considered while researching radiation propagation in such structures.

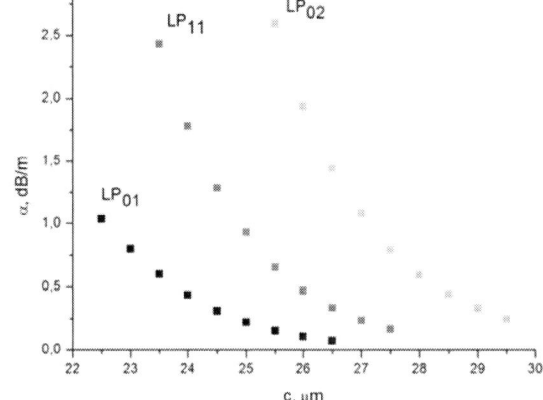

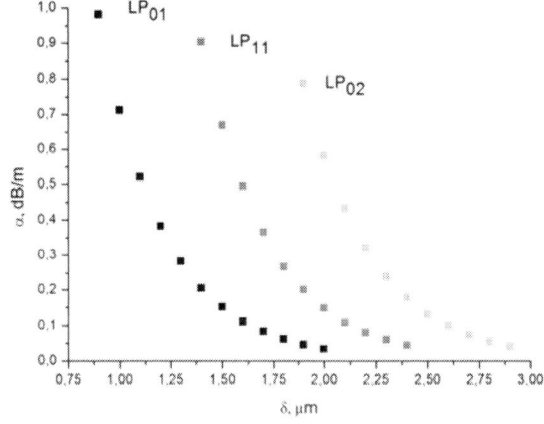

Fig. 1. Variation of the leakage loss with fiber parameters.

Optical fibers with a similar refractive index profile were fabricated. In these samples was carried out experimental research of mode content of the radiation. The experimental data with good accuracy coincided with the data obtained from computer modeling. Thus, it can be concluded that leaky modes contribute significantly to the properties of the radiation in W-fibers.

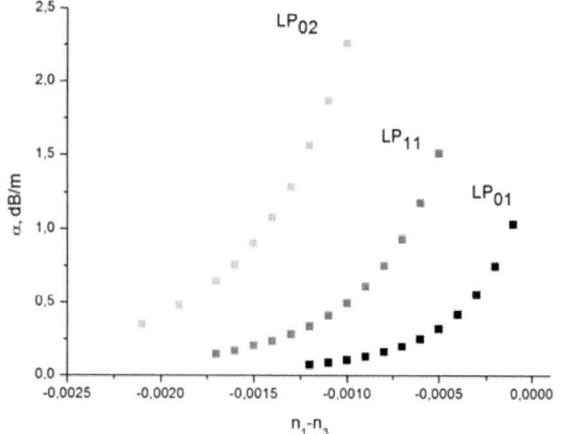

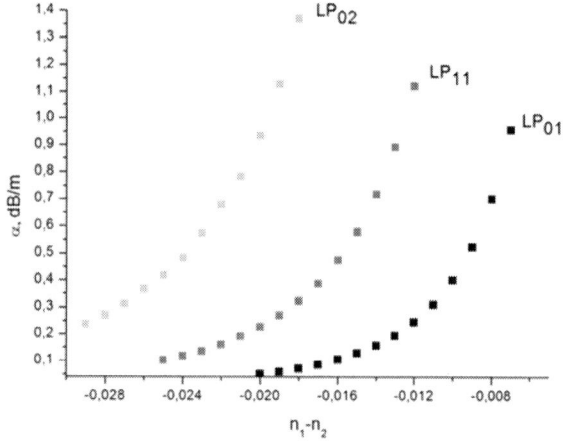

Fig. 2. Variation of the leakage loss with fiber parameters.

REFERENCES

[1] J. M. Fini, "Bend-resistant design of conventional and microstructure fibers with very large mode area," *Opt. Express,* vol. 14, no. 1, pp. 69-81, 2006.

[2] C. Liu, et al., "Effectively single-mode chirally-coupled core fiber," *Advanced Solid-State Photonics,* paper ME2, 2007.

[3] W. S. Wong, X. Peng, J. M. McLaughlin, and L. Dong, "Breaking the limit of maximum effective area for robust single-mode propagation in optical fibers," *Opt. Lett,.* vol. 30, no. 21, pp. 2855-2887, 2005.

[4] S. Yu. Otrokhov, Yu. K. Chamorovskiy, and A. D. Shatrov, "Leaking modes of guiding W-fibers with a large difference in the refractive index profiles," *Journal of Communications Technology and Electronics,* vol. 55, no. 10, pp. 1108-1115, 2010.

[5] С. Ю. Отрохов, Ю. К. Чаморовский и А. Д. Шатров, "Вытекающие моды W-световодов с большой разницей в профиле показателя преломления," *Радиотехника и электроника,* том 55, № 10, сс. 1185-1192, 2010.

[6] M. R. Shenoy, K. Thyagarajan, and A. K. Ghatak, "Numerical analysis of optical fibers using matrix approach," *Journal of Lightwave Technology,* vol. 6, no. 8, pp. 1285-1290, 1988.

Localized optical modes in photonic liquid crystals and low threshold DFB lasing

V.A.Belyakov

Landau Institute for Theoretical Physics, Russian Academy of Sciences, 142432, Chernogolovka, Moscow region

Abstract: A brief survey of the investigations on the localized optical modes (edge modes (EM) [1,2] and defect modes (DM) [1,3]) in chiral liquid crystals (CLC) and original results related to the analytic approach to the theory of the optical EM and DM and their applications in distributed feedback (DFB) lasing are presented.

The main experimental results which have to be understood and explained for lasing at localized optical modes are unusually low lasing threshold for DFB lasing [1] and existence of the anomalously strong absorption effect [4] in CLC at the localized mode frequency. The analytic study is facilitated by the choice of the problem parameters. Namely,

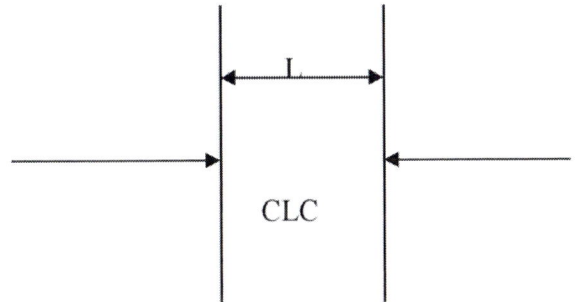

Fig.1

for DM a defect layer (with the averaged dielectric susceptibility equal to the average dielectric susceptibility of the CLC) sandwiched between two CLC layers inserted in an isotropic medium with the dielectric constant equal to the average dielectric constant of the CLC is studied. The chosen model allows one to get rid off the polarization mixing at the external surfaces of the localized mode structure (LMS) and to reduce the corresponding equations to the equations for light of circular diffracting in the CLC polarization only. The dispersion equations determining connection of the EM and DM frequency with the CLC layers parameters and other parameters of the LMS are obtained. Schematic of the boundary problem for EM is presented at Fig.1. Analytic expressions for the transmission and reflection coefficients of the LMS are presented and analyzed. For EM (Fig.1) amplitude reflection R(L) and transmission T(L) coefficients for light of diffracting circular polarization are:

$$R(L) = i\delta\sin qL / \{(q\tau/\kappa^2)\cos qL + i[(\tau/2\kappa)^2 + (q/\kappa)^2 - 1]\sin qL\} \quad (1)$$

$$T(L) = \exp[i\tau L/2](q\tau/\kappa^2) / \{(q\tau/\kappa^2)\cos qL + i[(\tau/2\kappa)^2 + (q/\kappa)^2 - 1]\sin qL\},$$

where $\kappa = \omega\varepsilon_0^{1/2}/c$, $q = \kappa\{1 + (\tau/2\kappa)^2 - [(\tau/\kappa)^2 + \delta^2]^{1/2}\}^{1/2}$, $\varepsilon_0 = (\varepsilon_{\shortparallel} + \varepsilon_\perp)/2$, $\delta = (\varepsilon_{\shortparallel} - \varepsilon_\perp)/(\varepsilon_{\shortparallel} + \varepsilon_\perp)$ is the dielectric anisotropy, and $\varepsilon_{\shortparallel}$, $\varepsilon_\perp$ are the local principal values of the LC dielectric tensor and $\tau = 4\pi/p$, where p is the cholesteric pitch, is the reciprocal lattice vector of the CLC spiral. The dispersion equation:

$$tgqL = i(q\tau/\kappa^2)/[(\tau/2\kappa)^2 + (q/\kappa)^2 - 1] \quad (2)$$

determines the discrete frequencies of the EM which are complex quantity for finite L (i.e. correspond to a EM finite life-time). Under the condition $(L\delta/pn) \gg 1$ the life-time of the EM (where n is the EM number) is given by

$$\tau_m \approx (1/4)(\varepsilon_0^{1/2}L/c)(L\delta/pn)^2. \quad (3)$$

Hence, for sufficiently thick CLC layers as their thickness L increases the EM life-time τ_m increases as the third power of the thickness and is inversely proportional to the square of the EM number n connected with the EM frequency by the relation $qL = n\pi$. The distribution of the EM energy in the layer for the n = 1,2,3 is shown at Fig.2 (L/p=200, the number of maxima at each curve coincides with n).

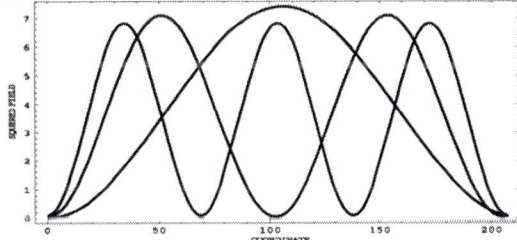

Fig.2

The studies of DM [3] for LMS presented at Fig.3 result in the following. To be specific as defect layers of thickness d were considered isotropic, birefringent, absorbing and amplifying layers in a perfect CLC structure.

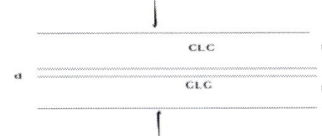

Fig.3

978-1-4799-0019-0/13 $31.00 © 2013 IEEE

The transmission $|T(d,L)|^2$ and reflection $|R(d,L)|^2$ coefficients for DM structure (Fig.3) with an isotropic defect layer for light of diffracting circular polarization are:

$$|T(d,L)|^2 = |[T_eT_d\exp(ikd)]/[1-\exp(2ikd)\,R_dR_u]|^2, \quad (4)$$

$$|R(d,L)|^2 = |\{R_e+R_uT_eT_u\exp(2ikd)/[1-\exp(2ikd)R_dR_u]\}|^2,$$

where $R_e(T_e)$, $R_u(T_u)$ and $R_d(T_d)$ are the amplitude reflection (transmission) coefficients of an individual CLC layer for the light incidence (Fig.3) at the outer (top) layer surface, for the light incidence at the inner top CLC layer surface from the inserted defect layer and for the light incidence at the inner bottom CLC layer surface from the inserted defect layer, respectively.

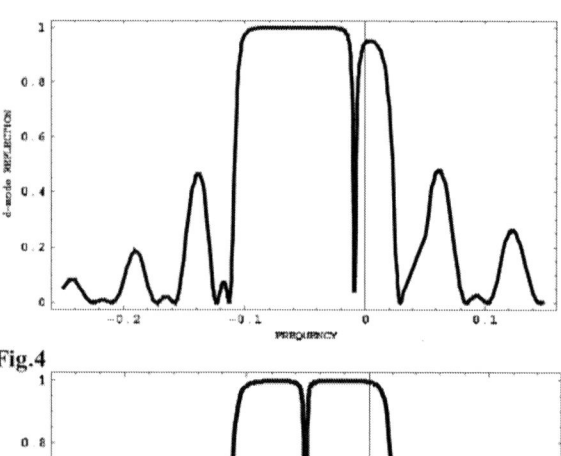

Fig.4

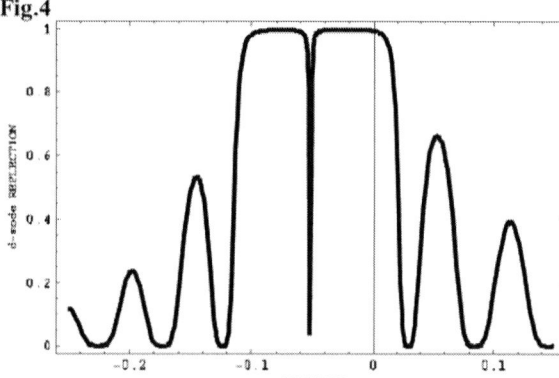

Fig.5

Calculations of reflection coefficient (δ=0.05, l=200, l=Lτ=2πN, where N is the director half-turn number at the CLC layer thickness L) for isotropic defect layer presented at Fig.4 (d/p=0.1) and Fig.5 (d/p=0.25) reveal a strong dependence of the DM frequency (corresponding to the reflection minimum) on d. At Fig.4 and in all figures below the parameters of LMS are the same as at Fig.4 and at the frequency axis the frequency deviation from the stop band edge is plotted (normalized by the Bragg frequency multiplied by δ).

The DM frequencies ω_D (which are complex quantity for finite L, i.e. correspond to a finite DM life-time) are determined by the following dispersion equation:

$$\{\exp(2ikd)\sin^2 qL-\exp(-i\tau L)[(\tau q/\kappa^2)\cos qL$$
$$+i((\tau/2\kappa)^2+(q/\kappa)^2-1)\sin qL]^2/\delta^2]\}=0. \quad (5)$$

The maximum for the DM life time τ_D corresponds to the location of the DM frequency just at the middle of the stop band. For thick CLC layers in the LMS at the condition $|q|L\gg1$ and the DM frequency at the middle of the stop band the life-time τ_D is:

$$\tau_D=(3\epsilon_0^{1/2}/4)(L/c)\exp[2\pi\delta L/p], \quad (6)$$

i.e. grows exponentially with the thickness increase. The coordinate distribution of DM intensity at DM frequency for infinitely thick CLC layers in the Fig.3 are presented at Fig.6 (with the distance normalized by p and δ =0.05, 0.1, 0.2 from the top to the bottom, respectively.

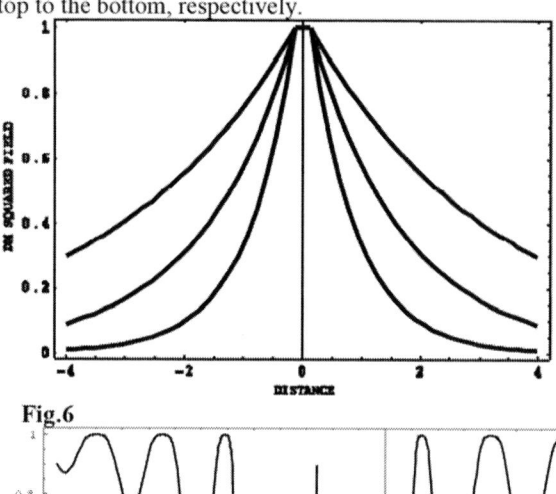

Fig.6

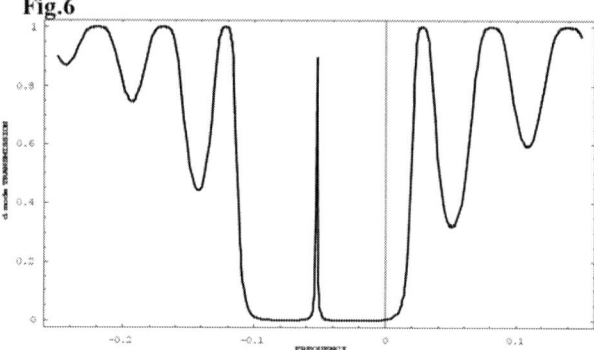

Fig.7

The localized EM and DM reveal themselves in a number of optical phenomena, the most known is DFB lasing. The lasing threshold for DFB lasing occurs to be lower than in the conventional lasing. The main corresponding results are as the following. An active media in lasing should be amplifying at the lasing frequency and absorbing at the pumping frequency. To take into account the absorption or amplification in the active media we define the ratio of the dielectric constant imaginary part to the real part of ϵ for the active media (CLC) as γ, i.e. $\epsilon=\epsilon_0(1+i\gamma)$ with positive and negative γ for absorbing and amplifying media, respectively. It happens that the lasing threshold for DFB lasing at the frequency of localized modes is decreasing with the CLC layers thickness increase. In a general case one has to solve numerically the dispersion equation (2) for EM and (5) for DM, respectively, relative to γ [2,3]. In the case of thick CLC layers γ corresponding to the lasing threshold for DFB lasing may be found analytically [2,3]:
$\gamma= -\delta(n\pi)^2/(\delta L\tau/4)^3$ for EM and

$$\gamma = -(4/3\pi)(p/L)\exp[-\pi\delta(L/p)] \text{ for DM.} \qquad (7)$$

The discussed here lowering of the lasing threshold is observed experimentally [1]. Another option to reduce the lasing threshold is connected with the anomalously strong absorption effect in CLC which reveals itself at the frequencies of localized modes in absorbing media [4]. The experimental observation of this option (reducing the lasing threshold) was reported in [5] where the pumping frequency was coinciding with the EM frequency at the short wave length edge of the stop band and the lasing frequency was coinciding with the EM frequency at the long wave length edge of the stop band. Under these circumstances a low threshold lasing at the EM frequency and an enhanced absorption (at another EM frequency) were ensured simultaneously resulting in a further lowering of the lasing threshold.

Another effects are related to the LMS presented at Fig.3 in the case of active (birefringent, absorbing or amplifying) defect layers. It is shown [6,7] that the layer birefringence reduces the DM life-time in comparison with the case of LMS with an isotropic defect layer and only at discrete values of LMS parameters it achieves the value of the corresponding LMS with an isotropic defect layer. The cause of it is the light polarization conversion in the defect layer resulting in transformation of the diffracting polarization light into light of nondiffracting polarization escaping from the LMS. In the case of a defect layer with a low birefringence the transmission $|T(d,L)|^2$ and reflection $|R(d,L)|^2$ coefficients for DM structure (Fig.3) with a birefringent defect layer for light of diffracting circular polarization are:

$$|T(d,L)|^2 = |[T_eT_d\exp[ikd]\cos(\Delta\varphi/2)/$$
$$[1-\exp[i2kd]\cos^2(\Delta\varphi/2)R_dR_u]|^2, \qquad (8)$$
$$|R(d,L)|^2 = |\{R_e+R_dT_eT_u\exp[i2kd]\cos^2(\Delta\varphi/2)/[1-$$
$$\exp[i2kd]\cos^2(\Delta\varphi/2)R_dR_u]\}|^2, \qquad (9)$$

where $\Delta\varphi$ is the phase difference of two beam component with different linear eigen polarization at propagation in the defect layer of thickness d. At small $\Delta\varphi$ the the spectral shape of transmission and reflection are the same as for an isotropic defect layer (see Figs.7).However, for larger $\Delta\varphi$ the spectral shape of transmission and reflection changes significantly (see Fig.8, Fig.9 for $\Delta\varphi = \pi/16$, $\pi/8$, respectively).

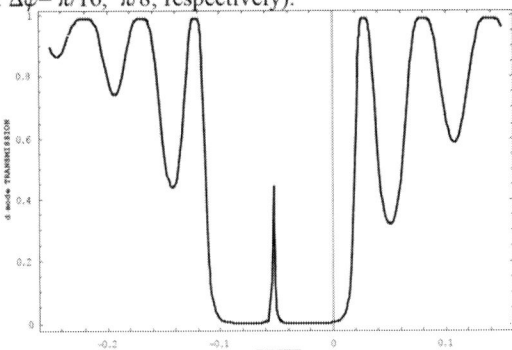

Fig.8

Correspondingly, the effect of anomalously strong light absorption (and amplification) at the defect mode frequency and, consequently, the lowering of the lasing threshold are not so pronounced for a birefringent defect layer as in the case of the LMS with an isotropic defect layer.

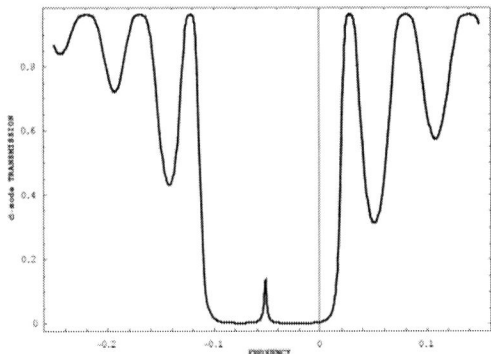

Fig.9

The effect of anomalously strong light absorption at the DM frequency for absorbing defect layer is discussed. It is shown that in DFB lasing at LMS with an amplifying defect layer an adjusting of the pumping frequency to the DM or EM frequency results in a significant lowering of the lasing threshold and the threshold gain is lowering with increase of defect layer thickness. The options of effectively influence at the DM properties by varying the defect layer parameters are discussed.

The work is supported by the RFBR grants 12-02-01016-a, 13-02-9045-Ucr-f-a and 13-02-92601-RS-a.

References:

[1] V. I .Kopp, Z.-Q. Zhang, and A. Z.Genack, "Lasing in chiral photonic structures",*Prog. Quant. Electron.* 2003. V.27, n 6, P.369.

[2] V.A.Belyakov, S.V.Semenov, "Optical edge modes in photonic LC," JETP., **109**, 687 (2009).

[3] V.A. Belyakov and S.V.Semenov, *Defect modes in photonic LC, JETP,* **112**, 694 (2011).

[4] V.A.Belyakov, A.A.Gevorgian, O.S.Eritsian and N.V.Shipov, " Abnormal optical absorption in chiral LC ," Zhurn.Tekhn.Fiz.,**57**, 1418(1987) [Sov. Phys. Technical Physics, **32** (n7), 843- 845 (1987), English translation]; Sov. Phys. Crystalography. **33** (n3), 337 (1988).

[5] Y.Matsuhisa, Y.Huang, Y.Zhou et al., "Low-threshold lasing," *Appl.Phys.Lett.* **90**, 091114 (2007).

[6] V.A. Belyakov, "Defect modes at birefringent defect layer," *MCLC*, 559, 50 (2012).

[7] V.A. Belyakov, "Defect modes at active defect layer," *MCLC*, **559**, 39 (2012).

The researches of optical fibers speckle at the presence of optical vortices

D.V. Kiesewetter, N.V. Ilin
Saint Petersburg State Polytechnical University, Saint Petersburg, Russia

Abstract: Investigated the interference pattern and the speckle structure of radiation of optical fibers near output end in the presence of optical vortices. The basic properties of these speckle structures experimentally and by numerical method defined.

The technical features of fiber-optic devices on base of multimode optical fibers (OF) depend from intensity distribution of the optical radiation, coming out of optical fiber. When use the coherent radiation on output of multimode OF appears speckles as a result of interference of the radiation of waveguide modes. From the parameter of speckles depends parameters of optical fiber sensors, features of the noise in photo receiving devices, performance attributes of devices for optical manipulate micro particles. For waveguide modes with flat wave front, which will hereinafter be identified as "usual", the main characteristic of speckle studied fairly well.

However, in optical fibers can also be distributed optical vortices. Properties of speckle structures formed optical vortices differ from the properties of speckle, formed by usual waveguide modes [1], that affects, for example, to the signal to noise ratio (SRS) at spatial filtering radiation [2]. Therefore, the study of distributions of intensity of radiation near the output end of the optical fibers is relevant and can be applied value.

The basic properties of the interference patterns, formed by optical vortices and waveguide modes, could be found if you treat interference of solitary waves, which have a flat wave front, and solitary vortices. It is known that the interference pattern of divergent beam with spiral-like wave front and the supporting plane wave is a spiral in the plane of observation. The interference pattern at the core of optical fiber usual waveguide modes and an optical vortex have the similar property, and the phase spiral changes along the fiber axis. In other words, the rotation of interference pattern takes place when moving plane surveillance along the fiber.

Similar properties have the speckle structure, formed in the result of interference of optical vortices and waveguide modes of optical fiber. In most theoretical and experimental works is investigated intensity distribution either in the near or far field of diffraction (NFD, FFD). Therefore, studying the properties of the interference patterns, formed separate waves and vortices, you can generalize a result to the properties of speckle structures. The study of such interference patterns it is possible to perform using the method of numerical simulation.

Investigate experimentally the interference pattern of the waves in the core of the OF along the axis of the fiber is not possible. In the far-field diffraction in any direction relative to

the axis of the waveguide phase of the interfering waves, caused by outgoing radiation as conventional waveguide modes, and optical vortices, change equally, depending on the distance to the output end.

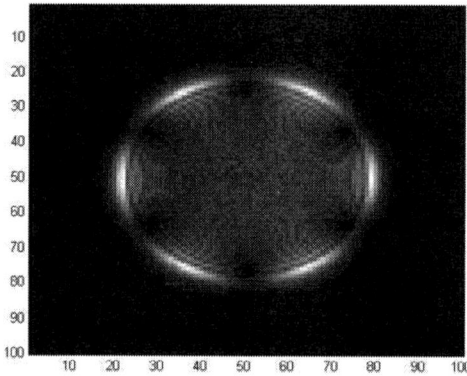

Fig. 1. Visualized intensity distribution at a distance of 20 micrometers from the output end of OF for $m=5$, $s=1$

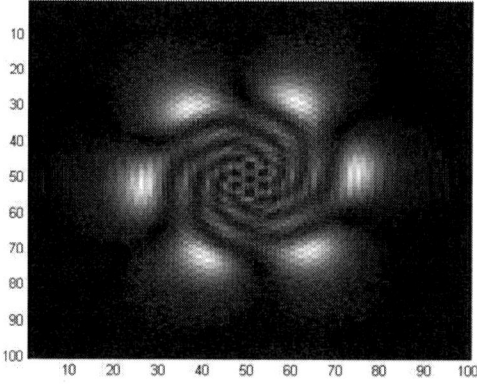

Fig. 2. Visualized intensity distribution at a distance of 400 micrometers from the output end of OF for $m=5$, $s=1$

Therefore, if you change the distance from the output end to the image plane of rotation of the speckle are not observed, and the linear dimensions of the interference pattern increases proportionally to the distance to the end, i.e. in FFD intensity angular distribution remains unchanged.

Currently remain poorly studied characteristics of speckle structures and the interference patterns that occur with the participation of optical vortices, transformation of the near field in a far field of diffraction. For this case, the calculations were made using the expression for the electromagnetic field in the approximation of weakly guided modes, forming a linearly polarized waveguide group in an optical fiber with a step index refractive profile.

Electromagnetic waves, the azimuthally dependence which is described by the function of sine or cosine ($\sin\varphi$, $\cos\varphi$) should be considered as standing waves to a coordinate φ. Such waves are common waveguide modes. The waves of the azimuthally dependence of which is described by a function of the type $\exp(-il\varphi)$, where l is the azimuth index, should be considered as a traveling wave to coordinate φ. Such a wave is called optical vortex. The movement of the wave front of such wave is spiral-like.

Calculation of diffraction for an arbitrary distance to the plane of observation was made by the numerical method [3] using the following formula

$$E(x', y') = \frac{ik}{4\pi}(1+\cos\theta) \int\limits_{(S)} E(x, y) \frac{\exp(-ik|r - r'|)}{|r - r'|} dxdy$$

where $k = 2\pi/\lambda$ (λ – is the wavelength, θ – is the angle between the axis Oz and the line passing through the point $(0, 0)$ in the plane of Oz (x', y') in the plane of observation $X'O'Y'$. The plane of butt is XOY, the coordinates of points in the plane – (x, y), the radius-vector – r; distribution of electric tension – $E(x, y)$. The distance counted the axis Oz. Point O' also lays on the axis Oz. The plane of observation is $X'O'Y'$, the point in this plane coordinates is (x', y'). The distance from the plane XOY to plane $X'O'Y'$, i.e. the distance to the plane of observation, we denote by L_p. S is the area of integration.

As an example, Fig. 1 and Fig. 2 show the distribution of intensity of the outgoing radiation of the optical fiber with step index profile of the refractive index. The core radius is 100 micrometers; numerical aperture is 0.22 at wavelength of 0.6328 micrometers. Azimuth index longitudinal projection of the vector of electric strength of the waves is $m=5$, radial index – is $s=1$. The calculations allow us to describe the main regularities of transformation of distributions of intensity of the interference patterns, formed by the optical vortex and a plane wave at the output end of the optical fiber. From the data obtained it follows that the interference patterns that have near-field diffraction radial distribution of the intensity of the transformation in a far field can acquire spiral type. The similar result is obtained for the different types of waveguide modes and optical vortices for multimode OF with any parameters.

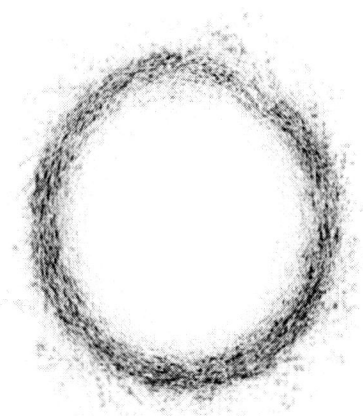

Fig. 3. Invert image of the speckle pattern of radiation of optical fiber in the presence of optical vortices with similar directions of the wave front.

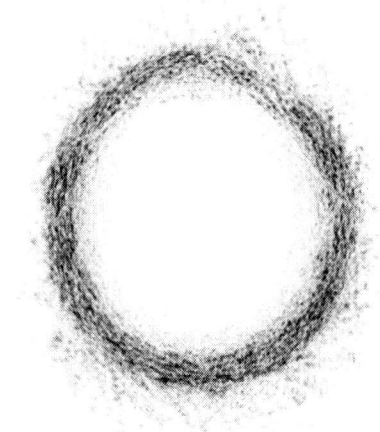

Fig. 4. Invert image of the speckle pattern of radiation of optical fiber, which differ from Fig. 3 the fact that the plane of observation is shifted to 50 micrometers along the fiber axis.

With increasing distance from the output end of the fiber to the plane of observation the rotation of interference pattern takes place, i.e. the rotation continues in free space. With increasing distance L_p angular velocity of rotation ($\Delta\varphi/\Delta L_p$) decreases and gradually decreases to zero in FFD. The angular velocity of rotation of the interference pattern depends on the indexes of modes and vortices' orders.

For excitation of optical vortices in optical fibers used experimental setup described in [4]. The essence of the applied method of excitation of optical vortices is to illuminate the end face of the optical fiber focused laser beam, the centre of which is shifted relative to the axis of the fiber on the value of l_s to the optical axis of the beam is inclined about the axis OF the angle γ_s in the plane perpendicular to the plane passing through the axis of the fiber and the center of the focused beam in the plane of the end OF.

Using the terminology of the ray approximation, we can say that the used device created in optical fiber oblique rays. The studies were performed OF a stepped profile of the refractive index with quartz core diameters of 200, 600 and 1000 micrometers and polymer reflective jacket. As a radiation source used gas, semiconductor and solid state lasers with wavelengths of radiation 632.8 nm, 650 nm and 532 nm. Registration of distributions of intensity of the output radiation was made of the matrix television camera with focusing micro lens located on the micro metric movable table.

As an example, Fig. 3, 4 presents the distribution of the intensity of the outgoing radiation in the plane perpendicular to the axis of the waveguide; plane shifted 50 microns along the axis OF. It is possible visually observe the rotation spots of speckles on some angle and elasticity of grains of speckles [1, 5].

To determine the angle of rotation of speckle the calculation of the cross correlation functions (CCF) distributions of intensity versus $\Delta\varphi$ (azimuth shift) at different distances from the output end of the following formula

$$F_{SC}(\Delta\varphi) = \frac{1}{2\pi(r_1 - r_2)} \int_{r_1}^{r_2} \int_0^{2\pi} (I_1(r,\varphi) - \bar{I}_1)(I_2(k_p r, \varphi - \Delta\varphi) - \bar{I}_2) r dr d\varphi$$

where I_1, I_2 – is intensity distribution at two different distances from the output end, k_p is the coefficient of proportionality, taking into account the radiation divergence, superscript "-" means averaging in the area of integration, r_1 and r_2 – the limits of integration on the radius. In the extreme case, we can assume $r \rightarrow 0$, and for the calculation of the CCF of the obtained images to use is the intensity of the corresponding pixel.

Example of the calculation of cross correlation and self correlation functions for image contained in fig. 3, 4, is shown in Fig. 5. The data obtained for OF with a core radius of 300 micrometers at wavelength of 532 nm; input angle of radiation into fiber - approximately 10^0, displacement of the centre of the laser beam from the axis of the fiber to the cladding - approximately 2/3 of the radius of the core. Designation of the angles and the other parameters listed above, similar to the scheme presented in [5-7]. Shift of the maximum of the CCF characterizes the image rotation approximately 1.6 degrees. Decrease the maximum value compared with the self correlation function (SCF) means that in addition to the rotation of the speckle structure is subject to change. To extend the function SCF compared with CCF indicates the different velocity of rotation of grains of speckle.

Thus, the paper show that the speckle structure of multimode optical fiber, formed by a group of optical vortices with the same direction of rotation of the wave front and waveguide modes subject tends to rotate in an open space near the surface of the output end when plane of surveillance is moving along the axis.

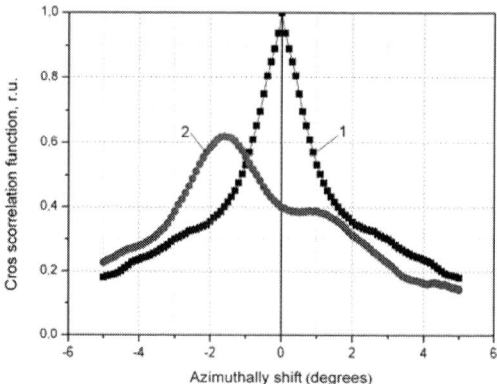

Fig. 5. Self correlation (1) and cross correlation (2) functions for images contained in fig. 3, 4.

REFERENCES

[1] D. V. Kiesewetter, "Numerical simulation of a speckle pattern formed by radiation of optical vortices in a multimode optical fibre", *Quantum Electron*, vol. 38. No 2. pp. 172–180, 2008.

[2] D. V. Kizevetter, "How defects of the end surfaces of a lightguide affect the mode-interference parameters when optical vortices are present", *Journal of Optical Technology*, vol. 80, No 1. pp. 7-11, 2013.

[3] N. V. Il'in, D. V. Kiesewetter, "The numerical simulation of the light intensity distribution in proximity to the output end of optical fiber given the optical vortices", *Nauchno Tech. Vedom. SPbSTU*, vol. 165. No 1. pp. 108-113, 2013.

[4] N. V. Il'in, D. V. Kiesewetter, "Method of optical vortices excitation in graded index optical fibers", *Nauchno Tech. Vedom. SPbSTU*, vol. 98. No 2. pp. 96-101, 2010.

[5] D. Kiesewetter. *Multimode optical fibers. Polarization and interference effects in multimode optical fibers.* LAP LAMBERT Academic Publishing Gmbh & Co KG, Leipzig, 232 p., 2011.

[6] D. V. Kizevetter, "Quasi beam description of the inter-mode interference of radiation of optical vortices in the short fibers", *Journal of Optical Technology*, vol. 75, No 1. pp. 80-82, 2008.

[7] D. V. Kiesewetter, "Characteristics of speckle of optical fibers", *Nauchno Tech. Vedom. SPbSTU*, vol. 64. No 3. pp. 72-80, 2008.

Spectral dependence of all-solid photonic bandgap fiber transmittance

A. Plastun, A. Konyukhov, E. Romanova[1],
T. Benson[2], *Senior Member, IEEE*, G. Athanasiou[2]
J. Lousteau, G. Scarpignato, E. Mura, N. Boetti, D. Milanese.[3]

[1] Saratov State University, Saratov, Russia
[2] University of Nottingham, Nottingham, UK
[3] Politecnico di Torino, Turin, Italy.

Abstract: The results of the numerical modeling of the transmittance of an all-solid photonic band gap fiber are presented. For the modeling, a vectorial "wide-angle" Fourier transform based beam propagation method is applied. The actual fiber structure, having morphological deformations of a hexagonal lattice, is considered. The coexistence of high-order total internal reflection modes of high-index rods and the fundamental photonic band gap mode is demonstrated.

I. INTRODUCTION

A photonic bandgap is a property of periodic dielectric structures, whereby light of certain frequencies is forbidden to propagate [1]. Bandgap-guiding optical fibers incorporating air [2], fluids [3] and contrasting glasses [4] have been reported, where the core is surrounded by a cladding with a 2-D bandgap [5]. The spectral selectivity of photonic bandgap fibers (PBF) offers great potential for their use in spectroscopic applications [5]. Solid-core PBF has low bending loss, high bandwidth and allows dispersion to be tailored [3, 4]. Fiber losses, and their spectral dependency, are critical in the operation of the PBF. In the PBF there are several main sources of loss, the principal ones being confinement loss, scattering loss and coupling to index guiding modes or surface modes. It is important to properly understand the losses which arise when propagating through the PBF.

In this paper, a solid-core low-contrast tellurite bandgap fiber is considered [6]. The manufacturing process, via a preform drawing approach, and the characterization process are described in [6].

To calculate fiber losses, the Fourier transform based beam propagation method [7] was used. In this method the transverse derivatives for the field and dielectric susceptibility are treated using Fourier series. The Fourier transform based beam propagation method [7] was modified using the "wide-angle" scheme. This numerical scheme allows us to take into account the effect of the second order derivatives with respect to the field propagation direction *z*.

The transmittance of PBF was calculated under excitation by a Gaussian beam focused into the fiber core.

II. DISPERSION DIAGRAM FOR SOLID-CORE PBF

In this section the bandgap properties of the PBF considered [6] are described.

The transverse distribution of the refractive index is shown in fig.1a. Here white regions correspond to the high-

refractive index rods with refractive index n_{high}. Black regions correspond to the host glass with refractive index n_{low}. The fiber exhibits some morphological deformations. The average diameter of the high refractive index rods is 2.25 μm. The average inter-core distance is 8.6 μm [6]. The refractive index of the rods and refractive index of the host glass are shown in fig.1b.

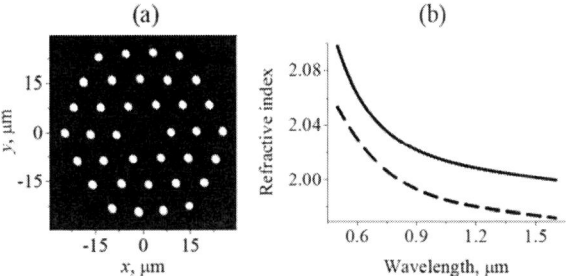

Fig.1. Solid-core PBF [6]. (a) Transverse distribution of the refractive index; (b) Refractive index of the glass of the PBF. Solid and dashed curves show n_{high} and n_{low} respectively.

Figure 2 shows the calculated dispersion diagram of the PBF. The diagram shows the spectral dependency of the effective refractive indices n_{eff} of the PBF modes. Effective refractive indices are calculated from the wave equation for the magnetic field vector [1]. A plane-wave expansion method [8] was used. The transverse distribution of the refractive index was calculated directly from a bitmap image of the PBF end-face. The dispersion properties of the high refractive index rods and the host glass are similar (fig. 1b). This fact allows us to calculate the dispersion diagram with medium dispersion accounted for [8].

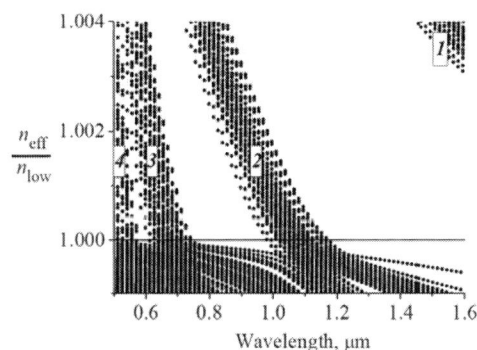

Fig. 2. Dispersion diagram of the PBF. Labels *1,2,3,4* indicate different groups of index guiding modes.

978-1-4799-0019-0/13 $31.00 © 2013 IEEE

The PBF has three bandgaps for the range of wavelengths considered (fig.2). The first bandgap can be found for wavelengths $\lambda > 1.2$ µm, the second one is localized between 0.75 µm and 1.0 µm and the third bandgap is localized around $\lambda = 0.6$ µm. At bandgap wavelengths the light can be confined by the PBF central area (photonic crystal defect). The confined light can be considered as a guiding bandgap mode. The experimental observations of the PBF guiding modes in [6] agree well with localization of bandgaps found by the plane-wave expansion method (fig.2).

Bandgap modes have $n_{eff}/n_{low} < 1$ while total internal reflection modes have $n_{eff}/n_{low} > 1$. Total internal reflection modes are guided mostly by the high index rods which act as step-index fibers. The wavelength $\lambda = 1.2$ µm (fig.2) corresponds to the cut-off frequency of high-order index guiding modes. In the wavelength range 0.5 µm $< \lambda <$ 1.2 µm the coexistence of high-order index guiding modes and a bandgap mode guided by the central low index area (defect) can be found. For $\lambda > 1.2$ µm only the fundamental index guiding mode propagates within the high index rods.

The cross-section of the high index rods of the experimental PBF is not circular (fig.1a) and their shape also varies from rod to rod. This leads to the removal of degeneracy of index guiding modes. Modes having similar symmetries in transverse field distribution are grouped within defined wavelength ranges (fig.2). The symmetry properties of index-guiding modes can be described using the notation for the modes of a step-index circular fiber. The first group corresponds to fundamental HE_{11}−like modes. The second group belongs to TE_{01}−, TM_{01}− and HE_{21}−like modes. The third group corresponds to the EH_{11} and HE_{12} modes of step-index fiber. The fourth group is formed by EH_{21}−, HE_{31}−like modes and other high-order modes.

III. MODELING OF THE FIELD PROPAGATION IN THE PBF

The vectorial propagation equations for the transverse components of magnetic field vector $\mathbf{H} = (H_x, H_y, H_z)$ are

$$\left(1 - \frac{i}{2\beta}\frac{\partial}{\partial z}\right)\frac{\partial H_x}{\partial z} - \frac{i}{2\beta}\left(P_{xx}H_x + P_{xy}H_y\right) = 0,$$
$$\left(1 - \frac{i}{2\beta}\frac{\partial}{\partial z}\right)\frac{\partial H_y}{\partial z} - \frac{i}{2\beta}\left(P_{yx}H_x + P_{yy}H_y\right) = 0,$$
(1)

where z is the propagation direction, (x, y) are transverse coordinates, β is the reference propagation constant and the operators $P_{xx}, P_{xy}, P_{yx}, P_{yy}$ are

$$P_{xx}H_x = \nabla_\perp^2 H_x + \left(k^2 n^2 - \beta^2\right)H_x - \frac{\partial \ln(n^2)}{\partial y}\frac{\partial H_x}{\partial y},$$
$$P_{yy}H_y = \nabla_\perp^2 H_y + \left(k^2 n^2 - \beta^2\right)H_y - \frac{\partial \ln(n^2)}{\partial x}\frac{\partial H_y}{\partial x},$$
$$P_{xy}H_y = \frac{\partial \ln(n^2)}{\partial y}\frac{\partial H_y}{\partial x}, \quad P_{yx}H_y = \frac{\partial \ln(n^2)}{\partial x}\frac{\partial H_y}{\partial y},$$
(2)

where $k = 2\pi/\lambda$, $n(x,y)$ is transverse distribution of refractive index. The refractive index is complex. The real part of the refractive index $n(x,y)$ corresponds to the distribution shown in fig 1a. The imaginary part of the refractive index is $\alpha\lambda(4\pi)^{-1}$, where α is the power absorption coefficient for bulk glass. Effective absorption coefficients $\alpha = 11$ m^{-1} (47.77 dB/m) for bulk glass and $\alpha = 400$ m^{-1} for absorbing boundaries were used.

Equations 1 are solved using two-dimensional Fourier series for magnetic field and complex refractive index profile. In calculations the (1,1) Padé approximant and Crank-Nicolson scheme were used. At each propagation step a system of linear equations was solved to update values of the transverse components of magnetic field vector. A Gaussian linearly polarized beam was used to excite the PBF. The beam radius $w_0 = 8$ µm.

Near field outputs after field propagation in a 60 mm long section of the PBF are shown in figure 3a-c. The fiber remains single-mode in a remarkably wide spectral range. Figures 3a and 3c show the field at wavelengths located within bandgaps. The transverse structure of the field (fig.3a, c) is formed by the PBF mode propagating in a central low index area and the field guided by high index rods which are step-index fibers. At $\lambda = 0.61$ µm (fig.3a) the PBF mode coexists with the high-order modes of high-index rods. Out of bandgap (fig.3b) the field is guided mainly by high-index rods.

We found strong field confinement for wavelengths $\lambda \geq 1.23$ µm (fig.3c). For this wavelength range the field propagates in the fundamental PBF mode and partially within six high index rods. As the wavelength $\lambda = 1.2$ µm corresponds to cut-off of high-order modes of separate high-index rods (fig.2) the field propagating within the rods corresponds to HE_{11}-like mode of circular step-index fiber.

The polarization of the central area is mainly linear, while the field inside high-index rods is polarized elliptically. Due to modal birefringence the ellipticity of the polarization varies during beam propagation.

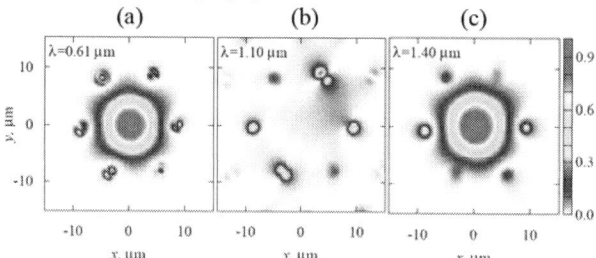

Fig.3. Near field output at $z=60$ mm. Transverse distribution of z-components of Pointing vector at three different wavelengths.

The power of the field propagating within the central area of the PBF is shown in fig.4. At a wavelength which corresponds to the photonic bandgap the power exhibits damping oscillations (fig.4, $\lambda = 1.4$ µm). After a transient process the field becomes propagating in the central region of PBF (fig.3c). At the bandgap edges a part of the radiation is confined by the high index rods. Interference between the PBF mode and high-index rod modes leads to oscillations of the beam power (fig.4, λ=1.0 μm). Out of the bandgap the power calculated for the field propagating within central area quickly goes down (fig.4, λ=1.1 μm). Correspondingly the PBF transmittance for this wavelength is quite low.

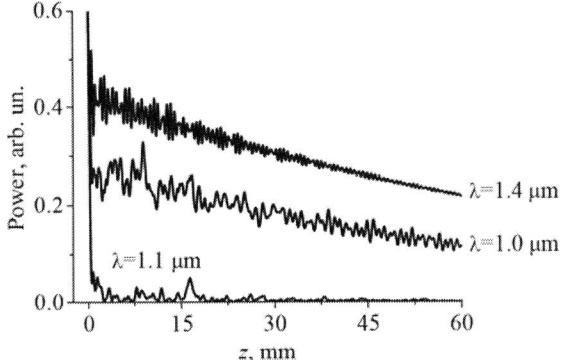

Fig. 4. The fractional power of the field propagating in the central area of PBF for three different wavelengths.

The transmittance spectrum of the PBF is shown in fig.5. The transmittance was calculated for a field propagation distance of 60 mm. There are several wavelength bands of "low-loss" guidance. At these wavelength bands the field is guided by the low refractive index PBF area, in other words a so-called PBF "defect" mode is excited. The attenuation spectrum agrees well with bandgap structure calculated by plane-wave expansion method (fig.2). The transmittance spectrum agrees with the experimental results [6]. The results show the existence of a robust single-mode structure in a wide range of wavelengths.

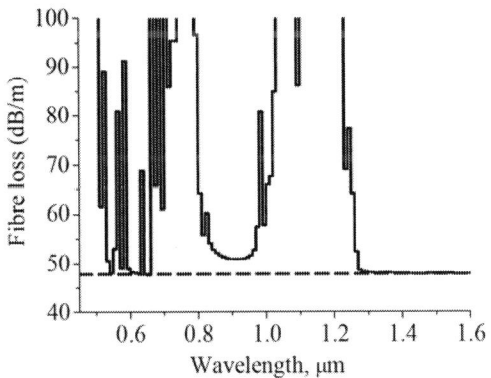

Fig. 5. Calculated transmission spectrum of PBF. The dashed line at 47.77 dB/m shows the attenuation of the bulk.

IV. CONCLUSION

The Fourier transform based beam propagation method with (1,1) Padé approximant and Crank-Nicolson scheme is used to model the transmittance spectrum of a solid-core photonic bandgap fiber. The actual fiber structure was considered using the transverse distribution of refractive index from the fiber end-face bitmap image. Our simulation allows us to calculate the field distribution at any wavelength and thus to calculate the fiber attenuation over the wide range of experimental wavelengths.

The attenuation of the solid-core PBF was analyzed. The fiber was excited by the Gaussian beam focused at the input face of the fiber. Low-loss guidance appears within the bandgap wavelengths. The PBF considered supports guiding only of the fundamental photonic bandgap mode. The mode coexists with high-order modes of the rods surrounded PBF core. At bandgaps wavelengths the energy concentrates in the PBF core rather than in surrounding waveguides.

ACKNOWLEDGMENT

A. Konyukhov is grateful for the grant of the Directorate of the Program of International Training Fellowships of the Saratov State University, supported by the Ministry of Education and Science, Russian Federation.

REFERENCES

[1] J.D. Joannopoulos, R.D. Meade, J.N. Winn, "*Photonic Crystals,*" Princeton University Press, 1995.
[2] R.F. Cregan, B.J. Managan, J.C. Knight, T.A. Birks, P. St. J. Russell, P.J. Roberts, D.C. Allen, "Single-mode photonic band gap guidance of light in air," *Science,* vol. 285, pp. 1537-1539, 1999.
[3] R.T. Bise, R.S. Windeler, K.S. Kranz, C. Kerbage, B.J. Eggleton, D.J. Trevor, "Tunable photonic band gap fiber," *Optical Fiber Communication, Opt. Soc. of Am.* vol. 70 , pp. 466-468, 2002
[4] F. Luan, A. K. George, T. D. Hedley, G. J. Pearce, D. M. Bird, J. C. Knight, P. St. J. Russell, "All-solid photonic band gap fiber," *Opt. Lett.* vol. 29, pp. 2369-2371, 2004.
[5] P. St. J. Russell, "Photonic crystal fibers," *Science,* vol. 299, pp. 358-362, 2003.
[6] J. Lousteau, G. Scarpignato, G.S. Athanasiou, E. Mura, N. Boetti, M. Olivero, T. Benson, P. Sewell, S. Abrate, D. Milanese, "Photonic bandgap confinement in an all-solid tellurite-glass photonic crystal fiber," *Opt. Lett.,* Vol. 37, pp. 4922-4924 2012
[7] J. M. López-Doña, J. G. Wangüemert-Pérez, I. Molina-Fernández, "Fast-Fourier-Based Three-Dimensional Full-Vectorial Beam Propagation Method," *IEEE Phot. Tech. Lett.,* vol. 17, No. 11, pp. 2319-2321, 2005
[8] L. Melnikov, I. Khromova, A. Scherbakov, N. Nikishin "Softglass hollow-core photonic crystal fibers," *Proc. SPIE,* Vol. 5950, *Photonic Crystals and Fibers,* 595012, (Sept. 2005).

Non-reciprocal beam behavior in two-dimensional photonic crystal slab

A.P. Ostroukh, R.A. Lymarenko, V.A. Kolyadenko, V.B. Taranenko
International Center "Institute of Applied Optics" of NASU, Kiev, Ukraine

Abstract: We simulate Gaussian beam propagation in two-dimensional photonic crystal slab (PCS). The found results show that character of beam propagation depends on the refractive index modulation on PCS boundaries. We observe a non-reciprocal beam behavior for PCS with different modulations on the boundaries. The non-reciprocity reaches its maximum at maximum of the refractive index modulation on the one boundary and minimum one on another boundary.

Photonic crystals are of great interest because they allow realization of various interesting and useful effects of light beam transformation such as focusing and negative refraction. We focus our attention on non-reciprocal behavior of a beam propagating in PCS.

The optical non-reciprocity can be obtained in some specifically constructed elements. Manipatruni et al showed the effect in optomechanical element with movable Bragg reflector [1]. The liquid crystal layer introduced between periodic structures with a different number of SiO_2/TiO_2 periodic cells was described in [2]. Horsley et al used cold atom Bragg mirrors in motion to demonstrate the non-reciprocity [3].

For beam propagation in medium modeling, we use the Narrow Angle Parabolic Wave Equation in the following form:

$$\frac{\partial E}{\partial z} = -\frac{i}{2kn}\frac{\partial^2 E}{\partial x^2} + i\frac{k}{2}mE$$

where k is the free space wave number, n is the refractive index of the medium, m is the refractive index modulation. The equation describes the forward narrow-angle beam propagation. Therefore, we limit the angle range to 10° because computational error dramatically increases at larger angles [4]. To create beam propagating model we use the Split Step Fourier Method.

The periodic structure in our model has form:

$$m = \cos[K(z/w + x * \sin(\varphi))]\ldots$$
$$+\cos[K(x - z/w * \sin(\varphi))]$$

where x, z are the transverse and longitudinal coordinates respectively, K is the lattice vector, w is the scaling coefficient for longitudinal direction, φ is responsible for the transverse shift of structure after every longitudinal period $a = 2\pi/K$. At $\varphi = \pi/2$, the grating transverse shift is half of period (Fig. 1). We limit this structure in longitudinal direction so that the obtained grating has different modulation amplitudes on opposite boundaries.

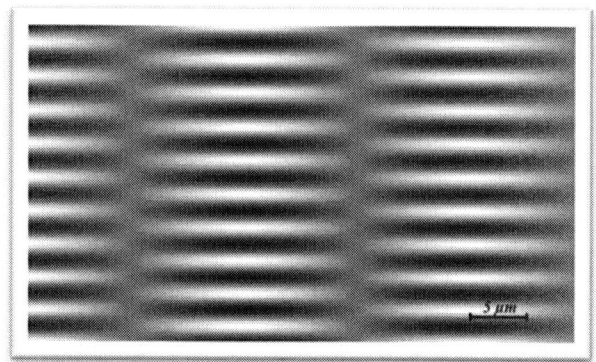

Fig. 1. Fragment of the periodic structure.

To simulate the beam propagation in PCS we use scheme shown in Fig. 2. The input Gaussian beam first propagates 100 μm in homogeneous medium ($n = 1$), then 890 μm in the slab ($n = 1.48$) and 546 μm in homogeneous medium ($n = 1$) again.

We use the Gaussian beam with wavelength 633 nm. The beam radius at which the field amplitude drops to 1/e is 50 μm.

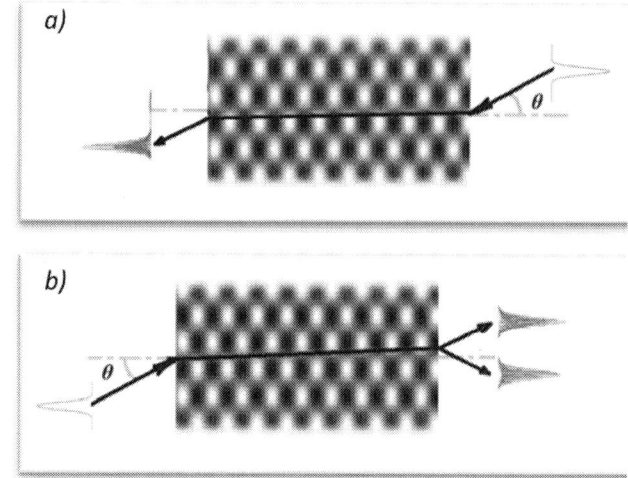

Fig. 2. Schematic presentation of calculation scheme.

First, consider the 512x890 μm PCS with transverse period 5 μm, longitudinal period 40 μm, average refractive index 1.48 and refractive index modulation 0.03. Modulation takes its maximum value on the one slab boundary and minimum value on another one. Introducing the Gaussian beam on the boundary with maximum of refractive index modulation and on structure's half Bragg angle for the transverse grating period, we observe one beam inside the slab and one beam outside in second diffraction order (Fig. 3a). When the incident beam introduce on the boundary with minimum modulation on the same angle, we observe two beams both inside and outside structure (Fig. 3b).

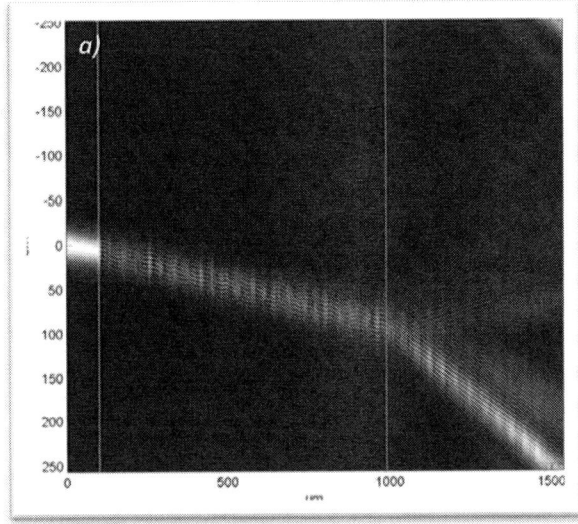

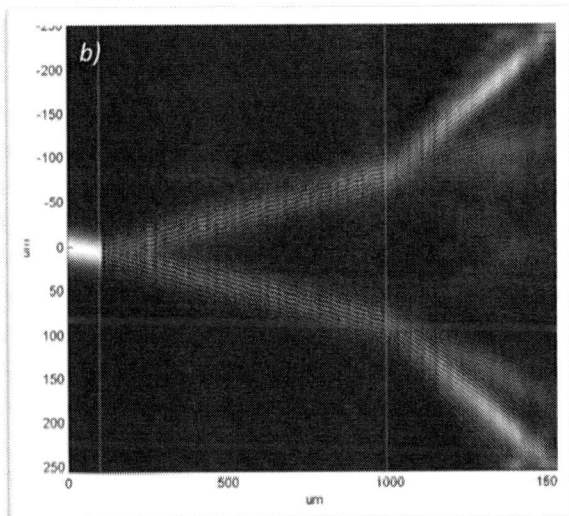

Fig. 3. Beam propagation for two opposite orientations of the PCS: (a) maximum refractive index modulation on the input boundary and minimum one on the opposite, (b) minimum refractive index modulation on the input boundary and maximum one on the opposite. Transverse period of grating is 5 μm, longitudinal period is 40 μm, refractive index modulation 0.03

Fig. 4 shows the dependence of transmitted beam intensity from the incident beam angle.

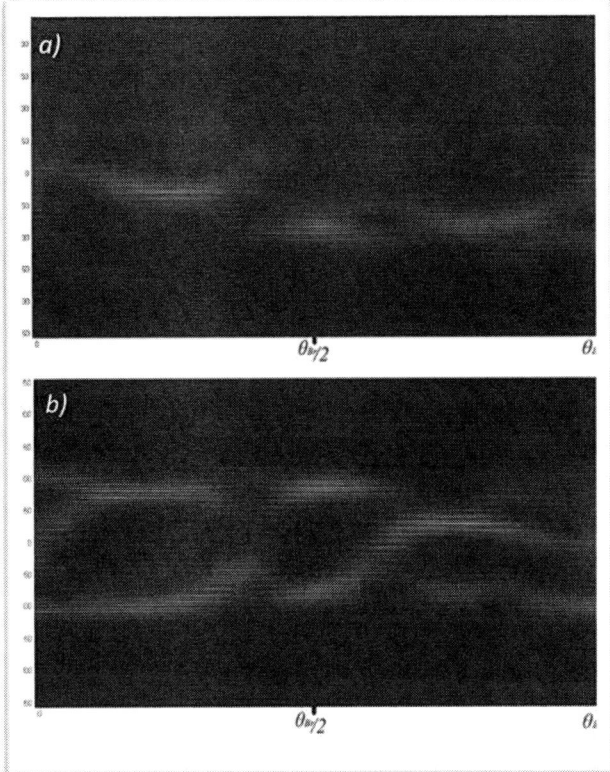

Fig. 4. Dependence of output beam intensity after PCS on the angle of incidence: a) maximum modulation on the input edge and minimum on the output, b) minimum on the input edge an maximum on the output; transverse period of grating 5 μm, longitudinal 40 μm, refractive index modulation 0.03

Now consider similar PCS with transverse period 5 μm, longitudinal period 55 μm, average refractive index 1.48 and its modulation 0.0165.

Introducing the Gaussian beam on the boundary with maximum of refractive index modulation and on structure's half Bragg angle, we observe one beam inside PCS but two beams out of it appeared from one point (Fig. 5a). Introducing the beam from the opposite boundary with minimum modulation, we see two beams split inside the slab. They do not split on the exit boundary so we have also two beams but displaced relative to first case (Fig. 5b).

The transmitted beam intensity dependence from the incident beam angle is similar to the first example (Fig. 6).

Simulation results show the non-reciprocal behavior of beam propagation in PCS with smooth modulation variations in one direction and half-period shift after every period of variation in the other. Two cases of non-reciprocal behavior are described: angular (Fig. 3) and spatial (Fig. 5) types. This feature is related to different diffraction conditions on boundary layers, thicknesses of which are full longitudinal period or half longitudinal period respectively. This behavior is observed in wide range of incident angles.

978-1-4799-0019-0/13 $31.00 © 2013 IEEE

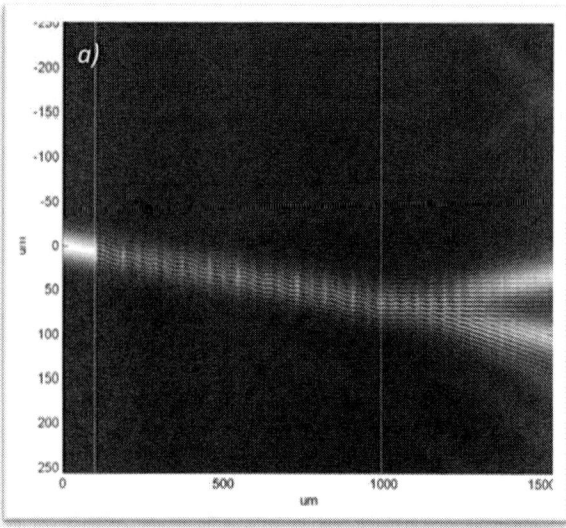

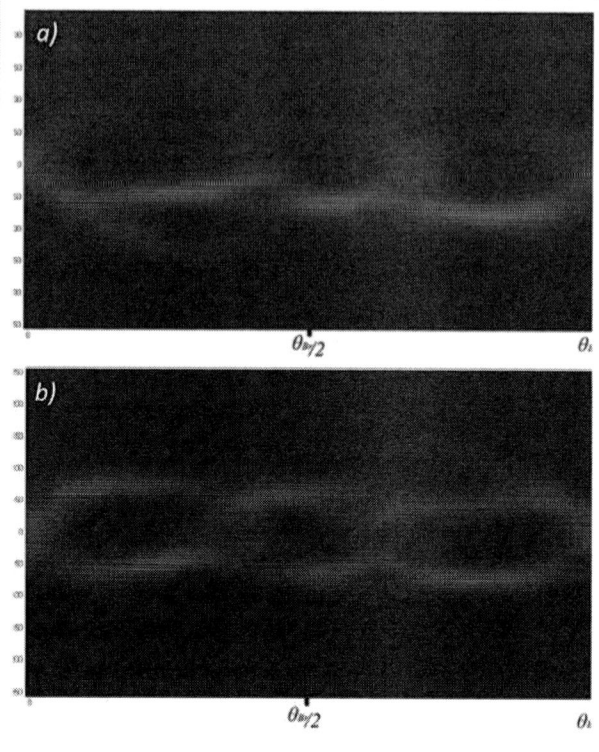

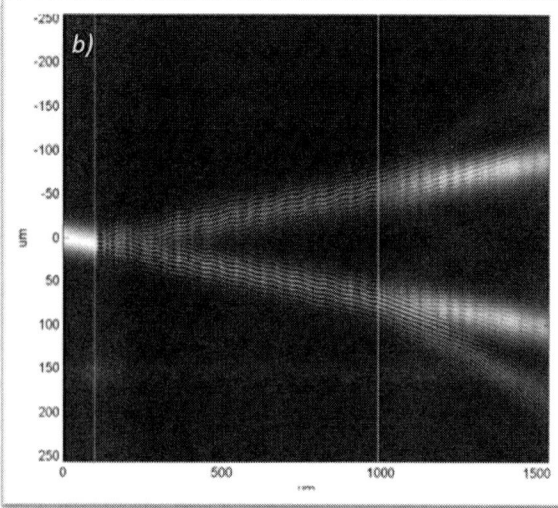

Fig. 6. Dependence of output beam intensity after PCS on the angle of incidence: a) maximum modulation on the input edge and minimum on the output, b) minimum on the input edge an maximum on the output; transverse period of grating 5 μm, longitudinal period 55 μm, refractive index modulation 0.0165

Fig. 5. Beam propagation for two opposite orientations of the PCS: a) maximum refractive index modulation on the input boundary and minimum one on the opposite, b) minimum refractive index modulation on the input boundary and maximum one on the opposite. Transverse period of grating is 5 μm, longitudinal period is 55 μm, refractive index modulation 0.0165

REFERENCES

[1]. Sasikanth Manipatruni, Jacob T. Robinson, Michal Lipson, "Optical Nonreciprocity in Optomechanical Structures", Phys. Rev. Lett. 102, 213903 (2009)

[2]. Andrey E. Miroshnichenko, Etienne Brasselet, Yuri S. Kivshar, "Reversible optical nonreciprocity in periodic structures with liquid crystals", Appl. Phys. Lett. 96, 063302 (2010)

[3]. S. A. R. Horsley, Jin-Hui Wu, M. Artoni, G. C. La Rocca, "Optical non-reciprocity of cold atom Bragg mirrors in motion", Phys. Rev. Lett. 110, 223602 (2013)

[4]. Merielle Levy "Parabolic equation methods for electromagnetic wave propagation", IEE Electromagnetic waves series 45, The Institution of Electrical Engineers, London, United Kingdom, 2000

Bragg resonance in multilayer structure with compound nonlinear layer

D.V. Sidorov and V.F. Borulko, *Senior Member, IEEE*
Oles Honchar Dnipropetrovsk National University, Dnipropetrovsk, Ukraine

Abstract: Transmission of plane monochromatic waves through the Bragg layered resonators have been considered. Iterative numeric solutions for complex amplitudes of reflected and transmitted waves are found by modified method of transmission matrices. Influences of Kerr nonlinearity of permittivity of some layers on resonance properties have been investigated. Hysteresis behavior is observed for case of combination of high quality factor and nonlinearity.

The emergence of powerful radiation sources in microwave, submillimeter, and optical frequency ranges stimulated intensive theoretical research and provided an opportunity to the experimental study of nonlinear phenomena.

There are a lot of effects caused by nonlinearities: focusing, defocusing, wave interaction, bistability and multistability, the generation of spatial structures, optical turbulence, etc. One of the most important technical applications of nonlinear material properties is to create a logical device, memory and information processing devices based on the phenomenon of hysteresis [1-4].

It is known that bistability (multistability) is characteristic property of nonlinear systems with feedback. The system can take two (or more) stable states at one and the same input power [5]. For the existence of multistable input-output characteristics in given intervals of the input variable, the values of feedback and nonlinearity must be in a specific ratio [6].

The simplest model of nonlinearity is the Kerr nonlinearity. Such a model quite well agrees with a lot of experimental results [1, 6], but does not take into account the generation of higher frequency harmonics. It assumes the permittivity dependence on the square modulus of the electric field with a proportionality coefficient α. The nonlinearity coefficient has a broad range of variation. For example, the sodium vapor has coefficient of nonlinearity $\alpha \sim 10^{-8}\ cm^2/kW$, the semiconductors have coefficient of nonlinearity $\alpha \sim 10^{-5}...10^{-2}\ cm^2/kW$, liquid crystals are characterized by "gigantic" nonlinearity $\alpha > 0.1\ cm^2/kW$. And in all cases, the physical mechanisms of nonlinearity are different, and the increasing of the value of α is accompanied by increasing inertial response of the nonlinear system [1, 8].

Using the Bragg reflectors for strong frequency-selective feedback in Fabry – Perot resonator allows us to observe multistability with less thickness of the nonlinear medium [8]. In most cases, the exact solution of problem of electromagnetic field distribution on the boundary of the layered Bragg structure with Kerr nonlinear layers can not be found, and we have to use approximate numerical methods [9].

This approximate numerical method takes into account changing the field amplitude on the thickness of the nonlinear layer. The sufficient accuracy of proposed method can be achieved by increasing the number of sublayers decomposition of the nonlinear resonance layers. The presence of several nonlinearities in the resonant layer of Bragg structure leads to a complex hysteretic behavior of the frequency characteristics due to the redistribution of the field between the parts of the resonance layer with different (decreasing and increasing) nonlinearity.

The characteristics of layered structure are strongly dependent on the location of the nonlinear lumped element inclusion [10]. The width of the hysteresis loops and the shift of the resonance frequency can be varied by moving of the nonlinear lumped element with respect to the antinode of the electric field.

In the paper the Bragg structures with compound nonlinear resonant layer with Kerr nonlinearity have been considered. The pseudoinverse method for calculation of the field at boundaries of the structures has been proposed.

The proposed method allows us to investigate the resonance properties of structures with different combinations of nonlinear layers, including Bragg structures containing layers with several types of Kerr nonlinearity.

In presented work we investigate nonlinear phenomena in Bragg layered structures with two nonlinear layers with of abnormal length.

In this paper we consider two type of nonlinearity – conventional Kerr nonlinearity [9] and "rational" Kerr nonlinearity of permittivity of nonlinear layers. In these cases values of permittivity as functions versus magnitude of electric field have the following forms:

$$\varepsilon_c = \alpha_c |E|^2 + \varepsilon_{C0}, \tag{1}$$

$$\varepsilon_M = \varepsilon_{M0}\left(1 + \xi|E|^2\right)/\left(1 + \eta|E|^2\right), \tag{2}$$

where ε_{C0} and ε_{M0} are the linear parts of permittivity and α_c, ξ, η are the nonlinearity coefficients. Dependence in the form of (2) has two nonlinearity parameters: ξ and η, which allows more properly approximation of permittivity dependence. Moreover, depending on the value of ξ/η with respect to the unit, the dependence (2) is either decreasing or increasing function of the squared modulus of the electric field.

In case of strong nonlinearity exact solution cannot be obtained. In Ref [11] the problem of diffraction of a plane wave by a transversely inhomogeneous isotropic nonmagnetic dielectric layer filled with a Kerr weakly nonlinear medium is reduced to a cubic-nonlinear integral equation of the second kind. This approach does not allow consideration of hysteresis phenomena.

In this paper we propose approximate numeric solution based on modification method of transmission matrices [12].

Further we will describe proposed method for layered structure with one nonlinear layer situated between two Bragg layered mirrors.

At the beginning for fixed magnitude A_t of transmitted wave we determine amplitudes of electric E_r and magnetic H_r fields on the right side of the structure

$$E_r = A_t \; ; \; H_r = A_t / Z_0 \qquad (3)$$

where Z_0 is the wave impedance of vacuum.

Amplitudes of electric E_{nr} and magnetic H_{nr} fields on the right side of nonlinear layer are found through transmission matrix $\mathbf{M}_r$ of the right Bragg mirror

$$\begin{pmatrix} E_{nr} \\ H_{nr} \end{pmatrix} = \mathbf{M}_r \begin{pmatrix} E_r \\ H_r \end{pmatrix}, \qquad (4)$$

The components of the transmission matrix $\mathbf{M}_r$ are uniquely determined by the parameters of layers [12]

We divide each thick nonlinear layer into some thin sublayers. Permittivity of each sublaeyer is supposed to be the same on its thickness and function versus the root-mean-square value of magnitudes of electric field on left and right sides of sublayer. We assume that electric and magnetic fields E_j and H_j on right surface of the j-th sublayer is known and fields E_{j+1} and H_{j+1} on the left surface of the sublayer and its permittivity must be found from system of nonlinear equations

$$\begin{pmatrix} E_{j+1} \\ H_{j+1} \end{pmatrix} = \mathbf{M}(\varepsilon_j) \begin{pmatrix} E_j \\ H_j \end{pmatrix}, \qquad (5)$$

$$\varepsilon_j = \varepsilon_{c0} + \alpha \left(|E_j|^2 + |E_{j+1}|^2 \right) / 2 , \qquad (6)$$

where $\mathbf{M}(\varepsilon_j)$ is the transmission matrix of sublayer, ε_{c0} is the linear part of permittivity, α_c is the nonlinearity coefficient in case of traditional Kerr nonlinearity.

Solving the system (5)-(6) was implemented by the Jacobi iterative method [9]. Preliminary values of initial estimate of sublayer permittivity ε_j were determined by electric

magnitude on right side of sublayer. Substantial improvement of initial estimate was achieved by linear and quadratic extrapolations from values of previous sublayers. Errors caused by subdivision of nonlinear layers into sublayers were verified by comparison with the results obtained with doubled number of sublayers.

We evaluate amplitudes of electric E_l and magnetic H_l fields on the left side of structure in terms of transmission matrix $\mathbf{M}_l$ of the left Bragg mirror

$$\begin{pmatrix} E_l \\ H_l \end{pmatrix} = \mathbf{M}_l \begin{pmatrix} E_{nl} \\ H_{nl} \end{pmatrix}. \qquad (7)$$

Complex amplitudes of incident A_i and reflected A_{rf} waves are found as follows:

$$A_i = (E_l + H_l Z_0)/2 , \qquad A_{rf} = (E_l - H_l Z_0)/2 \qquad (8)$$

Dependence of amplitude of the transmitted wave as function of amplitude of incident wave can be inverted numerically [9]. It is important that direct function can be nonmonotonic, so that inverse function will be multiple valued one. Presented approach can be easily generalized for cases of two or more nonlinear layers included between layered Bragg structures.

Very thin layers with high contrast of permittivity can be approximately described as parallel lumped inhomogeneities. The model of series element is applicable for thin layers with high contrast of permeability.

Figure 1 shows the frequency dependence of the field amplitude of the transmitted wave for the Bragg resonator with a compound central nonlinear layer. The left part of the compound layer has the traditional Kerr nonlinearity ($\xi_1 > 0$, $\eta_1 = 0$), and right part has "rational" decreasing one ($0 < \xi_2 / \eta_2 < 1$). The amplitude of the incident wave is $A_i = 2 \times 10^3 \, V/cm$.

The parameters of the left and right Bragg reflectors of resonator are follows:

$$\varepsilon_{2j-1} = 4 , \quad \varepsilon_{2j} = 2 , \quad \mu_{2j-1} = \mu_{2j} = 1 , \quad M = 15 .$$

Nonlinearity coefficients of layers have values:

$$\xi_1 = 1 \times 10^{-6} \, cm^2 / V^2 , \quad \eta_1 = 0 ,$$

$$\xi_2 = 1 \times 10^{-6} \, cm^2 / V^2 , \quad \eta_2 = 1.25 \times 10^{-6} \, cm^2 / V^2 ,$$

for the first and the second parts of the compound nonlinear layer, respectively. For given parameters of reflectors, the quality factor of the resonator is large enough. Thus, a small frequency variation leads to significant changes of the field amplitude and compound layer permittivity (Fig. 2, lines 1-3).

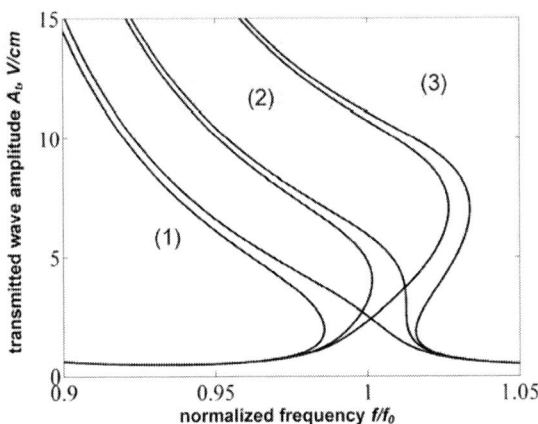

Fig. 1. Hysteresis phenomenon in the Bragg resonator with compound nonlinear layer. The transmitted wave amplitude dependence versus frequency for different thickness ratio of the left h_1 and right h_2 sides of resonance layer: $h_1 = 0.5h$ for line 1, $h_1 = 0.4h$ for line 1, $h_1 = 0.25h$ for line 3.

When $f/f_0 = 1.1$ (Fig. 2, line 1) electrical thicknesses of the resonant sublayers are nearly equal. If incident wave frequency closer to the frequency of the Bragg resonance (Fig. 2, line 2 and line 3) total electric thickness of compound resonance layer satisfies the condition of $h_{n1} + h_{n2} > \lambda_0/2$. In this case, the transmission maximum is shifted down with respect to the Bragg resonance frequency (Fig. 1, line 1). The resonant frequency does not extend beyond the Bragg reflection band wherein the Q-factor of the resonator is much lower.

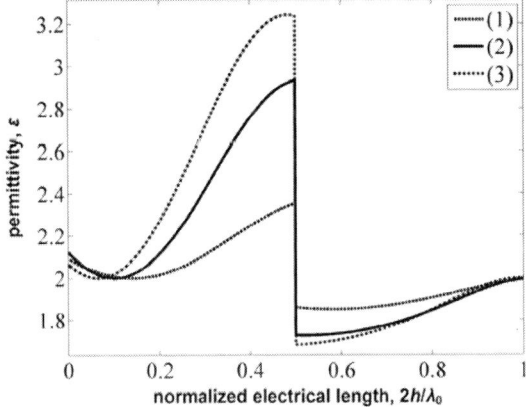

Fig. 2. The permittivity distribution in nonlinear central compound layer for different frequency of incident electromagnetic wave: $f/f_0 = 1.1$ for line 1, $f/f_0 = 1.05$ for line 2, $f/f_0 = 1.0$ for line 3.

In contrast to classical single-frequency nonlinear oscillations, the complex (Fig. 1, line 3) frequency hysteresis phenomenon can be observed for structures with composite nonlinear layer.

Fulfilled investigations show that combination of nonlinearity and high quality causes complicated hysteresis phenomena. Proposed modification of the method of transmission matrices allows analyzing structures with ambiguous dependence amplitudes of scattered waves against the amplitude and frequency of incident wave.

REFERENCES

[1] H.M. Gibbs *Optical Bistability: Controlling Light with Light*. – N.Y.: Academic, 1985.

[2] V.R. Tuz, S.L. Prosvirnin, S.V. Zhukovsky "Polarization switching and nonreciprocity in symmetric and asymmetric magnetophotonic multilayers with nonlinear defect," *Phys. Rev. A*. – 2012. – vol. 85, no. 4. – pp. 043822 (8).

[3] O.V. Shramkova, A.G. Schuchinsky "Gaussian pulse scattering by nonlinear Thue-Morse quasiperiodic multilayers," *International Conference on Mathematical Methods in Electromagnetic Theory*, – Kharkov, Aug. 28-30. – pp. 371-373, 2012.

[4] V.R. Tuz and S.L. Prosvirnin "Bistability, multistability, and nonreciprocity in a chiral photonic bandgap structure with nonlinear defect," *J. Opt. Soc. Am. B*. – vol. 28, no. 5. – pp. 1002-1008, 2011.

[5] D.A. Powell, I.V. Shadrivov, and Yu.S. Kivshar "Nonlinear electric metamaterials," *Appl. Phys. Lett.* –vol. 95, no. 8 – pp. 084102 (3), 2009.

[6] L. A. Lugiato "Theory of optical bistability". *In: Progress in Optics* / Ed. by E. Wolf. – Amsterdam: Elsevier. –vol. 21. – p. 69-216, 1984.

[7] P.W. Smith "On the physical limits of digital optical switching and logical elements," *Bell Syst. Tech. J.* – vol. 61, no. 8. – pp. 1975-1993, 1982.

[8] H.G. Winful, J.H. Marburger, E. Garmire "Theory of bistability in nonlinear distributed feedback structures," *Appl. Phys. Lett.* – vol. 35, no. 5. – pp. 379-381, 1979.

[9] V. Borulko, D. Sidorov "Nonlinear Resonances in Bragg layered structures with Kerr nonlinear layers," *International Conference on Laser and Fiber-Optical Networks Modeling (LFNM)*. – Kharkov, Sept. 5-8. – pp. 1-3, 2011.

[10] M.W. Feise, I.V. Shadrivov and Yu.S. Kivshar "Tunable transmission and bistability in left-handed bandgap structures," *Phys. Lett.* – vol. 85, no. 9. – pp. 1451–1453, 2004.

[11] Yu.V. Shestopalov and V.V. Yatsyk "Diffraction by a Kerr-type nonlinear dielectric layer," *PIERS online*. – vol. 3, no. 6. – pp. 759-763, 2007.

[12] M. Born and E. Wolf *Principles of Optics*. – Oxford: Pergamon Press, 1975.

Transmission and reflection spectra of photonic crystal with plasmonic defect layer

S. G. Moiseev[1,2], V. A. Ostatochnikov[1], D. I. Sementsov[1]

[1]Ulyanovsk State University, Ulyanovsk, Russia

[2]Kotel'nikov Institute of Radio Engineering and Electronics of RAS, Ulyanovsk, Russia

Abstract: The possibility of using plasmonic nanocomposite with nonspherical metal inclusions as a tuning defect in a photonic crystal is considered. Polarization-dependent features appearing when the electromagnetic wave is reflected from and transmitted through a photonic crystal with plasmonic defect are studied.

When used as inclusions in dielectric matrix, metal nanoparticles allow obtaining composite materials with optical properties vastly different from that usually found in natural materials. Absorptive and dispersion properties of such media are determined by the frequencies of the plasmon resonances, which depend on the shape of impurity metallic nanoparticles. Therefore the composite material with uniformly oriented elongated nanoparticles behaves like a dichroic crystal [1]. In the work [2] the relationship between the structural parameters of a plasmonic nanocomposite and the spectral properties of the photonic crystal structures was investigated. In this work the possibility of using plasmonic nanocomposite with nonspherical inclusions as a tuning defect in a photonic crystal is considered and the features of the behavior of electromagnetic eigenmodes in an artificial layered periodic structure with a finite number of periods and a plasmonic defect are analyzed.

We consider a symmetric microcavity photonic crystal structure in which a composite layer is sandwiched between two dielectric photonic crystal mirrors inverted relative to one another. The structure has a double defect: inversion and inserted layer. The inversion is due to the change in the stacking sequence of the layers in going from one part of the structure to the other. The transfer matrix of the photonic crystal under consideration, with an inserted defect layer and two photonic crystal mirrors, has the form $\hat{G} = \hat{N}^a \hat{D} \hat{\bar{N}}^a$, where $\hat{N}^a = \left(\hat{N}_1 \hat{N}_2\right)^a$ and $\hat{\bar{N}}^a = \left(\hat{N}_2 \hat{N}_1\right)^a$ are the transfer matrices of defect-free photonic crystal mirrors having a periods. The photonic crystal mirrors have a finite number of structural periods, each consisting of two layers of isotropic dielectrics having permittivity ε_j and thickness L_j (j=1,2). We neglect absorption in the frequency range of interest, so ε_j is real-valued and both $\hat{N}^a$ and $\hat{\bar{N}}^a$ are unimodular matrices with unity determinant.

The defect layer in the photonic crystal structure under consideration consists of a nanocomposite, which has the form of a dielectric material containing evenly distributed metallic nanoparticles in the shape of ellipsoids of revolution. The nanoparticles are aligned with their polar axis parallel to the x axis. The nanocomposite has properties of a uniaxial crystal, and its effective permittivity is represented in the major axes by a diagonal tensor with components $\varepsilon_x = \varepsilon_\parallel$ and $\varepsilon_y = \varepsilon_z = \varepsilon_\perp$ (the subscripts $\parallel$ and $\perp$ refer to two orientations of the electric field vector of the wave: along and across the optic axis of the nanocomposite). The dependence of optical properties of the plasmonic defect on the geometric (shape and concentration of inclusions) and material (permittivities of the matrix and metal nanoparticles) parameters of composite are calculated within the effective-medium approximation. Specifically, we apply the Maxwell–Garnett model [3], whose results for matrices with a moderate content of spheroidal inclusions are in fairly good agreement with the results of exact electrodynamic calculation.

Analytical Maxwell–Garnett formula for the dielectric function ε of mixture with spheroidal inclusions has the form

$$\varepsilon_{\perp,\parallel} = \varepsilon_m \left(1 + \frac{\eta(\varepsilon_p - \varepsilon_m)}{\varepsilon_m + (1-\eta)(\varepsilon_p - \varepsilon_m)L_{\perp,\parallel}} \right),$$

where ε_m and ε_p are the permittivities of the matrix and inclusions, respectively; η is the volume fraction of the inclusions; and $L_{\perp,\parallel}$ are geometric factors that take into account the effect of nanoparticle shape on the induced dipole moment of the nanoparticles. The permittivity of the metallic nanoparticles is given by

$$\varepsilon_p(\omega) = \varepsilon_0 - \frac{\omega_p^2}{\omega^2 + i\omega\gamma},$$

where ω_p is the plasmon frequency; ε_0 is the lattice contribution; and γ is the relaxation parameter. The geometric factor can be expressed through the ratio of the semi-polar axis to the semi-equatorial axis of the nanoparticles:

$$g_\parallel = \frac{1}{1-\xi^2}\left(1 - \xi \frac{\arcsin\sqrt{1-\xi^2}}{\sqrt{1-\xi^2}} \right),$$

$$g_\perp = (1 - g_\parallel)/2 .$$

The magnetic permeabilities of all the layers in the structure are taken to be unity.

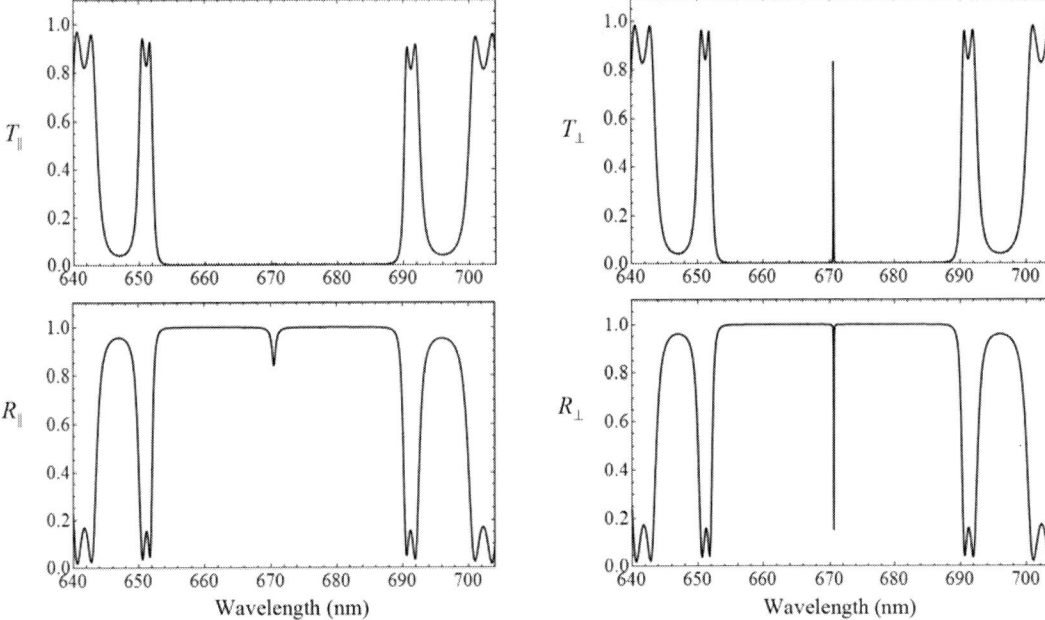

Fig. 1. Polarization-dependent spectral reflectance R and transmittance T of the defective layered periodic structure for the light polarized parallel to the long axis of spheroids (left plots) or to the short axis of spheroids (right plots). The computational parameters: volume concentration of nanoparticles equals $5 \cdot 10^{-5}$, the ratio of the length of polar semi-axis and equatorial semi-axis of spheroids equals 3.

The presence of a defective layer creates additional resolved levels in forbidden zones of the nondefective artificial layered periodic structures (by analogy with additional resolved levels in solids), which correspond to localized modes associated with the defect. By combining different types of defects and their positions in the structure as well as selecting the material, it is possible to efficiently control the optical properties of photonic crystal. The position and shape of the spectral line of the defect mode are analyzed as functions of the structure parameters and polarization-dependent features appearing when the electromagnetic wave is reflected from and transmitted through a periodic structure with plasmonic defect are studied.

The case is considered when two orthogonal polarizations of an incident wave correspond to different plasmon resonance frequencies of the nanocomposite. If one of the plasmon frequencies coincides with the defect mode frequency in one of the photonic bandgaps, complete suppression of the defect mode in the transmission spectrum is possible, which makes the spectra of such structures polarization-sensitive. And the case is possible, when orthogonal polarizations of waves correspond to defect mode suppression in different photonic bandgaps.

According to Fig. 1, reflection and transmission spectra of the defective layered periodic structure exhibits high polarization contrast. For radiation of the region near the spectral line of the defect mode, this photonic crystal absorbs the light polarized parallel to the long axis of spheroids, and for the light polarized parallel to the short axis of spheroids the structure is transparent.

In addition, plasmonic defect is considered to be formed by monolayers of uniformly oriented silver spheroids suspended in a transparent media. Our results show that the polarization-dependent optical characteristics of the photonic crystal structure can be archived using a single layer of silver nanoparticles.

The performed analysis and the revealed features of wave characteristics of the photonic crystal with plasmonic defect layer can be used in the development of various devices for the control of optical radiation on the basis of such structures.

This work was supported by the Russian Foundation for Basic Research and the Ministry of Education of the Russian Federation through project contracts within the framework of the Federal Target Program 'Science, Academic and Teaching Staff of Innovative Russia for 2009-2013'.

REFERENCES

[1] S. G. Moiseev, "Thin-film polarizer made of heterogeneous medium with uniformly oriented silver nanoparticles," *Appl. Phys. A.*, vol. 103, no. 3, pp. 775-777, 2011.

[2] S. Ya. Vetrov. A. Yu. Avdeeva and I. V. Timofeev, "Spectral properties of a one-dimensional photonic crystal with a resonant defect nanocomposite layer," *Journal of Experimental and Theoretical Physics* , vol. 113, pp. 755-761, 2011.

[3] C. F. Bohren, D. R. Huffman, *Absorption and Scattering of Light by Small Particles*, Wiley, New York, 1983.

Magneto-Optical Spectra of Microcavity One-Dimensional Magnetophotonic Crystals with Double Layer Bismuth-Substituted Iron Garnet

V. N. Berzhansky[1], A. N. Shaposhnikov[1], T. V. Mikhailova[1], A. R. Prokopov[1], A. V. Karavainikov[1],
Yu. M. Kharchenko[2], I. M. Lukienko[2], O. V. Miloslavskaya[2], M. F. Kharchenko[2]

[1]LIA LEMAC-LICS, Taurida National V.I. Vernadsky University, Simferopol, Ukraine
[2]Institute for Low Temperature Physics and Engineering of the NASU, Kharkov, Ukraine

Abstract – Optical transmittance and magneto-optical Faraday and Kerr rotation and magnetic circular dichroism spectra and its features for microcavity one-dimensional magnetophotonic crystals with double layer bismuth-substituted iron garnet films are considered.

I. INTRODUCTION

Within the last years the unique optical and magneto-optical (MO) properties of 1D magnetophotonic crystals (1D-MPCs) have attracted a great deal of attention of numerous research groups. The large Faraday and Kerr rotation angles and high transmissivity of 1D-MPCs in visible wavelength are very fruitful for applications in some MO devices. Typical 1D-MPC structures are formed from the layers of transparent magnetic and non-magnetic materials. One of the first proposed microcavity 1D-MPCs is structure $(Ta_2O_5/SiO_2)^m/Bi:YIG/(SiO_2/Ta_2O_5)^m$ on glass substrates [1]. Here Bi:YIG are polycrystalline Bi-substituted iron garnet films with Bi content of about 1 at./f.u. In our previous works [2, 3] double layer iron garnet films with a high Bi content (above 1.5 at./f.u.) and dioxides of TiO_2 and SiO_2 in Brag mirrors were used to improve the efficiency of this type of structures. Used non-magnetic dielectrics have the most favorable optical contrast and minimum absorption in the visible range of wavelength.

In present work the design, realization and properties of microcavity 1D-MPC with double magneto-active layer are discussed. The influence of the spectral characteristics (the differences in refractive index and absorption of right and left circularly polarized waves, i.e. Faraday rotation θ_F and magnetic circular dichroism $\Delta\alpha$) of the used double layer iron garnet film on the magneto-optical spectra of microcavity 1D-MPC is considered.

II. FABRICATION AND CALCULATION PROCEDURES

The microcavity structures with different repetition number of Bragg oxide mirrors m ($m = 4$ and 7) and double layer Bi:YIG films on the substrates of fused quartz: $KU-1/(TiO_2/SiO_2)^m/M1/M2/(SiO_2/TiO_2)^m$ were fabricated. Here M1 is sub-layer of content $Bi_{1.0}Y_{0.5}Gd_{1.5}Fe_{4.2}Al_{0.8}O_{12}$,

and M2 is the main magneto-active layer of content $Bi_{2.8}Y_{0.2}Fe_5O_{12}$. The selected compositions for double iron garnet layers showed a significant increase of the Faraday rotation angle and magneto-optical figure of merit Q of microcavity structures [3]. Bi:YIG films were fabricated by reactive ion beam sputtering of corresponding targets in argon-oxygen mixture [2], TiO_2 and SiO_2 mirrors were fabricated by electron beam evaporation. The targets for Bi:YIG films were produced by conventional ceramic technique. All magnetoactive layers were formed and crystallized separately at optimal $T_{cryst} = 680$ °C [2]. The first layer with a lower Bi content M1 was deposited on SiO_2 and after its crystallization the layer with a higher Bi content M2 was deposited. The optical thickness of the films was controlled by the nd monitor *in-situ*.

The thicknesses of TiO_2 d_{TiO2}, SiO_2 d_{SiO2}, M1 d_{M1} and M2 d_{M2} layers in MPC structures were defined as

$$d_{TiO2} = \lambda_0/(4 \cdot n_{TiO2}), \ d_{SiO2} = \lambda_0/(4 \cdot n_{SiO2}), \quad (1)$$

$$\lambda_0/2 < l_M < \lambda_0 \quad \text{or} \quad (2)$$

$$\lambda_0 < l_M < 3\lambda_0/2, \ l_M = n_{M1} \cdot d_{M1} + n_{M2} \cdot d_{M2} \quad (3)$$

where n_{TiO2}, n_{SiO2}, n_{M1} and n_{M2} are the refractive indexes of TiO_2, SiO_2, M1 and M2 films, respectively. At these conditions the photonic band gap (PBG) centered at wavelength λ_0 and with two resonant peaks corresponding to a magnetic defect in MPC at λ_{R1} and λ_{R2} is formed (see section III).

Theoretical calculations of optical and magnetooptical properties of multilayer structures are performed by conventional 4×4 transfer matrix method [1]. The method is based on the solution of Maxwell equations for MPC structure using the permittivity tensors of functional layers

$$\nabla \times \mathbf{E}(\mathbf{r},t) = i\omega\mu_0 \mathbf{H}(\mathbf{r},t), \ \nabla \times \mathbf{H}(\mathbf{r},t) = -i\omega\varepsilon_0 \hat{\varepsilon} \mathbf{E}(\mathbf{r},t), \ (4)$$

where $\mathbf{E}(\mathbf{r},t)$ and $\mathbf{H}(\mathbf{r},t)$ are magnetic and electric fields vectors of light wave respectively, $\mathbf{r}$ is radius-vector, t is time, μ_0 and ε_0 are vacuum permittivity and permeability respectively, $\hat{\varepsilon}$ is the permittivity tensor of corresponding

978-1-4799-0019-0/13 $31.00 © 2013 IEEE

layer. For optical frequencies $\mu = 1$. The permittivity tensor is written as

$$\hat{\varepsilon} = \begin{pmatrix} \varepsilon_1 & -i\varepsilon_2 & 0 \\ i\varepsilon_2 & \varepsilon_1 & 0 \\ 0 & 0 & \varepsilon_3 \end{pmatrix}, \qquad (5)$$

where tensor elements ε_1 and ε_2 are complex in the case of magnetic layers, $\varepsilon_j = \varepsilon_j' + i \cdot \varepsilon_j''$. Element ε_1 is determined by refractive index n and extinction coefficient κ.

$$\varepsilon_1' = n^2 - \kappa^2, \ \varepsilon_1'' = 2 \cdot n \cdot \kappa \qquad (6)$$

Gyrotropic effects are described by ε_2 [4]. The element ε_2 depends linearly on the magnetization $\mathbf{M}$, $\varepsilon_2(\mathbf{M}) = \varepsilon_2(-\mathbf{M})$. For magnetic layers ε_2 can be expressed using the experimentally determined quantities

$$\varepsilon_2' = (-\lambda/\pi) \cdot \left(n\theta_F + \tfrac{1}{4}\kappa\Delta\alpha \right), \varepsilon_2'' = (-\lambda/\pi) \cdot \left(\kappa\theta_F - \tfrac{1}{4}n\Delta\alpha \right), \quad (7)$$

where λ is wavelength of incident light, θ_F is specific Faraday rotation and $\Delta\alpha$ is the difference of absorption coefficients for right and left circularly polarized light. For nonmagnetic layers $\varepsilon_2 = 0$. Elements ε_1 and ε_2 of MPC layers were determined using the experimental data for n, κ, θ_F and $\Delta\alpha$ for single films.

The transmittance T and reflectance R, Faraday θ_F and Kerr θ_K rotation of MPCs were determined by computational solution of equation (4). It was considered that TM polarized light falls to the structure. The amplitudes of reflected and transmitted TM and TE, left and right circularly polarized modes were found. Magnetic circular dichroism (MCD) of MPCs is calculated as

$$\text{MCD in transmittance} = \left(T^+ - T^- \right) / \left(T^+ + T^- \right), \qquad (8)$$

$$\text{MCD in reflectance} = \left(R^+ - R^- \right) / \left(R^+ + R^- \right), \qquad (9)$$

where T^+ and R^+, T^- and R^- are transmittance and reflectance of left and right circularly polarized modes, respectively.

III. RESULTS AND DISCUSSION

The highest values of θ_F and Q were obtained for microcavity 1D-MPCs at λ_{R1}. For $m = 4$ at $\lambda_{R1} = 624$ nm $\theta_F = 13.6$ deg (43 deg/μm) and $Q = 15.1$ deg, for $m = 7$ at $\lambda_{R1} = 626$ nm $\theta_F = 20.6$ deg (66 deg/μm) and $Q = 8.05$ deg. Faraday rotation enhancement at λ_{R1} in comparison with single M2-layer are 10.5 and 16 for 1D-MPC with $m = 4$ and $m = 7$, respectively. For these structures the experimental (symbols) and calculated (solid lines) transmittance, Faraday and Kerr rotation spectra are present in Fig. 1 and 2. Optical transmittance and Faraday rotation spectra were measured in magnetic field of 16 kOe.

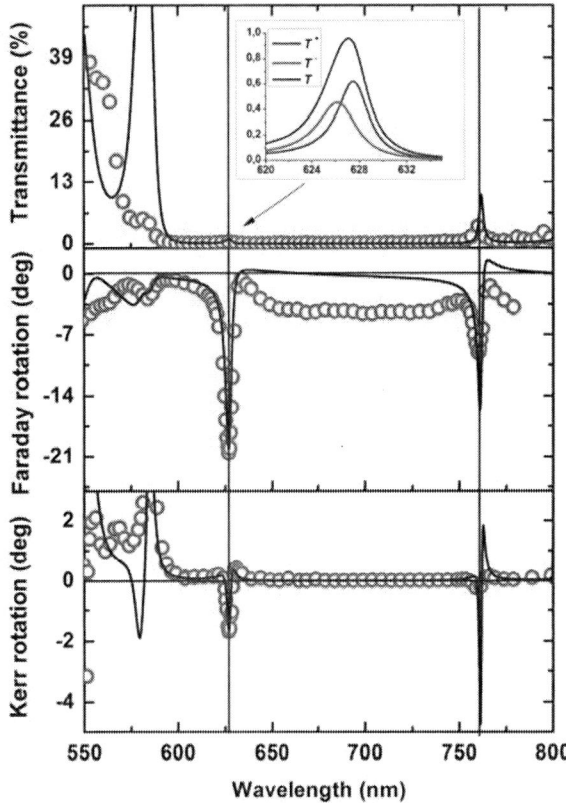

Fig. 1. Transmittance, Faraday and Kerr rotation spectra of microcavity 1D-MPC with $m = 7$ and $\lambda_0 < l_M < 3\lambda_0/2$. Transmission of left T^+ and right T^- circularly polarized modes is shown in the inset.

Fig. 3 shows the experimental (symbols) and calculated (solid lines) transmittance, Faraday rotation and MCD spectra for structure with $m = 4$ and a thinner magnetic layer ($\lambda_0/2 < l_M < \lambda_0$). MCD spectra were measured in magnetic field of 3.4 kOe.

Peculiarities of observed magneto-optical spectra are region with a negative Faraday rotation (Fig. 2) and a few peaks in Kerr and MCD spectra in the vicinity of the resonance wavelengths. It is caused by the difference in propagation of right and left circularly polarized waves in the structure.

In PC with magneto-active defect resonance conditions that defined by refractive index and thickness of the defect (2)-(3) for left and right circularly polarized waves are different. As a result, the resonant transmission peaks for right and left circularly polarized components are at different wavelengths. If the absorption of differently circularly polarized waves in the magnetic defect is present the corresponding peaks are differ in amplitude (inserts in Fig. 1-3). Hence, in accordance with (8) two opposite and symmetric or asymmetric peaks in MCD spectra in the vicinity of the resonant wavelengths are formed. Similarly, shape of peaks in Faraday rotation spectra can be as symmetric and asymmetric with negative region of values.

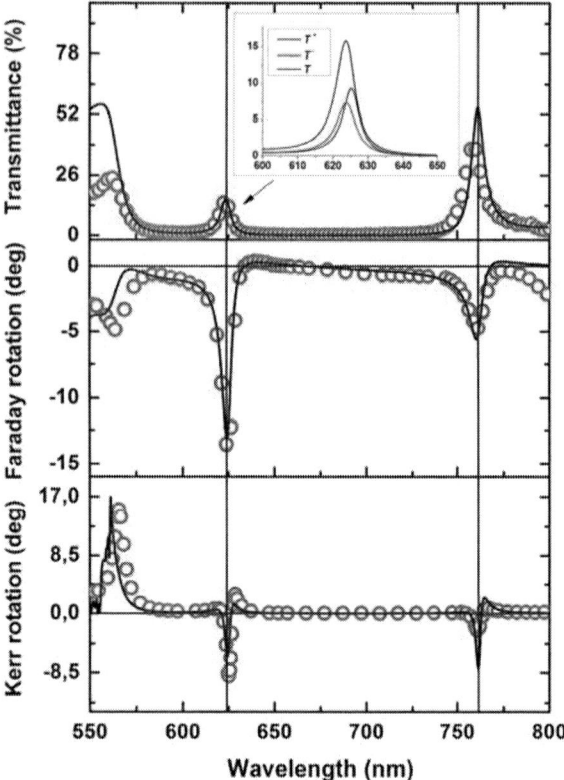

Fig. 2. Transmittance, Faraday and Kerr rotation spectra of microcavity 1D-MPC with $m = 4$ and $\lambda_0 < l_{\mathrm{M}} < 3\lambda_0/2$. Transmission of left T^+ and right T^- circularly polarized modes is shown in the inset.

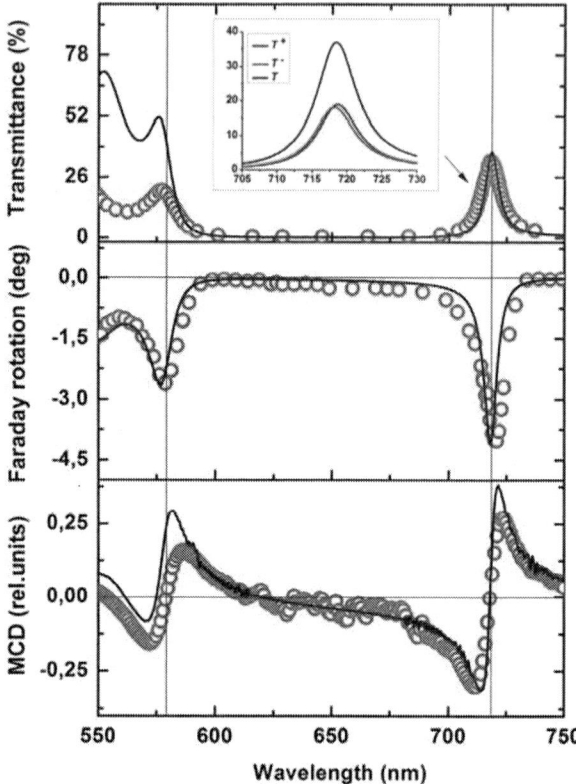

Fig. 3. Transmittance, Faraday rotation and MCD spectra of microcavity 1D-MPC with $m = 4$ and $\lambda_0/2 < l_{\mathrm{M}} < \lambda_0$. Transmission of left T^+ and right T^- circularly polarized modes is shown in the inset.

A similar is observed in reflection. Therefore, in Kerr rotation spectra two positive asymmetric peaks surrounding negative peak that coincided with resonance wavelength are presented. Negative Kerr rotation peak coincides to peak of TE mode, positive Kerr rotation peaks coincide to peaks of MCD in reflection.

It should be noted that the observed features connected to increasing contributions of both components of ε_2 (7) in the appearance of magneto-optical effects for microcavity 1D-MPC structure.

V. Conclusion

The microcavity 1D-MPCs with double magneto-active layer and different repetition number of Bragg oxide mirrors m ($m = 4$ and 7) are fabricated. The highest obtained values of Faraday rotation angle is 20.6 deg (66 deg/µm) at 626 nm and figure of merit is 15.1 deg at 624 nm.

The Faraday rotation and magnetic circular dichroism of the used double layer iron garnet film equally important influence on MCD, Faraday and Kerr rotation spectra of MPC. The presence of significant magnetic circular dichroism of the magneto-active films can significantly change the spectral characteristics of the devices based on microcavity 1D-MPC (as in bad and for the better).

References

[1] H. Kato, T. Matsushita, A. Takayama, M. Egawa, K. Nishimura, M. Inoue, Theoretical analysis of optical and magneto-optical properties of one-dimensional magnetophotonic crystals, *J. Appl. Phys.*, **93**, 3906 (2003).

[2] V. Berzhansky, A. Karavainikov, E.Milyukova, T. Mikhailova, A. Prokopov, A. Shaposhnikov, Synthesis and properties of substituted ferrite-garnet films for one-dimensional magnetophotonic crystals, *Functional Materials* **17**, 120 (2010).

[3] V. Berzhansky, T. Mikhailova, A. Karavainikov, A. Prokopov, A. Shaposhnikov, I. Lukienko, Yu. Kharchenko, O. Miloslavskaya, N. Kharchenko, Microcavity One-Dimensional Magnetophotonic Crystals with Double Layer Iron Garnet, *J. Magn. Soc. Jpn.* **36**, 42 (2012).

[4] A. K. Zvezdin, and V. A. Kotov, *Modern Magnetooptics and Magnetooptical Materials*, (Institute of Physics Publishing, Bristol and Philadelphia, IOP Publishing Ltd., 1997).

Research of Misalignments and Cross-Sectional Structure Influence on Optical Loss in Photonic Crystal Fibers Connections

A.I. Filipenko, O.V. Sychova

Kharkov National University of Radio Electronics, Kharkov, Ukraine

Abstract: In the report the factors of optical loss in PCF connecting are analyzed. The cross-sections of optical fibers several types are constructed and distributions of their fundamental mode field are received by computer simulation. Optical loss are calculated for several variants of fiber connections. Dependence of attenuation from displacements are obtained.

When using and installing of functional electronic elements on photonic crystal fibers (PCF) is necessary to carry out their connection with each other or with a standard optical fiber. As with the connection standard optical fibers, connections PCF can be detachable or non-detachable. This gives rise to displacement of optical fibers relative to one another. When welding fibers, besides collapsing air holes in the PCF cladding and their arrangement disturbed. It influences the distribution of mode fields. On fig. 1 presents the factors of optical signal loss for detachable and welded connection of PCF.

The greatest losses occur when the transverse, longitudinal and angular displacement. The basic requirement for quality fiber connection is low loss at the junction. But because of its complex periodic structure PCF cross-section, this process is much more difficult and needs to greater precision and control implementation than in the case of standard optical fiber connection. Therefore an important task is the control of optical-geometric parameters of the PCF connection during installation.

Purpose of this research is to identify depending loss of signal from optical-geometric deviations of connected fibers. For its achievement it is necessary to solve the following tasks:

- determine the dependence of the mode field distribution from PCF structure (size of holes in the cladding, the distance between them, the diameter of the core);

- research the dependence of the optical signal loss from the difference of fibers mode fields;

- calculate the signal loss in optical fiber connections of different types in transverse, longitudinal and angular displacement.

The characteristics of optical fibers used in the simulation are presented in the Table I. On Fig. 2 the PCF simulated cross section (a, b), standard optical fiber (c) and their field distributions of the fundamental mode are shown.

An indicator of the connection quality is the transfer coefficient of optical signal from one fiber to another. In the cases of absence displacement connection, it can be defined roughly so [1, 2]:

$$T \approx \frac{4 A_{eff\,1} \cdot A_{eff\,2}}{\left(A_{eff\,1} + A_{eff\,2}\right)^2}, \tag{1}$$

where $A_{eff\,1}, A_{eff\,2}$ - effective area of connection fibers mode fields:

$$A_{eff} = \frac{\left(\int\limits_{-\infty}^{+\infty}\int\limits_{-\infty}^{+\infty} |E(x,y)|^2\,dxdy\right)^2}{\int\limits_{-\infty}^{+\infty}\int\limits_{-\infty}^{+\infty} |E(x,y)|^4\,dxdy}, \tag{2}$$

where $E(x,y)$ - amplitude of the electric field.

The loss in this case can be defined by the formula

$$\alpha = -10lg(T). \tag{3}$$

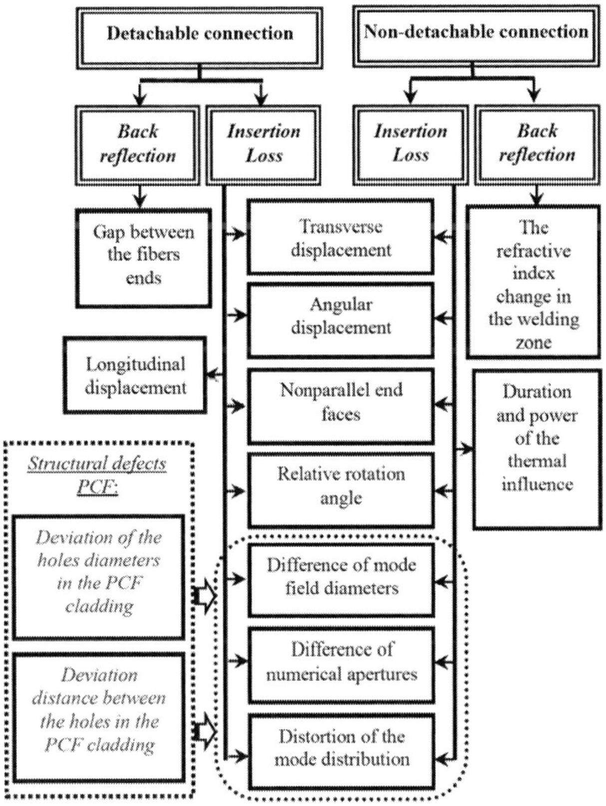

Fig. 1. Losses factors in PCF connections

The transfer coefficient in the ideal case is equal to one, and loss, respectively, is equal to zero when connect two identical fibers with the same distribution of mode field. But in practice these values are not achievable due to various factors that affect the quality of connection (fig. 1).

TABLE I
CHARACTERISTICS OF MODELED FIBERS

Image	Core diameter, μm	Holes diameter, μm	Distance between holes, μm	n_{clad}	n_{core}	Mode field radius, μm
a (PCF-10)	10	2.142	7.14	1.46	-	23
b (PCF-40)	39.4	6.402	13.2	1.446	-	32.5
c (SMF)	10	-	-	1.438	1.446	12.5

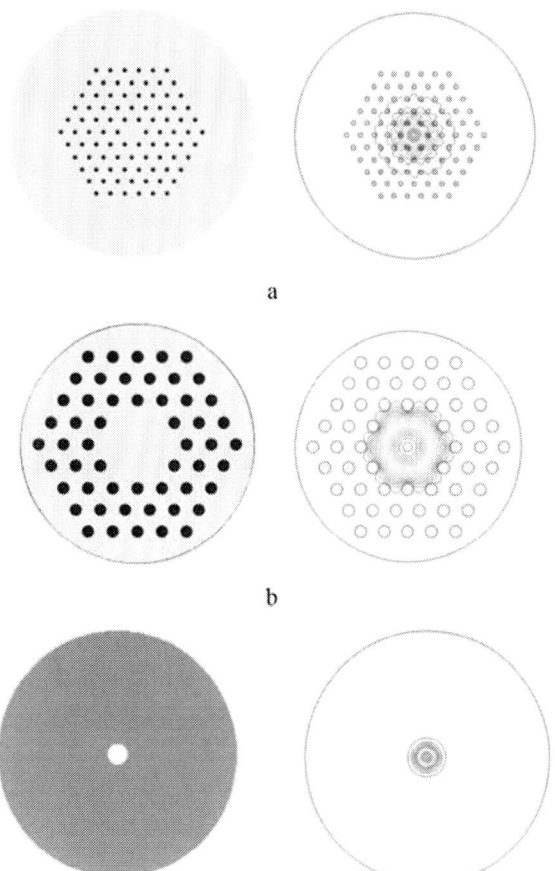

a

b

c

Fig. 2. Cross section of fibers and mode field distributions

Loss of transverse displacement fibers can be calculated by the formula

$$\alpha_t = -10\,lg\left(\frac{4w_1^2 w_2^2}{\left(w_1^2 + w_2^2\right)^2}\,exp\left(-\frac{2r_d^2}{w_1^2 + w_2^2}\right)\right), \qquad (4)$$

where $r_d = \left(d_x^2 + d_y^2\right)^{1/2}$ - transverse displacement of fibers; w_1, w_2 - fibers mode fields radiuses.

Losses of longitudinal displacement of fibers can be determined by

$$\alpha_l = -10\,lg\left(\frac{4w_1^2 w_2^2}{\frac{l^2}{k_0^2} + \left(w_1^2 + w_2^2\right)^2}\right), \qquad (5)$$

where l - longitudinal displacement of fibers; $k_0 = \frac{2\pi}{\lambda}$.

Losses of angular displacement determined by the formula

$$\alpha_\theta = -10\,lg\left(\frac{4w_1^2 w_2^2}{\left(w_1^2 + w_2^2\right)^2}\,exp\left(-\frac{k^2\theta^2 w_1^2 w_2^2}{2\left(w_1^2 + w_2^2\right)}\right)\right), \qquad (6)$$

where θ - angular displacement of fibers [3].

The results of calculations by formulas (4-6) for various types of optical fibers connection are presented in table II.

TABLE II
RESULTS OF CALCULATIONS

Connection	r_d, μm	α_t, dB	l, μm	α_l, dB	θ, deg.	α_θ, dB
PCF-10/SMF	0.5	3.516	0.5	3.508	0.2	3.513
	2.5	3.691	2.5	3.508	0.6	3.551
	4.5	4.099	4.5	3.508	0.8	3.585
PCF-10/PCF-10	0.5	0.005	0.5	$0.014 \cdot 10^{-5}$	0.4	0.012
	2.5	0.118	2.5	$0.340 \cdot 10^{-5}$	0.6	0.028
	4.5	0.383	4.5	$1.101 \cdot 10^{-5}$	0.8	0.050
PCF-40/PCF-40	0.5	0.005	0.5	$0.034 \cdot 10^{-6}$	0.4	0.006
	2.5	0.059	2.5	$0.852 \cdot 10^{-6}$	0.6	0.014
	4.5	0.192	4.5	$2.761 \cdot 10^{-6}$	0.8	0.025

By results of calculations were built graphics of optical signal loss dependencies from transverse misalignment (fig. 3), longitudinal (fig. 4) and angle (fig. 5) displacement.

After analyzing of the research results can conclude that the losses of transverse shifted two identical PCF will be higher for fibers with smaller diameter of mode field, but do not exceed 0.5 dB at displacements up to 5 μm. Connection loss of a PCF and a standard optical fiber increase and achieve values of up to 4.2 dB (for the case of PCF connection with a mode field radius 23 μm and standard optical fibers).

Longitudinal displacement losses of two PCF with identical structure even more insignificant, and when fibers with different mode field diameters are connected the attenuation is approximately as for the transverse displacement - up to 3.5 dB for gap size to 5 μm. The same dependence is observed for the angular displacement.

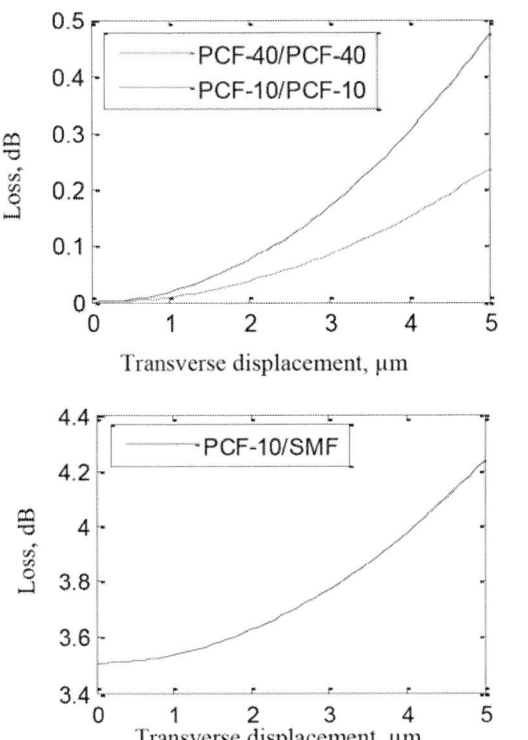

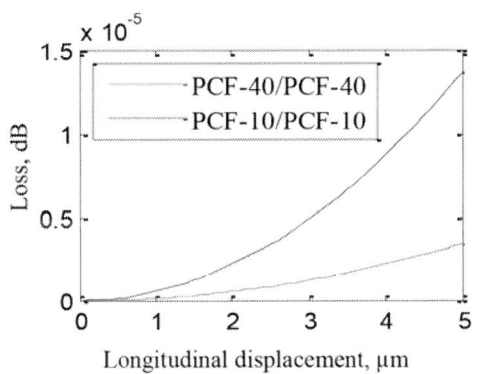

Fig.3. Dependence of the loss from transverse displacement

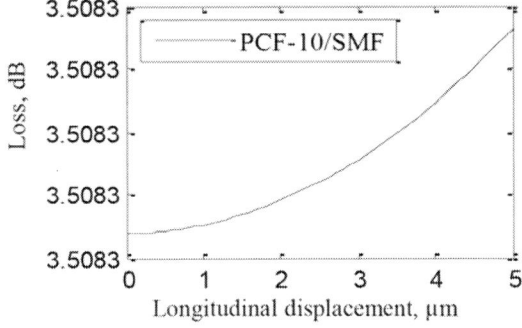

Fig.4. Dependence of the loss from the longitudinal displacement

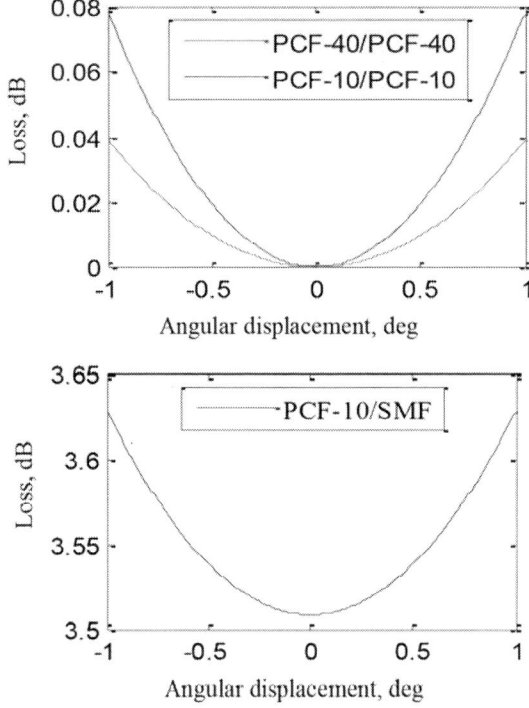

Fig.5. Dependence of the loss from angular of inclination

Further research aimed at studying of the influence of other optical-geometric parameters on transmission signal in PCF connection (such as relative angle of rotation, roughness and parallelism of the fiber end faces, difference of numerical aperture).

REFERENCES

[1] W. Zhi , R. Guobin, L. Shuqin, J. Shuisheng. Supercell lattice method for photonic crystal fibers // OPTICS EXPRESS, 5 May 2003, Vol.11, No9, pp.980-991.

[2] N. A. Mortensen. Effective area of photonic crystal fibers // OPTICS EXPRESS, 8 April 2002, Vol.10, No7, pp.341-348.

[3] G. S. Kliros, J. Konstantinidis, C. Thraskias. Prediction of Macrobending and splice losses for photonic crystal fibers based on the effective index method // Wseas transactions on communications, 2006, Vol.5, Issue 8, pp.1314-1321.

Influence of Water Medium and Flame on the Processes of Laser Materials Treatment

(Invited Paper)

A.F. Glova, S.V. Gvozdev, V.Yu. Dubrovskii, S.T. Durmanov,
A.G. Krasyukov, A.Yu. Lysikov, G.V. Smirnov

State Research Center of Russian Federation "Troitsk Institute for Innovation and Fusion Research",
142190, Moscow, Troitsk, Russia, e-mail: afglova@triniti.ru

Abstract-The conditions of drilling and cutting in water of 0.15-mm-thick titanium and stainless steel plates with the radiation of a repetitively pulsed Nd:YAG laser are studied experimentally in the absence of water and gas jets. Measurements of the absorption coefficient of radiation of a repetitively pulsed Nd:YAG laser and of a cw fiber Yb laser by the diffusion flame of aviation kerosene are performed. The radiation power scattered by the flame are measured under flame irradiation by Nd:YAG laser radiation.

I. Introduction

Laser cutting and welding of metal constructions in water has practical interest for many technological operations. To increase the efficiency of these processes, use is usually made of water or gas jets, targeted at the site of the focal spot localization on the surface. The jet effect leads either to elimination of the beam defocusing due to refraction on gas bubbles by removing the bubbles from the interaction volume (water jet) [1], or to creation of a locally dried volume (gas jet) [2]. If supplying a water or gas stream to the interaction site is not reasonable or hard to achieve, one has to perform laser processing of the constructions in water in the absence of the jet. In this case, due to the increase in the local heat extraction from the constructions submerged in water, one should expect the growth of energy consumption as compared with processing in the air.

Use of laser radiation for igniting gaseous and condensed media and for exerting influence upon a flame has recently a great interest. This is due to the possibility of investigating inflammation conditions [3] and the capability of controlling combustion regimes [4] with the help of a local energy source with controllable parameters, which may be formed remotely in an arbitrary region of the object under investigation. Another example of practical application of laser radiation in the presence of the flame is remote cutting of metal constructions on the emergency oil or gas borehole [5]. Under these conditions increase of the energy consumption also is possible due to absorption and scattering of laser radiation by the flame.

The aim of this work is to study the features of laser processing in water of thin metal plates in the absence of water and gas jets and measurements of the absorption and scattering of laser radiation by the diffusion flames.

II. Drilling and Cutting of Metal Plates in Water

Metal plates made of titanium and stainless steel (X18H10T) with the dimensions 1.5×3 cm and thickness $h = 0.13$ and 0.15 mm, respectively, were used in experiments. The velocity v of the sample movement did not exceed 0.2 mm/s. The water layer thickness, covering the front surface of the plate facing the laser beam, was ~2 mm, and the water layer thickness adjacent to the back side of the plate was 10 mm. A repetitively pulsed Nd :YAG laser with the FWHM pulse duration $\tau = 130$ μs and variable pulse repetition rate f and pump power was used. The size d of the focal spot, focused onto the plate surface, was equal to 0.22 or 0.3 mm, which corresponds to the condition of irradiation of thermally thin plates ($d/h > 1$).

Under irradiation in water of both motionless and moving plates practically in all exposure regimes, one could observe intense boiling process in the water volume adjacent to the focal spot, accompanied by the outcome of gas bubbles to the surface and their symmetric propagation in water in the radial direction off the beam axis.

A. Drilling of plates

It was found, that at a fixed mean laser power $<P>$ the drilling time t_0 in air increases with f increasing, for the stainless steel plate the drilling time is greater than for the titanium one, and in water the drilling time is considerably longer than in air.

In the case of drilling in air, the noticed features are related to a decrease in the metal vapour pressure under lowering the fluctuating temperature δT in the near-surface layer of the plate with increasing f and the heat conductivity coefficient of stainless steel as compared to titanium, having nearly the same plate thickness, approximately equal radiation absorption coefficient and unchanged $<P>$ [6].

The increase in t_0 when drilling in water may be explained by the increase in heat extraction in water. Let us compare two channels of laser radiation power loss, namely, the heat release from two opposite surface regions of the plate having the area $S \approx \pi d^2/4$, heated up to nearly the same mean temperature ΔT, $Q_1 \approx 2\alpha S\Delta T$ (here α is the heat-transfer coefficient), and the transverse heat conduction $Q_2 \approx 2\pi h\lambda\Delta T$, where λ is the heat conductivity coefficient. Under these assumptions, we obtain $Q_1/Q_2 \approx \alpha d^2/(4\lambda h)$. For a plate in air, the heat-transfer coefficient may be estimated using the expression $\alpha = \lambda_a(g/(d\eta^2))^{1/4}$ [7] (here λ_a, η are the heat-transfer coefficient and the kinematic viscosity coefficient for air, respectively; g is the free fall acceleration), then we get $Q_1/Q_2 \approx 3 \cdot 10^{-3}$. For film boiling in water, the heat release is also small ($Q_1/Q_2 \sim 10^{-2}$), in spite of almost triple increase in the heat-transfer coefficient [8]. One can suppose that the considerable increase in the plate drilling time in water compared with that in air occurs due to the mixed boiling regime, namely, the film boiling under the action of a radiation pulse, when the fluctuating temperature δT exceeds the critical temperature for the film boiling beginning, and the bubble boiling during the pause between the pulses, when the mean temperature is smaller than δT. The heat-transfer coefficient for bubble boiling is considerably higher than for film boiling, particularly under the conditions of pulsed heating, when the bubbles, located on the surface, have no time to tear off the surface [9]. To determine the mean heat-transfer coefficient in the case of the mixed boiling regime, supposed in these experiments, one has to perform special experimental and numerical studies.

B. Cutting of plates

Fig. 1 presents the power dependences of the maximal velocity of the target movement v^{max}, for which the stable cut still exists. From the dependences presented in Fig. 1 it is seen that the maximal values of the velocity at a given power and frequency are smaller for steel than for titanium, and become smaller with an increase in frequency.

The parameter $\beta = v^{max}/(<P>/h)$, equal to the cut area per unit consumed energy, provides a convenient quantitative criterion of the continuous cutting efficiency [6]. Thus, for titanium at $f = 50$ Hz the parameter β in air and in water is equal to ≈ 3.8 and ≈ 0.3 mm²/kJ, respectively, and weakly depends on $<P>$ due to the linear character of the dependences of v^{max} on $<P>$ in the ranges of power shown in Fig. 1. Note, that the value $\beta \approx 0.5$ mm²/kJ is typical for the technology of remote cutting of metals in air with the radiation of a cw CO_2 laser [6].

III. ABSORPTION AND SCATTERING OF LASER RADIATION BY THE DIFFUSION FLAMES

A. Experimental facility

Schematic of the unit in Fig. 2 is presented. The radiation of a repetitively pulsed (RP) Nd:YAG laser (radiation wavelength $\lambda = 1.06$ μm, $\tau = 130$ μs, $f = 1$–50 Hz, a highest pulse energy of 0.5 J) or a cw ytterbium fibre laser ($\lambda = 1.07$ mm, a peak power of 4 kW) (1) is collimated using lenses (2) and (3) into a laser beam, which may be treated as parallel over the length of the flame region (4). The beam diameter d can be varied. A power meter (7) served to measure the radiation power with and without the flame.

The scattered radiation power was measured with a photodiode (14) mounted, together with a set of elements (8–13), on an optical bench rotatable about O-point. Elements (9), (10) and the plate (6) are employed only in the calibration of the photodiode. The aperture stop (8) 4 mm in diameter is used to separate out the radiation scattered from a small flame segment. Translating the flame cell along the beam axis for a fixed angle of bench rotation yields the radiation power scattered by different flame segments.

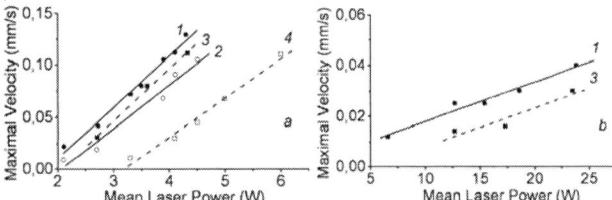

Fig. 1. Power dependences of the maximal velocity v^{max} of stable cut in air (a) and water (b) for titanium (1,2) and stainless steel (3,4). Focal spot diameter $d = 0.22$ (a) and 0.3 mm (b), pulse repetition rate $f = 50$ (1,3) and 100 Hz (2,4).

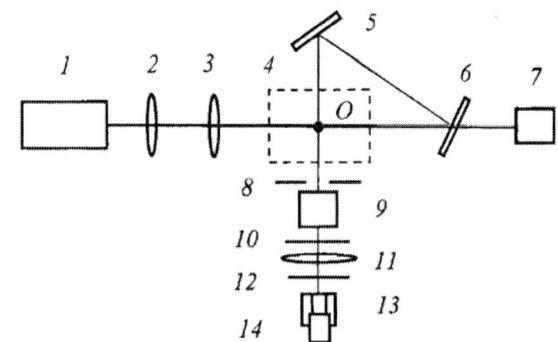

Fig. 2. Layout of the facility: (1) cw fibre or RP Nd:YAG laser; (2,3) collimator lenses; (4) flame region; (5, 6) radiation-attenuating plates; (7, 9) radiation detectors; (8) aperture stop; (10) set of neutral density light filters; (11) shaping lens; (12) IR light filter; (13) light shield; (14) photodiode.

B. Radiation scattering

Measurements of the radiation power scattered by the flame were made in its propagation through the yellow part of the aviation kerosene flame of length $l = 4.5$ cm for a distance $h = 2$ cm of the beam axis from the cell edge. As a radiation source a RP Nd:YAG laser with a pulse repetition rate $f = 15$ Hz was used; the laser beam diameter is equal to $d = 5.5$ mm. The average radiation power at the flame input, the pulsed input power and intensity were fixed and were equal to $<P_{in}> = 2.5$ W, $P_{in}^p = <P_{in}>/(f\tau) = 1280$ W, and $I_{in}^p = P_{in}^p/S \approx 4 \cdot 10^3$ W/cm², respectively (here, $S = \pi d^2/4$).

Under these values of the intensity and the spacing h, the radiation–flame interaction is attended with a glow of the beam path and the emergence of a characteristic sound with a pulse repetition rate. This may arise from additional heating and laser-field-induced intense combustion of the particles located in this part of the flame. Irrespective of their nature and production mechanism [4], the scattering of laser radiation by the flame is primarily due to these particles.

Measurements of the scattered radiation power were performed for different flame segments separated out by the aperture stop (8) in dependence on angle of bench rotation about O-point. It was found that the scattering is anisotropic as along flame length as well as on angle. To estimate the total scattered radiation power, we have assumed that the scattering is isotropic and similar to all flame segments regardless of the angle. For the total scattered radiation power W_Σ and its fraction δ relative to P_{in}^p we have $W_\Sigma = W_1 N(4\pi/\Omega) = 3$ W and $\delta = W_\Sigma/P_{in}^p = 0.23$ %, where $W_1 = 0.022$ mW is the average measured radiation power scattered by each segment of length $x = 5$ cm, $N = l/x$ is the number of segments; $W = 8.3 \cdot 10^{-4}$ sr is the solid angle at which the lens (11) (with an aperture of 2 cm and a lens–flame separation of 61.5 cm) is seen from the centre of a segment. Note, that it are upper estimates, since they do not take into account the increase in spectral energy brightness of the flame at $\lambda = 1$ μm arising from the temperature increase due to laser-induced particle heating.

C. Radiation absorption

A lowering of the average radiation power $<P_{out}>$ of the RP laser at the flame output caused the absorption and scattering of the radiation may be described by the expression $<P_{out}> = $

$<P_{in}>(1 - R)\exp(-\alpha l)$, where R is the effective scattered radiation fraction and α is the absorption coefficient averaged over the flame length l. Putting $R \approx \delta = 0.23\%$, we assume that $R \approx 0$ and for the average absorption coefficient of the RP laser radiation we have the expression $\alpha = -(1/l)\ln(<P_{out}>/<P_{in}>)$. The magnitude of R for the cw laser radiation is also assumed to be approximately equal to δ, and on replacing $<P_{in}>$ with P_{in} and $<P_{out}>$ with P_{out} we obtain a similar expression for the radiation absorption coefficient, where P_{in} and P_{out} are the input and output powers of the cw radiation, respectively.

Table 1 shows the absorption coefficients α in relation to l, h, and $<P_{in}>$ upon irradiation of kerosene flame by the RP Nd:YAG laser pulses. The radiation intensity I_{in}^{p} at the flame input was about the same for all irradiation regimes and was equal to $(2-5)\cdot 10^3$ W/cm^2. The input power was varied by changing the pulse repetition rate for a fixed pump power. The values of f and d corresponding to a given power $<P_{in}>$ are indicated in the heading of Table 1 in brackets.

The data collected in Table 1 possess three main features. First, the absorption coefficient is independent of the input power. This is due to the convective motion of the medium and the products of its interaction with the radiation, with the consequential complete renewal of a medium volume of size d during the interpulse period, so that the next radiation pulse interacts with the same medium as the previous pulse. Second, the absorption coefficient depends on the flame length, and with its increase it becomes stronger in the blue region of the flame ($h = 1$ cm) and shows a tendency for a decrease in the yellow region ($h = 2$ cm). Third, for equal lengths the absorption coefficient in the blue flame region is lower than in the yellow one. The two last named features are attributable to the variation of flame composition with height.

In the blue region, because of the low density of solid particles, the interaction takes place primarily with the gaseous medium. Here, it is possible to neglect the radiation power loss arising from the interaction with particles, and if there are no strong absorption bands at $\lambda = 1.06$ μm, the absorption coefficient will be lower than in the yellow region. The lowering of α in the yellow region of the flame with increase in its length is related to radiation absorption by the gaseous products of particle combustion: as the radiation is attenuated with length, the rate of additional product generation in the heating of particles under laser irradiation becomes lower, and therefore the length averaged absorption coefficient becomes smaller. To interpret the effect of the increase in α with length in the blue region of the flame we assume that the polarization of molecules in the laser radiation field results in their clustering, and these clusters may have absorption bands corresponding to the radiation wavelength. Since the total number of clusters rises with length and their structure changes owing to the lowering of the radiation field, α may also exhibit a change, including an increase.

We compare the measured α data for the RP Nd:YAG laser radiation and $h = 2$ cm with the data for the cw fibre laser and the same h value for kerosene flame, which are collected in Table 2. One can see from Table 2 that the absorption coefficient for the cw laser radiation, like for the RP laser radiation, lowers with increase in flame length. However, for close intensities ($I_{in} \sim I_{in}^{p} \sim 10^3$ W/cm^2) for both lasers and lengths it is

significantly higher than for the RP laser. The greater α value for the cw laser is attributable to the fact that the average temperature of the in-flame heated particles will be higher under the additional cw radiation heating than under RP radiation heating. As a consequence, the density of radiation-absorbing combustion products will be higher to result in a greater α.

We have measured also the absorption coefficient for ethanol and kerosene-ethanol mixture flames. The absorption coefficient of ethanol flame is independent of the laser operation regime and the radiation intensity. This is due to the absence of hot solid particles in the flame, as witnessed by the blue colour of the flame throughout its length. The absorption coefficient of ethanol flame is significantly lower than that of kerosene flame and is defined by water vapour [10]. The absorption coefficient rises with lowering the fraction of ethanol. Simultaneously changed is the flame appearance: the emerging yellow region with heated particles, which are responsible for the bulk of radiation absorption, becomes more pronounced.

TABLE 1
ABSORPTION COEFFICIENTS (CM⁻¹) UPON IRRADIATION OF KEROSENE FLAME BY RP Nd:YAG LASER PULSES

h, cm	l, cm	$<P_{in}>$, W (f, Hz; d, mm)			
		0.6 (5; 5.5)	2.4 (15; 5.5)	3.8 (25; 6.3)	6.2 (50; 7.5)
1	3.5-4	0.015	-	0.015	0.015
	11	0.029	-	0.026	0.026
2	2.5-3	0.064	0.066	-	0.062
	9-11	0.054	0.053	-	0.051

TABLE 2
ABSORPTION COEFFICIENTS (CM⁻¹) UPON IRRADIATION OF KEROSENE FLAME BY CW FIBRE LASER RADIATION ($h = 2$ CM)

P_{in}, W	d, mm	I_{in}, W/cm²	l, cm		
			3.5	8	13
100	3	1.4·10³	-	0.117	0.078
	8	1.9·10²	0.114	0.074	0.07
950	8	1.8·10³	0.23	0.132	0.121
	15	5.5·10²	0.145	0.13	0.116
3000	15	1.7·10³	-	0.32	0.24

ACKNOWLEDGMENT

This work was partially supported by Gazprom Gazobezopasnost.

REFERENCES

[1] M.I. Arzuov, Zh.I. Dzhumabekov, V.I. Konov, et al. *Fizika i khimiya obrabotki materialov*, n.3, pp. 136-140, 1989.

[2] X. Zhang, E. Ashida, S. Shono, F. Matsuda. *J. Mater. Proces. Technol.*, vol. 174, pp.34-41, 2006.

[3] O.V. Vysokomornaya, G.V. Kuznetsov, P.A. Strizhak P.A. *Teplovye Protsessy v Tekhnike*, vol. 3, pp. 113-117, 2011.

[4] Ch.B. Li, V. Li, K.Ch. O, et al. *Fizika Goreniya I Vzryva*, vol. 42, pp. 74-81, 2006.

[5] O.A. Blokhin, V.G. Vostrikov, V.D. Gavrilyuk, et al. *Khimicheskoe i Neftegazovoe Mashinostroenie*, n. 5, pp. 52-54, 2001.

[6] G.G. Gladush, S.V. Drobyazko, N.B. Rodionov, et al. *Quantum Electron.*, vol. 30, pp. 1072-1076, 2000.

[7] B. Gebhart, Y. Jamuria, R.L. Mahajan, B. Sammakia. *Buoyancy Induced Flows and Transport*. New York, Hemisphere, 1988.

[8] S.S. Kutateladze. *Osnovy teorii teploobmena*. Moscow, Atomizdat, 1979.

[9] L.I. Antonova, E.Yu. Afanas'eva, A.F. Glova, et al. Preprint of Troitsk Istitute for Innovation and Fusion Research, A-122, Moscow, 2005.

[10[R. Measures. *Laser Remote Sensing*. Moscow, Mir, 1987.

Adaptive system for high power (more than 100 kW) CW CO2 lasers.

(Invited Paper)

Alexis Kudryashov, Alex Alexandrov, Vadim Samarkin, Alexey Rukosuev
Moscow State Open University, Active Optics NightN Ltd.

Abstract: In this paper we consider the design and realization of wide aperture adaptive system to correct for high power CW CO2 laser radiation. Two new approaches for development of the wavefront sensors to measure 10 μ laser radiation are considered.

It is very well known that the wavefront of the radiation of most of high power lasers is highly aberrated. This does not allow to obtain a good focus and high concentration of the energy of laser beam. The reason for the wavefront distortions are first of all thermally induced aberrations in active elements and also some residual aberrations of various optical elements. In general the initial quality of each optical element is high enough (P-V about $\lambda/10$) but the whole optical setup consists sometimes of tens of such elements that altogether introduce sufficiently large aberration. So, in order to improve the quality of the input laser beam first one needs to be able to measure it and after correct for them. If we talk about some solid state lasers that generate in the range of 400 – 1100 nm there is a huge variety of wavefront sensors, interferometers that could be used for this application. Also coating technique for mirrors is very well developed – one can get up to 99.98% of reflectivity of the mirror. But as soon as we move to far infrared spectrum (10 μ radiation) the situation changes – there are almost no commercially available wavefront sensors that could be used in this range. On one hand it is understandable – there is a problem with a reliable infra red cameras (arrays of independent sensors) or single sensors. At the same time – most of industrial lasers and laser complexes are still based on high power CO_2 lasers with 10.6 μ output! And this kind of radiation need to be evaluated, moreover, in the real time and also corrected!

This paper represents two types of Hartmann wavefront sensors – one – based on baloometer IR camera produced by INO, Canada. Another – is based on Russian technology of thing film deposition on Si substrates.

Hardware and software of Hartmann sensor have some important advantages:

- Characterize response functions of bimorph mirror;
- Reconstruct wavefront and intensity distribution simultaneously;
- Possibility to measure both CW and pulse radiation and single shots;
- Ability to estimate a quality of laser radiation;
- Wide dynamic range;
- High precision;

- Fast measurements;
- Low sensitivity to mechanical noises and vibrations;
- Low-cost.

One of the demands of any optical system – is its reliability and ability to work not only in laboratory, but also in the real conditions, so that every student could use it and not break it. From this point of view Hartmann wavefront sensor is the most suitable one to be included in industrial laser. These kinds of sensors are widely used by astronomers or in medical research but in fact very rarely were applied to control for laser beam. One of the shortcoming of existing wavefront sensors – is their relatively high price, that varies from 25,000$ to 60,000$ or even to 200,000$ (it depends on the tasks and parameters of the system).

In this paper we present two types of Hartmann sensors – one based on IR commercial camera and another one – on thin film technology sensors

1. HARTMANN SENSOR BASED ON IR BOLOMETER CAMERA

Picture of such sensor is presented on Fig. 1. To build wavefront sensor on the base of commercial IR camera has several advantages, such as compactness; convenience to use it for various applications; it is like standard video camera – "plug and play"; possible high resolution. In our case we constructed this sensor based on INO (Canada) IR camera. Below we give the main parameters of the "bolometer" sensor.

Fig. 1. Hartmann sensor based on IR camera

Sensor INO160 - Microbolometer uncooled FPA bolometer; 160X120 pixels; 52 mm pitch
Video Output - Gigabit Ethernet Link RJ-45 connector; 16 bit raw data; 8 bit corrected data

Frame Rate - 30 Hz (with extension to 60 Hz)
Number of input diaphragms – 16x16, 20x20
Available Options - External trigger input (opto-isolated); TEC driver; Microshutter electronic driver; Serial interface; Thermistor interface (x2); Random access readout; Real time clock
Overall Dimensions - 65 mm(H) X 60 mm(W) X 105 mm(L)
Weight~230 g

This is a very nice device, but it also has some disadvantages: all bolometer cameras are too sensitive to input radiation (up to 2 W/cm^2). And as soon as you want to use them to measure the radiation of some industrial lasers you need to be very careful not to destroy the sensor and to provide a complicated set of filters. Another problem - input window has a limited size. So, if the input high power laser beam has large aperture a special telescope should be used to reduce the input beam size to fit the sensing window.

2. SENSOR BASED ON THING FILM TECHNOLOGY

Fig. 2 presents the idea of Hartmann sensor based on thin film technology. It consists of set of diaphragms and a sensing area - a number of quadrant thermo-electric converters (TEC) made on thin film technology. Main advantage of such wavefront sensor is it could be used with high CW power laser beams and plus large aperture beams might be easily detected. Parameters of proposed sensor are the following:

- Wavelength of measurements, 4 – 12 μ;
- Input beam size – ring Ø160 x 80 mm;
- Number of subapertures – 72;
- Range of measurements – 50 μ;
- Precision of measurements – < 0.6 μ;
- Frame frequency – 80 Hz;
- Interfacing port – USB.

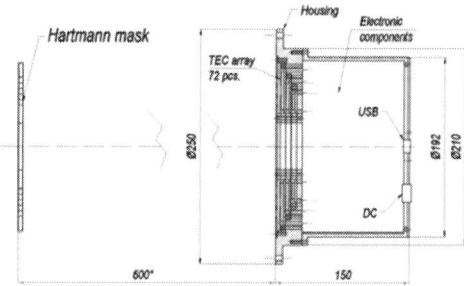

Fig. 2. Thin-film Hartmann sensor

As a thermo-electric converters we used the anisotropic quadrant elements shown on Fig. 3.

Properties of converter:

Material	Si
Thickness of layer	0.465 mm
Working wavelength	λ= 4 – 12 μAbsorption for

λ=10,6μ and α=(0-13)° < 10%

Sensitivity for λ=10,6μ,	> 0.5 mV/W
Response time	< 10^{-3}s
Resistance	< 100 Om
Resistance of isolation	> 10 kOm

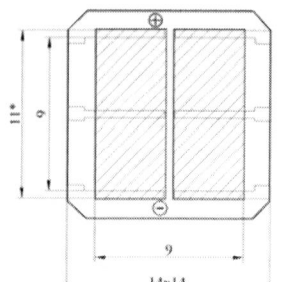

Fig. 3. Quadrant TECs

It is interesting to point out that main disadvantages of this kind of sensors are continuation of their advantages – they cannot measure small input beams and low CW power beams. At the same time – number of subapertures of such a sensors is also limited by its design. And they have a large size, so it is not very comfortable to install them.

3. DEFORMABLE MIRROR

As a wavefront corrector we suggest to use a watercooled bimorph deformable mirror (fig. 4). Main advantage of the use of such type of correctors is that they almost perfectly fit the aberrations (mainly thermally induced) of laser beam that need to be compensated. We either use wafer type water cooling system (fig. 4a)^2 or so called passive cooling system (fig. 4b).

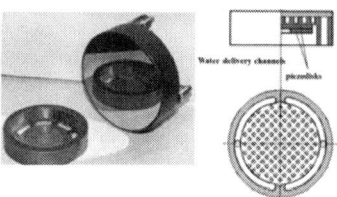

Fig. 8. Water-cooled deformable mirrors: left – waffle structure of cooling; right – "passive" cooling system.

CONCLUSION.

We demonstrated two types of wavefront sensors to measure the radiation of CO2 lasers and two types of deformable mirrors to correct for high power CO2 laser radiation. They both have advantages and problems. And only field of application determines which kind of mirror or sensor is the most appropriate.

REFERENCES

[1] J.V.Sheldakova, A.V.Kudryashov, V.Y.Zavalova, T.Y.Cherezova, "Beam quality measurements with Shack-Hartmann wavefront sensor and M2-sensor: comparison of two methods", *Proc. SPIE* **6452**, 207, (2007).

[2] A.V.Kudryashov, V.V.Samarkin, "Control of high power CO$_2$ laser beam by adaptive optical elements", *Opt. Comm.* **118**, pp. 317-322, 1995.

The control of energy, temporal and spatial characteristics a microchip laser with active output mirror

(Invited Paper)

V.V. Kiyko[1], S.V. Gagarsky[2], V.I. Kislov[1], V.A. Kondratyev[1], E.N. Ofitserov[1], A.N. Sergeev[2]

[1] Prokhorov General Physics Institute, Russian Academy of Science, Moscow, Russia
[2] Saint Petersburg National Research University of Information Technologies, Mechanics and Optics, Saint-Petersburg, Russia

Abstract: Presented a results of the theoretical and experimental investigation of the Nd^{+3}:YAG microchip laser with passive Q-switch modulation based on Cr^{+4}. The laser operation is analyzed on the base of operator model of a microchip laser with an active output mirror based on Fabry-Perot interferometer. The experimental and theoretical results are in a good agreement. The feasibility of control over temporal and spatial characteristics of micro-chip laser radiation is demonstrated.

The feasibility of control over temporal and spatial characteristics of micro-chip laser radiation is limited. Especially problematic is radiation control over such passive Q-switched lasers operating in a high frequency mode. Radiation temporal characteristics (repetition rate and pulse duration) control is realized by varying of pumping power. Simultaneously, owing to change of the thermal lens, pumping power density also inevitably occurs variation of spatial and temporal characteristics of output radiation. One of the ways for microchip laser output parameters control is based on application of an output mirror with controlled reflection coefficient [1, 2].

In this work presented is an operator model of a micro-chip laser with output mirror on the basis of Fabry-Perot interferometer with non-flat mirrors and variable gap.

This approach is especially preferred for application in small-size and microchip lasers with small diameter of the output beam.

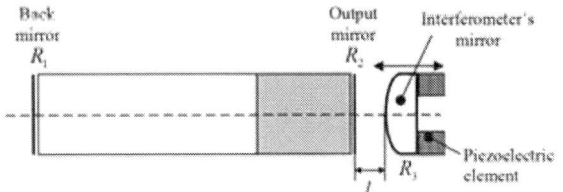

Fig.1 The scetch of the microchip laser with active output mirror.

In the modeling of such type of a mirror it is necessary to take into account beam diffraction at each pass through the interferometer meanwhile simplified geometrical models may results in essential errors. Applicability boundary of the geometrical model is determined by an expression

$N_{eff} = N_{int} / F \gg 1$ (where N_{eff} is effective Fresnel number, $N_{int} = d^2 / 8\lambda l$ is Fresnel number of the interferometer, d is beam diameter, λ is radiation wavelength, F is number of interfering beams, l is size of the gap). Comparative analysis of geometrical and diffraction models is conducted by means of developed operator approach to the interferometer description. Within the frame of this approach reflection coefficient across the field is determined by a relation

$$R_{int} = \left(R_2 - \left(R_3 |T_2|^2 H_{int}\right)/\left(E - R_2^* R_3 H_{int}\right)\right)\exp\left(i\Psi\right)$$

(where $R_{2,3}$ are of the first and the second interferometer's reflectors reflection coefficients taken into account along direction of falling onto the interferometer field, H_{int} is propagation operator per round trip, Ψ is phase addition, E is unit operator, T_2 is transparency of reflector 2).

Within the frame of the operator model of the laser eigenvalues γ and eigenfunctions U are being found from equation:

$$R_{int}H_{rez}U = \gamma U$$

where: U describes field distribution at the output mirror of the cavity; R_{int} is the interferometers reflection coefficient; H_{rez} is propagation operator per round trip for the resonator. In the calculations applied was matrix description of the operators.

The model enables to optimize characteristics of system resonator and mirror-interferometer taking into account thermal lens (TL) and requirements on output radiation characteristics. One of the cavity basic characteristics is squared modulus of eigenvalue $|\gamma|^2$ (EV) [1, 3]. Some results of calculations are presented in Fig. 2. Here the initial cavity is composed of flat mirrors and the interferometer is of output cavity mirror and spherical semi-transparent mirror (Fig.1). In Fig. 2 depicted are of $|\gamma|^2$ vs. gap mismatch in the range *(0-0,5)λ* for dominated mode (solid line) with azimuth index 0. Also the dependences for modes with azimuth indexes 1,2 is presented. The dominated structure of the field inside of the cavity is formed by modes with azimuth indexes 0 and 1. In the degeneracy points of the modules of eigenvalue the both modes with indexes 0 and 1 is existed. The increasing of the output power is possible at these points. It is demonstrated in experiment (Fig.3).

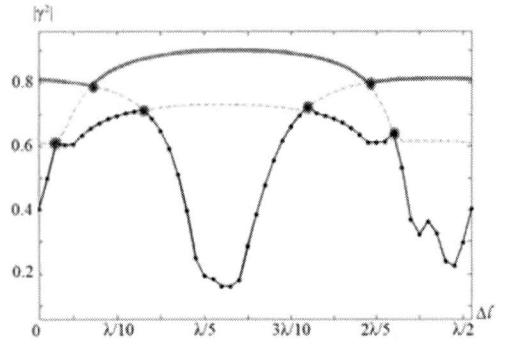

Fig.2 The modulus of eigenvalue vs. gap mismatch $\varDelta l$ (mirror 1 is flat; mirror 2 curvature radius is -0,06m;. $R_1 = R_2 = 0,65$; $\lambda = 1,064\mu m$)

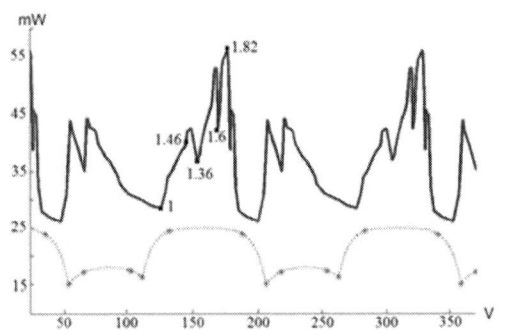

Fig.3 The average output power for microchip laser vs. control voltage (gap mismatch), solid curve - experiment , dashed curve – theory, numbers – M^2 values, points – the degeneracy points of eigenvalue modulus.

In the course of the experimental investigation was possibility of control not only by energy (power) and spatial characteristics of radiation, but also temporary is found.

The experimental investigation were developed on the experimental setup (Fig.4).

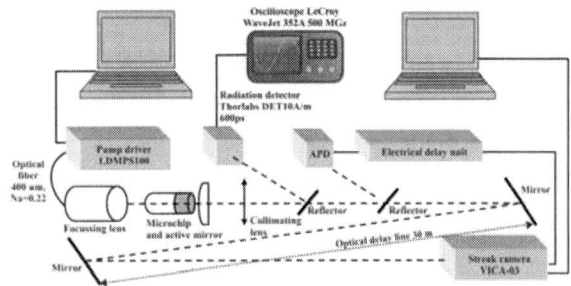

Fig. 4. The scheme of experimental setup.

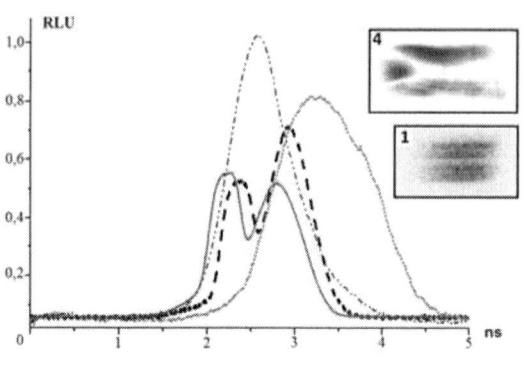

Fig. 5. Time dependences for pulses with different gap mismatch (control voltage in range 0-150V), 1- 0V, 2-50V, 3-100, 4 – 150. In the borders - streak photography samples for 1 and 4 regimes.

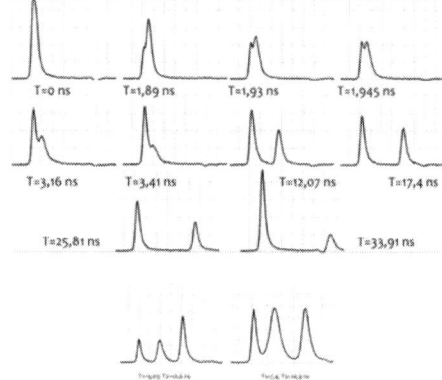

Fig.6 Time dependencies for different gap mismatch and pumping volume. T – time betveen pulses maximums.

Conducted numerical and experimental investigations of system of a microchip laser demonstrate that use of an active cavity output mirror based on Fabry-Perot interferometer enables to control not only mode composition and power of laser output radiation, but also time structure of the output irradiation in wide range.

References

[1] O. Svelto Principles of Lasers, 4-th ed., 604p., 1998.

[2] S. De Silvestri, P. Laporta, V. Magni, and O. Svelto *Optics Letters*, vol. 12, pp. 84-86, 1987.

[3] A.G Fox. and T. Li. B*ell Syst. Tech. J.,* vol. 40, p.453, 1961.

978-1-4799-0019-0/13 $31.00 © 2013 IEEE

Study of Stable Laser Cavity With Hole-coupling Output Mirror

(Invited Paper)

Z. S. Tian, Y. C. Zhang, Z. H. Sun, S. Y. Fu, Q. Wang

Information Optoelectronics Research Institute, Harbin Institute of Technology at Weihai, Weihai 264209, China

Abstract: A new type of CO_2 laser resonator with hole-coupling output is presented in this paper. The laser beam in the resonator is focused into the reflector assembly system composed of a reflector with a small central hole and an output mirror with a central coupling hole. According to the theory of diffraction optics, the laser optical field distribution in the cavity and the far-field distribution are analyzed by numerical calculation. In the experiment, the variations of the laser output power with continuous current discharged is measured. The maximum laser output power is 14.5W at the discharge current 12mA. The laser power distribution of the output beam at the distance of 1m is measured, and the result is in good accordance with numerical calculation. The laser cavity with hole-coupling can be used in intracavity pumped THz laser output and other types of intracavity pumped lasers.

I. INTRODUCTION

Optically pumped THz lasers have been widely applied in the fields of THz imaging, security, medical and so on [1-5]. Compared to other optical and electronical methods of generating THz wave, the method of optically pumped THz laser owns the advantages of high efficiency and output power. The optically pumped THz lasers are commercially available, but the average power of the laser is only hundreds of mW level. At present, optically pumped far infrared lasers are operated in external-cavity pumped mode. It consists of two parts: the CO_2 pumped laser and THz laser generator [6, 7]. In general, the laser intensity of pumped laser inside the cavity is much stronger than the intensity of the output laser, so operating in intracavity pumped mode is a good method to increase the pump laser power. At present, the optical lens and coating technology applied to both the THz band and CO_2 laser band are not mature, so it is difficult to find ideal optics to meet the intracavity pumped requirements.

In order to realize intracavity pumped THz laser, a new type of laser resonator is presented based on our previous study for optically pumped THz laser and the references about external-cavity pumped THz laser resonator with small hole-coupling [7-12]. The main feature of the laser is that the laser beam in the resonator is focused into the reflector assembly system composed of a reflector with a small central hole and an output mirror with a central coupling hole. The laser optical field distribution in the cavity and the far-field distribution are analyzed by numerical calculation. In the experiment, the maximum laser output power is 14.5W at the discharge current 12mA. The laser power distribution of the output beam at the distance of 1m is measured, and the result is in good accordance with numerical calculation.

This work was supported by the Fundamental Research Funds for the Central Universities (Grant No. HIT. BRET. 2010014), the Science and Technology Planning of Shandong Province, China (Grant No. 2011GHY11514)

II. THE EXPERIMENT SETUP

The schematic drawing of the resonator is shown in Fig. 1. The laser consists of two parts: gain medium area and reflector assembly system. The gain medium area is a modified industrial CO_2 laser tube with continues discharge current. The output mirror of the laser tube is replaced by a ZnSe window with antireflection coating both sides. The laser beam in the resonator is reflected by two reflectors inclined at 45° to the optical axis and focused into the reflector assembly system through a small hole by a ZnSe lens with focus length 185mm. The reflector assembly system is composed of two gold-coated mirrors with 40mm diameter. One mirror is a flat total reflector with a central coupling hole 2mm in diameter. The other mirror is a concave reflector with a central coupling hole 10mm in diameter, and the radius of curvature is 3m. The distance between the two mirrors is 1.5m. The focused laser beam through the reflector with small hole is expanded with a certain divergence angle to the concave mirror, and then the laser in the cavity is partially output from the coupling hole. Therefore, the concave reflector can be called hole-coupling output mirror.

The divergent laser reflected by concave mirror is converted into parallel beam, then transmit to the mirror with small hole. The reflected laser returns along the original direction. The parallel beam bounces twice at the concave mirror and is focused by concave mirror, and finally re-enter the area of gain medium through the ZnSe lens. Then the laser is amplified again, thereby the laser oscillation build up. The laser system is shown in Fig. 2. The glass tube is filled with CO_2 laser working gas, and the laser gain can be obtained by gas discharge with high voltage. The two mirrors with coupling hole compressing "O" rubber rings to the flanges of metal tube form the reflector assembly system.

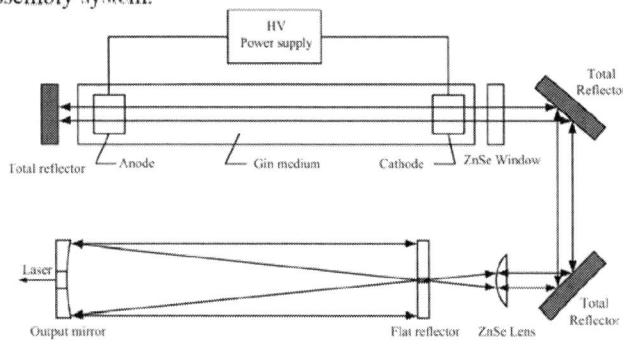

Fig.1. Schematic drawing of the laser resonator with hole-coupling

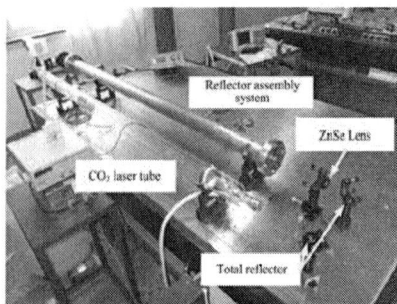

Fig.2. The photograph of the laser system

III. THEORETICAL ANALYSIS

According to optical diffraction theory, the beam propagation equation in the cavity (as shown in Fig.1) can be expressed in the form of being solved by Fast Fourier Transform iteration method. The intra cavity optical field distributions on the surface of the total reflector in the gain medium area and hole-coupling output mirror are calculated and the results are shown in Fig. 3. It can be seen that the optical field distribution on the surface of the total reflector is similar to the Gaussian function (Fig. 3a), and the optical field distribution on the surface of the output mirror is a dip ring. At the distance of 1m, the intensity distribution of the output laser is given by numerical simulation, and the result is shown in Fig. 4. Due to the diffraction effect of coupling hole, the output laser power distribution does not fit the Gaussian function, and there are a series of concentric circles at the cross section of laser beam.

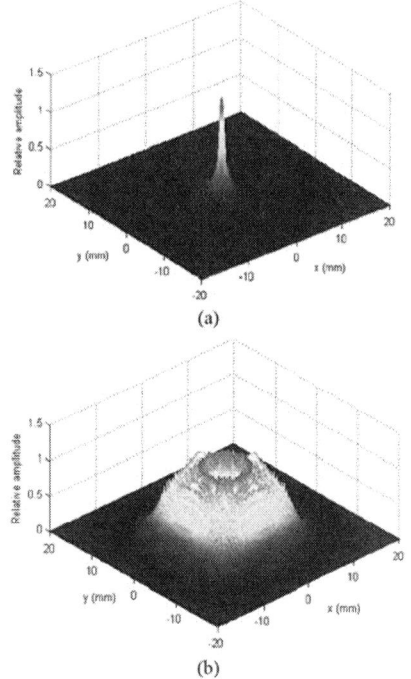

(a)

(b)

Fig.3. The optical field distributions of the cavity. (a) Optical field distribution on the total reflector (b) Optical field distribution on the output mirror

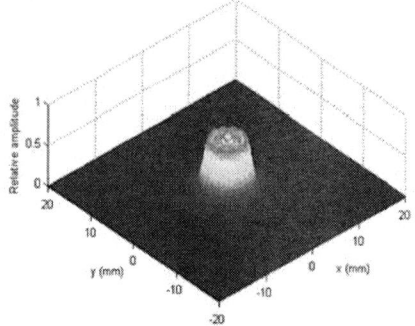

Fig.4. The far-field laser power distribution of the output laser beam

IV. RESULTS AND DISCUSSION

A. The output power CO_2 laser

In the experiment, the glass laser tube filled with CO_2 laser working gas is sealed by reflector and ZnSe window. The tube length is 1.6m and the discharge length in the tube is 1.4m. The laser cavity shown in Fig. 2 is adjusted precisely using a He-Ne laser. The laser output from the coupling hole was measured by the laser power meter. The variation of the laser output power at discharge current 5-20mA is shown in Fig. 5. It can be seen that the maximum output power of the laser is up to 14.5W at 12mA discharge current.

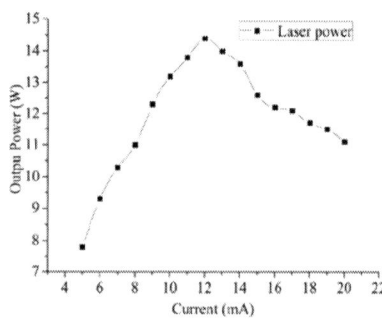

Fig.5. The variation of the laser output with discharge current

B. The intensity distribution of the laser

At the distance of 1m from the output mirror, the intensity distribution of the laser on the receiving screen is recorded by an infrared camera, as shown in Fig.6. It can be seen that the experimental results is basically consistent with the theoretical analysis.

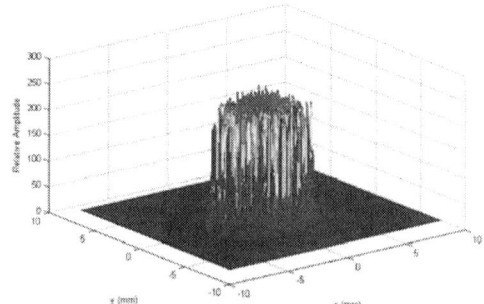

Fig.6. The far-field laser power distribution of the laser beam measured by infrared camera

V. CONCLUSION

In conclusion, a new stable laser resonator with hole-coupling is reported in this paper, and the maximum output power of 14.5W CO_2 laser is obtained. The optical field distribution of the output laser is analyzed by numerical calculation, and the theoretical analysis is consistent with the experimental results. Though the CO_2 output laser is not ideal Gaussian beam, it has little effect on the THz laser beam quality and output power while applied to intracavity pumped THz laser in the future. In addition, the optically pumped THz lasers usually require the pumped lasers with high spectral purity and stability. The laser we reported can satisfy the requirements by replacing the total reflector of the laser tube (area of the gain medium) to the grating with PZT and adding an appropriate stabilization circuit. The laser cavity presented in this paper can also be applied to other types of lasers. This new laser structure can provide a helpful reference for the development of the laser resonator technology.

REFERENCES

[1] L. Miao, D.L. Zuo, Z.X. Jiu, Z.H. Cheng, C.C. Qi and J.Z. Wu, "High energy optically pumped NH_3 terahertz laser with simple cavity," Chin. Opt. Lett., Vol. 8, pp. 411-413, April 2010.

[2] E. R. Mueller, R. Henschke, W. E. Robotham, Jr., L.A. Newman, L.M. Laughman, R.A. Hart, J. Kennedy and H.M. Pickett, "Terahertz local oscillator for the Microwave Limb Sounder on the Aura satellite," Appl. Opt., Vol.46, pp.4907-4915, January 2007.

[3] J. F. Federici, B. Schulkin, F. Huang D. Gary, R. Barat, F. Oliverira and D. Zimdars, "THz imaging and sensing for security applications—explosives, weapons and drugs," Semicond. Sci. Technol., Vol.20, pp. S 266- S 280, July 2005.

[4] T. Kleine-Ostmann and T. Nagatsuma, "A review on Terahertz communications research," J. Infrared Milli. Terahz. Waves, Vol. 32, p. 143, August 2011.

[5] Q. Li, S.H. Ding, Y.D. Li, K. Xue and Q. Wang, "Experimental research on resolution improvement in CW THz digital holography," Appl. Phy. B., Vol.107, pp.103-110, July 2012.

[6] L.J. Geng, D.M Ren, W.J. Zhao, Y.C. Qu, H.Y. Chen and J. Du, "An efficient, compact pulsed D_2O terahertz super-radiant laser pumped with a fundamental transverse mode transversely excited atmospheric pressure CO_2 laser," Laser Physics, Vol. 23, pp. 025001, July 2013.

[7] T.Y. Chang, T.J. Bridges and E.G. Burkhardt, "CW submillimeter laser action in optically pumped methyl fluoride, methyl alcohol, and vinyl chloride gases," App. Phys. Lett., Vol.17, pp.249-251 July 1970.

[8] Y.X. Qin, X.H. Tang, Y. Xiao, J. Liu, D. Wang, H. Peng, Q.S. Deng, X. Zhu and Z.J. Li, "Toric concave mirror laser resonator with a big Fresnel number," Opt. Lett., Vol.34, pp.1120-1122 March 2009.

[9] M. Endo, "Azimuthally polarized 1 kW CO_2 laser with a triple-axicon retroreflector optical resonator," Opt. Lett., Vol. 33, pp.1771-1773, July 2008.

[10] L.L. Wang, Z.S. Tian, Y.C. Zhang, J. Wang, S.Y. Fu, "Frequency stabilization of pulsed CO_2 laser using setup-time method," Chin. Opt. Lett., Vol.10, p.011402, August 2012.

[11] Z.S. Tian J. Wang, F. Fei, J.G. Yang, Y.C. Zhang, S.Y. Fu and Q. Wang. "Study of optically pumped all-metal terahertz laser," Chinese J. Lasers Vol.37, pp.2323-2326, September 2010.

[12] Y.D. Sun, S.Y. Fu, J. Wang, Z.H. Sun, Y.C. Zhang, Z.S. Tian, Q. Wang, "Optically pumped terahertz lasers with high pulse repetition frequency: theory and design," Chin. Opt. Lett., Vol.7, pp.127-129, August 2009.

The technique of measuring the velocity of melt removal in gas-laser cutting technology using multi-channel pyrometer

Y. N. Zavalov, A. V. Dubrov, V. D. Dubrov, N. G. Dubrovin, E. S. Makarova A. N. Antonov
Institute on Laser and Information Technology of Russia Academy of Science,
Shatura, Moscow Region, Russia

Abstract: Multichannel pyrometer was used to measure fluctuations of temperature on the front of action of laser radiation on steel in technology of laser cutting with an assisted gas. Time-resolved measurements were made of local temperature in the five areas with spot size of about 100 mkm each, spaced approximately 0.6 mm apart in the upper part of the front cut. The technique of measuring the velocity of melt removal using the cross-correlation of data of multichannel pyrometer is described. It is shown that the reject filtering allows emphasizing the speed of so-called "fast" and "slow" waves, as the interface velocity. in which they take into account the generation of surface waves in thin film blowing a jet of gas. Obtained experimental data of the velocity of the melt and film thickness are consistent with the results of the model, Chen (1999), which takes into account the gas- jet generation of surface waves on the film.

The technology of laser cutting is accustomed now, but means of on-line monitoring are required to improve the reliability of the operation of this process [1-2]. The diagnostics of processes during laser cutting is relevant, and the following pioneering works are distinguished [3-5]. Those control devices are designed for laser welding technology [6]. The understanding of processes at the front of laser radiation action on the material should be the basis of such monitoring devices.

The multi-channel pyrometer was used in experiments; it described in [7]: local pulsations of the brightness temperature is measured at five areas (each area is sized of ~∅100mkm), spaced about 0.6 mm apart at the top edge of cutting front. The sensor K1713-05 (Hamamatsu company) is used in each channel comprising bred under each other translucent Si diode and InGaAs-diode. The scheme of two-colour pyrometry is used: the temperature is defined from division of currents of Si diode and InGaAs-diode avoiding changes in the fluctuations of surface brightness associated with other causes than the brightness temperature of the melt, fig.1.

The spectrum of random fluctuations of the measured temperature displays two independent processes: the formation of the low frequency pulsation of transverse striations on the side of the cut surface, and high temperature fluctuations associated with the movement of the melt downstream. Autocorrelation function, fig.2a, of the data shown in Fig. 1 can be represented respectively as the sum of the autocorrelation functions of two independent autocorrelation functions.

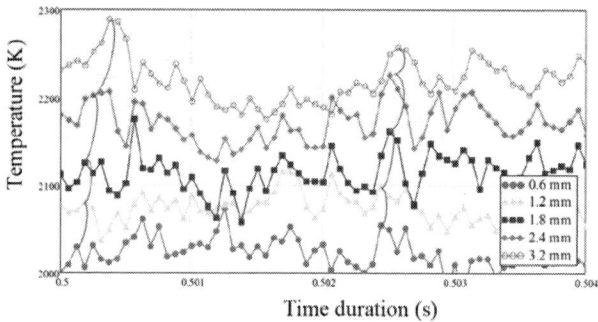

Fig. 1. Experimental data of temperature (K) along cutting front, mild steel, 6mm, v=33 mm/s, P (O₂)= 0.45 MPa.

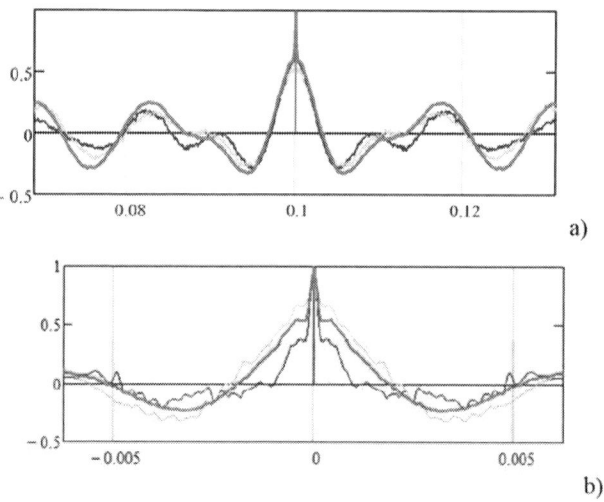

Fig. 2. Autocorrelation function of 4th channel: a) the function of unfiltered data; b) function of filtered data.

The delay of moving of the temperature inhomogeneity from channel to channel was defined by calculation index of the cross-correlation of adjacent channels. Thus the velocity of the temperature inhomogeneity was determined for given distance between channels (0.6 mm). The obtained velocity is about 15 m/s and lead too small values of melt's film thickness:

$$d_m \approx v_{cut} \cdot h / v_m \approx 10 \mu m \qquad (1)$$

If data is filtered in tight range [f_0-200 Hz, f_0+200 Hz], before computing the cross-correlation, then computations produce, fig. 3, the velocity of moving of temperature inhomogeneity depending on current f_0.

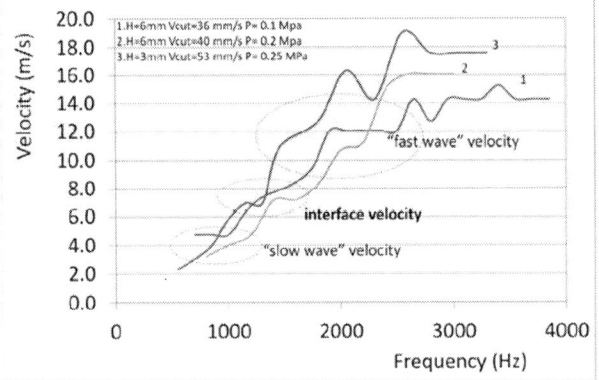

Fig. 3. Spectral distribution of velocity of moving of temperature inhomogeneity.

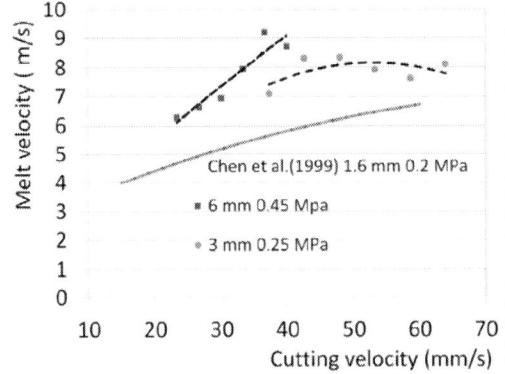

Fig. 4. The dependence of average velocity of melt removing on the cutting velocity for various steel sample parameters. Solid line is the calculation from [11].

It is known, [9], that the steady flow of melt layer under the action of the jet is unstable. Unsteady melt layer is stable [10], when some waves propagate in a range of speeds [c_0-c_1, c_0+c_1], where c_0 is the speed at the surface of the melt and c_1 is the limit of wave velocity relatively moving surface of melt film taking in account the second-order instability of film movement:

$$c_0 = v_m \cdot \Gamma \ ; \ c_1 \approx 0.5 \cdot c_0 , \text{ where} \qquad (2)$$

v_m is the average velocity of melt film;

Γ is form factor of depth profile of melt velocity,

$\Gamma = \dfrac{4}{3}$ for laminar profile of movement.

Thus the method enable to identify the speed of the "fast" and "slow" waves on the surface of the melt, as well as to determine the velocity of the surface of the melt. The dependence of average velocity of melt removing on the cutting velocity for various steel sample parameters is determined, fig.4. Solid line is the calculation from [11].

The technique of measuring the velocity of melt removal in gas-laser cutting technology is considered by computing the cross-correlation of data of multi-channel pyrometer.

ACKNOWLEDGMENT

The work has been supported by the Russian Foundation for Basic Research (grants № 13-08-00987-a).

REFERENCES

[1] Hacketta C.M., Gargb S. "Visualization of thermal cutting fluid flows", Eur. Phys. J. Special Topics, 182, pp.145–159, 2010.

[2] Zebala W., Matras A., Kowalczyk R. "Quality aspects of steel parts after laser cutting", Advances in Manufact. Sc. & Technology, Vol. 36(2), pp.5-13, 2012.

[3] Hansmann M., Decker I., Ruge J. "On-line control of the laser cutting process by monitoring the shower of the sparks", Power Beam Technology, Brighton, England, 10-12 Sept. 1986, p.440-445, 1986.

[4] Sforza P., Santecesaria V., "Analytical dependence of the roughness of the cut edge on the experimental parameters and process monitoring", Proceed. SPIE, v.2207, pp. 836-847, 1994.

[5] Arata Y., Maruo H., Miyamoto I., Takeuchi S. Dynamic Behavior in Laser Cutting of Mild Steel; Transactions of Japanese Welding Research Institute, V.8 (2), pp.15 – 26, 1979.

[6] Bardin F., Morgan S., Williams S. "Process control of laser conduction welding by thermal imaging measurement with a color camera" App. Optics, V. 44, pp. 6841-6848, 2005.

[7] Dubrov A. V., Dubrov V. D., Zavalov Y. N. "Application of optical pyrometry for on-line monitoring in laser-cutting technologies," Appl. Phys.B, Vol.105(3), 537–543, 2011.

[8] Dubrov A. V. , Dubrov V. D. , Zavalov Y. N., "Thermocapillary effects in CO_2 laser cutting of metals", In: 11th International Conference on Laser and Fiber-Optical Networks Modeling (LFNM), 5-9 Sept. 2011, Kharkov.

[9] Vicanek M., et al., "Hydrodynamic instability of melt flow in laser cutting" J. Phys. D: Appl. Phys. 20 (1987) 140-145.

[10] Asali J. C. and Hanratty T. J. "Ripples generated on a liquid film at high gas velocities" Int. J. Multiphase Flow Vol. 19, No. 2, pp. 229-243, 1993.

[11] Kai Chen, Y. Lawrence Yao "Striation Formation and Melt Removal", J. of Manufacturing Processes, Vol. 1/No. 1, 1999.

Intracavity singly-resonant optical parametric oscillator pumped by a semiconductor disk laser

Yu. A. Morozov, M. Yu. Morozov,
Kotelnikov Institute of RadioEngineering and Electronics (Saratov Branch),
Russian Academy of Sciences, Saratov, Russia

Abstract: A novel scheme of the compact intracavity singly-resonant optical parametric oscillator (OPO) pumped by a semiconductor disk laser is presented. The cavity of the device is shared between the laser's optical field (the pump of the OPO) and the signal radiation emitted by the OPO. The length of the cavity doesn't exceed 10 mm that should hopefully favor for a single mode operational regime without any frequency selective elements (e.g. Fabry-Perot etalons). By estimation, the idler wavelength could be tuned over a few percents in the mid-infrared range (~ 17 μm) as the primary pump power is changed.

Since 60-th of the last century, OPOs are well known as sources of widely tunable coherent radiation in near-and mid-infrared ranges. Among those devices, continuous-wave OPOs being the narrow linewidth ones are almost ideally suitable for high-resolution spectroscopic applications [1]. The principle of OPO's operation is based on dividing the energy of photon of highest energy (the pump) into two lower energy photons (denoted the signal and idler). Such a frequency down-conversion appears as one, two or all three involved optical fields are resonant and threshold of a parametric interaction is reached. Now doubly- and singly-resonant OPOs have only found a widespread implementation. The doubly-resonant realizations of the resonance conditions for both the signal and idler waves suffer from extreme perceptibility on mechanical mirror displacements coupled with complexity of smooth tuning, whilst show a low level of pumping threshold.

The singly-resonant OPOs (SROs) with one wave resonating (most usually, the signal wave) do not require very high mechanical stability and can be tuned without complex tuning schemes. The SROs, however, need tens-of-watt pumping lasers in order to overcome the threshold. One of the routes to diminish the threshold is to place an SRO within a high-finesse cavity of a pumping laser. In particular, such the intracavity SRO (ICSRO) was demonstrated with a neodymium laser (Nd:YAG) [2]. Unfortunately, ICSROs pumped by neodymium lasers appear to show pronounced relaxation oscillations of output wave intensity. To eliminate this disadvantage the pumping of the ICSRO by a semiconductor disk laser what is also referred to as a vertical external cavity surface-emitting laser (VECSEL) has been proposed recently [3].

In this paper we propose and analyze a novel type of a compact ICSRO pumped by a VECSEL (the scheme of the device is presented in Fig.1). One can see from the Figure that the device would look like the end-pumped VECSEL equipped with linear plano-concave cavity [4] as the nonlinear crystal

was removed from the cavity. The preference of using the primary end-pumping is motivated by compactness of the ICSRO with very short cavity (the total length of the cavity with a nonlinear crystal of quasi-phase-matched GaAs [5] inserted is supposed to be about 10 mm). This cavity is formed between a Bragg mirror grown in the laser chip and a curved mirror. Note that for avoidance of misunderstanding, one should distinguish between the primary pumping (by a diode laser at 808 nm) of the VECSEL and the pumping of the OPO by the optical field generated by the VECSEL.

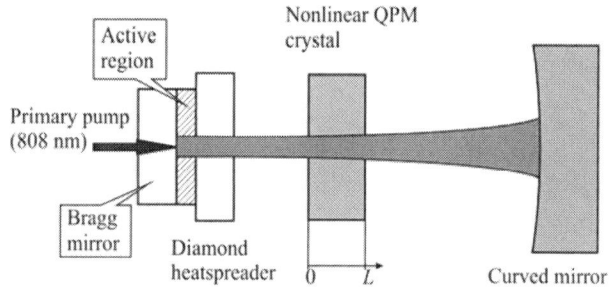

Fig. 1. Schematic view of the ICSRO.

Unlike to the ICSRO presented in [3], we have usage of the common cavity both for the pumping laser and the signal down-converted wave. When doing so we suppose that the normalized difference between the pump and the signal wavelengths doesn't exceed about ten percents. Besides, frequency selective elements, e.g. Fabry-Perot etalons or birefringent tuners have been removed from the cavity, thus taking advantage of maximal compactness and simplicity of the construction. According to the findings of A. Garnache et.al [6], a VECSEL itself shows single-frequency operation without intracavity etalons provided a cavity is quite short (about 10-20 mm). These observations allows us to expect a single-mode operation of the proposed ICSRO, while the point needs the additional study.

The active region of the ICSRO is composed of InGaAs quantum wells (QWs) separated by GaAs barriers. In order to take advantage of the resonant periodic gain, the QWs are located in the antinodes of the pumping optical field. For efficient heat removal from the active region an intracavity diamond heatspreader is used. The curved output mirror as well as the Bragg mirror are supposed to be highly reflective at the pump and signal wavelengths (λ_p and λ_s) and transparent for the idler radiation at wavelength $\lambda_i = \lambda_p \lambda_s / (\lambda_s - \lambda_p)$.

Fig. 2,*a* shows the intracavity pump (*1*) and signal (*2*) powers (left axis) as well as the power of idler radiation propagating in one direction (*3*, right axis) vs the primary pump power. Fig. 2,*b* displays the dependence of QWs' carrier density on the primary pump power. The following values of parameters have been used under the simulation: $\lambda_p = 1040$ nm, $\lambda_s = 1108$ nm, $\lambda_i = 16.95$ μm, the round-trip loss of the pump and signal optical fields $S_p = S_s = 0.025$, the nonlinear crystal length $L = 3$ mm, the waist radii of the pump and signal in the active region are approximately the same (about 50 μm) if the output mirror's radius of curvature is 15 mm and the total length of the cavity is 8.5 mm. The primary pumping of the QWs has been simulated according to [7].

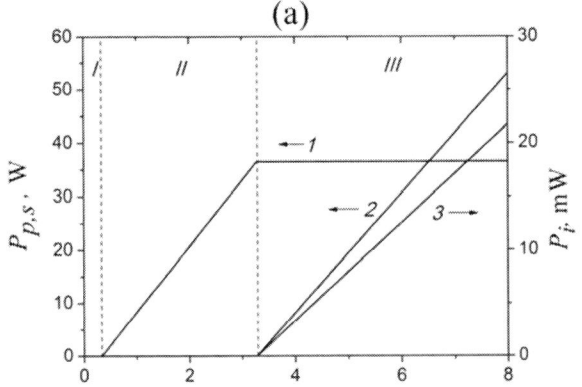

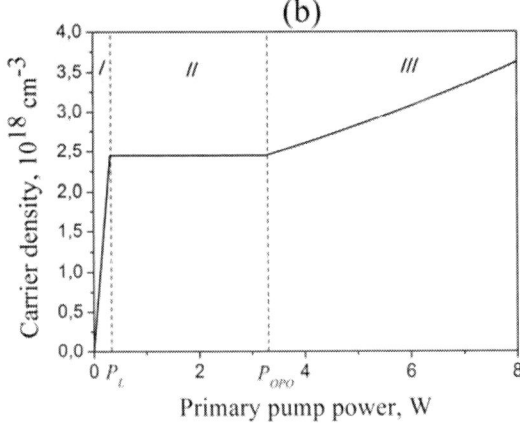

Fig. 2. Dependence of (a) the intracavity pump (*1*) and signal (*2*) power as well as the power of idler radiation (*3*) and (b) the quantum well carrier density on the primary pump power.

In the regions marked by *I* and *II* the signal radiation is not excited and the operational regime of the ICSRO doesn't sub-

stantially differ from the one of an end-pumped VECSEL. As the parametric gain (depending on the pump photon flux) equals the round-trip loss S_s, the stimulated parametric emission of signal and idler photons begins. Further increasing of the primary pump power results in pinning of the pump power at the threshold value (≈ 36 W), the power of signal and idler increases in line with primary pumping. Note that the power of idler radiation travelling in one direction reaches about 20 mW what is quite sufficient for most of applications. One can see from Fig. 2,*b* that the QW carrier concentration is not longer fixed in region *III*. Instead, it increases to compensate the loss of the pump photons arising from the nonlinear conversion. Since the spectral position of the maximal gain of the VECSEL radiation (and, consequently, λ_p) depends on the QW carrier density, the idler radiation wavelength can likely be tuned over a few percents under the primary pump power changes.

We believe the proposed ICSRO might be quite competitive with a quantum cascade laser counterpart regarding most if not all of operational characteristics.

REFERENCES

[1] F. Tittel, D. Richter, A. Fried, "Mid-infrared laser applications in spectroscopy", in *Solid-state mid-infrared laser sources*. Eds. T. Sorokina, K. L. Vodopyanov, Springer-Verlag Berlin Heidelberg, 2003.

[2] D. Stothard, M. Ebrahimzadeh, M. Dunn, "Low pump threshold, continuous-wave, singly resonant, optical parametric oscillator", *Opt. Lett.*, vol.23, pp. 1895-1897, 1998.

[3] D. Stothard, J. Hopkins, D. Burns, M. Dunn, "Stable, continuous-wave, intracavity, opical parametric oscillator pumped by a semiconductor disk laser (VECSEL)", *Opt. Express*, vol.17, pp. 10648-10658, 2009.

[4] J. Lee, J. Kim, S. Lee, J. Yoo, K. Kim, S. Cho, S. Lim, G. Kim, S. Hwang, T. Kim, Y. Park, "9.1-W high-efficient continuous-wave end-pumped vertical-external-cavity surface-emitting semiconductor laser", *IEEE Phot. Techn. Lett.*, vol.18, pp. 2117-2119, 2006.

[5] O. Levi, T. Pinguet, T. Skauli, L. Eyres, K. Parameswaran, J. Harris, M. Fejer, T. Kulp, S. Bisson, B. Gerard, E. Lallier, L. Becouarn, "Difference frequency generation of 8-μm radiation in orientation-patterned GaAs", *Opt. Lett.*, vol.27, pp. 2091-2093, 2002.

[6] A. Garnache, A. Ouvrard, D. Romanini, "Single-frequency operation of external-cavity VCSELs: nonlinear multimode temporal dynamics and quantum limit", *Opt. Express*, vol.15, pp. 9403-9417, 2007.

[7] Y. Morozov, T. Leinonen, M. Morozov, S. Ranta, M. Saarinen, V. Popov, M. Pessa, "Effect of pump reflections in vertical external cavity surface-emitting lasers", *New Journal of Physics*, vol.10, 063028, 2008

Infrared laser emission in a compact CW and quasi-CW diode pumped Nd^{3+}: GdLuCOB laser

C.A. Brandus[*], *Student Member, IEEE*, L. Gheorghe, T. Dascalu
National Institute for Laser, Plasma and Radiation Physics
Laboratory of Solid-State Quantum Electronics, Magurele-Ilfov, R-077125, Romania
[*] Doctoral School of Physics, University of Bucharest, Romania
E-mail: catalina.brandus@inflpr.ro

Abstract: Nd^{3+}: GdLuCOB laser crystals were grown by the Czochralski pulling technique. CW and quasi-CW laser emission at 1.06 μm has been achieved from an uncoated, 5-at. % Nd^{3+}: GdLuCOB crystal of 4.5-mm thickness, using end pumping at 811.9 nm and 812.3 nm, respectively, with fiber-coupled diode lasers. The maximum output powers for the CW and quasi-CW operations, were 167 mW and 178 mW for 1.31 W absorbed pump power, corresponding to the optical-to-optical efficiency with respect to the absorbed pump power of 0.13 and 0.24. Slope efficiency for both CW and quasi-CW operations were 0.45 and 0.31 and the absorbed pump powers at threshold 0.7 W, and 0.6 W, respectively.

Harmonic frequency conversion of the infrared laser radiation with generation of the radiation in the VIZ (green, blue) or UV spectral region has many applications in various domains, such as lithography, medicine (clinical surgery), biology, spectroscopy, communications, display, thin film deposition, and other. The simple way to produce such a visible laser in a compact and lower price system is to use a self frequency doubling crystal, or self sum-frequency mixing crystal, such as Nd: GdCOB. Experiments on this crystal have demonstrated by now conversion efficiencies of 20% together with slope efficiency for the laser emission of the fundamental wave of 45% when pumping by Ti: sapphire CW laser. [1]

By replacing Gd ions with Lu ions in the GdCOB matrix, the efficiency for the frequency doubling process is increased and noncritical phase matching at room temperature is achieved. [5]

Such as Nd: GdCOB, the Nd: GdLuCOB crystal is optically biaxial negative and belongs to spatial group C_m. Its crystalline structure is monoclinic and it poses low symmetry. Nd: GdLuCOB crystal has very good nonlinear coefficients, if compared to the BBO, KTP or LBO frequency doubling crystals.

The aim of the present work was to study the laser properties of the self-frequency doubling Nd: GdLuCOB crystal. The crystal was grown by Czochralski pulling technique in our laboratory.

Here we report for the first time, to our knowledge, emission of laser radiation in a compact laser system emitting at 1.06 μm based on self frequency doubling Nd:GdLuCOB single crystal, when is pumped by laser diode at 811.9 nm and 812.3 nm, in both CW and quasi-CW regime. Laser emission of the fundamental wave at 1060.5 nm, has been experimentally achieved for a solid state NdLuGdCOB laser, in a plano-plano cavity, when is end-pumped by a high power laser diode (DL LIMO35).

The crystal has the dimensions 4 x 4 x 4.5 mm^3 and the concentration in Nd^{3+} ions was 5 at%. The crystal is uncoated and is cut for doubling laser radiation at the fundamental wavelength of 1.06μm. The resonator cavity length was 13 mm. The transmission for the output couplers were 1,2,3 % for the CW wave pumping and 1,3, 5, 10, 20% for the quasi-CW pumping. The pumping wavelength in the CW regime was 811.9 nm. For the quasi-CW regime of pumping we have used an optimized wavelength of 812.3 nm, delivering an absorption of 82.46% (according to Lambert-Beer law), which is the maximum obtainable for this length of crystal and this doping level (5 at% Nd^{3+}).

In order to avoid damages caused by thermal heating during the CW pumping, we have cooled the crystal by mounting it in a cooper holder, and we have controlled and set the temperature of the crystal at 20^0 C. The cooling liquid was water. The cooling was done by an ensemble of thermistor-TEC element.

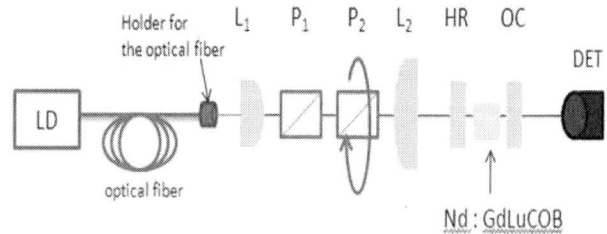

Fig.1. Experimental set-up for the Nd: GdLuCOB laser pumped by laser diode in the CW regime

The pumping beam from the laser diode is connected through an optical fiber with 100 μm diameter and 0.22, numerical aperture, and is imaged on the crystal with a pair of two near infrared achromatic doublets with focal length ratio 19mm : 30 mm. For controlling of the pumping intensity we have used also a pair of Glan Laser Polarizers designed to work in near IR at high power.

In the first experiment, when pumping in CW by diode lasers, at 811.9 nm and having 2.3 nm bandwidth, we obtained laser emission at 1.06 μm for three different output couplers (1%, 2%, and 3%).

The pumping beam waist diameter is 170 microns (determined by Knife-Edge method). The crystal absorbs in this configuration only 60% of the total pumping power. The slope efficiency rises to 31% for 3% transmission of the OC. The optical to optical conversion efficiency was 0.12, in respect to absorbed pump power, for transmission of 2% of the OC. The maximum laser output power at 1060.5 nm was 167 mW for an absorbed pump power of 1.3 W.

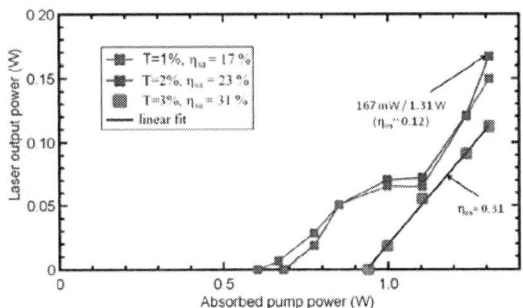

Fig.2. Laser emission at 1.06 μm in the case of CW laser diode pumping

For the second experiment, at the quasi-CW pumping we use almost the same simple experimental set-up but we insert after L1 and before P1, a chopper wheel in order to get higher peak pumping power and maintain the medium power at sufficient low level so we can avoid damaging of the crystal by thermal heating during the pumping.

In this experiment we have obtained 187 mW laser output peak power in the IR region, for 788 mW absorbed pump peak power, and the slope-efficiency is 30.3%, but in this case the OC had the transmission of 5%. Also we have calculated from linear fittings of the experimental data, slope efficiencies of 30% and 20% for the transmissions of 10% and 20% of the output coupler mirror. The optical to optical conversion efficiency rises to 0.24 (it is two times larger than in the case of CW pumping).

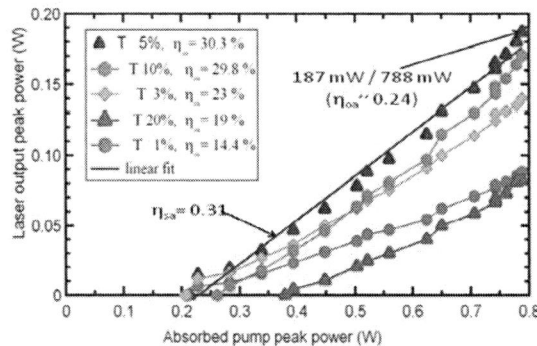

Fig.3. Laser emission at 1.06 μm in the case of quasi-CW laser diode pumping

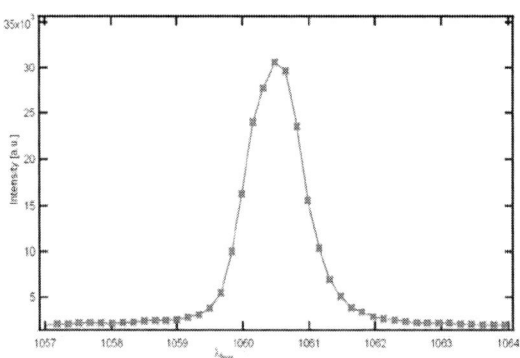

Fig.4. Laser emission line at 1.06 μm for the case of Nd: GdLuCOB laser

In further experiments we are interested in studying the self-frequency doubling characteristics for this compact solid state laser Nd^{3+}:GdLuCOB pumped by diode lasers. Experiments aiming improvements of the laser performances are under development

REFERENCES

[1] C.Chen et. al., "Nonlinear Optical Borate Crystals. Principles and Applications", Wiley Edition, 2012.

[2] Y.Shao et.al., A CW green laser emission by self-sum-frequency-mixing in Nd: GdCOB crystal", Y.Shao et.al. Laser Physics Letters 8, No. 10, 715-718, 2011.

[3] L.Leclin et. al., "Diode-pumped self-frequency-doubling Nd: GdCa4O (BO₃)₃ lasers: toward green microchip lasers", J.Opt.Soc.Am.B, Vol. 17, No. 9 , 2000.

[4] C.Gheorghe et.al, "Spectroscopic features and laser performance at 1.06 μm of Nd^{3+}-doped Gd$_{1-x}$Lu$_x$Ca₄O (BO₃)₃ single crystal", Journal of Appl. Phys. 111, 013102 , 2012.

Large aperture bimorph deformable mirror for extremely high power laser systems

Julia Sheldakova, Vadim Samarkin, Alexis Kudryashov, Alexey Rukosuev
Moscow State Open University, AKAoptics SAS

Abstract: In this work we present our recent results in modeling and fabrication of large aperture deformable mirrors first of all for laser beam correction but also for astronomical applications. Right now our company can produce bimorph mirrors with diameter from 200 to 500 mm. One of the main new advantages of our technology is that we can easily substitute damaged reflecting surface of our mirrors. Also, these wavefront correctors could be efficiently used in wide aperture high power lasers and laser systems.

It is well know now that exactly bimorph deformable mirrors are widely and most efficiently used in high power Tera and Peta Watt laser systems. Almost all newly designed high power lasers include bimorph deformable mirrors. And the reason is very simple – bimorph mirrors are rather inexpensive, they are perfectly suited to correct for thermally induced aberrations of optical elements of high power lasers, and any kind of coating could be easily put on their surface, so you can clean the surface of the mirror and might not think about any damage of such a corrector. Moreover nowadays the size of so called "standard" bimorph mirror rose to 200 mm. And is this presentation we report about new generation of bimorph wavefront correctors with active aperture up to 600 mm.

In general problem of bimorph mirrors is that it is almost impossible to make them small enough – less then 10 mm and also large. Last is due to the fact that ratio between diameter and thickness of mirror itself usually lays in the range of 30 – 40. And moreover with the increase of mirror diameter this parameter becomes even worse – 40 -50. So, mirror behaves like a thin sheet of paper and it is rather difficult to fix it in the housing and try to save the initial flatness. To overcome this effect we use a set of specially placed stacked actuators that support mirror both inside and at the edge (fig. 1). These actuators are only used to maintain initial mirror flatness and are almost not used in the whole process of wavefront correction.

Next sequence of pictures (fig. 2-4) clearly show the design and also the interferogramms of mirror surface before assembling and after correction of mirror surface with stacked actuators as well with some mirror electrodes.

Fig. 1. Design of wide-aperture deformable mirror.

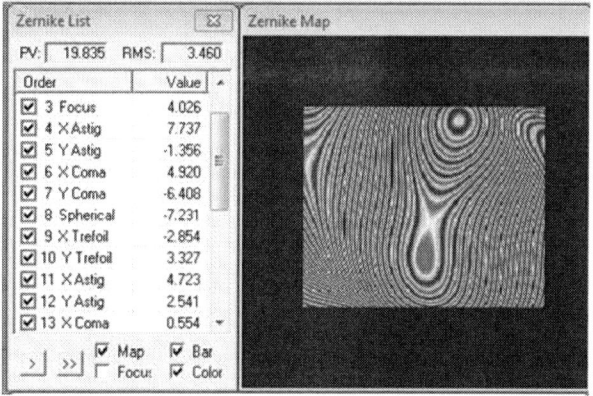

Fig. 2. Initial surface of DM with combined actuators
PV=19.835 µ, RMS = 3.460 µ

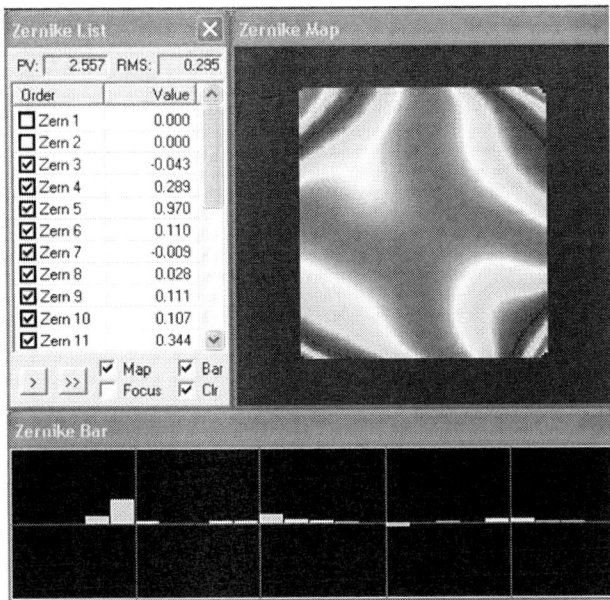

Fig 3. 1st step. Corrected surface 200x200 mm by 12 stack actuators and 4 defocus electrodes - PV=2.557 μ, RMS = 0.296 μ

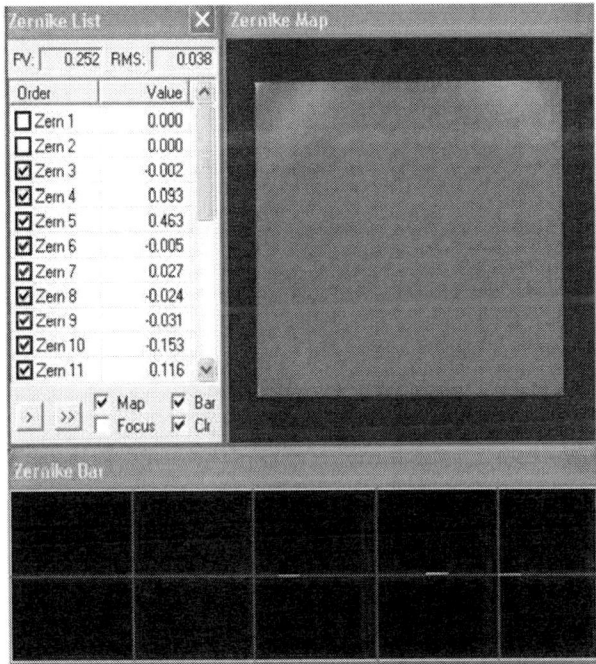

Fig. 4. 2nd step. Corrected surface by 60 bimorph electrodes PV=0.252 μ, RMS = 0.038 μ

Segments of Oversized Waveguides in Open Resonant Systems

I.K. Kuzmichev and A.Yu. Popkov

The A.Ya. Usikov's Institute for Radio Physics and Electronics of the NAS of Ukraine, Kharkov, Ukraine

Abstract – **The article discusses a hemispherical open resonator (OR) with a segment of circular or coaxial waveguide in the center of the plane mirror. Excitation efficiency for axially symmetric TE_{01} and TM_{01} modes in a circular and TEM and TE_{01} in coaxial waveguides with help of TEM_{01q}, TE_{01q} and TM_{01q} OR modes is analyzed. It shows that the efficiency of excitation of these waves can be reached 90% in case of the appropriate choice of geometric dimensions of resonator.**

I INTRODUCTION

In [1] is analyzed the excitation of TE_{01} mode in a circular waveguide which is located in the center of flat mirror of hemispherical open resonator (OR) with help of TEM_{01q} OR oscillation. In this case, the amplitude distribution of electric field components of the resonator and waveguide was considered. As it was turned out, the resonant system maintains the axially symmetric oscillation TE_{01q}. These oscillations are called "ox eye" [2]. As were shown by preliminary studies this resonant system has angular and frequency spectrum selection in the range of wavelength of the excited mode. However, we need to consider the vector nature of electric field components for these studies. In this paper, the excitation efficiency of axial-symmetric modes in the circular and coaxial waveguide using OR oscillations were considered, taking into account the vector nature of the electric field.

II CALCULATION RESULTS

As we know from the theory of reflector antennas [8, 9], it is necessary to provide fields matching in the focal plane of the reflector and the emitter aperture to obtain a high aperture efficiency (AE). From a physical point of view, it's the same as to field matching the mode of resonator with the field of waveguide which located in the centre of plane mirror of OR. So, we will use the ratio [8, 10], for excitation efficiency determination of various waves in the circular and coaxial waveguides via TEM_{01q}, TE_{01q} and TM_{01q} OR oscillations.

$$\eta = \frac{\left| \int\limits_{b}^{a} \int\limits_{0}^{2\pi} \vec{E}_e(\rho,\varphi) \vec{E}_w^*(\rho,\varphi) \rho \, d\rho d\varphi \right|^2}{\left\| \vec{E}_e(\rho,\varphi) \right\|^2 \left\| \vec{E}_w(\rho,\varphi) \right\|^2}. \quad (1)$$

The symbol * indicates the complex conjugate function.

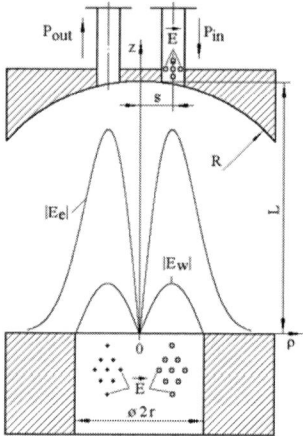

Fig. 1. The hemispherical OR with a segment of the circular waveguide

The values $\left\| \vec{E}_e(\rho,\varphi) \right\|^2$ and $\left\| \vec{E}_w(\rho,\varphi) \right\|^2$ in (1) – the squares of norms of exciting (in the resonator) and work (in the waveguide) fields. Note that vector of the electric field of the wave in the feeding TE_{10} rectangular waveguide (see Fig. 1) is perpendicular to the plane of the figure when excited the TE_{01} wave in a circular or coaxial waveguide with help of TEM_{01q} or TE_{01q} OR oscillation. Electric field vector of TE_{10} wave that propagating in feeding waveguide, lies in plane of the figure in the case of excitation of TM_{01} mode in the circular waveguide or TEM mode in the coaxial with help of TEM_{01q} or TM_{01q} OR oscillation. We neglect the reflection from the open ends of the circular and coaxial waveguides and assumed that the aperture mirrors are infinite.

Next, consider the efficiency of excitation the TE_{01} wave of circular waveguide using various oscillation of OR. To do this in the ratio (1) substitutes expressions describing the distribution of fields of TEM_{01q} and TE_{01q} ($E_e(\rho,\varphi)$ on fig.1) OR oscillations and equation for E_φ field component of TE_{01} wave of circular waveguide ($E_w(\rho,\varphi)$ on fig.1). The results of calculation by formula (1) was given on Fig. 2. Curve 1 corresponds to excitation of TE_{01} wave with help of axially asymmetric TEM_{01q} oscillation of the OR. Curve 2 – excitation the same wave using axially symmetric TE_{01q} oscillation. As can be seen from the figure the maximum efficiency of excitation the TE_{01} wave with help of TEM_{01q}

978-1-4799-0019-0/13 $31.00 © 2013 IEEE

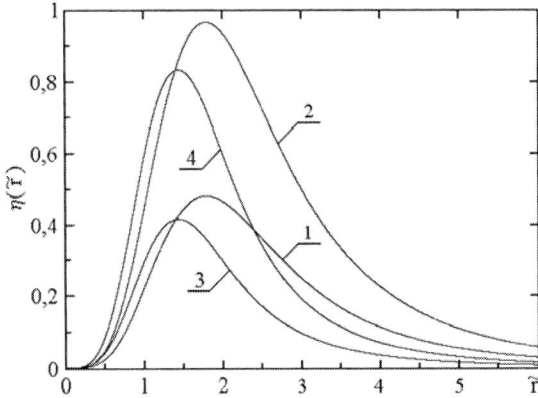

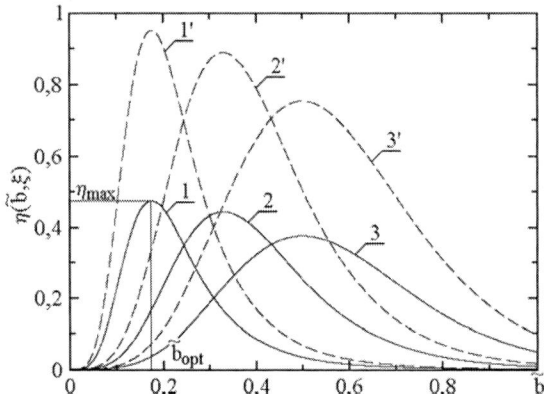

Fig. 2. The excitation efficiency for TE_{01} and TM_{01} modes in the circular waveguide with help of TEM_{01q} and TE_{01q} OR oscillations

Fig. 3. The excitation efficiency for TE_{01} and TM_{01} modes in the coaxial waveguide with help of TEM_{01q} and TE_{01q} OR oscillations

OR oscillation occurs when $\tilde{r} = r/w_0 = 1.773$ ($\eta = 0.484$). Here w_0 indicates the radius of field spot of the basic OR oscillation on the flat mirror where the segment of waveguide is made. At the same radius of the circular waveguide $\tilde{r} = 1.773$, the coefficient η has increased significantly in case TE_{01q} oscillation of OR. It maximum value is 0.967.

Now, the efficiency of excitation TM_{01} wave in a circular waveguide using the TEM_{01q} and TM_{01q} oscillations of resonator, we are analyzing. In that case, $E_w(\rho,\varphi)$ – the E_r-component of TM_{01} wave in the circular waveguide. The result of calculation is also shown on Fig.2. Curve 3 on figure corresponds to excitation of TM_{01} wave with help of TEM_{01q} OR oscillation, and curve 2 – the same wave excitation with help of TM_{01q} oscillation of resonator. In current case, as the previous one, takes place obviously expressed maximum of excitation efficiency of the wave in a circular waveguide using the TEM_{01q} oscillation of OR. Value η reaches a maximum which is equal to 0.418 when $\tilde{r} = 1.433$ (curve 3). In the case when the wave TM_{01} is excited by axially symmetric oscillation TM_{01q} the efficiency of excitation of the wave in waveguide is significantly higher. Thus, as in the previous case, curve 4 has obviously expressed maximum of $\eta = 0.835$ when $\tilde{r} = 1.433$.

Now consider the excitation of TE_{01} mode in a coaxial waveguide with help of TEM_{01q} oscillation of resonator. The waveguide is located in the center of the flat mirror of OR. $2b$ and $2a$ are diameters of inner and outer conductors of the coaxial. As in case of circular waveguide, in that case, the expression of OR TEM_{01q} oscillation and expression for E_φ-component of TE_{01} waveguide mode we substitute in the ratio (1). Fig. 3 shows the $\eta(\tilde{b})$ ($\tilde{b} = b/w_0$) for three values of $\xi = a/b$: $\xi = 10$ (curve 1), $\xi = 5$ (curve 2) and $\xi = 3$ (curve 3). Growth of ξ occurs by reducing the diameter of the central conductor of the coaxial waveguide $2b$ when a diameter the outer conductor is $2a = const$. With increasing of ξ the value η will seek to excitation efficiency of TE_{01}

wave in the circular waveguide using TEM_{01q} oscillation of OR. Dashed lines on the Fig. 3 indicates efficiencies of excitation of TE_{01} modes with help of TE_{01q} OR oscillation for same values of ξ: $\xi = 10$ (curve 1'), $\xi = 5$ (curve 2') and $\xi = 3$ (curve 3'). As it shown on the figure, the behaviors of these dependencies are similar to the case when the waveguide mode excited with help of TEM_{01q} OR oscillation, but the value of η is significantly higher. With increase of ξ, its maximum value seeks to the value of excitation efficiency for TE_{01} mode of the circular waveguide ($\eta_{max} = 0.967$) using of TE_{01q} oscillation of OR.

The practical interest is also the basic TEM mode of coaxial waveguide excitation using TEM_{01q} and TM_{01q} OR oscillation. The calculation results are presented on Fig. 4. There are shown dependencies η of $\tilde{b}$ for three values of ξ: $\xi = 3$ (curve 1); $\xi = 4.443$ (curve 2) and $\xi = 10$ (curve 3). The maximum value of η is 0.440 when the $\xi = 4.443$ and $\tilde{b}_{opt} = 0.399$ (curve 2). The dashed lines on the same figure are demonstrated dependences for excitation efficiency of TEM

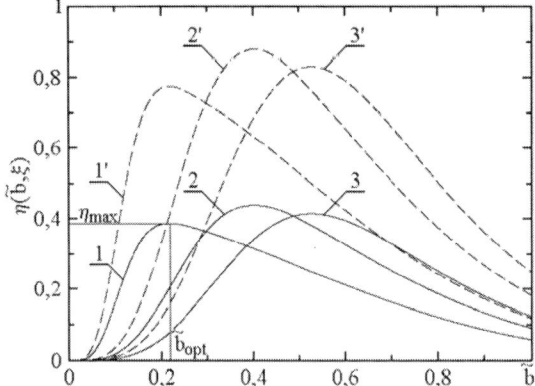

Fig. 4. The excitation efficiency of TEM mode in the coaxial waveguide with help of TEM_{01q} and TM_{01q} OR oscillations

mode in coaxial waveguide using the TM_{01q} OR oscillation. These dependences are calculated for same values of ξ: $\xi =3$ (curve 1'), $\xi =4,443$ (curve 2') and $\xi =10$ (curve 3'). As can be seen from the figure, the behaviors of these dependences are similar to previous one. However in these case, the value of η is increased. Thus the maximum value of η is 0.879 (curve 2') under the same parameters $\xi = 4.443$ and $\widetilde{b}_{opt} = 0.399$.

III. EXPERIMENTAL RESULTS

The experimental results confirmed that in these systems are excited resonant oscillation named the "ox eye". Analysis of spectral characteristics has showed that the OR with segment of waveguide has the unimodal resonance curve.

REFERENCES

[1] I.K. Kuzmichev, P.N. Melezhik, A.Ye. Poyedinchuk, "An open resonator for physical studies", International Journal of Infrared and Millimeter Waves, vol. 27, no. 6, pp. 857-869, 2006.

[2] D. Auston, R. Primich, R. Hayami. "The application of Fabri-Perot for diagnostic of plasma", Kvazioptika (in Russian), Moscow: Mir, pp. 387-423, 1966.

[3] R.C. Hansen, "Microwave scanning antennas", New York: Academic Press, 1966; Moscow: Sov. Radio, vol. I, 1968.

[4] R. Kühn, "Microwellenantennen", Berlin: Veb verlag Technik, 1964; Leningrad: Sudostroenie, 1967.

[5] I.K. Kuzmichev, "Exitation efficiency of quasioptical resonance systems", Telecommunications and Radio Engineering, vol. 68, no. 1, pp. 49-63, 2009.

Degree of paraxiality for monocromatic beams

A.B. Katrich

Kharkov National University, Kharkov, Ukraine

Abstract: The degree of paraxiality is one of the important parameters to characterize beam propagation. It shows the difference between the real beam and some reference beam that is the same beam without divergence. Such a beam is nonphysical and we propose to expand the beam propagation into the full space and use the point source as a reference beam.

Propagation of nonparaxial beams has been extensively investigated by different methods but the standard approach is substitution of known or assumed field distribution or its Fourier spectrum in some plane into one of diffraction integrals [1-5]. To simplify calculations these integrals are often replaced by their paraxial analogues. Accuracy of results substantially depends on the difference between propagation of the real beam and some reference beam which propagation is exactly predicted by used paraxial approximation of diffraction integral. To characterize such an error quantitatively two new parameters has been recently introduced: the paraxial estimator C_{PE} [3] and the degree of paraxiality C [4]. Both parameters are the ratio of the total power and integrated intensity. In both of them only the field distribution in the initial plane is used but in C_{PE} this distribution is the cross-section of the solution of paraxial wave equation whereas in C its origin doesn't matter. The importance of the parameter that characterizes paraxiality is obvious because the calculation error itself may be negligible but to estimate reliability of results the level of approximation error must be known and acceptable. Note, then the ratio C is also known as radiation efficiency and has been used to characterize the difference between partially and fully coherent beams [6].

Both mentioned parameters are equal to 1.0 when the beam is fully paraxial. C is equal to 0 when the beam is totally nonparaxial, e.g. Bessel-Gauss beam which width tends to zero [4]. But there is only one true nonparaxial beam – radiation of the point source which far-field angular distribution is uniform. Since any distribution of zero width is delta-function [7], its Fourier spectrum must be uniform and according known expression for the far-field zone [8] must have the cosine angular distribution. This contradiction follows from the fact that C is defined only in a half-space. However correct C must be defined in a full space because it has to characterize radiation not only for the diffraction by apertures in infinite metal screen but also radiation of real sources and scatterers that radiate into the full space. In that case the far-field angular distribution is an analytical function and cannot be equal to zero in any finite solid angle [9] and propagation necessarily must be considered in the full space.

When the width of the beam tends to zero C_{PE} becomes negative [3,10] but definitions of C_{PE} and C are the same and C_{PE} must be equal to zero as well. At least both definitions as the reference beam use such that does not diverge but has the same intensity distribution as the beam under study. Such a true paraxial beam is nonphysical since their quality parameter M^2 must be equal to zero but really it cannot be less then 1.0. The goal of this paper is to analyze proposed parameters and to find more appropriate one.

The degree of paraxiality C by definition [4] is the ratio of the total power and integrated intensity

$$C = \frac{P_0}{N} = \frac{\int\limits_0^{2\pi} d\varphi \int\limits_0^\infty S_z(r,\varphi) r\, dr}{\int\limits_0^{2\pi} d\varphi \int\limits_0^\infty |U_0(r,\varphi)|^2 r\, dr} \quad (1)$$

where $S_z(r,\varphi)$ is the component of the Poynting vector normal to some plane and $U_0(r,\varphi)$ is the cross-section of the scalar monochromatic field by the same plane; $r^2 = x^2 + y^2$. To calculate $S_z(r,\varphi)$ the field and its normal derivative must be known. To obtain the last value is usually a difficult problem. Using Parseval's theorem Eq. (1) can be transformed into (Eq. (6) in [4] is in Cartesian coordinates, here are used polar for convenience)

$$C = \frac{\int\limits_0^{2\pi} d\psi \int\limits_0^1 |A_0(\rho,\psi)|^2 m\, \rho d\rho}{\int\limits_0^{2\pi} d\psi \int\limits_0^\infty |A_0(\rho,\psi)|^2 \rho d\rho} \quad (2)$$

where $A_0(\rho,\psi)$ is the Fourier spectrum of the boundary value $U_0(r,\varphi)$; $m = (1-\rho^2)^{1/2}$; $\rho^2 = p^2 + q^2$; p and q are normalized by k_0 spatial frequency variables; $\rho < 1$ corresponds to propagating waves and $m = i(\rho^2-1)^{1/2}$ for evanescent waves when $\rho^2 > 1$. Eq.(2) can be also obtained from Eq.(1) using Fubini's theorem and governing equation of the angular spectrum representation method [8]

$$U(\mathbf{r}) = \int\limits_{-\infty}^\infty \int\limits_{-\infty}^\infty \left[A_0(p,q) e^{ik_0 mz} \right] e^{ik_0(px+qy)} dp dq \quad (3)$$

If $U_0(r,\varphi)$ is absolutely integrable then above transformations are valid. But if the angular distribution in the initial plane $D(\pi/2,\varphi)$ is not equal to zero for all φ then $U_0(r,\varphi)$ is proportional to r^{-1} far from the origin and the boundary value is no more absolutely integrable. It

978-1-4799-0019-0/13 $31.00 © 2013 IEEE

results in discontinuity of its spectrum at the circle $\rho=1$. The typical case is the spectrum of the free-space Green's function [8]. However if the function is not absolutely integrable then Fourier integral formula cannot be proved and Fubini's and Parseval's theorems cannot be applied.

To characterize any physical object by comparing with some ideal object the last must be physical object as well or must be a limiting case of it. The reference beam in C_{PE} and C is $U(\mathbf{r})=U(x,y)\exp(ik_0z)$ that corresponds to $m=1$ in Eq.(3). From known expression [8] for the field in the far-field zone

$$U(\mathbf{r}) = -\frac{2\pi i}{k_0}\frac{z}{R} A_0\left(\frac{x}{R},\frac{y}{R}\right)\frac{e^{ik_0R}}{R} =$$
$$= D(\theta,\varphi)\frac{e^{ik_0R}}{R}, \quad R\to\infty \tag{5}$$

follows that $D(\theta,\varphi)=-2\pi i/k_0 A_0(0,0)$, i.e. the angular width is equal to zero but the width in the waist is finite. This is impossible physically and in fact denominator in Eq.(1) is just the normalization constant. The only physical object that can be a reference is the totally nonparaxial beam, i.e. the radiation of the point source. The parameters to be compared are clear: it must be divergences because just they characterize directivity of radiation. The same in fact is proposed in [10] where C_{PE} using Fresnel approximation of Eq.(1) is reduced to

$$C_{PE} = 1 - \left(\sigma_{\infty x}^2 + \sigma_{\infty y}^2\right) \tag{6}$$

where σ_{Ax} and σ_{Ay} are normalized second moments or variances of the spectrum $A_0(x,y)$. From Eq.(6) it follows that σ_{Ax} and σ_{Ay} are variances of the angular distribution divided by the factor $z/R=\cos(\theta)$. Note, that in the far-field evanescent waves are absent and the support of $A_0(\rho,\psi)$ is bounded by the circle $\rho\leq1$.

Thus, the degree of paraxiality that is defined in the full space and uses as the reference the limiting case of any nonparaxial beam is

$$C = 1 - \frac{w_b}{w_0} \tag{7}$$

where w_b is the divergence of the beam and w_0 is the divergence that corresponds to reference beam. The value of w_0 depends on the chosen space. When divergence is calculated as the solid angle $w_0=4\pi$ and in plane $w_0=2\pi$. Methods of measurements and calculation of w_b are stated in the standard ISO11146 [11].

REFERENCES

[1] P.C.Chaumet, "Fully vectorial highly nonparaxial beam close to the waist." *J. Opt. Soc. Am. A* 23 (2006) 3197-3202.

[2] C.J.R.Sheppard, "High aperture beams". *J. Opt. Soc. Am. A* 18, pp. 1579-1587 (2001).

[3] P.Vaveliuk, B.Ruiz, A.Lencina, "Limits of the paraxial approximation in laser beams". *Opt. Letters* 32 (2007) 927-929.

[4] O.E.Gawhary, S.Severini, "Degree of paraxiality for monochromatic light beams". *Opt. Letters* 33 (2008) 1360-1362.

[5] S. R. Seshadri, „Dynamics of the linearly polarized fundamental Gaussian light wave," *J. Opt. Soc. Am. A* **24**, 482-492 (2007).

[6] P.Ostlund, A.T.Friberg, "Radiation efficiency of partially coherent electromagnetic beams". *J. Opt. Soc. Am. A*, 18, 1696-1703 (2001).

[7] G.A. Korn, T.M. Korn, *Mathematical Handbook*. 2nd ed. (McGrow-Hill, New York, 1968).

[8] L. Mandel, E. Wolf, *Optical Coherence and Quantum Optics*. (Cambrige University Press, New York, 1995).

[9] E. Wolf, M.Nieto-Vesperinas, "Analyticity of the angular spectrum amplitude of scattered field and some of its consequences," *J. Opt. Soc. Am. A* **2**, 886-890, (1985).

[10] P.Vaveliuk, "Quantifying the paraxiality for laser beams from the M^2 factor," *Opt. Letters* 34 (2009) 340-342.

[11] ISO DIN 11146: "*Optics and optical instruments, test methods for laser beam parameters: Beam width, divergence angle and beam propagation factor.*"

Narrow-band DFB laser based on a dye-doped volume Bragg grating

T.N. Smirnova[1], O.V Sakhno[2], V.M. Fitio[3], J. Stumpe[2]

[1]Institute of Physics NAS of Ukraine, prospect Nauki, 46, 03068 Kiev, Ukraine
[2]Fraunhofer Institute for Applied Polymer Research IAP, Geiselbergstr. 69, 14476, Potsdam, Germany
[3]Lviv Polytechnic National University, 12 Bandera Str., 79013 Lviv, Ukraine

Abstract - **The distributed feedback (DFB) lasers based on a waveguide with volume grating providing feedback through second order Bragg diffraction were experimentally and theoretically investigated and optimized. The gratings are readily inscribed holographically in a layer of a dye-doped organic nanocomposite containing high refractive index inorganic nanoparticles. It is shown that the improvement of the parameters of waveguide and grating provides narrow-band lasing emission with a beam profile close to Gaussian intensity distribution.**

I. INTRODUCTION

Distributed feedback based on periodic structures is widely used to provide feedback in laser sources. The investigations of DFB structures, started more than 40 years ago, are of great interest up to nowadays. Recent development of new organic and composite materials and low-cost methods for fabrication of periodic micro- and nanostructures resulted in cost-efficient, maintenance-free tunable microlasers based on permanent distributed feedback. Those can be integrated into optical and fluidic microchips usable for medical and biosensing.

Polymer DFB laser under study is a planar waveguide containing a volume grating doped with a laser dye. In this work we experimentally investigated and theoretically analyzed the influence of the waveguide and grating parameters on spectral and energy characteristics of this kind of DFB laser in order to obtain a single-mode narrow-band emission of low divergence.

II. EXPERIMENTAL RESULTS

The gratings were formed by holographic structuring of the photopolymerizable nanocomposite consisting of a mixture of acrylate monomers, an initiator of radical polymerization, the $LaPO_4$ nanoparticles (NP) (REN-X green, Nanosolutions GmbH) and activating with a Pyrromethene 567 (PM567) laser dye. The grating periods were matched to the second-order operation. The grating fabrication procedure and measurement of the laser characteristics were published elsewhere [1, 2]. In Fig. 1 the scheme of the laser structure and the directions of the pump beam and the output emission are shown. The devices are pumped with a frequency-doubled Nd:YAG laser which emits 500 ps long pulses at a wavelength of 532 nm [2]. The spectral characteristics and the output angles (θ_{ex}) of the emitted radiation as well as the beam divergence and the profile were measured.

Changing the concentration of NP, C_{NP}, and, consequently, the refractive index (RI) of an active waveguide layer, n_1, the parameters of the waveguide can be tuned (RI of the glass substrate $n_0 = n_2 = 1.515$, in the case of an asymmetric waveguide $n_0 = 0$). The number of the waveguide modes for a chosen wavelength λ depends on the thickness of the guiding layer, d, and the values of n_1, n_2 and n_0.

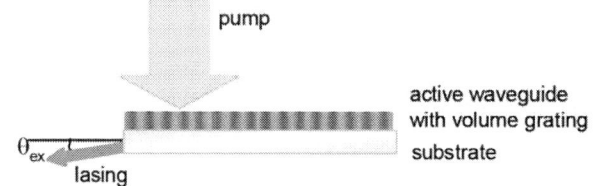

Fig. 1. Scheme of the DFB laser structure under study (asymmetric waveguide with grating).

At low difference between the RIs of the waveguide layer and the substrates only several guided modes with slightly different propagation constants β_j will propagate. Decreasing the number of guided modes leads to the decrease in the line-width of laser emission.

General features of the emission spectra obtained in our experiments are the following. Varying the C_{NP} in the range of 20–26 wt. %, the emission line-width (FWHM, full width at a half maximum) was measured to be about 0.05 nm, limited by the spectral resolution of the spectrometer used (Fig. 2). The following optimized parameters for the volume DFB laser were found: $n_1 = 1.5177$, optical density of an active layer at 532 nm is $D \cong 0.4$, the RI modulation amplitude of the grating $\tilde{n}_1 \geq 0.007$, d=10 μm. Varying the grating period the laser emission was obtained in the range of 568 – 620 nm. The lowest lasing threshold energy of 0.12 – 0.16 μJ/pulse (6 – 8 μJ/cm^2) has been measured in the tuning range of 594-610 nm. The FWHM of the output emission does not exceed 0.05 nm and remains constant even at a pump energy 5-times higher, than the corresponding laser threshold.

978-1-4799-0019-0/13 $31.00 © 2013 IEEE

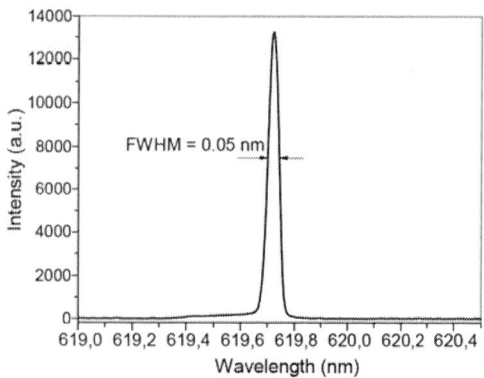

Fig. 2. Emission spectrum of the sample with optimized parameters

The output emission from the volume DFB laser (a symmetric waveguide) is demonstrated in Fig. 3. It was found that

Fig.3. Emission output from the optimized DFB-laser

the lasing emission propagates inside the substrate(s) and then couples out at the angle θ_{ex} with respect to substrate's edge. The output angle θ_{ex} was found to be $5.6\pm0.2°$ for the improved DFB laser. Observed out-coupling through the substrates provides lower beam divergence compared to the out-coupling through the edge of the active layer that typically has a thickness of 5 - 10μm. The far-field beam divergence of the DFB laser, whose emission spectrum is presented in Fig. 2, was ~ 5 mrad.

The intensity profile of the output beam was found to be very close to a Gaussian distribution. In order to explain the out-coupling of the lasing emission through the substrate a theoretical model, presented below, was developed.

III. THEORETICAL MODEL

The theory of propagation of light in a planar dielectric waveguide with periodic surface corrugation, providing distributed feedback, was developed in [3]. The waveguide mode was considered as a plane wave that propagates in a guiding layer due to total internal reflection and diffracts by the grating generating secondary plane waves. We develop this approach for the case of an active waveguide with permanent volume grating. It was found that interaction of a waveguide mode with a diffraction grating operating in the second Bragg order generates not only a reflected Bragg wave, but also other waves, propagating almost perpendicularly and at a grazing angle with respect to the waveguide surface. A part of the waveguide with a volume grating is shown in Fig. 4.

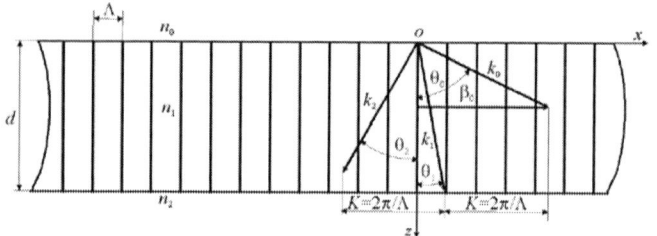

Fig. 4. Propagation of the waves in the waveguide with a volume grating

At the second-order Bragg condition the modulus of the vector of an inverse grating K and the propagation constant β are connected as:

$$K = \frac{2\pi}{\Lambda} \cong \beta \qquad (1)$$

We consider a symmetric waveguide for which $n_0 = n_2 < n_1$, $n_1 - n_2 << n_1$ and the thickness of a waveguide layer is in the range of 5-10 μm. The calculations show that at each wavelength λ several modes can be guided. Let us consider a plane wave with a wave vector modulus $k_0 = 2\pi n_1 / \lambda$ that falls on the volume grating. The propagation constant of the fundamental TE mode corresponding to the wavelength λ, is equal to β_0.

With the knowledge of k_0 и β_0 the incidence angle of a plane wave on the grating θ_0 can be expressed as:

$$\theta_0 = \arcsin(\beta_0 / k_0), \ \beta_0 = k_0 \sin\theta_0 = k_{0,x} \qquad (2)$$

The moduli of the wave vectors of the waves, diffracted at the angles θ_1 (first order diffraction) and θ_2 (second order diffraction), satisfy the relation:

$$k_0 = k_1 = k_2 = 2\pi n_1 / \lambda. \qquad (3)$$

According to the diffraction theory, the x-components of the wave vectors are expressed as:

$$k_{1,x} = k_1 \sin\theta_1 = k_0 \sin\theta_0 - 2\pi / \Lambda,$$
$$k_{2,x} = k_2 \sin\theta_2 = k_0 \sin\theta_0 - 4\pi / \Lambda. \qquad (4)$$

From equations (4) one can find the respective diffraction angles of the first, θ_1, and the second, θ_2, Bragg orders:

$$\theta_1 = \arcsin\left(\sin\theta_0 - \frac{\lambda}{n_1\Lambda}\right), \qquad (5)$$

$$\theta_2 = \arcsin\left(\sin\theta_0 - \frac{2\lambda}{n_1\Lambda}\right). \qquad (6)$$

Knowing the diffraction angles in the waveguide layer with a grating and using the law of refraction the propagation angles of the diffracted waves in the adjacent media (Fig.4) can be found.

The out-coupling efficiency of the laser emission will be determined by the efficiencies of the first and the second order diffraction (η_1 and η_2) of a plane wave by the grating and by the transmittance T of the waveguide-substrate interface. In order to calculate η_1 and η_2 the equations from [4] were used:

978-1-4799-0019-0/13 $31.00 © 2013 IEEE

$$\eta_1 = \left| \frac{b\left[\exp\left(i\Delta_1 d\right) - 1\right]}{\Delta_1 \sqrt{\cos\theta_0 \cos\theta_1}} \right|^2 \cdot T , \tag{7}$$

where $b = \pi\tilde{n}_1 / \lambda$, $\Delta_1 = k_1\left(\cos\theta_0 - \cos\theta_1\right)$.

$$\eta_2 = \frac{\chi^2 \sin^2\sqrt{\left(\tfrac{1}{4}\Delta_2^2 + \chi^2\right)d^2}}{\tfrac{1}{4}\Delta_2^2 + \chi^2} \cdot T , \tag{8}$$

where $\chi = \dfrac{\pi\left(\tilde{n}_2 - \dfrac{\tilde{n}_1^2}{4n_1 \sin^2 \alpha/2}\right)}{\lambda\sqrt{\cos\theta_0 \cos\theta_2}}$ is the coefficient of coupling

between an incident and a second order diffracted wave, $\tilde{n}_2$ is the RI modulation amplitude of the grating with period of $\Lambda/2$, formed due to a nonlinear response of the nanocomposite that provides a non-sinusoidal RI modulation;

$$\alpha = \arcsin\left(\frac{\lambda}{n_1\Lambda}\right); \ \Delta_2 = k_1\left(\cos\theta_0 - \cos\theta_2\right);$$

T is the optical transmittance of the boundary between the active layer (n_1) and the substrate (n_2) (Fig.4). T was calculated using the Fresnel formulas for a plane wave incident on the boundary at the non-zero angle. If θ_1 is close to zero, transmission coefficient, T, approaches unity (Eq.7). We also omit $\tilde{n}_2$ in Eq.7 because its contribution is negligible.

As an example, we determined the parameters mentioned above for the sample which emission spectrum is shown in Fig. 2. The calculations were done for fundamental TE mode possessing the lowest excitation threshold. The waveguide of this laser is characterized by three propagation constants: β_0=15.3884 μm^{-1}, β_1=15.3821 μm^{-1}, β_2=15.3722 μm^{-1}. The angles of diffraction for the fundamental mode β_0 are the following: θ_0 =1.554 rad, θ_1 = 0.002 rad, θ_2 = 1.487 rad. The corresponding angle θ_{ex} is equal to 5.15^0 that is in a good agreement with the experimental value. As it was expected a plane wave guided in the waveguide layer at the angle θ_0 will not propagate into the substrate. The dependence of the out-coupling efficiency on the waveguide thickness, the RIs of the active grating and the substrates and the values of $\tilde{n}_1, \tilde{n}_2$ were calculated using the obtained equations.

CONCLUSIONS

In this work, we report the enhancement of a polymer DFB laser comprising a dye-doped volume grating. Optimization of various parameters of both the waveguide and the grating leads to the DFB laser with an emission line-width less than 0.05 nm. The shape and the line-width of the lasing peak remain constant even at pump energy that exceeds 5 times the lasing threshold within the 568 – 620 nm spectral range. The lowest pump threshold energy of about 0.12–0.16 μJ/pulse in the emission range of 594-610 nm has been obtained. The lasing emission is transferred from an active layer into the substrate and goes out from the substrate edge. The spatial intensity distribution in the output beam has an almost Gaussian profile and the beam divergence was found to be ~ 5 mrad. The emission out-coupling is provided by the second order Bragg diffraction of the waveguide mode by the grating. The out-coupling efficiency varies periodically with the increase of waveguide thickness, achieving its maximum value of 0.16 at d = 4.5 and 8.7 μm.

A constructive contribution of the first-order Bragg diffraction by the second harmonic of RI modulation significantly increases the efficiency of the light output. The presence of even a minor nonlinearity ($\tilde{n}_2$ is about one order less than $\tilde{n}_1$) results in more than two orders increase of the out-coupling efficiency.

ACKNOWLEDGMENT

The work was supported by the National Target Scientific and Engineering Program "Nanotechnologies and Nanomaterials" (Project 1.1.4.13/13-H-25) and by the Cooperation-Project between Germany and Ukraine UKR 10/043 (Federal Ministry of Education and Research, Germany).

REFERENCES

[1] O.V. Sakhno, T.N. Smirnova, L.M. Goldenberg, J. Stumpe, "Holographic patterning of luminescent photopolymer nanocomposites", *Mater. Sci. Eng. C*, vol. 28, pp.28-35, 2008

[2] O. V. Sakhno, T. N. Smirnova, J. Stumpe, "Distributed feedback dye laser holographically induced in improved organic–inorganic photocurable nanocomposites", *Appl. Phys. B*, vol. 103, pp. 907-916, 2011

[3] R.F. Kazarinov, Z.N. Sokolova, R.A. Suris, "To the theory of planar waveguides with distributed feedback", *J.Tech.Phys.*, vol. 66, pp. 229-239, 1976.

[4] V.M. Fitjo, O.V. Sakhno, T.N. Smirnova, "Analysis of the diffraction by the gratings generated in materials with a nonlinear respons", *Optik*, v.19, pp. 236-2462, 2008.

Broadband superluminescent diodes of NIR range with quasi-Gaussian spectra

E. V. Andreeva [1], S. N. Il'chenko [1], Yu. O. Kostin [1], M. A. Ladugin [2], P. I. Lapin [1], A. A. Marmalyuk [2], and S. D. Yakubovich [3], *Senior Member, IEEE*

[1] SUPERLUM Ltd., 117454 Moscow, PO Box 70, Russia;
e-mail: ilchenko@superlumdiodes.com
[2] JSC R&D Inst. POLYUS, 117342 Moscow, Vvedenskogo 3, Russia;
[3] MSTU MIREA, 119454 Moscow, Vernadskogo prosp. 78, Russia.

Abstract: Quantum-well (QW) superluminescent diodes (SLDs) with an extremely thin (AlGa)As and (InGa)As active layers and central wavelengths of 810 nm, 840 nm, 860 nm and 880 nm are studied. Their emission spectra possess quasi-gaussian shape with FWHM of 30 – 60 nm depending on active channel length and pumping level. In CW operation mode light-emitting modules based on these SLDs exhibit optical power of 1.0 – 25 mW ex SMF and high enough life time exceeding 30 000 h. The prototypes of combined light sources of BroadLighter series based on these SLDs with bell-shaped spectra of up to 100nm FWHM are realized.

Key words: nanoheterostructure, quantum-well superluminescent diode, optical coherence tomography.

I. INTRODUCTION

In quantum-well SLDs with an extremely (several nm) thin active layers the excited subbands of energy spectrum are shifted to higher energies. In this case, the excited subband at reasonable pumping levels is not filled and the superluminescent spectrum is determined by the quantum transitions only from the fundamental state. SLD spectrum possess quasi-gaussian shape and its FWHM heavily relies on active channel length L_a and pumping level and can exceed 60nm. The present paper is devoted to the study of light sources based on the SLDs of such type, which are of great interest for optical coherence tomography (OCT) [1].

II. EXPERIMENTAL RESULTS

Fig.1 shows a typical for the bulk and one of the QW SLD investigated in this work L-C characteristics and FWHM of the emission spectrum as a function of the injection current. The samples have exactly the same structure and composition of heterolayers except the active layer with a thickness of 28 nm (GaAs) and 6.5 nm $(In_{0.04}Ga_{0.96})$As, respectively. Their superluminescencent spectra strikingly differ in both the width and the nature of its dependence on the pump level. Both samples possess quasi-gaussian shape. The inset on Fig.1b shows the approximation of the quantum-well SLD spectrum

by the Gaussian function. The slight asymmetry has almost no effect on the "purity" of the coherence function central peak.

The most popular for SLD-modules used in OCT systems is the wavelength of about 840 nm. In the present work was grown a series of similar single quantum-well (SQW) separate confinement double heterostructures (SC DHS) with different thicknesses of the active layer, which magnitude λ_m maintained at a specified value due to changes in the chemical composition of the solid solution (concentrations of In and Ga). Values of output power and FWHM of the spectrum for SLDs with various L_a, made from these heterostructures at typical injection current density were measured. The main result is a very weak dependence of the main parameters of the SLDs from thickness of active layer d_a in the range of 4.0-10.0 nm. Later in the mass production of the devices under consideration it is expected to stay at a d_a value of about 6.0 nm.

Changes in the chemical composition of the SC DHS active layer with its constant thickness can widely vary median wavelength λm of the produced SLDs. The displacement to the shorter wavelengths leads to the narrowing of the spectrum width Δλ while the rest of the output characteristics remain close. Tab.1 shows the main characteristics of the SLDs with active channel width W=4 μm and La values of 700, 900, 1100 and 1300 μm. The selected operation points ensure the PSM values of about 1.0, 3.0, 10, and 20 mW. Low-power "short" devices have extremely wide Δλ of about 60 nm. Lifetime tests showed high reliability of the developed SLDs (Fig.2). Median wavelength λ_m and FWHM Δλ changes are not more than 0.3% and 2.0% respectively. Approximate lifetime exceeds 35000 hours.

978-1-4799-0019-0/13 $31.00 © 2013 IEEE

Table 1. Characteristics of the SLDs with a thickness of active layer about 6.0 nm (L_a - the length of the active channel; I - injection current; J - injection current density; P_{FS} - output power in a free space; P_{SM} - output power via single-mode fiber (SMF); λ_m - median wavelength; $\Delta\lambda$ - FWHM; L_c - the coherence length; TE / TM - polarization ratio).

The active layer composition	L_a (μm)	I (mA)	J (kA/sm²)	P_{FS} (mW)	P_{SM} (mW)	λ_m (nm)	$\Delta\lambda$ (nm)	L_C (μm)	TE/TM
Type I $Al_{0.02}Ga_{0.98}As$	700	150	5.4	2.1	0.9	806	43	15	3
	900	200	5.5	6.5	3.0	806.5	35.5	18	4
	1100	250	5.7	18	10	807.5	33	20	6
	1300	260	5.0	32	19	809	27	24	10
Type II $In_{0.05}Ga_{0.95}As$	700	160	5.7	3.2	1.3	835	57	12	5
	900	180	5.0	8.5	4.1	840	46	15	7
	1100	240	5.5	24	13	844	42	17	15
	1300	280	5.4	46	24	847	34	21	25
Type III $In_{0.09}Ga_{0.91}As$	700	170	6.1	2.5	1.2	856	58	13	6
	900	195	5.4	7.9	4.0	856	49	15	10
	1100	240	5.5	17	10	860	42	18	22
	1300	280	5.4	38	24	861	36	20.5	32
Type IV $In_{0.13}Ga_{0.87}As$	700	170	6.1	2.8	1.4	872	63	12	13
	900	180	5.0	6.2	3.1	875	50	15	16
	1100	200	4.5	18	9.5	876	48	16	28
	1300	240	4.6	34.5	19	879	44	17.5	34

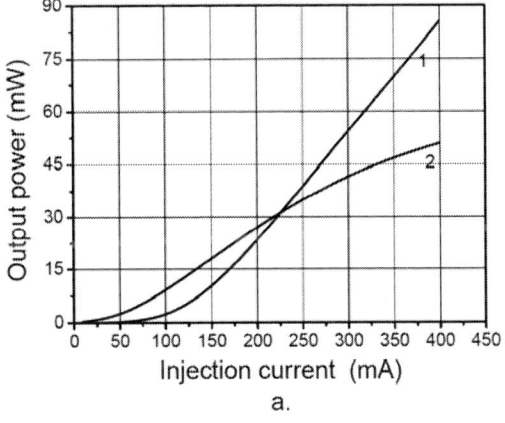 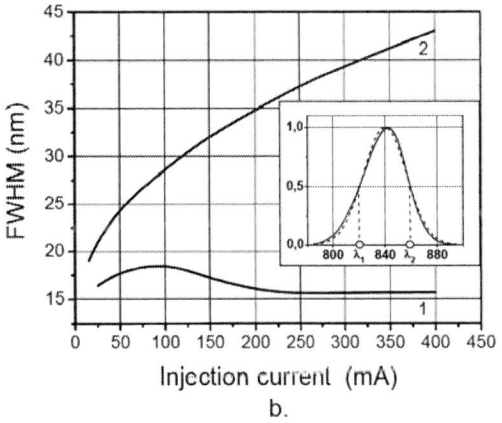

a.

b.

Fig. 1 Typical L-C characteristics(a) and spectrum FWHM as a function of the injection current (b) of the bulk (1) and the quantum-well (2) SLDs (d_{Bulk} = 28nm, d_{SQW} = 6.5nm, L_a = 1200 μm, W = 4 μm). The insert – an approximation of the quantum-well SLD spectrum by Gaussian function.

III. COMBINED LIGHT SOURSES WITH THE BELL-SHAPED SPECTUM

Along with light-emitting SLD modules in OCT systems and in optical metrology are widely used combined light sources, in which the optical outputs of two or more broadband SLD-modules with offset spectra are combined with fiber optic couplers [2,3].

As is known, the superposition of two shifted spectra of Gaussian shape with close FWHM values allows, under certain conditions, to obtain the bell-shaped spectrum, whose FWHM is close to the sum of the composed spectra widths. The form of this spectrum differs little from the Gaussian shape, but the "pedestal" of coherence function is almost absent. This approach was used in the study of prototypes of new combined light sources based on the developed SLD-modules with output radiation through the SMF. Tab.2 shows the main characteristics of the implemented prototypes. Such combinations of parameters together with the bell-shaped spectrum are realized for the first time.

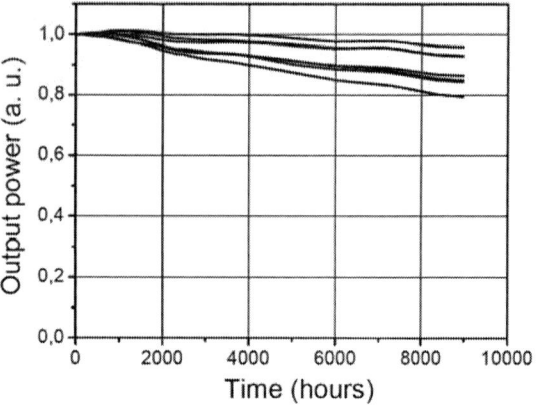

Fig.2 Lifetime test chronogram of the Type II SLDs (L_a = 1400 µm, P_{FS}~35 mW).

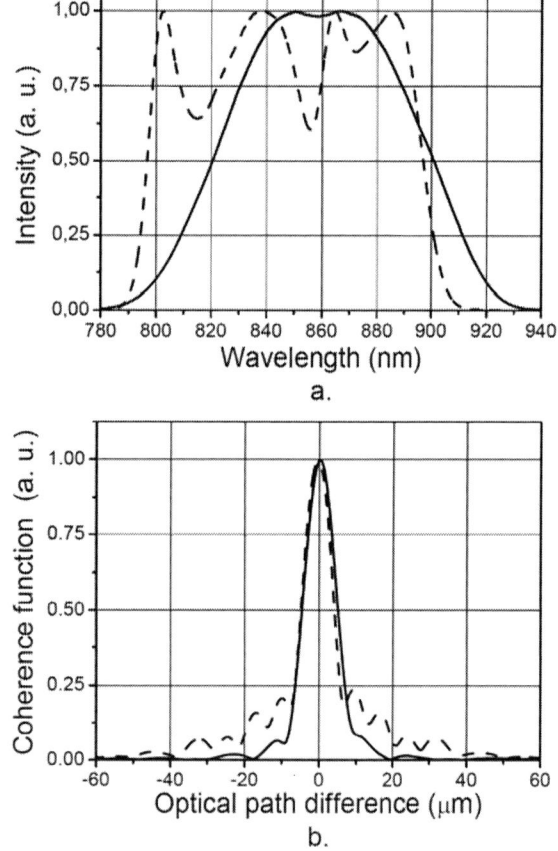

a.

b.

Fig.3 Output emission spectra (a) and the coherence function central peaks (b) of the serial source BroadLighter D-840 (dashed line) and one of the new models.

Fig.3 shows the emission spectra and coherence function of the most popular device (model D-840-HP) and the new model (preliminary name - D-860-G). There is every reason to believe that the new device will be preferred for OCT systems.

Table 2. Examples of the main combined light sources characteristics.

SLD Types	L_a (µm)	P_{SM} (mW)	λ_m (nm)	$\Delta\lambda$ (nm)	L_C (µm)
I + II	900	3.0	830	70	9.8
	1100	8.0	830	65	10.5
I + III	700	1.0	830	100	6.9
II + IV	1100	7.0	860	80	9.2
	1300	20	860	70	10.5

IV. CONCLUSION

The research and development of SLD-modules of spectral range 800-900 nm on the basis of QW heterostructures with the thickness of the active layer of several nm were carried out. They possess quasi-gaussian shape of the spectrum with FWHM 2-3 times higher than that of the bulk SLDs of same spectral range. The high reliability of these devices was demonstrated. Prototypes of new broadband combined light sources with a bell-shaped spectrum were realized.

ACKNOWLEDGEMENTS

The authors thank A.T.Semenov for initiating of this research. This work was partially supported by a grant of the Federal Target Program №14.B37.21.0756.

REFERENCES

[1] W.Drexler, J.G.Fujimoto, Optical coherence tomography, Springer-Verlag Berlin Heidelberg (2008).

[2] T.H. Ko, D.C. Adler, J.G. Fujimoto, D.S. Mamedov, V.V. Prokhorov, V.R. Shidlovski, and S.D. Yakubovich, Optics Express, 12 (10), 2112-2119 (2004).

[3] D.C. Adler, T.H. Ko, A.K. Konorev, D.S. Mamedov, V.V. Prokhorov, J.G. Fujimoto, S.D. Yakubovich, Quantum Electronics, 34 (10), 915 (2004).

Numerical analysis of the performance of AlGaAs/GaAs multi-quantum well Superluminescent diodes

A. Asgari[1-3,*], P. Navaeipour[1]

[1]Research Institute for Applied Physics & Astronomy, University of Tabriz, Tabriz 51665-163, Iran
[2]School of Electrical, Electronic and Computer Engineering, The University of Western Australia, Crawley, WA 6009, Australia
[3]Excellence Center for Photonics, University of Tabriz, Tabriz, Iran
*Corresponding author: asgari@tabrizu.ac.ir

Abstract: In this paper we have investigated numerically the performance of AlGaAs/GaAs multi-quantum well superluminescent diodes. In this device the dependence of optical gain, output power on the cavity length and the density states have been analyzed. It is observed that the optical gain and its FWHM bandwidth increase with increasing the density state. Also, the output power increases with increasing the cavity length, whereas the FWHM bandwidth decreases..

Superluminescent diodes (SLDs) are highly desired for applications including optical coherence tomography (OCT), fibers optic gyroscopes, fiber optic sensors and optical testing [1-4]. It is known that the main feature of SLDs is a combination of the high radiation brightness, comparable to that of laser diodes, with their broad emission spectrum (low coherence) [5]. To expand the band width of SLDs, researchers have used quantum – well active regions exhibiting simultaneous emission from both the first and second quantized states [6,7]. An important challenge to achieve superluminescent operation with low spectral modulation is to minimize the optical feedback and, therefore, avoid lasing. Most well-known techniques of reducing optical reflection from facets include utilizing antireflection (AR) coating, tilted, bent or trapped waveguides, buried facets and integrated absorption regions [8,9].

In this paper we have tried to analyze numerically the performance of multi-quantum well (MQW) superluminescent diodes. We make investigated in to the effects of the structural performance like cavity's length and operating condition (density state) on the SLD performance (e.g., output power spectrum and optical gain profile).

The model presented here employs a MQW structure with non- identical quantum wells (QWs), on a GaAs substrate, emitting in range of 750 nm to 850 nm. The active layer is composed of 6, 8, and 15- nm QWs separated by 15 nm barriers. To suppress lasing, the ridge waveguide oriented at ☐ respect to the crystal end facet, as shown in Fig.1. The ridge angle effectively suppresses facet reflections and undesired laser action or spectral narrowing. In practice, the facet reflection can be reduced dramatically by AR coating combined with tilted waveguide structures [10].

Under sufficient current injection the semiconductor material becomes optically active and exhibits optical gain (also called material gain). Gain is, therefore, a function of density of carriers inside the active regions. For QW active regions, gain can be calculated as [11]:

$$
g(E) = \frac{\pi \hbar e^2}{n_r c \varepsilon_0 m_0^2 E} \sum_n \sum_m \left| I_{hm}^{en} \right|^2 \int_0^\infty \rho_r^{2D}(E_t) \left| M_{if} \right|^2
$$
$$
\times \frac{\Gamma/2\pi}{(E_{hm}^{en} + E_t - E)^2 + (\Gamma/2)^2} [f_e^n(E_t) - f_v^m(E_t)] dE_t
$$

(1)

Double summation in this relation ensures contributions of transitions between all subbands considered. n counts the conduction subbands. Similarly, m is used to count the valence subbands. I_{hm}^{en} is the overlap integral of the conduction and valence band envelope functions. The two dimensional reduced density of states function is given as $\rho_r^{2D} = \frac{m_r}{\pi \hbar^2 I_x}$ with L_x the width of quantum well. f_c^n, f_v^m represent quasi-Fermi levels, and F_c, F_v are Fermi-Dirac distribution functions for conduction and valence sub bands respectively.

The total SLD output power is, given by [12, 13],

$$
P_L(\omega) = \frac{\hbar \omega}{2\pi} n_{sp} \Delta \omega (e^{g_n L} - 1) \frac{(1 - r_1^2)(r_2^2 e^{-i\omega t} + 1)}{(1 - r_1 r_2 e^{-i\omega t})^2}
$$

(2)

g_n is the modal gain, r_1, r_2 are the reflectivity of the facets, and n_{sp} is the spontaneous emission factor [13].

By simulation of band diagram, it is possible to determine allowable optical transition in QWs, In our structure, there is

978-1-4799-0019-0/13 $31.00 © 2013 IEEE

only one transition $(e_1 \rightarrow h_1)$ in 6, 8 nm QWs and two transitions $(e_1 \rightarrow h_1, e_2 \rightarrow h_2)$ in 15 nm QW.

The optical gain is presented on Fig 2, which is in five sheet carrier densities for the SLDs with 6 and 15 nm single QWs. It can be seen that 15 nm- QW structures has two peaks around 860 and 825 nm coming from $(e_1 \rightarrow h_1)$, and $(e_2 \rightarrow h_2)$ transitions, respectively. The 6 nm- QW has one peak at 833.89 nm related to the $(e_1 \rightarrow h_1)$ transition. Furthermore, it has shown that the 6 nm-QW has much stronger gain in comparison with 15 nm- QW at the same sheet density. Also, there are not any broadening for the peak of gain.

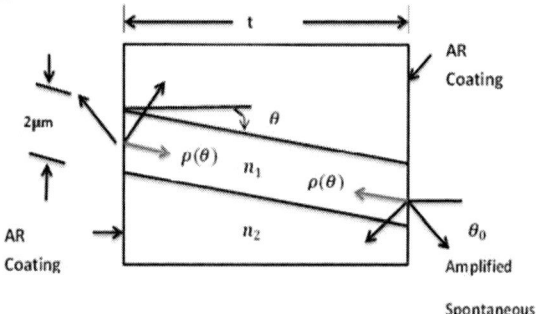

Fig. 1. Incline stripe SLD - top view.

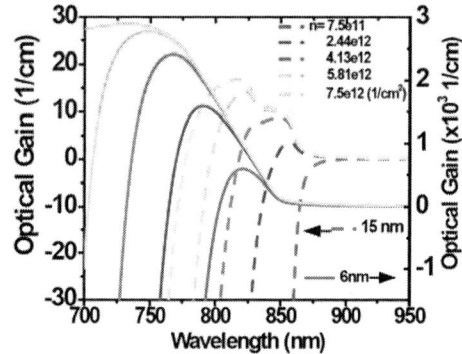

Fig. 2. Material gain for 6nm- and 15nm- QWs at five sheet charge densities.

Fig 3 indicates the optical gain spectrum for the SLDs with MQW structures for various density of states. With increasing the density the broadening of gain peak increases. At low density, the optical gain shows only one peak at about 860 nm corresponding to the first transition of 15 nm- QW. As the density increases, the second peak appears at 840 nm due to the first transition of 8nm- QW and 6nm- QWs. By increasing the density, other transitions contributes and make the gain spectrum broader.

In Fig 4, the output power spectrum of the SLD with MQWs for different density states are shown. As the output power depends exponentially on the optical gain and the optical gain increases with increasing the density. So with increasing the density the output power increases. The broadening of the output power with increasing the density state represents the optical transitions related to quantum wells with various sizes.

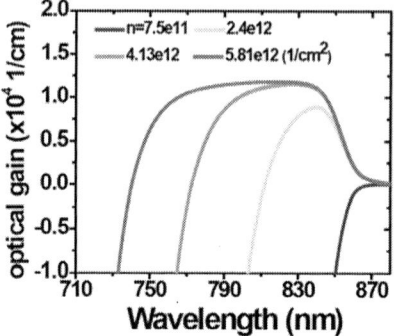

Fig. 3. Material gain for SLD at various density states.

To find out more about the widening of the output power spectrum with density state, the Full width at half maximum (FWHM) bandwidth is presented in Fig 5 in which the increasing of density state increases it.

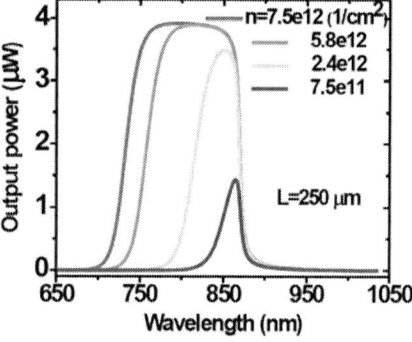

Fig. 4. Output power for SLD at various density states.

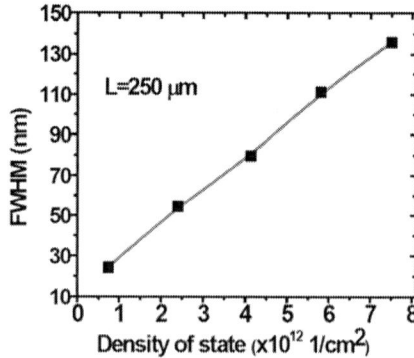

Fig. 5. FWHM bandwidth for different density states at L=250 μm.

The output power spectra for the SLDs with different cavity lengths and for $n = 2.41 \times 10^{16} (1/m^2)$ are shown in Fig 6, and its FWHM bandwidth as a function of cavity length is shown in Fig. 7. It is observed that the output power increases with increasing the cavity length, whereas the FWHM bandwidth decreases. By increasing the cavity length, the photon lifetime in the cavity increases and spontaneous emission amplifies. That means narrowing of output power bandwidth with increasing cavity length.

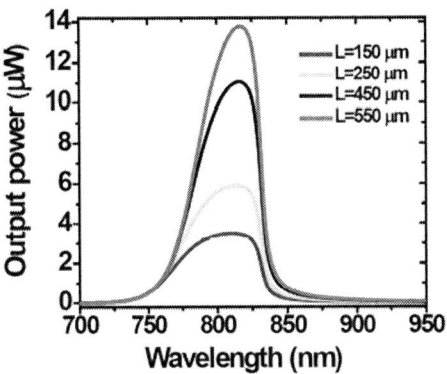

Fig. 6. Output power for SLD at different cavity length.

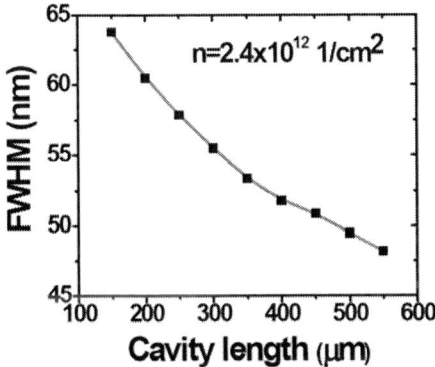

Fig. 7. The FWHM bandwidth as a function of cavity's length.

In summery at this paper we have investigated the dependence of the optical gain, the output power on the cavity length and the density states. From the results it can be seen that the increasing of output power and optical gain with increasing density states. Also, the broadening of the output power spectrum is observable with increasing the density states.

REFERENCES

[1]   T. Ko, D. Adler, J. Fujimoto, D. Mamedov, V. Prokhorov, V. Shidlovsky and S. Yakubovich, "

Ultra high resolution optical coherence tomography imaging with a broadband superluminescent diode light source, " Optics Express, Vol.12, pp.2112-2119, (2004).

[2]   W. K. Burns, C. L. Chen and R. P. Moeller, " Fiber-Optic gyroscopes with broad band source, " OSA/IEEE J. Lightw. Technol., Vol.1, p.98, (1983).

[3]   S. Martin-Lopez, M. Gonzalez-Herraez, A. Carrasco-Sanz, " broadband spectrally flat and high power density light source for fiber sensing purposes, " Measurement science and technology, Vol.17, pp.1014-1019, (2006).

[4]   A. F. Brooks, T. L. Kelly, P. L. Veitch and J. Munch, " Ultra- Sensitive wavefront measurement using a hartmann sensor, " Optics Express, Vol. 15, pp.10370-10375, (2007).

[5]   B. D. Paterson, J. E. Epler, B. Graf, H.W. Lehmann, H. C. Sigg, " Superluminescent diodes at 1.3 □m with very low spectral modulation, " IEEE J. Quantum Electron., Vol.30, pp. 703-712, (1994).

[6]   B. R. Wu, C. F. Lin, L. W. Laih, and T. T. Shih, "Extremely broad band InGaAsP/InP superluminescent diodes, " Electron. Lett, Vol.36, pp.2093-2095, (2000).

[7]   C. F. Lin, B. L. Lee, and P. C. Lin, " Broad band superluminescent diodes fabricated on a substrate with asymmetric dual quantum wells, " IEEE Photon. Technol. Lett., Vol.8, pp. 1456-1458, (1996).

[8]   F. Causaa, L. Burrow, " Ripple- Free High- Power superluminescent diode arrays, " IEEE j. Quantum Electronics, Vol.43, pp. 1055-1059, (2007).

[9]   N. S. K. Kung, K. Y.Lau, N. Bar-Chaim, I. Ury and K.J.Lee, " High power, high efficiency window buried heterostructure GaAlAs superluminescent diode with an integrated absorber," Appl. Phys. Lett., Vol.51, p.1879, (1987).

[10] G. A. Alphonse, D. B. Gilbert, M. G. Harvey, M. Ettenberg, " High-Power superluminescent diodes," IEEE J. Quantum Electronics, Vol.24, pp.2454-2457, (1988).

[11] S. L. Chuang, Physics of Optoelectronic Devices. New York: Willey- Interscience, (1995).

[12] C. H. Henry, " Theory of Spontaneous emission noise in open resonators and its application to lasers and optical amplifiers, " J. Lightw. Technol., Vol.4, pp.288-297, (1996).

[13] Z. Q. Li, Z. M.Simon Li, " Comprehensive Modeling of superluminescent Light- Emitting Diodes, " IEEE J. Quantum Electronics, Vol.46, pp.454-461, (2010).

Modeling of XeCl excilamps with barrier discharge in frequency regime of work

S. S. Anufrik, A. P. Volodenkov, K. F. Znosko

Yanka Kupala State University of Grodno, 22, Ozheshko Street, 230023, Grodno, Belarus,
Fax: +375 (152) 73-19-10; e-mail: a.volodenkov@grsu.by

Abstract: The results of modeling of XeCl excilamps with barrier discharge in frequency regime are considered. Model of XeCl excilamp taking into account process of Cl_2 regeneration is developed. Excilamp efficiency can reach 4 % at use of small values of storage capacity and at that pulse energy is equal 0,032 J and power of radiation is 3200 W.

SIMULATION METHOD

Modeling of electro-discharge XeCl excilamps is discussed. Generally the computer model includes the following modules.

1) The module of the solving of Boltzmann equation for the electron energy distribution function (EEDF) [1]. This module on composition of a mixture, on value of a degree of ionization and set E/N (E - intensity of an electric field in an interelectrode gap; N - full concentration of particles) allows to find EEDF and accordingly to define rates of plasma-chemical reactions with participation of electrons, and also to define electron mobility.

Rate factors of reactions with participation of electrons were obtained by averaging on EEDF expressions of next type.

$$k = \left\langle \sigma(\varepsilon) \cdot \sqrt{2\varepsilon \cdot /m} \right\rangle \qquad (1a)$$

Rate coefficients are determined as tables in which their dependence on mixture of the active medium, electron concentration n_e, the resulted strength of electric field U/Pd (U - a voltage on an interelectrode gap; d - distance between electrodes; P - the total pressure of a gas mix) is submitted. Calculations have been executed for a discrete file of points. Extrapolation on intermediate values is carried out by means of cubic splines.

For the solving ready program Bolsig + [2], which automatically calculates rate factors of reactions, is used.

2) The module of the solving of system of the equations of plasma-chemical reactions [3]. Models of XeCl-excilamp (halogenide Cl_2) are investigated and analyzed. On the base of these models the program module for solving of system of the equations of plasma-chemical reactions in electro-discharge XeCl-excilamp was developed. In simplest case the next plasma-chemical reactions must be taken into account.

Ionization and excitation

$$Xe + e \rightarrow Xe+ + e + e; \ (ki);$$
$$Xe + e \rightarrow Xe* + e; \ (ke);$$
$$Xe* + e \rightarrow Xe+ e + e; \ (ks);$$
$$Xe^* + Xe^* \rightarrow Xe^+ + e + Xe \ (k_{Pen})$$

Dissociative attachment

$$Cl_2 + e \rightarrow Cl^- + Cl; \ (ka,);$$

Production of XeCl-molecules

$$Xe* + C_2 \rightarrow XeCl* + Cl \ (k_{harp}) ;$$
$$Xe^+ + Cl^- + M \rightarrow XeCl* + M; \ (\beta_{rec})$$

Quenching of XeCl- molecules

$$XeCl* + e \rightarrow Xe + Cl + e; \ (kq);$$
$$XeCl* + N \rightarrow Xe + Cl + N; \ (\tau q)$$

Emission of XeCl- molecules

$$XeCl* \rightarrow Xe + Cl + h\nu; \ (\tau_{sp});$$

The power density of radiation of the excilamps (the power, obtained from unit of volume) is determined by next expression.

$$P = \left[XeCl^* \right] \cdot h\nu \ / \ \tau_{sp} \qquad (1b)$$

$[XeCl*]$ is concentration of excimer molecules; τ_{sp} - time constant of spontaneous emission.

3) The module of the solving of the equations of an electric circuit [4]. This module describes work of system of excitation of the volume discharge in active medium. On the total resistance of plasma this module allows to define time dependence of E/P, formed by excitation system in active medium. For excitation of the discharge the system of excitation on the basis of a LC-contour on the lumped capacities has been used (Fig. 1.)

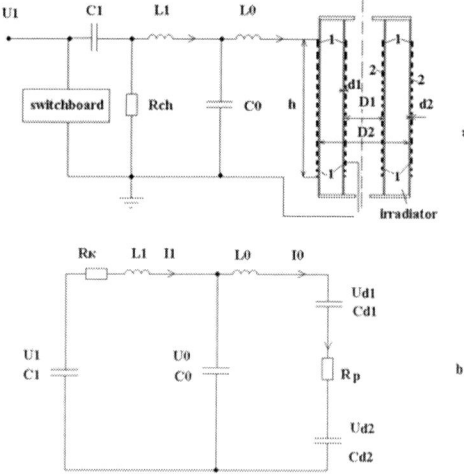

Fig. 1. The system of excitation.

The following designations were used: $C1$ – the storage capacity; $L1$ – inductance of recharging of the storage capacity on peak capacity $C0$; $Rk(t)$ –switchboard resistance; $Rp(t)$ - discharge resistance; $L0$ - inductance of recharging of $C0$ on discharge; $Cd1$, $Cd2$ – dielectric capacity. The length of electrodes h=15 cm. The coaxial radiator was used, The value of $d1$ and $d2$ was 0,2 cm. Distance between electrodes $D2$-$D1$=1 cm ($D1$=6 cm), it was considered, that the area of cross-section of the discharge is equal S. Value S was considered as parameter which accepts values in a range 280 cm^2. It has been used charging voltage Uch=$U1$=30 kV. Value of storage capacity varied in limits $C1$=1 - 10 nF.

Thus at carrying out of concrete calculations it is used the equivalent electric circuit in Fig. 1b.

RESULTS OF COMPUTER MODELING

In Fig. 2 dependences of energy of a pulse of radiation from composition of a mix which have been received under the various conditions specified in signatures to figure are submitted. From dependences, submitted in Fig. 2, follows, the optimum composition of a mix which provides the greatest energy of a pulse of radiation, depends on partial pressure of Cl_2 molecules. So at value of these parameters $C1$=2 nF, $C0$=0,2 nF S=0,25 cm^2 optimum mix is Cl_2:Xe:He=1,5:10:290.
:

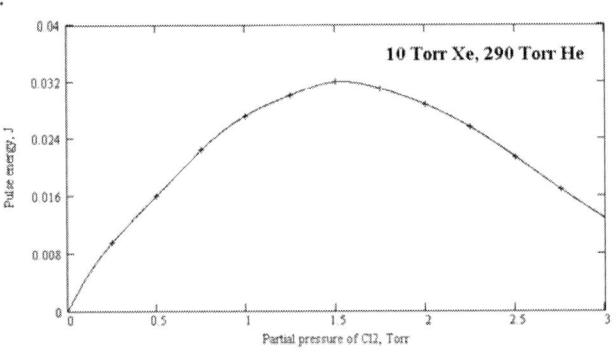

$C1$=2 nF, $C0$=0,2 nF, S=280cm^2;
Fig. 2.Dependence of pulse energy from mix.

These data have been used for definition of power of radiation of excilamp depending on pulse repetition frequency.

At work of excilamp in a pulse-periodic mode during action of a pulse of excitation there is a disintegration of molecules of chlorine, and in an interval between pulses of excitation there is a regeneration of molecules of chlorine. It results to that through some number of pulses in excilamp some fixed partial pressure of Cl_2 is established for the moments of time corresponding to the beginning of a pulse of excitation.

Process of regeneration of Cl_2 molecules is taken into account as follows. Molecule Cl_2 is formed owing to three-partial association by reaction [5].

$$2Cl+He=Cl_2+He \quad (kCl_2=5\ 10^{-33}\ cm^6/s) \quad (2)$$

On the basis (1) the following system of the kinetic equations has been written (Xe is used as buffer gas):

$$\frac{d[Cl_2]}{dt} = kCl_2 \cdot [Xe][Cl][Cl], \quad \frac{d[Cl]}{dt} = -2kCl_2 \cdot [Xe][Cl][Cl] \quad (3)$$

The decision of the equations system (3), provided that during the initial moment of time $t=0$ concentration of atoms of chlorine is equal $[Cl]$ (0) and concentration of molecules of chlorine is equal $[Cl_2]$ (0) has the following kind.

$$[Cl](t) = \frac{[Cl](0)}{1+2\cdot kCl_2 \cdot [Xe]\cdot [Cl](0)\cdot t} \quad (4a)$$

$$[Cl_2](t) = [Cl_2](0) + \frac{[Cl](0)}{2}\left[1 - \frac{1}{1+2kCl_2[Xe][Cl](0)t}\right] \quad (4b)$$

The equations (4) have a constant of time equal to the following value.

$$\tau_{ac} = \frac{1}{2\cdot kCl_2 \cdot [Xe]\cdot [Cl](0)} \quad (5)$$

If initial concentrations of Cl_2 is 6,6 10^{16} 1/cm^3 (2 Torr) and 50 % of Cl2 has dissociated during discharge, then τ_{ac}=160 microsecond. Then for pulse repetition cycle T>> τ_{ac} we have mono-pulse regime and for T<<τ_{ac} we have frequency regime of work. If the established value of partial pressure at the moment of the beginning of a pulse of excitation is known, it is possible to define energy of a pulse of radiation under the conditions submitted in Fig. 2. In Fig. 3 dependence of power of radiation on frequency of recurrence of the pulses, received for various initial partial pressure of molecules of chlorine is submitted.

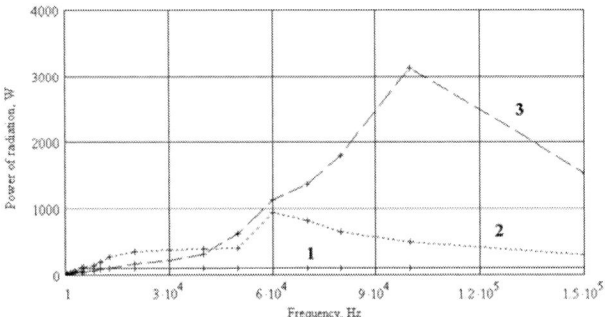

1 – Initial partial pressure Cl_2 1 Torr; 2 -2 Torr; 3- 3 Torr
Fig. 3. Dependences of power of radiation from frequency of pulses.

From the submitted dependences follows, that at the fixed initial partial pressure of Cl_2 power of radiation increase with grows of frequency. At some frequency power of radiation achieves a maximum. At the further increase in frequency power of radiation falls and leaves on some level which corresponds to the continuous glow discharge.

The maximum of power is achieved for such value of frequency of recurrence of pulses at which partial pressure Cl_2 for the moments of time corresponding to the beginning of a pulse of excitation, approximately coincides with optimum value of partial pressure for a mono-pulse operating mode. In Fig. 2 dependence of energy of a pulse from initial partial pressure of molecules of chlorine which corresponds to conditions at which dependences in Fig. 3 is calculated. The optimum value of partial pressure is equal 1,5 Torr in Fig. 2. All maximums of power of radiation are received for such value of established partial pressure of Cl_2 for the moments of time corresponding to the beginning of a pulse of excitation, is approximately equal 1,5 Torr.

At the further increase in frequency of following of pulses there is a falling power of radiation which is caused by burning out of molecules of chlorine and falling of energy of separate pulses. As it follows from the dependences submitted in Fig. 3 power of radiation increases at increase initial partial pressure of molecules of chlorine. So at increase initial partial pressure of molecules of chlorine with 2 Torr up to 3 Torr the maximal power of radiation grows from 1000 W up to 3200 W. This growth is caused by that at the greater initial maintenance of molecules of chlorine their burning out up to the optimum value corresponding to a maximum of energy for mono-pulse work, occurs at the greater frequency of recurrence of pulses.

Dependence 1, which is submitted in Fig. 3, has no maximum of power of radiation depending on frequency of following of pulses. Therefore the established value of partial pressure at the moment of the beginning of a pulse of excitation turns out much less, than the optimum size corresponding to a maximum of energy for mono-pulse work.

CONCLUSIONS

On the basis of results of modelling of XeCl-excilamps it is possible to draw the following conclusions.

1. The basic channel of formation of XeCl-molecules is harpoon reaction.

$$Xe^* + Cl_2 \rightarrow XeCl^* + Cl$$

At the given total pressure the channel ion - ionic recombination can be not taken into account.

$$Xe^+ + Cl^- + M \rightarrow XeCl^* + M$$

It is connected by that harpoon reaction is two-partial, and ion - ionic recombination three-partial and its speed depends on the total pressure of a mix. In that case when at modelling this channel was not taken into account the same dependence for change of concentration of XeCl-molecules turned out depending on time.

Model of XeCl excilamp taking into account process of Cl_2 regeneration is developed. Excilamp efficiency can reach 4 % at use of small values of storage capacity and at that pulse energy is equal 0,032 J and power of radiation is 3200 W.

It is established, that at small values of storage capacity power of radiation increases at increase in frequency of following of pulses up to some value. And the maximum of power is reached for such value of frequency of recurrence of pulses at which partial pressure Cl_2 for the moments of time corresponding to the beginning of a pulse of excitation, approximately coincides with optimum value of partial pressure for a mono-pulse operating mode. At the further increase in frequency there will be the reduction of power caused by burning out of halogen donor. At the big values of storage capacity power of radiation at increase in frequency leaves on some constant level.

The developed techniques of modeling allow to optimize more purposefully issue characteristics of pulse sources of radiation on the basis of the discharge in mixes of inert gases with halogens.

REFERENCES

[1] G. J. M. Hagelaar, L. C. Pitchford. "Solving the Boltzmann equation to obtain electron transport coefficients and rate coefficients for fluid models," *Plasma Sources Sci. Technol.*, vol.14, no. 1, pp.1-12, 2005.

[2] http://www.codiciel.fr/plateforme/plasma/bolsig/bolsig.php.

[3] A. N. Panchenko, A. S. Polyakevich, E. A. Sosnin, V. F. Tarasenko. "Glow discharge in excilamps of low pressure", *Proceedings of institutes of higher education. Physics,* vol. 42, no. 6, pp. 50-66, 1999.

[4] S. S. Anufrik, V. O. Shkleinik, A. P. Volodenkov, K. F. Znosko. "XeCl-excilamps computer modeling," *Proceedings of the VII symposium of Belarus and Serbia on physics and diagnostics of laboratory and astrophysical plasmas (PDP`2008), September 22-26, 2008, Minsk, Belarus*, pp. 118-121, 2008.

[5] D. L. Donohoue, D. Bauer, A. J. Hynes. "Temperature and pressure dependent rate coefficients for the reaction of Hg with Cl and the reaction of Cl with Cl: A pulsed laser photolysis-pulsed laser induced fluorescence study," *J. Phys. Chem. A.*, vol. 109, pp. 7732 – 7741, 2005.

Broadband semiconductor optical amplifiers of NIR range based on nanoheterostructures

E.V.Andreeva,[1] S.N.Il'chenko,[1] M.A.Ladugin,[2] A.A.Lobintsov,[1] A.A.Marmalyuk,[2] M.V.Shramenko,[1] and
S.D.Yakubovich,[3] *Senior Member, IEEE*

[1]SUPERLUM DIODES Ltd. P.O.Box-70, Moscow 119454 Russia
[2]JSC R&D Inst. "POLYUS", Vvedenskogo 3, Moscow 117342 Russia
[3]MSTU MIREA, Vernadskogo prosp.78, Moscow 119454 Russia
E-mail: yakubovich@superlumdiodes.com

Abstract: The series of travelling-wave semiconductor optical amplifiers (SOAs) based on QW-heterostructures used for the production of broadband superluminescent diodes (SLDs) is developed. Small-signal fiber-to-fiber gain of SOA-modules is about 25dB. They possess spectral gain bands of 70-125 nm at 10 dB level. Together they cover the IR-range of 750-1100 nm. Their high reliability at CW output optical power of up to 50 mW ex SMF was demonstrated. An example of the application of one of the developed SOA-modules as an active element of high-performance tunable laser is presented.
Key words: semiconductor optical amplifier, QW heterostructure, tunable laser.

INTRODUCTION

It is well-known that QW semiconductor heterostructures permit to realize in NIR range optical gain spectra with bandwidth of the order of 100 nm. Serial superluminescent diodes (SLDs) based on such heterostructures exhibit spectral FWHM of 50 nm, 70 nm, 95 nm and 110 nm in NIR bands of 750-800 nm, 800-900 nm, 900-1000 nm and 1000-1100 nm respectively [1-4]. The present work was aimed to the development of broadband traveling-wave semiconductor optical amplifiers (SOAs) operable in these NIR bands. High-performance tunable laser with one of such SOAs as active element was realized.

EXPERIMENTAL RESULTS

The studied samples were based on six different QW SCDHs grown by MO CVD. SOAs had traditional design with straight tilted 4μm-width active ridge waveguide. SOA-modules were assembled in Butterfly packages with input and output SMF pigtails. Several technological improvements ensured their reliable operation in temperature range from -55°C to +70°C at maximum CW output optical power from 20 mW to 50 mW. The stationary characteristics of SOA-modules were investigated by methods described in [5] using tunable semiconductor lasers as input signal sources. The parameters of active layers of the developed SOAs and their main technical characteristics are shown in Table 1.The values in the last column correspond to the output powers at which modules passed the reliability tests. Small-signal fiber-to-fiber gain spectra at maximum FWHM

of the studied samples are shown on Fig.1. Our experience shows that the "pure" gain of about 10 dB is enough to fulfill the threshold conditions of the laser with an external fiber cavity with reasonable insertion loss level, in which SOA-module is used as an active element. In accordance with this estimation the developed SOA-modules possess effective gain bands from 70 nm (Type I) to 125 nm (Type V).

TABLE I
MAIN PARAMETERS OF THE DEVELOPED SOA-MODULES.

Type	QW composition and thickness	Active channel length, μm	10 dB gain bandwidth, nm	Max gain, dB	Max output power, mW
I (SQW)	$Al_{0.1}Ga_{0.9}As$ 10 нм	1100	755-825	26	25
II (SQW)	GaAs 9.0 нм	1200	795-875	25	50
III (SQW)	$In_{0.02}Ga_{0.98}As$ 11 нм	1200	825-915	27	30
IV (SQW)	$In_{0.2}Ga_{0.8}As$ 6 нм	1200	885-990	28	30
V (DQW)	$In_{0.3}Ga_{0.7}As$ 2x5.5 нм	1000	955-1080	25	20
VI (DQW)	$In_{0.35}Ga_{0.65}As$ 2x7.0 нм	1200	1010-1110	24	20

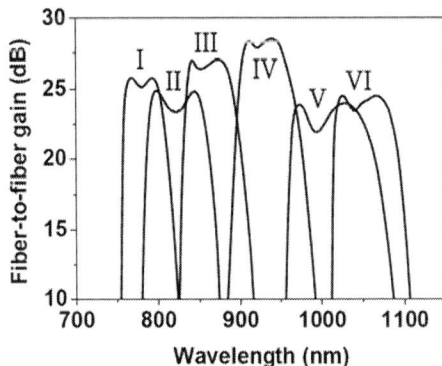

Fig.1. Small-signal gain spectra of the novel broadband SOAs of "nearest" IR range (Max FWHM).

The example of the usage of SOA-module (Type II) as the power booster of narrowband ($\delta\lambda < 0.05$ nm) input signal is illustrated by Fig.2-4. The value SMS corresponds to the excess of the amplified signal spectral density over ASE "pedestal".

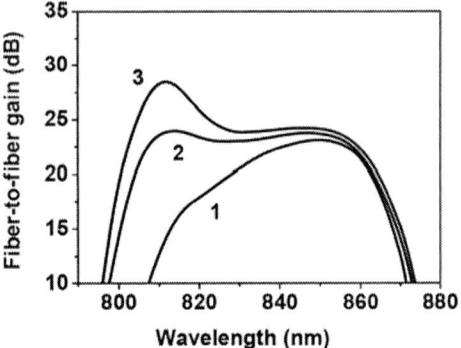

Fig.2. Small-signal gain spectrum evolution of SOA (Type II): 1 - I_{SOA}= 100mA; 2 - I_{SOA}= 150mA; 3 - I_{SOA}= 200mA.

This study has made it possible to start the serial production of new six models of SOA-modules, that to our knowledge have no analogs at photonics market.

As was mentioned above the pigtailed SOA-modules may be used in tunable and single-frequency lasers with an external fiber cavity. The prototype of such laser based on the module of Type VI with tuning range of 1010-1110 nm was studied in present work. Its ring cavity based on PMF Corning PANDA-980 contained SOA-module, optical isolator (OI), that ensured unidirectional lasing, and an acousto-optic tunable filter (AOTF) used as selective element. The main advantages of such design that does not contain any moving parts are high precision and reproducibility of spectral tuning. In comparison with linear external cavities [5,6] the ring cavity with OI ensures higher output power. Two different optical schemes that differ by the position of the output fiber splitter were investigated. In the scheme of Type 1 the splitter is placed after AOTF, in the scheme of Type 2 – just after SOA. The advantage of the first scheme is the "pure" spectrum of the output emission (SMS > 55 dB). The second scheme ensures higher external efficiency, but the output spectrum contains ASE "pedestal" whose relative level changes in the range from -30 dB to -50 dB together with wavelength tuning. Typical output spectra and tuning curves of the both types of lasers are shown on Fig.5 and Fig.6. The original controller ensured thermostabilization of SOA and AOTF and manual spectral tuning with the accuracy of 0.05 nm or linear sweeping between any two wavelengths within tuning range with the rate of up to 10^4 nm/s and an instantaneous linewidth below 0.01 nm in APC-mode.

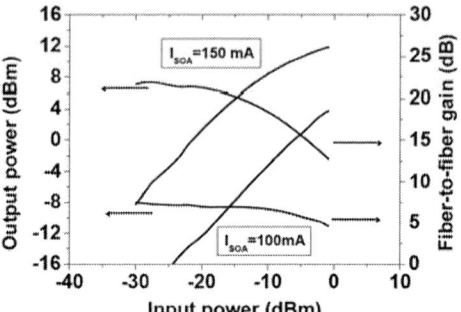

a.

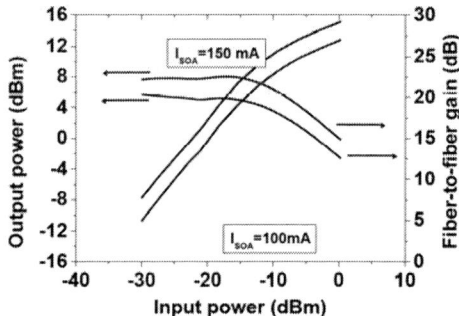

b.

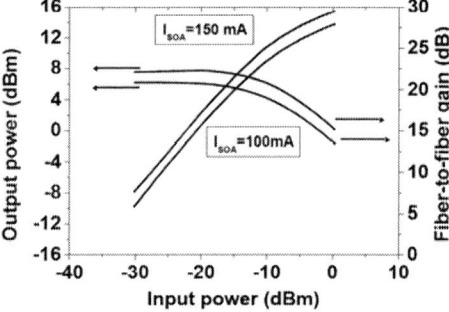

c.

Fig.3. Transmission characteristics at different I_{SOA} and input signal wavelengths of 805nm (a), 830nm (b) and 860nm (c).

CONCLUSION

New six models of SOA-modules with fiber-to-fiber gain of about 25 dB based on heterostructures with (AlGaAs), (GaAs) and (InGaAs) QW active layers were developed. They possess optical gain bands of the order of 100 nm and cover together the NIR-range of 750-1100 nm. The possibility of the effective usage of such SOAs as active elements of broadband tunable lasers was demonstrated.

The authors are grateful to A.T.Semenov for the initiation of this study. The work was partially supported by the grant of Federal target program № 14.B37.21.0756. The executed study will permit to improve significantly the main technical parameters of the earlier developed tunable laser of this spectral range (Model BroadSweeper 1060-01).

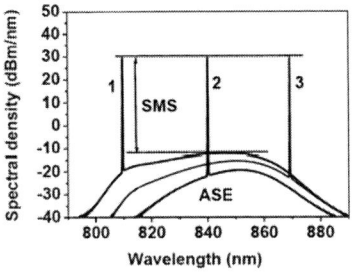

a.

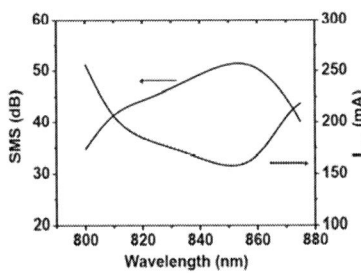

b.

Fig.4 a. Output emission spectra at CW output power of 50mW (P_{input}= 2mW at 810nm(1), 840nm(2) and 870nm(3); b. SMS and I_{SOA} vs input signal wavelength.

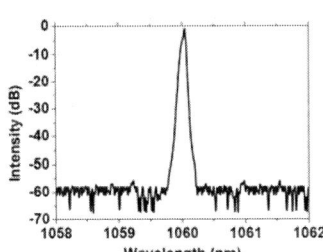

a.

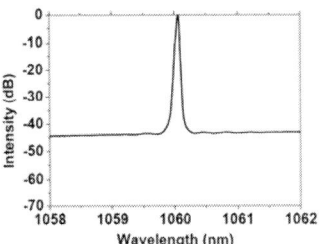

b.

Fig.5. Output emission spectra at 1060nm (a. – Type 1; b. – Type 2).

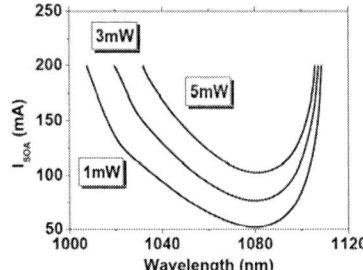

a.

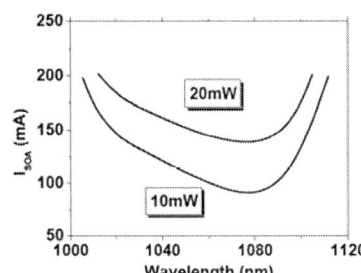

b.

Fig.6. Tuning curves at different output powers (APC-mode: a. – Type 1; b. – Type 2).

REFERENCES

[1] S.N.Il'chenko, Yu.O.Kostin, I.A.Kukushkin, M.A.Ladugin, P.I.Lapin, A.A.Lobintsov, A.A.Marmalyuk, S.D.Yakubovich, "Broadband SLDs and SOAs for the spectral range 750-800nm," *Quantum Electronics*, vol. 41, no. 8, pp. 677-680, 2011.

[2] Yu.O.Kostin, P.I.Lapin, V.R.Shidlovski, S.D.Yakubovich, "Towards 100nm-band NIR SLDs," *Proc. SPIE*, vol. 7139, pp. 713905:1-7, 2008.

[3] D.S.Mamedov, V.V.Prokhorov, S.D.Yakubovich, "Extremely broadband high-power SLDs at 920nm," *Quantum Electronics*, vol. 33, no. 6, pp. 471-473, 2003.

[4] P.I.Lapin, D.S.Mamedov, A.A.Marmalyuk, A.A.Padalitsa, S.D.Yakubovich, "High-power broadband SLDs emitting in 1000-1100nm spectral range," *Quantum Electronics*, vol. 36, no. 4, pp. 315-318, 2006.

[5] A.A.Lobintsov, M.V.Shramenko, S.D.Yakubovich, "SOAs for the 1000-1100nm spectral range," *Quantum Electronics*, vol. 38, no. 7, pp. 661-664, 2008.

[6] E.V.Andreeva, D.S.Mamedov, V.R.Shidlovski, M.V.Shramenko, S.D.Yakubovich, "NIR semiconductor laser with fast broadband tuning," *Proc. SPIE*, vol. 6079, pp. 275-282, 2006.

Dependence of efficiency of generation of the DYE laser from a wave length of microsecond pumping

S.S.Anufrik, V.Yu. Kurstak, V.V. Tarkovsky

Yanka Kupala Grodno State University, Grodno, Belarus

Abstract: Dependence of efficiency of generation of the dye laser from a wave length of a pumping radiation of microsecond duration is explored. It is shown, that the given effect is caused by effect of the thermal lens induced by a pumping radiation in the active medium of the laser.

The executed experimental researches of efficiency of generation of the dye laser (DL) depending on a wave length of an exciting radiation of microsecond duration have allowed to establish presence of some decrease of efficiency of generation near to a maximum of absorption of dye [1]. The precise information on the concrete physical mechanism, underlying in the basis of the given dependence, is represented important not only at use of dye as active mediums of high energy laser systems, but also in general as the systems providing conversion of radiation spectrum.

For interpretation of the given dependence some effects in the active medium of DL as TT-absorption, losses in the channel of the singlet levels, thermal heterogeneity induced by a pumping radiation, interaction of generated radiation with products of photo-destruction of dye, induced by pumping, can be involved.

The executed preliminary analysis of kinetics of generation of DL on the basis of a method of the velocity equations at duration of pumping 1 mks has specified on feeble action on investigated effect of TT-absorption and, at the same time, particular action of a relaxation of energy of electronic excitation in the higher excited singlet levels. It is necessary to take into account also, that the spectral maximum of the induced SS-absorption corresponds to a maximum of usual absorption.

The energy-level diagram, allowing to describe processes in the dye laser in view of the induced absorption in channels of S-S- and T-T energy levels is given on fig. 1. For reception of dependence of efficiency of generation of radiation at change of a wave length of pumping of microsecond duration spectral dependences of section of absorption in channels of S_0-S_1 levels - $\sigma_P(\lambda)$, sections of absorption in channels of S_1.S_2 levels $\sigma_S(\lambda)$ have been taken into account. Effect of spectral change of T-T absorption section $\sigma_T(\lambda)$ shows. The sistem of equations (1), featuring generation of the dye laser in view of the specified processes looks like:

$$\frac{dn_1}{dt} = \frac{\sigma_e c}{n_p}(n_3 - n_1)q + \frac{n_3}{\tau_{31}} + d_{31}n_3 - n_1 d_{10};$$

$$\frac{dn_2}{dt} = P_{32}n_3 + d_{42}n_4 - \frac{\sigma_{TT} c}{n_p}(n_2 - n_4)q - P_{20}n_2;$$

$$\frac{dn_3}{dt} = I_p(t)\sigma_p(\lambda_P)(N - n_1 - n_2 - 2n_3 - n_4 - n_5 - n_6) +$$
$$d_{53}n_5 - \frac{n_3}{\tau_{31}} - d_{31}n_3 - \frac{\sigma_e c}{n_p}(n_3 - n_1)q - P_{32}n_3 +$$
$$-I_p(t)\sigma_S(\lambda_P)(n_3 - n_5) - \frac{\sigma_S(\lambda_\Gamma)c}{n_p}(n_3 - n_5)q$$

$$\frac{dn_4}{dt} = \frac{\sigma_{TT} c}{n_p}(n_2 - n_4)q - d_{42}n_4; \qquad (1)$$

$$\frac{dn_5}{dt} = I_p(t)\sigma_S(\lambda_P)(n_3 - n_5) + \frac{\sigma_S(\lambda_\Gamma)c}{n_p}(n_3 - n_5)q - d_{53}n_5 - P_{56}n_5$$

$$\frac{dn_6}{dt} = P_{56}n_5 - \frac{\sigma_a c}{n_p}n_6 q;$$

$$\frac{dq}{dt} = \frac{\sigma_e c}{n_p}(n_3 - n_1)q - \frac{\sigma_S(\lambda_\Gamma)c}{n_p}(n_3 - n_5)q - \frac{\sigma_{TT} c}{n_p}(n_2 - n_4)q -$$
$$\frac{q}{\tau_c} + \Omega\frac{n_3}{\tau_{31}} - \frac{\sigma_a c}{n_p}n_6 q$$

$$\tau_c = \frac{-2 \cdot l}{c \cdot \ln\left(R_1 R_2 \cdot (1 - T_{df})^2\right)},$$

where N - concentration of particles, n_i - concentration of particles in the relevant excited states, q_i - an volume density of generated quantums for radiation with a wave length λ_g, σ_e - section of an induced radiation, $I_P(t)$ - intensity of quantums of pumping of the Gaussian form. Coefficients d_{ij} are responsible for nonradiative relaxation, P_{ij} - for intercombination conversion. Value Ω defines a part of the spontaneous radiation, falling on a spectral range of generation and spreading in the same corporal corner, as generation, n_P - refractive index of a dye solution, τ_c is responsible for interaction of radiation with the resonator of the dye laser.

At powerful pumping in the higher singlet levels there can be a convertible and irreversible photodeblooming, and also photodecay of molecules of dye. Yields of the given phototransmutations will be to interreact in the greater or smaller measure with generated radiation. At modeling of the dye laser, excited by radiation of microsecond duration

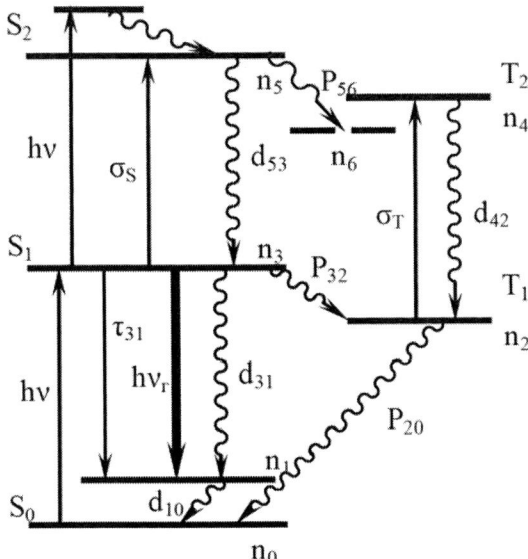

Fig. 1. The energy-level diagram of dye used in accounts. S_0, S_1, S_2 - singlet levels of energy, T_1, T_2 - triplet levels of energy, d_{ij} - probabilities of nonradiative transitions, P_{ij} - probabilities of intercombination conversion.

parameters of laser colouring agent Rodamin 6G, frequently used in experimental researches, are used: $\sigma_e = 2,0 \cdot 10^{-16}$ cm^2, $\sigma_{TT} = 6,1 \cdot 10^{-17}$ cm^2, $\tau_n = 1$ mks, $\tau_{31} = 5,9$ ns, $1 = 5$ cm, $\lambda_g = 5,6 \cdot 10^{-5}$ cm, $R_1 = 1$, $R_2 = 0,6$, $P_{32} = 0,5 \cdot 10^{-3} \cdot$ns^{-1}, $d_{31} = 0,01 \cdot$ns^{-1}, $d_{42} = 0,005 \cdot$ns^{-1}, $n_p = 1,36$, $N = 1,2 \cdot 10^{17}$ cm^{-3}, $\sigma_p = 4,0 \cdot 10^{-16}$ cm^2 (in a maximum), $\sigma_s = 5 \cdot 10^{-17}$ cm^2.

Spectral dependences of section of the absorption, the induced absorption and the amplifications used in accounts, have been received with the help of interpolation of the data available in the literature [2,3].

Research of kinetics of generation depending on pumping level $\gamma = E_p/E_{th}$, where E_p - energy of pumping pulses, E_{th} - a threshold energy, has shown, that maximums of pumping pulses and generations coincide, and increase of γ gives in increment of power and duration of pulses.

The generated pulse with growth of an excitation level is extended symmetrically concerning the maximum that is well compounded with the data of experimental researches.

That generation finishes on the same pumping level specifies on absence of accumulation of losses, bound with upper laser level S_1.

The executed numerical researches of operation of the dye laser at its excitation by a coherent radiation by pulses of 1 mks duration have shown, that decrease of efficiency of generation observed experementally when the wave length of pumping is near to a maximum of absorption of colouring agent is not bound to processes in the higher singlet levels of dye molecules.

At studying of dependence of efficiency of generation of the dye laser from a wave length of an exciting radiation at

microsecond duration of pumping by all means it is necessary to take into account, that change of absorption at rebuilding of a wave length of pumping will give in respective alteration of a threshold power density of excitation. As the given value changes expediently to find out effect of a power density of excitation on efficiency of generation at the fixed wave length of pumping that should clear substance of the phenomenon. The specified dependence is given on fig. 2. That fact, that this dependence has a maximum at $\gamma = 5$, and then smoothly decreases, specifies on presence of losses in the active medium, which are considerably incremented with growth of a power density of pumping.

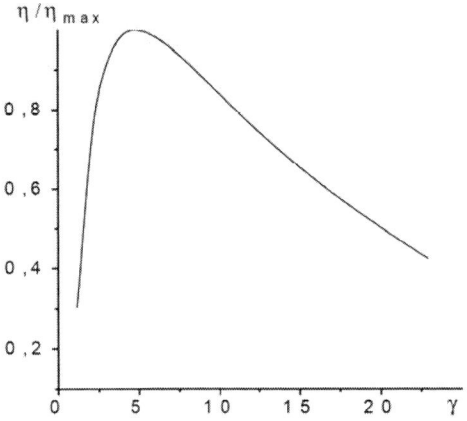

Fig. 2. Dependence of efficiency of generation on a power density of excitation, normalized on threshold. A wave length of pumping - 532 nm.

The question is that part of energy of pumping which during operation of the laser is converted in heat and gives in origination thermo-optical distortions of the resonator. Joule heat losses of pumping are proportional to its value when pumping will exceed some particular meaning of energy E_{min} for the given operating conditions of the laser. In this case action of the negative thermal lens, induced by a pumping radiation, increments divergency of generated radiation above the diffraction limit.

The trend of a curve, given on fig. 2, is attained due to effect of specified thermo-optical distortions, incipient in the active medium, on generated radiation.

Dependence of efficiency of generation on a wave length of the exciting radiation, calculated at various meanings of the normalized pumping, is given on fig. 3. At value of the normalized pumping which are not exceeding 5 (fig. 3,a), dependence of efficiency of generation on a wave length of pumping has one maximum. To major meanings of pumping also the greater meaning of efficiency corresponds. In this case the loss coefficient because of thermo-optical distortions is not high. At power density of the normalized pumping which are in limits from 8 up to 17 dependence of efficiency of

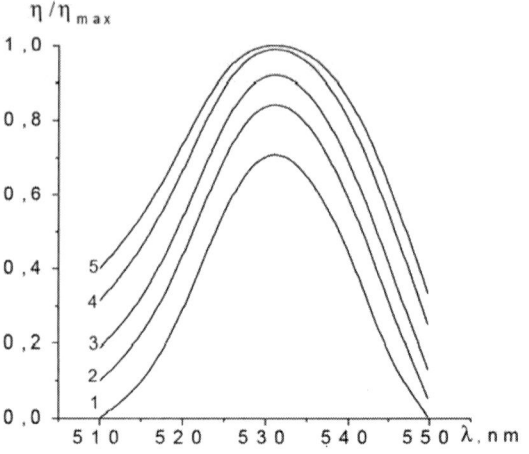

a)

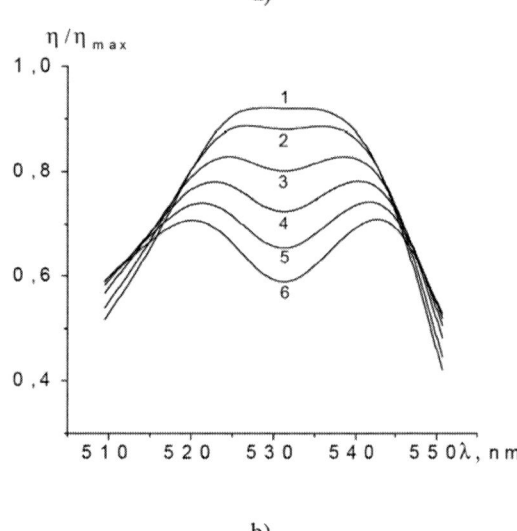

b)

Fig. 3. Dependence of efficiency of generation on a wave length of pumping at various pumping levels γ. a): $1 - \gamma = 2$; $2 - \gamma = 2,5$; $3 - \gamma = 3$; $4 - \gamma = 4$; $5 - \gamma = 5$; b): $1 - \gamma = 8$; $2 - \gamma = 9$; $3 - \gamma = 11$; $4 - \gamma = 13$; $5 - \gamma = 15$; $6 - \gamma = 17$.

generation from a wave length of a pumping radiation is given on fig. 3, b).

The increase of pumping results in reduction of efficiency, curves thus are displaced downwards. At curves the dip near to a maximum of absorption in this case is formed, and size of a dip is incremented with growth of value of pumping. In this case the loss coefficient because of thermo-optical distortions is high enough, and its greatest value is attained in a maximum of absorption as on the given wave length the greatest absorbed energy is spent on thermo-optical distortions of the active medium.

REFERENCES

[1] В. В. Тарковский, В. Ю. Курстак, С. С. Ануфрик Зависимость эффективности генерации от частоты возбуждения при накачке раствора родамина 6Ж лазерными импульсами микросекундной длительности // ЖПС. 2002. Т.69, №6. С.747-750.

[2] Копылов С.М., Лысой Б.Г., Серегин С.Л., Чередниченко О.Б. Перестраиваемые лазеры на красителях и их применение. // М.: Радио и связь, 1991.

[3] Magde D., Gaffney T., Campbell B. T. Excited Singlet absorption in Blue Laser Dyes: Measurement by Picosecond Flash Photolysis // IEEE J. Quant. Electron. −1981. Vol. QE-17, № 4. P.489 − 495.

Spectral-luminescent and lasing properties of coumarin dyes

N.Kh. Ibrayev[1], E.V. Seliverstova[1], T.N. Kopylova[2], R.M. Gadirov[2], V.I. Alekseeva[3], L.E. Marinina[3], L.P. Savvina[3]

[1] E.A. Buketov Karaganda State University, Karaganda, Kazakhstan
[2] Tomsk State University, Tomsk, Russia
[3] State Scientific Center of the Russian Federation (NIOPiK), Moscow, Russia

Abstract: Spectral-luminescent and lasing properties of new derivatives of Coumarin-7 were studied. Induced emission generation of studied dyes was observed in the red wave-lenghth region of spectrum. The efficiency of lasing was equal to 40%. The threshold of generation of dyes was observed upon excitation with the power density less than 0.01 MW/cm². It was revealed that studied dyes have a high photostability.

Organic dyes due to its unique photophysical properties are the important and widely-used in technique class of compounds with the complex molecular structure.

The class of coumarin dyes can be distinguished from the large variety of organic dyes. Interest to this dyes growing due to the fact that they are effectively used as indicators for defectoscopy, in fluorimetry, in laser technology as a laser active media, in biology and medicine for staining of biological specimens and fluorescent probes. Also this dyes have nonlinear optical properties [1,2].

The results of studies of spectral-luminescent and lasing properties of new synthesized derivatives of Coumarin-7 are presented. Structure of studied dyes is shown in fig. 1.

Fig. 1. Structure of coumarine dyes

Absorption and fluorescence spectra were measured on spectrophotometer CM 2203 and Cary Eclipse Varian correspondingly. Quantum yields of fluorescence was measured with comparative method. Ethanol solution of Rhodamine C was chosen as standard. Calculated values of fluorescence quantum yield of dyes were equal to 0.99 and 0.75 for ethanol solutions of chromene-3 and chromene-13 correspondingly.

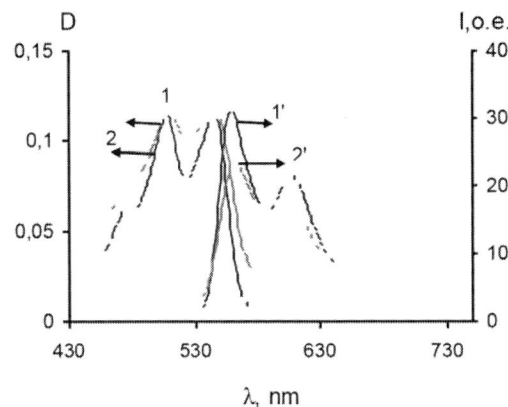

Fig. 2. Absorption (1,2) and fluorescence (1',2') spectra of chromene-3 (1,1') and chromene-13 (2,2')

Absorption spectra of ethanol solution of dyes (fig. 2) appears as wide band in the region of 450-600 nm with two maxima at 551 and 517 nm – for chromene-3 and 550 and 517 nm – for chromene-13. Fluorescence spectra of dyes in ethanol have a maximum at 573 nm and weakly evidenced shoulder at 615-630 nm.

Spectral measurements showed that the changes of shape of bands and shift of maxima of absorption and fluorescence spectra were observed upon changes in the polarity of medium. For example, in solutions of low polarity the spectrum of dyes are the blue-shifted.

Lasing parameters were studied upon excitation of samples with the second harmonic of Nd:YAG laser (SOLAR Q 129, λ_{gen}=532 nm, Eimp=100 mJ) in cross-section direction. Resonator was formed between the opaque mirror and side of cell. The pump emission was focused onto front side of cell in the form of line with the area of 0.1 cm². Power density of pumping was varied by neutral filters and was equal to 0.1-55 MW/cm². Spectra and shape of pulse generation were measured with laser spectrometer «Real» and digital oscilloscope Tektronix TDS 224. Energetic parameters were measured with measuring instrument of optical energy Gentec E DUO and OPHIR NOVA II. Relative errors of energetically measurements were equal to 1.5%, spectral and luminescent – 1% and 3% correspondingly. At the measurements of lasing properties of dyes its concentration in toluene solution was equal to 10^{-3} mol/l.

It was found that for both dyes the generation of the band of induced emission located in the long-wavelength band of fluorescence spectrum. The maxima of the lasing band is appears at the 609 nm – for dye chromene-3 and at 605 nm – for chromene-13.

The generation threshold for chromene-3 was observed upon excitation with the power density equal to 0.01 MW/cm^2. Whereas for chromene-3 this parameter was equal to 0.25 mJ. The efficiency of lasing of coumarin dyes was equal to 40 and 30% for chromene-3 and chromene-13 correspondingly.

The measurements showed that the maximum of the lasing spectrum in ethanol solutions is shifted to shorter wavelengths and is located in the short-wavelength of fluorescence band. This shift is due to the peculiarities of the interaction of dye molecules with molecules of the solvent.

The effect of power density of second harmonic of Nd:Yag laser on efficiency of lasing of coumarin dyes was studied. Measurements are showed, that maximal efficiency of lasing corresponds to W=10 MW/cm^2. Further increasing of the power of excitation do not leads to growing of lasing efficiency of dyes (fig. 3).

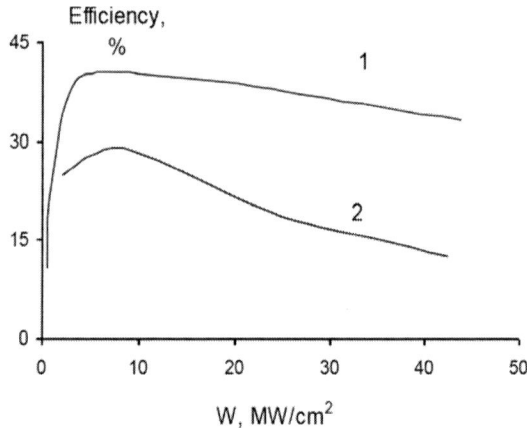

Fig. 3. The effect of the power dencuty of pumping of dyes on the efficiency of lasing:1 - chromene-3; 2 – chromene-13

Photostability of dyes in solutions and in the matrix of PSS was studied. The scheme of experiment is similar to that above mentioned. The matrix of PSS was excited by laser with the power equal to 9 MW/cm^2.

The photostability in liquid and solid matter was estimated on dependence of maximal efficiency of lasing of dyes on the time of excitation. It was found that intensity if lasing of both dyes decreased on 20 % upon excitation of polymer film during 20 min. Further occurs the decreasing of intensity over 50 %.

The full time of excitation was equal to 33 min. The studying of photostability of dyes in PSS shows that lifetime of active medium based on the coumarin dyes equal to 2000 pulse.

Thus, in present work spectral-luminescent properties of new coumarin dyes were studied. Performed measurements shown that medium polarity have an essential influence on the shape and position of maxima of absorption and flurescence spectra. It was shown, that new compounds have high fluorescence quantum yield. Generation of induced emission of studied dyes was observed in the red region of spectrum. New dyes have high value of lasing efficiency (40%) and low lasing threshold, and good photostability.

REFERENCES

[1] B.M. Krasovitsky, B.M. Bolotin, *Organic luminophores,* Chemistry, 1984. (in Russian)

[2] V.I. Zemsky, U.L. Kolesnikov, I.K. Meshkovsky. Physics and techniques of pulse dye lasers, ITMO, 2005. (in Russian)

Super/Subradiant Frequency Doubling by Quantum Wells Coupled to External Resonator

Gennady A. Koganov
Physics Department
Ben-Gurion University of the Negev
P.O.B. 653, Beer Sheva 84105, Israel
Email: quant@bgu.ac.il

Reuben Shuker
Physics Department
Ben-Gurion University of the Negev
P.O.B. 653, Beer Sheva 84105, Israel
Email: shuker@bgu.ac.il

Abstract— A scheme for active frequency doubling is suggested. The system comprises N semiconductor quantum wells situated into resonant cavity. Theoretical model for such systems is based on three-level virtual atoms in ladder configuration. It is found that the system can lase in either superradiant or subradiant regime, depending on the number of atoms N. When N passes some critical value the transition from the super to subradiance occurs in a phase-transition-like manner. Stability study of the steady state supports this conclusion.

The first model of statioanry superradiance, a superradiant laser was described by Haake et. al. [1], who considered a model of three-level atoms placed inside a resonant cavity, and pumped with a classical external electromagnetic field. In addition, another "passive" cavity mode was used to coherently couple one of the non-lasing atomic transition. This "passive" cavity mode, being adiabatically eliminated, results in nonlinear collective decay. The steady state laser intensity calculated for this model scales as N^2 typical for superradiance. However, the superradiance obtained in Ref. [1] essentially differs from the Dicke superradiance [2] by: (i) it is stationary rather than transient, (ii) the linewidth of the superradiant laser scales as $1/N^2$ which is extremely small compared to the spectral width of Dicke superradiant pulse, (iii) intensity fluctuations of the superradiant laser are essentially squeezed, while those of superfluorescence pulse are close to fluctuations of a coherent state. Since then many theoretical works have been devoted to study this model as well as some other models [3]–[10].

The key mechanism responsible for stationary superradiance in such lasers is collective nonlinear spontaneous decay of one of the atomic states. Such a cooperative decay is provided by incorporating an additional "passive" resonator to couple the non-lasing atomic transition [1], [3], or to couple two non-lasing transitions in a four-level scheme [6]. Additionally, in order to reach the effect of super-radiance, when the laser intensity scales as N^2, the pumping field strength was taken proportional to N. However, as we have shown in a recent work [11], the two aforementioned requirements are not necessary to obtain superradiant lasing. We proposed a model of superradiant laser based on N three-level systems in ladder configuration, driven by two pumping lasers. All spontaneous decay processes are linear, i.e. no correlation in spontaneous emission is introduced. We called this new type

of superradiance Field Driven Superradiance since it stems from simultaneous coherent interaction of the atomic system with two driving laser fields. Utilizing semiclassical treatment, we solved optical MaxwellBloch equations both numerically and, in some particular cases, analytically. It was found that in the steady state, the generated laser field phase is locked to those of the driving lasers, and at a strong enough pump, the number of photons inside the resonator scales as N^2. It was also found that this system exhibits subradiant behavior manifested in a departure from N^2 scaling, in appropriate conditions.

In this article we suggest a scheme of active super/subradiant frequency doubling in semiconductor quantum wells (SQW) [12]–[14]. Within the scope of this article we consider N^2 dependence of the laser intensity as an indication of stationary superradiance, whereas saturation with N as indication of subradiance. It is shown that the radiation field generated by the system can exhibit both superradiance and subradiance, depending on the number of atoms, which is strong manifestation of collective behavior. The phase of the generated field is locked to that of the pumping laser. We have found that in the vicinity of some critical value of the pump Rabi frequency the system behaves in a phase-transition-like manner, changing from super to subradiant regime.

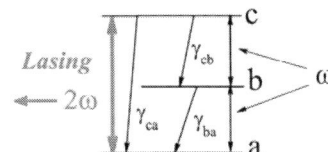

Fig. 1. (Color online) Schematic diagram of three-level system.

Consider a three-level system shown in Fig. 1. The two transitions $|a\rangle \to |b\rangle$ and $|b\rangle \to |c\rangle$ have closed enough frequencies, i.e. $\omega_{ab} \approx \omega_{bc}$, to the extent that both transitions can be driven by a single pumping laser. The transition $|a\rangle \to |c\rangle$ is in resonance with the laser resonator, therefore the frequency of the generated laser field will be twice the frequency of the

978-1-4799-0019-0/13 $31.00 © 2013 IEEE

driving laser. Relaxation constants γ_{ij} ($i, j = a, b$) describe the linear spontaneous decay of the corresponding atomic states, where the events of spontaneous emission of different atoms are independent of each other. Equations of motion for the atom+field density matrix, in the rotating wave approximation, have the following form:

$$\dot{\rho}_{cc} = -\left(\gamma_{cb} + \gamma_{ca}\right)\rho_{cc} + i\Omega(\rho_{cb}e^{-i\varphi} - \rho_{bc}e^{i\varphi})$$
$$+ig^*b\rho_{ca} - igb^\dagger\rho_{ac} \tag{1}$$

$$\dot{\rho}_{bb} = \gamma_{cb}\rho_{cc} - \gamma_{ba}\rho_{bb} + i\Omega(\rho_{bc}e^{i\varphi} - \rho_{cb}e^{-i\varphi} - \rho_{ab}e^{-i\varphi} + \rho_{ba}e^{i\varphi}) \tag{2}$$

$$\dot{\rho}_{bc} = -\frac{\gamma_{ca}+\gamma_{cb}+\gamma_{ba}}{2}\rho_{bc} - i\Omega[(\rho_{cc} - \rho_{bb})e^{i\varphi} + \rho_{ac}e^{-i\varphi}] + ig^*b\rho_{ba} \tag{3}$$

$$\dot{\rho}_{ac} = -\frac{\gamma_{ca}+\gamma_{cb}}{2}\rho_{ac} - ig^*b(\rho_{cc}-\rho_{aa}) + i\Omega(\rho_{ab}e^{i\varphi} - \rho_{bc}e^{-i\varphi}) \tag{4}$$

$$\dot{\rho}_{ab} = -\frac{\gamma_{ba}}{2}\rho_{ab} - i\Omega[(\rho_{bb} - \rho_{aa})e^{i\varphi} - \rho_{ac}e^{-i\varphi}] - ig^*b\rho_{cb} \tag{5}$$

$$\dot{b} = -\kappa b + i\sum_j^N g\rho_{ca} \tag{6}$$

$$\dot{b}^\dagger = -\kappa b^\dagger - i\sum_j^N g^*\rho_{ac} \tag{7}$$

where N is the number of atoms. This is a standard set of optical Maxwell-Bloch equations where irreversible processes of spontaneous emission and resonator losses are governed by linear terms with atomic relaxation constants γ_{ij} and the field decay rate κ in the cavity. $\Omega = \hbar\vec{d}\cdot\vec{E}$ is Rabi frequency of the driving field, $\vec{d}$ is the relevant dipole matrix element, $\vec{E}$ is the pumping field amplitude, g is the coupling constant and $\hat{b}$ is the photon annihilation operator. In semiclassical approximation the field operators can be replaced with c-numbers, so we put $b = \sqrt{n}e^{i\Psi}$, where n is the number of photons in the lasing mode, Ψ is the laser field phase. No assumption is made regarding the initial atomic cooperativity such as nonlinear collective relaxation imposed by the presence of a "passive" resonator, as in [1], [3], [6].

Steady state solution of Eqs. (1)-(7) can be obtained analytically. For the sake of simplicity we put three atomic relaxation rates equal, i.e. $\gamma_{ij} = \gamma$. Among nine formal solutions for the photon number n the two physically acceptable solutions (positive photon number n) can be roughly approximated, using $g \ll \Omega$, as follows:

$$n_1 \approx \frac{4g^2\gamma^4\Omega^4}{\kappa^2(\gamma^4 + 12\gamma^2\Omega^2 + 12\Omega^4)^2}N^2, \tag{8}$$

$$n_2 \approx \frac{4\Omega^2 - 3\gamma^2}{4g^2}. \tag{9}$$

Expression (8) is valid at $N \ll N_c$, while expression (9) is valid for $N > N_c$, where

$$N_c \approx \frac{3\kappa\Omega^2}{\gamma g^2}. \tag{10}$$

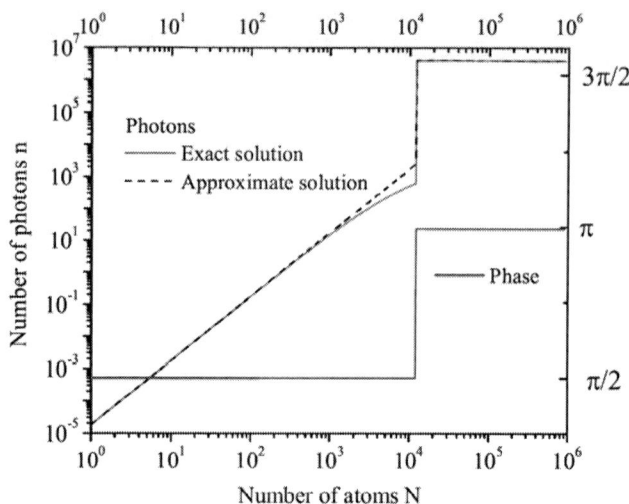

Fig. 2. (Color online) Number of photons and the field phase as a function of the number of atoms.

is a critical number of atoms that separates the superradiant regime from the subradiant one, as will be seen in the following. The field phase $\Psi \approx 2\varphi + \pi/2$ for $N < N_c$ and $\Psi \approx 2\varphi + \gamma/\Omega$ for $N > N_c$. In Fig. 2 the number of photons and the field phase are plotted as a function of the number of atoms. One can see that approximate analytical Eqs. (8) and (9) describe quite well the system behavior, except the vicinity of the critical point $N \sim N_c$, where it differs from the numerical solution.

When N approaches its critical value N_c defined by Eq. (10), both the number of photons and the field phase abruptly change their values, which indicates to the presence of phase transition. At $N \ll N_c$ the photon number scales as N^2 in agreement with Eq. (8) - a signature of superradiance. At $N = N_c$ the photon number n sharply increases by several orders of magnitude, and at $N > N_c$ the photon number remains asymptotically constant. We call this regime subradiant as the photon number does not depend on N. Figure 3 shows a "phase diagram" of the system where the solid (red online) line separates two different areas of parameters corresponding to superradiant (above the line) and subradiant (below the line) regimes.

Analysis of time behavior and the stability of the steady state solutions of Eqs. (1)-(7) shows that at $N < N_c$ the superradiant solution is stable, while the subradiant one is unstable. At $N > N_c$ the situation is opposite: the subradiant solution becomes stable and the superradiant one becomes unstable. Thus the critical point $N = N_c$ is the instability point, which gives another indication to the presence of phase transition at this point. Figures 4 and 5 illustrate different stability properties of the super and subradiant regimes. Phase portrait of the system at $N < N_c$ presented in Fig. 4 shows that in superradiant regime there is a single stationary point/attractor which is a stable node. In this regime the system gradually approaches its steady state that is stable.

978-1-4799-0019-0/13 $31.00 © 2013 IEEE

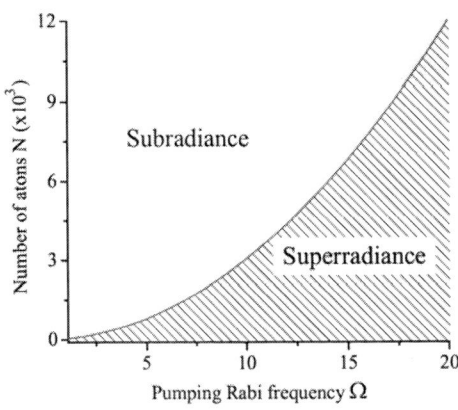

Fig. 3. (Color online) Phase diagram N vs Ω.

$\gamma = 1,\ g = 0.01,\ \kappa = 0.001,\ N = 1000,\ \Omega = 10$

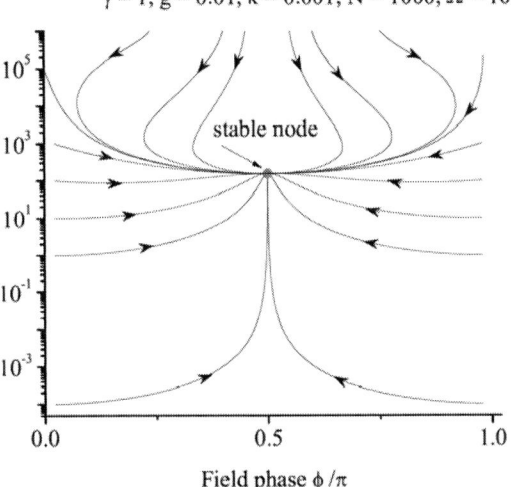

Fig. 4. (Color online) Superradiant regime: phase portrait of $n(t)$ vs $\phi(t)$.

$\gamma = 1,\ g = 0.01,\ \kappa = 0.001,\ N = 4000,\ \Omega = 10$

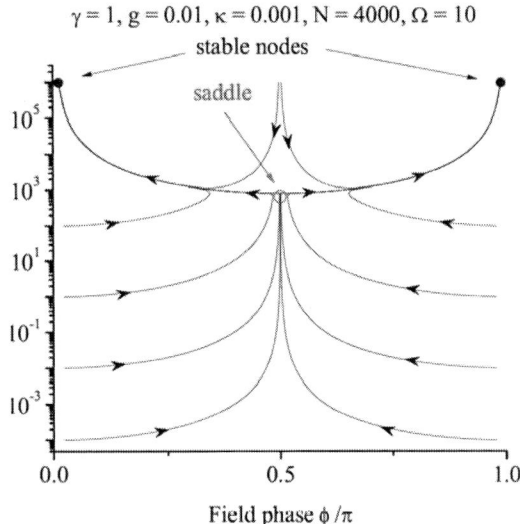

Fig. 5. (Color online) Subradiant regime: phase portrait of $n(t)$ vs $\phi(t)$..

state is established also in phase-transition-like manner, via critical slowing down. The lasing field phase is locked to the phase of the driving laser. Details of the time evolution of this system will be published elsewhere.

Essentially different behavior is seen in Fig. 5 showing phase portrait in the subradiant regime. The node at $\phi = \pi/2$ has lost its stability and transformed into a saddle point that repels slowly approaching phase trajectories, which then quickly reach the stable steady state defined by Eqs. (9) and $\Psi \approx 2\varphi + 3\gamma/8\Omega$.

Analysis of time dependence of the laser field in the subradiant regime shows that first the field reaches the long-lived metastable state, and then, at some critical time, grows abruptly to stable steady state. In theory of phase transitions such a behavior is know as a phenomenon of critical slowing down.

In summary, semiconductor quantum wells with equally separated virtual levels, coupled to an external resonator, can double the frequency of the pumping laser in superradiant or subradiant manner, depending in the ratio between the number of virtual atoms and intensity of the pumping laser. Transition from one regime to another occurs abruptly, in phase-transition-like manner. In subradiant regime the steady

REFERENCES

[1] Fritz Haake, Mikhail I. Kolobov, Claude Fabre, Elisabeth Giacobino, and Serge Reynaud. Superradiant laser. *Phys. Rev. Lett.*, 71(7):995–998, Aug 1993.
[2] R. H. Dicke. Coherence in spontaneous radiation processes. *Phys. Rev.*, 93(1):99, Jan 1954.
[3] Fritz Haake, Mikhail I. Kolobov, Carsten Seeger, Claude Fabre, Elisabeth Giacobino, and Serge Reynaud. Quantum noise reduction in stationary superradiance. *Phys. Rev. A*, 54(2):1625–1637, Aug 1996.
[4] Carsten Seeger, Mikhail I. Kolobov, Marek Kuś, and Fritz Haake. Superradiant laser with partial atomic cooperativity. *Phys. Rev. A*, 54(5):4440–4452, Nov 1996.
[5] Yu. M. Golubev. Kinetic theory of a super-radiating laser. *Theoretical and Mathematical Physics*, 109:1437–1452, 1996.
[6] Deshui Yu and Jingbiao Chen. Four-level superradiant laser with full atomic cooperativity. *Phys. Rev. A*, 81(5):053809, May 2010.
[7] D. Meiser and M. J. Holland. Intensity fluctuations in steady-state superradiance. *Phys. Rev. A*, 81:063827, Jun 2010.
[8] M. Vogl, G. Schaller, and T. Brandes. Counting statistics of collective photon transmissions. *Annals of Physics*, 326(10):2827 – 2833, 2011.
[9] A Auffves, D Gerace, S Portolan, A Drezet, and M Frana Santos. Few emitters in a cavity: from cooperative emission to individualization. *New Journal of Physics*, 13(9):093020, 2011.
[10] J.G. Bohnet, Z. Chen, J.M. Weiner, D. Meiser, M.J. Holland, and J.K. Thompson. A steady-state superradiant laser with less than one intracavity photon. *Nature*, 484(7392):78–81, 2012.
[11] Gennady A. Koganov, Boris Shif, and Reuben Shuker. Field-driven super/subradiant lasing without inversion in three-level ladder scheme. *Opt. Lett.*, 36(15):2779–2781, Aug 2011.
[12] E. Rosencher and Ph. Bois. Model system for optical nonlinearities: Asymmetric quantum wells. *Phys. Rev. B*, 44:11315–11327, Nov 1991.
[13] G. B. Serapiglia, E. Paspalakis, C. Sirtori, K. L. Vodopyanov, and C. C. Phillips. Laser-induced quantum coherence in a semiconductor quantum well. *Phys. Rev. Lett.*, 84:1019–1022, Jan 2000.
[14] C. Gmachl. Semiconductors: Quantum optics by design. *Nature materials*, 5:170, 2006.

Manifestations of optical transient nutation in frequency- modulated cw laser beams

I. L. Plastun, *Member, IEEE*, A. G. Misurin
Saratov State Technical University, Saratov, Russia

Abstract: On the basis of numerical simulations an optical transient nutation in frequency-modulated cw laser beams is investigated. This effect is developed on high modulation amplitude of frequency - modulated cw laser beam propagating in resonance conditions. At modulation periods comparable with the atomic relaxation times the time dependence of the output intensity exhibits the combined manifestations of optical nutation and resonance self-action.

In the present paper we focus our attention at the beams initially modulated in frequency. Passing through resonant absorbing medium, the beam gradually acquires intensity modulation, which is simply caused by different absorption at different frequencies. This effect is neither nonlinear, nor transient, it takes place even at low intensity and low frequency of modulation. Note, that in this case it is not affected by the lensing properties of the medium, since they are due to saturation and not to frequency-dependent refraction as such. First, we study how the saturation of absorption and refraction affects the transmission of modulation from phase at the input to the intensity at the output. We take the frequency of modulation low enough to consider the saturation as adiabatically following the field variation in time. The most prominent manifestations of the induced lens can be seen in moderate saturation regime. In this regime we vary the modulation frequency to indicate the transient effects. We observe the modulation phase shift and amplitude decrease due to the inertia of the medium response.

On the other hand, we investigate a various modulation regimes and found that if the modulation frequency is greater than the width of the absorption line, begins to manifest effect of optical transient nutation. The effect of optical nutation predicted and experimentally observed in the two-level non-linear optical systems with a homogeneously broadened transition with a sharp turn on or off the resonant excitation field (see, e.g., [1]). It consists in the fact that, under certain conditions, when the Rabi frequency exceeds the rate of relaxation of the polarization, the transmission coefficient of excited medium approaches the steady state, passing stage damped oscillations, which leads to damped oscillations of the exciting laser signal. Usually the experimental observation of optical nutation is based on technology of pulse Stark frequency switching, suggested Brewer and Shoemaker [2], which is due to the Stark shift of certain groups of molecules resonate with the exciting field or leave it, which causes periodic damped fluctuations at different frequencies. Similar results can be obtained, as noted in [1], in the case of the rapid change of the laser frequency. Recently, interest to the development of optical nutation has grown significantly: the optical nutation experimentally observed for example, in nitrous oxide [3] in cold atoms [4] and in gases [5].

In present investigation optical transient nutation observed in case of a frequency-modulated laser signal propagation in a nonlinear optical two-level system with a homogeneous broadening of the transition when the modulation frequency is comparable with the relaxation rate of medium polarization and population levels, and the modulation amplitude is several times higher than the spectral width of the absorption line. Under these conditions the resonant self-action effects are manifested and may affect optical nutation.

Numerical model is based on the set of normalized Maxwell-Bloch equations

$$2ik\left(\frac{\partial E}{\partial z}+\frac{1}{c}\frac{\partial E}{\partial t}\right)+\left(\frac{\partial^2}{\partial r^2}+\frac{1}{r}\frac{\partial}{\partial r}\right)E=gP \qquad (1)$$

$$\frac{\partial D}{\partial t}=-\gamma\left[D-1+i\left(E^* P - E P^*\right)\right] \qquad (2)$$

$$\frac{\partial P}{\partial t}=-(\Gamma+i\Delta)P-\frac{i}{2}\Gamma DE, \qquad (3)$$

where g is the unit length absorption, γ, Γ are the population and polarization decay rates, respectively, $D(z,\rho\varphi,t)$ is the population difference normalized to its non-saturated value, $E(z,\rho\varphi,t)$ and $P(z,\rho\varphi,t)$ are the slow-varying amplitudes of the electric field and the polarization, respectively, Δ is the detuning of the carrier frequency of the field from the atomic transition frequency. The unit field amplitude corresponds to the CW saturation level of $D=0.5$. The axial coordinate z is scaled to the diffraction length, while the radial coordinate r is scaled to the typical beam radius.

Eqs. (1)-(3) should be solved under the initial conditions: $E(z=0,\rho\varphi,t)=E^0(z,\rho\varphi,t)$; $E(z,\rho\varphi,t=0)=0$; $D(z,\rho\varphi,t=0)=1$; $P(z,\rho\varphi,t=0)=0$.

To solve Eqs. (1)-(3) we use the second-order scheme [6] based on the decomposition of the transverse field pattern in terms of Gauss-Laguerre (GL) modes.

We investigate an initially Gaussian beam. The frequency of the beam at the input of the medium is taken to be harmonically modulated in time, $\omega=\omega_0+\omega_1 sin\Omega t$, , where ω_0 is the carrier laser frequency, ω_1 is the amplitude of the frequency modulation, Ω is the modulation frequency. Taking into account that the instantaneous frequency of the field is the time derivative of its phase, we can write the input field complex amplitude in the form:

$$E(0,\rho,\varphi,t)=E_0\exp(-\frac{\rho^2}{2a^2})\exp[i\frac{\omega_1}{\Omega}\cos(\Omega t)] \tag{4}$$

The initial beam radius a in the examples considered below was taken equal to 1. The natural time and frequency scale is provided by the decay rate, for simplicity we took $\gamma=\Gamma=1$. We considered the central carrier frequency ω_0 to be equal to the atomic transition frequency, so that $\Delta=0$ in Eq. (3) and two regimes when carrier frequency is tuned to one half- linewidth in the focusing and defocusing side ($\Delta=+1;-1$). In resonant case the modulated field frequency oscillates symmetrically with respect to the exact resonance value. The amplitude of the frequency modulation was $\omega_1=1$, i.e., the deviation of the field frequency from resonance is one half-linewidth without optical nutation. In this case one can expect considerable modulation of the output intensity due to substantial variation of absorption. In case when optical nutation is occur the amplitude of the frequency modulation was several times higher than the half-width of the absorption line: $\omega_1=5, 10$. We consider a slow modulation regime: $\Omega=0.5$ when the transient effects in the medium response are negligible. The propagation length $Z=z_{max}=6$ is large enough, so that the free beam diverges significantly. The linear absorption in our examples is taken to be $g=1$, and, therefore, the output field intensity is small because of both absorption and divergence.

resonance cause the same change in linear absorption. In saturation regime (Fig. 1b, $E_0=5$) the half- periods of modulation become non-equivalent, since the induced lens in positive above the frequency of atomic transition and negative below it. Hence the growth of intensity below the resonance due to lower absorption is decreased by the defocusing. In the focusing region the intensity peaks are larger. When the saturation is strong (Fig. 1c, $E_0=10$) due to saturation broadening the effective amplitude of the frequency modulation is smaller than in linear regime. That is why the relative amplitude of the output intensity modulation is much smaller than in the previous cases. The same reasons explain why during the defocusing half-periods there is only a decrease of intensity due to the beam divergence (and no peak due to the decrease of absorption), while the positive induced lens, naturally, causes the growth of intensity. In the focusing area intensity peak are increased (Fig.1b). If carrier frequency is detuning in defocusing area affect of self-focusing is fully compensated, and at offset into the focusing region smoothing the influence of defocusing.

When the amplitude of frequency modulation to a value outside of the absorption line begins to manifest itself transient optical nutation, which is a consequence of the delayed response of the medium to the acting field (Fig.2).

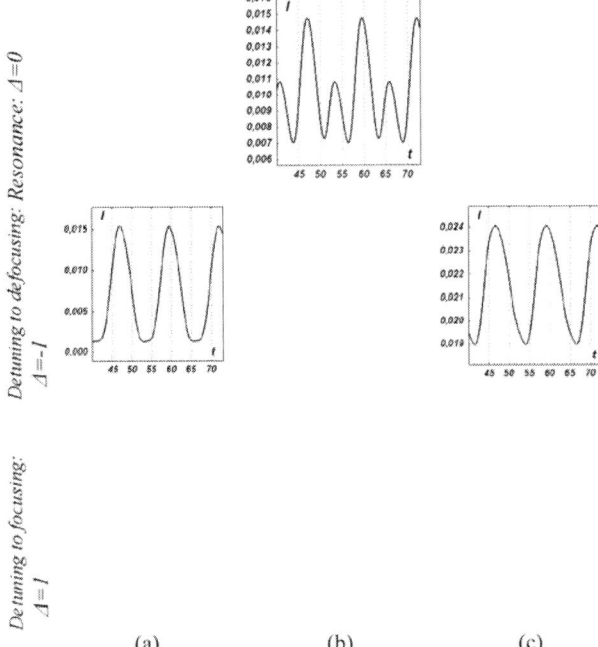

(a)   (b)   (c)

Fig.1. Output on-axis intensity I versus time at low modulation frequency $\Omega=0.5$ and low modulation amplitude $\omega_1=1$: (a) – no saturation ($E_0=0.1$), (b) – moderate saturation ($E_0=5$), (c) – strong saturation ($E_0=10$).

In linear regime (Fig. 1a, $E_0=0.1$) the modulation of the output intensity is harmonic-like, its frequency being twice the modulation frequency. This follows from the fact that the symmetric shifts of the laser frequency to both sides of the

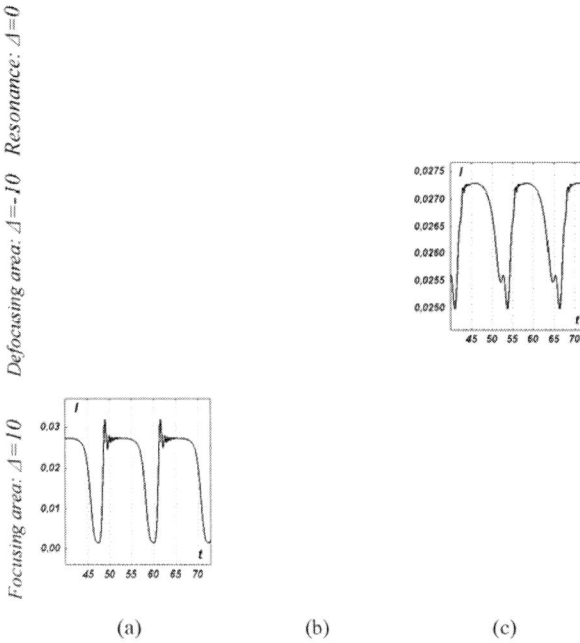

(a)   (b)   (c)

Fig. 2. Manifestation of optical transient nutation at high modulation amplitude $\omega_1=10$: (a) – no saturation ($E_0=0.1$), (b) – strong saturation ($E_0=10$), (c) – extra strong saturation ($E_0=20$).

Figure 2 demonstrate the optical nutation manifestations in exact resonance with the carrier frequency and its offset by the value of the modulation amplitude $\omega_1=10$. In this case the resonance point is passed only once. Thus, at the exact resonance the laser beam frequency twice crosses the line absorption, causing not only a doubling of the frequency modulation, but more distinct optical nutation signal (Fig.2, top). When the carrier frequency detuning from resonance so that the modulated beam frequency is in resonance only once, there is a significant decrease in the amplitude of the oscillations, which can be explained by less sharp transition to a region of strong absorption followed by a return to resonance (Fig.2, middle and bottom).

The nature of the damped oscillations is explained the oscillations of the polarization P (Fig.3b,d,f) caused by transition of particles from one level to another. In case of strong saturation (E_0=20), when it is possible to achieve the alignment level populations, displays optical nutation are smoothed by reducing the amount of absorption that promotes smoother fluctuations in intensity (Fig. 3c) with the constant of polarization behavior (Fig. 3f). We see that the nature of the oscillations is independent of the level of saturation - change only the value of the population difference jumps.

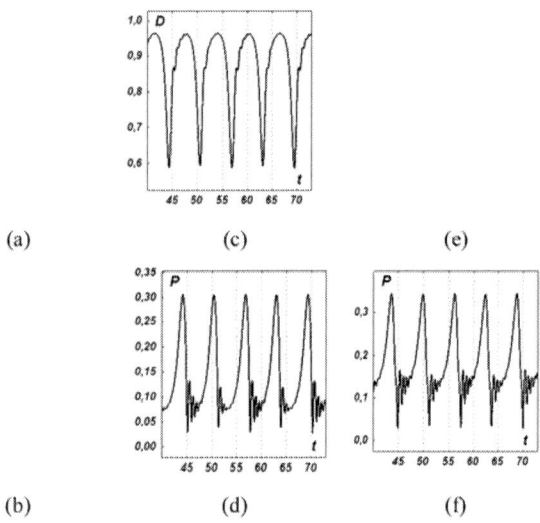

Fig.4. Oscillations of population difference D and polarization P caused optical nutation at high modulation amplitude ω_1=10: (a,b) – no saturation (E_0=0.1), (c,d) – strong saturation (E_0=10), (e,f) – extra strong saturation (E_0=20).

Thus, as a result of the study it can be concluded that the nature of the optical transient nutation related primarily with a sharp jump of the populations, causing a jump in absorption occurring at the approach to the resonance frequency of the laser beam variable. Increased intensity of the acting field leads to an increase of the saturation effect, which gradually reduces the absorption at frequencies near the resonance. In turn, a lower level of absorption reduces the manifestations of optical nutation with strong saturation, causing the population inversion levels, which occurs when the input amplitude of the field exceeds the saturation threshold is more than 10 times.

REFERENCES

[1] Y.R. Shen *The principles of nonlinear optics* A Wiley-Interscience Publication, John Wiley&Sons, Inc. N.Y., 1984

[2] R.G. Brewer, R.I. Shoemaker "Optical Free Induction Decay" *Phys. Rev.A.,* vol.6, no.6, pp.2001-2007, 1972.

[3] G. Duxbury, JF. Kelly, T.A. Blake, N. Langford "Observation of infrared free-induction decay and optical nutation signals from nitrous oxide using a current modulated quantum cascade laser". *J Chem Phys,.* vol.136, no.17 pp.174-181, 2012.

[4] V. Shim, S.B. Cahn, A. Kumarakrishnan, T. Sleator, J.-T. Kim "Optical Nutation in Cold 85Rb Atoms" *Jpn.J.Appl.Phys,* vol.44, no.1A pp.168-173, 2005.

[5] N. N. Rubtsova, E. B. Khvorostov, D. A. Vorona, V. A. Reshetov "Coherent Control over Optical Nonstationary Processes in Gases" *Laser Phys.,* vol.15, pp.763-768, 2005.

[6] V.L. Derbov, I.L. Plastun, "Resonant self-focusing of periodically modulated laser beams", *Proc. Of SPIE,* vol. 4706, pp.82-87, 2002.

Propagation of Frequency-Modulated CW Laser Beams in Coherent Population Trapping Conditions

A.N. Bokarev, I.L. Plastun

Saratov State Technical University, Saratov, Russia

Abstract: The effect of coherent population trapping is investigated using the numerical simulation. Numerical model includes nine equations. Resolving the equations provides the description of the spatial and temporal dynamics of the two laser beams in a three-level system.

Coherent population trapping (CPT) in three-level Λ-systems is known to produce narrow resonances under the condition of equal frequency detuning of two laser waves from two corresponding adjacent transitions. These resonances appear as sharp dips in the population of the upper state usually monitored via laser-induced fluorescence intensity measurements.

Solution of the nonlinear wave equation [1] is usually the most important part of the modeling:

$$\left[\nabla \times (\nabla \times) + \frac{1}{c^2}\frac{\partial^2}{\partial t^2} \right] \vec{E}(\vec{r},t) = -\frac{4\pi}{c^2}\frac{\partial^2}{\partial t^2}\vec{P}(\vec{r},t), \quad (1)$$

where $\vec{E}(\vec{r},t)$ is the electric field vector, $\vec{P}(\vec{r},t)$ - the polarization vector, $\vec{r}$ - a spatial variable, t - time, c - the speed of light, and ∇ is the del operator. To determinate the polarization on the right-hand side of this equation we need to calculate the density matrix elements ρ_{ij}, which are the solutions to the motion equations for ρ.

In the case of a nonstationary media response we need to resolve the system of three equations: for the pump field, the probe field and the third field, which results from the interaction of the first two fields. To calculate the response we need to resolve the system of six equations for density matrix elements [1]. These procedures require a large amount of computational time and resources. So far such investigations have not been conducted, which makes the given research particularly relevant.

The aim of the work is investigation of coherent population trapping in frequency-modulated CW laser beams which propagate in a nonlinear optical three-level system under the resonance conditions.

We consider the three-level system with energy levels $E_1 < E_2 < E_3$ (figure 1). This system is excited by two laser fields at two frequencies: $\omega_p \approx \omega_{31}$ and $\omega(t) = \omega_s + \omega_1\sin\Omega t \approx \omega_{32}$, where $\omega_{ij} = (E_i - E_j)/\hbar$ is the transition frequency $i \rightarrow j$, ω_s - the carrier frequency of a frequency modulated laser, ω_1 - the amplitude of the modulation, and Ω - the modulation frequency.

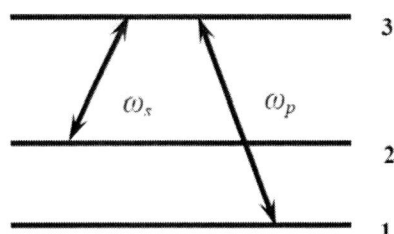

Fig. 1. Λ – level scheme.

To get the description of the spatial and temporal dynamics of the two laser beams in the three-level system we need to resolve nine equations. Three wave equations for E: for the pump field E_p (2), for the probe field E_s (3) and for the induced field E_{p+s} (4), which results from the interaction of the first two fields, and six equations for the density matrix (5)-(10). Dimensionless wave equations (2-4) are obtained from the equation (1) based on the usage of the slowly varying amplitudes approximation [1]:

$$2i\frac{\omega_p}{c}\left(\frac{\partial E_p}{\partial z} + \frac{1}{c}\frac{\partial E_p}{\partial t}\right) + \left(\frac{\partial^2}{\partial r^2} + \frac{1}{r}\frac{\partial}{\partial r}\right)E_p = 4\pi\frac{N_0}{c^2}\left(-2i\omega_p\frac{\partial\tilde{\rho}_{21}}{\partial t}d_{12} - \omega_p^2\tilde{\rho}_{21}d_{12}\right); \quad (2)$$

$$2i\frac{\omega_s}{c}\left(\frac{\partial E_s}{\partial z} + \frac{1}{c}\frac{\partial E_s}{\partial t}\right) + \left(\frac{\partial^2}{\partial r^2} + \frac{1}{r}\frac{\partial}{\partial r}\right)E_s = 4\pi\frac{N_0}{c^2}\left(-2i\omega_s\frac{\partial\tilde{\rho}_{32}}{\partial t}d_{23} - \omega_s^2\tilde{\rho}_{32}d_{23}\right); \quad (3)$$

$$2i\frac{\omega_{p+s}}{c}\left(\frac{\partial E_{p+s}}{\partial z} + \frac{1}{c}\frac{\partial E_{p+s}}{\partial t}\right) + \left(\frac{\partial^2}{\partial r^2} + \frac{1}{r}\frac{\partial}{\partial r}\right)E_{p+s} = 4\pi\frac{N_0}{c^2}\left(-2i\omega_{p+s}\frac{\partial\tilde{\rho}_{31}}{\partial t}d_{13} - \omega_p^2\tilde{\rho}_{31}d_{13}\right), \quad (4)$$

where r is a transverse coordinate, z - a longitudinal coordinate, N_0 - the particles number density, ω_{p+s} - the frequency of the induced field, and c is the speed of beam propagation («the speed of light»).

The state of the three-level quantum system interacting with an external electromagnetic field is described using the system of equations for density matrix elements ρ_{ij} in the approximation of the rotating field [1]:

$$\frac{\partial\tilde{\rho}_{21}}{\partial t} = \frac{1}{i}(\omega_{21} - (\omega_p - \omega_s) - i\gamma_{12})\tilde{\rho}_{21} + \frac{1}{\hbar}((\tilde{\rho}_{31}(V_s^-)_{23} - \tilde{\rho}_{32}^*(V_p^+)_{31}); \quad (5)$$

$$\frac{\partial\tilde{\rho}_{31}}{\partial t} = \frac{1}{i}(\omega_{31} - \omega_p - i\gamma_{13})\tilde{\rho}_{31} + \frac{1}{\hbar}((\rho_{11} - \rho_{33})(V_p^+)_{21} + \tilde{\rho}_{21}(V_s^+)_{32}); \quad (6)$$

$$\frac{\partial\tilde{\rho}_{32}}{\partial t} = \frac{1}{i}(\omega_{32} - \omega_s - i\gamma_{32})\tilde{\rho}_{32} + \frac{1}{\hbar}((\rho_{22} - \rho_{33})(V_s^+)_{32} + \tilde{\rho}_{21}^*(V_p^+)_{31}); \quad (7)$$

$$\frac{\partial\rho_{11}}{\partial t} = \frac{1}{\hbar i}((V_p^-)_{13}\tilde{\rho}_{31} - (V_p^-)_{13}^*\tilde{\rho}_{31}^*) + W_{21}\rho_{22} + W_{31}\rho_{33}; \quad (8)$$

$$\frac{\partial\rho_{22}}{\partial t} = \frac{1}{\hbar i}(\tilde{\rho}_{32}(V_s^-)_{23} - c.c.) + W_{32}\rho_{33} + W_{21}\rho_{22}; \quad (9)$$

$$\frac{\partial\rho_{33}}{\partial t} = \frac{1}{\hbar i}((-\tilde{\rho}_{31}(V_p^-)_{13} + c.c. - \tilde{\rho}_{32}(V_s^-)_{23} + c.c.) - W_{32}\rho_{33} - W_{31}\rho_{33}, \quad (10)$$

where ρ_{ij} is the density matrix element, ω_{ij} - the transition frequency $i \rightarrow j$, γ_{ij} - the transition linewidth $i \rightarrow j$, W_{ij} - the relaxation transition probability $i \rightarrow j$ per unit time, ω_p - the frequency of the pump field, ω_s - the frequency of the probe field, $\hbar$ - reduced Planck constant, c.c. - a complex conjugate of the previous term; $(V_{p,s}^+)_{ij} = -d_{ij} E_{p,s}^0$, $(V_{p,s}^-)_{ij} = -d_{ij} E_{p,s}^{0*}$ are the slowly varying envelopes of interaction operators, where d_{ij} is the dipole moment of $i \rightarrow j$ transition, $E_{p,s}^0$ are the complex amplitudes of the fields; $\tilde{\rho}_{21} = \rho_{21}/\exp(-i(\omega_p - \omega_s)t)$, $\tilde{\rho}_{31} = \rho_{31}/\exp(-i\omega_p t)$, $\tilde{\rho}_{32} = \rho_{32}/\exp(-i\omega_s t)$, and t is time. The given system includes all the transition linewidths associated with the transverse relaxation time, and all the probabilities of spontaneous transitions, which are responsible for the relaxation of the populations. All the units of frequency in the equations (2-10) are normalized to the rate of relaxation.

For the numerical solution of the equations we used the implicit difference scheme of the second order, based on the method of splitting with the expansion of the field in the transverse coordinate by the Gauss-Laguerre modes [2]. The equations (5)-(10) were calculated using the Fourth-order Runge-Kutta method. As a result of resolving the equations (5)-(10) we have the diagonal elements of the density matrix ρ_{11}, ρ_{22}, ρ_{33}, which present the populations of levels 1, 2 and 3, and the nondiagonal elements ρ_{21}, ρ_{31}, ρ_{32}, which present the coherence contributing to the polarization of the media P.

We have conducted numerical experiments referring the effect of coherent population trapping (CPT) with the following input parameters: $E_p = 0,5$, $\omega_{21} = 4$, $\omega_{31} = 6$, $\omega_{32} = 2$, $d_{ij} = 1$, $\gamma_{31} = \gamma_{32} = 1$, $\gamma_{21} = 0,0001$, $W_{31} = W_{32} = 0,5$, $W_{21} = 0,0001$, $\omega_p = 6$, $\omega_s = 2$. The results of the numerical experiments are shown in Figures 2 and 3. The population $\rho_{33}(t)$ sharply falls when the frequency detuning passes through zero and gradually decreases between zeros. This shows the CPT effect (Figure 2).

In the case of saturation ($E_p = 5$) and an exact resonance half-cycles of modulation become unequal because the induced lens is positive with a frequency above the atomic transition and negative with a frequency below this transition. Therefore, increasing of intensity due to weak absorption at a frequency below a resonance is smoothed out by defocusing (Figure 3).

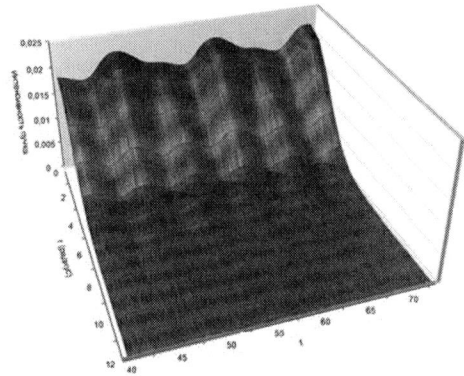

Fig. 3. Transverse profile evolution of the probe field in the case of strong saturation and an exact resonance of the carrier frequency.

With increasing of the frequency modulation, the effect of coherent population trapping gradually decreases and disappears, while the oscillation of the polarization becomes smaller.

References

[1] Shen, Y.R. [The principles of nonlinear optics] A Wiley-Interscience Publication, John Wiley&Sons, Inc. N.Y., (1984)

[2] Plastun I.L., Misurin A.G. Resonance Self-Action and Optical Transient Nutation in Frequency-modulated cw Laser Beams //Proceedings SPIE, Vol.8699, 869915-1-869915-2 (2013).

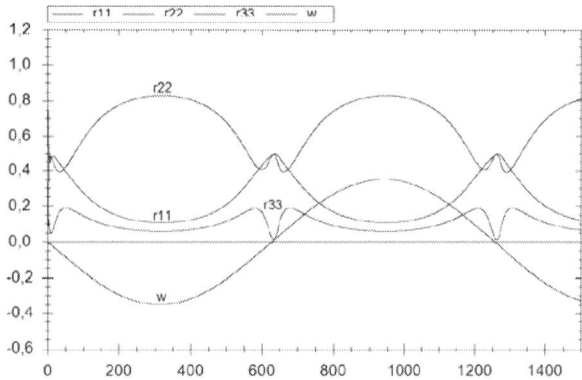

Fig. 2. Time dependence of the elements ρ_{11}, ρ_{22}, ρ_{33}.

The new laser media of dipyrromethene complexes with boron fluoride

[1]R.T., Kuznetsova, [1]Yu.V.Aksenova, [1]T.A.Solodova, [1]T.N.Kopylova, [1]E.N.Telminov, [1]G.V.Mayer,
[2]M.B.Berezin, [3]A.S.Semeikin, [4]S.M.Arabei, [5]T.A.Pavich, [5]K.N.Soloviov

[1]National Research Tomsk State University, 634050, Tomsk
[2]Institute of Solution Chemistry RAS, 153045, Ivanovo
[3]Ivanovo State University of Chemical Technology, 153000, Ivanovo
[4]Belorussian State Agricultural Technical University, 220023, Minsk
[5]Stepanov Physics Institute NANB, 220072, Minsk

Abstract: The characteristics of new active media for tuning lasers are presented in liquid and solid state.

I. Introduction

Tuning lasers on the base of organic compounds, which have broad band lasing spectra (up to 15-20 nm) are using by creation of optical devices and as sources of excitation at the spectroscopic instrument-making industry [1]. Boron fluorides of dipyrromethenes (BODIPY) exhibited intensive fluorescence on the visible region and good photostability, therefore some assortment is produced by commercial firms (Aldrich, Molecular Probes etc). However at last years assortment of BODIPY is expanded by syntheses of new structure [2, 3] for purposeful creation of optical devices with specify characteristics.

II. Objects and Methods

The photonics and lasing properties of new derivatives of BODIPY, which were syntheses in Ivanovo State University of Chemical Technology and Institute of Solution Chemistry with the known from the literature [4, 5] and by own methods [6, 7] were researched. The structures of researched compounds and their notation are presented on the fig.1. These notations are not universal adopted and applied only in this paper. Ethanol, ethylacetate (polar) and cyclohexene (nonpolar) were used as solvents. For some compounds were syntheses solid polymer films in corporation with scientists from Belarus on silica base doped by BODIPYs as shown for other dyes in [8]. Unfortunately, the quality of films is not high, is necessary to increase the homogeneity of films in order to improve emission characteristics.

Spectral-luminescent properties were studied with spectrometers CM2203 (SOLAR, Belarussia) and Cary Eclipse (Varian). Lasing characteristics were determined under excitation the second (532 nm, τ_{puls}=15 ns, E=70 mJ/pulse) and third (355 nm, τ_{puls}=9 ns, E=40 mJ/pulse) harmonics of Nd:YAG laser (SOLAR, Belarus). The energy characteristics were measured by OPHIR NOVA (Israil') and Gentec E100 (Canada), lasing spectra – with spectrometer AVANTES. The transverse scheme of excitation was used with length of cavity 2 cm. The BODIPY coloured thin films have lasing without external cavity. Emission get off 2 directions, efficiency was estimated as relation of lasing energy to absorbed pump energy without transmission. The variation of pump energy was produced with neutral light filters. Characteristics of photostability were determined by methods developed of authors [7, 8].

III. Results and discussion

Fluorescence: The results of spectral-luminescent, photochemical and lasing characteristics investigations are presented at the Table. There dates are shown that increase of number of alkylsubstitutes at 1–3 and 5–7 positions –it is BODIPY №1– BODIPY №5 - the slight long-wave shift of spectra and rising of quantum yields of fluorescence are produced. Skewness of methylsubstitution structure slightly violates this regularity. The most long-wave shift and efficiency fluorescence was observed for tetraphenyl substitution at 1, 3, 5, 7-positions (BODIPY №6) which has fluorescence with maximum 600 nm and quantum yield γ=1 (Table).

BODIPY №8 has the doubled chromophore of alkylsubstituted BODIPY which has at 6-position the spacer with CF$_3$-group. The spectral range is the same almost, however absorption intensity is increase.

Substitution BODIPYs at 4-position exhibits the short-wave spectral shifts. The significance of shift is determined of the substitute structure and is increase in row: BODIPY №3<BODIPY №4 < BODIPY №7. For example, the absorption maximum of BODIPY №3 is 528 nm, of BODIPY №4 – 522 nm and of BODIPY №7 –406 nm in ethanol [9], the efficiency of fluorescence is high (Table). The fluorescence exhibits the similar shifts. These great shifts are determined on the absent of alkylsubstitution in 1-3, 5–7 positions and introduction of propargylamino-group at 4-position. The reasons of such influence of substituted amino-group at 4-positions are ambiguous and require future researches, since introduction of amino-group at other positions of BODIPY-core do not produce short-wave shifts. As to practical applications, these results show that different substituted BODIPY are effective fluorophores for broad visible region 450-620 nm.

Coming from polar solvent (ethanol, ethylacetate) to nonpolar cyclohexene and to solid matrix show long-wave shift of absorption and fluorescence (up to 17 nm). The small Stokes

978-1-4799-0019-0/13 $31.00 © 2013 IEEE

shifts $\Delta\nu_{ss}$ =200-400 cm $^{-1}$ are tipical for alkyl-derivatives of BODIPY.

Very high influence of solvent on photonics of *bis*-BODIPY is observed. The quantum yield of its fluorescence is decreased by coming from cyclohexene to ethanol on 16 times however shifts are small. At frozen ethanol efficiency is increased to 1. This property is tipical for all complexes with dipyrromethenes and may be used for creation of differ optical devices: ethanol solutions of *bis*-BODIPY are recommend for limiters of power pulse UV-radiation and temperature marks on region 250-100K [7]. Cyclohexene solutions are good laser media (Tabl.).

Laser properties: Compound №3–№6 and №8 have good absorbance on 532 nm therefore were excited by radiation of the second harmonic at S_0-S_1 transition and in some cases of the third harmonic of Nd:YAG-laser, which excites these BODIPYs at S_2 and S_3 states.

The lasing properties of BODIPY 5 in cyclohexene were determined by excitation on 532 nm also (Tabl). This compound has lasing with maximum on 556 nm and exceed on efficiency of commercial BODIPYs, which are presented in works of A.Costela. BODIPY №5 solution has lasing with low threshold <<1 MW/cm^2, under intensity excitation =1 MW/cm^2 lasing efficiency consists 43 %. By introduction of BODIPY №5 in polymer material on the base of nanostructured silica and formation of film with thickness to 40 µm by spin-coated method lasing spectra is shifted at long-wave region on 12-15 nm (Tabl). This compound has high lasing efficiency (up to 17% in films and 76 % in solution), high photostability: quantum yields of phototransformations in cyclohexene consist $3\ 10^{-5}$ and high resource of solid thin-film laser: R_{90}=4.2 KJ/cm^3 under 40 MW/cm^2 pumping. It is mean, that before decrease of initial efficiency on 10 % (up to Eff/Eff$_0$=90%) as results of phototransformations laser media absorbed 4.2 KJ/cm^3 pumping radiation (532 nm). This parameter for PM567 solutions consist 570 J/cm^3 [8] and for BODIPY №5 in cyclohexene=650 J/cm^3. It is explained of the possibility of intermediated photoproduct geminate recombination in solid matrix which is absent in modify under excitation solvent core in solution. In solid matrix most excited molecules return to the ground state with possibility future participation in emission. It is necessary to point that decrease of pump intensity from 40 up to 10 MW/cm^2 increase the number of lasing pulses, however do not increase the summary absorbed pump energy which characterized the same resource. These results allow to recommend BODIPY №5 as the base for thin-film lasers active media for region 560-575 nm with high efficiency and resource. The increase of the quality of films must be to improve laser properties.

BODIPY №6 is the most long-wave dye from lasing BODIPYs (λ_{las}=601nm) and has the same feature in polymer films (Tabl). The more low efficiency compare with other BODIPYs is connected with transient absorption in lasing region, probably.

BODIPY №1, BODIPY №2 and BODIPY №7 have not absorption on 532 nm, therefore they were excited by the third harmonics only. The most short-wavelength laser dyes from

reserched BODIPYs is propargyl-amino substituted BODIPY №7: $\lambda_{las}{}^{max}$=476 (tabl). The efficiency consist 38% under W=10-20 MW/cm^2 and exceeds this parameter got by Costela. This compound has lasing in ethanol and ethylacetate solution and in thin film (tabl), however photostability and resource characteristics are essentially lower compare with other derivatives of BODIPY which are excited by the second harmonics (532 nm).

BODIPY №1 and BODIPY №2 have lasing at 535-550 nm under third harmonics excitation also (Tabl). These compound have more high photostability and resource compare with BODIPY №7, in spite of they are excited on S_2 state.

The presented results demonstrated the possibility of creation of active media for tuning lasers at the region 470-620 nm with the differ derivatives of BODIPY. The exhibited properties required future investigations for determination of phototransformation mechanism and improve of resource characteristics especially for BODIPY №7.

IV. ACKNOWLEDGMENT

This work is supported by RFBR (grant № 02-12-90008_Bel -a)

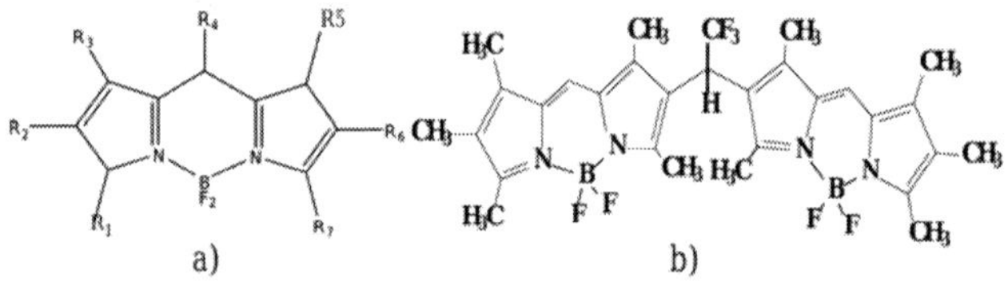

Fig.1. Structure formulae and notation of researched BODIPY:
a): BODIPY №1: $R_1=R_2=R_3=R_4=H$; $R_5=R_6=R_7=CH_3$; **BODIPY №2:** $R_1=R_3=R_5=R_7=CH_3$; $R_2=R_4=R_6=H$;
BODIPY №3: $R_1=R_3=R_5=R_7=CH_3$; $R_4=H$; $R_2=R_6=C_2H_5$;
BODIPY №4: $R_1=R_3=R_5=R_7=CH_3$; $R_4=Ph$; $R_2=R6=C_2H_5$;
BODIPY №5: $R_1=R_3=R_5=R_7=CH_3$; $R_4=H$; $R_2=R_6=CH_2-Ph$;
BODIPY №6: $R1=R_3=R_5=R_7=Ph$; $R_2=R_4=R_6=H$; **BODIPY №7:** $R_1=R2=R_3=R_5=R_6=R_7=H$; $R_4=NH-CH_2-C\equiv CH$; **b): BODIPY №8:** *bis*-BODIPY

Table. Spectral-luminescent, lasing and photochemical properties of researched BODIPY

Compound, solvent	$\lambda_{abs}{}^{S0-S1}$, nm (ε,M^{-1}cm^{-1})	λ_{fl}, nm (λ_{ex})	$\gamma_{fl}\pm10\%$ (λ_{ex} nm)	λ_{las}, nm (λ_{ex},W$_{pump}$ MW/cm^2)
BODIPY №1, ethanol	504 (57400)	518	0.7 (480)	547 (355, 15)
BODIPY №2, cyclohexen	509 (70000)	516	1.0 (460)	
BODIPY №3, ethanol	528 (57400)	545 (475)	0.82 (475)	560 (532, 25)
BODIPY №3, filmPOSS-poly	535	544 (470)		562 (532, 15)
BODIPY №4, ethanol	522 (72650)	538 (470)	0.83(470) 0.18(330)	551 (532, 25)
BODIPY №5, cyclohexen	531 (119380)	539 (500)	0.98(500)	557 (532)
BODIPY №5, filmPOSS-pol	534	556		569 (532,46)

978-1-4799-0019-0/13 $31.00 © 2013 IEEE

Compound, solvent	λ_{abs}^{S0-S1}, nm (ε, M^{-1}cm^{-1})	λ_{fl}, nm (λ_{ex})	$\gamma_{fl}\pm10\%$ (λ_{ex} nm)	λ_{las}, nm (λ_{ex}, W$_{pump}$ MW/cm^2)
BODIPY №6, ethanol	565 (17515)	599 (520)	1 (520)	601 (532, 40)
BODIPY №6, cyclohexen	568 (28500)	601 (550)	0.9 (520)	601 (532, 40)
BODIPY №6, filmPOSS-poly	570	605 (550)		605, 612 (532,40)
BODIPY №7, ethylacetate	409 (37390)	470 (370)	0.9 (370)	477 (355,10)
BODIPY №7 filmPOSS-poly	416	490 (370)		493 (355,15)
bis-BODIPY, ethanol	537 (101000)	547	0.06(480)	no lasing
bis-BODIPY, cyclohexen	542 (78500)	546 (480)	0.97(480)	567(532)
bis-BODIPY,f TEOS+VTEOS	539	548		no lasing

Compound, solvent	Eff$_{las}$%, (W$_{pump}$, MW/cm^2)	$\varphi_{phot}\times10^{-5}$ ($\lambda_{ex,nm}$) [R$_{98}$J/cm^3]
BODIPY №1, ethanol	26 (355, 15)	<6(355) [112]
BODIPY №2, cyclohexen	15(355, 20)	22(355) [15]
BODIPY №3, ethanol	74 (532, 25)	7 (532)[500]
BODIPY №3, filmPOSS-poly		
BODIPY №4, ethanol	56 (532, 25)	4(532) [1800]
BODIPY №5, cyclohexen	76 (532,40)	4 (532)[650]
BODIPY №5, filmPOSS-poly	17(45)	[4200]
BODIPY №6, ethanol	8.5 (532, 40)	
BODIPY №6, cyclohexen	7.8 (532, 40)	
BODIPY №6, filmPOSS-poly		
BODIPY №7, ethylacetate	38 (10)	<450 (355)[7]
BODIPY №7 filmPOSS-poly		
bis-BODIPY, ethanol		47(355) 0.035(532
bis-BODIPY, cyclohexen	1(355) 14(532)	0.26(532)
bis-BODIPY, film TEOS+ VTEOS		

V. References

1. Shankarling G.S., Jarag K.L. "Laser Dyes" *RESONANCE*. 2010. September. P.804-818.

2. Benstead M., Mehl G.H., Boyle R.W. "4,4'Difluoro-4-bora-3a,4a-diaza-s-indacenes (BODIPYs) as components of novel active materials" *Tetrahedron* (2011) doi:1016/j.tet.2011.03.028.

3. Loudet A., Burgess K. "BODIPY dyes and their derivatives: syntheses and spectroscopic properties" *Chem.Rev.* 2007. V.107. P.4891-4932.

4. Teets Thomas S., Partyka David V., Updegraff James B. III, Gray Thomas G. "Homoleptic, Four-Coordinate Azadipyrromethene Complexes of d^{10} Zinc and Mercury" *Inorg. Chem.* 2008. Vol. 47. N. 7. P. 2338.

5. Gomez-DuranC.F.A., Garcia-Moreno I., Costela A., Martin V., Sastre R., Banuelos J., Lopez-Arbeloa F., Lopez-Arbeloa I., Pena-Cabrera E. "8-Propargylamino BODIPY: unprecedented blue-emitting pyrromethene dye. Synthesis, photophysical and laser properties" *Chem.Com.* 2010. V.46. P.5103-5105.

6. Berezin M.B., Semeikin A.S., Antina E.V., Pashanova N.A., Lebedeva N.Sh., Bukushina G.B. "Synthesis and physics-chemical properties of alkylsubstitutute hydrobromides of dipyrrometgenes" *Journ. Organic Chemistry (Russia)* 1999. V.69. №12. P.2040-2041.

7. Kuznetsova R.T., Aksenova Yu.V., Orlovskaya O.O., Kopylova T.N., Telminov E.N., Mayer G.V., Antina E.V., Yutanova S.L., Berezin M.B., Guseva G.B., Antina L.A. Semeikin A.S. "The study of photoprocesses at coordinational compounds of Zn(II) and B(III) with openly-chain olygopyrroles for using in optical devices" *High Energy Chemistry (Russia)* 2012. V.46. №6. P.464-475.

8. Kuznetsova R.T., Mayer G.V., Manekina Yu.A., Telminov E.N., Arabei S.M., Pavich T.A., Soloviov K.N. "The active media on the base of silica matrices for tuning solid-state lasers" *Quantum Electronics*. 2007. V.37. №8. P.760-765.

9. Banuelos J., Martin V., Gomez-Duran C.F.A., Cordoba I.J.A, Pena-CabreraE., Garcia-Moreno I., Costela A., Perez-Ojeda M.E., Arbeloa T., Lopez-Arbeloa I. "New 8-amino-BODIPY derivatives: suprassinglaser dyes at blue-edge wavelengths" *Chem.Eur.J.* 2011. DOI:10.1002/chem.201003689

Photophysical properties of laser active elements based on dyes in new aliphatic polyurethane matrix

T.V. Bezrodna[1], L.F. Kosyanchuk[2], A.M. Negryiko[1], M.S. Stratilat[2], T.T. Todosiichuk[2]

[1]Institute of Physics NAS of Ukraine, Kyiv, Ukraine

[2] Institute of Macromolecular Chemistry NAS of Ukraine, Kyiv, Ukraine

Abstract: A new polymer matrix based on aliphatic polyurethane was proposed and investigated for the development of laser active media for the tunable dye lasers. The production method of the laser active media based on xanthene and pyrromethene dyes was described; their spectral, photophysical and generation characteristics were investigated. The main reason for the high photostability and operation lifetime of laser polyurethane-based active elements was shown to be the absence of reactive radicals.

To develop the dye-based laser elements, various solid-state matrices, such as polymers, porous glasses, polymer-doped porous glasses, and sol-gel materials, are used [1-5]. Among the listed matrices, the pure polymer ones are especially distinguished by their diversity, availability, cheapness, ability for easy dye doping, and material homogeneity. Polymer matrix should provide sufficient beam strength, transparency, high solubility of dyes, and dye stability during storage and operation. In particular, high laser damage threshold is observed for the polyurethane (PUA) elastic matrix [6, 7].

The aim of the present work is investigations of the probability for the aliphatic polyurethane (APU) to be used as a polymer matrix in the development of dye-doped active laser media. This study presents the data for the main operation parameters of active elements, produced on a base of similar by physical properties polymer matrices, APU and PUA.

This work deals with the laser dyes (Exciton, Inc.), which are known as perspective in different polymer matrices, namely, widely used Rhodamine 6G (Rh6G), and also efficient pyrromethene dyes, PM567 and PM597.

Aliphatic polyurethane matrix was synthesized by the prepolymer method. The prepolymer, produced from hexamethylene diisocyanate and oligodiethyleneglycol adipinate of 800 molecular mass, was hardened by trimethylolpropane, followed by a formation of the APU with the following chemical structure:

The samples under study were triplexes (glass – polymer – glass); a dye-doped polymer layer was located between optical substrates, made of K8 glass or KB quartz. The analysis of transmission spectra recorded for polymer samples of 1 mm thickness has been shown that APU possesses high transparency in a wide spectral region 260 – 2200 nm, covering nearly the whole range of dye laser operation. The short-wavelength absorption edge, presented in Fig. 1(1) indicates a possibility for a use of APU-based matrices in the production of dye-doped active elements, pumped by classical coherent sources. The APU testing on the beam strength has showed that a threshold for its laser single-pulse damage, E_d, at pulse duration of 10 ns and an irradiation zone diameter of 200 μm is determined to be above 18 J/cm^2, which is two times larger than that of PUA. The increased beam strength of APU is caused by high flexibility of the polymer macromolecules due to hexamethylene chains of APU diisocyanate and oligoester parts.

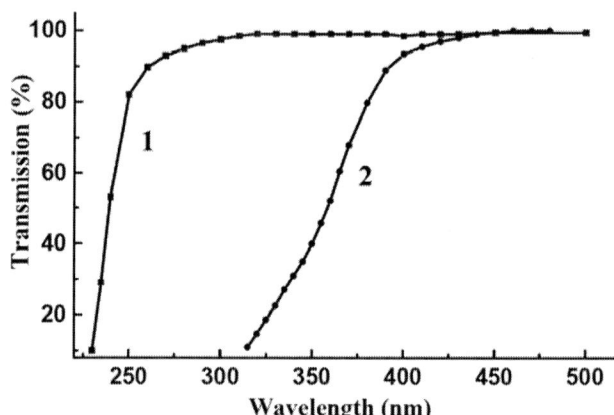

Fig. 1. The short-wavelength region of absorption spectra for the polymer matrices: 1 – APU, 2 – PUA.

Absorption and luminescence spectra of Rh6G, PM567 and PM597 dyes in the APU polymer matrix are presented in Fig. 2. The absorption and luminescence are practically the same as in the case of alcohols, except a bathochromic shift, typical of solid-state matrices. Luminescence spectra do not significantly broaden, which indicates no aggregation of dye molecules.

978-1-4799-0019-0/13 $31.00 © 2013 IEEE

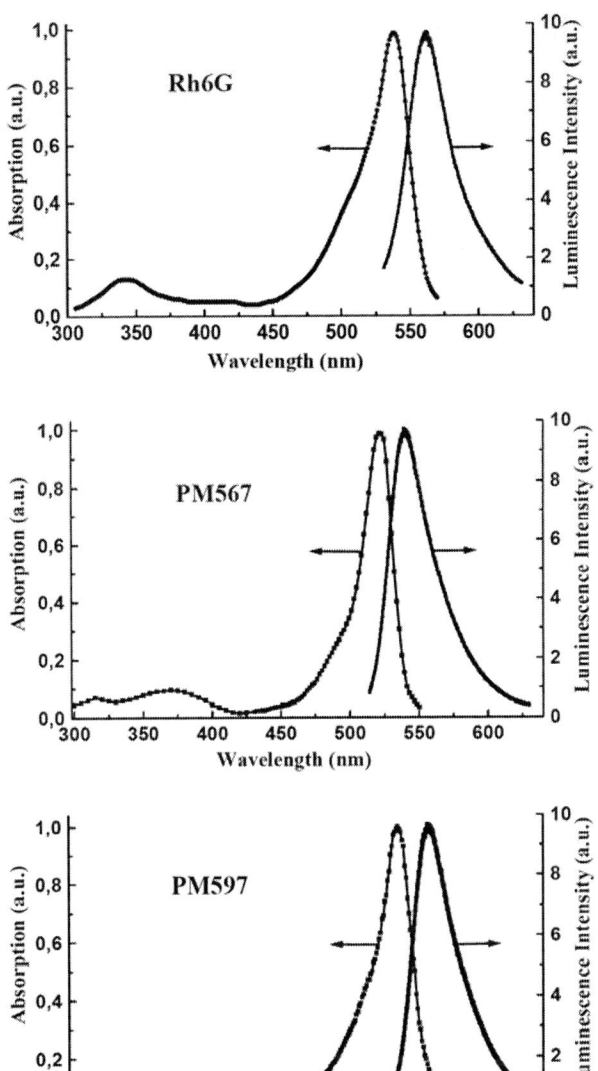

Fig. 2. Absorption and luminescence spectra of the dyes: Rh6G, PM567, PM597 in aliphatic polyurethane.

Photostability is an important operational parameter. To investigate photostability mainly for the $S_0 \rightarrow S_2$, S_3 transitions, the triplexes have been produced with substrates made of quartz glass. The thickness of dye-activated APU and PUA polymer films is 300 μm in the photostability experiments. Optical density, D, in the maximum of a basic transition is equal to about 1 for all samples. Irradiation of the samples is carried out by the light from an arc mercury lamp DRK-120 of ultrahigh pressure. The generation spectrum of such lamps has a nearly continuous character with high intensity in the spectral range $300 – 450$ nm, containing $S_0 \rightarrow S_2$, S_3 transitions. Quantitative data on photostability are listed in Table 1.

Table 1. Photostability of dyes in PUA, OUA and APU matrices.

Dye-doped samples	Irradiation dose by a DRK-120 lamp, causing decrease by two times of D_0, kJ/cm^2	Irradiation dose by a continuous laser (λ=532 nm), causing decrease by two times of D_0, kJ/cm^2	Decrease in dye initial concentration at the polymerization reaction, %
Rh6G in PUA	1.01	0.63	12
Rh6G in OUA	2.89		
Rh6G in APU	8.60	2.00	0
PM567 in PUA	0.06	3.20	19
PM567 in OUA	7.14		
PM567 in APU	12.43	12.28	0
PM597 in PUA	0.12	1.51	17
PM597 in OUA	21.23		
PM597 in APU	25.19	6.17	0

The photodecay of PM567 and PM597 in PUA is seen to occur quite rapidly. Photochemical instability of these dyes, and also Rh6G is resulted from a reaction between electron-excited dye molecules and radicals (macroradicals) of the matrix. To prove this statement, the photostability investigations of dyes in oligourethane acrylate (OUA) have been carried out under the same conditions and irradiation by a DRK-120 lamp. The samples were produced as previously, except the photoinducer adding. The multiple increase of dye photostability is observed in the dye-doped samples based on OUA, compared to those ones with PUA. These investigations have demonstrated a crucial role of a presence in the first case or an absence in the second one of the radicals in the polymer matrix.

Irradiation by a DRK-120 lamp of the polymer samples based on aliphatic polyurethane shows a very low photodegradation degree of Rh6G, PM567 and PM597 dyes, as in the case of the OUA matrix.

Photostability has been also investigated on irradiation by second harmonic continuous generation of a neodymium laser (532 nm) solely in the basic electron transition $S_0 \rightarrow S_1$ (532нм). Quantitative values of the irradiation dose, causing a decrease in two times of the initial optical density, D_0, are presented in Table 1. These results demonstrate a systematic significant growth of the dye photostability at a transition from PUA to APU.

During the synthesis process of the active elements, based on the PUA polymer matrix, radical polymerization is found out to decrease initial concentration of the dye (before solidification) on a value, listed in Table 1.

The regular increase in photostability, observed for the dyes in APU in the comparison with PUA is resulted from the absence of dye molecule decay in APU. Free radicals, formed during the PUA synthesis process induce further activation of radical reactions at the interactions with light, and finally, lead to the destruction of dye molecules. The advantage of APU is not only in the method of its production, but also in its high polarity. Since the dielectrical permittivity of APU is much higher (ε=8.8), than that of PUA (ε=4.1), ionic dyes dissociate into APU much better. Detachment of the counterions in APU is facilitated by a presence of a large amount of nucleophilic functional groups. These groups realize nucleophilic solvation, which decreases the probability of the contact ion pair formation, in the comparison with the polymers of low ε. This is correct not only for the ionic Rh6G, but also for the intraionic dyes, such as PM567 and PM597. Pyrromethenes consist of chemically bonded contact pairs, which cannot dissociate at any polarity of the polymer. But the charges in the mentioned pairs are quite well divided due to large electro-negativeness of fluorine atoms, and an intermolecular phototransition of the electron from boron to nitrogen is more complicated, than in the chemically non-bonded contact pair of the cationic dye Rh6G, where the counterions can approach or move away from each other in a dependence on the polarity and solvation of the medium.

Thus, a transition from the polymer matrix, obtained by radical polymerization to the one, formed by means of polycondensation reaction results in the significant increase of photostability. High dielectrical permittivity of the new polymer matrix increases the mentioned parameters, complicating the formation of various associates of organic molecules.

To conclude, the highly elastic polymer matrix based on aliphatic polyurethane has been developed, investigated and tested for the laser dye-doped active elements. This polymer matrix is synthesized by the prepolymer method. The prepolymer, obtained from hexamethylene diisocyanate and oligodiethyleneglycol adipinate of 800 molecular mass, was hardened by trimethylolpropane. The spectral properties of xanthene and pyrromethene dyes in this matrix are the same as in the liquid media (methanol, ethanol). Laser active elements based on APU significantly exceed the ones made of other polymer matrices, which are produced by the radical polymerization method and widely used for such purposes by photostability. The results of this work allow to make a conclusion about good prospects for the applications of the active laser media based on aliphatic polyurethane in the development of tunable dye lasers.

REFERENCES

[1] B.H. Soffer and B.B. McFarland, "Continuously tunable, narrow-band organic dye lasers," *Appl. Phys. Lett.*, vol. 10, no. 10, pp. 266–267, 1967.

[2] O.G. Peterson and B.B. Snavely, "Stimulated emission from flashlamp-excited organic dyes in polymethyl methacrylate," *Appl. Phys. Lett.*, vol. 12, no. 7, pp. 238–240, 1968.

[3] G.B. Altshuler, E.G. Dulneva, I.K. Meshkovskii, L.I. Krilov "Solid-state active media based on dye," *J. Appl. Spectr. USSR,* vol. 36, no. 4, pp. 592-599, 1982.

[4] M.D. Rahn and T.A. King, "Comparison of laser performance of dye molecules in solgel, polycom, ormosil, and polymethyl methacrylate! host media," *Appl. Opt.,* vol. 34, no. 36, pp. 8260–8271, 1995.

[5] A. Weissbeck, H. Langhoff and A. Beck, "Lasing and fluorescence properties of dye-doped xerogel," *Appl. Phys. B,* vol. 61, no. 3, pp. 253– 255, 1995.

[6] M.V. Bondar, O.V. Przhonskaya and E.A. Tikhonov, "Photodecomposition of dyes in a polymer matrix under lasing conditions," *Quantum Electron.*, vol. 19, no. 11, pp. 1415–1418, 1989.

[7] V.I. Bezrodnyi, V.P. Yashchuk, O.A. Prygodjuk "Multiple scattering effect on luminescence of the dyed polymer matrix", *Semiconductor Physics. Quantum Electronics & Optoelectronics*, vol. 7, no. 1, pp. 77-81, 2004.

Influence of the electric field on laser material processing

A.Yu.Ivanov, S.V.Vasiliev

Grodno State University, Ozheshko 22, 230023 Grodno, Belarus

Abstract: It is shown that by varying the external electric field with different polarity from 0 to 10^6 V m^{-1} in the course of laser processing with the mean radiation flux density ~106 W cm^{-2} the change in the evolution features of the plasma torch at the surface of some metals at early stages is quantitative rather than qualitative. At the same time the characteristic size of the target material droplets, carried out from the irradiated zone, becomes essentially (by several times) smaller as the amplitude of the external electric field strength grows, independently of its polarity.

Our aim was to study the influence of electric fields of different strength (from 0 to 10^6 V m^{-1}) on the spatial and temporal evolution of the laser plasma arising under the action of laser pulses with the average power density range ($10^6 - 10^7$ W/cm^2) at the surface of metals and on the mechanisms of formation of the surface relief of the irradiated samples.

The radiation of the GOR-100M ruby laser operating in the free oscillation regime (pulse duration τ ~ 1.2 ms) or rhodamine laser (pulse duration τ ~ 20 μs) passed through the focusing system and was directed through the hole in the electrode onto the sample that served as the second electrode and was mounted in air at a pressure of 10^5 Pa. The radiation spot diameter with sharp edges on the sample was varied in the course of the experiments from 1 to 2 mm. The energy of the laser pulses varied from 5 to 60 J. The voltage was applied to the electrodes from the source wich allowed the voltage variation within 25 kV and its stabilization in the course of the experiment. To study the spatial and temporal evolution of the laser plasma plume in the course of laser radiation action on the sample, we used the method of high-speed holographic motion-picture recording. The interelectrode gap was placed in one of the arms of a Mach-Zehnder interferometer, which was illuminated with the radiation of the ruby laser (λ = 0.694 μm) operating in the free oscillation regime. The pulse duration of the radiation amounted to ~ 400 μs. The transverse mode selection in the probing laser was accomplished using the aperture, placed in the cavity, and the longitudinal mode selection was provided by the Fabry-Perot cavity standard used as the output mirror. The probing radiation after the collimator was a parallel light beam with the diameter up to 3 cm, which allowed observation of the steam-plasma cloud development. The interferometer was attached to the SFR-1 M high-speed recording camera, in which the plane of the film was conjugate with the meridian section of the laser beam, acting on the sample, by means of the objective. The high-speed

camera operated in the time magnifier regime. The described setup allowed recording of time-resolved holograms of the focused image of the laser plasma plume. Separate holographic frames provided temporal resolution no worse than 0.8 μs (the single frame exposure time) and the spatial resolution in the object field ~ 50 μm. The error in the determination of the electron density was ~ 10% and it was governed by the precision with which the shifts of the fringes could be determined in the photographically developed interference patterns. The diffraction efficiency of the holograms allowed one to reconstruct and record interference and shadow pictures of the studied process under the stationary conditions. The shadow method was most sensitive to grad n, so that the nature of the motion of the front of a shock wave outside the laser plasma and of the motion of the plasma jet could be determined from the reconstructed shadow patterns. This gave information on the motion of the shock front and the laser plasma front generated at the surfaces of metal samples. It was found that the nature of the motion of the shock wave front was practically independent of the target material and was governed primarily by the average power density of the laser radiation. The reconstructed interference patterns were used to determine the spatial and temporal distributions of the electron density in a laser plasma plume. The reliability of the results obtained by the method of fast holographic cinematography was checked by determination of the velocity of the front of a luminous plasma jet by a traditional method using slit scans recorded with a second SFR-1M streak camera. To study the surface shape of the crater that appears on the plate, we used the fringe projection method [1].

The experimental results have shown that at any polarity of the applied voltage with positive or negative potential at the irradiated sample with respect to the electrode the topography of the crater is practically identical and is determined by the energy distribution over the focusing spot of the laser radiation.

Figure 1 presents the time dependences of the plasma plume front motion velocity at different directions of the external electric field strength vector, calculated by using the information, obtained by analyzing the temporal variation of the interferograms. It is seen that even when the plasma front reaches the electrode, its velocity not only does not decrease (which is typical for late stages of the laser plasma torch existence [2]), but even increases; this happens both in the presence of the external electric field of any orientation and in the absence of the field. As already mentioned, this is due to the permanent and significant increase in the mass of the material, carried out under the action of laser radiation on the irradiated sample, as well as to the secondary ionization of plasma by laser radiation.

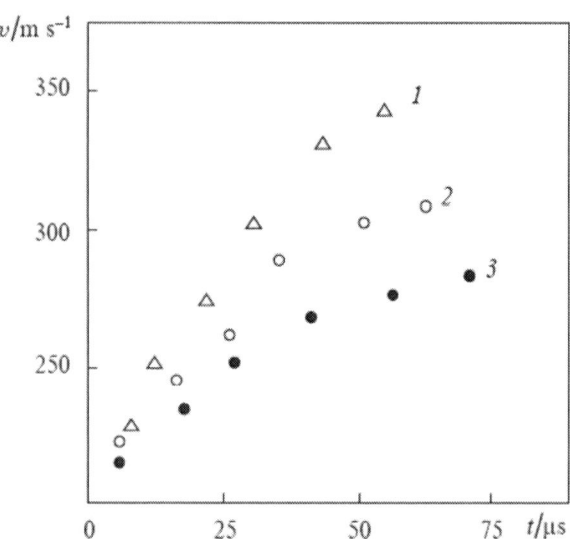

Fig. 1. Time dependences of the velocity of the plasma torch front motion at the negative target potential *(1)*, in the absence of the field *(2)*, and at the positive target potential *(3)*.

The experimental results have shown that at any polarity of the applied voltage with positive or negative potential at the irradiated sample with respect to the electrode the topography of the crater is practically identical and is determined by the energy distribution over the focusing spot of the laser radiation.

Figure 1 presents the time dependences of the plasma plume front motion velocity at different directions of the external electric field strength vector, calculated by using the information, obtained by analyzing the temporal variation of the interferograms. It is seen that even when the plasma front reaches the electrode, its velocity not only does not decrease (which is typical for late stages of the laser plasma torch existence [2]), but even increases; this happens both in the presence of the external electric field of any orientation and in the absence of the field. As already mentioned, this is due to the permanent and significant increase in the mass of the material, carried out under the action of laser radiation on the irradiated sample, as well as to the secondary ionization of plasma by laser radiation.

The maximal expansion velocity of the plasma torch amounted to 350 m/s for the negative voltage at the target, 310 m/s in the absence of the external electric field, and 270 m/s for the positive voltage applied to the target.

Our investigation showed that the time evolution of the leading edge of a luminous plasma moving away from the surface of a sample, deduced from the slit scans, differed from the time evolution of the front of the plasma jet, which was recorded by the shadow method. This allowed us to conclude that the concentration of the heavy particles, responsible for the radiation emitted by the plasma, was low at the front of the laser plasma jet, whereas the electron density was sufficient for reliable determination of the contribution of electrons to refraction in a hologram.

The distribution of the density of cold air was determined and the electron density distribution was refined by two-wavelength holographic cinematography. We supplemented the system described with a second probe laser and an SFR-1M camera which recorded holograms at the wavelength of the radiation emitted by this laser. The second source of probe radiation was a laser utilizing a rhodamine 6G solution excited by a coaxial flashlamp. The use of a standard power supply system from a GOR-100M laser made it possible to generate output radiation pulses of $30 - 40$ μs duration. The line width was reduced employing a plane-parallel Fabry-Perot interferometer. This made it possible to obtain scan holograms of the process at $\lambda_1 = 0.69$ μm and $\lambda_2 = 0.58$ μm, and to separate the contribution of electrons from that of heavy particles to the refraction of a plasma jet.

This two-wavelength holographic cinematography method was used to determine the radial distributions of the electron density and of the heavy-particle concentration at different moments in time and for different sections of laser plasma near the irradiated surface of an irradiated sample

At distances of $10 - 15$ mm from the surface of a sample it was found that heavy particles ("hot" atoms and ions) of metals and molecules of atmospheric gases made only a small contribution to refraction. At large distances (where there were no "heated" luminous particles) the contribution of the cold dense air became significant. This was due to the pushing out of air by a plasma cloud.

When either positive or negative potential is applied to the sample, many small droplets appear on its surface after the laser action. In particular, at the laser pulse energy 20 J, the diameter of the focusing spot 2 mm, and the electric field strength 10^6 V cm^{-1} we observed ejection of droplets having the mean characteristic size less than 0.1 mm to the distance up to 2 cm from the crater centre. The maximal characteristic size of the droplets was 0.4 mm. In the absence of the external electric field the mean size of the droplets was ~ 0.4 mm. The droplets were seen at the distance up to ~ 1 cm from the crater centre.

In accordance with the results presented above, the dynamics of the processes on the surface of a sample, placed in an external electric field with the strength from 0 to 10^6 V m^{-1} and subject to the action of the pulsed laser radiation with the parameters mentioned above, is thought to be the following. The primary plasma formation and the initial stage of the laser torch development, in principle, do not differ from those observed in the absence of the external electric field. The metal is melted and evaporated. As a result of local formation of steam and plasma [3, 4], the erosion torch begins to form with the fine-dispersed liquid-droplets phase. Note, that the bulk evaporation is promoted by the gases, diluted in the metal, and by the spatiotemporal nonuniformity of the laser radiation [4]. At a radiation flux density $10^6 - 10^7$ W cm^2 the bulk evaporation is typical of all metals used in the experiments [4]. Obviously, the presence of the external electric field affects (increases or decreases depending on the direction of the field strength vector) the velocity of motion of the plasma front and causes some distortion of the plasma cloud shape. It is essential that the mentioned differences (at the considered parameters of laser radiation) are observed only at the initial stage of the laser plume development, because after the steam-plasma cloud reaches the electrode an electric breakdown (short-circuit) occurs, and the external field in the interelectrode gap disappears.

Consider now the motion of the molten metal droplets in the steam-plasma cloud. In our opinion, the significant difference in the characteristic size of droplets, observed on the surface of the irradiated sample in the presence of the external electric field (independent of the direction of the field strength vector) and in the absence of the field, is a manifestation of the following mechanism of droplet formation. It is known that at the surface of a liquid (including a liquid metal) the formation of gravity-capillary waves is possible under the action of various perturbations. Using the method presented in [5], one can show that the dispersion equation for the gravity-capillary waves takes the form

$$\omega^2 = \frac{\alpha k^3}{\rho} + g k - \frac{k^2 E_0^2}{4\pi\rho} .$$

where α is the surface tension coefficient of the molten metal; ρ is the metal density; g is the free fall acceleration; k is the magnitude of the wave vector of the gravity-capillary wave. Because the frequency of the gravity-capillary waves ω is determined by the temporal characteristics of the abovementioned perturbations and, therefore, does not depend on the strength of the electric field E_0, the growth of the magnitude E_0 (independent of the direction of the vector $\vec{E}_0$) should cause the increase in the magnitude of the wave vector $k = \frac{2\pi}{\Lambda}$ and the decrease in the wavelength Λ of the gravity-capillary wave. If we assume that the droplets are 'torn away' by the plasma flow from the 'tops' of the gravity-capillary wave and, therefore, their characteristic size is proportional to Λ, then it becomes clear why in the presence of the external electric field (of any direction) the observed mean size of the droplets becomes essentially reduced.

The escaped droplets possess the charge of the same sign as the sample. That is why the droplets begin to move with acceleration towards the second electrode.

However, since the maximal initial velocity of the outgoing droplets under the analogous conditions [6] is ~ 45 m s^{-1}, i.e., an order of magnitude smaller than the velocity of steam-plasma cloud spreading, the droplets do not reach the electrode (3) before the moment of the breakdown in the interelectrode gap. In what follows (in the absence of the external electric field) the droplets move under the action of the same forces as in [6] and, therefore, in the way, described in [8]. In this case, having acquired at the stage of accelerated motion in the electric field the velocity, exceeding the initial one, the droplets may fly to a greater distance along the surface of the irradiated sample than in the absence of the electric field, which is observed in the experiment. Moreover, having moved to a greater distance from the sample surface and, therefore, being affected by the plasma for longer time before returning to the surface, the droplets may be split into finer parts than in the absence of the external field.

REFERENCES

[1] NA. Bosak, S.V. Vasil'ev, A.Yu. Ivanov, L.Ya. Min'ko, V.I. Nedolugov, A.N. Chumakov "Characteristic Features of the Formation of a Crater on the Surface of a Metal irradiated with repeated Laser Pulses", Kvantovaya Elektron., vol. 27, no. 1, pp. 69 – 72, 1999 [Quantum Elelctron., 29, 69 (1999)].

[2] B.A. Barikhin., A.Yu. Ivanov, V.I. Nedolugov "Fast Holographic Cinematography of a Laser Plasma", Kvantovaya Elektron., vol. 17, no. 11, pp. 1477 – 1480, 1990 [Soy. J. Quantum Electron., 20, 1386 (1990)].

[3] V.K. Goncharov, V.L. Kontsovoy, M.V. Puzyrev "Interaction of Laser Radiation with Metals included into alloys based on iron" Inzhenerno- Fizicheskii zhurnal, vol. 66, no. 5, pp. 585 – 589, 1994 [Journal of Engineering Physics and Thermophysics, 66, 588 (1994)].

[4] V.K. Goncharov, V.L. Kontsevoy, M.V. Puzyrev "Influence of different Factors on the Dinamics of Laser Plumes of Vetal Targets", Kvantovaya Elektron., vol. 22, no. 2, pp. 249 – 232, 1995 [Quantum Elelctron., 25, 232 (1995)].

[5] M.I. Rabinovich, D.I. Trubetskov "Vvedenie v teoriyu kolebanii i voln" (Introduction to the Theory of Oscillations and Waves), Moscow: Nauka, 1984.

[6] S.V. Vasil'ev, A.Yu. Ivanov, A.M. Lyalikov "Topography of Crater formed by the Action of a Laser Pulse on the Surface of a Metal", Kvantovaya Elektron., 22, 830 (1995) [Quantum Elelctron., 25, 799 (1995)].

978-1-4799-0019-0/13 $31.00 © 2013 IEEE

Modeling of XeCl excilamps with glow discharge in frequency regime of work

S. S. Anufrik, A. P. Volodenkov, K. F. Znosko

Yanka Kupala State University of Grodno, 22, Ozheshko Street, 230023, Grodno, Belarus,
Fax: +375 (152) 73-19-10; e-mail: a.volodenkov@grsu.by

Abstract: The results of modeling of XeCl excilamps with barrier discharge in frequency regime are considered. Model of XeCl excilamp taking into account process of Cl_2 regeneration is developed. Excilamp efficiency can reach 5 % at use of small values of storage capacity and power of radiation is 3130 W.

SIMULATION METHOD

Modeling of electro-discharge XeCl excilamps is discussed. Generally the computer model includes the following modules.

1) The module of the solving of Boltzmann equation for the electron energy distribution function (EEDF) [1]. This module on composition of a mixture, on value of a degree of ionization and set E/N (E - intensity of an electric field in an interelectrode gap; N - full concentration of particles) allows to find EEDF and accordingly to define rates of plasma-chemical reactions with participation of electrons, and also to define electron mobility.

Rate factors of reactions with participation of electrons were obtained by averaging on EEDF expressions of next type.

$$k = \left\langle \sigma(\varepsilon) \cdot \sqrt{2\varepsilon \cdot /m} \right\rangle \qquad (1a)$$

Rate coefficients are determined as tables in which their dependence on mixture of the active medium, electron concentration n_e, the resulted strength of electric field U/Pd (U - a voltage on an interelectrode gap; d - distance between electrodes; P - the total pressure of a gas mix) is submitted. Calculations have been executed for a discrete file of points. Extrapolation on intermediate values is carried out by means of cubic splines.

For the solving ready program Bolsig + [2], which automatically calculates rate factors of reactions, is used.

2) The module of the solving of system of the equations of plasma-chemical reactions [3]. Models of XeCl-excilamp (halogenide Cl_2) are investigated and analyzed. On the base of these models the program module for solving of system of the equations of plasma-chemical reactions in electro-discharge XeCl-excilamp was developed. In simplest case the next plasma-chemical reactions must be taken into account.

Ionization and excitation

$$Xe + e \rightarrow Xe+ + e + e; \; (ki);$$
$$Xe + e \rightarrow Xe^* + e; \; (ke);$$
$$Xe^* + e \rightarrow Xe+ e + e; \; (ks);$$
$$Xe^* + Xe^* \rightarrow Xe^+ + e + Xe \; (k_{Pen})$$

Dissociative attachment

$$Cl_2 + e \rightarrow Cl^- + Cl; \; (ka,);$$

Production of XeCl-molecules

$$Xe^* + C_2 \rightarrow XeCl^* + Cl \; (k_{harp}) \;;$$
$$Xe^+ + Cl^- + M \rightarrow XeCl^* + M; \; (\beta_{rec})$$

Quenching of XeCl- molecules

$$XeCl^* + e \rightarrow Xe + Cl + e; \; (kq);$$
$$XeCl^* + N \rightarrow Xe + Cl + N; \; (\tau q)$$

Emission of XeCl- molecules

$$XeCl^* \rightarrow Xe + Cl + h\nu; \; (\tau_{sp});$$

The power density of radiation of the excilamps (the power, obtained from unit of volume) is determined by next expression.

$$P = \left| XeCl^* \right| \cdot h\nu / \tau_{sp} \qquad (1b)$$

[XeCl*] is concentration of excimer molecules; τ_{sp} - time constant of spontaneous emission.

3) The module of the solving of the equations of an electric circuit [4]. This module describes work of system of excitation of the volume discharge in active medium. On the total resistance of plasma this module allows to define time dependence of E/P, formed by excitation system in active medium. For excitation of the discharge the system of excitation on the basis of a LC-contour on the lumped capacities has been used (Fig. 1.)

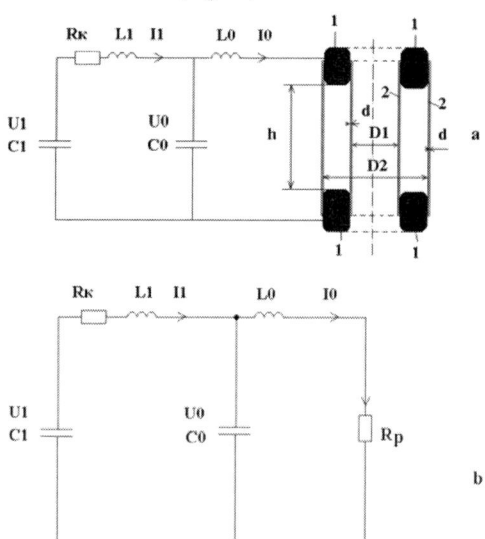

Fig. 1. The system of excitation.

978-1-4799-0019-0/13 $31.00 © 2013 IEEE   148

The following designations were used: $C1$ – the storage capacity; $L1$ – inductance of recharging of the storage capacity on peak capacity $C0$; $Rk(t)$ –switchboard resistance; $Rp(t)$ - discharge resistance; $L0$ - inductance of recharging of $C0$ on discharge. The length of electrodes h=30 cm. The coaxial radiator was used, The value of 1 was 0,2 cm. Distance between electrodes $D1$=6,4 cm $D2$=6,6 cm, it was considered, that the area of cross-section of the discharge is equal S. Value S was considered as parameter which accepts values in a range 2 cm^2. It has been used charging voltage $Uch=U1$=30 kV. Value of storage capacity varied in limits $C1$=0,5 - 1 nF.

Thus at carrying out of concrete calculations it is used the equivalent electric circuit in Fig. 1b.

RESULTS OF COMPUTER MODELING

In Fig. 2 dependences of energy of a pulse of radiation from composition of a mix which have been received under the various conditions specified in signatures to figure are submitted. From dependences, submitted in Fig. 2, follows, the optimum composition of a mix which provides the greatest energy of a pulse of radiation, depends on partial pressure of Cl_2 molecules. So at value of these parameters $C1$=1 nF, $C0$=0,143 nF S=2 cm^2 optimum mix is Cl_2:Xe=1:9, pressure 10 Torr.
:

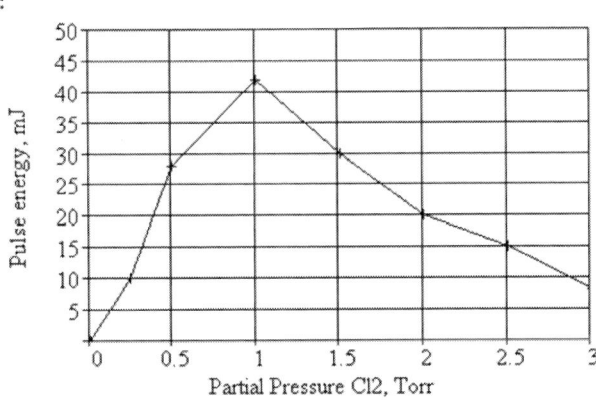

$C1$=1 nF, $C0$=0,143 nF, S=2cm^2;
Fig. 2.Dependence of pulse energy from mix.

These data have been used for definition of power of radiation of excilamp depending on pulse repetition frequency.

At work of excilamp in a pulse-periodic mode during action of a pulse of excitation there is a disintegration of molecules of chlorine, and in an interval between pulses of excitation there is a regeneration of molecules of chlorine. It results to that through some number of pulses in excilamp some fixed partial pressure of Cl_2 is established for the moments of time corresponding to the beginning of a pulse of excitation.

Process of regeneration of Cl_2 molecules is taken into account as follows. Molecule Cl_2 is formed owing to three-partial association by reaction [5].

$$2Cl+Xe=Cl_2+Xe \quad (kCl_2=5 \ 10^{-33} \ cm^6/s) \quad (2)$$

On the basis (1) the following system of the kinetic equations has been written (Xe is used as buffer gas):

$$\frac{d[Cl_2]}{dt}=kCl_2 \cdot [Xe][Cl][Cl], \quad \frac{d[Cl]}{dt}=-2kCl_2 \cdot [Xe][Cl][Cl] \quad (3)$$

The decision of the equations system (3), provided that during the initial moment of time t=0 concentration of atoms of chlorine is equal $[Cl]$ (0) and concentration of molecules of chlorine is equal $[Cl_2]$ (0) has the following kind.

$$[Cl](t)=\frac{[Cl](0)}{1+2 \cdot kCl_2 \cdot [Xe] \cdot [Cl](0) \cdot t} \quad (4a)$$

$$[Cl_2](t)=[Cl_2](0)+\frac{[Cl](0)}{2}\left[1-\frac{1}{1+2kCl_2[Xe][Cl](0)t}\right] \quad (4b)$$

The equations (4) have a constant of time equal to the following value.

$$\tau_{ac}=\frac{1}{2 \cdot kCl_2 \cdot [Xe] \cdot [Cl](0)} \quad (5)$$

If initial concentrations of Cl_2 is 6,6 10^{16} 1/cm^3 (2 Torr) and 50 % of Cl2 has dissociated during discharge, then τ_{ac}=4500 microsecond. Then for pulse repetition cycle T>> τ_{ac} we have mono-pulse regime and for T<< τ_{ac} we have frequency regime of work. If the established value of partial pressure at the moment of the beginning of a pulse of excitation is known, it is possible to define energy of a pulse of radiation under the conditions submitted in Fig. 2. In Fig. 3 dependence of power of radiation on frequency of recurrence of the pulses, received for various initial partial pressure of molecules of chlorine is submitted.

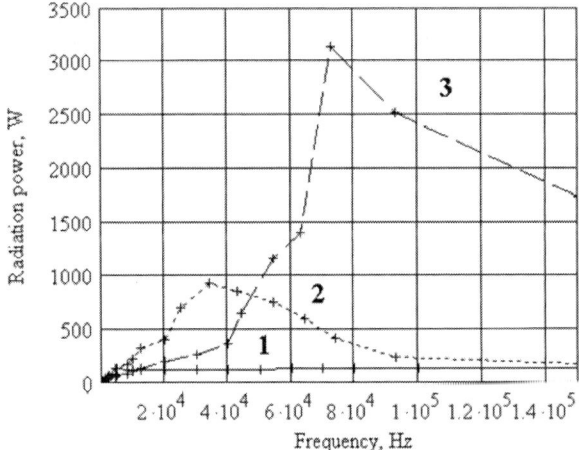

1 – Initial partial pressure Cl_2 1 Torr; 2 -2 Torr; 3- 3 Torr
Fig. 3. Dependences of power of radiation from frequency of pulses.

From the submitted dependences follows, that at the fixed initial partial pressure of Cl_2 power of radiation increase with grows of frequency. At some frequency power of radiation achieves a maximum. At the further increase in frequency power of radiation falls and leaves on some level which corresponds to the continuous glow discharge.

The maximum of power is achieved for such value of frequency of recurrence of pulses at which partial pressure Cl_2 for the moments of time corresponding to the beginning of a pulse of excitation, approximately coincides with optimum value of partial pressure for a mono-pulse operating mode. In Fig. 2 dependence of energy of a pulse from initial partial pressure of molecules of chlorine which corresponds to conditions at which dependences in Fig. 3 is calculated. The optimum value of partial pressure is equal 1 Torr in Fig. 2. All maximums of power of radiation are received for such value of established partial pressure of Cl_2 for the moments of time corresponding to the beginning of a pulse of excitation, is approximately equal 1 Torr.

At the further increase in frequency of following of pulses there is a falling power of radiation which is caused by burning out of molecules of chlorine and falling of energy of separate pulses. As it follows from the dependences submitted in Fig. 3 power of radiation increases at increase initial partial pressure of molecules of chlorine. So at increase initial partial pressure of molecules of chlorine with 2 Torr up to 3 Torr the maximal power of radiation grows from 1000 W up to 3130 W. This growth is caused by that at the greater initial maintenance of molecules of chlorine their burning out up to the optimum value corresponding to a maximum of energy for mono-pulse work, occurs at the greater frequency of recurrence of pulses.

Dependence 1, which is submitted in Fig. 3, has no maximum of power of radiation depending on frequency of following of pulses. Therefore the established value of partial pressure at the moment of the beginning of a pulse of excitation turns out much less, than the optimum size corresponding to a maximum of energy for mono-pulse work.

CONCLUSIONS

On the basis of results of modelling of XeCl-excilamps it is possible to draw the following conclusions.

1. The basic channel of formation of XeCl-molecules is harpoon reaction.

$$Xe^* + Cl_2 \rightarrow XeCl^* + Cl$$

At the given total pressure the channel ion - ionic recombination can be not taken into account.

$$Xe^+ + Cl^- + M \rightarrow XeCl^* + M$$

It is connected by that harpoon reaction is two-partial, and ion - ionic recombination three-partial and its speed depends on the total pressure of a mix. In that case when at modelling this channel was not taken into account the same dependence for change of concentration of XeCl-molecules turned out depending on time.

Model of XeCl excilamp taking into account process of Cl_2 regeneration is developed. Excilamp efficiency can reach 5 % at use of small values of storage capacity and at that pulse energy is equal 0,042 J and power of radiation is 3130 W.

It is established, that at small values of storage capacity power of radiation increases at increase in frequency of following of pulses up to some value. And the maximum of power is reached for such value of frequency of recurrence of pulses at which partial pressure Cl_2 for the moments of time corresponding to the beginning of a pulse of excitation, approximately coincides with optimum value of partial pressure for a mono-pulse operating mode. At the further increase in frequency there will be the reduction of power caused by burning out of halogen donor. At the big values of storage capacity power of radiation at increase in frequency leaves on some constant level.

The developed techniques of modeling allow to optimize more purposefully issue characteristics of pulse sources of radiation on the basis of the discharge in mixes of inert gases with halogens.

REFERENCES

[1] G. J. M. Hagelaar, L. C. Pitchford. "Solving the Boltzmann equation to obtain electron transport coefficients and rate coefficients for fluid models," *Plasma Sources Sci. Technol.*, vol.14, no. 1, pp.1-12, 2005.

[2] http://www.codiciel.fr/plateforme/plasma/bolsig/bolsig.php.

[3] A. N. Panchenko, A. S. Polyakevich, E. A. Sosnin, V. F. Tarasenko. "Glow discharge in excilamps of low pressure", *Proceedings of institutes of higher education. Physics,* vol. 42, no. 6, pp. 50-66, 1999.

[4] S. S. Anufrik, V. O. Shkleinik, A. P. Volodenkov, K. F. Znosko. "XeCl-excilamps computer modeling," *Proceedings of the VII symposium of Belarus and Serbia on physics and diagnostics of laboratory and astrophysical plasmas (PDP`2008), September 22-26, 2008, Minsk, Belarus*, pp. 118-121, 2008.

[5] D. L. Donohoue, D. Bauer, A. J. Hynes. "Temperature and pressure dependent rate coefficients for the reaction of Hg with Cl and the reaction of Cl with Cl: A pulsed laser photolysis-pulsed laser induced fluorescence study," *J. Phys. Chem. A.*, vol. 109, pp. 7732 – 7741, 2005.

Spectral features of some red and NIR laser dyes in silica matrices

I. M. Pritula[1], O. N. Bezkrovnaya[1], V. M. Puzikov[1], V.V. Maslov[2], *Senior Member, IEEE,*
A. G. Plaksiy[1], A.V. Lopin[1], Yu. A. Gurkalenko[1]

[1]Institute for Single Crystals, SSI "Institute for Single Crystals", National Academy of Sciences of Ukraine, Kharkiv, Ukraine

[2]O. Ya. Usikov Institute for Radiophysics and Electronics, National Academy of Sciences of Ukraine, Kharkiv, Ukraine

Abstract: We synthesized a series of sol–gel silica matrices doped with laser dyes DCM, LK678, LD1, LD2 which alcohol solutions have efficient lasing in the red spectral region and Rhodamine 800 – in the near infrared (NIR) one. Absorption, fluorescence, and laser properties of the dyes in silica samples prepared by sol–gel process were investigated. It has found some peculiarities in preparation process of the laser matrices and in spectral characteristics of the dyes in these matrices.

I. INTRODUCTION

The lasers tuned in the red and the near infrared (NIR) regions of the spectrum and dyes-sensitizers for this spectral range are widely applied for diagnostics of different biological and medical objects. For instance, an oxazine dye [1] is used for early diagnostics of Alzheimer's disease. The urgency of laser sources for this region is due to the fact that the penetration depth of the radiation into living tissue for them is about a centimeter. Furthermore study of new photosensitizers with the silica coating for photodynamic therapy at infrared excitation [2] necessitates quest of new radiation sources.

Dye lasers developed specially may be used for this purpose. However, for many practical applications solid-state dye lasers are the most challenging because of they are not required bulky dye-flow systems, and no toxic and inflammable solvents [3]. Matrices prepared by the sol-gel process, polymeric ones or porous glasses are used in these lasers.

The matrices prepared with the sol-gel method have higher thermal conductivity and lower temperature coefficient of the change of the refractive index than polymeric ones [4]. In addition the photostability of Rhodamine 6G in sol-gel matrix by 2.7 times exceeds the same value for poly (methyl methacrylate) host media [3]. Moreover, the silica gel matrices have higher laser damage threshold as against the polymer media [3], due to higher thermal conductivity of silica gel than of polymer materials [4].

For fabrication laser media based on SiO_2 matrices for the spectral range above mentioned we used laser dyes with high efficiency in alcohol solutions studied early [5, 6] and two standard laser dyes DCM and Rhodamine 800 (Rh800) [7]. The spectral and laser characteristics of these matrices were studied. The influence of the molar ratio of the reaction components on the matrix transparency was investigated.

II. EXPERIMENTAL

The silica gel were synthesized using tetraethoxy- and tetramethoxysilane (TEOS and TMOS respectively; Aldrich), additionally purified ethyl alcohol, formamide (FA) chemically pure, and twice distilled water, as well as laser dyes LD1 and LD2 (synthesized at the V. Karazin Kharkiv National University [5]), LK678 (synthesized at the D. Mendeleev Moscow Chemical Technology University [8]), DCM (Aldrich) and Rh800 (Aldrich). Presented below are the structural formulas of some dyes:

SiO_2 matrices were synthesized using sol-gel method by the hydrolysis of TEOS (or TMOS) with addition of nitric acid as a reaction catalyst [9, 10]. The alcohol solutions of the dyes (LD1, LK678, Rh800 in ethanol; LD2 in ethanol with added 1 mM HNO_3; DCM in methanol) were introduced after 30 min. mixing of alkoxysilane in ethanol (or in methanol for the matrices based on TMOS). Then we added twice distilled water, FA as drying control chemical additive (DCCA). The resulting mixture was stirred during 2 h. and then pyridine was added to it. The synthesized sol was placed into plastic cuvette, hermetically sealed and stored till the gel was formed. Afterwards the plastic cuvette were opened and dried for 3 – 4 weeks at room temperature and 7 –10 days at 60^0C. The density of obtained SiO_2 matrices was 1.4 – 1.6 g·cm^{-3}.

978-1-4799-0019-0/13 $31.00 © 2013 IEEE

The absorption spectra of the samples were recorded by spectrophotometer Lambda 35 (Perkin–Elmer, USA). The measurement of the luminescence spectra were carried out by the fluorimeter FluoroMax-4 (Horiba Jobin Yuon, USA). The luminescence of the dyes was excited near their absorption band maxima.

The lasing characteristics of the sol–gel matrices were studied using the laser [11] with non-selective cavity (LNSC) formed by two dielectric wideband mirrors with the reflection coefficient $R_1 \approx 99$ %, $R_2 \approx 60$ %. The samples placed in a quartz cuvette containing immersion liquid (ethylene glycol) were put into the laser cavity. The excitation of the samples was performed on the transverse scheme by flashlamp-pumped dye laser (FLPDL) with the output energy ≤ 230 mJ, the pulse duration ~ 1.5 µs, and the spectrum half-width ~ 3 nm. Its radiation was focused on the cuvette by a cylindrical lens with F= 110 mm in a strip with height ~ 1 mm. The matrices with LD1, LD2, and DCM dyes were pumped by the FLPDL with ethanol solution of coumarin 314 at $\lambda_p = 507$ nm. Those with Rh800 were pumped by the FLPDL with ethanol solution of oxazine 17 at $\lambda_p = 667$ nm, and matrix with LK678 – by the FLPDL with ethanol solution of Rhodamine 6G at $\lambda_p = 588$ nm. The output laser energy of the FLPDL and LNSC was measured by a device of IMO-2 type. The laser spectra were registered using a spectrograph based on UF-90 chamber with 1,200 line/mm diffraction grating, and then photographed by an EOS 400D DIGITAL camera.

III. RESULTS AND DISCUSSION

We synthesized transparent silica gel matrices with incorporated molecules of the mentioned laser dyes (Fig. 1).

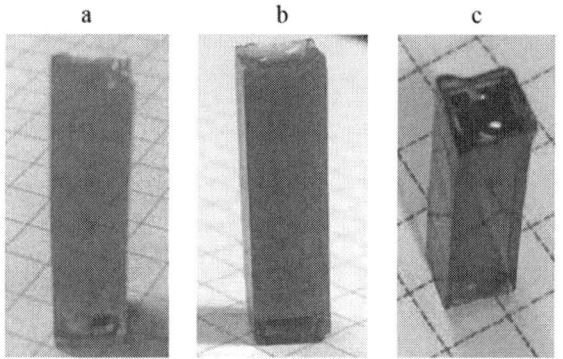

Fig. 1. SiO₂ matrices with the DCM (a), LK678 (b), and Rh800 (c) dyes; matrices were synthesized with FA.

The synthesis of SiO₂ matrices comprises the formation of the sol with subsequent aggregation of primary SiO₂ nanoparticles into spatial nets leading to gelation. The sol–gel transition is accompanied with the formation of the nanoparticles spatial structure which size may increase with time. The ageing and drying of the gel leads to the formation of amorphous xerogel with nanometric pores (5–20 nm) and 3–8 nm SiO₂ nanoparticles [12, 10]. In the process of ageing the solvent is removed from silica gels under the influence of

capillary forces [12]. The dye molecules are incorporated in the pores of the three-dimensional SiO₂ xerogel network. The gelation process is influenced by the DCCA [12, 10] (in our case – FA). In the samples prepared with FA the rate of the solidification and drying was by several times higher than for ones without FA.

There was established an essential influence of micro-environment on the spectral properties of the dye molecules incorporated in the matrix. In the gel the absorption and luminescence maxima of the LD1 dye correspond to 564 and 586 nm, respectively (Fig. 2, Table 1), but after drying of the matrix at 60⁰C the luminescence maximum was shifted to 582 nm.

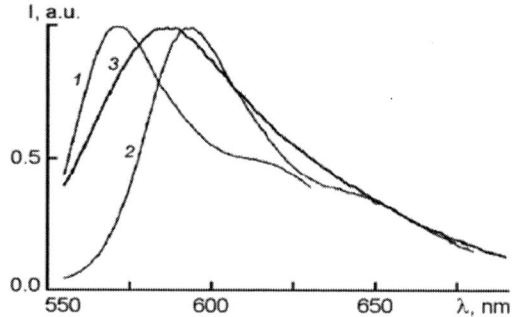

Fig. 2. Luminescence spectra of LD1 in: ethanol (1), ethanol + 0.44 mM HNO₃ (2), and SiO₂ matrix (3).

The LD2 luminescence maximum was shifted from 598 nm (in the gel) to 592 nm and 587 nm while the sample was dried at 20⁰C and 60⁰C respectively (Fig. 3, Table 1).

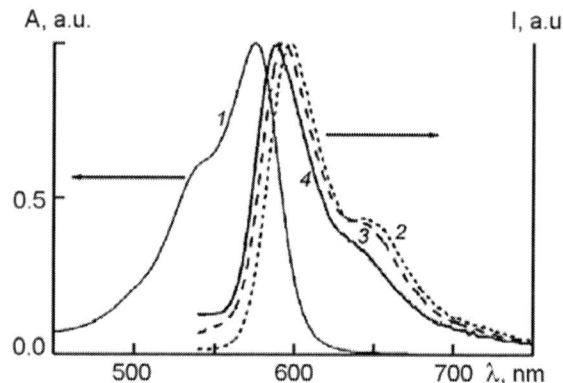

Fig. 3. Absorption (1) and luminescence (2-4) spectra of LD2 (0.35 mM LD2/1 M TEOS) in: SiO₂ matrix: gel (2), xerogel dried at 20⁰C (3), and 60⁰C (4).

The position of the absorption and luminescence maxima of the LD1 and LD2 dyes are caused by influence of solvation shell of the solvent molecules in the liquid-impregnated silica gel, and are prone to protonation in the presence of acid. In the process of drying the solvent and water are practically removed from the matrices and the dye molecules enter the silica gel pores. Moreover, is obvious that the molecules of LD1 and LD2 dyes undergo partial deprotonation while interacting with the surface of the SiO$_2$ nanoparticles which form the silica gel skeleton.

TABLE 1. Spectral characteristics* of LD1 and LD2 dyes.

Medium	C (HNO$_3$), mM	λ_a, nm	λ_l, nm
LD1			
Ethanol	–	549	572
Ethanol	0.44	570	592
SiO$_2$-gel	1.33	564	586
SiO$_2$-matrix (60^{0}C)	–	566	582
LD2			
Ethanol	–	567	586
SiO$_2$-gel	1.33	–	598
SiO$_2$-matrix (60^{0}C)	–	575	587

* λ_a – absorption maximum wavelength, λ_l – luminescence maximum wavelength.

As we have established, the process of drying is accompanied with bleaching of LD1 and LD2 in SiO$_2$ matrices. The observed effect seems to be caused by protonation of nitrogen in the 7th position of benzopyran base of the dyes. It is more noticeable for LD1 in which diethylaminogroup in this position may leave the plain of the molecular skeleton at protonation, thus sharply diminishing the probability of S$_1$→S$_2$ transaction and leading to luminescence quenching [5]. Therefore, while fabricating the active media on the base of silica matrices we added pyridine to reduce the acidity in sol synthesis process. However, we did not manage to obtain lasing of the matrices with the LD1 and LD2 because of their essential bleaching.

We observed the complete bleaching of the LK687 dye at the matrix synthesis and low luminescence intensity of DCM in the matrices when not using pyridine. In this case the last do not lasing.

The laser radiation of SiO$_2$ matrices activated with DCM and LK678 dyes was obtained (Fig. 4, 5) when at their synthesis a small dose of pyridine was added for reducing the acidity of the medium. In the process of this synthesis we used both TEOS and TMOS. We founded that the matrices with these dyes are more transparent when TMOS is used as a precursor compared with TEOS.

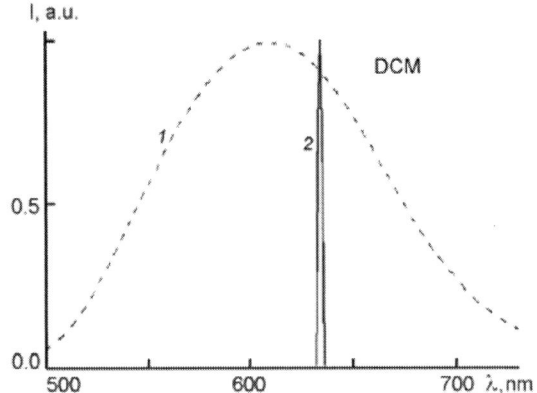

Fig. 4. Luminescence (1) and laser emission (2) spectra of DCM in SiO$_2$ matrices.

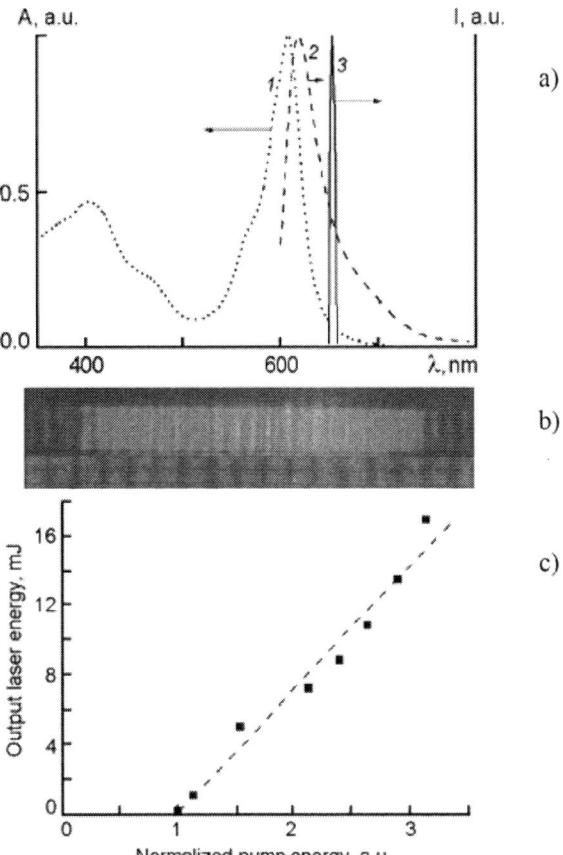

Fig. 5. Characteristics of the LK678 in SiO$_2$ matrices: a) absorption (1), luminescence (2) and laser emission (3) spectra; b) lasing spectrogram (the reciprocal dispersion of spectrograph ~ 6.2 Å/mm); c) dependence of the output laser energy on the pumping one (λ_p=588 nm).

On the TMOS matrices doped with the LK678 dye we obtained effective laser radiation at central wavelength λ_{las}= 654 nm with half-width of spectrum band – 4 nm (Fig. 5a, b). The dependence of output laser energy of this matrix on normalized pump energy (the last is normalized to the threshold pump energy E$_{th}$ ~ 70 mJ) is depicted in Fig. 5c. It is necessary to note that for this matrix (C$_{LK678}$ = 1.08 mM) under pumping energy ~200mJ output laser energy was by 1.5 higher than for matrix doped with 1.9 mM Rh6G.

Our studies of influence of the molar ratio of water and formamide – n (H$_2$O/FA) on transparency of the silica matrices have enabled to fabricate those activated with Rh800 and obtain laser emission on them in the NIR spectral range. We synthesized these matrices at n (H$_2$O/FA) = 8.8 and here pyridine was not used.

Spectral characteristics of Rh800 dye in different media are shown in Fig. 6. The absorption and luminescence spectra of it in SiO$_2$ matrix are shifted to long-wavelength side relative to ethanol solution. This shift indicates that the intermolecular interactions of the dye molecule with micro-environment in matrix become stronger then in the alcohol solution. The laser spectra of this dye in SiO_2 matrix were located at the long-wavelength slope of the luminescence band because of strong overlapping of the absorption and luminescence spectra and as a result the self-absorption in the active medium.

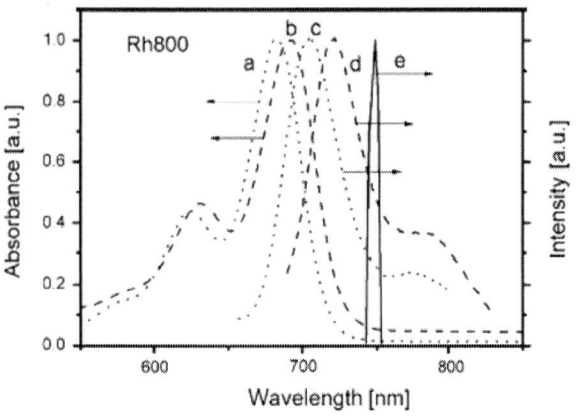

Fig. 6. Absorption (a, b) and luminescence (c, d) spectra of Rh800 in ethanol solution (a, c), in SiO_2 matrices (b, d), and laser spectrum of 1.1 9 mM Rh800 in SiO_2 matrix (e).

The main spectral parameters of the DCM, LK678, and Rh800 dyes in studied media are presented in Table 2 (λ_{las} – central laser wavelength, $\Delta\lambda_{las}$ – half-width of laser spectrum.). One can see that the Stokes shift $\Delta\nu^{St}$ between the absorption and luminescence maxima of DCM in solvents is more about by an order than ones for LK678 and Rh800. As a result the lasing of the last was at the long-wavelength wing of the luminescence spectrum but DCM laser emission was near its maximum.

TABLE 2. Spectral parameters of the dyes.

Medium	λ_a, nm	λ_l, nm	$\Delta\nu^{St}$, cm^{-1}	$\lambda_{las}\pm\Delta\lambda_{las}$, nm
DCM				
Acetonitrile	463	622	5500	632±6
Methanol	472	630	5280	635±4
SiO_2 matrix		610		634±2
LK678				
Methanol	609	626	450	650±10
SiO_2 matrix	608	620	320	654±4
Rh800				
Ethanol	683	705	460	725±3
SiO_2 matrix	690	720	600	747±4

It is necessary to note that the lasing spectrogram with an explicitly pronounced line structure shown in Fig. 5b is typical and for all other lasing matrices. This points to the fact that for these matrices the lifetime of photons in the laser cavity resonator is sufficient for forming the high-quality laser emission.

IV. CONCLUSIONS

Silica matrices activated with series of laser dyes for 600 – 750 nm spectral range of lasing were synthesized and their spectral characteristics were determinated. The necessity to employ acidity reduction additive in sol-gel process for some of the dyes was shown. For the matrices synthesized on the basis of TMOS with pyridine as such adding and doped with DCM and LK678 dyes there was obtained laser emission, whereas for those with LD1 and LD2 it was not achieved because of their essential bleaching in xerogel, connected probably with protonation of the nitrogen in 7th position of their benzopyran basis. The output laser energy of LK678 dye in the matrix was by 1.5 higher than of Rh6G at the equal conditions. A well-defined structure of the lasing matrix spectra testifies to their relatively high optical quality.

REFERENCES

[1] M. Hintersteiner, A. Enz, P. Frey, et al., "In vivo detection of amyloid-β deposits by near-infrared imaging using an oxazine-derivative probe." *Nature Biotechnology,* vol. 23, no 5, pp. 577 – 583, 2005.

[2] P. Zhang, W. Steelant, M. Kumar, and M. Scholfield "Versatile photosensitizers for photodynamic therapy at infrared excitation," *J. Am. Chem. Soc.,* vol. 129, no. 15, pp. 4526-4527, 2007.

[3] M. D. Rahn and T. A. King. "Comparison of laser performance of dye molecules in sol-gel, polycom, ormosil, and poly (methyl methacrylate) host media" *Appl. Opt.,* vol. 34, no. 36. pp. 8260-8271, 1995.

[4] W. Koechner, *Solid-State Laser Engineering,* 4th edition, Springer-Verlag, 1996, pp. 400 – 403.

[5] V.V. Maslov, M.I. Dzyubenko, S.N. Kovalenko, V.M. Nikitchenko, and A.I. Novikov. "New efficient dyes for the red part of the lasing spectrum," *Sov. J. Quantum Electron.,* vol. 17, no. 8 pp. 998 - 1001, 1987.

[6] V.V. Maslov, "Spectral and fluorescence characteristics of laser dyes for 650-800 nm range," *Function. Mater.,* vol. 13, no 3, pp. 419 –422, 2006.

[7] U. Brackmann, *Laser Dyes,* 3rd edition, Lambda Physik AG, Goettingen, 2000.

[8] B.I. Stepanov, N.N. Bychkov, V.G. Nikiforov, et al., "New generation of dyes for 660–860 nm spectral region for flashlamp-pumped lasers", *Lett. in Sov. J. Technic. Physics,* vol. 14, no. 7, pp. 650 –653, 1988, (in Russian).

[9] F. Salin, G. Le Saux, P. Georges, and A. Brun, "Efficient tunable solid-state laser near 630 nm using sulforhodamine 640-doped silica gel", *Opt. Lett.,* vol. 14, no. 15, pp. 785 – 787, 1989.

[10] O.N. Bezkrovnaya, I.M. Pritula, V.V. Maslov, et al., "Luminescent properties of rhodamine 6G dye in silica sol-gel matrices", *Functional Materials,* vol. 17, no 4, pp. 433–437, 2010.

[11] M.I. Dzyubenko, V.V. Maslov, V.P. Pelipenko, V.V. Shevchenko, and E.A. Kupko, "Laser tunable source for the red and near UV regions of the spectrum", *J. Appl. Spectrosc.,* vol. 71, no. 3, pp. 435–440, 2004.

[12] N.N. Khimich. "Synthesis of silica gels and organic–inorganic hybrids on their base," *Glass. Phys. and Chem.,* vol. 30, no. 5, pp. 430-442, 2004.

Faraday rotator for the optical switch of a two-wave light flux in fiber-optic networks

G. D. Basiladze, V. N. Berzhansky, A. I. Dolgov

Taurida National V.I. Vernadsky University

Abstract - In the work is shown the possibility of creation of the Faraday rotator based on two planar-oriented magneto-optical film elements for modulating the planes of polarization of the two light beams with wavelengths of 1310 and 1550 nm with an angular amplitude of ±45°.

I. INTRODUCTION

Solution of the problem of optical switches for all-optical WDM networks on the nearest future is associated with the development of optical switching technologies with utilizing active optical elements, which provide passing of a light signal through switching system. One of such technologies utilizes magneto-optical (MO) material, having Faraday effect, as an active optical element. With the annex to it a pulse magnetic field parallel to a beam of light, there is a modulation of the plane of the polarization of the linearly polarized light that applied to the input of the MO element. Switching of a light flux in such switches is carried out by polarization beam splitters, that adjacent to the output of the MO element [1-3]. Faraday rotators (in this case MO elements with coils of the solenoid magnetizing them) in the mentioned works contained one MO element, which, owing to spectral dependence of coefficient of Faraday provided operation of the switch only on one of lengths of waves of telecommunication spectral range.

The aim of the present work is to check the possibility of practical use of Faraday rotator (FR) based on two planar-oriented to its optical axis MO elements, each of which acts on its own wavelength. The main essence of a problem consists in receiving out of grown on a 3 inch substrate epitaxial ferrite-garnet film (EFGF) of the MO elements that capable to produce the equal angle rotation of the plane of polarization of light beams with specifically designated for MO elements wavelengths at joint magnetization by one inductor of a magnetic field.

II. STRUCTURE OF THE FARADAY ROTATOR AND ONE OF POSSIBLE SCHEMES OF THE SWITCH ON ITS BASIS

One of possible optical schemes of the switch of light fluxes with lengths of waves of 1310 and 1550 nm, that can be realized with the help FR represented in this work is shown in Fig. 1.

Basic elements of the FR are the inductor of a magnetic field 5 in the form of the solenoid and the planar MO elements 6, 7 focused to its axis, which are made of EFGF.

To the inputs of MO elements is summed polarization maintaining optical fibers 1, 2 of the "Panda". The fibers are connected to the MO elements via by planar glass ferrule 3, 4. From exits of MO elements by means of similar ferrules 3, 4 light fluxes with wavelengths of 1310 and 1550 nm, through polarization maintaining fibers, are supplied in polarization beam splitters PBS_{1310} and PBS_{1550}. The optical signals arriving to their entrances splits on orthogonally polarized S and P components and come to the inputs of the multiplexers M_1, M_2. Where S components of the both of wavelengths are sent to the M_1 multiplexer, and P components are sent to the M_2 multiplexer.

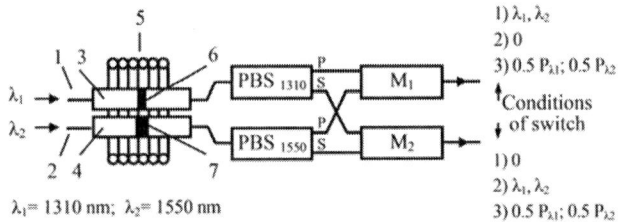

$\lambda_1 = 1310$ nm; $\lambda_2 = 1550$ nm

Fig. 1. Scheme of optical switch.

Distribution of the signals with the wavelengths of 1310 and 1550 nm at the exits of these multiplexers i.e. the switch depends on polarization state of light at the exits of FR. Therefore, the size of MO elements L_{1310} and L_{1550} for the corresponding wavelengths in the direction of light propagation must be such that, the magnetic saturation field H_S provides rotation of the planes of polarization of either 90°, if unipolar magnetizing device is used, or to ±45°, if bipolar magnetizing device is used. The need of polarization modulation with such angular amplitude is dictated by the physical mechanism of action of the polarization beam splitters, where one of conditions of complete separation applied to their inputs linearly polarized signals is strictly orthogonal plane of polarization of the signals.

In this work, FR with bipolar magnetizing device was simulated and investigated. The switch with such rotator can be in three conditions and works as follows. The fluxes of light, with wavelengths of 1310 and 1550 nm with the equal orientation of the polarization planes, through fiber ports are supplied to the corresponding to them MO elements of the FR.

When in the solenoid there is no electric current, the planes of polarization at the outputs of the MO elements are oriented at an angle of 45° to the axis of birefringence of the input fibers of the splitters PBS_{1310} and PBS_{1550}.

In this state, the light fluxes entering the polarization splitters by each are divided into two fluxes and signals are sent to both wavelengths of both multiplexer. Thus, in the both output ports of the switch are present the optical signals with both wavelengths, and the operation of the switch is implemented in broadcast mode. Distribution of signals corresponding to this condition between two output ports of the switch is labeled 1 in Fig. 1.

When the magnetic field is of unipolarity, ie $+H_S$, or $-H_S$, then the Faraday rotation angle, respectively, $+45°$ or $–45°$. Plane of polarization of the light beams at the output MO elements are oriented, respectively, along one or another of the axes of birefringence fibers. In one of these states of the switch, polarizing splitters directs fluxes of the lights with their respective wavelengths to the respective inputs one of a multiplexers and both flux of the lights come out via its output port, or the corresponding output of the switch. In the second of these states, due to the polarization plane rotation by 90° in the reverse direction, the fluxes of the light via polarization splitters will be directed to the second spectral multiplexer and they come out from second output port of the switch. Corresponding to these states distribution of the signals between the two output ports of the switch in Fig. 1 indicated by numbers 2 and 3.

III. CALCULATION AND FABRICATION OF MO ELEMENTS FOR THE FARADAY ROTATOR

As it was noted above, in our case, switching between output ports of the switch is carried out by magnetic fields of $+H_S$ and $-H_S$ which provide EFGF magnetization before saturation, alternately, in the direction and against the direction of light propagation inside the films. Thus for each of the operating wavelengths the optical path length $L(\lambda)$ in EFGF layer must be such that the Faraday effect can provide rotation of polarization of light by an angle of $\theta = \pm45°$.

These the longs of the optical paths or in other words the longitudinal dimensions of MO elements are calculated by the formula:

$$L(\lambda) = \frac{45°}{\theta_F(\lambda)} \quad \text{cm,} \qquad (1)$$

where $\theta_F(\lambda)$ – Faraday's coefficients for the used EFGF on operating of wavelengths of the switch. They were determined by us by a technique described in the work [4], which allows to investigate Faraday rotation on different lengths of waves of near infrared range when light passes into the films plane.

The starting material for the manufacture of MO elements of FR was EFGF with the thickness of 12 µm, consisted of $(BiLuCa)_3(FeGaV)_5O_{12}$, grown the method of liquid phase epitaxy from solution-melt on GGG substrate with a crystal orientation of 111, thickness of about 500 µm end diameter of 76 mm. The specific Faraday rotation of the EFGF at 1310 and 1550 nm was, respectively, 93.4 and 60.9

deg/mm. EFGF has a magnetic anisotropy of the "angle phase" with the saturation fields along the axis of easy and hard magnetization intensity of 1 and 30 Oe, respectively.

Based on this EFGF with the help of the method, briefly described in [5] for each of the operating wavelengths were manufactured optical modules OF-EFGF. Optical module OF-EFGF, which is, united in a single design with MO element and optical fiber "panda". The input and output edges of a films of optical modules are plane parallel, end optically polished so that light extended along an axis of its easy magnetization.

The longitudinal sizes of the MO elements L_{1310} and L_{1550} in optical modules had to satisfy to the values calculated on above given formula, respectively 0.48 and 0.74 mm. These values serve as a guideline for selecting the longitudinal sizes of blanks of EFGF for elements with a stock of length necessary for carrying out technological operations of polishing film edges. It is necessary to note that this technology in our case includes the final polishing operation to minimize the broken layers of EFGF, which can worsen the work of MO element in the device.

In the present study, FR was made without adjacent to the outputs MO elements of optical fibers (on the right) and the following them remaining components of the switching circuit. Its main feature is the fact that both modules FO-EFGF and magnetizing device are assembled into a single structure with rigidly fixed elements in it. This ensured the permanence conditions of magnetization of both MO elements in the research process. As the magnetizing device, the inductor, in the form of the solenoid coil from a copper wire with a diameter of 0.67 mm. The number of turns in the coil 16, the packing factor is 0.74, along the length of the coil is 11 mm, an inner diameter is 10 mm.

OF-EFGF modules are arranged inside the inductor, so that their optical axes are parallel to the axis of the solenoid coil.

IV. RESULTS OF RESEARCH OF THE FARADAY ROTATOR AND THEIR DISCUSSION

During the study, we measured the angles of rotation of the polarization plane on the outputs of these modules the magnetization of the films in the forward and backward direction of the light propagation, the degree of polarization in such a magnetization. Moreover, the response of FV to magnetic reversal by impulses of a magnetic field with a fixed frequency of 125 kHz was investigated.

The optical signal from the polarization analyzer installed for measurements at the exit of FR, gets on the germanium photodetector connected to the amplifier of photocurrent. The amplified electrical signal proportional to the intensity of optical signal on an entrance of a photodetector, was fed to the input of the digital voltmeter in static measurements or vertical channel oscilloscope for dynamic measurements.

When measuring the rotation angle and the degree of polarization of the light intensity of the magnetic field it was possible continuously change from -20 to +20 Oe in proportion to change of the current, transiting through in the inductor from -1.5 to +1.5 A.

Measurements of angles of Faraday rotation yielded the following results: in the optical channel with a length of wave of 1310 nm the angle of rotation in both directions of magnetization of MO element made average $\theta_{1310} = \pm(45.7 \pm 0.5)°$, in the optical channel with a length of wave of 1550 nm $- \theta_{1550} = \pm(45.2 \pm 0.5)°$. The degree of polarization of the light output of both optical channels was not less than the value 0.98.

At studying of a response of FR on remagnetizing magnetic field the source of a magnetic field was connected to the pulse power unit which gave out a variable voltage from -5 to $+5$ V with a fixed frequency of impulses of 125 kHz. At such voltages the inductor of a magnetic field provided magnetization of saturation of EFGF alternately in the direction and against light propagation in a film. The signal from the photo diode of the photoreception module arrived on a measuring entrance of an oscillograph.

When you turn on the source of the magnetic field at the output of each of the optical channels of FR observed modulation of the plane of polarization of light. The analyzer thus remained in the position of fixing 45-degree rotation of the polarization plane of the previous measurements.

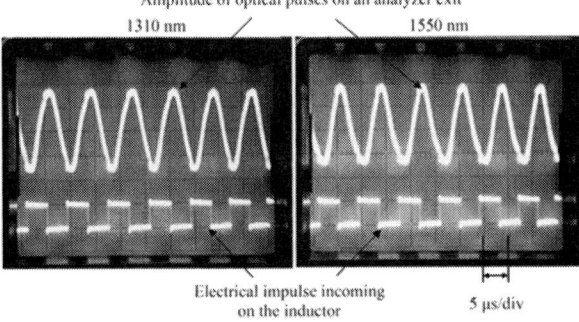

Fig 2. Oscillograms of the electric signals incoming on the inductor (below) and optical signals on output ports of the switch (above) for wavelengths of 1310 and 1550 nm.

Fig. 2 shows the waveform of pulses applied to the solenoid, and the waveform of pulses registered on each of two outputs FR. The voltage pulses supplied to the inductor of magnetic field have amplitude of ±2 V, and repetition frequency of 125 kHz. Oscilloscope sweep was 5 μs/div.

The received oscillograms show that time of increase and signal falling for each of exits of FR on two lengths of waves of 1310 and 1550 nm makes 4 μs.

That is, per such a time there is a switching the plane of polarization in both optical channels of FR. Consequently, FR in the submitted version is able to provide the speed switch of 4 μs.

Some inclination, which is observed on oscillograms at fronts of impulses of the response of the FR on given impulses of voltage, is explained by existence of a quite considerable inductance of the used solenoid. When using

inductors of a magnetic field of other type, for example, the strip line, this time can be reduced considerably. So in work [6] with such inductor of a magnetic field on a surface of MO film was achieved modulation of polarization, at which the time between maxima of intensity made about 1 ns.

V. CONCLUSION

Is shown the possibility to practical realization of the two-wave Faraday rotator, which based on the two planar by oriented on its optical axis MO elements, each of which operates at its wavelength of light.

Such MO rotator, being embedded in a fiber-optical circuit of spectral multiplexers and polarization beam splitters can be used for switching of two-wave light flux in fiber-optical networks [7].

REFERENCES

[1] Weng Zihua, Chen Zhimin, Huang Yuanqing et al., "High-speed all-fiber magneto-optic switch and its integration", SPIE, Vol. 6021, p.p. 725-736, 2005.

[2] R. Bahuguna, M. Mina, J.W. Tioh, R. J. Weber, "Magneto–optic–based fiber switch for optical communications", IEEE Trans. Magn, Vol. 42, No 10, p.p. 3099–3101, 2006.

[3] Jianjian Ruan, Zihua Weng, Shaohan Lin, "High-speed magneto-optic switch for optical communication", Proc. of SPIE, Vol. 7509, 75090B-1 – 75090B-9, 2009.

[4] G.D. Basiladze, V.N. Berzhansky, A.I. Dolgov, "Influence of spectroscopic dependence of Faraday coefficient on characteristics of radiation transmitted the planar magnetoactive element of the fiber-optic switch", Scientific", Notes of Taurida National V.I. Vernadsky University. Series: Physics and Mathematics Sciences, Vol. 25 (64), No 1, p.p. 160-169, 2012.

[5] G.D. Basiladze, V.N. Berzhansky, A.I. Dolgov, "Design of Faraday rotator for the optical switch", 10th Intern. Conf. on Laser & Fiber-Optical Networks Modeling (LENM'2010), Sevastopol, Ukraine, Proc. – [S. l.], p.p. 158-160, 12-14 Sept. 2010.

[6] S.E. Irvine, A.Y. Elezzabi, "A miniature broadband bismuth-substituted yttrium iron garnet magneto-optic modulator", J. Phys. D: Appl. Phys, Vol. 36, p.p. 2218–2221, 2003.

[7] G.D. Basiladze, A.I. Dolgov, V.N. Berzhansky, "Fiber-optic magnetooptical switch", UA Patent, No 75525, 04 Apr. 2012.

Realization of two-electron mechanism of two charged ions creation upon multiphoton ionization of barium atoms by infrared laser radiation

I. I. Bondar', V. V. Suran

Department of Physics, Uzhgorod National University, ul. Voloshina 54, Uzhgorod, 88000 Ukraine

Abstract: Our experimental investigations show that neutral Ba atoms serve as target upon the creation of the doubly charged ions in the infrared spectral range. This fact indicate that the two-electron mechanism of the creation of the Ba^{2+} ions is implemented upon the ionization of the Ba atoms by the IR radiation.

The creation of the doubly charged ions (A^{2+}) upon the multiphoton ionization of atoms was observed for the first time in [1]. Two mechanisms (cascade and two-electron) were proposed in to account for the creation of doubly charged ions upon the multiphoton ionization of the alkaline-earth elements. In the cascade mechanism, doubly charged ions are created owing to the multiphoton ionization of singly charged ions that appear in the presence of the same laser pulse due to the multiphoton ionization of atoms. The two-electron mechanism leading to the creation of doubly charged ions involves the simultaneous detachment of two outer electrons from neutral atom. So, the neutral atoms serve as targets upon the creation of the doubly charged ions in case of two-electron mechanism realization.

Our experimental investigations was aimed at the analysis of the effect of atomic concentration on the creation of the two-charged ions in the infrared spectral range (IR). In the experiments, the concentration of neutral atoms was varied both over the duration of the laser pulse needed for the creation of the Ba^{2+} ions and prior to the pulse.

In the first case, a variation in the atomic concentration results from an increase in the atomic ionization probability due to the additional excitation. The Ba^{2+} ions are created owing to the IR irradiation with the fixed frequency $\omega_1 = 9395$ cm^{-1} (the radiation of the laser on yttrium-aluminum garnet (LYAG)) or $\omega_1 = 8620$ cm^{-1} (the radiation of the laser on colour centre (CCL)). The dye-laser radiation (DL) with the frequency tuning in the range $\omega_2 = 17860\text{-}18160$ cm^{-1} is employed for the additional excitation of the Ba atoms. The action of the DL radiation on the Ba atoms in the ground $6s^2\,^1S_0$ state leads to either one-photon excitation of the $6s6p\,^1P_1^\circ$ state ($\omega_2 = 18060$ cm^{-1}) or two-photon excitation of the $6s7d\,^3D_2$ state ($\omega_2 = 17881$ cm^{-1}) and $5d6d\,^3D_2$ state ($\omega_2 = 18100$ cm^{-1}). We compared the yields of the Ba^+ and Ba^{2+} ions created under the simultaneous and separate action of the IR and DL beams on the Ba atomic beam. Figure 1 demonstrates the results obtained in the case when the creation of the ions Ba^{2+} is induced by the LYAG radiation.

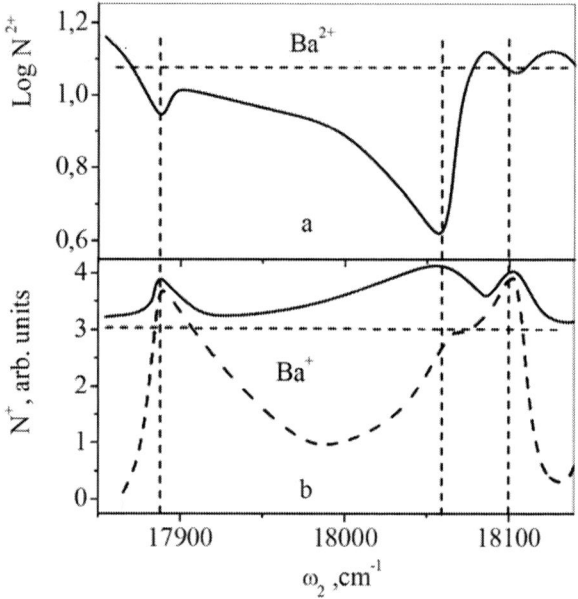

Fig. 1. Plots of Ba^{2+}(a) and Ba^+ (b) ion yields vs. the DL frequency for the ionization process involving the simultaneous action of the LYAG and DL radiations (solid lines), the ionization induced by the DL radiation (dashed lines), and the ionization induced by the LYAG radiation (dashed horizontal lines), The vertical dashed lines indicates the DL-radiation frequencies that correspond to the resonant transitions in the spectrum of the Ba atom .

In contrast to the Ba^+ ion yield, the Ba^{2+} ion yield is lower in the case of the simultaneous action of the two laser beams. Note that the minimum Ba^{2+} ion yields are observed in the vicinity of the DL frequency, which corresponds to the excitation of neutral atoms. The strongest decrease in the Ba^{2+} ion yield is observed in the vicinity of the DL frequency $\omega = 18060$ cm^{-1}. Recall that the maximum increase in the Ba^+ ion yield is also observed in the vicinity of this frequency. Such simultaneous decrease in the Ba^{2+} ion yield and increase in the Ba^+ ion yield cannot be interpreted in the framework of the cascade model of the creation of doubly charged ions. The above behavior of the ion yields caused by the additional excitation of the Ba atoms indicates that neutral Ba atoms serve as targets upon the creation of the Ba^{2+} ions in the presence of the LYAG irradiation. In particular, an increase in the ionization probability of the Ba atoms under the saturation conditions for this process can lead to a significant decrease in the concentration of the Ba atoms in the interaction region. A decrease in the Ba atomic concentration during the laser pulse prior to the moment at which the effective creation of the Ba^{2+} ions is started must result in a decrease in the ion yield. Such a scenario is in agreement with the experimental data.

We have described the effect of the additional excitation of the Ba atoms on the creation of the Ba^{2+} ions under the action of the LYAG radiation. Note that the same results were obtained for the creation of the Ba^{2+} ions under the action of the CCL radiation with the frequency $\omega = 8620$ cm^{-1}.

Thus, the results show that neutral Ba atoms serve as targets upon the creation of the Ba^{2+} ions induced by the IR radiation.

We also performed experiments in which a variation in the concentration of neutral atoms in the beam preceded the arrival of the laser pulse creating the doubly charged ions. Under such conditions, we studied the creation of the Ba^{2+} ions under the action of the linearly polarized LYAG radiation. A variation in the atomic concentration is implemented with the aid of an additional laser beam. The LYAG radiation serves as the additional radiation whose parameters are chosen in such a way that the ionization of the Ba atoms in the beams occurs under the saturation conditions or close to the saturation. For the correctness of the experiments, we employ the circularly polarized additional radiation, since the Ba^{2+} ion yield is strongly suppressed in the presence of the circularly polarized LYAG radiation (in comparison with the case of the linearly polarized radiation). The additional radiation pulse acts upon the Ba atomic beam earlier (by 4×10^{-8} s) than the main LYAG pulse that induces the creation of the Ba^{2+} ions.

The action of the additional radiation prior to the action of the main radiation pulse makes it possible to substantially increase the Ba^+ ion concentration and to decrease the concentration of the neutral Ba atoms. In the experiments, the intensity of the additional radiation is varied and the intensity of the main radiation remains constant. We determine the yields of ions created due to the action of the additional and main beams. Figure 2 shows the results. It is seen that an increase in the intensity of the additional radiation leads to an increase in the amount of the Ba^+ ions in the interaction region. Note a decrease in the yields of the Ba^+ and Ba^{2+} ions created by the main radiation pulse delayed relative to the additional pulse.

Such a behavior of the Ba^{2+} ion yield cannot be interpreted in the model of the cascade creation of the doubly charged ions and can easily be explained using the two-electron model. Under the conditions for the cascade process, the preliminary increase in the Ba^+ ion concentration in the interaction region due to the action of the additional radiation must lead to an increase in the amount of the Ba^{2+} ions created by the main radiation. At least, the amount of the Ba^{2+} ions must not decrease.

In the case of the two-electron process (when neutral atoms serve as targets upon the creation of the doubly charged ions), a decrease in the concentration of the neutral Ba atoms in the interaction region must result in a decrease in the Ba^{2+} ion yield. This scenario is in agreement with the experimental data.

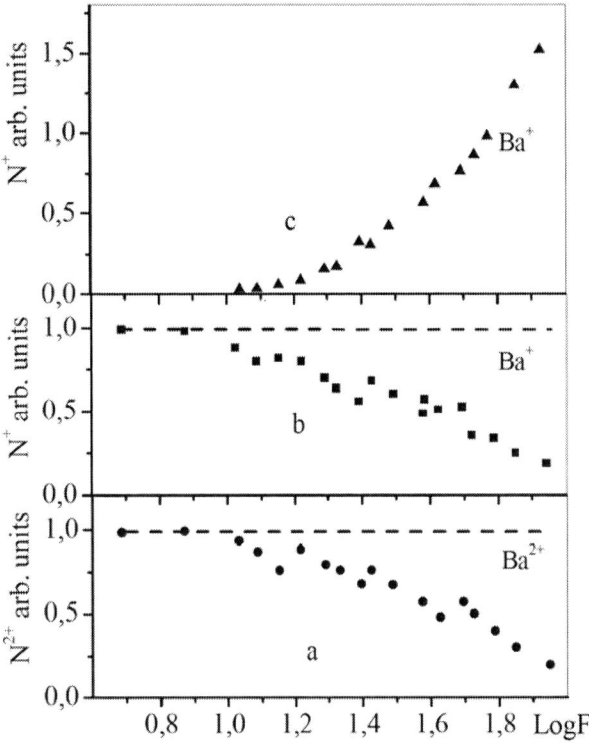

Fig. 2. Plots of the Ba^{2+} ion yield resulting from the action of the LYAG radiation in the case when the LYAG radiation pulse is delayed relative to the pulse of additional radiation (a), the Ba^+ ion yield resulting from the action of the fundamental LYAG radiation in the case when the LYAG radiation pulse is delayed relative to the pulse of additional radiation (b), and the Ba^+ ion yield resulting from the action of the additional LYAG radiation (c). The dashed horizontal lines show the mean Ba^+ and Ba^{2+} ion yields resulting from the action of the fundamental LYAG radiation in the absence of the preliminary irradiation of the Ba atoms using the additional radiation.

Thus, the experimental results show that neutral atoms serve as targets upon the creation of the doubly charged ions at multiphoton ionization of Ba atoms in the IR spectral range. Therefore the two-electron mechanism of the creation of the Ba^{2+} ions is realized upon the ionization of the Ba atoms by the IR radiation.

REFERENCES

[1]   V.V.Suran and I.P.Zapesochnyi, *Sov. Tech. Phys. Lett.*, vol. 1, p. 420, 1975.

Formation of a quasi-uniform output beam in the waveguide CO_2 laser

O.V.Gurin, A.V.Degtyarev, V.A.Maslov, *Member IEEE*, V.A.Svich, *Member IEEE*, A.N.Topkov, T.F.Ruban

V.N.Karazin Kharkiv National University, Kharkiv, Ukraine

Abstract: An experimental sample of a waveguide CO_2 laser with a quasi-uniform profile of the output radiation intensity is designed on the basis of a waveguide quasi-optical cavity of a new type comprising the generic confocal cavity with a nonuniform mirror and the hollow waveguide with the dimensions satisfying the conditions for self-imaging the quasi-uniform field. The surface of the mirror has the discrete large-scale absorbing nonuniformities. Results of theoretical and experimental investigations of spatial-energy characteristics of the laser in using uniform or amplitude-stepped reflecting mirrors are presented.

Waveguide gas lasers are widely applied in technology, medicine, spectroscopy, and space communication [1, 2]. Problems in these fields are optimally solved by employing the beams with a quasi-uniform cross-section distribution of the radiation intensity.

The present work is aimed at development, fabrication, and study of an experimental sample of the waveguide CO_2 laser with a quasi-uniform output beam.

1. THEORETICAL RELATIONS

Theoretical consideration is based on the methods of Fourier optics and eigenoscillations [3, 4]. The process of formation of cavity oscillation types is interpreted as the result of various diffraction interactions of the waves with the optical elements comprised in the cavity. Due to these interactions, the eigenwaves recover the relative spatial amplitude and phase distributions and the polarisation state in a cavity cross-section in each successive transit. The lens corrector and mirror nonuniformities we will describe by the amplitude-phase correction function [4]. The scheme of the cavity under study is

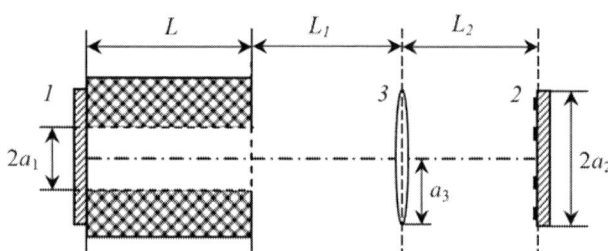

Fig. 1. Schematic diagram of a waveguide quasi-optical cavity.

given in Fig. 1

One arm of the cavity near the output semitransparent reflector (*1*) has a circular dielectric waveguide. The waveguide dimensions should correspond to the selfimaging conditions for the radiation beams with the super-Gaussian field amplitude distribution in hollow waveguides. These conditions have been obtained in [5]. We denote the diameter of the output mirror and waveguide by $2a_1$ and the waveguide length by L. At a distance L_1 from the waveguide end, the thin lens corrector (*3*) of radius a_3 and focal length F is placed, which performs Fourier transformation of the field at the waveguide output. At a distance L_2 from the phase corrector, there is the nonhomogeneous mirror (*2*) of diameter $2a_2$ with a spatial filter characterised by the amplitude correction function $T(\rho_2)$ ($\rho_2 = r_2/a_2$ is the dimensionless radial coordinate for mirror (*2*)). The transversal dimensions of the cavity elements are assumed to satisfy the conditions of quasi-optical approximation $(ka_i)^2 >> 1$ ($i = 1, 2, 3$), where $k = 2\pi/\lambda$ (λ is the wavelength), and paraxial conditions $k_\parallel >> k_\perp$ (the longitudinal wavenumber is much greater than the transverse one).

The method for calculating the characteristics of lower cavity modes is based on the quasi-stationary condition for the transversal structure of the mode field in the cavity [6]. In this case, the field in the waveguide is presented as a superposition of eigenwaves and in the open parts of the cavity – as the diffraction integral in the Fresnel approximation [3]. Finally, the problem of eigenoscillations in the considered cavity is reduced to the following system of linear algebraic equations:

$$\mu C_k = e^{i\gamma_k L} \sum_{m=1}^{M} C_m e^{i\gamma_m L} \int_0^1 V_m(\rho_1) Q_k(\rho_1)\rho_1 \, d\rho_1 \quad , \quad (1)$$

where

$$Q_k(\rho_1) = \int_0^1 Q^0(\rho_1, \rho_1^{'}) \, V_k(\rho_1^{'}) \, \rho_1^{'} \, d\rho_1^{'};$$

$$Q^0(\rho_1, \rho_1^{'}) = \frac{N_1 N_2}{(1-G_1)\,(1-G_2)} \int_0^1 Q(\rho_1, \rho_2) Q(\rho_2, \rho_1^{'}) \, T(\rho_2)\rho_2 \, d\rho_2;$$

$$Q(\rho_p, \rho_n) = -4\pi^2 \, N_0 \, e^{ik(L_1+L_2)} \, e^{i\pi(N_1\,\rho_p^2 + N_2\rho_n^2)}$$

$$\times \int_0^1 e^{i\pi N_0 Z \rho_3^2} J_0(2\pi N_1 \xi_1 \rho_p \rho_3) J_0(2\pi N_2 \xi_2 \rho_n \rho_3)\rho_3;$$

ρ_1 and ρ_1' are the dimensionless (normalised to a_1) radial coordinates on the mirror (1) at the start and finish of a round trip in the cavity; ρ_3 is the dimensionless radial coordinate (normalised to a_3) on the phase corrector (3); J_0 is the zero-order Bessel function of the first kind; $p = 1, 2$ is the mirror number; $n = 3 - p$; $k = m = 1,...,M$; M is the number of modes for the waveguide laser tube; and γ are waveguide wave propagation constants [7]:

$$N_{1(2)} = \frac{a_{1(2)}^2}{\lambda L_{1(2)}}; \quad N_0 = \frac{a_3^2}{\lambda F}; \quad \xi_{1(2)} = \frac{a_3}{a_{1(2)}}; \quad G_{1(2)} = 1 - \frac{L_{1(2)}}{F}; \quad Z = \frac{1 - G_1 G_2}{(1 - G_1)(1 - G_2)}.$$

The solution to system (1) yields M eigenvalues μ, determining the relative energy losses for cavity modes and their additional phase intrusion over the cavity round trip and the same number of eigenvectors C_k whose components determine the profiles of the corresponding transversal cavity modes. The relative energy losses, which include the mode energy losses in the waveguide and in free space segments per round trip, are defined, along with the phase shift of the modes, by the expressions

$$\delta_r = 1 - |\mu|^2; \qquad \qquad \Phi = \text{Arg}\,\mu. \qquad (2)$$

Assume that the distribution of a complex amplitude of field component at output mirror (1) of the waveguide quasi-optical cavity and, correspondingly, at the end of the multiharmonic waveguide is described, at the corresponding waveguide length, by the circular function

$$\text{circ}(\rho_1) = \begin{cases} 1, & \rho_1 \le 1, \\ 0, & \rho_1 > 1. \end{cases} \qquad (3)$$

In the approximation of infinite aperture of the phase corrector, the Fourier-Bessel transformation of this function, with the accuracy to an insignificant constant factor, has the form

$$\text{somb}(\Theta) = \frac{2J_1(\pi\Theta)}{\pi\Theta}, \qquad (4)$$

where $Q = 2N_{12}\rho_2$; $N_{12} = a_1 a_2 / [\lambda F(1 - G_1 G_2)]$ is the GCC Fresnel number.

By placing the absorbing elements on the nonuniform mirror (2) in such a way that $\rho_{2\chi} = v_{1\chi}/2\pi N_{12}$, where $v_{1\chi}$ are roots of function J_1, $\chi = 1, 2, 3$, and taking into account possible selection of transversal modes by the above mentioned elements one may hope that a solution to system (1) will be formed from functions close to analytical forms (3), (4). In this case, the transverse dimensions of the uniform domains, in whose boundaries the matter constants experience discontinuity, should be noticeably greater than the wavelength.

System (1) can only be solved with the help of a computer. It was performed by the matrix method by using the modified Rutishauser algorithm. There are three independent kinds of solutions of system (1): for hybrid (EH_{nm}-), transversal electric (TE_{0m}-) and transversal magnetic (TM_{0m}-) modes, where n and m are the azimuthal and radial mode indices, respectively. The results of calculations presented below refer to the practically important modes from the class of axisymmetric EH_{1m}-modes, which have a linear polarisation of the field. Their complex amplitudes are described by a complete system of orthonormalised functions $V_m(\rho_1) = \sqrt{2}J_0(Y_m\rho_1)/J_1(Y_m)$, where J_0, J_1 are Bessel functions of the first kind, Y_m are roots of equation $J_0(Y_m) = 0$. The propagation constants for these modes are [7]

$$\gamma_m \approx k\left[1 - \frac{1}{2}\left(\frac{Y_m\lambda}{2\pi a_1}\right)^2\left(1 - \frac{iv_1\lambda}{\pi a_1}\right)\right],$$

where $v_1 = 0.5(v^2 + 1)/(v^2 - 1)$ and v is the refractive index of waveguide wall.

3. EXPERIMENTAL SETUP

The design of an experimental sample of a waveguide CO_2 laser is schematically shown in Fig. 2. The laser operated in the regime of slow gas mixture pumping ($CO_2 : N_2 : He : Xe = 1 :$

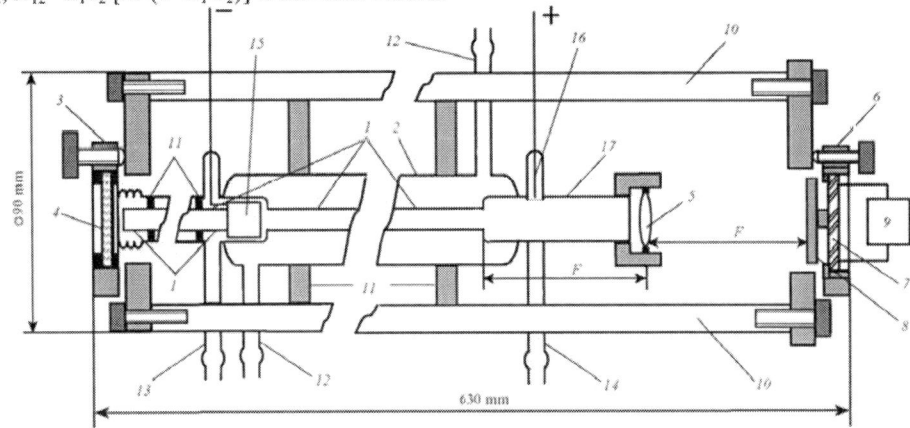

Fig. 2. Construction of a waveguide CO_2 laser: (1) waveguide; (2) water heat exchanger jacket; (3) adjustment unit of a semitransparent mirror; (4) semitransparent mirror; (5) ZnSe lens; (6) adjustment unit of a nonuniform mirror; (7) KP-1 piezoelectric corrector; (8) nonuniform mirror; (9) DC source; (10) invar rods; (11) centring rings; (12) fittings of the water cooling system; (13) exhaust fitting; (14) working mixture puffing fitting; (15) cathode; (16) anode; (17) glass tube.

1 : 5 : 0.25). A longitudinal glow discharge was maintained by the DC power supply with a voltage of up to 26 kV. The discharge chamber was cooled by flowing water passing through fittings (*12*) into the water heat exchanger jacket (*2*).

The laser cavity is formed by two plane circular mirrors (*4*) and (*8*) with the diameter of 20 mm, the lens (*5*) is made of ZnSe (the phase corrector serving as the Fourier transformation element in the simulation model of the laser cavity) with the diameter of 19 mm, focal length of 76 mm, and a fraction of the hollow dielectric waveguide (*1*) with the diameter of 4 mm and length of 460 mm. The waveguide length $L \approx 1.2 a_1^2 / \lambda$ was chosen to satisfy the self-imaging conditions in a hollow dielectric waveguide for the radiation beams with a super-Gaussian field amplitude distribution. The gap in the waveguide (the length of approximately 10 mm) did not affect the character of field replication in the waveguide and provided space for placing a hollow cylindrical cathode in the considered laser construction for obtaining a stable glow DC discharge. The discharge gap length was 370 mm.

The distance between the phase corrector and the nonuniform mirror from one side and the waveguide from the other side was chosen equal to the focal length. The inner diameter of the glass tube (*17*) is 15 mm and does not affect the field formation in the cavity. The radiation beam in this section propagates as in free space. In the experiments, for the mirror (*8*) we used the plane mirror made from stainless steel with gold deposited in vacuum, or the plane nonuniform amplitude-stepped mirror (ASM) made by the method of photolithography – by depositing thin annular strips from an absorbing material to a similar substrate. Parameters of the nonuniform mirror were preliminarily calculated according to (1). After fabricating the mirror, the widths of the reflecting and absorbing rings were measured and substituted again into (1) for recalculating the mode characteristics of the real cavity model.

The measured diameter of the central reflecting circle of the ASM was 0.743 ± 0.005 mm ($70.11\lambda \pm 0.51\lambda$). The widths of the reflecting and corresponding absorbing rings

The radiation of laser was extracted through the plane semi-transparent germanium mirror (*4*) with the transmission coefficient of ~10 %. The mirror was arranged at a distance not longer than 1 mm from the waveguide end. The highly reflecting mirror (*8*) and semi-transparent mirror (*4*) were mounted in adjustment units (*3, 6*), which provided exact adjustment of the cavity by the radiation of a He-Ne laser. The cavity was supported by three invar rods (*10*), which provided the long-term stability of the cavity length. The construction was made rigid due to supports (*11*), which also made it possible to align the system comprising the waveguide, discharge chamber, phase corrector, ASM, and semitransparent mirror. The highly reflecting mirror was mounted on the piezoelectric corrector (*7*) of type KP-1, which served to remotely change the cavity length in the limits ± 5 µm.

4. COMPARISON OF EXPERIMENTAL AND NUMERICAL RESULTS

Fig. 3 presents the calculated relative transversal distributions of the intensity and phase of the field on the output mirror for the experimental model of laser at the amplitude spatial filter. The absolute measure Π of distinct between the etalon circular function $\text{circ}\rho_1$ and the profile of field intensity $I(\rho_1)$ at the output mirror is defined, for the case of non-uniform mirror in the cavity, as

$$\Pi = \frac{1}{S} \sum_{s=1}^{S} |1 - I(\rho_{1s})|,$$

where S is the number of points of the discrete set of the field intensity on the mirror between the maximal and $1/e^2$ intensity levels. The value of Π does not exceed 30 %. The value of field quasi-uniformity is comparable to that obtained by other methods.

The measurements were performed at the total pressure of the working mixture in the discharge tube of 18 mm Hg and the discharge current of 10 mA. The laser operated in a single-mode regime. The radiation power of the laser measured by a calorimetric power meter of type IMO-2N was 2.5 W for the uniform mirror and 1.9 W for the amplitude-stepped mirror. The fall of power in the case of amplitude-stepped mirror is explained by greater waveguide and diffraction losses.

Due to constructive features of CO_2 laser it was impossible to measure the radiation intensity profile directly on the output mirror; thus, the experimental distributions of output beam intensity were recorded at various distances from the semi-transparent mirror in the near field. Experimental and the corresponding calculated intensity profiles are presented in Fig. 4. The experimental transversal intensity distributions of output radiation were recorded by a scanning pyroelectric detector (with the spatial resolution of 0.2 mm) in the plane normal to the radiation propagation direction. One can see that the profiles slightly differ. There are characteristic oscillations at the peak of both experimental and calculated transverse

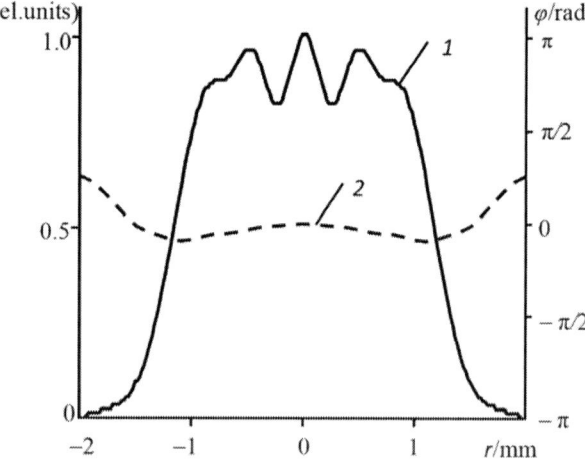

Fig. 3. Calculated radial distributions of (*1*) intensity and (*2*) phase of the field on laser output mirror.

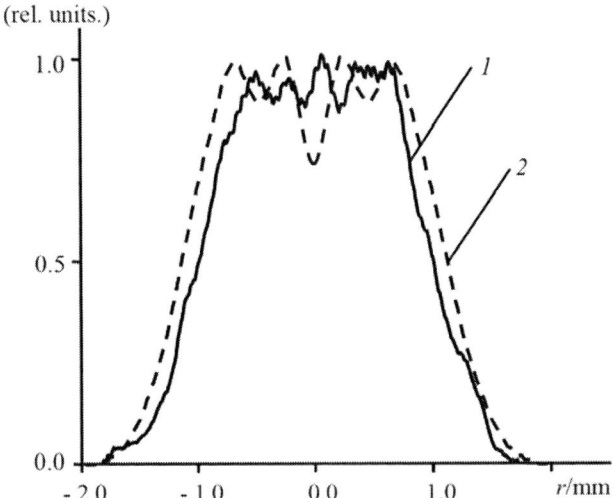

Fig. 4. (*1*) Experimental and (*2*) calculated radial radiation intensity distributions of the laser with an amplitude-stepped mirror at a distance of 20 mm from the output mirror.

intensity distributions. A fall of the laser output power by 10 % in detuning the cavity did not result in a noticeable change in the output field distribution. Calculations show that at the same parameters of the nonuniform mirror, the absolute distinct measure is changed by at most 5 % for the two wavelengths corresponding to the neighbouring laser generation lines 10P(20) and 10P(22).

For testifying the close-to-uniform radial distribution of field intensity at the output of the waveguide CO_2-laser we investigated the intensity distribution at a focus of the positive lens with $f = 130$ mm. The latter was placed so that one of its focal planes coincided with the plane of laser output mirror. In this case the beam intensity distribution can be recorded in the other focal plane, which will be the Fourier-transform of the

function, which describes the distribution of the field formed on the semi-transparent mirror. In our case it should be a curve close to the 'sombrero' function. The obtained experimental and calculated curves are presented in Fig. 5. A comparison confirms that there are no qualitative discrepancies as well. The amplitude of side lobes in the experimental curve is comparable to the noise amplitude of the pyroelectric detector.

Experimental and calculated relative radial distributions of radiation intensity for the laser with a uniform mirror are shown in Fig. 6 in the same cross-section as in Fig. 4. One can see that the radiation profile corresponds to a transversal distribution of the waveguide mode EH_{11}. Comparison of distributions in Figs 4 and 6 reveals that employment of the amplitude-stepped mirror makes it possible to obtain a quasi-uniform output beam in a waveguide laser.

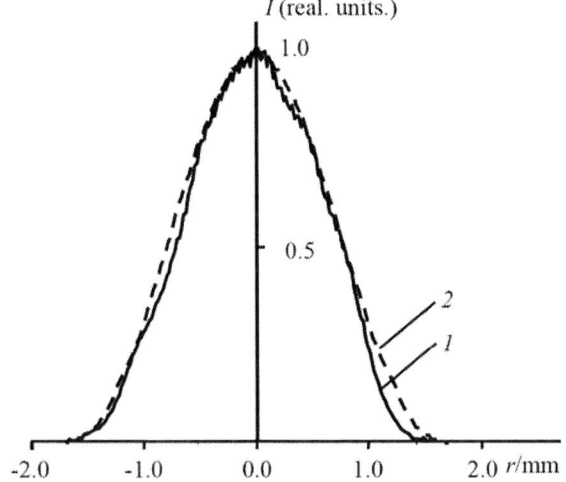

Fig. 6. (*1*) Experimental and (*2*) calculated radial distributions of the radiation intensity of the laser with a uniform mirror at a distance of 20 mm from the output mirror.

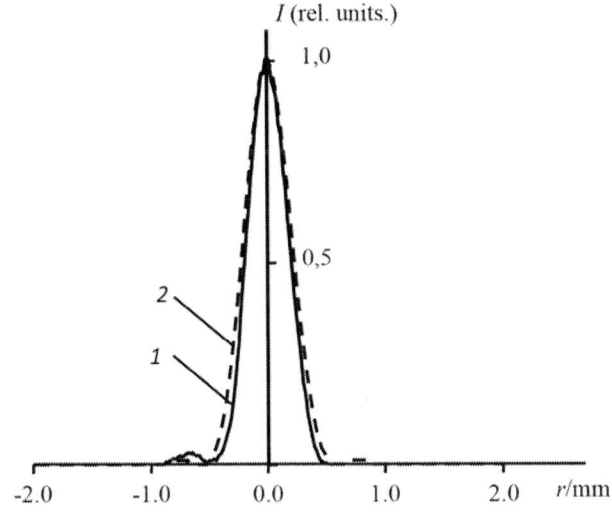

Fig. 5. (*1*) Experimental and (*2*) calculated radial distributions of the radiation intensity at the lens focus.

REFERENCES

[1] N. Hodgson, H. Weber, *Laser Resonators and Beam Propagation: Fundamentals, Advanced Concepts and Applications*, New York: Springer, 2005.

[2] V.N. Ochkin, *Waveguide Gas Lasers*, Moscow: Znanie, 1988.

[3] J.W. Goodman, *Introduction to Fourier Optics*, New York: McGraw Hill, 1968.

[4] B.Z. Katsenelenbaum, *High-Frequency Electrodynamics*, Moscow: Nauka, 1966.

[5] O.V. Gurin, V.A. Maslov, V.A. Svich et al., in Radiotekhnika, Kharkov: Kharkov Nat. Univ. Radioelectr., 2001, Issue 121, pp.117-120.

[6] V.P. Bykov, O.O. Silichev, *Laser Cavities*, Moscow: Fizmatlit, 2003.

[7] E.A.J. Marcatili, R.A. Schmeltzer, Bell. Syst. Techn. J., vol.43, no. 4, pp.1783-1809, 1964.

CONICAL 90^0 MIRRORS IN A TERAHERTZ GAS-DISCHARGE LASER

V.P. Radionov[1], V.K. Kiseliov[1,2],

[1] A. Ya. Usikov Institute of Radiophysics and Electronics of National Academy of Sciences of Ukraine,
12 Ac. Proskury Street, Kharkov, 61085, Ukraine.
[2] V.N. Karazin Kharkov National University, 4 Svoboda Square, 61022 Kharkov, Ukraine.

Tel: 380 57 7203335, E-mail: kiseliov@ire.kharkov.ua

Abstract: For the purpose of improvement of economic and consumer qualities of terahertz lasers test of conical 90^0-th mirrors was carried out. Our experiments showed that a conical mirror does not require precise alignment. Use of such mirrors gives the chance to increase stability of radiation, considerably to simplify a design of the laser and to reduce labor input of its service.

Gas lasers are among the most common and available generators in the mid and high parts of the terahertz range of frequencies (from ~1 to 10 THz). However, essential shortcomings of such lasers are their constructional complexity and labor input of operation. High qualified personnel are required even for the maintenance of gas-discharge lasers, which are relatively simple. One of the most responsible and difficult operations which should be carried out during using such lasers – an adjustment of mirrors of the resonator. For ensuring stability of radiation it is necessary to support an adjustment of mirrors in the course of operation of the laser. Various systems of stabilization of elements of fastening of the mirrors, considerably complicating a laser design are used for this purpose. Development of the laser resonators which mirrors do not require an adjustment, is an actual task.

The application of conical mirrors in lasers resonator is known [1, 2]. Features of the alignment process of conical 90^0 mirrors are of great interest. This paper presents the results of the test terahertz gas-discharge HCN - laser with a conical 90^0 mirrors. The laser generates a laser beam with a wavelength of 337 μm. It was found that the conical 90^0 mirrors require no alignment.

The resonator consists of a discharge tube – waveguide 1 and two mirrors 2, 3. The length of resonator is 150cm, diameter of the discharge tube is 50mm. The mirror 2 is made in the form of a conical reflective surface with an apex angle of about 90^0. A mirror 2 is equipped with a mechanical device 4 that moves a mirror 2 along the resonator axis. Setting the length of the resonant cavity is produced with the help of this mechanism 4. The exit mirror 3 is made in the form of a conical reflective surface with an apex angle of about 90^0 (a little less). In the center of a mirror 3 is made a hole, through which radiation is emitted from the resonator. The diameter of the hole is 4mm. It should be noted that optimization of coefficient of a transmission of an output mirror wasn't performed. The direction of propagation of radiation in the resonator is shown by arrows 5. The active substance is synthesized in the discharge tube under the impact of pump energy. The laser light is generated and amplified in the active substance as a result of multiple reflections from the mirrors 2, 3. The radiation which has got to area of the central opening in a mirror is brought out of the resonator. Laser radiation gets into the central hole through the diffraction divergence of the laser beam. The fact that a conical mirror 3 has an apex angle of a little less than 90^0 displaces laser radiation to the resonator axis.

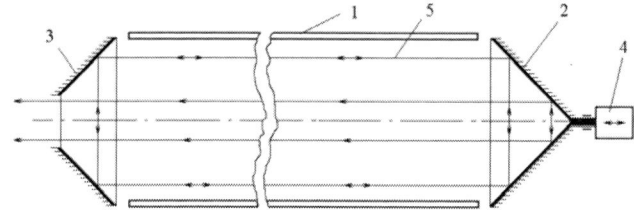

Fig. 1. The scheme of the laser resonator with two conical 90^0 mirrors.

These two conical 90^0 mirror laser resonator shown in Fig. 2.

Fig. 2. Conical 90^0 mirrors.

The experimental results demonstrated that the skewing conical mirror is more than an order of magnitude smaller impact on generating process than skewing the flat mirror. Comparison of the influence of skewing the flat mirror and conical mirror was tested in a resonator formed by a flat mirror and a conical mirror (Fig. 3).

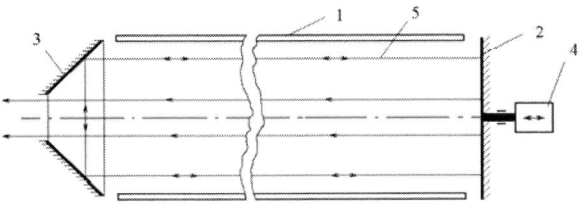

Fig. 3. The scheme of the laser resonator with a flat mirror and a conical 90^0 mirror.

The resonator consists of a discharge tube – waveguide 1 and two mirrors 2, 3. The length of resonators is 150 cm, diameter of the discharge tube is 50mm. The mirror 2 is made of metal and has a flat surface. This mirror 2 is equipped with a mechanical device 4 that moves the mirror 2 along the resonator axis. The exit mirror 3 is made in the form of a conical reflective surface with an apex angle of about 90^0. In the center of a mirror 3 is made a hole, through which radiation is emitted from the resonator. The diameter of the hole is 4mm. The direction of propagation of radiation in the resonator is shown by arrows 5. The active substance is synthesized in the discharge tube under the impact of pump energy. The laser light is generated and amplified in the active substance as a result of multiple reflections from the mirrors 2, 3. Laser radiation, hitting into the central hole in the mirror 3, comes out of the resonator. Laser radiation gets into the central hole through the diffraction divergence of the laser beam. The fact that a conical mirror 3 has an apex angle of a little less than 90^0 displaces laser radiation to the resonator axis.

The fig. 4 shows the dependence of the laser power on the angle skewing mirrors (1 - skewing flat mirror, 2 - skewing conical mirror).

Our experiments allow making a conclusion that conical 90^0 mirrors do not require precise alignment. Consequently, complex and expensive adjustment mechanisms of these mirrors can be excluded from the laser. In addition, the use of such mirrors can significantly simplify the laser. In principle, it is possible to exclude a system of thermally stabilized rods, on which are installed the mirror from a laser design. Thus mirrors can be fixed directly on a resonator wave guide. Thermal and mechanical effects have little influence on alignment of this resonator. Stability of the generation can be provided by a an auto-adjusting system in the cavity length [3]. Therefore, application of conical mirrors gives the chance to simplify considerably a design of the laser and to reduce its cost, and also, to simplify laser service. However, it is necessary to consider that conical 90^0 mirrors bring a little big losses in the resonator, at the expense of double reflection.

However, it should be noted that the conical mirrors makes some big losses in the resonator due to the double reflection. Furthermore, the conical 90^0 mirror are difficult to manufacture.

REFERENCES

[1] 1. Yurij N. Parkhomenko, Boris Spektor, Joseph Shamir "Mode Selection in Resonators With Conical Reflectors", IEEE JOURNAL OF QUANTUM ELECTRONICS, Vol. 44, No. 5, pp. 456-461, MAY 2008.

[2] 2. Y.E. Kamenev, A.M. Korobov, V.P. Radionov "Laser", Author's certificate USSR №1829832, 1992.

[3] S.V. Mizrakhi and V.K. Kiseliov. Automated Submillimeter Laser Power Stabilizing // Telecommunications and Radio Engineering – 2007, Vol. 66, No 1, pp. 89-96.

Fig. 4. Dependence of the lasing power P from the angle mirrors for a gas-discharge HCN laser. **1** - skewing flat mirror, **2** - skewing conical mirror.

High efficient vertical LED with pattern surface texture

M. H. Mustary, V. V. Lysak

Chonbuk National University, Chonju, Republic of Korea

Abstract: A quantitative investigation of light extraction efficiency for hemispherical and pyramidal pattern surface textured GaN based vertical light emitting diode (VLED) was reported by various distance and size ratios. There is a significant increase in light extraction efficiency because of increase number of random scattering of textured surface. Hemisphere pattern surface always gives the 30% better enhancement than pyramid type pattern. We also showed the efficiency dependency on the reflectivity of GaN/metal interface. Furthermore, a linear diffraction grating with certain grating period can eliminate the distance and size ratio dependency on output power which can be considered as an integrated surface texture.

GaN based light emitting diodes have been developed rapidly for many applications such as full color display, solid state lighting, backlighting in liquid crystal displays [1, 2]. In recent years, blue GaN LED together with yellow phosphor is one of the key component for high efficiency white LED. Although LED market is growing fast, further improvement of light output power is required for future demand. Only 4 % of emitted light from active layer can escape from LED because of low escape cone (24°) at the GaN-air interface [3]. In this case texture the surface is an efficient way to improve the light output power because of multiple chances of light escape from the surface. However, vertical LED on metal substrate shows many excellent performances over lateral LED such as vertical current path for low operation voltage, better current spreading, higher heat dissipation, better light extraction, higher driving current density, flexible chip size scaling and good reliability [4, 5, 6]. In addition, various types of texture can be easily formed on the comparatively thick n-side up VLED. Periodically pattern micro structures such as hemisphere holes [7], pillar [8], air prism array [9, 10], photonic crystal [11], and hexagonal cone [12] have been successfully demonstrated in producing improved light output. In addition to one-step texture two-step surface patterning, combines with micro and sub micro structures also investigated [13, 14]. Considering these, optimization of surface textured VLED is an indispensable issue for best light extraction.

In this work, we represent a simulation analysis on texture surface VLED of different distance to size (D/S) ratios. Extraction efficiency of VLED with hemisphere and pyramid type texture surface was analyzed at three different sizes 0.1, 0.5, 2.0 μm respectively. Results showed that for more efficiency more compact pattern is required. There is no dependency of output power on the size of pattern. We also took this investigation considering different reflectivity of

GaN/metal interfaces 100% and 80% respectively because reflectivity becomes more important in textured surface VLED. Furthermore, applying a linear grating on hemisphere pattern surface with a certain nano scale grating period can solve the problem of making compact pattern.

Simulation was realized by Light Tools™ commercial package which is a 3D ray tracing tools required for optical design and engineering [15]. It allows you to set up, view, modify and analyze optical system graphically.

To analyze the LED, geometrical structure for vertical LED is presented in table 1. As conductive metal substrate is used for VLED, Si metal is used as substrate with refractive index 3.42 and dimension $1000 \times 1000 \times 100 \mu m^2$.

Table 1 shows the structural parameters of GaN based LED with thickness 4 μm. As the thickness of the GaN, AlGaN, N-GaN, MQW, P-GaN is very small compared to the surface size of LED we can ignore the photons that escape from four sides. Furthermore, total internal reflection also can be ignored as the indices of these layers are similar.

Table 1: Parameter of each layer of the simulated VLED. Chip size: $1000 \times 1000 \mu m^2$

Layer	Thickness(μm)	Refractive Index
Silicon	100	3.42
P-GaN	0.05	2.45
MQW	0.1	2.54
N-GaN	2.0	2.42
AlGaN	0.05	2.40
GaN	2.0	2.40

The surface source with lambertian intensity distribution is used as an emitter without the need of any additional features. The size of the source matches the size of the LED. After simulation the extraction efficiency of planar vertical LED was found to be 21.63%.

In the vertical injection LED, power efficiency mainly controlled by two interfaces, one is GaN/air interface and another is GaN/metal interface. The main reason for low extraction efficiency of LED is total internal reflection (TIR) of light at the semiconductor-air interface which results from the high refractive index difference between the semiconductor and air. According to Snell's law, critical angle [$\theta_c = \sin^{-1}(n_{air} / n_{GaN})$] in case of GaN (n_{GaN}=2.5) and air (n_{air}=1.0) interface is 24° [16]. So any light ray incident from MQW with an angle more than 24° on the GaN-air interface will be trapped inside the semiconductor.

978-1-4799-0019-0/13 $31.00 © 2013 IEEE

Surface texture is an important method in enhancing light extraction efficiency. The enhancement in light extraction is attributed to the multiple chances for lights to escape from the cone through surface texture. However, surface texturing is difficult for lateral LED because the top p-GaN layer is too thin for texturing and sensitivity of p-GaN to electrical deterioration and plasma damage. Any engineering on thin p-GaN layer can affect the underneath MQW. In case of VLED, all the surface engineering occurs on n-GaN layer ($>2\ \mu m$) which is much more thicker than p-GaN. So texture engineering on n-GaN layer will not affect the underneath MQW.

The reflectivity of metal/GaN layer plays an important role to enhance the extraction efficiency of light. The reflectivity at the metal/semiconductor interface can be calculated by using (1) [17] for reflection of a wave perpendicularly incident from media 1 onto the plane boundary of a solid with refractive index n. The ratio R of reflected-to-incident irradiance is given by Fresnel expression

$$R = \frac{(n_s - n_m)^2 + k_m^{\,2}}{(n_s + n_m)^2 + k_m^{\,2}} \qquad (1)$$

where n_s the refractive index of semiconductor, n_m is the refractive index of metal, and k_m is extinction coefficient of metal. Greater reflectivity at the semiconductor/metal interfaces gives more light output power of LED.

Fig. 1 shows the change of extraction efficiency (EE) with distance/size parameter of textured VLED with R=80% at the GaN/metal interfaces at the same radius (hemisphere) and side width (pyramid) as in fig. 3. In this case, EE increase up to 44% for hemisphere and 30% for pyramid whereas in case of planar surface it was only 15.4%. Therefore, we can conclude that for better EE it is essential to increase the reflectivity of GaN/metal interfaces because 20% increase in reflectivity results in more than 2 times increase in EE.

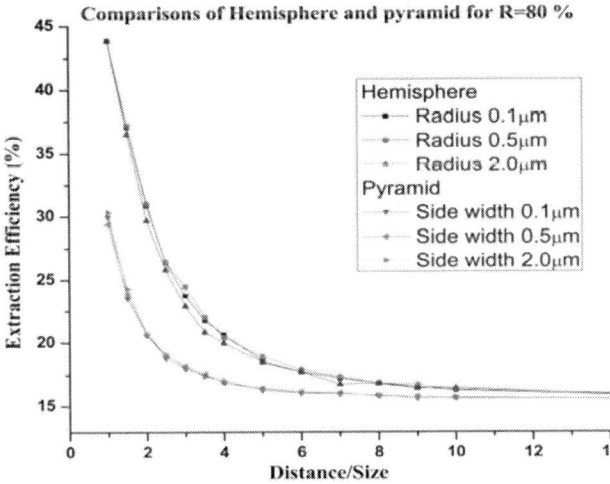

Fig. 1 Extraction efficiency of hemisphere and pyramidal pattern surface at different sizes as a function of distance/size of the texture at reflectivity R=80%

Results show that there is a weak dependence of light extraction efficiency on sizes of hemisphere and pyramid but strong dependence on the variation of distance/size parameter. For better efficiency compact pattern is required. This is because as the D/S increase, number of hemisphere/pyramid pattern decrease. Thus random scattering probability decreases.

Fig. 2 shows the efficiency of hemisphere pattern surface as a function of D/S by applying an additional linear grating on it. Analysis was taken on different grating period at 100nm, 200nm and 400 nm respectively. General grating equation is often referred to as

$$\sin\theta_d - \sin\theta_i = m\frac{\lambda/n}{\Lambda} \qquad (2)$$

where, θ_i is angle of propagated light incident on the texture surface and θ_d is the angle of diffracted light from the surface normal, m is the diffraction order, λ is free space wavelength, n is the refractive index of medium and Λ is spatial period of the scattering sites i.e. grating period. The value of diffraction order can be m= $0, \pm1, \pm2, \dots$.

If $\Lambda \gg \lambda$, many diffraction order m are possible and thus efficiency can increase. However, maximum propagating order is limited to < 90°. When this angle is exceed diffraction angle is no longer possible.

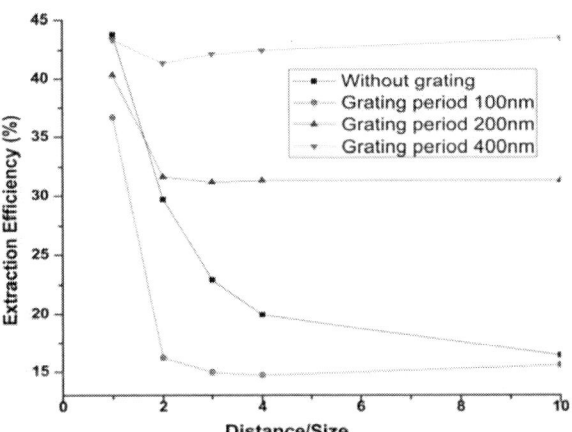

Fig. 2. Effect of grating period on the efficiency of hemisphere texture surface as a function of distance/size parameter

Above result indicate that, for grating period 100 nm, EE is even less than without grating because in this case only m=0 is worked. For grating period 400 nm, larger m is propagated that is why we get better efficiency. Moreover, in this case D/S dependency does no longer exist and we can get almost constant efficiency as a function of D/S.

In this work, surface textured GaN based vertical light emitting diode was designed and simulated. Texturing surface with hemispherical and pyramidal pattern at different placement between patterns was analyzed separately. We also investigate this structure at two different reflectivity 100% and 80% of GaN/metal interfaces and three different sizes of the pattern. Results shows that extraction efficiency depends strongly on the placement between the pattern and it is independent on the size of the pattern. There is an exponential decrease in efficiency as distance/size ratio increase. Texture surface gives 2.8 times and 2.0 times enhancement of efficiency for hemisphere and pyramid pattern respectively with R=80%. Hemispherical pattern always gives 30% more enhancement than pyramid. In addition to this applying a linear grating at a certain grating period can solve problem about the dependence of light output power on distance/size ratio.

REFERENCES

[1] T. Mukai, H. Narimatsu, and S. Nakamura" Amber InGaN-Based Light-Emitting Diodes Operable at High Ambient Temperatures," *Jpn. J. Appl. Phys.*, Part 2 vol. 37, pp. L479-481, 1998.

[2] E. F. Schubert, Light Emitting Diode (Cambridge University Press, Cambridge, England, 2003.

[3] A. I. Zhmakin "Enhancement of light extraction from light emitting diodes," Phys. Rep. Vol 498 no. 4-5, pp. 189-241, 2011.

[4] C. A. Tran, C. F. Chu, C. C. Cheng, W. H. Liu, J. Y. Chu, H. C. Cheng, F. H. Fan, J. K. Yen, Trung, Doan "High brightness GaN vertical light emitting diodes on metal alloyed substrate for general lighting application," J. Crystal Growth vol. 298, pp. 722-724, 2007.

[5] J.-T. Chu, H.-W. Huang, C.-C. Kao, W.-D. Liang, F.-I. Lai, C.-F. Chu, H.-C. Kuo, and S.-C. Wang, "Fabrication of large-area GaN-based light-emitting diodes on Cu substrate," Jpn. J. Appl. Phys. Vol. 44, no. 4B, pp. 2509–2511, 2005.

[6] W. J. Liu, X. L. Hu, J. Y. Zhang, G. E. Weng, X. Q. Lv, H. J. Huang, M. Chen, X. M. Cai, L. Y. Ying, B. P. Zhang "Low-temperature bonding technique for fabrication of high-power GaN-based blue vertical light-emitting diodes," Opt. Mater. Vol. 34 no. 8, pp. 1327-1329, 2012.

[7] J. Lee, J. Oh, S. Choi, Y. Kim, H. Cho and J. Lee, "Enhancement of InGaN-based vertical LED with concavely patterned surface using patterned sapphire substrate," IEEE Photon. Technol. Lett., vol. 20, no. 5, pp. 345-347, 2008.

[8] W. K. Wang, S. Y. Huang, S. H. Huang, K. S. Wen, D. S. Wuu, and R. H. Horng "Fabrication and efficiency improvement of micropillar

InGaN/Cu light-emitting diodes with vertical electrodes," Appl. Phys. Lett. Vol. 88, no. 18, pp. 181113 1-3, 2006.

[9] J. H. Kang, H. G. Kim, S. Chandramohan, H. K. Kim, H. Y. Kim, J. H. Ryu, Y. J. Park, Y. S. Beak, J. S. Lee, J. S. Park, V. V. Lysak, C.-H. Hong "Improving the optical performance of InGaN light-emitting diodes by altering light reflection and refraction with triangular air prism arrays," Optics Letters, vol. 37, no. 1, pp. pp. 88-90, 2012.

[10] V.V. Lysak, J. H. Kang, C.-H. Hong "Conical air prism arrays as an embedded reflector for high efficient InGaN/GaN light emitting diodes," Appl. Phys. Lett., vol. 102, no. 6, pp. 061114, 2013.

[11] I.V. Guryev, O. V. Shulika, I.A. Sukhoivanov, O.V. Mashoshina "Improvement of characterization accuracy of the nonlinear photonic crystals using finite elements-iterative method," Appl. Phys. B, Vol. 84, Nr 1-2, pp. 83 – 87, 2006.

[12] K. M. Uang, S. J. Wang, S. L. Chen, Y. C. Yang, T. M. Chen, and B. W. Liou "Effect of Surface Treatment on the Performance of Vertical-Structure GaN-Based High-Power Light-Emitting Diodes with Electroplated Metallic Substrates," Jpn. J. Appl. Phys. Part 1 vol. 45, pp. 3436-3441, 2006.

[13] W. C. Lee, S. J. Wang, K. M. Uang, T. M. Chen, D. M. Kuo, P. R. Wang, and P. H. Wang, "Enhanced light output of GaN-based vertical-structured light-emitting diodes with two-step surface roughening using KrF laser and chemical wet etching," IEEE Photon. Technol. Lett. Vol. 22, no. 17, pp. 1318–1320, 2010.

[14] H. Kim, K. K. Choi, K. K. Kim, J. Cho, S. N. Lee, Y. Park, J. S. Kwak, and T. Y. Seong "Light-extraction enhancement of vertical-injection GaN-based light-emitting diodes fabricated with highly integrated surface textures," *Opt. Lett.*, vol. **33** no. 11, pp. 1273-1275, 2008.

[15] *Light Tools, Core Module User's Guide Version 7.0*, Optical research associates, 2010.

[16] H. A. Macleod, *Thin Film Optical Filter*, 3rd ed. Bristol, U.K.: Institute of Physics Pub. 2001.

[17] H. W. Jang, S. W. Ryu, H. K. Yu, S. Lee and J. L. Lee "The role of reflective p-contacts in the enhancement of light extraction in nanotextured vertical InGaN light-emitting diodes," Nanotechnology, vol. 21, no. 2, pp. 025203, 2010.

978-1-4799-0019-0/13 $31.00 © 2013 IEEE

End diode pumped solid state mini-lasers passive Q-switched by colored polymeric matrix

A.O. Yaskovets,[1] P.V. Shpak,[2] V.I. Bezrodnyi,[1] A.M. Negriyko,[1] V.A. Orlovich,[2]

[1]Institute of Physics, Nat. Acad. of Sci. of Ukraine (46, Nauky Prosp., Kyiv 03028, Ukraine)

[2]B.I. Stepanov Institute of Physics, Nat. Acad. of Sci. of Belarus (68, Nezalezhnasty Ave., Minsk 220072, Belarus)

Abstract: We studied the generation characteristics of compact passively q-switched diode pumped Nd:YAG lasers. Lasing of 1064 nm with a pulse repetition of up to 100 kHz, a pulse duration of 2-27 ns and an average output power of 520 mW was realized with the use of a polymer q-switcher on the basis of polyurethane matrix doped with BDN (bis-(4-dimethylaminodithiobenzyl)-nickel) dye.

Diode pumped solid state mini-lasers which can produce high peak power and nanosecond duration pulses are very useful in the fields of nonlinear optics, spectroscopy, biology, environment sensing, etc. Q-switching is the technique of modulation Q-factor of laser cavity and allows the production of short light pulses. There are two main types of Q-switching: active and passive. Under active Q-switching, the modulator (usually an acousto-optic or an electro-optic device) is externally controlled by trigger event, typically an electric signal. In case of passive Q-switching, the Q-switcher is a saturable absorber, a material whose transmission increases when the light intensity exceeds some threshold. The passive Q-switch is initiated by the laser intensity in the resonator itself. Therefore, passive Q-switching technique compared to an active simplifies the operation and the alignment, improves compactness and reduces the costs of laser source. The most widely used passive Q-switchers (PQS) for Nd lasers operating near 1064 nm are Cr^{4+}: YAG, the films of irradiated lithium fluoride (LiF: F_2^-), semiconductor absorption mirrors (SESAMs) and bleachable dye incorporated in a polymeric matrix.

Polymeric PQS may be produced in the form of triplex – dye doped polymeric film placed between two glass substrates [1]. This allows avoiding processes of polishing and grinding which present in fabrication of crystalline elements. The high polymer adhesion gives mechanical strength and increases lifetime as compared to the brittle crystal PQS one. Such a polymeric PQS can be easily made under laboratory conditions and this process is quite cheap.

We have investigated the generation dynamics of Q-switched solid-state mini-lasers. The minimal laser cavity length of 18 mm was used in our experiments. Pump duration was 10 ms with repetition rate of 10 Hz. Under the quasi-periodical pumping regime laser produces train of the pulses (Fig.1). Pulse energy and duration depends on initial transmission of the Q-modulator (see Table 1.). We obtained the value of pulse energy ten times more compared to those, obtained by authors in work [2] for diode pumped mini-laser q-switched by

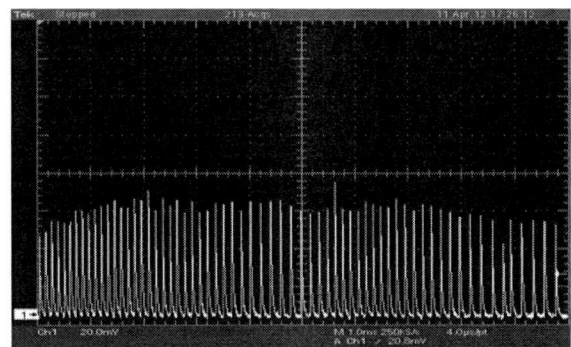

Fig.1 An oscilloscope pulse train from the q switched minilaser

Table 1. Generation characteristics of mini-lasers passively q-switched by polymeric saturable absorbers

Initial transmission of saturable absorber, %	FWHM, ns	Pulse energy, µJ	The number of laser pulses, N
94,5	27	1	$5 \cdot 10^6$
89	8	5	$3,3 \cdot 10^7$
65	3,2	7	$1,4 \cdot 10^7$
47	2,15	15	$1,1 \cdot 10^7$

polymeric PQS. We have tested our PQS in continuous pumping regime also.

It should be noticed, that not any additional cooling of triplex was applied. The heat from polymeric film dissipated in air through glass substrates. The number of pulse generation within laser radiation parameters, such as pulse duration, pulse energy, remain unchanged, are shown in Table 1. It was established that the q-switcher with smaller initial transmission have better parameters reproducibility as compared with absorbers with larger transmission one, because under the same pumping conditions the first operate at low pulse repetition rate.

So, laser generation in diode pumped solid state mini-laser passive q-switched by polyurethane triplex was realized and characteristics, and lifetime of polymeric PQS were measured.

REFERENCES

[1] V.I. Bezrodnyi, L.V. Vovk, N.A. Derevyanko, A.A. Ishchenko, L.V. Karabanova, and I.L. Mushkalo, Kvant. Elektron. 22, 245 (1995).

[2] A. Inoue, J. Hayashi, T. Komikado, and S. Umegaki, Opt.Lett. 32, 2807 (2007).

Hyperbolic Airy beams

(Invited Paper)

V.V. Kotlyar, A.A. Kovalev

Laser Measurements Laboratory, Image Processing Systems Institute of the Russian Academy of Sciences, 151
Molodogvardejskaya street, Samara 443001, Russia

Abstract - We discuss finite-energy and infinite-energy Airy beams of the second kind. Similarly to the well-known paraxial Airy beams, these are accelerating beams that can be analytically described in the Fresnel diffraction zone but propagate along a hyperbolic rather than parabolic path over a certain length, with their "gravity center" shifting linearly with distance. The Airy beams of the second kind can be generated near a phase transparency with cubic phase modulation.

The solution to the paraxial equation of propagation in the form of Airy functions was first offered in [1-3]. The Airy beams discussed in Refs. [1-3] possess an infinite energy. In [4,5] finite-energy Airy beams in optics were analyzed. The solution of the equation

$$2i\frac{\partial E}{\partial \xi} + \frac{\partial^2 E}{\partial s^2} = 0 \qquad (1)$$

has been found in the form of a function [4]

$$E(s,\xi) = \text{Ai}\left(s - \xi^2/4 + ia\xi\right) \times$$
$$\times \exp\left(is\xi/2 + ia^2\xi/2 - i\xi^3/12 - a\xi^2/2 + as\right), \qquad (2)$$

which takes the form of the exponentially apodized Airy function at the input, $\xi = 0$:

$$E(s,\xi = 0) = \text{Ai}(s)\exp(as), \quad a > 0, \qquad (3)$$

where $s = x/x_0$, $\xi = z/(kx_0^2)$ are the dimensionless transverse and longitudinal Cartesian coordinates, $k = 2\pi/\lambda$ is the wavenumber, x_0 is an arbitrary transverse scale, Ai(x) is the Airy function, and a is constant. The Airy beam can be produced by passing a Gaussian beam through a phase transparency with cubic dispersion on the transverse coordinate, followed by implementing the Fourier transform with a spherical lens. This can be inferred from the Fourier image of the input field (3):

$$\tilde{E}(t) = \exp\left(-at^2\right)\exp\left(it^3/3 - ia^2t + a^3/3\right). \qquad (4)$$

The key peculiarity of the Airy beam is that its major lobe propagates along a curved trajectory, which has the form of a parabola. Analytic expressions for different types of the accelerating beams, including the Airy beams in the ABCD optical system have been obtained [6, 7]. Propagating beams with their rays forming a desired caustic curve have been examined [8, 9]. Accelerating beams that propagate on a circular and elliptic trajectory have also been studied [10, 11]. To our knowledge there have been no publications handling the Airy beams accelerating along a hyperbolic trajectory. In this work, we show that in an optical setup conventionally employed to generate the Airy laser beams of Eq. (2) there is a path section found immediately behind the transparency of Eq. (4) on which the Airy beam is propagating with acceleration on a hyperbola. We have termed such a beam as the Hyperbolic Airy (HA) beam. While lacking the property of being diffraction-free upon propagation on the hyperbolic path, the HA beam shows a number of other notable properties, such as having high acceleration (though rapidly decaying with distance due to highly curved path) and preserving its shape up to a scale, i.e. showing the linear divergence. To analyze a beam which is generated in the Fresnel diffraction zone of the phase transparency (4), let us find the complex amplitude of the Gaussian beam directly behind the phase transparency:

$$E(x,0) = \exp\left[-(x/w)^2 + i\alpha(x/x_0)^3 + i\beta(x/x_0)\right], \qquad (5)$$

where w is the Gaussian beam's waist radius and λ and β are dimensionless parameters of the phase transparency. Then, the paraxial approximation of the light field amplitude at distance z is given by

$$E(x,z) = \sqrt{\frac{-i2\pi k}{z}}\frac{x_0}{\sqrt[3]{3\alpha}} \times$$

$$\exp\left[\frac{1}{3\alpha}\left(\frac{x_0}{w}\right)^2\left(\beta - \frac{kx_0 x}{z}\right) + \frac{2}{27\alpha^2}\left(\frac{x_0}{w}\right)^6\left(1 - \frac{3z_0^2}{z^2}\right)\right] \times$$

$$\exp\left[\frac{ikx^2}{2z} - \frac{iz_0}{z}\frac{1}{3\alpha}\left(\frac{x_0}{w}\right)^2\left(\beta - \frac{kx_0 x}{z}\right)\right] \times$$

$$\exp\left[-\frac{2i}{27\alpha^2}\left(\frac{x_0}{w}\right)^6\left(3\frac{z_0}{z} - \frac{z_0^3}{z^3}\right) + ikz\right] \times \qquad (6)$$

$$\text{Ai}\left\{\frac{1}{\sqrt[3]{3\alpha}}\left[\beta - \frac{kx_0 x}{z} + \frac{1}{3\alpha}\left(\frac{x_0}{w}\right)^4\left(1 - \frac{iz_0}{z}\right)^2\right]\right\},$$

where $z_0 = kw^2/2$ is the Rayleigh range. The expression in Eq. (6) describes an HA beam. It is noteworthy that while the propagation of the Airy beam in an ABCD optical system was discussed in Ref. [6], a relationship similar to Eq. (6) was not deduced and the possibility of the beam acquiring the acceleration upon propagation on a hyperbolic path was not discussed. The relationship in Eq. (6) suggests that unlike the linear phase of the Airy beam (2), the HA beam has a quadratic phase, thus experiencing the divergence upon propagation. Besides, similarly to Eq. (2), the argument of the Airy function in Eq. (6) is complex, although the z-dependence is different: the value of the Airy function argument is in direct proportion to z^2 in Eq. (2) and in inverse proportion to z in Eq. (6). It is possible to obtain the infinite energy Airy beams if the cubic phase transparency is illuminated by a plane wave ($w \rightarrow \infty$) rather than a Gaussian beam. Then, we obtain, instead of Eq. (6):

$$E(x,z) = \sqrt{\frac{-i2\pi k}{z}} \frac{x_0}{\sqrt[3]{3\alpha}} \times$$
$$Ai\left[\frac{1}{\sqrt[3]{3\alpha}}\left(\beta - \frac{kx_0 x}{z} - \frac{k^2 x_0^4}{12\alpha z^2}\right)\right] \times \qquad (7)$$
$$\exp\left[\frac{ik}{2z}\left(x^2 + \frac{kx_0^3 x}{3\alpha z} - \frac{\beta x_0^2}{3\alpha} + \frac{k^2 x_0^6}{54\alpha^2 z^2}\right) + ikz\right].$$

Because the phase relation remains quadratic, the beam in Eq. (7) will diverge upon propagation. As opposed to the beam in Eq. (6), the Airy function argument has become real, making it possible to deduce an equation for the beam path. Putting the Airy function's argument equal to values y_m at which the function has local maxima, we find an explicit expression of the trajectory of the HA beam maximum:

$$x = \frac{\left(\beta - y_m \sqrt[3]{3\alpha}\right)z}{kx_0} - \frac{kx_0^3}{12\alpha z}. \qquad (8)$$

Unlike a parabolic trajectory of the beam (2), the HA beam propagates along a hyperbolic path (8). Unlike the beams (2), the acceleration is not uniform and decreases as z^{-3}. Let us analyze the following parameters: $x_0 = \lambda = 532$ nm, $\alpha = -1$, $\beta = 10$, $m = 0$, $y_0 = -1.01879$. In this case, the condition (9) is satisfied, so that the trajectory has the acceleration at $z > z_1 \approx 330$ nm. The intensity pattern of Eq. (7) is depicted in Fig. 1a. Fig. 2 depicts the intensity profiles in the planes (a) $z = \lambda$, (b) 2λ, and (c) 4λ.

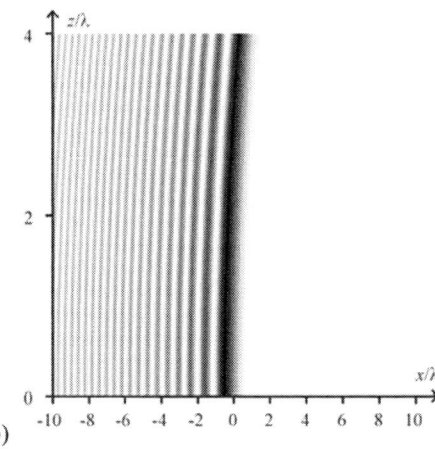

b)

Fig. 1. Intensity pattern of the infinite energy accelerating HA(a) and Airy (b) beam in the xz-plane.

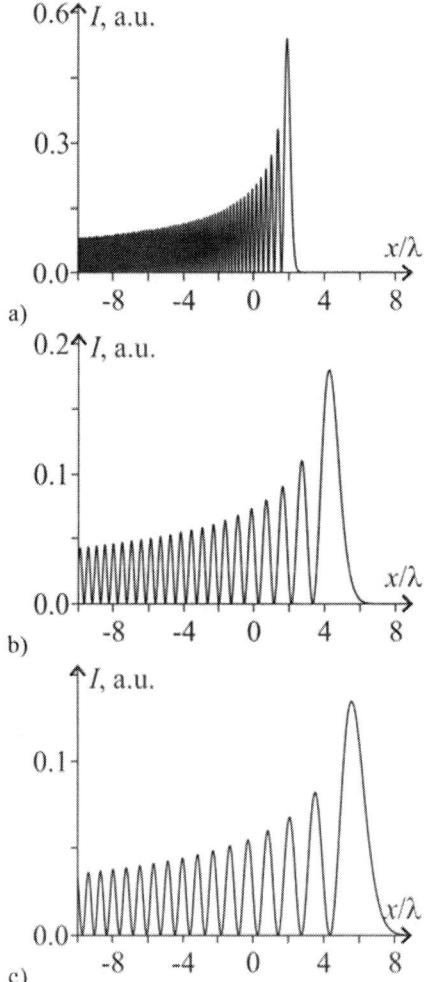

a)

b)

c)

Fig. 2. Intensity profiles for the HA beam in the planes (a) $z = \lambda$, (b) 2λ, and (c) 4λ.

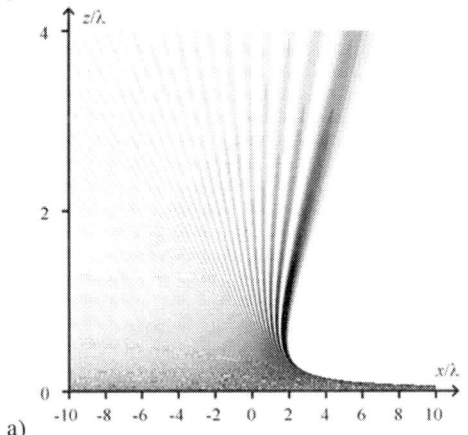

a)

For comparison, let us examine the Airy beam of Eq. (2) at $a = 0$. Putting the Airy function's argument in Eq. (2) equal to values of y_m at which it has local maxima, the trajectory of the Airy beam maximum can be explicitly expressed as $x = x_0 y_m + z^2/(4k^2 x_0^3)$. From the equation, the beam is seen to have a uniform acceleration equal to $1/(2k^2 x_0^3)$, while the Airy beam of Eq. (2) is seen to be diffraction–free, because $x_1 - x_2 = x_0(y_m - y_n)$ is independent of z. Meanwhile for the HA beam, from Eq. (7) it follows that $x_1 - x_2 = (3\alpha)^{1/3} z(y_m - y_n)/(kx_0)$ and the beam shows a linear divergence with increasing z (see Fig. 2). Figure 1b shows the intensity pattern from the field (2) for the following parameters: $\lambda = 532$ nm, $x_0 = \lambda/2$. The acceleration of the beam (2) at $a = 0$ equals $1/(\pi^2\lambda)$, whereas for the HA beam of Eq. (8) shown in Fig. 1, the acceleration at $z = z_1 \approx 330$ nm is about $19.87/(\pi^2\lambda)$. This is the reason why the beam's trajectory in Fig. 1a is more curved. Note that both beams in Fig. 1a and Fig. 1b have been computed under the same conditions.

The work was partially funded by the Federal Program "Research and Academic Cadres of Innovative Russia" (Agreement # 8027), the RF President's grants for Support of Leading Scientific Schools (NSh-4128.2012.9) and Young Doctors of Sciences (MD-1929.2013.2), as well as the Russian Foundation for Basic Researches grants (## 12-07-00269, 12-07-31117, 13-07-97008).

REFERENCES

[1] E.G. Kalnins, W. Miller Jr., "Lie theory and separation of variables," J. Math. Phys. 15, 1728–1737 (1974).

[2] M.V. Berry, N.L. Balazs, "Nonspreading wave packets," Am. J. Phys. 47, 264–267 (1979).

[3] I.M. Besieris, A.M. Shaarawi, R.W. Ziolkowski, "Nondispersive accelerating wave packets," Am. J. Phys. 62, 519–521 (1994).

[4] G.A. Siviloglou, D.N. Christodoulides, "Accelerating finite energy Airy beams," Opt. Lett. 32, 979–981 (2007).

[5] G.A. Siviloglou, J. Broky, A. Dogariu, D.N. Christodoulides, Phys. Rev. Lett. 99, 213901 (2007).

[6] M.A. Bandres, J.C. Gutierrez-Vega, "Airy-Gauss beams and their transformation by paraxial optical systems," Opt. Express 15, 16719–16728 (2007).

[7] M.A. Bandres, "Accelerating parabolic beams," Opt. Lett. 33, 1678–1680 (2008).

[8] E. Greenfield, M. Segev, W. Walasik, O. Raz, "Accelerating light beams along arbitrary convex trajectories," Phys. Rev. Lett. 106, 213902 (2011).

[9] D.M. Cottrell, J.A. Devis, T.M. Hazard, "Direct generation of accelerating Airy beams using a 3/2 phase-only pattern," Opt. Lett. 34, 2634–2636 (2009).

[10] I. Kaminer, R. Bekenstein, J. Nemirovsky, M. Segev, "Nondiffracting accelerating wave packets of Maxwell's equations," Phys. Rev. Lett. 108, 163901 (2012).

[11] P. Aleahmad, M. Miri, M.S. Mills, I. Kaminer, "Fully vectoral accelerating diffraction-free Helmholtz beams," Phys. Rev. Lett. 109, 203902 (2012).

Dielectric properties of some practical-use materials in the low-frequency part of the terahertz band

V. V. Meriakri, E. E. Chigryai, I. P. Nikitin

Kotel'nikov Institute of Radio Engineering and Electronics, Russian Academy of Sciences, Fryazino, Moscow region, Russia

(Invited Paper)

Abstract: Dielectric properties of practical use- materials are measured in the low-frequency part of the terahertz band using beam-waveguide spectroscopy methods. These properties are important for the study of propagation of these waves in urban and indoor conditionstons, as well as for the problems of introscopy and radio imaging.

The growing application of terahertz waves stimulates interest in the study of the absorbing, transmitting, and reflecting properties of many practical-use materials in the terahertz-wave band, which are crucially important for the problems of propagation of terahertz waves in urban and indoor conditions, as well as in the problems of introscopy and imaging. These characteristics are determined by the complex refractive index $n + ik$ and loss tangent $tan\delta$ of materials. We investigated the dielectric properties of the above-mentioned materials in the range of wavelengths λ from 2 to 0.6 mm (frequencies f = 0.15–0.5THz) using the beam-waveguide quasi-optical spectroscopy methods [1].

For wavelengths λ longer than approximately 6÷8 mm, there exist effective waveguide techniques for measuring the dielectric properties of materials. On the other hand, for wavelengths shorter than 0.5 mm, the Fourier spectroscopy, laser, and time domain spectroscopy methods based on geometrical optics principals prove to be efficient.

However, there are some difficulties to use these methods in the wavelength interval from 5÷4 mm to 0.6÷0.5 mm. due to a decrease in the waveguide dimensions and insufficient accuracy of geometrical optics formulas due to diffraction effects [1, 2].

Therefore the best way for the measurement of material properties in this frequency band is to use a quasi-optical lens beam waveguide [3]. This waveguide transmits only the fundamental low-loss Gaussian-type mode with small enough cross-sectional dimensions of a wave beam and high losses for higher order modes. In this case, the wave structures incident on the plane-parallel specimen and the wave on the receiving aperture are the same, and it is possible to estimate the measurement errors due to a thickness l of a plane-parallel specimen and cross-sectional dimensions of the specimen a and receiving aperture b [4]. Thus, for a beam waveguide consisting of lenses with $a \approx b > 10\lambda$, $l \leq 1$ cm, and the refractive index is

$n < 5$, the error in power transmission $|t|^2$ and reflection $|r|^2$ coefficients are less than $5 \cdot 10^{-3}$ [4], whereas, in conventional free space measurements, these values may be about or greater than 10^{-1} [2]. Using this quasi-optical beam waveguide set up (Fig.1), we measured the complex refractive indexes $n + ik$ in a wide interval of n from 1.05 to 10, $|t|^2$ from approximately 1 to 10^{-6}, $\tan\delta = \dfrac{2n\kappa}{n^2 - \kappa^2}$ from 10^{-4} to 1, and $|r|^2$ from ≈ 1 to 10^{-3}.

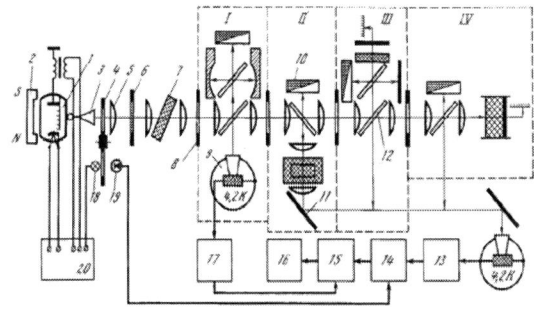

Fig. 1. Schematic diagram of the measurement setup; I – resonator for low loss material properties measurement, II – transmission measuring circuit, III– Michelson or Max–Zehnder interferometer, IV – reflectometer; 1 – backward wave oscillator, 2 – magnet, 3 – horn, 4 – modulator, 5 – lens, 6 – polarizer, 7 – attenuator, 8 – iris, 9 – receiver, 10 – absorber, 11 – mirror, 12 – beam splitter, 13 – amplifier, 14 – synchronous detector, 15 – digital voltmeter, 16 – storage unit, 17 – volt- meter, 18 – light source, 19 – LED, 20 – power supply.

The experimental setup I for measuring very low-loss samples is based on a hemispherical Fabry-Perot resonator The quality factor of this resonator is more then 50000. Set ups II, IV allow one to measure the $|t|^2$, $|r|^2$, and interferometer III allows to measure complex refractive index of materials, including birefringence and dichroism of anisotropic materials.

In many cases we must also take in account moisture of these materials due to very high absorption of water at THz frequencies.

Table 1 presents absorption coefficients of water in the millimeter and submillimeter range.

Table 1. Absorption of water, (T =20^0 C)

λ, mm	8	4	2	1	0.6
α, dB/mm	18.0	29.5	41.0	48.0	64.5

Table 1 shows that absorption α increases with decreasing wavelength λ and therefore moist media with free water content of a few percentage become practically opaque for electromagnetic waves with frequencies more than 0.15- 0.20 THz.

On the other hand Table 2 gives the properties of nonpolar liquids with minimal loss in the low frequency part of THz range. These liquids are of interest for many applications.

Table 2 Dielectric properties of low- loss liquids. (T=20°C)

liquid	n	$\tan\delta \times 10^3$	λ, mm
cyklohexane	1.424	0.50	0.63
octane	1.396	0.74	0.63
decane	1.407	0.83	0.63
nonane	1.405	0.93	0.63
transdecalin	1.461	0.6	1.0
crude oil	1.470-1.570	0.8-1.4	2.0

Some results of investigation of polymers are summarized in Table 3 [4]. The measurements of n and $\tan\delta$ were carried out at approximately 20°C on polymers with the highest commercial purity. Unsintered PTFE features the lowest losses..

Table 3. Dielectric properties of polymers

material	$n \pm 0.5\%$	$\tan\delta \times 10^3 \pm 10\%$	λ, mm
PTFE, teflon,	1.43	0.7	1.3
Polyethylene	1.52	0.6	0.63
Polypropylene	1.51	0.6	0.63
TPX	1.45	0.7	1.0
Rexolite	1.59	1.0	2.2
Duroid	1.48	1.0	3.0
Polystyrene	1.59	2.9	1.0
teflon-4MB	1.42	1.2	0.63

Tables 4 presents the characteristics of the most interesting low-loss materials with $n > 1.6$: ceramics and glasses [4, 5. In the frequency range 100÷400

GHz, $\tan\delta$ for materials presented in Table 4 increases depending on f as f^γ ($\gamma = 0.5 \div 1.0$).

Some results of investigation of many types of glass cloths and resins used for preparing glass plastics were investigated in [6]. Thus, cloths based on non-alkaline and quartz glass fibers have $n =1.8 \div 2.5$ and $\tan\delta$ ranging from $(0.2 \div 2.0) \times 10^{-3}$ at frequencies 30÷35 GHz to $(0.3 \div 4.0) \times 10^{-3}$ at frequencies 300÷350 GHz

Table 4 Dielectric properties of low-loss materials

Material	$n \pm 0.5\%$	$\tan\delta \times 10^3 \pm 10\%$	λ, mm
BN	1.73	1.0	0.95
BN	2.07	0.64	1.22
SiO$_2$	1.92	2.0. 0.67	0.65 0.26
MgF$_2$	2.16	0.9 0.6	1.0 1.2
BeO	2.63	1.2	1.0
AlN	2.88	4.0 7.0	3.0 1.4
Al$_2$O$_3$	3.10	0.26 1.3	2.18 0.95
BaTiO$_3$+TiO$_2$ (50%)	6.1	1.1	0.95
SrTiO$_3$	15.1	35	1.0
TiO$_2$	9.4	10	1.4
MgAl$_2$O$_4$	3.14 2.90	1.5 0.6	1.0 2.5
La$_{7/12}$Na$_{1/4}$TiO$_3$	9.3	12	2.2
ZnS	2.89	1.9	3.0
ZnSc	3.02	2.2	3.0
Na$_2$O 6Al$_2$O$_3$	3.6	2.0	2.3
SiO$_2$ (fused)	1.95	1.4	0.85

Resins used for glass plastics (epoxy and silicon-bounded types) have $n = 1.6 \div 1.8$ and $\tan\delta$ from $(1.2 \div 2.5) \times 10^{-2}$ at frequencies 30÷35 GHz to $(2.5 \div 3.5) 10^{-2}$ at frequencies 300÷350 GHz. The refractive index for each material is practically constant at frequencies 10÷350 GHz. Antenna cover materials based on porous SiO$_2$ and Al$_2$O$_3$ have $n =1.1 \div 1.9$ and $\tan\delta$ from $(1 \div 5) \times 10^{-3}$ at frequencies of 150÷300 GHz.

Table 5 shows the properties of natural, building and common-use materials which are of great interest for communications applications, as well as for instruments for non-destructive test of materials [5, 7]. Here the smaller value of $\tan\delta$ for pine tree wood corresponds to the orientation of the electric field perpendicular to the wood fibers. The water content in wood materials is less than 7%.

978-1-4799-0019-0/13 $31.00 © 2013 IEEE   174

Table 6 presents the properties of cloth materials [7]. Here $|t|^2$ and $|r|^2$ are the power transmission and reflection coefficients, d is the material thickness. The frequency dependence of n for materials in Tables 5 and 6 is weak, but $\tan\delta$ and $|t|^2$ vary considerably as frequency changes. For example, $\tan\delta$ of brick at $\lambda = 6 \div 7$ mm is only 10^{-2}, and $\tan\delta$ of concrete is 6×10^{-3}.

Table 5. Dielectric properties of building and common-use materials ($\lambda = 1.7$–2 mm)

material	$n \pm 1\%$	$\tan\delta \times 10^2 \pm 10\%$	ρ, g/cm^3
brick (red)	1.78	3.5	1.5
brick (silic.)	1.82	4.2	1.8
Concrete	2.40	5.5	1.7
Asphalt	1.50	8.0	1.3
Sand	1.55	2.5	1.8
Soil	1.60	3.8	1.5
Snow, T= - 36 C	1.12	0.65	0.23
pine tree wood	1.4	3.4/2.0	0.5
glass, window	1.45	5.0	
organic glass	1.60	1.5	
Marble	1.50	1.0	
Ebonite, λ= 0.6 mm	1.67	1.1	
Cardboard	1.80	6.0	
Cautchuck	1.66	30	
Glues	1.57-1.72	1.0-2.0	
Phenolone	1.8	3.9	
Polymyde	1.66	1.8	
Veneer, λ=7.6 mm	1.5	10	
Plaster	1.7	0.7	

Table 6. Dielectric properties of cloth materials ($\lambda = 1.6$ mm)

material	$|t|^2$, %	$|r|^2$, %	d, mm
cloths for tents	82–94	≤ 1.3	0.3–0.5
cloths for coat, wool	77–84	≤ 3.0	2–4
cloths for suits, wool	85–98	≤ 1.0	0.5–1.0
Silk	89–93	≤ 1.0	0.15–0.25
leather, natural	79–85	≤ 5.0	0.9–1.5
leather, artificial	75–89	≤ 5.0	0.7–0.8
fur, artificial	75–89	≤ 3.0	4.0–12
Astrakhan	71	≤ 1.0	35
cloths for shirts	92–95	≤ 5.0	0.2–0.3

All materials in Table 6 exhibit strong dependence of transmission on frequency and moisture. Table 4 shows that for dry materials $|t|^2 = 98 \div 75\%$ in the temperature interval $T = 5°\div20°C$, whereas, for moisture of $W = 28\%$, $|t|^2$ reduces to $80\div25\%$. At higher frequencies transmission decreases much stronger. For example, the transparency of dry clothes decreases as frequency increases from $98\div75\%$ at $\lambda = 1.6$ mm to $76\div30\%$ at $\lambda = 0.5$ mm.

This review illustrates the state of the art in the study of dielectric properties of materials for applications in the low-frequency part of the terahertz band. These properties are important for the study of propagation of these waves in urban and indoor conditions, as well as for solving the problems of introscopy and radio imaging.

REFERENCES

1. Apletalin V.N., Meriakri V.V., and Chigrai E.E., Quasi-Optical Techniques of Studying Liquid and Solid Dielectrics at Submillimeter Wavelengths, Proc. Symposium on Submillimeter Waves, N.-Y., USA, pp. 631–641, 1970.

2. J.R. Birch, G.J. Simons, M.N. Afsar et. al., An intercomparsion of Measurement Techniques for the Determination of the Dielectric Properties of Solids at Near Millimetre Wavelengths, NPL Report Des 115, UK, October 1991.

3. Goubau G., Shwering F., On the guided propagation of electromagnetic beam waves, IRE Trans. AP, v. AP-9, pp. 248–256, 1961.

4. Apletalin V.N., Meriakri V.V., Kopnin A.N., et. al., Submillimeter Beam Wavequide Spectroscopy and Its Applications, in Kotel'nikov, V.A., (ed.) Problems of Modern Radio Engineering and Electronics, Nauka Publishers, Moscow, pp. 179–197, 1985.

5. Meriakri V.V., Chigrai E.E., Parkhomenko M.P., Beam Waveguide Spectroscopy of Materials in Millimeter and Submillimeter Waves Ranges, 6th International Conference on Broadband Dielectric Spectroscopy and Its Applications, Madrid, p. 163, 2010.

6. Chigrai E.E., Meriakri V.V. Millimeter Wave Characteristics of Glass Plastics for Antenna Covers, Nordic Antenna Symposium Antenn 03, Kalmar, Sweden, pp. 131–134, 2003.

7. Meriakri V.V., Chigrai E.E, Nikitin I.P., Dielectric properties of building and common-use materials in the low-frequency part of the terahertz band, International THz Conference, book,ISBN 978-3-85403-287-8, pp. 143–147, Villach, Austria, 2011.

Gyrotropic Metamaterials and Polarization Experiment in the Millimeter Waveband

(Invited Paper)

S.I. Tarapov[1], *Member IEEE*, S.Yu. Polevoy[1]

[1]O.Ya. Usikov Institute for Radiophysics and Electronics of NAS of Ukraine, Kharkov, Ukraine

Abstract: The experimental research of polarization rotation effects, which take place in artificial materials (volumetric chiral structures, photonic crystals) is presented. Three main types of such effect have been considered. The rotation due to electrodynamical Tamm state (surface state), the rotation due to volumetric features of rosette-type structure and rotation due to mixing of both reasons are under analysis.

The investigation of the gyrotropic media is of strong interest now both from fundamental and applied points of view. Namely, these media may be extremely promising for the study of the artificial materials (namely, bi-isotropic media) electrodynamics and for the development of the compact magnetically-controlled microwave devices.

The chiral medium is a special case of gyrotropic medium. Note that the electromagnetic waves propagation in the natural gyrotropic media today is quite thoroughly investigated [1-4]. However, the physics of artificial media of such type, where the gyrotropy effects are quite pronounced, does not have sufficient experimental confirmation yet.

In the framework of research the following phenomena were study:

1 – The gain of polarization properties (increasing of the rotation angle of the polarization) of the wave in

(1a) the bulk metamaterial (3D array of rosettes) that simulates the gyrotropic media with off-diagonal components of the dielectric permittivity matrix;

(1b) the bulk metamaterial (2D array of rosettes on magnetically controllable ferrite substrate) [5];

2 – The gain of the Faraday effect in the bounded photonic crystal, where the surface oscillation are occurred.

During the experiments the technique that described in [5, 6] (Fig. 1) was applied. The test object has been irradiated by 20-40 GHz microwave field; and the static magnetic field was applied along the wavevector.

In the first round we investigated the interaction of electromagnetic waves with the metamaterial formed by so-called "electric" gyrotropic material, formed by the bulk rosette structure (Fig. 2) [5]. These metamaterials are of great interest, because they can be used to realize the rotation of the plane wave polarization plane at large angles in mm waveband. In our experiments we construct the experimental model of the such bulk metamaterial based on the array of planar chiral structures with which we can realize the rotation of the polarization plane of the transmitted wave at the angle of 90° in the millimeter waveband [7] (Fig. 2).

In such structure, the frequency dependence of the polarization plane rotation angle and the values of the chirality parameter were measured, the possibility to control the maximum of frequency dispersion of the rotation angle and the chirality parameter by changing the thickness (distance between elements) of the structure was showing (Fig. 3). Besides, the structure shows the possibility of exciting the left-handed mode (LH mode) for the left-circularly polarized wave. Its likely origin and conditions of its exciting are under discussion. This so-called "magnetic mode" first detected apparently in [7, 8], has negative dispersion: i.e., when the distance between the layers of the chiral structure growth the mode eigen frequency increases.

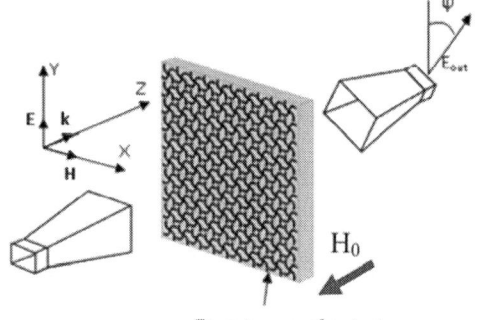

Structure under test

Fig. 1. The scheme of experimental test bench for study of polarization properties of the metamaterials with the longitudinal magnetization.

 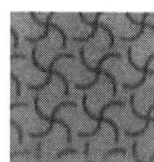

Fig. 2. The chiral structures (a) under study; the fragment (b) of the structure.

However, for the wave with right-circularly polarization has a traditional so-called "electric mode" (RH mode), with positive dispersion. A possible reason of the LH mode appearance is most likely lies in the fact that it arises from the interaction of the electromagnetic field with effective magnetic dipoles (not electric, as in the RH mode!) induced by currents flowing along the curved paths.

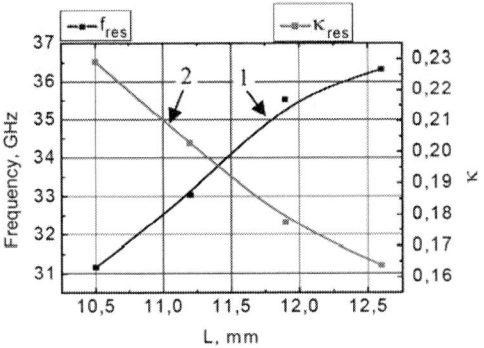

Fig. 3. Ehe frequency region of maximum angle of the polarization plane (1) and the resonant chirality parameter (2) versus the thickness of structure.

In the second round the experimental and theoretical studies was carried out for the chiral structure formed by 2D-metamaterial of the rosettes placed on the ferrite (magnetogyrotropic) slab of finite thickness.

It has been shown that using such a structure one can achieve conditions when the gyrotropy effects called as by the magnetic gyrotropy of the ferrite as well as the gyrotropy effects caused by the 2D array of rosettes lead to the significant increasing of the polarizing properties of such metamaterial. In particular, Figure 4 shows the experimental results (the rigorous calculation method and experiment technique see in [5]). It can be seen that there is a range (marked by the arrow)of frequencies and external magnetic field, for which this material can provide the polarization rotation angle up to 40°.

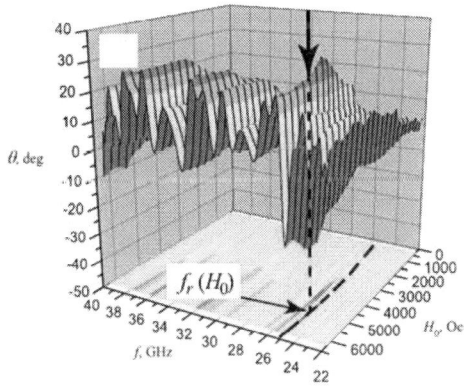

Fig. 4. Experimental dependence of the rotation angle θ of the polarization plane on the frequency and magnetic field for bulk chiral metamaterial (2D array of rosettes plus the ferrite slab).

The third round of studies is devoted to the gyrotropy in metamaterials with the band nature of the frequency spectrum

[9, 10]. The well-known demonstration of the natural gyrotropy is the Faraday effect, which is the rotation of the polarization plane of linearly polarized wave propagating throw the longitudinally magnetized magnet (ferromagnetic as a rule). A significant enhancement of the Faraday effect can be obtained by insertion such magnetically active media in the resonant multilayer structures. Attractive examples of such structures are axially-symmetrical photonic crystals with the defect or photonic crystals, bounded by the medium that can be assigned the negative constitutive parameter. The electrodynamical Tamm surface states [10] (or the surface oscillations) can be formed there. Features of such phenomena are described in details in [9], where the investigated structure, which is the axially-symmetrical photonic crystal (air/teflon/quartz), loaded with ferrite and/or thin-metal layer placed to the metal waveguide (Fig. 5) of circular cross-section. The static magnetic field (H_0) was applied along to the wavevector.

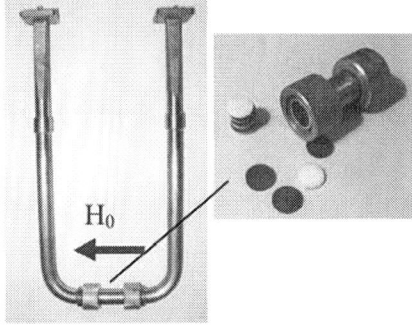

Fig. 5. Unit for the study of gyrotropic photonic crystals (PC).

From the transmission spectrum (Fig. 6) of the photonic crystal and photonic crystal loaded with ferrite and metal layers (serially placed), see the following. When the photonic crystal excited by the TE mode, the mode of the surface oscillation is formed in the band gap of the photonic crystal spectrum (the arrow in Fig. 6). This mode is the TM mode, which arises as a result of amplification of the Faraday effect in the ferrite layer.

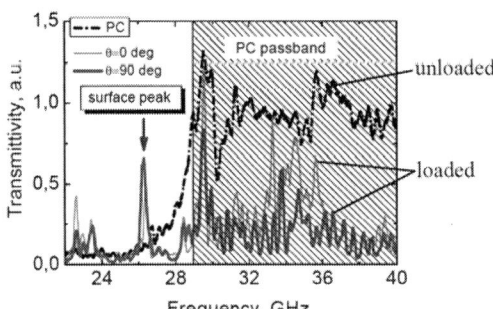

Fig. 6. The transmission spectrum of gyrotropic photonic crystal: unloaded PC and PC, loaded by the negative permittivity boundary.

With the purpose of detailed studying the properties of this mode (which display itself as Tamm peak or the Surface peak [10], Fig. 6) the dependence of its intensity on the rotation angle (θ) of the receiving waveguide (Fig. 7a) and the dependence of its frequency on the applied magnetic field at θ = 90° (Fig. 7b) have been analyzed. A satisfactory agreement between experiment and the results of numerical calculation [9] was obtained in [9]. Here we note only, that the maximum intensity of Tamm-peak occurs at θ=90°, that confirms the theoretical assumptions about the origin of TM mode in this structure at θ = 90° at the frequency of Tamm peak.

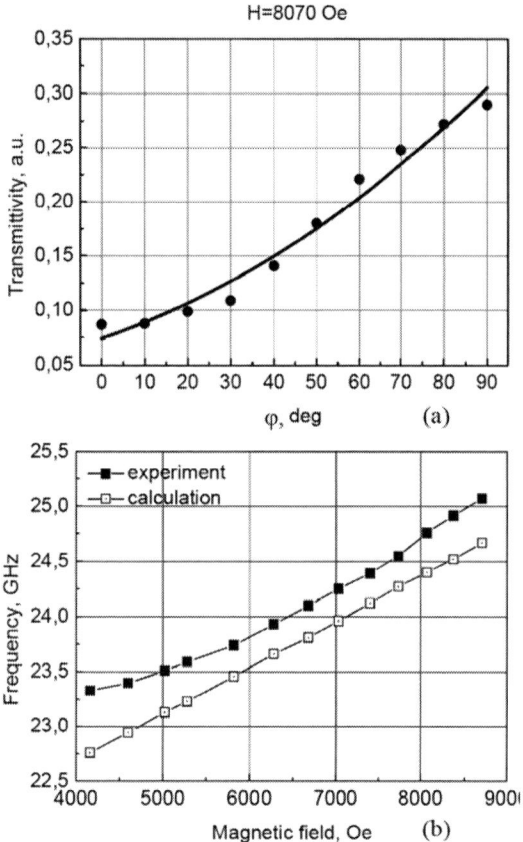

Fig. 7. (a) Tamm-peak frequency versus the polarization angle (θ); (b) experimental and calculated dependence of Tamm peak frequency on the magnetic field at θ = 90°.

In summary, we can conclude that the possibility of performing the artificial gyrotropic media for the millimeter waveband intended for developing both fundamental and application tasks is demonstrated experimentally.

REFERENCES

[1]   V.I. Agranovich, V.L. Ginzburg, Crystal optics with spatial dispersion and theory of excitons. - Moscow: Nauka, 1979 (rus).

[2]   F. I. Fedorov, The theory of gyrotropy, Nauka i technika, Minsk, 1976 (rus).

[3]   A. G. Gurevich "Ferrites in microwave frequencies", Fizmatgiz, Moscow, 1960 (rus).

[4]   O. V. Ivanov, "Electromagnetic wave propagation in anisotropic and bianisotropic layered structures", UlSTU, Ulyanovsk, 2010 (rus).

[5]   S.Y. Polevoy, S.L. Prosvirnin, S.I. Tarapov, V.R. Tuz, "Resonant features of planar Faraday metamaterial with high structural symmetry", The European Physical Journal – Applied Physics, Vol. 61, Iss. 3, pp. 30501, 2013.

[6]   S.Y. Polevoy, S.L. Prosvirnin, S.I. Tarapov, "Resonance properties of planar metamaterial, formed by array of gammadions on unmagnetized ferrodielectric substrate", Radiophysics and electronics, Vol. 4, Iss. 18, No. 1, pp. 42-46, 2013.

[7]   Plum E. Metamaterial with negative index due to chirality / E. Plum, J. Zhou, J. Dong, V.A. Fedotov, T. Koschny, C.M. Soukoulis, N.I. Zheludev // Phys. Rev. B. – 2009. – Vol. 79, Iss. 3. – p. 035407.

[8]   J. Dong, J. Zhou, T. Koschny, C. Soukoulis, "Bi-layer cross chiral structure with strong optical activity and negative refractive index", Optics Express, Vol. 17, Iss. 16, pp. 14172-14179, 2009.

[9]   Experimental Study of the Faraday Effect in 1D-Photonic Crystal in Millimeter Waveband, A.A. Girich, S.Y. Polevoy, Sergey I. Tarapov, A.M. Merzlikin, A.B. Granovsky, D.P. Belozorov, Solid State Phenomena, 2012, v.190, p.365-368.

[10]   Microwaves in Dispersive Magnetic Composite Media (Review Article), S.I. Tarapov and D.P. Belozorov, Low Temperature Physics, 2012, v.38, N.7, p.766-792

Electromagnetic wave diffraction by periodic structures with nonlinear inclusions

(Invited Paper)

Vyacheslav V. Khardikov[1,2], Pavel L. Mladyonov[1], Sergey L. Prosvirnin[1,2], and Vladimir R. Tuz[1,2]

[1]Institute of Radio Astronomy of National Academy of Sciences of Ukraine, 4, Krasnoznamennaya st., Kharkiv 61002, Ukraine
[2]School of Radio Physics, Karazin Kharkiv National University, 4, Svobody Square, Kharkiv 61077, Ukraine.

Abstract— The results of the study of the electromagnetic waves resonant reflection from and transmission through planar metamaterials based on periodic structures, which are composed of some elements consisted of nonlinear materials are presented.

I. INTRODUCTION

Modern nanotechnologies effort an opportunity to structure optically thin layers of materials with a periodic subwave-length pattern in order to produce planar metamaterials. The planar metamaterials are an impressive contemporary object, which is driven by certain fascinating facilities both mostly well known over the decade and quite novel.

Latest works aim to develop an electromagnetic framework for the design of tiny controlling optical and THz devices by using novel resonant planar metamaterials with active media inclusions. These artificial open high-Q periodic structures are promising to achieve a strong localization and enhancement of internal field which are necessary to produce amplifiers, generators, and tunable metamaterials. Strong intensity of electromagnetic field inside these novel planar metamaterials is achieved by their designing to bear a resonant regime of excitation of trapped modes [1], [2].

Recent papers report results of aggregating laser materials with these planar metamaterials to develop gaining or lasing devices such as the spaser [3], [4], [5]. Another high promised phenomenon is an optical bistability or multistability of a transmission response of nonlinear planar metamaterials. This effect is employed to construct optical switches. diodes, transistors, logical elements, and systems of optical storage [6], [7], [8], [9].

In this paper we report our recent results on the theoretical study of trapped mode planar metamaterials with nonlinear inclusions.

II. SATURATION EFFECT IN ACTIVE METAMATERIALS

We have proposed a simple design of all-dielectric silicon-based planar metamaterial [10] manifested an extremely sharp resonant reflection and transmission in the wavelength of about 1550 nm due to both low dissipative losses and involving a trapped mode operating method. The quality factor of the resonance exceeds in tens times the quality factor of resonances in known plasmonic structures. The designed metamaterial is envisioned for aggregating with a pumped gain medium

to achieve an enhancement of luminescence and to produce an all-dielectric analog of a "lasing spaser". We report that an essential enhancement (more than 500 times) of luminescence of in a layer contained pumped quantum dots (QD) may be achieved by using the designed metamaterial. This value exceeds manyfold the known the value of luminescence enhancement by in known plasmonic planar metamaterials.

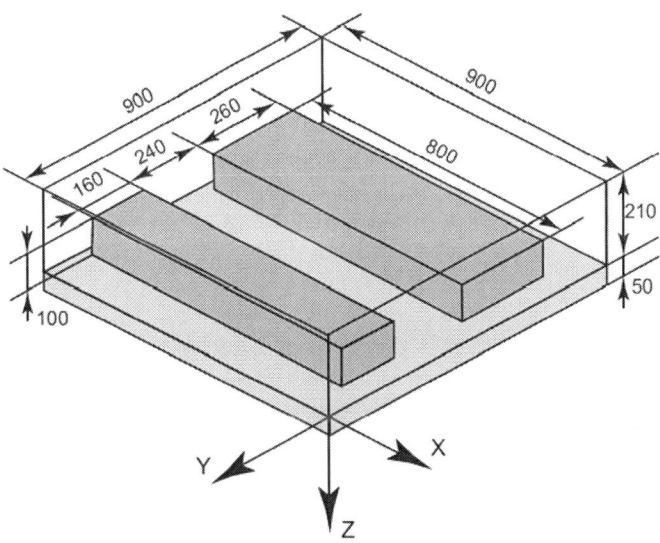

Fig. 1. A sketch of the unite cell of the double-periodic planar structure. The all-dielectric array composed of two dielectric bars per a periodic cell is immersed into the QD layer. All dimension are in nm.

We use the model of gain nonlinear medium by introducing negative frequency dependent conductivity

$$\sigma(\omega) = \frac{1}{1 + I/I_s} \frac{\sigma_0(1 + i\omega\tau)}{(1 + \omega_0^2\tau^2) + 2i\omega\tau - \omega^2\tau^2}$$

where $\omega_0 = 1.26 \cdot 10^{15}$ s^{-1} which corresponds to wavelength $\lambda_0 = 1550$ nm; $\tau = 4.85 \cdot 10^{-15}$ s; $\epsilon_{QD} = 2.19$ which corresponds to refractive index $n_{QD} = 1.48$ of non-pumped quantum dot laser medium, and $\sigma_0 = -500$ Sm/m corresponding to an emission factor $\tan\delta_e = -0.021$ on the analogy of a lossy factor of media. Small value of τ results in a wide-band QD spectral line and it enables to exclude from consideration the effects caused by displacement of meta-material dissipation peak and maximum of exciton emission

978-1-4799-0019-0/13 $31.00 © 2013 IEEE

line of QDs. Let us notice that the pump level (parameter σ_0) is in one order less than it was needed in the case of plasmonic metamaterials because of low losses of all-dielectric array. The factor $(1 + I/I_s)^{-1}$ allows considering the effect on luminescence enhancement of the gain saturation effect inherent in gain media. Here the parameter I_s is proportional to the saturation intensity and it displays effect of inversion population reducing in the gain medium by simulated emission which is proportional to the maximum of the internal field (I). The saturation factor $((1 + I/I_s)^{-1})$ is calculated separately for each point of the spatial grid that allows considering the inhomogeneous of the QD layer. We should note that the small value of an emission factor results in independence of the QD refractive index from the saturation factor. Thus the effect of saturation on the luminescence enhancement of QD layer hybridized with all-dielectric metamaterial can be considered under this model.

The diffraction approach proposed in [11] was used to calculate the luminescence enhancement for QD layer hybridized with all-dielectric metamaterial. This approach consists of evaluation of considered structure luminescence through the difference of energy dissipation in the passive and gain structure. The dissipation energy is calculated from the solution of corresponding diffraction problem for plane wave. The luminescence enhancement equals to the ratio of the luminescence of the hybrid structure to the luminescence of 210 nm homogeneous QD layer placed on 50 nm silica substrate. The wavelength dependencies of the luminescence enhancement of QD layer hybridized with all-dielectric metametrial for different value of the saturation intensity are shown in Fig. 2. The reducing of the luminescence enhancement with decreasing of the saturation intensity may be explained by exciting strong local field in the hybrid structure which results in decreasing of saturation factor. The distribution of the saturation factor in cross section ($z = -155$ nm) is shown in Fig. 3. One can see the burning hole appearance in the distribution (see dark blue areas in Fig. 3). The energy of optical pumping within these hole was completely used by simulated emission. The effect of gain saturation dose not strongly influence on the photoluminescence in the system but it needs to be taken into account for modeling of the optical amplifier and lasing spacer.

III. BISTABILITY AND MULTISTABILITY OF RESPONSE OF NONLINEAR PLANAR METAMATERIALS

The phenomenon of optical bistability or multistability is a common property of nonlinear optical systems with feedback, which means that there are two or more stable states of the system corresponding to different amplitudes or polarizations of the field [12].

A. Magnetically controllable array on nonlinear antiferromagnetic substrate

The structure under study is shown in Fig. 4. The doubling-ring (DR) array is placed on a substrate made of antiferromagnetic film (AF). The metamaterial parameters are $d = d_x = d_y = 0.3$ mm $a_1 = 0.11$ mm and $a_2 = 0.09$ mm,

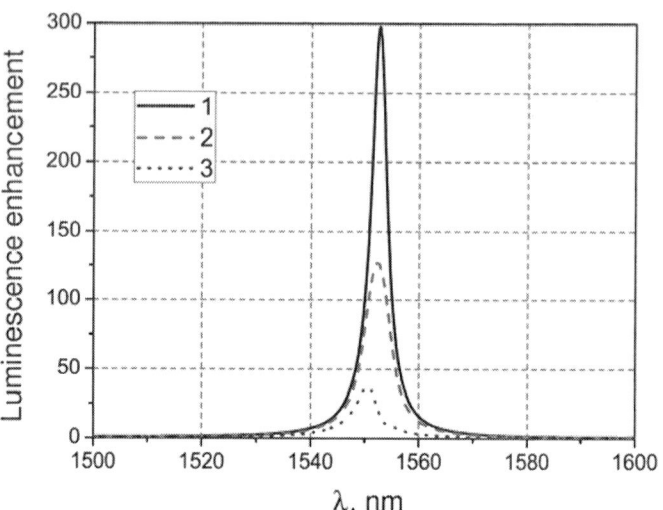

Fig. 2. The wavelength dependencies of the enhancement of the luminescence of the QD layer hybridized with all-dielectric metamaterial for different value of I_s. Line 1 - $I_s = 2.0$, line 2 - $I_s = 0.4$, line 3 - $I_s = 0.04$. The saturation intensities present in arbitrary units.

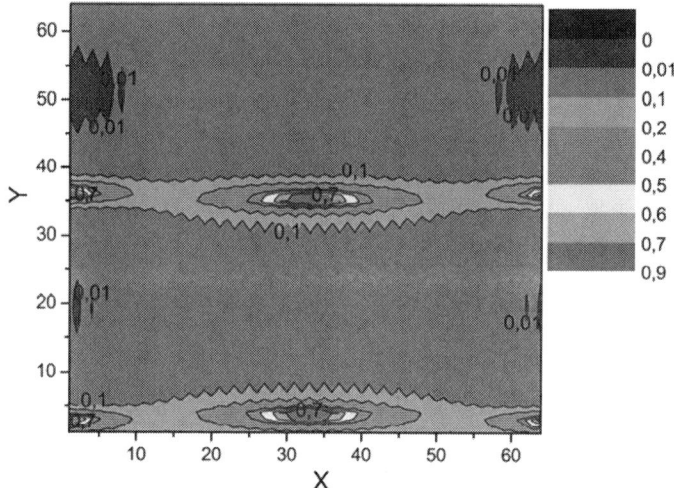

Fig. 3. The distribution of saturation factor in the cross section $z = -155$ nm. $\lambda = 1553$ nm, $I_s = 0.4$.

$2w = 0.004$ mm, and $h = 0.05$ mm. As a material for the substrate, MnF_2 antiferromagnetic film is considered [13], [14]. The external static magnetic field (ESMF) is applied to the system in the Faraday geometry.

For the fixed external magnetic field $H_0 = 1.0$ kG, in the dispersion dependences of the permeability tensor components there are two resonant frequencies, $\text{æ}_1 = 0.262$ ($f = 0.2618$ THz) and $\text{æ}_2 = 0.268$ ($f = 0.2678$ THz) where $\text{æ} = d_x/\lambda$.

Under the action of intense light the dynamical magnetization in AF media is coupled nonlinearly with the wave magnetic field which leads to the magnetic optical nonlinearity. Remarkably, when the magnetic field strength inside the AF film increases both the real and imaginary parts of magnetic permeability undergo changes. Under a certain threshold level of the input light intensity, this can lead to dispersion-

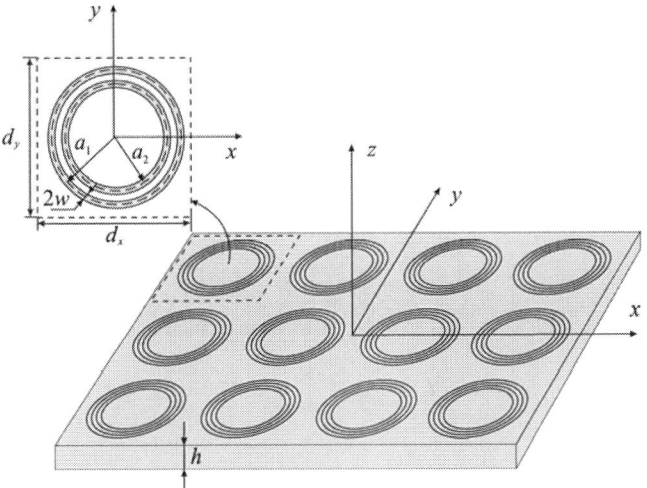

Fig. 4. Fragment of the planar metamaterial and its elementary unit cell.

absorption bistable behavior of the system.

Here the way to obtain large magnitude of the magnetic field strength in the AF film lies in the choice of the parameters of the DR array of the metamaterial to tune the frequency of the trapped-mode excitation to be close to the frequency of antiferromagnetic resonance. Such a situation is illustrated in Fig. 5 where the optical response of the metamaterial in the case when the intensity of input light is small (linear regime) is presented.

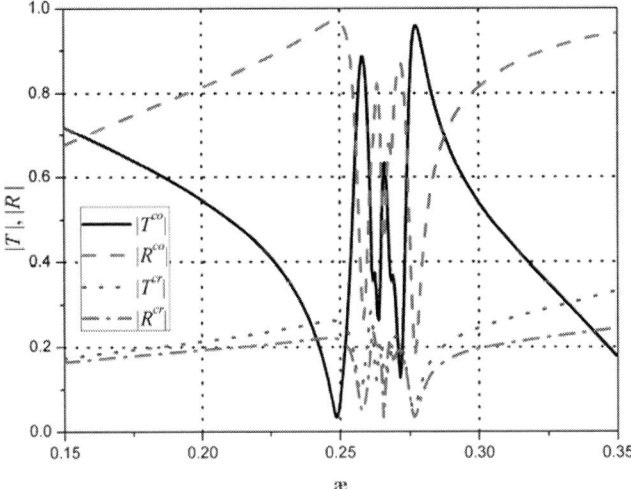

Fig. 5. Reflection and transmission spectra of the array placed on an antiferromagnetic substrate, linear regime.

The behavior of the dispersion characteristics of the AF permeability leads to appearing alternate bands of high transmission and absorption in the spectra of metamaterial. In the frequency band of the trapped-mode excitation there is a peak of current magnitude, but its frequency dependence has a form of alternating maxima and minima due to the strong absorption in the substrate in the vicinity of the AF resonances nearly $æ_1$

and $æ_2$.

Since the magnetic field strength is proportional to the current and the current magnitude obviously increases when the intensity of the incident field rises, under a certain intensity of the incident field the magnetic properties of substrate can change.

Our calculations [15] show that in case of the nonlinear permeability of substrate, dependences of the inner field intensity versus the incident field intensity have a form of hysteresis (Fig. 6). Such form of curves is studied quite well and is explained by the nonlinear phase-shift and nonlinear attenuation which appear in the nonlinear system. As the incident field intensity increases, the nonlinear phase shift rapidly raises and the attenuation decreases that guarantee the presence of obvious bistable switching.

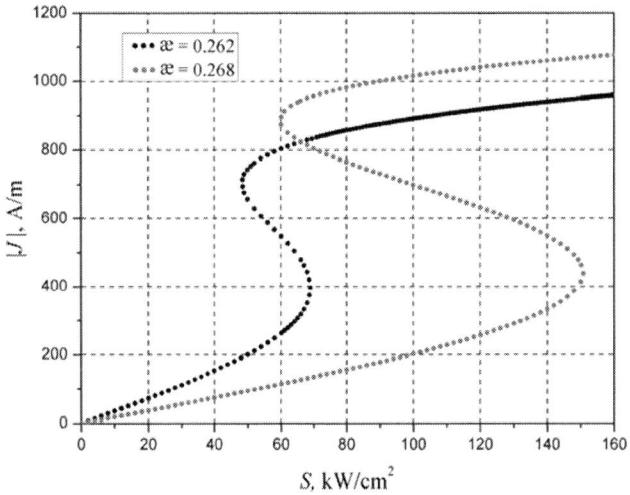

Fig. 6. Inner field intensity versus incident field intencity.

The frequency dependences of the transmission coefficient magnitudes also manifest discontinuous switching with frequency changing (Fig. 7). Since the dispersion curves of nonlinear susceptibility have the bands of grow and decay, the real and imaginary parts of the permeability tensor coefficients also increase and decrease with frequency. The bistable transmission occurs exactly in these frequency bands and is manifested in the ambiguity of the transmission coefficient magnitudes at the leading and trailing edges of the resonance.

Thus, the all-optical switching can be realized due to the capability of a planar DR metamaterial provides the sufficient field localization inside the thin nonlinear substrate at the frequency of the trapped-mode excitation.

B. Double array consisted of wavy strips on nonlinear substrate

A studied structure consists of two gratings of planar perfectly conducting infinite strips placed on a dielectric slab with thickness h (see Fig. 8). We assume that this slab is a Kerr-type nonlinear dielectric which permittivity ε linearly depends on the intensity $|E|^2$ of the electric field. The gratings

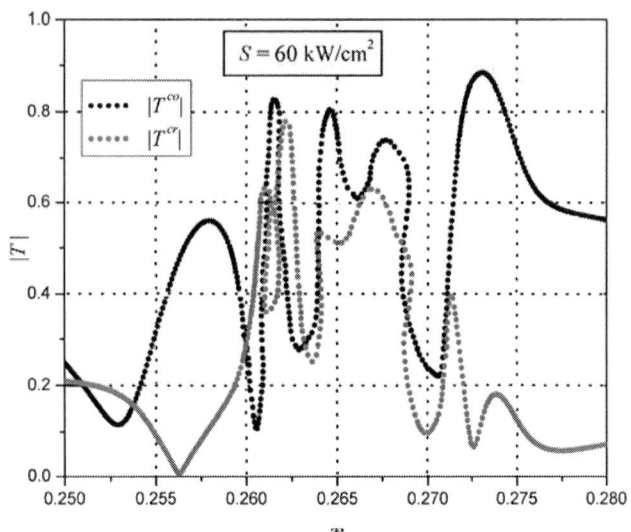

Fig. 7. Transmission spectrum of the array placed on the antiferromagnetic substrate, nonlinear regime.

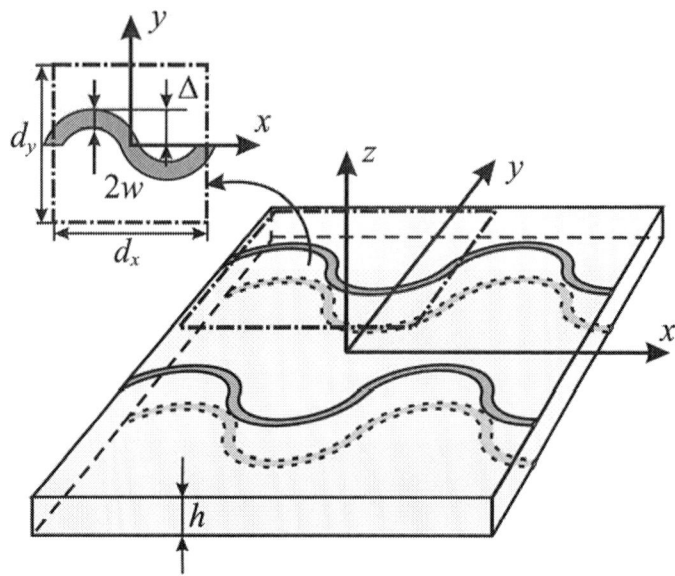

Fig. 8. Fragment of a bilayer planar fish-scale metamaterial and its unit cell.

with wavy-line strips are located on the both sides of the slab at planes $z = 0$ and $z = -h$. The elementary translation cell of the structure under study is a square with sides $d = d_x = d_y$. The full length of the strip within the elementary translation cell is S. Suppose that the thickness h and size d are much less then the wavelength λ of the incident electromagnetic radiation. The width of the metal strips and their deviation from the straight line, respectively, are $2w$ and Δ. Assume that the normally incident field is a plane monochromatic wave polarized parallel to the strips (x-polarization), and the magnitude of the primary field is A. We suppose that the intensity of the incident field is enough for the nonlinearity to become apparent, i.e., it is about 1 kW/cm^2.

In our publications [16], [17] the detailed description of the numerical method based on the Method of Moments (MoM) is given to study electromagnetic properties of planar fish-scale metamaterials in both single- and bilayer configurations when the intensity of the incident field is small (linear regime). It involves solving the integral equation related to the surface currents which are induced in the metallic pattern by the field of the incident wave. Remarkably that in the bilayer configuration the method of solution rigorously takes into account a coupling between two gratings via evanescent partial spatial waves. Obtained solution allows us calculating the magnitude and distribution of the current J along the strips, the reflection R and transmission T coefficients as functions of frequency ω, permittivity ε and other parameters of the structure.

When the studied structure is under an action of the intense light (in the nonlinear regime), permittivity of the substrate ε depends on the intensity of the electromagnetic field inside it, $\varepsilon = \varepsilon(I_{in})$. In [18], [19] an approximate treatment was proposed to solve the nonlinear problem.

The origin of trapped-mode resonances is the opposite directed but being almost equal currents which appear in two closely spaced metallic wires. The scattered fields produced in this situation is very weak, and, as a consequence, the coupling of the metamaterial array to free space is small and therefore its radiation losses are reduced, which ensures a high-Q resonant response.

The first distribution is the antiphased current oscillations near point of inflection wavy-line of each grating (see Fig. 9a). The observers structure can be considered as a system of two coupled resonators which work on the same frequency because the gratings are identical. Obviously that the distance h between the gratings will strongly effect on the resonant frequency position since this parameter define the electromagnetic coupling degree. The Q-factor of this resonance is higher in the bilayer structure in comparison with a single-layered one but their similarity is in the fact that the current magnitude in the metallic pattern depends relatively weakly on the thickness and permittivity of the substrate.

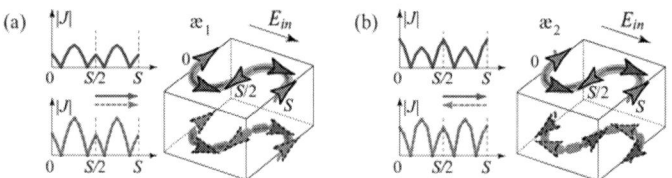

Fig. 9. The resonant current distribution along the strips in the case of bilayer structure composed of gratings with wavy-shaped strips. The resonant frequencies are (a) æ$_1 = 0.7855$, (b)æ$_2 = 0.82$ (æ $= d/\lambda$).

The second distribution is the antiphased current oscillations between two adjacent gratings (see Fig. 9b). It is well known that the closer are the interacting metallic elements, the higher is the Q-factor of the trapped-mode resonance. Thus varying the distance between the gratings or the substrate permittivity

978-1-4799-0019-0/13 $31.00 © 2013 IEEE

changes the trapped-mode resonant conditions and this changing manifests itself in the current magnitude. Remarkably that in this type of current distribution the field is localized between the gratings, i.e. directly in the substrate, which can sufficiently enhance the nonlinear effects if the substrate is made of intensity dependent material.

This circumstance is depicted in Fig. 10 where typical curves of the inner field intensity and the transmission coefficient magnitude are shown as functions of the frequency and incident field intensity in the nonlinear regime.

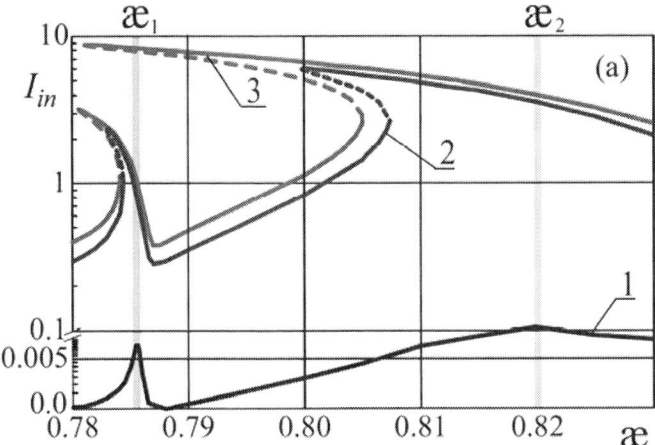

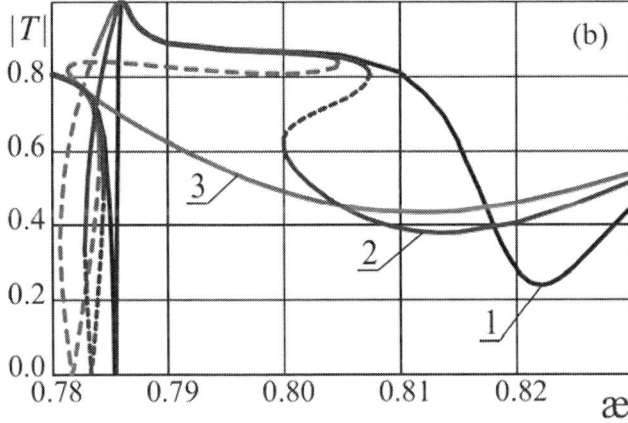

Fig. 10. The frequency dependences of the inner field intensity (on the logarithmic scale) (a) and the transmission coefficient magnitude (b) versus the incident field intensity in the case of the nonlinear permittivity ($\varepsilon = \varepsilon^l + \varepsilon^{nl} I_{in}$, dimension of I_{in} is in kW/cm^2); $\varepsilon^l = 3.0$, $\varepsilon^{nl} = 5 \times 10^{-3}$ cm^2/kW; curve 1 - $A = 1$ kW/cm^2, curve 2 - $A = 200$ kW/cm^2, curve 3 - $A = 300$ kW/cm^2.

For the nonlinear conditions the curves of the transmission coefficient magnitude experience different distortion nearly the trapped-mode resonance frequencies. At the frequency $\text{æ} \approx 0.78$ the antiphased current oscillations localize in area of each grating and are weakly affected on dielectric substrate. In such case the resonances curve transforms into a closed loop that is typical for the sharp nonlinear Fano-shape resonance. The second resonance $\text{æ} \approx 0.82$ is smooth but the current oscillations between two adjacent gratings has lad to

greater concentration of field in dielectric substrate. For the nonlinear conditions the resonance near $\text{æ} \approx 0.82$ undergoes more distortion in the wider frequency band, and at a certain incident field intensity this resonance can overlap the first one (Fig. 10b). Evidently that in this case the transmission coefficient has more than two stable states, i.e. the effect of the multistability occurs.

C. Polarization bistability in magnetophotonic structures

Magnetophotonic crystals (MPCs) are periodic structures that contain magnetic materials and therefore exhibit interesting physics arising from the interplay between photonic band gap (PBG) phenomena and magnetooptical effects [20], [21]. Examples include external magnetic tunability of PBGs and strong enhancement of Faraday polarization rotation. It is even more interesting to consider an MPC in presence of optical nonlinearity. In particular, the interplay between the Faraday effect (which is associated with optical nonreciprocity) and the Kerr effect (which is known to result in direction-sensitive optical bistability [22], [23], [24]) can give rise to new types of asymmetric or unidirectional light transmission, and the effects of this interplay on the polarization of transmitted and reflected light still remains to be investigated.

A one-dimensional magnetophotonic crystal with a Kerr-type nonlinear defect placed either symmetrically or asymmetrically inside the structure is considered (see Fig. 11). If such a system is under the longitudinal action of an external static magnetic field, the simultaneous effects of time-reversal nonreciprocity and nonlinear spatial asymmetry take a place. These effects manifest themself in the bistable response accompanied by abrupt polarization switching between two circular or elliptical polarizations for transmitted and reflected waves [25].

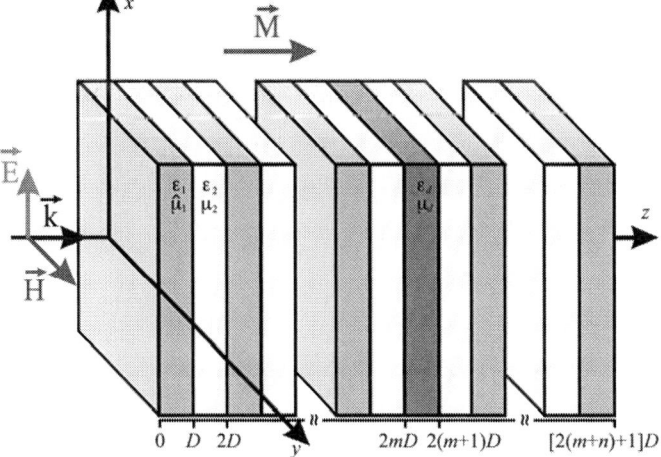

Fig. 11. A magnetophotonic crystal structure with a nonlinear cavity.

In the underlying structure, as a result of the action of an external static magnetic field, the defect resonances split into Zeeman-like doublets (see the black and red lines in Fig. 12 and Fig. 13).

978-1-4799-0019-0/13 $31.00 © 2013 IEEE

If the incident field is linearly polarized, the reflected and transmitted fields become elliptically polarized due to the Faraday rotation and circular dichroism. Directly at the Zeeman-like resonances the fields are circularly polarized. Therefore, in the nonlinear regime, it is possible to use multistability to switch between two distinct polarization states in the transmitted and/or reflected light (see the dark cyan and purple lines in Fig. 12 and Fig. 13).

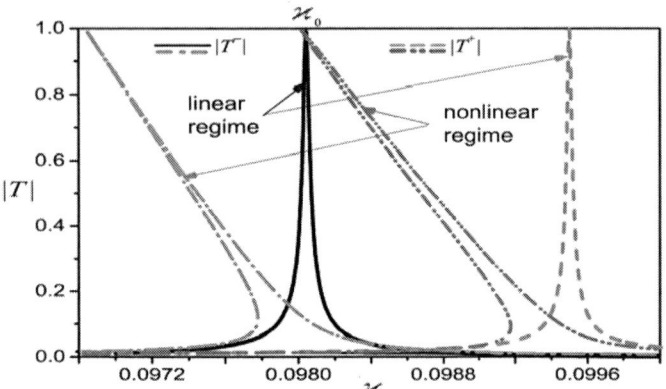

Fig. 12. Frequency dependences ($æ = D/\lambda$) of the transmission coefficient (T) of the LCP ($-$) and RCP ($+$) waves for $m = n = 5$.

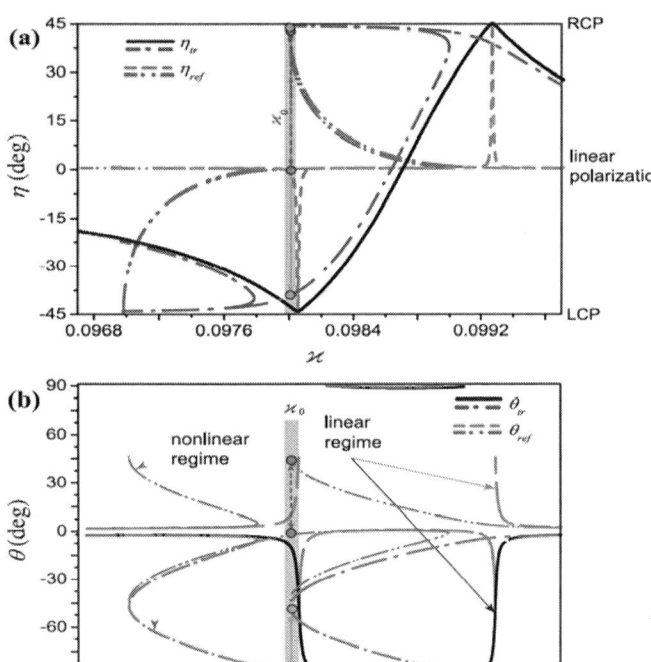

Fig. 13. Frequency dependences ($æ = D/\lambda$) of (a) the ellipticity angle η and (b) the polarization azimuth θ of the transmitted and reflected fields for $m = n = 5$. The incident light is linearly polarized. The vertical line marks the bistable polarization switching at $æ_0$.

In conclusion, metamaterials, which bears so-called trapped-mode resonance operation is a promising technique to design all-optical control devices. From our numerical calculations it seems reasonable to conclude that the bistable response can be obtained at the incident power densities of 10–100 kW/cm^2 with available materials in the considered structure configuration.

ACKNOWLEDGMENT

This work was supported by the Ukrainian State Foundation for Basic Research, the Project no. $\Phi 54.1/004$.

REFERENCES

[1] S. Prosvirnin and S. Zouhdi, "Resonances of closed modes in thin arrays of complex particles," in *Advances in Electromagnetics of Complex Media and Metamaterials*, S. Zouhdi and et al., Eds. Printed in the Netherlands: Kluwer Academic Publishers, 2003, pp. 281–290.

[2] V. A. Fedotov, M. Rose, S. L. Prosvirnin, N. Papasimakis, and N. I. Zheludev, "Sharp trapped-mode resonances in planar metamaterials with a broken structural symmetry," *Phys. Rev. Lett.*, vol. 99, no. 14, p. 147401, 2007.

[3] D. J. Bergman and M. I. Stockman, "Surface plasmon amplification by stimulated emission of radiation: Quantum generation of coherent surface plasmons in nanosystems," *Phys. Rev. Lett.*, vol. 90, no. 2, p. 027402, January 17 2003.

[4] N. I. Zheludev, S. L. Prosvirnin, N. Papasimakis, and V. A. Fedotov, "Lasing spaser," *Nature Photonics*, vol. 2, no. 6, pp. 351–354, June 2008.

[5] S. D. Campbella and R. W. Ziolkowski, "Impact of strong localization of the incident power density on the nano-amplifier characteristics of active coated nano-particles," *Optics Communications*, vol. 285, no. 16, pp. 3341–3352, 15 July 2012.

[6] S. Pereiraa, P. Chakb, J. E. Sipeb, L. Tkeshelashvilic, and K. Busch, "All-optical diode in an asymmetrically apodized kerr nonlinear microresonator system," *Photonics Nanostruct. Fundam. Appl.*, vol. 2, no. 3, pp. 181–190, 2004.

[7] A. Alberucci and G. Assanto, "All-optical isolation by directional coupling," *Opt. Lett.*, vol. 33, pp. 1641–1643, 2008.

[8] H. Zou, G. Q. Liang, and H. Z. Wang, "Efficient all-optical dual-channel switches, logic gates, half-adder, and half-subtracter in a one-dimensional photonic heterostructure," *J. Opt. Soc. Am. B*, vol. 25, no. 3, pp. 351–360, 2008.

[9] C.-H. Chen, S. Matsuo, K. Nozaki, A. Shinya, T. Sato, Y. Kawaguchi, H. Sumikura, and M. Notomi, "All-optical memory based on injection-locking bistability in photonic crystal lasers," *Opt. Express*, vol. 19, pp. 3387–3395, 2011.

[10] V. V. Khardikov and S. L. Prosvirnin, "New type high-q planar dielectric metamaterial," in *NATO Advanced Research Workshop on Detection of Explosives and CBRN (Using Terahertz) (TERA-MIR 2012)*, Cesme, Izmir, Turkey, November 2012, pp. 58–59.

[11] ——, "Enhancement of quantum dot luminescence in all-dielectric metamaterial," *arXiv:1210.4146 [physics.optics]*, p. 16, October 2012.

[12] H. M. Gibbs, *Optical bistability: controlling light with light*. Academic Press, Orlando, Fla., 1985.

[13] S.-C. Lim, J. Osman, and D. Tilley, "Calculation of nonlinear magnetic susceptibility tensors for a uniaxial antiferromagnet," *Journal of Physics D: Applied Physics*, vol. 33, pp. 2899–2910, 2000.

[14] Y. Zhao, S.-F. Fu, H. Li, and X.-Z. Wang, "Bistable transmission of antiferromagnetic fabri-perot resonator," *Journal of Applied Physics*, vol. 110, p. 023512, 2011.

[15] V. Dmitriev, S. Prosvirnin, V. Tuz, and M. N. Kawakatsu, "Electromagnetic controllable surfaces based on trapped-mode effect," *Advanced Electromagnetics*, vol. 1, no. 2, pp. 89–95, August 2012.

[16] S. L. Prosvirnin, S. A. Tretyakov, and P. L. Mladyonov, "Electromagnetic wave diffraction by planar periodic gratings of wavy metal strips," *J. Electromagnetic Waves and Applications*, vol. 16, no. 3, pp. 421–435, 2002.

[17] P. L. Mladyonov and S. L. Prosvirnin, "Wave diffraction by double-periodic gratings of continuous curvilinear metal strips placed on both sides of a dielectric layer," *Radio Physics and Radio Astronomy*, vol. 1, no. 4, pp. 309–320, 2010.

[18] V. R. Tuz, S. L. Prosvirnin, and L. A. Kochetova, "Optical bistability involving planar metamaterials with broken structural symmetry," *Phys. Rev. B*, vol. 82, p. 233402(4), 2010.

[19] V. R. Tuz and S. L. Prosvirnin, "All-optical switching in planar metamaterial with high structural symmetry," *Eur. Phys. J. Appl. Phys.*, vol. 56, no. 3, p. 30401, 2011.

[20] I. L. Lyubchanskii, N. N. Dadoenkova, M. I. Lyubchanskii, E. A. Shapovalov, and T. Rasing, "Magnetic photonic crystala," *J. Phys. D: Appl. Phys.*, vol. 36, pp. R277–R287, 2003.

[21] M. Inoue, R. Fujikawa, A. Baryshev, A. Khanikaev, P. B. Lim, H. Uchida, O. Aktsipetrov, A. Fedyanin, T. Murzina, and A. Granovsky, "Magnetophotonic crystals," *J. Phys. D: Appl. Phys.*, vol. 39, pp. R151–R161, 2006.

[22] V. S. C. M. Rao, S. D. Gupta, and G. S. Agarwal, "Study of asymmetric multilayered structures by means of nonreciprocity in phases," *J. Opt. B: Quantum Semiclass. Opt.*, vol. 6, pp. 555–562, 2004.

[23] V. Grigoriev and F. Biancalana, "Bistability, multistability and nonreciprocal light propagation in thue-morse multilayered structures," *New J. Phys.*, vol. 12, p. 053041, 2010.

[24] A. E. Miroshnichenko, E. Brasselet, and Y. S. Kivshar, "Reversible optical nonreciprocity in periodic structures with liquid crystals," *Appl. Phys. Lett.*, vol. 96, p. 063302, 2010.

[25] V. R. Tuz, S. L. Prosvirnin, and S. V. Zhukovsky, "Polarization switching and nonreciprocity in symmetric and asymmetric magnetophotonic multilayers with nonlinear defect," *Phys. Rev. A*, vol. 85, p. 043822, April 2012.

Frequency and wave-vector dispersion of the microwave mobility of drifting electron gas in GaN

V. V. Korotyeyev, G. I. Syngayivska and V. A. Kochelap

V. Lashkaryov Institute of Semiconductor Physics, National Academy of Sciences of Ukraine, 03028 Kyiv, Ukraine

Abstract: We have studied the high-frequency response of drifting electron gas on time- and spatial-dependent harmonic perturbation in frame of the exact solution of Boltzmann transport equation using Monte-Carlo method. It was demonstrated that results obtained by Monte-Carlo for the case of low-density electron gas differ from ones that are given by conventional hydrodynamic approach. It was found the region of frequencies and wave-vectors where negative microwave mobility is realized. The appearance of the relatively low-frequency region can be interpreted as a manifestation of the Cherenkov-like effect and higher-frequency regions as a manifestation of well-known optical phonon transit-time resonance.

Introduction

The development of compact, tunable and solid-state emitters and detectors operating in THz frequency range is one of the hottest problem of modern opto- and microelectronics. One of the possible approaches to realize such systems is based on the excitation of electron plasma oscillations in the channel of Grating-Gate Field-Effect-Transistor. The subwavelength metallic grating ensures the spatial inhomogeneity that gives rise to the effective coupling of short- wavelength plasmons and relatively long- wavelength THz electromagnetic waves. Generally speaking, description of nanoscale devices with built- in spatial inhomogeneities, including plasmonic structure requires taking into account both time- and spatial dispersion of electric characteristics. For example, to calculate plasmons dispersion law it is necessary to know the frequency and wave-vector dispersion of microwave mobility which determines the electron response on time- and spatial- dependent electric field, $E(\vec{r},t) = E_{\omega,q} \exp(I\vec{q}\vec{r} - I\omega t)$ with frequency ω and wave-vector q.

Usually, the physics of plasmon excitation and basic principles of plasmonic emission [1,2] and detection[3,4] of THz radiation in plasmonic structures are considered in the frames of very simplified hydrodynamic model. In the hydrodynamic model, microwave mobility of the drifting electron gas is given by the following expression [1]:

$$\mu_{\omega,q} = \frac{e}{m} \frac{\omega\tau}{[\omega - V_d q]} \frac{1}{[1 - I(\omega\tau - V_d q\tau)]}, \quad (1)$$

where $\vec{V}_d$ is the drift velocity, τ is the phenomenologically introduced electron relaxation time, e and m are the elementary charge and effective mass, respectively. At the absence of electron drift latter formula has a form of standard Drude-Lorentz model. At finite values of drift velocity, the real part of microwave mobility becomes negative at $\omega < \vec{V}_d \vec{q}$ and, consequently, plasmons with phase velocity less than drift velocity can be amplified. Such Cherenkov-like mechanisms of plasma instability was proposed by Mikhailov [1,2] for the realization of electrical – pumped and electrical- tunable solid-state THz emitters.

However, the applicability of the hydrodynamic approach and Eq. (1) is strongly limited. This approach can well describe the weakly-damped electron gas in the case of very low temperatures when details of electron distribution function become inessential or in the case of dense electron gas when the electron-electron scattering is the dominating.

The aim of this paper is to study features of $\mu_{\omega,q}$ in the frames of more general kinetic theory, particularly, based on the numerical solution of Boltzmann transport equation by Monte-Carlo method. This powerful method provides the detailed analysis of the hot carrier kinetics taking into account the appropriate electron relaxation mechanisms, their parameters and character of carrier distribution in the momentum space. The Monte-Carlo calculations of frequency and wave-vector dispersion of microwave mobility are applied to the GaN sample. It is remarkable that due to a strong electron–optical-phonon interaction, nitride materials possess small electron relaxation time that provides large electron response up to the THz frequency range [5]. Moreover, in the case of low-density electron gas in GaN, in the range of moderate strength of applied steady-state electric fields (typically 1-10 kV/cm) another type of current instability can exist. Latter one is associated with a formation of the specific streaming transport regime at which the strongly anisotropic electron distribution function is taken place in the momentum space. It leads to electric field-induced resonances in microwave mobility even for spatially uniform ($q = 0$) case (it is so-called optical phonon transit - time resonance (OPTTR) [6-7].

Transport model

Our model of electron transport includes the following key points. We consider a thin film of crystalline GaN of cubic modification with thickness, d (other dimensions are assumed to be infinitive). We suppose that d is much greater than the de Broglie wavelength of electrons, and, consequently, electron transport in the film can be treated as three-dimensional. But plasmons in such sample can be treated as a two-dimensional if the thickness of the film is essentially smaller than the plasmon wavelength. For the simulation of the electron transport we take into account three main scattering mechanisms: scattering by ionized impurities, acoustic phonons and polar-optical phonons. To exclude the quenching effect on the OPTTR by electron–electron scattering we suppose that the sample is compensated, thus the electron concentration is much smaller than impurity concentration. For ionized impurity scattering, we exploit the 'mixed' scattering model that unifies both Brooks–Herring and Conwell–Weisskopf approaches. This approach is more appropriate for the analysis of compensated materials. Throughout the calculations we set impurity concentration 10^{16} cm^{-3} and electron concentration 10^{15} cm^{-3}. At such parameters, the steady-state characteristics of GaN and frequency spectra of microwave motility, $\mu_{\omega,0}$, including magnetic field effects have been analyzed previously in ref. [8,9]

Results and Discussions

Results of Monte-Carlo calculations of the spectrum of microwave mobility, $\mu_{\omega,q}$ of drifting electron gas in THz frequency range is shown in Fig.1 (here, vectors $\vec{q}$ and $\vec{V}_d$ are assumed to be parallel). It is seen that the frequency dispersion of $\mu_{\omega,q}$ at finite q has essential asymmetry in respect to zero frequency. It is associated with the different response of electron gas on perturbations propagating along and against the drift.

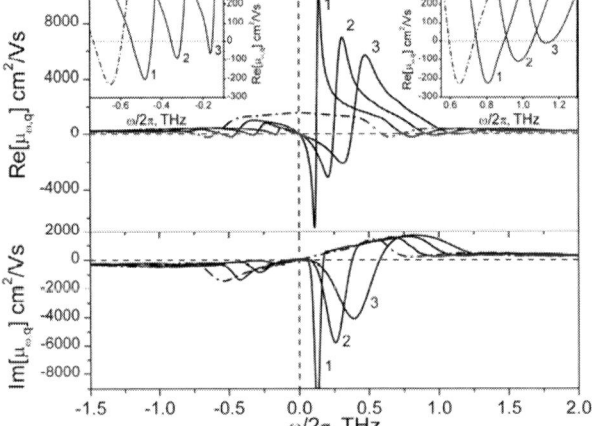

Fig.1. Frequency dispersion of real and imaginary parts of $\mu_{\omega,q}$ calculated at three different wave-vectors $q = 0.5, 1, 1.5 \times 10^5$ cm^{-1} (curves 1, 2, 3, respectively). Dash-dotted curves are $\mu_{\omega,0}$. The special features of the spectra relates to the OPTTR are shown in detail in the inserts. Temperature $T = 30\,K$, magnitude of steady-state field is $3\,kV/cm$. At such field, $V_d = 1.6 \times 10^7\ cm/s$.

Also, there are several frequency bands where real part of microwave mobility becomes a negative. It means the perturbation with such frequencies and appropriate wave-vector will be unstable and amplified.

The appearance of such frequency bands has clear physical explanations. The low-frequency band with relatively large values of negative microwave mobility is the manifestation of Cherenkov-like effect. The existence of "Cherenkov" band is also predicted by hydrodynamic approach (see Eq.1), however, in the contrast of the latter, Monte-Carlo calculation gives the finite values of $\mu_{\omega,q}$ at characteristic frequency $\omega = V_d q$, that reflects statistical nature of electron gas, particularly, thermal velocity spread.

The high-frequency bands of negative microwave mobility are associated with the effect of OPTTR. The small values of negative microwave mobility are typical for conditions of the streaming regime, when the drift velocity saturates and the differential mobility tends to zero. Note, at finite q OPTTR frequencies are shifted by factor $V_d q$ (see inserts in Fig. 1). The imaginary part of $\mu_{\omega,q}$ behaves like a dielectric function near narrow resonance: change the sign when the real part have maximum.

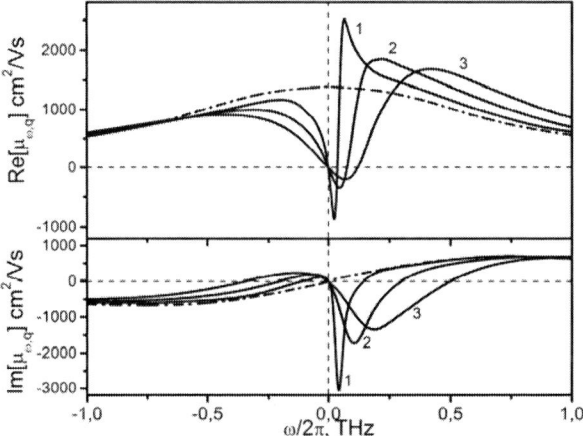

Fig.2. The same for the temperature $T = 300$ K

It should be noted that in the case of III-nitride compounds the simultaneous existence of both instability frequency bands (Cherenkov and OPTTR bands) is possible only at low temperatures; typically, it is nitrogen temperature, 77 K.

The similar behavior of $\mu_{\omega,q}$ at low temperatures were obtained analytically in ref. [10] using Barraf approximation for the solution of Boltzmann transport equation

At room temperature, 300 K, streaming transport regime is destroyed due to strong inelastic mechanism of the absorption of optical phonons, as a result the effect of OPTTR does not occur (see Fig. 2). In turn, the Cherenkov resonance is less sensitive to the electron distribution function and exists even at room temperature, but it is less pronounced.

Calculating the microwave mobility at different frequencies and wave vectors we can determine in $\{\omega, q\}$ plane the *amplification* region ($\text{Re}[\mu_{\omega,q}] < 0$). Fig. 3 provides the set of isolines each of them corresponds to the certain negative values of $\mu_{\omega,q}$. The OPTTR amplification region is shown by closed isoline 1 in upper panel. As can seen the negative microwave mobility occurs in wide frequency range from $0.1\,THz$ to $1.2\,THz$ varying the wave vector from -1.5×10^5 to $1.5 \times 10^5\,cm^{-1}$. Note, in the case of spatially-uniform perturbation ($q = 0$), amplification due to OPTT resonance is possible only in the narrow frequency range of $0.6 - 0.73\,THz$, with absolute minimum of $\text{Re}[\mu_{\omega,q}] = -230\,cm^2/Vs$ at frequency $0.65\,THz$.

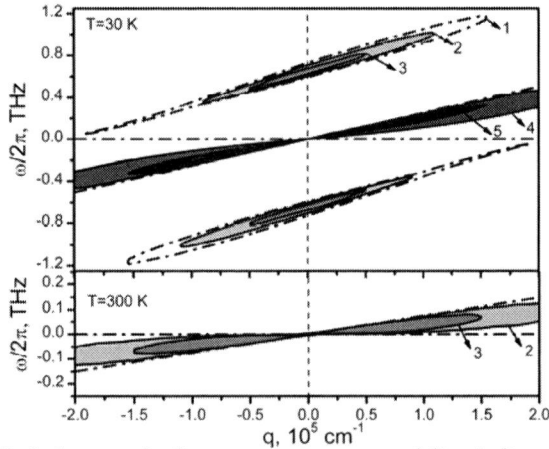

Fig.3. Contour plot for constant microwave mobility. Isolines 1, 2, 3, 4 and 5 restrict the regions of $\{\omega, q\}$ at which $\text{Re}[\mu_{\omega,q}] < 0, -100, -200, -1000, -2000\,cm^2/Vs$, respectively. Steady-state field is $3\,kV/cm$.

In the case of *Cherenkov* effect, the amplification occurs within the sector $\{\omega = 0, \omega = V_d q\}$ and covers the frequency range of $0 - 0.45\,THz$. Here the magnitudes of negative microwave mobility can reach the values of several thousands cm^2/Vs. At room temperatures (bottom panel), the amplitudes of negative microwave mobility are appeared to be of several hundreds cm^2/Vs and amplification is restricted by the frequencies less than $0.15\,THz$. This is in the agreement with the obtained value of drift velocity. At room temperature the calculations give, $V_d = 0.5 \times 10^7\,cm/s$ at the steady-state field of $3\,kV/cm$.

In summary, we developed Monte-Carlo procedure of the calculation of frequency and wave-vector dispersion of the microwave mobility. For the case of the compensated GaN sample it was found the regions of the frequencies and wave vectors where amplification of external time- and spatial-dependent signal is possible. We suggest that present work can improve the existing theory of the plasmonic grating structures and facilitate elaboration of field-controlled devices for THz optoelectronics, including THz emitters, amplifiers and detectors.

REFERENCES

[1] S. A. Mikhailov, "Plasma instability and amplification of electromagnetic waves in low-dimensional electron systems," *Phys. Rev. B*, vol. 58, no. 2, pp.1517-1532, 1998. S. A. Mikhailov, "Tunable solid-state far-infrared sources: New ideas and prospects," *Recent Res Devel. Applied Phys.*, vol. 2, pp. 65-108, 1999;

[2] S. A. Mikhailov "Graphene-based voltage-tunable coherent terahertz emitter," *Phys. Rev B* vol. 87, 115405, 2013.

[3] V. V. Popov, "Plasmon Excitation and Plasmonic Detection of Terahertz Radiation in the Grating-Gate Field-Effect-Transistor Structures," *J. Infrared Milli Terahertz Waves*, vol. 32, pp. 1178-1191, 2011.

[4] T. Otsuji , H. Karasawa, T. Watanabe, T. Suemitsu, M. Suemitsu , E. Sano , Wojciech Knap, Victor Ryzhii "Emission of terahertz radiation from two-dimensional electron systems in semiconductor nano-heterostructures," *C.R. Physique* vol. **11**, pp. 421-432, 2010.

[5] V. N. Sokolov, K. W. Kim, V. A. Kochelap and D. L.Woolard, "High-frequency small-signal conductivity of hot electrons in nitride semiconductors," *Appl. Phys. Lett.* vol. **84**, no. 18, pp. 3630-3632, 2004.

[6] Korotyeyev V.V., Kochelap V.A., Klimov A.A., Kim K.W., Woolard D.L. "Tunable terahertz-frequency resonances and negative dynamic conductivity of two-dimensional electrons in group-III nitrides." *J. Appl. Phys.* vol. **96**, no. 11, pp.6488-6491, 2004.

[7] P. Shiktorov, E. Starikov, V. Gružinskis, L. Varani, C. Palermo, J.-F. Millithaler, L. Reggiani "Frequency limits of terahertz radiation generated by optical-phonon transit-time resonance in quantum wells and heterolayers" *Phys. Rev B* vol. **76**, 045333, 2007.

[8] G.I. Syngayivska and V.V. Korotyeyev "Electrical and high-frequency properties of compensated GaN under electron streaming conditions," *Ukrainian Journal of Physics* vol. **58**, no.1, pp. 40-55 2013.

[9] G I Syngayivska, V V Korotyeyev and V A Kochelap "High-frequency response of GaN in moderate electric and magnetic fields: interplay between cyclotron and optical phonon transient time resonances," *Semicond. Sci. Technol.* vol. **28** 035007, 2013.

[10] V. V. Korotyeyev, V. A. Kochelap, and L. Varani. "Wave excitations of drifting two-dimensional electron gas under strong inelastic scattering," *J. Appl. Phys.* vol. **112**, 083721 (2012)

Zero reflection phenomenon in terahertz crystalline spectra

S. G. Felinskyi[1], G. S. Felinskyi [2]

Taras Shevchenko Kyiv National University, Ukraine, 03127, Kyiv, Glushkov Ave 4,

e-mail:[1]felinskyi.sg@gmail.com, [2]felinskyi.gs@gmail.com

Abstract: Research results of deep minima reflectance of terahertz (far-infrared) radiation one may observe experimentally in a number of crystalline media are firstly presented in the paper. Analysis of physical conditions for the almost complete disappearance of the electromagnetic waves reflection from the interface with crystal is given. We based on single oscillatory model of crystal and it allows us to derive the quantitative expressions for the frequency corresponds to minimum value of the reflection coefficient R_{min}. It is shown the frequencies where the reflection intensity riches extremely low values are always greater than the appropriate frequencies of longitudinal vibrations, and $R_{min} = 0$ for an ideal case of damping absence. Two models of the crystalline medium are considered: (i) the ideal case without damping and (ii) real model corresponds to the case of nonzero phonon damping. The applicability of both models for describing the phenomenon of zero reflection of terahertz waves is established by comparing the calculations results together and by its matching with actual experimental data.

Prospects and the possibility of creating anti reflection covers (invisible) throughout the electromagnetic spectrum is associated with the synthesis of metamaterials [1], which should have a negative refractive index resulting from the presence of negative dielectric permittivity (DP) ε and magnetic permeability μ. However anti reflection features may be found in real crystals as we will show in our paper. The basic conditions of the abnormally low reflection coefficient R formation, which is observed in the real crystalline media, were considered in this work. The areas with almost zero R are located outside of the area with negative permittivity [2], as it is shown in our paper.

Negative DP is located within the T-L frequency splitting interval, as recently was showed in our works [3,4], and explore this area using standard IR techniques. The traditional processing of the crystals reflection spectra in the far infrared (terahertz) region is usually restricted by analysis of polar oscillation parameters, which are formed near the reflection peak and just it don't paid enough attention to the reflection minimum. Moreover, these gaps in the spectrum are often hidden by noises of spectral equipment.

Even at superficial analysis of reflection spectra of a large number of crystalline media in terahertz range one can identify specific frequency bands with anomalously low reflection coefficient R. The experimental reflection spectra of TbMnO$_3$ crystal [5] for the three polarizations of incident radiation are shown on Fig. 1 as example. In some areas we marked on the spectra by dashed arrows in Fig. 1, the reflection coefficient drops to such small values as the air to

crystal intersection becomes virtually indistinguishable to electromagnetic radiation.

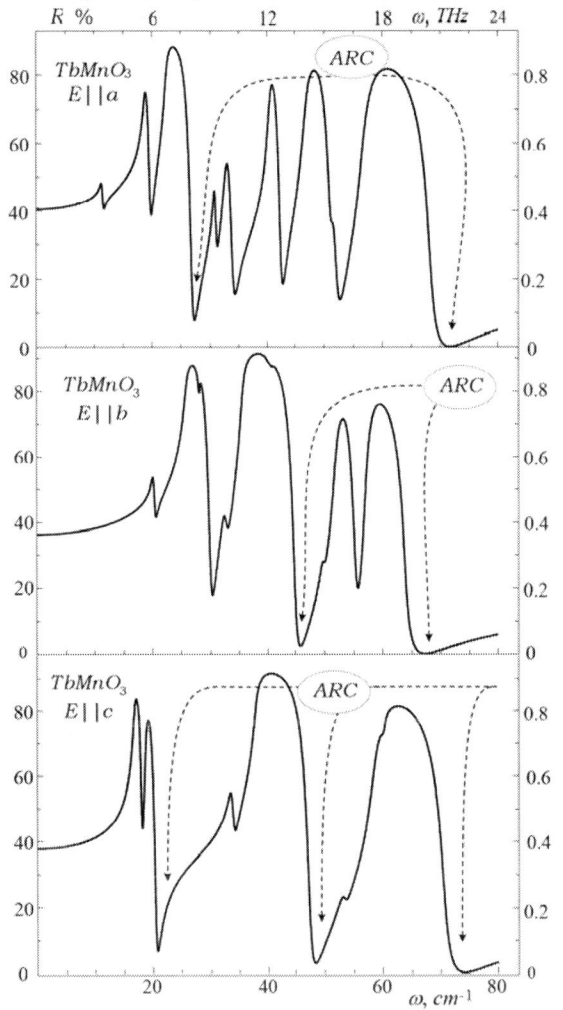

Fig. 1. A typical example of the deep minima occurrence in the infrared reflectance spectra (0-24 THz) for crystal TbMnO$_3$, taken from [5]. We point out by arrows the areas with abnormally low reflection of radiation in comparison to reflection both in radio frequency range ($\omega \to 0$) and the optical transparency frequencies ($\omega \to \infty$).

Firstly paid attention to areas of abnormally low reflection, we have detected the specific bands with a nearly zero R value at minimum and it really is many times smaller than a reflection on both sides of the resonance region.

These circumstances allows us to introduce the term "anti-reflection channel" (ARC) for refer to areas with abnormally low reflection. Next we consider the physical reasons of ARC appearance and equations for the main channel parameters (center frequency and minimum value of the reflection coefficient) were obtained in this paper.

Our modeling results for basic ARC parameters such as minimum frequency $\omega_{\min}$ and the value of minimum reflection coefficient $R_{\min}$ for the crystal with one oscillation based on the dispersion analysis (DA) method. Simulation consists of two parts of single oscillation model depends on the value of the damping constant Γ:

• ideal or case without damping ($\Gamma = 0$);
• real or case with no zero damping ($\Gamma \neq 0$).

Equivalency of both models together for small values of the damping constant Γ is also demonstrated in the article.

The reflection coefficient reaches the absolute minimum at a frequency $\omega_{\min}$ in the ideal case ($\Gamma = 0$) and its value can be found as:

$$\omega_{\min} = \omega_T \sqrt{1 + \frac{S}{\varepsilon_\infty - 1}} = \omega_T \sqrt{\frac{\varepsilon_0 - 1}{\varepsilon_\infty - 1}} = \omega_L \sqrt{\frac{1 - 1/\varepsilon_0}{1 - 1/\varepsilon_\infty}}, \quad (1)$$

where ε_0 and ε_∞ are DPs at low and high frequencies, respectively, S is oscillator strength, ω_T and ω_L are frequencies of transverse and longitudinal vibrations, respectively.

The inequality $\omega_{\min} > \omega_T > \omega_L$ is directly resulted from the equation (1). This means that the frequencies of zero reflection are not located within the T-L frequency splitting of the crystal. Therefore zero reflection should be outside of the negative DP region [6].

The minimum value of the reflection coefficient for an ideal case is:

$$R_{\min} = R(\omega_{\min}) = 0, \quad (2)$$

Thus, for arbitrary crystalline media there is the frequency in the vicinity of single polar vibration and incident radiation doesn't reflect from the crystal at this frequency in case of phonon damping absence. However, this doesn't mean that the media becomes completely invisible, since the crystal is opaque. So we can't see anything that is located behind of crystal. Conversely, at minimum frequency, so to speak, the medium behaves as a "black hole" that all electromagnetic radiation that falls on the crystal irreversibly lost in it.

For a real case of no zero damping minimum frequency has a bulky look, unlike (1):

$$\omega_{\min} = \frac{1}{\sqrt{2\left(\varepsilon^\infty - 1\right)}} \left(\begin{array}{c} \Gamma^2\left(1 - \varepsilon^\infty\right) + \omega_L^2 \varepsilon^\infty + \omega_T^2\left(\varepsilon^\infty - 2\right) + \\ + \left(\Gamma^4\left(\varepsilon^\infty - 1\right)^2 + \varepsilon^{\infty 2}\left(\omega_L^2 - \omega_T^2\right)^2 + \\ + 2\Gamma^2\left(1 - \varepsilon^\infty\right)\left(\omega_L^2 \varepsilon^\infty + \omega_T^2\left(\varepsilon^\infty - 2\right)\right)\right)^{1/2} \end{array} \right)^{1/2} \quad (3)$$

The minimum value of the reflection coefficient ($R_{\min}$) isn't zero in the real case and its finite value can be determined as follows:

$$R_{\min} = \frac{\left(P + 2\left(\dfrac{\Gamma \varepsilon^\infty \omega_{\min}\left(\omega_L^2 - \omega_T^2\right)}{\sqrt{2P}\left(\Gamma^2 \omega_{\min}^2 + \left(\omega_{\min}^2 - \omega_T^2\right)^2\right)} - 1 \right) \right)^2}{\left(P + 2\left(\dfrac{\Gamma \varepsilon^\infty \omega_{\min}\left(\omega_L^2 - \omega_T^2\right)}{\sqrt{2P}\left(\Gamma^2 \omega_{\min}^2 + \left(\omega_{\min}^2 - \omega_T^2\right)^2\right)} + 1 \right) \right)^2}, \quad (4)$$

where

$$P = \varepsilon^\infty \sqrt{\frac{\Gamma^2 \omega_{\min}^2 + \left(\omega_{\min}^2 - \omega_L^2\right)^2}{\Gamma^2 \omega_{\min}^2 + \left(\omega_{\min}^2 - \omega_T^2\right)^2}} - \varepsilon'\left(\omega_{\min}\right)$$

If damping constant direct to zero in (3) and (4) than we obtain the expressions (1) and (2) respectively.

Table 1. ARC options of MnF$_2$ crystal for temperatures 300 and 5 K.

Structure	MnF$_2$ (300 K)	MnF$_2$ (5 K)
ω_T, THz	8.67	8.78
ω_L, THz	14.49	14.69
Γ, THz	0.48	0.13
$\omega_{\min}$, THz ($\Gamma \neq 0$)	17.81	18.08
$R_{\min}$, % ($\Gamma \neq 0$)	$1.22 \cdot 10^{-2}$	$8.78 \cdot 10^{-4}$
$\omega_{\min}$, THz ($\Gamma = 0$)	17.82	17.86

We have used additional experimental data for a crystal MnF$_2$ from [7] by way of our results approbation because a single pronounced oscillation is really observed in C polarization of its terahertz reflectivity spectrum. The IR reflection spectra of crystal MnF$_2$ are shown on Fig. 2 for two temperatures 5 K and 300 K.

The quantitative parameters for real and ideal models gathered with spectroscopic ARC parameters of the crystal MnF$_2$ are presented in Table 1. The maximum value of the reflection coefficient in resonance by three orders more than $R_{\min}$ at 300 K and it is by five orders greater than minimum reflection at 5 K. The channel frequency is shifted relative the longitudinal phonon frequency on about of 4 THz for both temperatures. Minimum frequencies are calculated using the expressions (3) and (1) and they are almost equal for both temperatures because damping constants Γ are close to zero. So the ideal model of ARC can be used for small values of Γ as it is mathematically simpler. The value of $R_{\min}$ for temperature 5 K on two orders less than for 300 K (see Table 1) whereas the damping constant Γ at 5 K is almost on four times less than at 300 K.

The characteristic curves for R_{min} as function of damping parameter Γ in MnF_2 crystal for two temperatures 300 K and 5 K are shown on Fig. 3. The minimum value of the reflection coefficient increases with increasing damping, however, still remains at level <1% for oscillations with a high degree of damping.

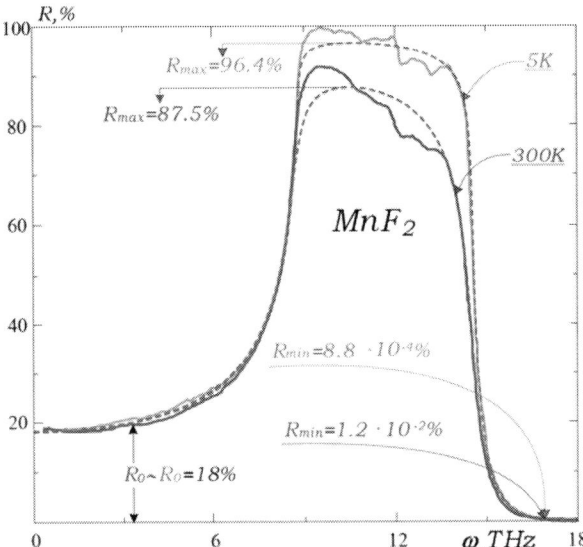

Fig. 2 Analysis of the ARC in the crystal reflection spectra of MnF_2, based on the data of [7] for temperatures 300 K and 5 K. The solid line shows the experimental data, dotted calculation of the parameters in Table 1. The calculated value of R_{min} is less by three orders for 300 K and over four orders for the 5 K with respect to R_0 in non-resonant region.

Reflection spectra of crystals (Fig. 1 and Fig. 2) show the electromagnetic anomalies in terahertz range: the disappearance of the crystal interface at a certain frequency. These effects aren't drawn enough attention until now. Formation ARC, i.e. the emergence of regions with an extremely low electromagnetic reflection should be observed in any crystalline spectrum with at least one pronounced polar mode and low phonon damping. The sharp drop reflection of electromagnetic wave power from the crystal interface is observed at a definite frequency $\omega_{min} > \omega_T > \omega_L$ and such gap deals with resonant interaction between radiation and crystal polar vibrations.

The minimum reflectance can be reduced by several orders compared to the resonance region. Effect of ambient temperature is to add reducing of the damping constant Γ, and it reduces the value of R_{min} by two orders for the crystal MnF_2.

Thus, full incident intensity of electromagnetic wave is absorbed on ARC frequencies in the crystal as it is shown by our calculations. This effect demonstrates that the main channel of heat transfer from the environment into crystal isn't located at the resonant frequencies of transverse and/or longitudinal phonons but it is shifted toward higher frequency.

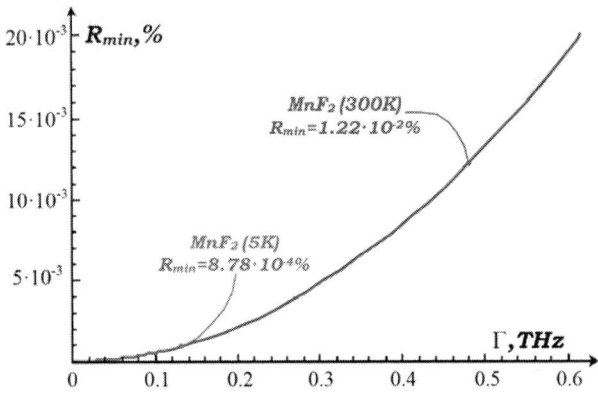

Fig. 3 The minimum reflection coefficient as function of damping parameter Γ (dots point out the measured parameters in crystal MnF_2 for temperatures 300 K and 5 K). The minimum value of the reflection coefficient is increased as damping, however, it still remains <<1% at level even for polar modes with a high damping degree.

REFERENCES

[1] V.G. Veselago, "Energy, momentum and mass transfer by an electromagnetic wave in a negative refraction medium," Physics – Uspekhi, vol.**179**, no.6, pp.689-694, 2009.

[2] P. A. Korotkov, and G. S. Felinskyi, "Research of negative dielectric permeability area in the media without inversion center," Bull. Kyiv Univ.: ser. Phys. & Math., no. 2, pp. 162-171, 2008.

[3] Felinskyi S. G., Korotkov P. A. Felinskyi G. S. Negative dielectric permittivity of nonmagnetic crystals in the terahertz waveband / Semiconductor Physics, Quantum Electronics & Optoelectronics, 2012. V. 15, N 1. P. 83-88.

[4] S. G. Felinskyi, and G. S. Felinskyi, "Criterion for the appearance of negative dielectric areas in crystals," *in* Proc. 10th Int. Conf. LFNM'2010, Sevastopol, Ukraine, pp.58-59.

[5] R. Schleck, R. L. Moreira, H. Sakata, and R. P. S. M. Lobo, Infrared reflectivity of the phonon spectra in multiferroic $TbMnO_3$ / PHYSICAL REVIEW B **82**, 144309 (2010), DOI: 10.1103/PhysRevB.82.144309

[6] S. G. Felinskyi, P. A. Korotkov, and G. S. Felinskyi, "Negative dielectric function appearance areas on the frequencies of polar vibrations in crystals," Bull. Kyiv Univ.: ser. Phys.&Math., no. 1, pp. 191-196, 2010.

[7] R. Schleck, Y. Nahas, and R. P. S. M. Lobo Elastic and magnetic effects on the infrared phonon spectra of MnF_2 / PHYSICAL REVIEW B **82**, 054412 (2010), DOI: 10.1103/PhysRevB.82.054412

Amplification of terahertz radiation by plasmons in graphene with a planar Bragg grating

O. V. Polischuk[1,2], V. V. Popov[1,2], S.A. Nikitov[2,3], V. Ryzhii[4], T. Otsuji[4], M. S. Shur[5]

[1]Kotelnikov Institute of Radio Engineering and Electronics (Saratov Branch),Russian Academy of Sciences, Saratov 410019, Russia

[2]Saratov State University, Saratov 410012, Russia

[3]Kotelnikov Institute of Radio Engineering and Electronics, Russian Academy of Sciences, Mokhovaya 11-7, Moscow, 125009, Russia

[4]Research Institute for Electrical Communication, Tohoku University, Sendai 980-8577, Japan

[5]Department of Electrical, Computer, and Systems Engineering, Rensselaer Polytechnic Institute, Troy, New York 12180, USA

Abstract: The giant amplification of terahertz radiation due to stimulated generation of plasmons in graphene with a planar Bragg grating strongly coupled to terahertz radiation is predicted. The amplification of terahertz wave at the plasmon resonance frequencies is several orders of magnitude stronger than away from the resonances. It is shown that the dynamic and frequency ranges of terahertz graphene amplifier can be strongly enhanced in the structure with a narrow-slit Bragg grating and/or thin barrier layer.

Graphene, a two-dimensional monolayer of graphite, has received a great deal of interest recently due to its unique electronic properties stemming from a linear (Dirac-type) gapless carrier energy spectrum $\mathcal{E} = \pm V_F |\mathbf{p}|$ (see the inset in Fig. 1), where $\mathcal{E}$ and $\mathbf{p}$ are the electron (hole) energy and momentum, respectively, $V_F \approx 10^8$ cm/s is the Fermi velocity, which is a constant for graphene, and upper and lower signs refer to the conduction and valence bands, respectively [1, 2]. Nanostructured graphene exhibits strong plasmonic response at terahertz (THz) frequencies due to both high density and small "relativistic" effective mass of free carriers [3]. As compared with the stimulated emission of the electromagnetic modes (photons), the stimulated emission of plasmons by the interband transitions in the population inverted graphene exhibits a much higher gain due to a small group velocity of the plasmons in graphene and strong confinement of the plasmon field in the vicinity of the graphene layer [4, 5]. However, a large plasmon gain in graphene leads to strong dephasing of the plasmon mode, thus preventing THz lasing. Also, strong coupling between plasmons in graphene and electromagnetic radiation can hinder THz lasing. Therefore, neither resonant amplification of THz radiation nor THz plasmonic lasing in graphene have been reported so far although stimulated emission of near-infrared and THz photons from population inverted graphene was recently observed [6, 7].

In this paper, we study theoretically and simulate numerically the amplification of a THz wave by the stimulated generation of resonant plasmons in a graphene sheet with the Bragg grating (see Fig. 1).

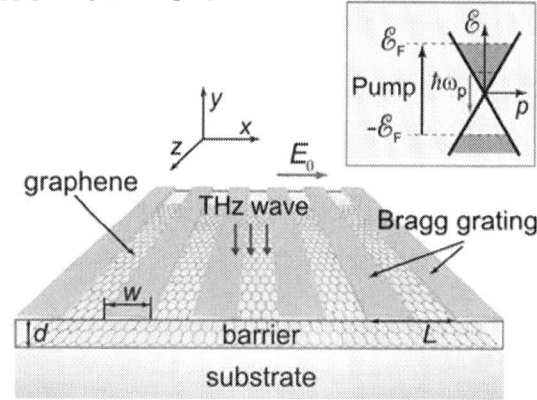

Fig. 1. Schematic view of the graphene sheet with the Bragg grating and the coordinate system. Incoming electromagnetic wave is incident from the top at normal direction to the structure plane with the polarization of the electric field across the metal grating contacts. The inset schematically shows the energy band structure of pumped graphene and the stimulated emission of the plasmon quantum $\hbar\omega_p$ in graphene.

Incoming THz wave of frequency ω is incident upon the array at normal direction to its plane and strongly couples to the plasmon modes in the graphene plane. Graphene is described by its dynamic conductivity $\sigma(\omega)$:

$$\sigma_{Gr}(\omega) = \sigma_{intra}(\omega) + \sigma_{inter}(\omega),$$

where

$$\sigma_{intra}(\omega) = \frac{e^2 8 k_B T \tau}{4\pi\hbar^2 (1 - i\omega\tau)} \ln\left[1 + \exp\left(\frac{\mathcal{E}_F}{k_B T}\right)\right]$$

and

$$\sigma_{\text{inter}}(\omega) = \frac{e^2}{4\hbar} \tanh\left(\frac{\hbar\omega - 2\mathcal{E}_F}{4 k_B T}\right) -$$
$$- \frac{e^2}{4\hbar} \frac{4\hbar\omega}{i\pi} \int_0^\infty \frac{G(\mathcal{E}, \mathcal{E}_F) - G(\hbar\omega/2, \mathcal{E}_F)}{(\hbar\omega)^2 - 4\mathcal{E}^2} d\mathcal{E}.$$

Here ω is the frequency of the incoming electromagnetic wave, e is the electron charge, $\hbar$ is the reduced Planck constant, k_B is the Boltzmann constant, and

$$G(\mathcal{E}, \mathcal{E}') = \frac{\sinh(\mathcal{E}/k_B T)}{\cosh(\mathcal{E}/k_B T) + \cosh(\mathcal{E}'/k_B T)}.$$

The conductivity $\sigma_{\text{intra}}(\omega)$ describes a Drude-like response involving the intraband processes with the phenomenological electron and hole scattering time τ, which can be estimated from the measured dc carrier mobility: $\tau = \mu \mathcal{E}_F / e V_F^2$. The temperature independent carrier mobility $\mu > 250000$ cm^2/V·s observed recently in multilayer epitaxial graphene on 4H-SiC substrate [9, 10] corresponds to $\tau \approx 10^{-12}$ s for $\mathcal{E}_F = 40$ meV at room temperature [8].

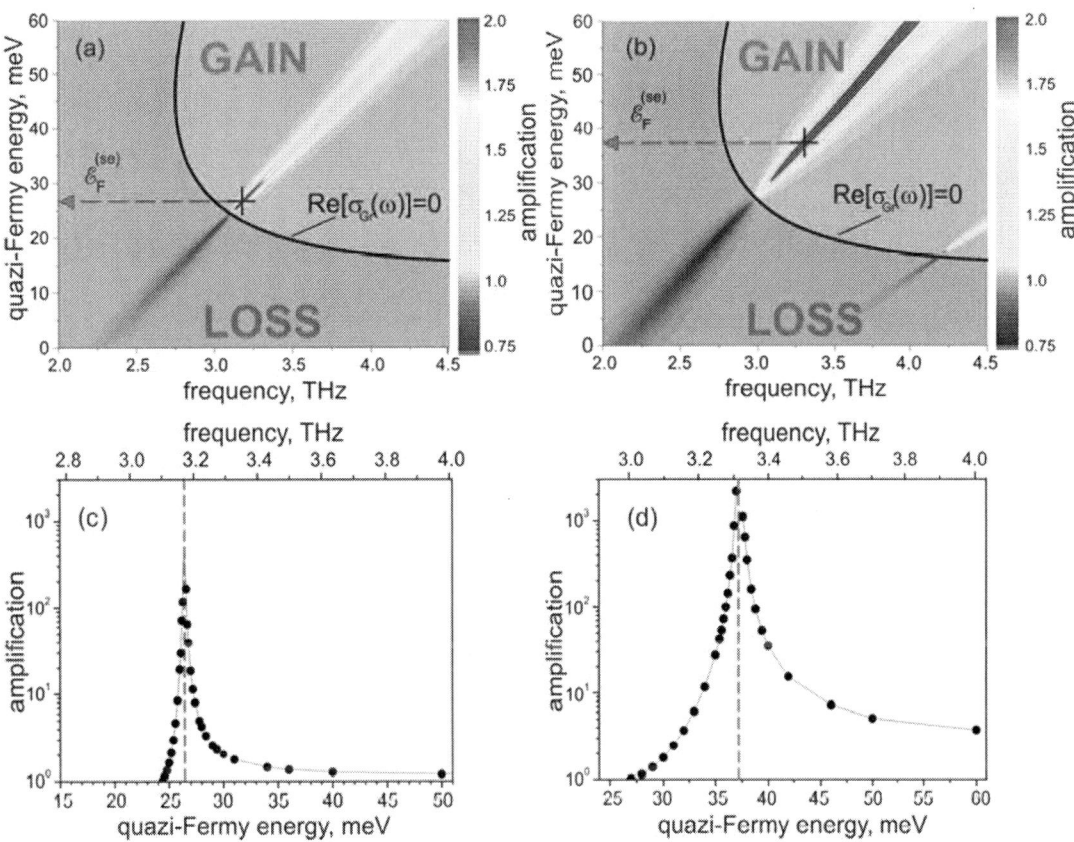

Fig. 2 (a, b) Contour maps of the power amplification coefficient as a function of the quasi-Fermi energy and the frequency of incoming THz wave for the Bragg grating aspect ratio $a/L = 0.5$ (panel (a)) and $a/L = 0.8$ (panel (b)). The period of the Bragg grating is L=500 nm and the barrier-slab thickness is $d = 50$nm. The electron scattering time in graphene is $\tau = 10^{-12}$ s. Red arrows in panels (a) and (b) mark the quasi-Fermi energies for the THz lasing regimes in the fundamental plasmon resonance. (c,d) The variation of the power amplification coefficient along the fundamental plasmon resonance lobe around the self-excitation regime marked by the vertical dashed lines in panels (c) and (d).

Under optical or injection pumping, the electrons and holes in graphene are characterized by different quasi-Fermi energies $\pm \mathcal{E}_F$ of the same absolute value but the opposite signs in respect to the energy of the Dirac point (see the inset in Fig. 1) that corresponds to the carrier population inversion. For sufficiently strong pumping (large absolute value of the quasi-Fermi energy), a real part of its conductivity becomes negative, Re[$\sigma(\omega)$]<0, that corresponds to the energy gain (negative absorption) in graphene.

The Bragg grating plays a two-fold role. First, it forms a distributed planar resonator for the plasmon modes excited or generated in graphene by the incident THz radiation or stimulated plasmon emission, respectively. Second, the Bragg grating serves as an effective coupler between the plasmons in graphene and THz radiation (incident upon or emitted from the structure).

Fig. 3 shows the calculated power amplification coefficient as a function of the quasi-Fermi energy (which corresponds to the pumping strength) and frequency for the Bragg grating of period 500 nm and the grating aspect ratio $a/L = 0.5$ (Fig. 3(a)) and $a/L = 0.8$ (Fig. 3(b)). The fundamental plasmon resonance is exhibited as colored lobes in Figs. 3(a) and 3(b). The value of $\text{Re}[\sigma_{Gr}(\omega)]$ is negative above the solid black line in Figs. 3(a) and 3(b) corresponding to $\text{Re}[\sigma_{Gr}(\omega)] = 0$ (i.e., to transparent graphene). The amplification coefficient is greater than unity above this boundary line while less then unity below this line at all frequencies and pumping strengths. Correspondingly, the plasmon absorption resonances below the $\text{Re}[\sigma_{Gr}(\omega)] = 0$ line give way to the plasmon amplification resonances above that line.

Above the graphene transparency line $\text{Re}[\sigma_{Gr}(\omega)] = 0$, the THz wave amplification at the plasmon resonance frequency (see Fig. 3(c)) is several orders of magnitude stronger than away from the resonances. The amplification coefficient at the coefficient along the lobe of the fundamental plasmon amplification resonance around the self-excitation regime is shown in Figs. 3(c) and 3(b).

The radiative damping of plasmons can be tuned by varying the width of the slits between the conductive strips of the distributed Bragg grating. Much higher plasmon radiative damping can be achieved for narrow slits [11]. As a result, much higher energy gain (more negative real part of the graphene conductivity) is required to meet the self-excitation condition in the structure with a narrow-slit Bragg grating (cf. the red arrow positions in Figs. 3(a) and 3(b). However, the plasmon radiative damping not only determines the self-excitation (lasing) condition but also controls the electromagnetic power emitted from the structure [12]. Therefore, much broader frequency and dynamic range of THz amplification around the self-excitation regime can be realized for greater plasmon radiative damping achieved for a narrow-slit Brag grating, cf. Figs. 3(c) and 3(d).

Acknowledgment

The work was supported by the Russian Foundation for Basic Research (Grant ## 11-02-92101 and 12-02-93105) and by the Russian Academy of Sciences Program "Technological Fundamentals of Nanostructures and Nanomaterials." The work at RPI was supported by the US NSF under the auspices of I/UCRC "CONNECTION ONE", NSF I-Corp, and by the NSF EAGER program. This work was financially supported in part by NPRP grant # NPRP 09-1211-2-475 to RPI from the Qatar National Research Fund and by the Grant-in-Aid for Specially Promoting Research (#23000008), Japan, by the JSPS-RFBR Japan-Russian Collaborative Research Program, and by JST-CREST, Japan.

References

[1] K. S. Novoselov, A. K. Geim, S. V. Morozov, D. Jiang, M. I. Katsnelson, I. V. Grigorieva, S. V. Dubonos, and A. A. Firsov, "Two-dimensional gas of massless Dirac fermions in graphene," *Nature*, vol. 438, no. 10, pp. 197–200, 2005.

[2] A. H. Castro Neto, F. Guinea, N. M. R. Peres, K. S. Novoselov, and A. K. Geim, "The electronic properties of graphene," *Rev. Mod. Phys.*, vol. 81, no. 1, pp. 109–162, 2009.

[3] L. Ju, B. Geng, J. Horng, C. Girit, M. Martin, Z. Hao, H. A. Bechtel, X. Liang, A. Zettl, Y. R. Shen, and F. Wang, "Graphene plasmonics for tunable terahertz metamaterials," *Nature Nanotech.*, vol. 6, no. 10, pp. 630–634, 2011.

[4] F. Rana, "Graphene terahertz plasmon oscillators," *IEEE Trans. Nanotechnol.*, vol. 7, no. 1, pp. 91–99, Jan. 2008.

[5] A. A. Dubinov, V. Ya. Aleshkin, V. Mitin, T. Otsuji, and V. Ryzhii, "Terahertz surface plasmons in optically pumped graphene structures," *J. Phys.: Condens. Matter*, vol. 23, no. 14, pp. 145302–145302-8, 2011.

[6] T. Li, L. Luo, M. Hupalo, J. Zhang, M. C. Tringides, J. Schmalian, and J. Wang, "Femtosecond population inversion and stimulated emission of dense Dirac fermions in graphene," *Phys. Rev. Lett.*, vol. 108, no.16, pp. 167401-1–167401-5, 2012.

[7] S. Boubanga-Tombet, S. Chan, T. Watanabe, A. Satou, V. Ryzhii, and T. Otsuji, "Ultrafast carrier dynamics and terahertz emission in optically pumped graphene at room temperature," *Phys. Rev. B*, vol. 85, no. 3, pp. 035443-1– 035443-6, 2012.

[8] J. Chen et al, "Optical nano-imaging of gate-tunable graphene plasmons", *Nature*, vol. 487, pp.77–81, 2012.

[9] M. Orlita et al, "Approaching the Dirac point in high-mobility multilayer epitaxial grapheme", *Phys. Rev. Lett.*, vol. 101, pp. 267601, 2008.

[10] M. Sprinkle et al, "First direct observation of a nearly ideal graphene band structure", *Phys. Rev. Lett.*, vol. 103, pp. 226803, 2009.

[11] V. V. Popov, "Plasmon excitation and plasmonic detection of terahertz radiation in the grating-gate field-effect-transistor structures", *J. Infrared Millim. Terahertz Waves*, vol. 32, pp. 1178–91, 2011.

[12] V V Popov, O V Polischuk, A R Davoyan, V Ryzhii, T Otsuji and M S Shur, "Plasmonic terahertz lasing in an array of graphene nanocavities", *Phys. Rev. B*, vol. 86, pp. 195437, 2012.

Modeling of functional optical coatings based on plasmonic nanocomposites

S. G. Moiseev[1,2]

[1]Ulyanovsk State University, Ulyanovsk, Russia
[2]Kotel'nikov Institute of Radio Engineering and Electronics of RAS, Ulyanovsk, Russia

Abstract: Computer modeling of the optical properties and optimization of the structure of nanocomposite films with different distribution of inclusions for the anti-reflective and ultrathin polarizing coatings are presented.

Nanocomposite materials (metamaterials) with extraordinary electromagnetic properties can find a variety of applications, in particular for manufacturing superlenses, nonreflecting (absorptive) materials, as well as controlling the optical beam intensity and propagation direction, etc. In recent years, a large number of structures have been proposed and theoretically and experimentally investigated, which possess negative, high or small refractive indices, selective absorption of or transparency to optical light. In this work, a more detailed investigation of optical properties of a matrix metal-dielectric medium with metal inclusions is performed, and the possibility to realize plasmonic structures with beneficial effects in the visible region is considered.

The dependence of optical properties of the plasmonic nanocomposite with metal nanoparticles randomly distributed over the whole matrix volume on the geometric (shape and concentration of inclusions) and material (permittivities of the matrix and metal nanoparticles) parameters are calculated within the effective-medium approximation. Specifically, we apply the Maxwell–Garnett model [1], whose results for matrices with a moderate content of spheroidal inclusions are in fairly good agreement with the results of exact electrodynamic calculation.

The data obtained with the help of Maxwell-Garnett effective medium model show that a heterogeneous medium with plasmonic impurities, such as silver nanoparticles with a concentration of about 10^{21} per m^3, is an interesting object of research with many perspectives for applications. Our results show that such plasmonic medium can be used as a low-loss anti-reflection coating [2], weakly reflecting light-absorbing filter, or polarizing beam splitter with high performance in transmission and reflection [3].

For example, in Fig. 1 we plot the optical properties of the proposed anti-reflection composite film with random distribution of silver inclusions. The dependencies show that coating of dielectric surface by a composite layer has a positive effect in a wide ($\Delta\lambda \approx 100\,nm$) spectral range. One can see that the total intensity of reflected light decreases, and the minimum reflectance of the composite coating is lower than that of the substrate. The refracted wave intensity increases

only slightly in this case, and even decreases in comparison with the initial dielectric at $\lambda{>}500$ nm. The latter circumstance can be explained as follows: some part of the light wave energy spent on excitation of free-electron oscillations in composite nanoparticles is transformed into heat.

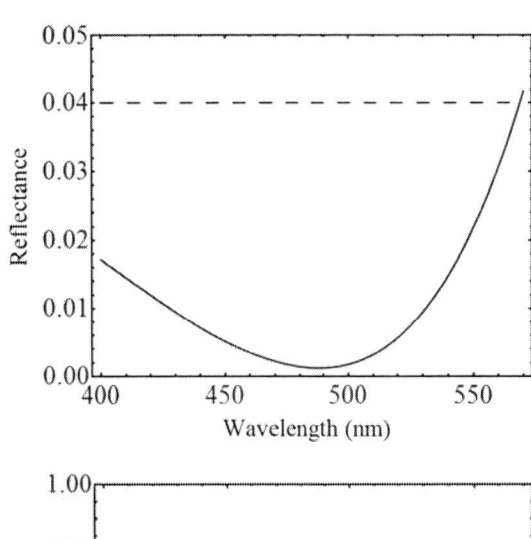

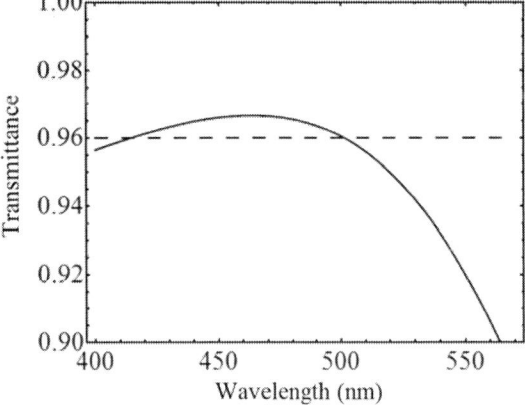

Fig. 1. Calculated reflectance and transmittance of the composite coating with silver nanoparticles for normal incidence of light, calculated within the effective-medium model. The case of prolate silver nanoparticles is considered. The refractive index of dielectric coating material and substrate is equal to 1.5. The volume fraction of silver nanoparticles is 0.05. The reflectance and transmittance of the clean dielectric surface are shown by dashed line.

978-1-4799-0019-0/13 $31.00 © 2013 IEEE

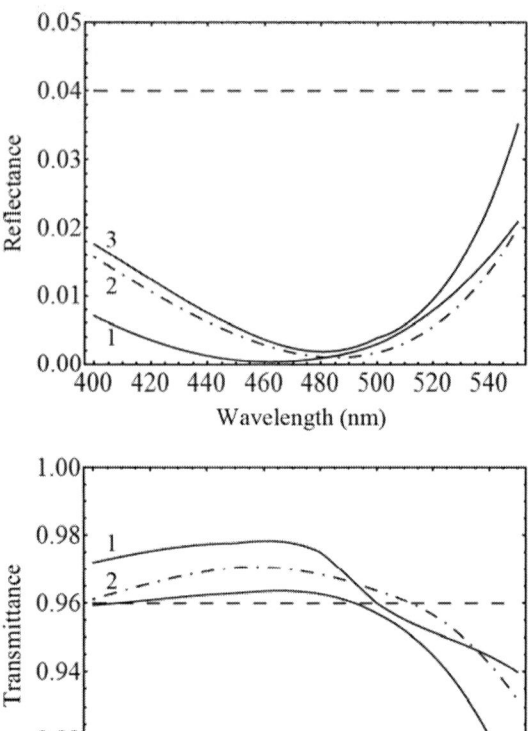

Fig. 2. Calculated reflectance and transmittance of the discontinuous composite coating with silver nanoparticles for normal incidence of light. The numbers in the pictures correspond to the quantity of monolayers of nanoparticles inside coating material. The refractive index of dielectric coating material and substrate is equal to 1.5. The reflectance and transmittance of the clean dielectric surface are shown by dashed line.

Nanocomposite functional structures are considered also to be formed by one, two, or more monolayers of uniformly oriented metal nanoparticles suspended in a transparent media. Monolayers are oriented parallel to the substrate boundary. In order to calculate the coefficients of the direct light transmission and specular reflection for a stack made of monolayers, we combine the quasi-crystalline approximation applied for calculations of the transmission properties of individual monolayers, with the transfer-matrix technique used for subsequent calculations of the transmission properties of multilayer structures [4,5].

Within the T-matrix method the field amplitudes in the left and right sides of a multilayer structure are related by the system T-matrix, which is a product of T-matrices of individual elements, i.e. nanoparticle monolayers ($\hat{T}_i^p$) and dielectric films ($\hat{T}_i^f$). Elements of the transfer matrix of each monolayer are determined by the amplitudes of its coherent transmission and reflection coefficients. For a nanocomposite coating made of N different monolayers separated by arbitrary dielectric films the system transfer matrix is

$$\hat{T}^{nc} = \hat{T}_0^f \left(\prod_{i=1}^{N-1} \hat{T}_i^p \hat{T}_i^f \right) \hat{T}_N^p \hat{T}_N^f ,$$

where $\hat{T}_0^f$ and $\hat{T}_N^f$ are transfer matrixes of the dielectric media outside metal monolayers. Using the T-matrix method one can calculate transmittance and reflectance of the whole multilayer structure.

Our results show that the same or better optical characteristics of the antireflection coating and the polarizing beam splitter can be obtained using discontinuous plasmonic nanocomposite. In Fig. 2 we plot the spectral characteristics of three antireflection coatings, each consisting of monolayers of spheroidal silver nanoparticles. It should be noted that considered cases correspond to the same volume fraction of inclusions, but the structure corresponding to the line 1 in Fig. 2 has both larger transmittance and smaller reflectance. In this case the antireflection coating contains only one monolayer of nanoparticles so the absorption of light is reduced.

This work was supported by the Russian Foundation for Basic Research and the Ministry of Education of the Russian Federation through project contracts within the framework of the Federal Target Program 'Science, Academic and Teaching Staff of Innovative Russia for 2009-2013'.

REFERENCES

[1] C. F. Bohren, D. R. Huffman, *Absorption and Scattering of Light by Small Particles*, Wiley, New York, 1983.

[2] S. G. Moiseev, "Composite medium with silver nanoparticles as an anti-reflection optical coating," *Appl. Phys. A.*, vol. 103, no. 3, pp. 619-622, 2011.

[3] S. G. Moiseev, "Nanocomposite-based ultrathin polarization beamsplitter," *Optics and Spectroscopy*, vol. 111, no. 2, pp. 233-240, 2011.

[4] C. C. Katsidis, D. I. Siapkas, "General Transfer-Matrix Method for Optical Multilayer Systems with Coherent, Partially Coherent, and Incoherent Interference," *Appl. Opt.*, vol. 41, pp. 3978-3987, 2002.

[5] S. M. Kachan, O. Stenzel, A. N. Ponyavina, "High-absorbing gradient multilayer coatings with silver nanoparticles," *Appl. Phys. B*, vol. 84, pp. 281-287, 2006.

Left-Handed Photonic Crystal Waveguide Sensors

D. El-Amassi and M. M. Shabat

Physics Department, Islamic University, Gaza, P.O. Box 108, Palestinian Authority

shabat@iugaza.edu.ps

Abstract: In this work, we examine analytically the propagation of TE polarized wave in a multilayer one dimensional photonic crystal consisting of alternate right-handed material and left-handed materials. The sensitivity of above-mentioned structure optical waveguide sensor is derived and investigated with various physical parameters of the structure.

Veselago first proposed in 1968 the possibility of electromagnetic wave propagation in a medium with simultaneously negative permittivity (ε) and permeability (μ) [1]. Such a medium came to be known as left-handed medium (LHM) or metamaterials [2, 3]. Photonic crystals (PhCs) also attracted intensive studies in the last decade due to their unique electromagnetic properties and possible applications. PhCs are novel class of optical media represented by natural or artificial structures with periodic modulation of the refractive index [4]. Optical waveguide wave sensors have been widely used for various application such as chemical sensing, humidity sensing, biochemical sensing, and biosensing. The effective refractive index of the propagating mode depends on the structure parameters, e.g., the guiding layer thickness and dielectric permittivity and magnetic permeability of the media constituting the waveguide. So, any change in the refractive index of the covering medium leads to a change in the effective refractive index of the guiding mode. The sensing concept of the planer waveguide sensor is to determine the change in the effective refractive index of the covering medium [5, 6].

In this paper, we examine analytically the propagation of TE polarized wave in a multilayer one dimensional photonic crystal consisting of alternate right-handed material and left-handed material, which refer to as left-handed photonic crystal. The proposed photonic crystal structure has been investigated for sensing applications. The sensitivity is derived for various physical parameters, analyzed and discussed taking into account the negative refractive index of the photonic crystals. It has been found that the sensitivity is increasing due to the left-handed materials.

We consider the propagation of electromagnetic wave along z-axis normal to the interface of a multilayer one-dimensional PhC and this layers composed of two different materials with a refractive index n_1 (RHM) and n_2 (LHM) and layer thickness d_1 and d_2, which is characterized by an electric permittivity ε_2 and a magnetic permeability μ_2 such that [8]

$$\varepsilon_2(\omega) = 1 - \frac{\omega_p^2}{\omega^2 + i\gamma\omega} \quad \text{and} \quad \mu_2(\omega) = 1 - \frac{F\omega^2}{\omega^2 - \omega_o^2 + i\gamma\omega}, \quad (1)$$

where ω_p is the plasma frequency, ω_o is the resonance frequency, γ is the electron scattering rate, and F is the fractional area of the unit cell occupied by the split ring.

Following the notation and approach in [9], we get the dispersion relation:

$$\cos(K_B d) = \cos(k_1 d_1)\cos(k_2 d_2) \\ - \frac{1}{2}\left(\frac{\mu_1 k_2}{\mu_2 k_1} + \frac{\mu_2 k_1}{\mu_1 k_2}\right)\sin(k_1 d_1)\sin(k_2 d_2) \quad (2)$$

where $k_i = \left[\left(\frac{n_i \omega}{c}\right)^2 - \beta^2\right]^{1/2}$, $i = 1, 2$, $\beta = \frac{\omega}{c}N$ and the constant K_B is known as the Bloch wave number. Note, K_B lies within the first Brillouin zone, the existence of Bloch requires that K_B is real, such as: $|\cos(K_B d)| < 1$.

The sensitivity of the sensor is calculated as the change of the effective refractive index with respect to the change of the cladding refractive index. Differentiating Eq. (2) with respect to N, we obtain:

$$S = \frac{\partial N}{\partial n_1}$$

$$= \frac{n_1\left[\frac{-2d_1}{k_1}u - \frac{d_1}{k_1}\zeta v + \frac{1}{k_1^2}\eta w\right]}{N\left[-2\left(\frac{d_1}{k_1}u + \frac{d_2}{k_2}v\right) - \left(\frac{d_1}{k_1}v + \frac{d_2}{k_2}u\right)\zeta + \left(\frac{k_2^2 - k_1^2}{k_1^2 k_2^2}\right)\eta w\right]} \quad (3)$$

Where $u = \sin(k_1 d_1)\cos(k_2 d_2)$, $v = \cos(k_1 d_1)\sin(k_2 d_2)$, $w = \sin(k_1 d_1)\sin(k_2 d_2)$, $\zeta = \left(\frac{\mu_1 k_2}{\mu_2 k_1} + \frac{\mu_2 k_1}{\mu_1 k_2}\right)$, and $\eta = \left(\frac{\mu_1 k_2}{\mu_2 k_1} - \frac{\mu_2 k_1}{\mu_1 k_2}\right)$.

In our calculations below, we have assumed the wavelength of Helium-neon laser ($\lambda = 632.8$ nm), the RHM to be glass with $n_1 = 1.5$ ($\varepsilon_1 = 2.25$ and $\mu_1 = 1$) and the thickness with $d_1 = 250$ nm, the LHM which is characterized by ε_2 and μ_2 given by Eq. (1) with $\omega_p = 2\omega$, $\omega_o = 0.4\omega_p$. we solve numerically the dispersion relation given by Eq. (2) for the effective refractive index N. To investigate the behavior of the effective refractive index and the sensitivity of the waveguide under consideration, the thickness of the LHM is changed from 250 nm to 600 nm and all these parameters were calculated and plotted with the thickness of the LHM layer for different K_B, γ, and F.

Fig. 1 shows the real part of the sensitivity of the proposed sensor versus the thickness of the LHM layer for the different values of the Bloch wavevector. It can be seen that, at $K_B = \pi/2d$ the Re(S) is increasing then reaching the stability with increasing of the LHM layer thickness (d_2). At $K_B = \pi/3d$ and $K_B = \pi/4d$ the Re(S) is decreasing with increasing range of d_2. The Re(S) has a peak at a specific value of the LHM layer thickness (d_2). The sensitivity exhibits different behaviors with K_B.

Fig. 2 shows the variation of the real part of the sensitivity of the proposed sensor with the thickness of the LHM layer and the electron scattering rate γ. The Re(S) shows similar behavior with γ. A little enhancement could be achieved by increasing γ.

Fig. 3, the real part of the sensitivity are plotted versus the thickness of the LHM layer for different values of the fractional area of the unit cell occupied by the split ring. As can be seen from the figure, Re(S) exhibits different behaviors with F. The Re(S) is increasing with increasing F.

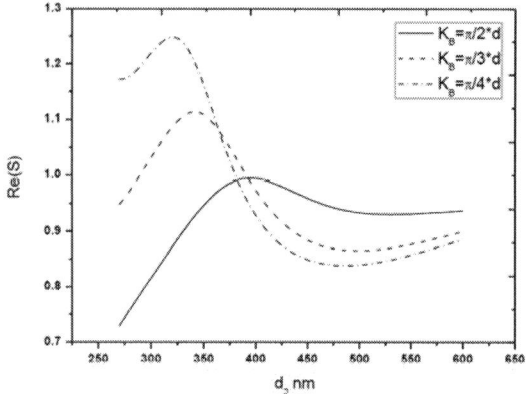

Fig. 1. The real part of the sensitivity of the proposed sensor versus the thickness of the LHM layer for different values of the Bloch wave vector for $\lambda = 632.8\ nm$, $\varepsilon_1 = 2.25$, $\mu_1 = 1$, $d_1 = 250\ nm$, $\gamma = 0.012\omega_p$ and $F = 0.56$.

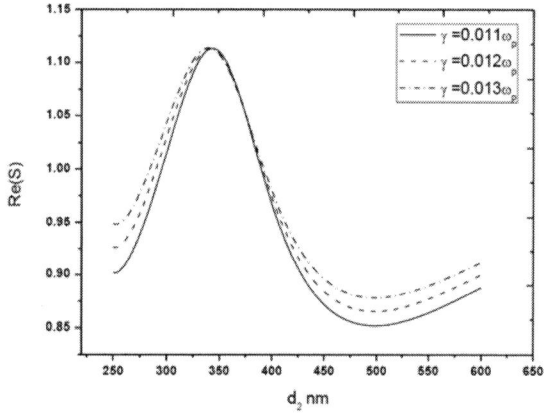

Fig. 2. The real part of the effective refractive index of the proposed sensor versus the thickness of the LHM layer for different values of the electron scattering rate for $\lambda = 632.8\ nm$, $\varepsilon_1 = 2.25$, $\mu_1 = 1$, $d_1 = 250\ nm$, $F = 0.56$ and $K_B = \pi/3d$.

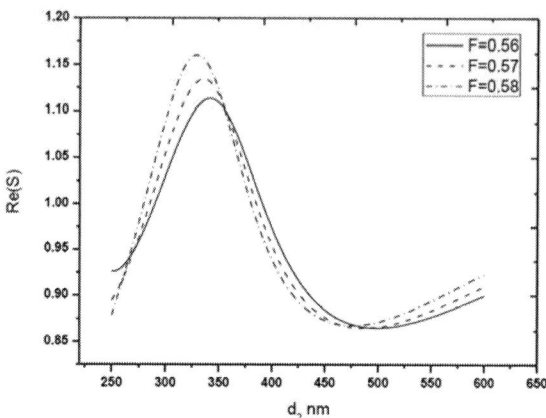

Fig. 3. The real part of the sensitivity of the proposed sensor versus the thickness of the LHM layer for different values of the electron scattering rate for $\lambda = 632.8\ nm$, $\varepsilon 1 = 2.25$, $\mu 1 = 1$, $d1 = 250\ nm$, $F = 0.56$ and $KB = \pi/3d$.

All the above figures show some kind of sensitivity with higher values than the sensitivity of the conventional sensors investigated before [5, 6].

REFERENCES

[1] V. Veselago, "The electrodynamics of substance with simultaneously negative index values of ε and μ," Sov. Phys. Usp., Vol. 10, pp. 509-514, 1968.

[2] A. Grbic and G. V. Eleftheriades, "Growing evanescent waves in negative-refractive-index transmission-line media, " J. Appl. Phys.,Vol. 92, pp. 5930-5933, 2002.

[3] I. V. Shadrivov, A. A. Zharov, Yu. S. Kivshar, "Giant Goos-Hanchen effect at the reflection from left-handed metamaterials," Appl. Phys. Lett., Vol. 83, pp. 2713-2715, 2003.

[4] Igor A. Sukhoivanov and Igor V. Guryev, "Photonic Crystals, Physics and Practical Modeling, " 1st edition, Springer-Verlag Berlin Heidelberg, 2009.

[5] S. Taya and M. Shabat, "Sensitivity enhancement in optical waveguide sensors using metamaterials," Appl. A: Phys. Vol. 103, pp. 611-614, 2011.

[6] S. Taya, M. Shabat, H. Khalil, and D. Jäger, "Theoretical Analysis of TM nonlinear asymmetrical waveguide optical sensors," Sensors and Actuators, A: Phys. Vol. 147, pp. 137-141, 2008.

[7] K. Park, B. Lee, C. Fu, and Z. Zhang, "Study of the surface and bulk polaritons with a negative index metamaterial," J. Opt. Soc. Am. B, Vol. 22(5), pp. 1016-1023, 2005.

[8] P. Yeh, A. Yariv, and A. Y. Cho, "Optical waves in layered media," Wiley, New York, 1998.

Fluid pumping cell of photonic - plasmonic microcavity sensor for biomedical application

Vladimir A. Saetchnikov[1], *Member, OSA*, Elina A. Tcherniavskaia[1], Anton V. Saetchnikov[1],
Gustav Schweiger[2], *Member, IEEE*, Andreas Ostendorf[2], *Member, IEEE*
[1]Belarusian State University, Minsk, Belarus
[2]Ruhr-Universitaet, Bochum, Germany

Abstract: Fluid pumping cell for plasmonic – photonic microcavity sensor for label-free biomolecule detection and identification has been developed and tested with drug and gold nanoparticle solutions including additional gold layer. Resonant spectra parameters have being analyzed.

Label-free biomolecule detection in sensing systems based on evanescent wave optical sensors is recently under very intensive development [1, 2]. New opportunity to increase a sensetivity of label-free detection down to single viruses based on nanoparticle (NP) plasmon resonance have been found [3–5]. On the other hand new recently developed tools for data processing can realize real-time identification of biological agents [6-9]. So combining advantages of plasmon enhancing optical microcavity resonance with identification tools can give a new platform for ulta sensitive label-free biomedical sensor.

We are developing and testing several schemes of sensor cell for reliable detection. The optimized technique is the following. Standard biocompartible polymer microspheres are used as sensitive elements. They are fixed in the solution flow by thin adhesive layer on the surface being in the field of evanescence wave. Compact spin-coater system with digital dosage was used to put and dry previously a thin film of adhesive on the surface of substrate or directly on the coupling element (Fig.1). After that, microspheres were superimposed on the surface of adhesive layer and final drying procedure by free solvent evaporation during 12 hours followed.

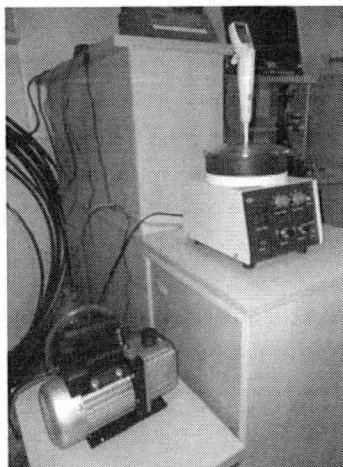

Fig. 1. Compact spin-coating system

Under optimized experimentally parameters of the process microspheres were reliably fixed as it was tested (Fig.2). The most part of their surface appeared in contact with tested solution and so can react to solution. To combine optical with plasmon resonance gold nanoparticles injected directly into solution or thin film gold layers deposited on the substrate before adhesive have been used.

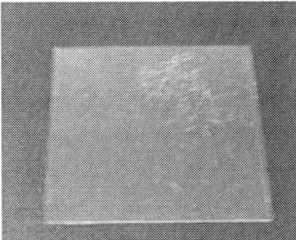

a

b

Fig.2.Aadhesive layer with fixed microspheres (a) and the same - 1 with optical coupling element - 2 (b).

The spheres used in these experiments were 50 ÷ 120 micron in diameter. The light from a tuneable diode laser (New Focus, 680 nm) is coupled into the microsphere through a prism (Fig.3). Laser beam was sharply focussed on the single microsphere to increase the contrast and intensity of the resonance scattering signal and decrease a power. Due to these only few microwatt of CW laser power was enough to register resonance spectra. Rectangular or equilateral prisms with refractive index 1.51 and 1.72 have been used as coupling element. Micromechanical system for adjustment of laser excitation to meet requirements of both optical whispering

978-1-4799-0019-0/13 $31.00 © 2013 IEEE   199

gallery mode (WGM) and plasmon resonances has been developed. The microsphere is submerged into a fluidic cell and brought into contact with the prism. The cell contains initially de-ionized (DI) water or physiological solution. To vary the refractive index, a solution of ethanol and water is incrementally added to the fluidic cell with a digital syringe. Following each injection, the WGM modes are monitored until equilibrium is reached, and then, the subsequent injection is made. To observe the WGM, the laser repeatedly scans across a spectral range of approximately 150 pm at a frequency of about 0.1 Hz. Light scattered by the micro sphere is collected through a microscope by a CMOS camera and monitored with a data acquisition card and computer (Fig.2b). While tuning the laser wavelength images were recorded by CMOS camera as avi-file. All sequences were broken into single frames and the location of the resonance was allocated in each frame. The image was filtered for noise reduction and integrated over two coordinates for evaluation of integrated energy of a measured signal. As input data the following parameters were used: normalized by free spectral range resonance shift of WGM and a relative efficiency of WGM excitation.

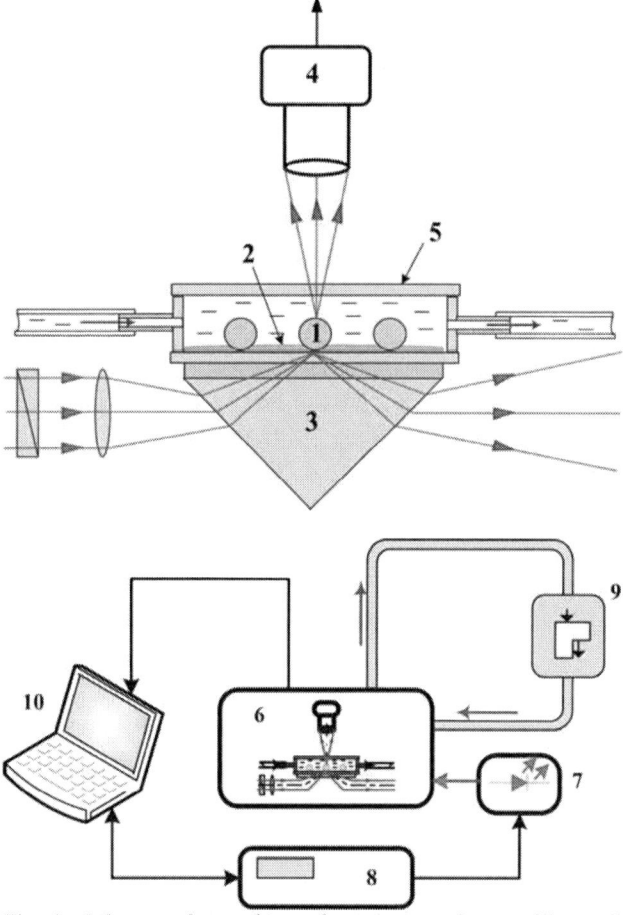

Fig. 3. Scheme of experimental geometry of a sensitive sell with a detector (a) and a complete set-up (b). 1 – microsphere, 2 – adhesive layer, 3 – coupling element (prism), 4 – CMOS camera, 5 – fluid pumping chamber, 6 – sensitive sell/detector

unit, 7- tuneable laser, 8 – laser power supply, 9 – digital dosage system, 10 – computer.

Laser beam was sharply focussed on the single microsphere to increase the contrast and intensity of the resonance scattering signal. When the wavelength of the tuneable laser corresponds to a resonance of the sphere, the power of the light scattered by the micro sphere increases, and a spectral maximum indicating the WGM spectral position is recorded. The width of such a resonance after filtering is used to estimate the resonance quality. The sensitivity of the scheme was tested to determine refractive index variation by monitoring the magnitude of the whispering gallery modes (WGM) spectral shift as in [16–19]. General overview of the experimental set up is represented on the Fig. 4.

Fig. 4. General overview of the experimental set up.

The data were obtained in the form of the video file form CMOS camera in a format *.avi. All sequence was broken into frames where the area of a resonance was allocated. The image was filtered for noise reduction and integrated on two coordinates for evaluation of integrated energy of a measured signal. As the entrance data following signal parameters were used: relative (to a free spectral range) spectral shift of frequency of WGM optical resonance in microsphere and relative efficiency of WGM excitation obtained within a free spectral range which depended on both type concentration of investigated agents. Last parameter was defined as normalized to an integrated resonant spectrum within a free spectral range. Then we broke the data set into two subsets – training and tested (randomly). The data before submitting on a network input ware preprocessed (normalized and standardized).

The network topology was designed: a number of the hidden layers of multilayered perceptron, a number of neurons in each of layers, a method of training of a neural network, activation functions of layers, type and size of a deviation of the received values from required values. For a network training the method of the back propagation error in various modifications has been used. Input vectors correspond to 6 classes of biological substances under investigation. The result of classification was considered as positive when each of the region, representing a certain substance in a space: relative spectral shift of an optical resonance maxima - relative efficiency of excitation of WGM, was singly connected. General window of an interface for the developed data processing software is represented on the Fig. 5.

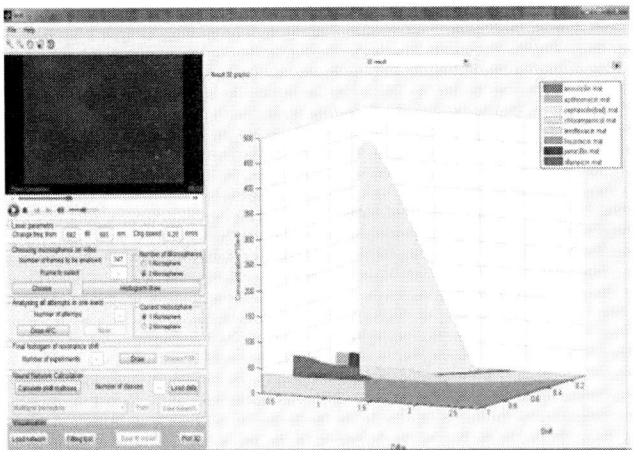

Fig. 5. General window of an interface for the data processing software.

The cell contains initially de-ionized water or physiological solution. Solutions of antibiotics of several generation or gold NP gels have been incrementally added to the fluidic cell with a digital syringe. Light scattered by the micro sphere is collected through a microscope by a CMOS camera and monitored with a data acquisition card and computer like in [6-9]. Both biological agents and NP injection was obtained caused WGM shift. But their influence looked competitive. Because the presence of drug decreased WGM shift due to NP and vice versa. WGM resonance in mictospheres fixed on substrate with gold layer was also observed under optimized layer thickness with higher intensity then without gold layer. More detailed investigation in this field are under development.

REFERENCES

[1] F. Vollmer and S. Arnold, "Whispering-gallery-mode biosensing: label-free detection down to single molecules," Nat. Meth., vol. **5**, pp. 591–596, 2008.

[2] X. D. Fan, I. M. White, S. I. Shopoua, H. Y. Zhu, J. D. Suter, and Y. Z. Sun, "Sensitive optical biosensors for unlabeled targets: A review," Anal. Chim. Acta, vol. 620, pp. 8–26, 2008.

[3] M. A. Santiago-Cordoba, M. Cetinkaya, S. V. Boriskina, F. Vollmer, M. C. Demirel, "Ultrasensitive detection of a protein by optical trapping in a photonic-plasmonic microcavity," J. Biophotonics, vol. 5, pp. 629–638, 2012.

[4] V. R. Dantham, S. Holler, V. Kolchenko, Z. Wan and S. Arnold, "Taking whispering gallery-mode single virus detection and sizing to the limit," Appl. Phys. Lett., vol. 101, pp. 043704-1–043704-4, 2012.

[5] M. Baaske and F. Vollmer, "Optical Resonator Biosensors: Molecular Diagnostic and Nanoparticle Detection on an Integrated Platform," Chem. Phys. Chem., vol. 13, pp. 427 – 436, 2012.

[6] E.A. Tcherniavskaia, V.A. Saetchnikov,"Application of neural networks for classification of biological compounds from the characteristics of whispering-gallery-mode optical resonance," Journal of Applied Spectroscopy, vol. 78, no 3, pp. 457-460, 2011.

[7] V.A. Saetchnikov, E.A. Tcherniavskaia, G. Schweiger, A. Ostendorf and A.V. Saetchnikov, "Neural Network analysis of the resonance whispering gallery mode characteristics of biological agents," Nonlinear Phenomena in Complex Systems, vol. 14, no 3, pp. 253–263, 2011.

[8] V.A. Saetchnikov, E.A. Tcherniavskaia, G. Schweiger and A. Ostendorf," Classification of the micro and nanoparticles and biological agents by neural network analysis of the parameters of optical resonance of whispering gallery mode in dielectric microspheres,", Proceeding of the SPIE, vol. 8090, - pp. 80900R1- 80900R11, 2011.

[9] V. A. Saetchnikov, E. A. Tcherniavskaia, G. Schweiger and A. Ostendorf, „Classification of antibiotics by neural network analysis of optical resonance data of whispering gallery modes in dielectric microsphere", Nanophotonics IV, Proceeding of the SPIE, vol. 8424, pp. 345-356, 2012.

Interaction wave at diffraction on slit

R.A. Lymarenko

International Center "Institute of Applied Optics" NASU, Kyiv, Ukraine

Abstract: New representation of field components of electromagnetic wave diffraction on a slit is proposed. It should be denoted that our solution of slit diffraction problem consist of two parts: diffraction field based of Fresnel complex integral derived from rigorous Sommerfield's solution and interaction wave described the peculiarity of diffraction process on subwavelength apertures. It is described the properties and peculiarities of the introduced component of diffraction field that corresponds to boundary condition in a Dirichlet-type screen.

Based on earlier interpretation of Sommerfeld's solution [1,2] of the problem of plane wave diffraction by a half-plane screen, consider case of normal incidence of linear polarized plane wave on ideal conducting slit. For half-plane diffraction without any additional approximations the diffracted field was represented as a superposition of four wave components. Two of them are ordinary plane waves with amplitudes half that of the incident wave, and other two are edge dislocation waves (EDW) [3].

The approximation of slit diffraction is more correct then parabolic and can be written in form [4]:

$$E(x,z) = \frac{1-i}{\sqrt{2\pi}}\left[\begin{array}{l} e^{-ikr}\left(\frac{1}{2} + \int_0^{U(x,z)} e^{i\mu^2} d\mu\right)... \\ ... \pm e^{ikr}\left(\frac{1}{2} + \int_0^{U(x,-z)} e^{i\mu^2} d\mu\right) \end{array}\right] \quad (1)$$

where $U(x,z) = \pm\sqrt{k}\sqrt{\sqrt{x^2 + z^2} - z}$.

In this case, field E does not satisfy the Dirichlet-type boundary conditions on the screen surface. Oscillations of EDW lead to the existence of non-zero field on the boundary of screen with amplitude decreased moving away from the slit edges.

The expression for the diffraction field (1) satisfies the boundary conditions with some error ΔE, which rapidly tends to zero when a slit width becomes much more wavelength (Fig. 2). The "interaction" slit edges converging leads to additional field appearing. This "interaction" field can be analyzed using the idea presented in the paper [5]. Thus, we consider the elementary solution of the Helmholtz wave equation:

$$\varepsilon(r,\varphi) = \frac{\exp(ikr)}{\sqrt{kr}}\sin\left(\frac{\varphi}{2}\right) \quad (2)$$

Similarly to [6] and taking into account that $\varepsilon(x-a,0) \equiv 0$ for $x < a$, we compose elementary waves (2) into wave D that satisfies the boundary condition:

$$D(x) = \int_{-\infty}^{-a} p(\xi)\varepsilon(r,\varphi-\pi)d\xi + \int_a^{\infty} p(-\xi)\varepsilon(r,\varphi)d\xi \quad (3)$$

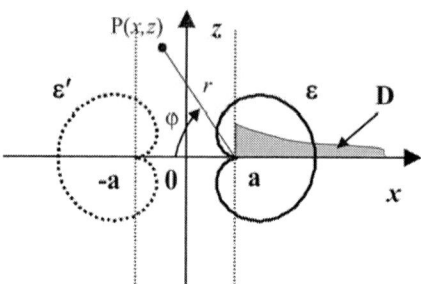

Fig. 1. Scheme of interaction wave definition.

Fig. 1 schematically shows the position of additional elementary waves ε in screen plane and composed solution part $D(x)$. The wave ε does not violate the boundary conditions on the opposite slit edge, as we can see from (2). So the waves ε and ε' may be consider separately.

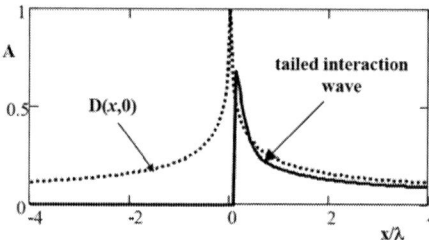

Fig. 2. Uncompensated part of diffraction field in expression (1) and interaction wave $D(x,0)$ in screen plane.

In the extreme case where the width of the slit width tends to zero, these waves create "evanescent" part of EDW [3] (Fig. 2).

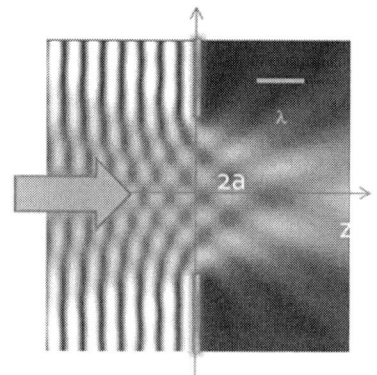

Fig. 3. Amplitude distribution of plane wave diffraction on slit.

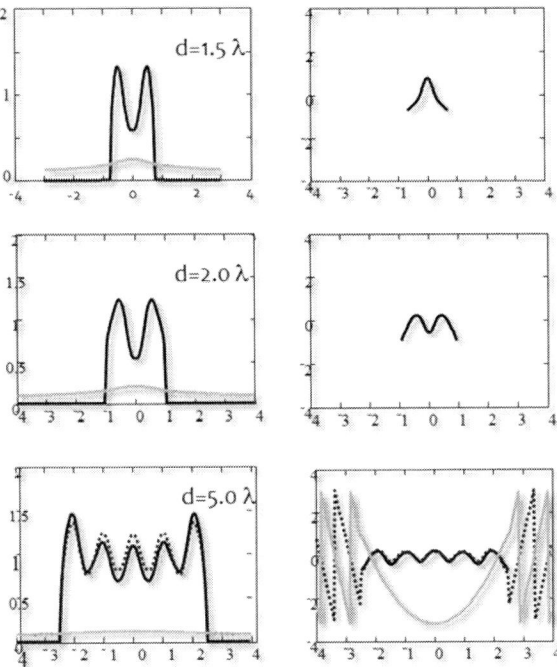

Fig. 4. Amplitude and phase distribution in screen plane for various slit width d.

In proposed model, uncompensated part is written in form:

$$D(x,0) = 2\frac{1-i}{\sqrt{2\pi}}\int_0^{\sqrt{k(x\pm a)}} e^{i\mu^2}d\mu - 1 \quad \text{for } |x| > a \quad (4)$$

We can find the representation of $D(x,0)$ as series of (2):

$$D(x,0) = \int_a^x p(\xi)\frac{\exp(ik(x-\xi))}{\sqrt{k(x-\xi)}} \quad (5)$$

Expression (4) can be approximate as:

$$D(x,0) \approx \frac{2}{\pi}\frac{1}{\sqrt{k(x\pm a)}}\exp\left(-ik(x\pm a) + \pi + \frac{\pi}{4}\right) \quad (6)$$

where after substitution $k(x-\xi) = \mu^2$ we obtain the distribution of elementary waves on slit plane:

$$p(kx - \mu^2) - \frac{1}{2}\exp\left[-\sqrt{(x-a)\frac{k}{2\pi}}\ \mu^2\right]\exp[-i(k(x-a)+\pi)] \quad (7)$$

Approximate analytical expression (7) illustrates the typical behavior of the interaction wave.

We obtain the integral equation for the interaction wave in conjunction with the main part (1) satisfies the boundary conditions on the surface of the screen. The interaction wave is expressed through elementary solutions of two-dimensional wave equation (2). The obtained result of field distribution is coincide with the FDTD method.

Consideration of additional wave components near the screen allows calculating field in the aperture more accurately. It is important for considering sub-wavelength holes [7] or going to the microwave range.

This approach can be generalized to the case of arbitrary incidence of the plane wave and to the case of H-polarization. Similarly, the additional field for plane wave diffraction on a strip can be obtained.

We also propose experimental scheme for pure interaction wave observation (Fig. 5).

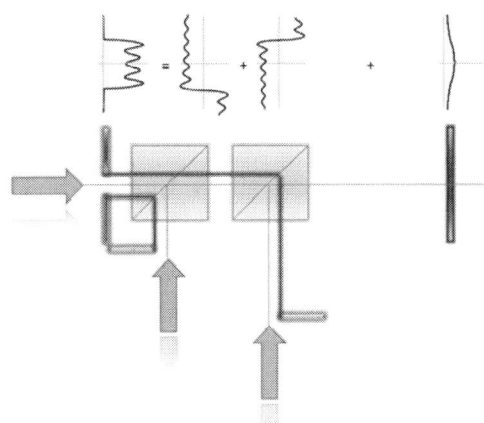

Fig. 5. Proposed experimental scheme for interaction wave observation.

REFERENCES

[1] Sommerfeld A. Mathematische Theorie der Diffraction. Math. Ann., vol. 47. pp. 317–374, 1896.

[2] M. Born and E. Wolf, Principles of Optics, 6th ed. Pergamon, Oxford, 1991.

[3] A.I. Khizhnyak et all. "The structure of edge-dislocation wave originated in plane-wave diffraction by a half-plane", *J. Opt. Soc. Am. A*, vol. 17, no. 12, pp. 2199-2207, 2000.

[4] Anokhov S.P., Lymarenko R.A., Khizhnyak A.I. "Arbitrary Tilted Plane Wave Diffraction by Slit" *Ukr. J. Phys.*, vol. 46, no 1, pp. 62-64, 2001.

[5] Gori F. "Diffraction from a half-plane. A new derivation of the Sommerfeld solution", *Opt. Comm.*, vol. 48, no 2. pp. 67-70, 1983.

[6] Gori F. "Diffraction by a half-plane. An elementary derivation of the rigorous solution", *Estrado da: Atti della fondazione Giorgio Ronchi Anno XXXVIII,* no. 5-6, pp. 593-605, 1983.

[7] Yann Gravel and Yunlong Sheng, "Rigorous solution for optical diffraction of a sub-wavelength real-metal slit", *Optics express,* vol. 20, no. 3, 2149, 2012.

Interaction of optical pulses in nonlinear plasmonic systems

D. O. Ignatyeva, A. P. Sukhorukov

Lomonosov Moscow State University, Moscow, Russia

Abstract— We investigate the collision of the two surface plasmon polariton pulses interacting at the interface between noble metal and dielectric with Kerr nonlinearity. The dynamics of the signal and pump SPP pulse was analyzed using both spectral and trajectory approaches. We derived the conditions under which the reflection of the weak signal SPP pulse from the intense pump SPP can be observed. Using such interaction one can control the propagation dynamics of the signal pulse via the modulation of the intensity of the pump pulse.

I. INTRODUCTION

This work is devoted to the analysis of the nonlinear interaction of two surface plasmon polariton (SPP) pulses at the interface between a noble metal and a dielectric with cubic nonlinearity. The intensity of the pump SPP pulse determines the propagation dynamics of the signal SPP pulse. Similar method of light-by-light control based on the phenomenon analogous to the total internal reflection of light implemented in the time domain was proposed earlier for the bulk optical pulses [1].

The idea of the method is the following. High-power pump pulse induces the change of the refractive index in the dielectric (which can occur due to the different nonlinear mechanisms). Weak signal pulse of a different frequency propagating with a different velocity due to the dispersion reaches the induced inhomogeneity. If the refractive index variation is enough the signal pulse can be ¡¡reflected¿¿ from the inhomogeneity that means it can slow down and continue travelling behind the pump pulse (or vice versa, depending on the sign of the group velocity dispersion). Therefore the relative position of the two pulses and the delay of the signal pulse can be controlled by the pump intensity.

The advantages of such light-by-light control method is that it allows to manipulate the signal pulse in a rather easy way. This can be done in the media with quadratic, cubic, photorefractive or thermal nonlinearity (see, for example, [2], [3]). No phase matching conditions are required to be satisfied. At the same time for the experimental realization of optical switching very fast nonlinear response together with a rather low critical value of the pump intensity is needed. However nonlinear mechanisms are either slow (for example, thermal or photorefractive nonlinearity), or require rather high pump intensities (for example, cubic or quadratic nonlinearities). In present work we discuss peculiarities of the implementation of such method in plasmonic systems.

II. PLASMONIC INTERFACE AS A SYSTEM FOR NONLINEAR INTERACTION

Let us discuss the features of plasmonic interface used for a nonlinear interaction of optical pulses. The main reason for using surface plasmon polariton pulses instead of the bulk waves is the following. Due to the high energy concentration in SPP wave near the metal-dielectric interface plasmonic systems can provide more efficient methods of light control in contrast to ordinary crystals or optical fibers [4]. Therefore the efficiency of the various light-matter interactions increases that leads to the increase of different linear (e.g. [5]–[7]) and nonlinear effects (e.g. [8]).

The bulk radiation can be transformed into the surface wave propagating along the metal-dielectric interface in several ways [9], [10]. First of all, one can use a coupling prism in a scheme of attenuated total internal reflection (ATR) with Otto or Kretschmann geometry. Second way supposed to be more convenient for SPP pulse management is to use the diffraction grating. The efficiency of the transformation from bulk to surface wave is considered to be 10% that is of the order of a the typical value. However if only 10% of energy is transformed into the surface wave its intensity is several orders higher than the intensity of the bulk wave due to the high localization near the interface that is about $1\mu m$ in optical frequency range.

Plasmonic systems are also characterized by high frequency dispersion. For a smooth metal-dielectric interface the SPP propagation constant β is:

$$\beta(\omega) = \frac{\omega}{c} \sqrt{\frac{\varepsilon_m(\omega)\varepsilon_d(\omega)}{\varepsilon_m(\omega) + \varepsilon_d(\omega)}}, \qquad (1)$$

where ω denotes frequency, c is the light speed in vacuum, $\varepsilon_{d,m}$ is the dielectric permittivity of metal or dielectric (index m or d, correspondingly). The experimental data on frequency dispersion of metal permittivity [11] can be approximated with a Drude-Lorentz model:

$$\varepsilon_m = \varepsilon_\infty + \frac{\omega_p^2}{\omega^2 - i\gamma\omega} \qquad (2)$$

The dispersion of the dielectric is usually much smaller than the dispersion of metal and can be neglected. For example, for silica glass which is a material with cubic nonlinearity one can use Sellmeier equation to describe its permittivity from 0,21 to 6,7 μm with a rather good accuracy [12]. In the region from 0,5 to 2 μm its permittivity changes very slightly in comparison with metallic dispersion. As a result plasmonic systems have much stronger dispersion than the bulk crystals.

Very strong dissipation should be referred to the negative but intrinsic features of a plasmonic system. The imaginary part of the propagation constant obtained from Eq. 1 can be used to estimate the propagation length of SPP:

$$l_{prop} = \frac{1}{2 \operatorname{Im} \beta}, \qquad (3)$$

which gives a little bit higher values than the experimental data [13], [14]. Eq. 3 shows that the propagation lengths increases with the increase of the wavelength. Typical values of the propagation lengths is assumed to be about $100 \ \mu m$ for a wavelength of $1 \ \mu m$. Therefore input and output gratings should be placed at this distance in order to allow the detection of signal after the interaction.

III. THEORETICAL APPROACH TO SPP PULSE INTERACTION

We consider SPP pulse propagation along the interface between metal (e.g. gold) and dielectric with cubic nonlinearity (e.g. SiO_2) taking into account dispersion, nonlinearity and energy dissipation. For a weak signal SPP pulse the nonlinearity can be neglected while the impact of the pump SPP on the dielectric permittivity should be taken into account (and vice versa for the pump SPP). Therefore the dynamics of SPP pulse can be described with the following equation:

$$\Delta \vec{E} - \frac{1}{c^2} \frac{\partial^2 \vec{E}}{\partial t^2} = \frac{4\pi}{c^2} \frac{\partial^2 \vec{P}}{\partial t^2}, \qquad (4)$$

where $\vec{P} = \vec{P}_L + \vec{P}_N$ — is the polarization, $\vec{P}_L$ is its linear part while $\vec{P}_N = \vec{P}_{nl} + \vec{P}_{ext}$ corresponds to the component arising due to the nonlinearity and external impact (of pump SPP). The dispersion of metal ($z < 0$) is considered in a usual way while the dispersion of the dielectric is neglected for the simplicity.

For the description of the SPP pulse dynamics with the Eq. (4) one should find its solutions for the two media matched by the boundary conditions. For the simplification of the analysis several simplifications can be made.

First of all, we consider the addition to the linear polarization term $\vec{P}_N$ to be small in comparison with the linear one $\vec{P}_{ext}, \vec{P}_{nl} \ll \vec{P}_L$. Secondly, we consider the nonlinear response of the Kerr dielectric to be nearly instantaneous (characteristic relaxation time of the Kerr dielectrics are about several femtoseconds) and the external permittivity change to be slow on the scale of one period. Next, we focus our attention on femtosecond pulses of about $\sim 50 fs$ duration so that their spectrum with the central frequency ω_0 has the width $\Delta \omega \ll \omega_0$. Finally, we neglect changes of SPP transversal structure and its polarization due to the nonlinear impact during the propagation.

The electric field of the SPP therefore can be found in the following form:

$$\vec{E}(\vec{r}, t) = \frac{1}{2} \left[\vec{S}(\vec{r}, t) \exp(i\omega_0 t) + \text{c.c.} \right], \qquad (5)$$

where $\vec{S}(\vec{r}, t)$ is the slowly varying on the scale of one SPP pulse period function. Polarization vector $\vec{P}$ can be found in a similar way. We use spectral representation for the calculation of the SPP pulse dynamics: $\tilde{\vec{S}}(\vec{r}, \omega - \omega_0) = \int_{-\infty}^{+\infty} \vec{S}(\vec{r}, t) \exp[i(\omega - \omega_0)t] dt$ and look for the soultion in the following form:

$$\vec{S}(x, z, \omega - \omega_0) = \vec{f}(z, \omega - \omega_0) A(x, \omega - \omega_0) \exp(i\beta_0 x), \quad (6)$$

where $\vec{f}(z, \omega - \omega_0)$ — describes the polarization and structure of SPP, $A(x, \omega - \omega_0)$ — is the slowly varying function of x, and β_0 — is the propagation constant at ω_0 frequency corresponding to the linear case. The profile and polarization of SPP for metal-dielectric interface whitout any perturbation or nonlinearity is well-known:

$$\vec{f}(z) = F_0 \exp(-\gamma_j^{(0)}|z|) \left\{ 1, 0, i\beta_0/\gamma_j \right\}. \qquad (7)$$

So the first order of the perturbation theory ($\beta = \beta^{(0)}(\omega) + \beta^{(1)}(\omega)$) gives us the change of the propagation constant:

$$\beta^{(1)} = \frac{k_0^2}{\beta_0} \int_{-\infty}^{+\infty} \Delta \varepsilon_j(z) |f(z)|^2 dz / \int_{-\infty}^{+\infty} |f(z)|^2 dz. \qquad (8)$$

Since the isolated metal-dielectric interface supports only one mode type on the given frequency SPP profile $\vec{f}$ has no corrections in the first order of perturbation theory.

Making the expansion of propagation constant:

$$\beta = \beta_0 + \frac{\partial \beta}{\partial \omega}(\omega - \omega_0) + \frac{1}{2} \frac{\partial^2 \beta}{\partial \omega^2}(\omega - \omega_0)^2 + \beta^{(1)}(\omega_0), \quad (9)$$

we obtain the equation describing SPP dynamics that in the time-domain representation of $\tilde{A}$ has the form:

$$\frac{\partial A}{\partial x} + \nu \frac{\partial A}{\partial \tau} + i D_{disp} \frac{\partial^2 A}{\partial \tau^2} + \beta^{(1)} A = 0, \qquad (10)$$

where time coordinate $\tau = t - z/v_{gr\ p}$ is associated with the group velocity of the pump pulse $v_{gr} = \partial \omega / \partial \beta|_{\omega = \omega_0}$, dispersion coefficient $D_{disp} = 0.5 \partial^2 \beta / \partial \omega^2|_{\omega = \omega_0}$ and $\nu = 1/v_{gr2} - 1/v_{gr1}$ is the group velocity detuning of the signal SPP pulse (for the pump SPP pulse this term is omitted).

Further analysis reveals that for the pump SPP pulse the phenomenon of self-action due to the cubic nonlinearity and dispersion spreading can be neglected since the nonlinear and dispersion lengths are much larger than the propagation distance of about $100 \ \mu m$. Therefore we can consider the induced by the pump SPP pulse inhomogeneity to have unchanged profile. Losses that are present due to the metallic absorption lead to the intensity decay of 40% at the middle of the interface between gratings.

IV. SPP PULSE REFLECTION FROM THE INDUCED INHOMOGENEITY

In order to reveal the phenomenon of signal SPP pulse reflection from the inhomogeneity induced by the pump SPP pulse we apply the eikonal method and trajectory approach to the description of signal SPP propagation [1].

978-1-4799-0019-0/13 $31.00 © 2013 IEEE

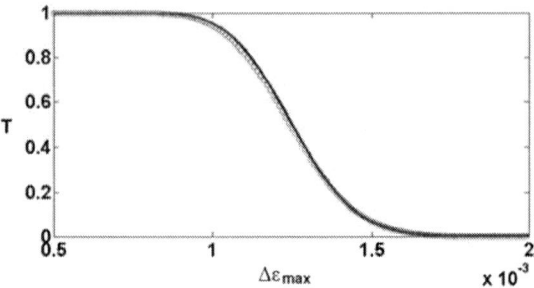

Fig. 1. Transmission of the signal SPP through the pump for different values of the pump-induced inhomogeneity (black line – theory, red line – numerical calculations).

The reflection of the signal SPP occurs if the induced inhomogeneity causes propagation constant variation:

$$\beta^{(1)} > \frac{\nu^2}{4D_{disp}}. \qquad (11)$$

For the interaction at the scale of $100 \ \mu m$ pulse duration of about $30 fs$ and group velocity detuning of about $\nu \sim 10^{-9} s/m$ is required. This can be achieved if the central frequency of signal SPP pulse is shifted from the pump at a distance of about $0.2 \ \mu m$ (for the range of 1-2 μm SPP frequency). The typical values of dispersion coefficient D_disp in plasmonic structure have the order of $10^{-19} s^2/m^2$ so to reach the required values of the permittivity change the intensity of the pump SPP should have the order of $10^11 W/cm^2$ (this evaluation was made for the nonlinearity of the silica glass). Note that the fluency of this pulse is three orders lower than the fluency causing thermal damage of the metallic structure [15].

Note that the SPP pulse of $30 fs$ duration has rather large spectral width so the whole signal SPP pulse is reflected from the inhomogeneity only if the condition (11) is satisfied for all spectral components. Therefore the switch between the propagation of the signal SPP through the pump and its reflection from it is rather smooth. In Fig. 1 the results of the calculations of the transmittance of the signal SPP through the pump are presented. If the value of the induced inhomogeneity is close to the critical value determined by the strict equality in (11) the pulse is divided into two sub-pulses one of which is transmitted through the inhomogeneity while the other is reflected from it.

Notice that the reflection from the inhomogeneity physically means the change of the signal SPP group velocity, so that the reflected signal SPP experiences a frequency shift. As far as signal SPP is reflected the difference between the group velocities of the signal and pump SPP changes its sign so that the group velocity detuning after the collision $\tilde{\nu} = -\nu$ that determines the spectral shift of a signal SPP. This shift occurs due to the movement of the inhomogeneity with the velocity close to the light speed and is analogous to the Doppler spectral shift.

V. CONCLUSION

We analyzed the collision of two SPP femtosecond pulses propagating along the metal-dielectric interface. Using both spectral and trajectory approach we revealed the conditions under which the weak signal SPP pulse is reflected from the moving inhomogeneity of the permittivity induced in the dielectric with cubic nonlinearity by pump SPP pulse. The switch between the regimes of the propagation through the inhomogeneity and the reflection from it is rather smooth and is determined by the spectrum of the signal SPP pulse. The spectral shift of the reflected signal SPP pulse is revealed.

ACKNOWLEDGMENT

The authors would like to thank Russian Foundation for Basic Research (grants No. 12-02-31396, 12-02-33100, 12-02-31298, 13-02-91334), the FTP Scientific and scientific-pedagogical personnel of the innovative Russia of The Ministry of Education and Science of the Russian Federation, and Dynasty Foundation for financial support.

REFERENCES

[1] V. E. Lobanov and A. P. Sukhorukov, "Total reflection, frequency, and velocity tuning in optical pulse collision in nonlinear dispersive media," *Phys. Rev. A*, vol. 82, p. 033809, 2010.

[2] V. E. Lobanov and A. P. Sukhorukov, "Repulsion and total reflection with mismatched three-wave interaction of noncollinear optical beams in quadratic media," *Physical Review A*, vol. 84, p. 023821(7), 2011.

[3] V. E. Lobanov, A. A. Kalinovich, A. P. Sukhorukov, F. Bennet, and D. Neshev, "Nonlinear reflection of optical beams in the media with a thermal nonlinearity," *Laser Physics*, vol. 19, no. 5, pp. 1112–1116, 2009.

[4] S. A. Maier and H. A. Atwater, "Plasmonics: Localization and guiding of electromagnetic energy in metal/dielectric structures," *Journal of Applied Physics*, vol. 98, no. 1, p. 011101, 2005.

[5] V. Belotelov, D. Bykov, L. Doskolovich, A. Kalish, and A. Zvezdin, "Giant transversal kerr effect in magneto-plasmonic heterostructures: The scattering-matrix method," *Journal of Experimental and Theoretical Physics*, vol. 110, pp. 816–824, 2010.

[6] D. O. Ignatyeva, A. N. Kalish, G. Y. Levkina, and A. P. Sukhorukov, "Surface plasmon polaritons at gyrotropic interfaces," *Physical Review A*, vol. 85, p. 043804, 2012.

[7] A. P. Sukhorukov, D. O. Ignatyeva, and A. N. Kalish, "Terahertz and infrared surface wave beams and pulses on gyrotropic, nonlinear and metamaterial interfaces," *Journal of Infrared, Millimeter, and Terahertz Waves*, vol. 32, no. 10, pp. 1223–1235, 2011.

[8] D. O. Ignatyeva and A. P. Sukhorukov, "Plasmon beams interaction at interface between metal and dielectric with saturable kerr nonlinearity," *Applied Physics A*, vol. 109, pp. 813–818, 2012.

[9] S. A. Maier, *Plasmonics: Fundamentals and Applications*. Springer Science+Business Media LLC, 2007.

[10] H. Raether, *Surface Plasmons*. Springer, Berlin, 1988.

[11] E. D. Palik, *Handbook of optical constants of solids*. Academic Press, 1998.

[12] R. Kitamura, L. Pilon, and M. Jonasz, "Optical constants of silica glass from extreme ultraviolet to far infrared at near room temperature," *Appl. Opt.*, vol. 46, no. 33, pp. 8118–8133, 2007.

[13] P. Dawson, B. A. F. Puygranier, and J.-P. Goudonnet, "Surface plasmon polariton propagation length: a direct comparison using photon scanning tunneling microscopy and attenuated total reflection," *Phys. Rev. B*, vol. 63, p. 205410, 2001.

[14] W. L. Barnes, "Surface plasmon polariton length scales: a route to subwavelength optics," *Journal of Optics A: Pure and Applied Optics*, vol. 8, no. 4, p. S87, 2006.

[15] P. Pronko, S. Dutta, D. Du, and R. Singh, "Thermophysical effects in laser processing of materials with picosecond and femtosecond pulses," *Journal of applied physics*, vol. 78, no. 10, pp. 6233–6240, 1995.

Laser irradiation effect on ZnO nanoparticles

W.A. Farooq[1,*], Walid Tawfik[1,2], A. Fatehmulla[1], S. M. Ali[1], M. Aslam[1]

[1]Department of Physics and Astronomy College of Science P. O. Box 2455 King Saud University Riyadh 11451 Saudi Arabia
[2]Department of Environmental Applications, NILES National Institute of Laser, Cairo University
Cairo, Egypt.
*Corresponding author email:wafarooq@hotmail.com

Abstract: Effect of laser irradiation on ZnO nanoparticles synthesized by co-precipitation technique with the post-oxidation annealing in air atmosphere is presented. We have observed agglomeration, enlargement and deformation of the ZnO nanoparticles after laser irradiation at 355 nm using SEM, XRD analysis..

1.Introduction

Nanoparticles have wide range of applications in many fields of science and technology because of tremendous increase in the surface area to volume ratio [1-3]. Large surface area is an important factor in chemical reactions. In the nano scale materials, interaction on atomic scale dominates and exhibits quantum mechanical behavior as compared to bulk materials. Behavior of light interaction with materials also changes drastically as we go down to the nano size of the materials. This is due to decrease in the dimensions below the critical wavelength of light [4]. There are many methods for nanoparticle generation [5-8]. We have synthesized the ZnO nanoparticles with co-precipitation technique and studied the effect of laser irradiation on the structural morphology at wavelength of 355 nm from Nd:YAG laser system.

2. Syntheses and irradiation of ZnO nanoparticles

ZnO nanoparticles were prepared by the co-precipitation technique with the post-oxidation annealing in air atmosphere. Zinc Chloride and NaOH powders of analytical grade purity were used as chemical reagents in the preparation of ZnO nanoparticles. 0.2 M Zinc Chloride solution was prepared in 20 ml ethanol by constantly stirring at 60°C for about half an hour and then mixed with separately prepared 0.1M NaOH solution at 60°C. After 1hr of constant stirring, the product was centrifuged and washed 10 times with deionized water and two times with ethanol to remove the by-products. The final filtered product was then dried into solid powder at low temperature, grinded and finally annealed at 450 °C to get ZnO nanoparticles. The nanoparticles were irradiated with one shot of laser beam of 6 ns pulse and 25 mJ energy from Nd:YAG laser system at 355 nm wavelength.

3. Results and Discussions

SEM images of the scanning electrom microscope of the synthesized ZnO nanoparticles before and after laser irradiation are depicted in fig. 1 and fig. 2 respectively. It can be observed from these figures that after laser irradiation, size of the particles are increased and agglomerated in different shapes. The XRD pattern shown in fig.3 suggests that the ZnO nanoparticles crystallize in Wurtzite structure (Hexogonal phase) and all the peaks match well with JCPDF file No. 36-1451, a = 0.3249 nm, c = 0.5206 nm. Increase in size of the particles is also observed as calculated from Scherrer relation before and after laser irradiation. Usually, the crystallite size calculated through Scherrer formula is smaller than the actual value as seen in SEM images. This is attributed to the widening of the XRD peak due to internal stress and defects [9]. The intensities of first three characteristic peaks (100), (002) and (101) almost reduced to half with substantial decrease in the broadening (FWHM) in case of laser irradiated samples indicating the improvement in the crystallinity or crystallite size. Similar observation of improvement in crystallite size has been reported in literature when the ZnO nanoparticles were thermally annealed [10]. The expansion and deformation of the sample are attributed to the thermal effect caused by the localized surface Plasmon resonances [4].

Compressed pulse after reflection from chirped mirror under normal light incidence. Pulse amplitude is increased and duration is reduced. Compression relative to FWHM is nearly 90%. But compression is not complete and time-bandwidth product still far from that for transform-limited pulse. This is due to limited negative dispersion in the mirror as well as due to group delay oscillations and contributions of high-order dispersion.

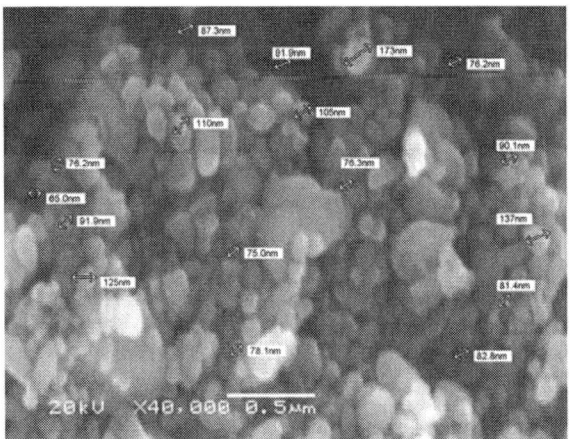

Fig. 1. SEM image of ZnO before laser shot

Particle size of the crystallites size of the crystals in the sample before and after laser irradiation is calculated from XRD patterns using following well-known Scherrer's formula [8]. The calculated values are given in table A.

$$D = \frac{0.9 X \lambda}{\beta Cos\theta}$$

Where D is Crystallite size, λ= 1.5405Å and β is the broadening of diffraction line measured at FWHM in radian and θ is the angle of diffraction.

Table. A: Calculated mean crystallite size of the sample from XRD patterns

	FWHM	Crystallite size (Å)
ZnO NP without laser	0.27096	304.69
ZnO NP with laser	0.22962	359.29

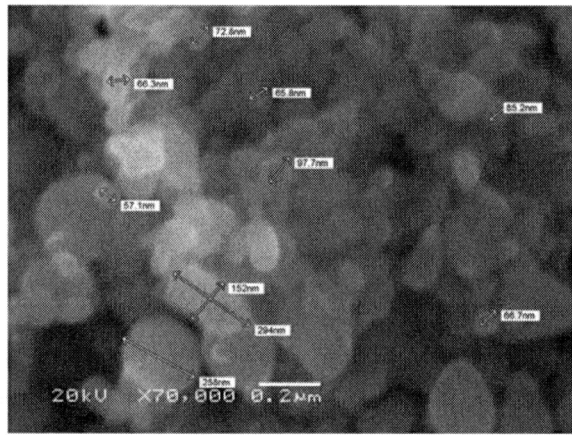

Fig. 2. SEM image of ZnO after laser shot

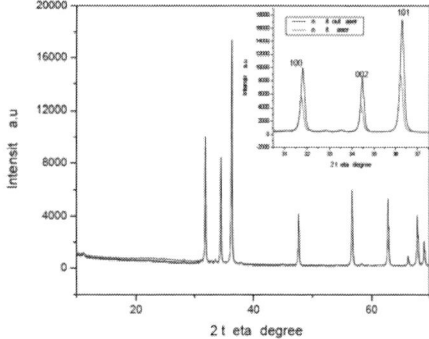

Fig. 3. XRD before and after laser shot.

4. Conclusion

Nanoparticles of ZnO powder synthesizes with co-precipitation technique are irradiated with UV laser. Expansion and deformation in the nano dimensions are observed after laser irradiation.

References

1. Y. C. Lee, W. Y. Lin, F. T. Jeng, Development of Controllable a Nanoparticle Generation Technique. Journal of Occupational Safety and Health 17: 2009, 163-174.
2. C. Darnault, K. Rockne, A. Stevens, G. A. Mansoori, and N. Sturchio, Water. Environ. Res, 177, (2005), 2576.
3. O.V. Makarova, T. Rajh, M.C. Thurnauer, A. Martin, P.A. Kemme, and D.Cropek, Environ. Sci. Technol. 34, 2000, 4797
4. Jing-Liang Li and Min Gu, Gold-Nanoparticle-Enhanced Cancer Photothermal Therapy, IEEE Journal of selected topics in quantum electronics,. 16, 2010, 4
5. K, Landfester, The Generation of Nanoparticles in Miniemulsions, Adv. Mater. 10, 2001, 13
6. N. S. Tabrizi , M. Ullmann, V. A. Vons. U. Lafont, A. Schmidt-Ott, Generation of nanoparticles by spark discharge, J Nanopart Res 11, 2009, 315–332
7. U. R. Kortshagen1, U. V. Bhandarkar1, M. T. Swihart and S. L. Girshick, Generation and growth of nanoparticles in low-pressure plasmas, Pure Appl. Chem., 71, 1999, 1871-1877
8. R. D. Glover, J. M. Miller, J. E. Hutchison, Generation of Metal Nanoparticles from Silver and Copper Objects: Nanoparticle Dynamics on Surfaces and Potential Sources of Nanoparticles in the Environmen, *ACS Nano*, *5* (11) 2011, 8950–8957.
9. Ashour A, Kaid MA, El-Sayed NZ, Ibrahim AA. Physical properties of ZnO thin films deposited by spray pyrolysis technique. Appl Surf Sci, 2006; 252:7844-8.
10. Davood Raoufi, Synthesis and microstructural properties of ZnO nanoparticles prepared by precipitation method, Renewable Energy, 50 (2013) 932-937.

Signal forming of chromophores secondary emission near noble metals plasmon films

N. D. Strekal, V. F. Askirka, A.E. German, S.A. Maskevich
Yanka Kupala Hrodna State Univeristy, Hrodna, Belarus

Abstract: Plasmon films of noble metals (gold PGF and silver PSF) evaporated onto quartz in vacuum shows possibility of selective exciting of fluorescence or SERS of molecules. We discovered plasmon-assisted fluorescence from PGF versus surface enhanced fluorescence from PSF.

Plasmon gold films (PGF) and plasmon silver films (PSF) possess several unique properties and usability despite of non-periodic and nonhomogeneous surface morphology. The main feature of mentioned substrates is possibility of smooth building of their spectral properties (optical density spectrum) to resonance conditions with adsorbed chromophores. It allows investigating different classes of chromophores within wide spectral range.

One more unique feature of plasmon films is possibility of selective exciting of fluorescence or SERS of the same molecules applying the same excitation changing only location of band of resonance excitation of localized plasmons (LP).

Mitoxantrone, anthraquinone derivative was chosen as test chromophore because of its practical interest (used in tumor therapy [1]) and it absorbs light in region of resonance exciting LP in PGF.

Technology of preparation of mentioned substrates (noble metal films evaporated in vacuum onto quartz) presented in [2].

We have determined main factor conditioning different kinds of secondary emission as dimension-spectral factor. According to this factor there is possibility to select substrates with specific dimensions of metal particles and frequency of resonance excitation of LP in maximum of absorbance band ν_{max}. Changing particle dimension allows changing ratio of scattering and absorbance cross-sections in total extinction cross-section of plasmon film. The 0–0-transition frequency ν_{00} of adsorbed chromophore is chosen as characteristic (resonant to plasmon film) parameter.

AFM images analysis shows that characteristic average dimension of particles on PGF annealed at 240 °C (PGF-240) is $d\approx200$nm and at 340 °C (PGF-340) − $d\approx70$ nm. At the same time spectral detuning $\Delta\nu$ for PGF-240 less than for PGF-340 (Fig. 1). Because of this if mitoxantrone is deposited directly to the surface of PGF-240 plasmon-assisted fluorescence is registered (Fig. 2, spectrum 1). At the same time if mitoxantrone is deposited onto PGF-340 surface high-resolved SERRS spectrum appears (Fig. 2, spectrum 2).

We have shown that surface enhanced fluorescence of mitoxantrone can be registered if molecule is deposited onto surface of PSF. For those substrates typical spectral detuning is

about $\Delta\nu$ =1000 meV versus 243 meV and 336 meV for PGF-240 and PGF-340 respectively.

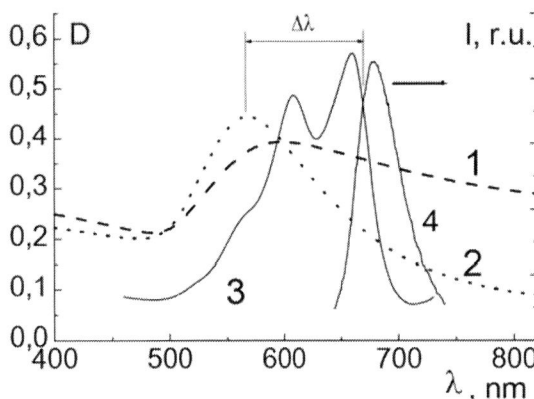

Fig. 1. Optical density spectra of PGF-240 (spectrum 1), PGF-340 (spectrum 2), water solution of mitoxantrone (spectrum 3) and fluorescence of mitoxantron (spectrum 4) in water solution

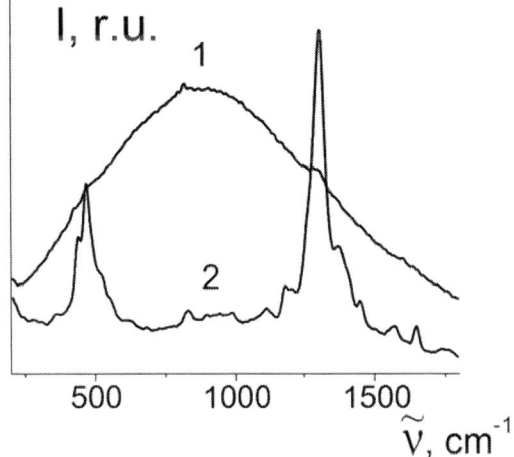

Fig. 2. Secondary emission spectra of mitoxantrone, deposited onto surface of PGF-240 (spectrum 1) and PGF-340 (spectrum 2), λ_{ex} = 632,8 nm

We propose spectral-dimension criteria for obtaining different kinds of secondary emission of adsorbed molecules. For registration of most intensive plasmon-assisted fluorescence of molecule plasmon films with minimal spectral detuning $\Delta\nu$ <240 meV (it provides resonance interactions between plasmons and fluorophore in the near filed) and maximum large metal particles on the surface should be chosen. Those conditions provide high emission in the far field. For registration of most intensive SERS or SERRS signals plasmon films with $\Delta\nu$ >240 meV and in average 30-50 nm of particles dimension should be chosen.

Despite of very similar shape of plasmon-assisted and surface enhanced fluorescence bands (Fig. 3 and Fig. 4), angular and polarization dependencies in excitation of mentioned kinds of secondary emission differ fundamentally. Plasmon-assisted fluorescence is mainly emission of molecule in opposite to surface enhanced fluorescence which is emission of plasmon-chromophore system in resonance conditions.

Polarization and angular dependencies of plasmon-assisted fluorescence from PGFs presented in Fig. 5 shows qualitative similarity to selective photo effect. It seems that our giant dipole resonance model is more approximate to model of induced resonance [3] among other models of enhancement of Raman and fluorescence signals. In the frame of mentioned model key issue is transition of electron between adsorbed molecule and metal.

We propose model of giant dipole resonance – short-lived dipole including surface plasmons and excited chromophore near surface of plasmon film. This model allows explaining of plasmon-assisted fluorescence.

We propose mechanism of transition of electron excitation energy between adsorbed chromophore and metal as explanation for surface enhanced fluorescence.

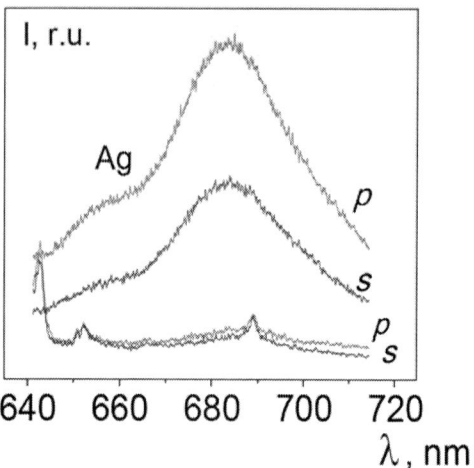

Fig. 4. Spectra of surface enhanced fluorescence of mitoxantrone on PSFs

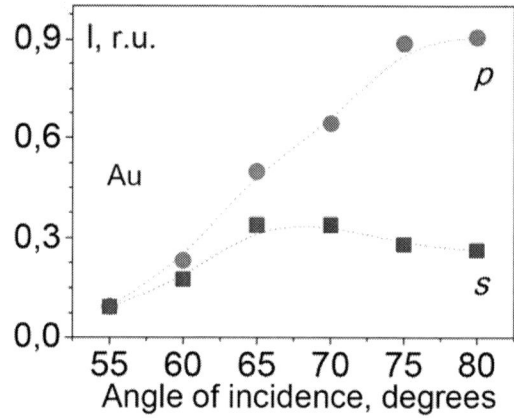

Fig. 5. Dependencies of plasmon-assisted fluorescence of mitoxantrone on excitation beam angles of incidence

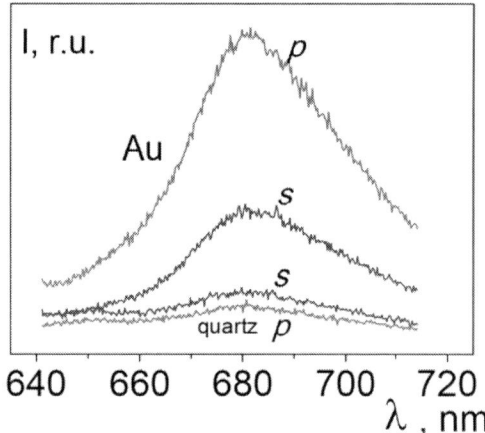

Fig. 3. Spectra of plasmon-assisted fluorescence of mitoxantron on PGFs

REFERENCES

[1] Kapuscinski J., and Z. Darzynkiewicz. "Interactions of antitumor agents ametatrone and mitoxantrone (Novatrone) with double-stranded DNA", Clin. Pharmacol. 34:4203-4213, 1985.

[2] A. Feofanov et al. "Nondisturbing and Stable SERS-Active Substrates with Increased Contribution of Long-Range Component of Raman Enhancement Created by High-Temperature Annealing of Thick Metal Films", Anal Chem, 69, 3731-3740, 1997.

[3] R.K. Chang, T.E. Furtak "Surface Enhanced Raman Scattering", Eds., Plenum Press, New York, 1982.

Effect of excitation intensity and surface morphology on the photoluminescence of ZnO films under the influence of surface plasmon resonance

S.I.Rumyantsev[1], V.M.Markushev[1], M.V.Ryzhkov[1], A.P.Tarasov[1], Ch.M Briskina[1], A.A.Lotin[2], O.A.Novodvorsky[2], V. L. Lyaskovskii[3]

[1]Kotel'nikov Institute of Radio engineering and Electronics of Russian Academy of Sciences (IRE RAS), 11 Mokhovaya St., 125009, Moscow, Russia

[2]The Institute on Laser and Information Technologies of the Russian Academy of Sciences (ILIT RAS), 1 Svyatoozerskaya St., 140700 Shatura, Moscow Region, Russia

[3]RF Research Institute for Optical and Physical Measurements, Ozernaya Str. 46, Moscow 119361, Russia

Author e-mail address: (rumyantsev@cplire.ru)

Abstract: It was observed that luminescence enhancement in Ag-treated ZnO polycrystalline (friable) films occurs under low pumping, while luminescence quenching plays the main role under high pumping intensity. For the explanation of this phenomenon an assumption about the occurrence of losses in the system was made. At the same time it was observed that on the Ag-coated epitaxial (dense) ZnO films only luminescence quenching can be seen independently of pumping level that evidently indicates that the roughness of ZnO films is required for the luminescence enhancement of ZnO films treated with Ag.

At present surface plasmon resonance (SPR) and its various manifestations are being studied intensively. Much attention is being paid to the investigation of the enhancement of excitonic photoluminescence under the influence of SPR, particularly on ZnO films (see for example [1-3]). However, the role of excitation intensity, i.e. pumping intensity, has not been studied so far. Vast majority of investigations have been carried out on polycrystalline ZnO films using He-Cd laser pumping (continuous radiation of 325 nm, power - 20-60 mW) i.e at a low excitation intensity.

Here the investigations were carried out on the sample of ZnO prepared by thermal deposition (Fig.1a) and treated with Ag (Fig.1b,c). From Fig. 1 it can be seen that Ag coating does not result in continuous layer but forms standalone almost spherical Ag nanoparticles embedded in ZnO polycrystalline film elements (Ag layer thicknesses indicated at caption to Fig.1 (10 nm, 20 nm) are conditioned values calculated on the base of suggestion that Ag used for treatment of ZnO samples forms continuous and uniform layers).

It was demonstrated that instead of excitonic luminescence enhancement with He-Cd laser pumping of ZnO films, luminescence quenching was observed in case of Nd:YAG laser pumping (third harmonic, 355 nm, pulse duration ~ 10 ns, pulse energy ~several mJ). However, the decrease of Nd:YAG laser pumping intensity to the level of micro joules

allows to achieve the enhancement of ZnO films luminescence due to Ag treatment -

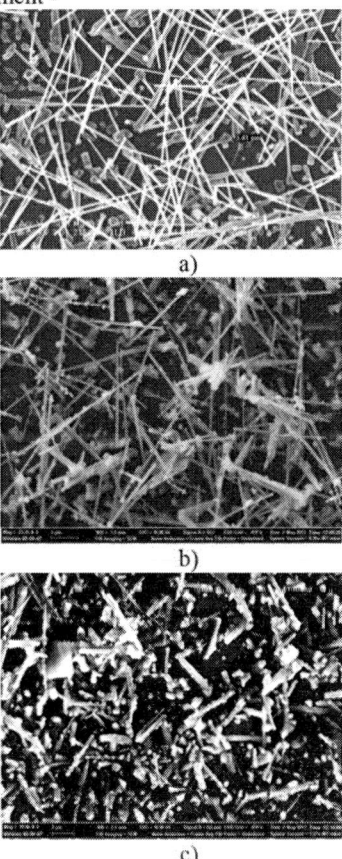

Fig.1 SEM images of ZnO film investigated: a) without Ag; b) covered with 10 nm Ag layer; c) covered with 20 nm Ag layer

- surface plasmon resonance effect. Results presented in Fig.2 convincingly demonstrate the dependence of luminescence enhancement under the influence of SPR on pumping level. With the increase of pumping intensity the enhancement lessens and converts into the quenching. Thus, the occurrence of such dependence accounts for the quenching but not the enhancement of luminescence under the influence of SPR in case of Nd:YAG laser without deliberate lessening of pumping intensity.

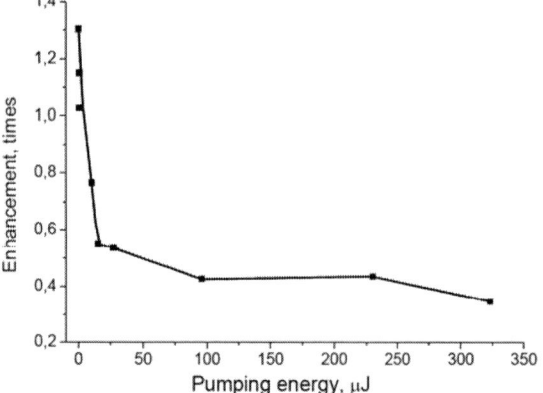

Fig. 2. Dependence of ratio of luminescence intensity of Ag treated and untreated parts of the sample (enhancement) on pumping level.

In addition, epitaxial ZnO film was prepared by pulsed laser deposition (PLD), covered with 5 nm Ag layer. The annealing of this sample leads to characteristic granular features on its surface (Fig.3).

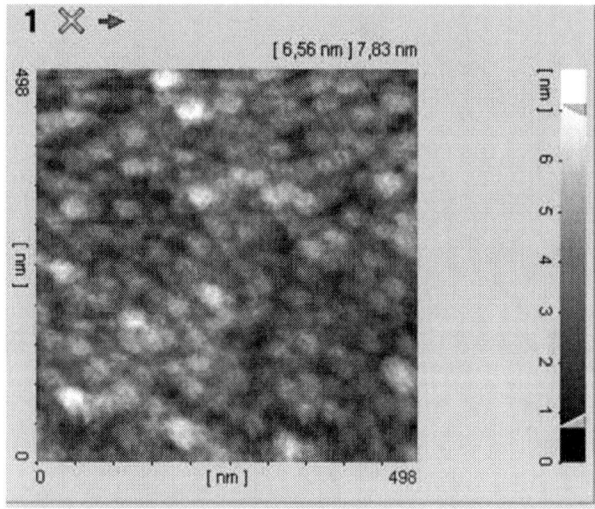

Fig. 3 AFM image of ZnO epitaxial film covered with 5 nm Ag layer and then annealed

However, only luminescence quenching was observed on such sample independent on pumping level (Fig.4).

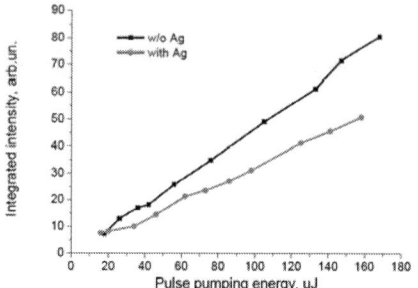

Fig. 4 Dependence of integrated luminescence intensity on pumping level for annealed epitaxial ZnO film (with and w/o Ag layer)

To the best of our knowledge dependence of photoluminescence enhancement under the influence of SPR on pumping level was observed only in [4] but without any interpretation. The study of this dependence for ZnO and the investigation of the nature of this dependence are of great interest both from the practical and scientific points of view. The first natural assumption about the origin of this effect is in its connection with losses in ZnO-Ag system due to plasmons. As plasmons are modes of collective electron vibration in the metal it is possible that they cause Joule's losses that efficiently provides additional non-radiative exciton recombination with the rate proportional to pumping intensity (P). The correctness of this assumption ($\Gamma_{nonrad} = \Gamma_o + bP$) can be checked by analyzing the dependence of photoluminescence intensity on pumping.

In the simplified variant for the number of excitons N following rate equation is true:

$$\frac{dN}{dt} = -\frac{N}{\tau} + P \qquad (1)$$

Here P is total pumping rate, $1/\tau = \Gamma_{rad} + \Gamma_{nonrad}$ is total rate of exciton emission decay, Γ_{rad} – is the rate of radiative recombination, Γ_{nonrad} - is the rate of nonradiative recombination.

In our experiments pulse duration for Nd:YAG laser was ~10 ns while lifetime of excitonic radiation was ~100 ps. This makes it possible to do one additional simplification and use stationary solution of equation (1).

Then: $N = P\tau$, and radiation intensity (I):

$$I = \Gamma_{rad}N = \frac{P\Gamma_{rad}}{\Gamma_{rad} + \Gamma_{nonrad}}$$

Let's denote: $\Gamma_{rad} = A$ and $\Gamma_{rad} + \Gamma_0 = a$.
This gives:

$$I = \frac{AP}{a + bP} = \frac{R_1 P}{R_2 + P} \qquad (2)$$

It was shown experimentally that for a non-coated ZnO film the dependence of luminescence intensity on the pumping level is linear but for Ag - coated film this dependence can be approximated by expression (2). Small pumping at (2) gives dependence close to linear with the slope higher than clean film has. When the pumping rises the bending of this dependence appears. As a result, under the pumping strong enough luminescence intensity of non-coated film appears to be higher than the intensity of Ag-treated film (Fig.5).

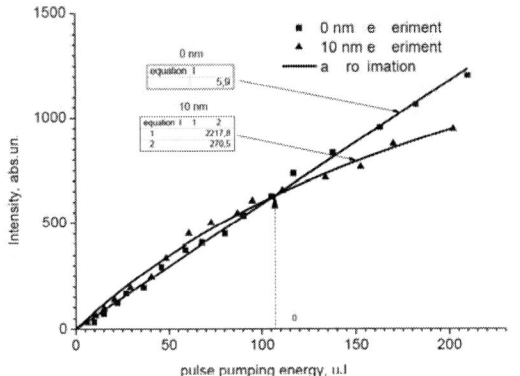

Fig.5. Intersection of luminescence intensity dependences on pumping level for non-treated ZnO film and ZnO film treated with Ag layer of 10 nm.

Thus, we have investigated the dependence of photoluminescence intensity of polycrystalline and epitaxial ZnO films covered with Ag layer on pumping level. Experimental dependences were compared with theoretical ones that were calculated on the assumption about the existence of non-radiative losses linearly dependent on the pumping level. It was proven that for the polycrystalline ZnO films investigated here experimental and theoretical dependences are rather close to each other. Such a result can be considered as the confirmation of this assumption at least for the films investigated. At the same time it was shown that on the Ag-coated epitaxial (dense) ZnO films only luminescence quenching can be observed independently on pumping level that evidently indicates that the roughness of ZnO films is required for the luminescence enhancement that was also shown in other investigations, for example in [1].

References

[1] A.P. Abiyasa, S. F. Yu, S. P. Lau, E. S. P. Leong, and H. Y. Yang "Enhancement of ultraviolet lasing from Ag-coated highly disordered ZnO films by surface-plasmon resonance" *Appl.Phys.Lett.*, vol. 90, no. 23, 231106, 2007.

[2] M. Liu, S. W. Qu, W. W. Yu, S. Y. Bao, C. Y. Ma, Q. Y. Zhang ,J. He, J. C. Jiang, E. I. Meletis, and C. L. Chen "Photoluminescence and extinction enhancement from ZnO films embedded with Ag nanoparticles" *Appl.Phys.Lett*. vol. 97, no. 23, 231906, 2010.

[3] P. Cheng, D. Li. Z. Yuan, P. Chen, and D. Yangb "Enhancement of ZnO light emission via coupling with localized surface plasmon of Ag island film" *Appl.Phys.Lett.*, vol. 92, no. 4, 041119, 2008.

[4] A. I. Dragan and C. D. Geddes "Excitation volumetric effects (EVE) in metal-enhanced fluorescence" *Phys. Chem. Chem. Phys* ., 2011, vol. 13, pp. 3831–3838

Finite Comb-Like Silver Nanostrip Grating in the Optical Range: Interplay of Resonances

Olga V. Shapoval[1], *Student Member, IEEE* and Jiří Čtyroký[2], *Senior Member, IEEE*

[1]Laboratory of Micro and Nano Optics, Institute of Radio-Physics and Electronics NASU, Kharkiv, Ukraine

[2]Institute of Photonics and Electronics AS CR, v.v.i., Chaberská 57, 18251 Prague 8, Czech Republic

Abstract: We analyze the interplay of several types of resonances in the scattering and absorption of light by optical nanoannennas shaped as finite comb-like nanostrip gratings illuminated with an H-polarized plane wave. Our numerical model is based on the combined use of the two-side generalized boundary conditions (GBC) imposed on the strip median lines, the coupled integral equations (IEs), and the rapidly convergent Nystrom-type algorithm of their discretization. We observe the expected excitation of the surface plasmons and the periodicity-induced grating resonances. Besides of that, we investigate specific cavity modes caused by the optical interaction between the adjacent strips.

Today's nanotechnologies available for nanoscale fabrication have lead to considerable progress in the understanding of the optical properties of metals on nanometer scale. In this context, thin noble-metal strips and wires are attractive object of research possessing fascinating physical properties for potential applications in nanoscale optics and optoelectronics. Indeed, they can be easily manufactured and serve as building blocks of optical nanoantennas and sensors with unique geometry-dependent optical properties. This is because they display intensive localized surface-plasmon resonances in the visible and far-infrared ranges that lead to near- and far-field enhancement effects [1-3]. Besides, optical interaction between adjacent strips or wires leads to so-called cavity resonances (a.k.a. gap resonances) [2]. Still besides, periodic gratings of sub-wavelength metal strips and wires display extraordinarily large reflection, absorption, and near-field enhancement at certain wavelengths. These phenomena are caused by the so-called grating (a.k.a. lattice, collective, and geometrical) resonances which appear due to periodicity [4-12].

Here we consider the 2-D scattering of the H-polarized plane wave by a comb-like finite grating of N identical thin (thinner than the free-space wavelength) silver nanostrips (see Fig. 1) of the width d, the thickness h and the period p. For a nanostrip with a thickness down to units of nanometers one can use the experimental data of the bulk silver from [13]. As a reliable instrument for the modeling of the scattering and the resonance effects we use developed in [OS] median-line IE method based on the two-side GBC [14,15] and the Nystrom-type interpolation discretization [16]. In the core of the GBC lays the fact that if the material layer thickness makes a small fraction of the optical wavelength ($h \ll \lambda$) then the electromagnetic analysis can be simplified by neglecting the field inside the layer and considering only the external field limiting values. If used in the thin-strip grating scattering

analysis, GBC lead to a set of the coupled singular IEs with the strip median lines being the intervals of integration. Such IEs can be solved by various numerical techniques. We use a Nystrom-type method based on the interpolation polynomials for the unknown effective-current functions and the quadrature formulas which guarantee the convergence of computations.

Using the above explained approach, we have investigated the spectral dependences of the scattering and absorption cross-sections of finite comb-type gratings of N silver nanostrips illuminated by the H-polarized plane wave - see Fig. 1 (the angle of incidence β is counted from the grating plane).

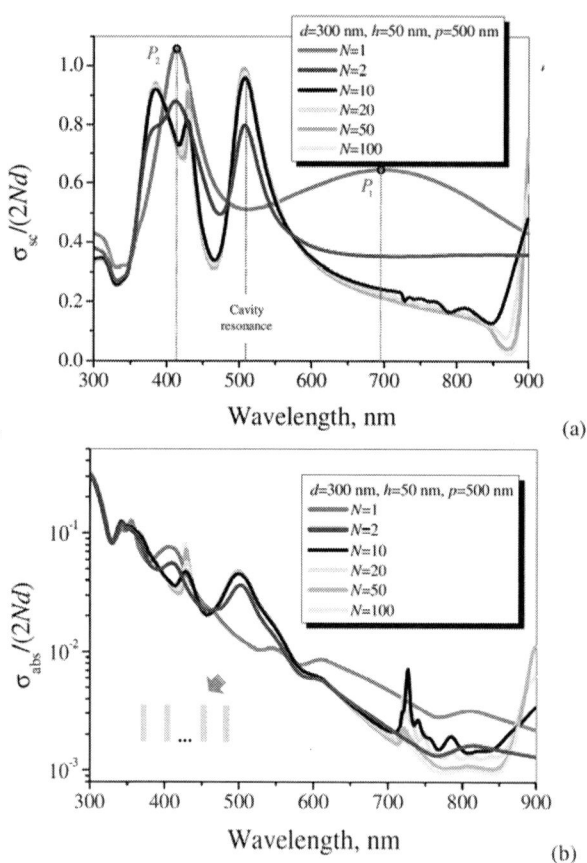

Fig. 1. Normalized scattering (a) and absorption (b) cross-section spectra for the comb-like gratings of N silver strips at the inclined incidence of the H-polarized plane wave ($\beta = \pi/4$).

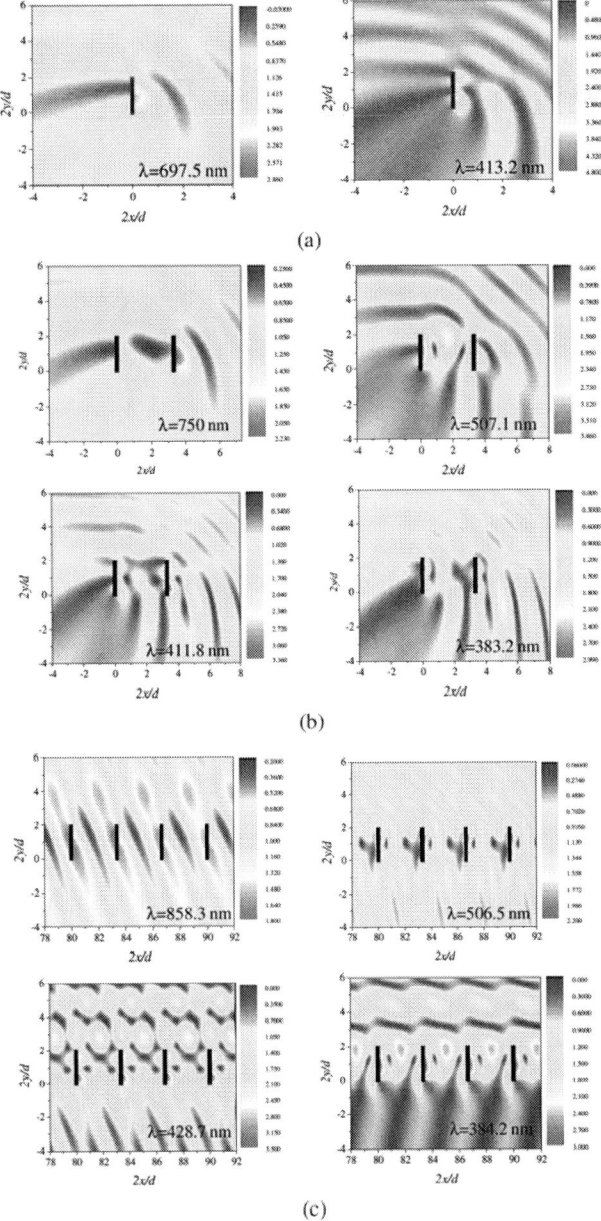

Fig. 2. Total magnetic near-field pattern for (a) $N = 1$ in the resonances on the plasmon modes P_1: $\lambda = 697.5$ nm and P_2: $\lambda = 413.2$ nm; (b) $N = 2$ at $\lambda = 750$ nm, $\lambda = 507.1$ nm, $\lambda = 411.8$ nm, and $\lambda = 383.2$ nm; (c) $N = 50$ around four central strips at $\lambda = 858.3$ nm, $\lambda = 506.5$ nm, $\lambda = 428.7$ nm and $\lambda = 384.2$ nm.

As one can see, for a stand-alone strip ($N = 1$) two plasmon modes of the 1st and 2nd orders are observed in the visible band (red curves) at the resonance wavelengths $\lambda_{P_1} = 697.5$ nm and $\lambda_{P_2} = 413.2$ nm. For better understanding of the nature of these resonances, we present the corresponding near-field patterns in Fig. 2 (a). If the strips are two ($N = 2$), the scattering and absorption spectra keep certain features characteristic for the single strip: the 2nd order plasmon resonance is only slightly

deformed and the 1st order plasmon resonance becomes less visible (twice lower in amplitude) because of the shadowing of the left strip by the right one. Still the spectra for $N = 2$ display something new: a strong resonance peak at the wavelength of $\lambda = 507$ nm. This new feature is kept almost intact if the number N is taken larger. The corresponding near field is visualized in the corresponding panel in Fig. 2 (b). It allows to conclude that this resonance is associated with the so-called "cavity mode" or "gap plasmon" mode [2].

Further getting more strips into a comb-like grating with $p = 500$ nm brings into the visible range. As known [4-13], the periodicity leads to specific resonances on the grating modes (a.k.a. lattice and geometrical modes). Their wavelengths are up-shifted by small values from the Rayleigh anomalies $\lambda_{\pm m}^{R.A.} = (p/m)(1 \pm \cos \beta)$, $m = 1, 2, 3...$ of the associated infinite gratings. Note that, if $\beta = \pi/4$, even the grating of $N = 10$ strips displays quite pronounced grating resonance around $\lambda \approx 428$ nm (Fig. 1). This is the vicinity of the +2nd Rayleigh anomaly at $\beta = \pi/4$. Similarly the extreme-red resonance at $\lambda \approx 890$ nm corresponds to the +1st Rayleigh anomaly. The near-field patterns in Fig. 2 (c) show the area around four central strips in an $N = 50$ grating; they support this explanation.

Finally, to bring together all the above discussed resonance phenomena, in Fig. 3 we show the relief of the total scattering cross-section (normalized by $2Nd$) as a function of two parameters: the wavelength and the incidence angle for the grating of $N = 50$ strips of $d = 300$ nm, $h = 50$ nm and $p = 500$ nm. White lines show the values of the Rayleigh anomalies, which are the functions of the anomaly index, m, and angle β.

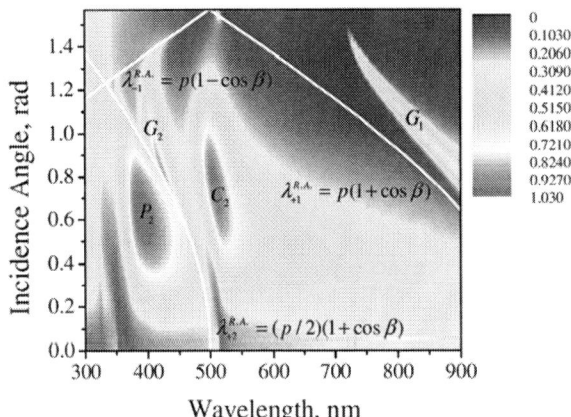

Fig. 3. The relief of the total scattering cross-section normalized per one strip versus the wavelength and the incidence angle for the comb-like silver-strip grating with parameters $d = 300$ nm, $h = 50$ nm, $p = 500$ nm, and $N = 50$.

As one can see, there are several broad and narrow bright "ridges" in the visible range while β varies from 0 (grazing incidence) to $\pi/2$ (normal incidence). Note that some of the ridges extinct at $\beta = 0$ or $\pi/2$ as their "parent" eigenmodes are not excited being orthogonal (of the opposite symmetry) to the incident plane wave coming along the *x*-axis or *y*-axis, respectively.

The largest "mountain" (i.e. bright spot) in the vicinity of 375 nm to 450 nm corresponds to the 2nd order surface plasmon mode on each strip of the comb-like grating. The other large bright spot around 510 nm corresponds to the mentioned above cavity mode between each pair of adjacent strips. Note that the both resonances in the scattering cross-section die off at $\beta = 0$ and $\pi / 2$ (i.e. turn from bright to dark resonances) because of the above-mentioned orthogonality considerations.

Besides of them, one can observe two narrow "ridges" stretching approximately along the Raylegh-anomaly curves $\lambda_{+2}^{R.A.} = (p / 2)(1 + \cos \beta)$ and $\lambda_{\pm1}^{R.A.} = p(1 \pm \cos \beta)$, to the right of them. Each "ridge" corresponds to the high-Q at $N = 50$ grating resonance associated with either the combination of the +2nd and -1st Floquet harmonics, or with the -1st Floquet harmonic, respectively, passing over horizon.

We have demonstrated some results of numerical modeling of the scattering and absorption characteristics of a comb-like finite grating made of silver nanostrips, in the optical range. The gratings are sparse and their periods are comparable to the wavelength. Tin all, three different types of resonances have been observed and analyzed: surface plasmon resonances, on the strips, cavity resonances between adjacent strips, and periodicity-induced grating resonances in the vicinities of the corresponding Rayleigh anomalies. Their interplay depends on the angle of incidence, period of the grating, and the width and thickness of each strip. Choosing these parameters in optimal manner may help design periodic sensors, absorbers, and SERS substrates with improved characteristics.

This work was supported, in part, by the National Academy of Sciences of Ukraine via the State Target Program "Nanotechnologies and Nanomaterials" and the International Visegrad Fund via a Ph.D. Scholarship to the first author. The authors thank A.I. Nosich for many fruitful discussions.

REFERENCES

[1] V. Giannini and J. A. Sànchez-Gil, "Calculations of light scattering from isolated and interacting metallic nanowires of arbitrary cross section by means of Green's theorem surface IEs in parametric form," *J. Opt. Soc. Am. A*, vol. 24, no. 9, pp. 241-248, 2007.

[2] T. Søndergaard and S. J. Bozhevolnyi, "Strip and gap plasmon polariton optical resonators," *Phys. Stat. Sol. (b)*, vol. 245, no. 1, pp. 9-19, 2008.

[3] A. Christ, T. Zentgraf, J. Kuhl, S. G. Tikhodeev, N. A. Gippius, and H. Giessen, "Optical properties of planar metallic photonic crystal structures: experiment and theory," *Phys. Rev. B*, vol. 70, no. 12, pp. 125113-125128, 2004.

[4] R. Gomez-Medina, M. Laroche, and J. J. Saenz, "Extraordinary optical reflection from sub-wavelength cylinder arrays," *Opt. Ex.*, vol. 14, no 9, pp. 3730-3737, 2006.

[5] D.M. Natarov, V.O. Byelobrov, R. Sauleau, T.M. Benson, and A.I. Nosich, "Periodicity-induced effects in the scattering and absorption of light by infinite and finite gratings of circular silver nanowires," *Opt. Ex.*, vol. 19, no 22, pp. 22176-22190, 2011.

[6] P. Ghenuche, G. Vincent, M. Laroche, N. Bardou, R. Haidar, J.-L. Pelouard, and S. Collin, "Optical extinction in a single layer of nanorods," *Phys. Rev. Lett.*, vol. 109, pp. 143903/5, 2012.

[7] T.V. Teperik and A. Degiron, "Design strategies to tailor the narrow plasmon-photonic resonances in arrays of metallic nanoparticles," *Phys. Rev. B*, vol. 86, pp. 245425/5, 2012.

[8] S.R.K. Rodriguez, M.C. Schaafsma, A. Berrier, and J. Gomez Rivas, "Collective resonances in plasmonic crystals: size matters," *Phys. B*, vol. 407, pp. 4081-4085, 2012.

[9] D.M. Natarov, R. Sauleau, and A.I. Nosich, "Periodicity-enhanced plasmon resonances in the scattering of light by sparse finite gratings of circular silver nanowires," *IEEE Photonics Technol. Lett.*, vol. 24, no 1, pp. 43-45, 2012.

[10] V.O. Byelobrov, T.M. Benson, and A.I. Nosich, "Binary grating of sub-wavelength silver and quantum wires as a photonic-plasmonic lasing platform with nanoscale elements," *IEEE J. Sel. Topics Quant. Electron.*, vol. 18, no 6, pp. 1839-1846, 2012

[11] T. L. Zinenko, M. Marciniak, and A.I. Nosich, "Accurate analysis of light scattering and absorption by an infinite flat grating of thin silver nanostrips in free space using the method of analytical regularization," *IEEE J. Sel. Topics Quant. Electron.*, vol. 19, no 3, 2013.

[12] O.V. Shapoval and A.I. Nosich, "Finite gratings of many thin silver nanostrips: optical resonances and role of periodicity," *AIP Advances*, vol. 3, no 4, pp. 042120/13, 2013.

[13] P. B. Johnson and R. W. Christy, "Optical constants of the noble metals," *Phys. Rev. B*, vol. 6, pp. 4370 – 4378, 1972.

[14] K. M. Mitzner, "Effective boundary conditions for reflection and transmission by an absorbing shell of arbitrary shape," *IEEE Trans. Antennas Propag.*, vol. 16, no. 6, pp. 706–712, 1968.

[15] E. Bleszynski, M. Bleszynski, and T. Jaroszewicz, "Surface-integral equations for electromagnetic scattering from impenetrable and penetrable sheets," *IEEE Antennas Propag. Mag.*, vol. 35, pp. 14-25, 1993.

[16] M.V. Balaban, E.I. Smotrova, O.V. Shapoval, V.S. Bulygin, A.I. Nosich, "Nystrom-type techniques for solving electromagnetics integral equations with smooth and singular kernels," *Int. J. Numerical Modeling: Electronic Networks, Devices and Fields.*, vol. 25, no 5-6, pp. 490-511, 2012.

Photonics of new metal–alkanoate composites contained semiconductor nanoparticles

A. Lyashchova[1], D. Fedorenko[1], G. Klimusheva[1], I. Dmitruk[2], S. Bugaychuk[1], T. Mirnaya[3]

[1] Institute of Physics of National Academy of Sciences of Ukraine, 46 Prospect Nauki, 03028 Kiev, Ukraine

[2] Taras Shevchenko National University, Physical Department , 4 Prospect Glushkova, 03127 Kiev, Ukraine

[3] V.I. Vernadskii Institute of General and Inorganic Chemistry, National Academy of Sciences of Ukraine, 32/34 Prospect Palladina, 03142 Kyiv, Ukraine

Abstract – The absorption and photoluminescence spectra of semiconductor CdS and CdSe nanoparticles (NPs) in cadmium alkanoates anisotropic glasses have been studied. NPs are chemically synthesized in thermotropic ionic liquid crystalline (TILC) smectic A phase of cadmium alkanoates which are used as nanoreactors. The glassy nanocomposites are obtained by rapid cooling the TILC to room temperature. In the new cadmium alkanoates matrices, the CdS and CdSe NPs have a small dispersion of their sizes, their shape is nearly spherical, they are stable over time, and they are ordered in glassy smectic A matrix. The thermo-optical nonlinearity of the cadmium octanoate nanocomposites contained CdSe NPs is characterized by extremely large values of the nonlinear refractive index, n_2, under relatively low-powered laser pulses.

I. INTRODUCTION

Class of metal-alkanoate compounds has attracted considerable attention due to their unique chemical and physical properties. Metal-alkanoate salts can form both lyotropic and thermotropic ionic liquid crystals (TILC). Formation of anisotropic glasses after super-cooling TILC to room temperature is another feature of these compounds [1]. Such an anisotropic glassy state is characterized by space ordering, which is similar to the mesomorphic smectic A structure. As have been shown in papers [2-4], the optical and nonlinear optical properties of pure and composite metal-alkanoates are very promising for applications in photonics.

Thermotropic ionic liquid crystals of metal-alkanoates can be used as nanoreactors for the synthesis, stabilization and ordering of semiconductor and metal nanoparticles, whose shape and size can be controlled [5]. We investigate the structural parameters of the cadmium alkanoate matrices, the size of the semiconductor CdS and CdSe NPs in the volume of the matrices by applying different physical methods, as well as absorption and photoluminescence spectra of the NPs in anisotropic glassy samples. All phases of the cadmium alkanoates with CdS NPs were studied by using small-angle X-ray scattering (XRS) technique namely the polycrystalline powder, the thermotropic mesophase, the melt and the glass. For visual confirmation of the size distribution of the CdS and CdSe NPs in the volume of the polycrystalline powder we made the TEM images. The nonlinear optical characteristics of the nanocomposites with CdSe NPs are studied by z-scan method.

II. ABSORPTION AND PHOTOLUMINESCENCE SPECTRA OF CdS QUANTUM DOTS COMPRISING IN CADMIUM ALKANOATE SMECTIC GLASSES

For synthesis of CdS quantum dots (QDs) are used cadmium capronate matrix or cadmium octanoate matrix and their binary mixture. Our results show that in the new cadmium alkanoates matrices, the CdS QDs have a small dispersion of their sizes, their shape is nearly spherical, they are stable over time, and they are ordered in a layered smectic matrix. CdS QDs in cadmium octanoate and in cadmium capronate matrices have the dominant characteristic sizes of 2.7 nm and 2.8 nm accordingly. In the binary mixture, the QDs have two dominant characteristic sizes of 2.7 nm and 3.6 nm. The glassy nanocomposites show spectra both of absorption and of photoluminescence CdS QDs in near-ultraviolet and blue visible spectral range.

The absorption spectra of the CdS NPs in metal alkanoate composites have narrow bands (Fig.1). The pure matrices of the cadmium alkanoates with QDs do not absorb light in this spectrum range. The observed absorption bands are due to excitation of free excitons in the CdS QDs. They are shifted in the UV spectrum range relative to the energy band gap of a CdS bulk crystal (E_g = 2.583 eV [6]). The nanocomposite of the binary mixture has two absorption bands. The presence of two maxima indicates the absorption of CdS QDs that have two different characteristic sizes, which are formed in the mesophase of the binary mixture of the cadmium alkanoates during the synthesis and retained in the glassy mesophase at room temperature. From the absorption spectra we estimated the size of the CdS QDs.

The luminescence spectra of the CdS NPs in different nanocomposites are obtained (Fig.2). The luminescence bands, which overlaps with the absorption bands of the nanocomposites are attributed to the recombination of free excitons in the volume of NPs. Intense next luminescence bands can be associated with others channels of recombination. The energy states, which form these bands, may appear due to structure defects within the QDs, as well as due to localized states at the surface of the QDs. Since the CdS QDs are synthesized directly in the smectic phase of the cadmium alkanoate, this spectral region

978-1-4799-0019-0/13 $31.00 © 2013 IEEE

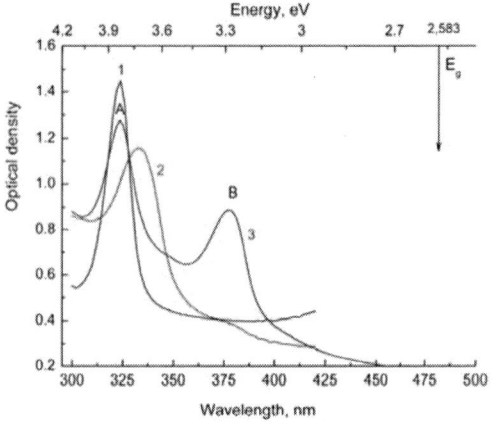

Fig. 1. The absorption spectra of CdS QDs of the anisotropic glass of the nanocomposite materials: 1 - CdC8 + CdS, 2 - CdC6 + CdS and 3 - CdC6/CdC8 + CdS. The arrow marks the band gap E_g of bulk crystal CdS.

of the luminescence can be associated with recombination of the excitons on the vacancies of cadmium either sulfur ions at the surface of the QDs, which create donor and acceptor levels for trapping of charge carriers. Excitons can be captured into such deeper levels, thus the luminescence band is broaden at the long-wave region of the spectrum.

The cadmium alkanoate anisotropic glasses with CdS NPs are new perspective materials for many applications including lasers and sensors of near-ultraviolet and blue visible spectral range.

III. THERMAL OPTICAL NONLINEARITY OF CADMIUM OCTANOATE CONTAINED CdSe NPs

The absorption spectra of glassy smectic nanocomposites with CdSe NPs synthesized at different temperatures are shown in the Fig. 3, curves 1 (the size of NP is 1.9 nm) and 2 (the size of NP is 2.3 nm). The CdC_8 matrix doesn't absorb the light in visible spectral region (Fig. 3, curve 3). Two well-resolved absorption bands of the CdSe NPs are presented in the spectra of the both samples.

Nonlinear optical properties of cadmium octanoate composites with CdSe nanoparticles (NPs) have been studied by using laser scanning technique (Z-scan). Typical normalized on-axis transmittance dependence on the sample position z/z_0 is obtained (Fig.4). The observed peak-valley dependence means that the process of self-defocusing is observed and the nonlinear refractive index of samples has negative value. The large difference between peak and valley values of the normalized light transmittance indicates the large nonlinear phase shift. For analysis of the experimental data in the case of large nonlinear phase shift we applied the model proposed by paper [7]. In this case nonlinear response of the samples is modeled as photoinduced thin lens (PhTL). By fitting experimental data, the photoinduced lens parameters and the nonlinear refraction coefficient, n_2, of nanocomposites were estimated.

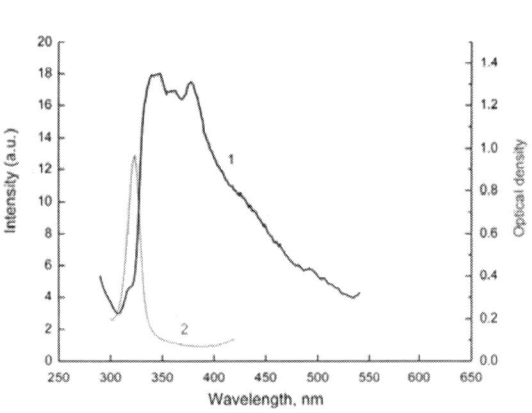

Fig. 2. The luminescence spectrum 1 and the absorption spectrum 2 of CdS QDs in the nanocomposite (I) (CdC8 + CdS).

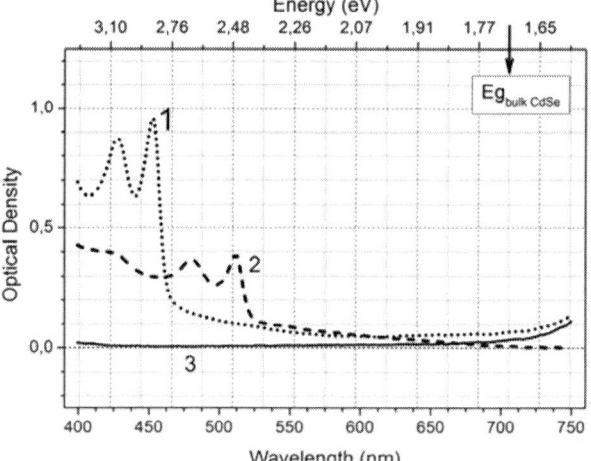

Fig. 3. The absorption spectra of the nanocomposites with CdSe NPs of the different characteristic sizes (curves 1, 2); the absorption spectrum of the pure matrix (curve 3).

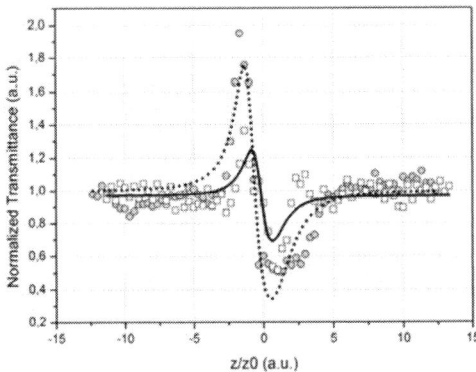

Fig. 4. Typical normalized on-axis transmittance dependence on the dimensionless sample position z/z_0 depicted by squares for samples (1) and circles for sample (2). The solid and dash curves corresponds to the fit by the model of photoinduced thin lens (PhTL).

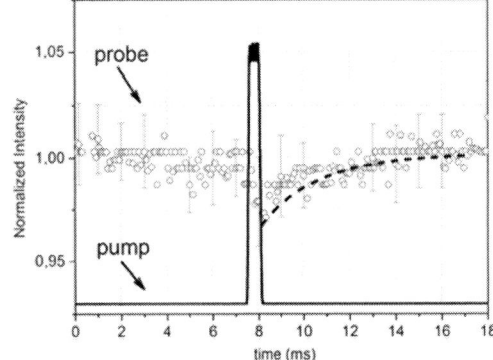

Fig. 5. The time-dependent normalized intensity of the probe pulse (open circles) and pump pulse (solid line). Fitting of the probe signal relaxation with $y = y_0 + A\exp(x/\tau)$ is shown as dashed line ($\tau = 2{,}9$ ms).

The thermo-optical nonlinearity of the new nanocomposites is characterized by extremely large values of the nonlinear refractive index, n_2, ($n_2 = -1{,}2 \times 10^{-2}$ esu in sample (1) and $n_2 = -1{,}9 \times 10^{-2}$ esu in sample (2)) under relatively low-powered laser pulses. Large values of the nonlinear optical coefficients n_2 of the nanocomposites can be explained as a result of the local heating due to the efficient exciton absorption of the CdSe NPs and following thermal dissipation which, in turn, produces the photoelastic tensions in the glassy smectic matrix.

These values are of the same order of magnitude as those reported for liquid crystals (LCs) and LC based mixtures. However, the lens development and nonlinear optic response are much shorter then characteristic times of the LCs orientational nonlinearity.

Large values of optical nonlinearity parameters and fast response times, together with the excellent photo- and thermo-stability of nanocomposites make them extremely promising for optical processing applications.

We studied kinetics of the thermo–optical processes by using the pump-probe z-scan technique. The beam of the He-Ne laser operated at $\lambda = 632{,}8$ nm is used as a probe.

The spot diameter of the probe beam is much smaller than the spot diameter of the pump beam. Both laser beams propagate coaxially. We estimated the time of the nonlinear lens disappearance. The typical kinetics of the normalized transmittance relaxation at the fixed position $z > 0$ is shown in the Figure 5 (open circles). The solid curve indicates the result of the fitting of the experimental data with the exponential grow function. Thus, the characteristic time of the thermal dissipation in CdC_8 matrix is $\tau = 2{,}9$ ms. The local media heating has been dissipating fully during intervals between pulses (time between pulses is 18 ms).

REFERENCES

[1] T.A.Mirnaya, S.V.Volkov, "Ionic liquid crystals as universal matrics (solvents).Main criteria for ionic for ionic mesogenicity", in *Green Industrial Applications of Ionic Liquids*, R.D. Rogers, K.R. Seddon, S.V. Volkov, Eds. London: Kluwer Academic Publishers, 2002, p. 439.

[2] A.B.Bordyuh, Yu.A.Garbovskiy, S.A.Bugaychuk, G.V.Klimusheva, T.A.Mirnaya, G.G.Yaremchuk, A.P.Polishchuk, "Dynamic grating recording in lyotropic ionic smectics of metal alkanoates doped with electrochromic impurities", *Opical Materials*, vol. 31, pp.1109 –1114, 2009.

[3] G.Klimusheva, S.Bugaychuk, Yu.Garbovskii, O.Kolesnik, T.Mirnaya, A.Ishchenko, "Fast dynamic holographic recording based on conductivity ionic metal–alkanoate liquid crystals and smectic glasses", *Optics Letters*, vol .31, pp. 235-237, 2006.

[4] Yu.A.Garbovskiy, A.V.Gridyakina, G.V.Klimusheva, A.S.Tolochko, I.I.Tokmenko, T.A.Mirnaya, "Tunable optical and nonlinear optical response of smectic glasses based on cobalt alkanoates", *Liquid Ctystals*, vol..37, No.11, pp. 1411-1418, 2010.

[5] T.A.Mirnaya, V.N.Asaula, S.V.Volkov, A.S.Tolochko, D.A.Melnik, G.V.Klimusheva, "Synthesis and optical properties of liquid crystalline nanocomposites of cadmium octanoate with CdS quantum dots", *Journal of Physics and Chemistry of Solid State*, vol. 13, pp. 131 -135, 2012

[6] P.A.Kurian, C.Vijayan, K.Sathiyamoorthy, C.S.Suchand Sandeep, R.Philip, "Excitonic transitions and off-resonant optical limiting in CdS quantum dots stabilized in a synthetic glue matrix", *Nanoscale Res. Lett.*, vol. 2, pp. 561-568, 2007.

[7] E.Reynoso Lara, Z.Navarrete Meza, M.D.Iturbe Castillo, C.G.Treviño Palacios, E.Marti Panameño, and M.L.Arroyo Carrasco, "Influence of the photoinduced focal length of a thin nonlinear material in the Z-scan technique', *Opt. Express*, vol. 15, pp. 2517-2529, 2007.

Selection rules for angular-resolved photoionization of single P donor atom in silicon

M.V. Klymenko, F. Remacle

Department of Chemistry, B6c, University of Liege, B4000 Liege, Belgium,
Email: mklymenko@ulg.ac.be

Abstract— In this work we analytically derive the expression of the dipole matrix elements for the photoionization process of a singe phosphorus donor atom embedded in silicon upon irradiation by linearly polarized optical pulses at different incidence angles relative to the bulk crystallographic axes. The derived expressions evidence a drastic dependence of the photoelectron angular distribution on the orientation of the polarization axis relative to the crystallographic axes of silicon. The narrowest angular distribution is obtained for a polarization directed along [001], [010] and [100] crystallographic axes.

Recent experiments that demonstrate that positioning dopants can be achieved at the atomic scale in Si [1] open the possibility of engineering single-atom transistors [2], [3] and designing new low-power computing architectures [4], [5]. One promising mechanism for classical and/or quantum control of single or few atom devices is based on optoelectrical addressing, i.e., on the photoionization and optically-induced coherent or incoherent electron transport within a single atom, between several atoms or between an atom and the leads of the semiconductor device. In order to design the devices based on the optical control, the optical selection rules must be known. In this work, we derive analytical expressions for the dipole matrix element for the photoionization of a single phosphorus donor atom in a silicon crystal lattice irradiated by a linear-polarized laser beam at different angles relative to the crystallographic axes.

The mono-crystalline silicon is characterized by six conduction band minima forming six valleys placed along directions [100], [010] and [001] in the k-space (see Fig. 1 a) [6]. As has been established in earlier works [7], [8], the wave functions corresponding to the bound state of the donor atom in silicon are affected by the valley-orbit coupling effect [7] and can be expressed quite generally as a superposition of the bulk states belonging to different valleys [9]:

$$\langle r|j\rangle = \sum_{\mathbf{k}_0} \alpha_{j,\mathbf{k}_0} f_{j,\mathbf{k}_0}(\mathbf{r}) u_{\mathbf{k}_0}(\mathbf{r}) e^{i\mathbf{k}_0 \mathbf{r}} \quad (1)$$

where $\alpha_{j,\mathbf{k}_0}$ is an expansion coefficient, $f_{j,\mathbf{k}_0}(\mathbf{r})$ is the slow-varying envelope function, $u_{\mathbf{k}_0}(\mathbf{r})$ is the periodic Bloch function, $\mathbf{k}_0$ is the wave vector corresponding to one of six conduction band minima and j is the quantum number of the donor atom's bound state.

In the simplest case of a single donor atom embedded in a perfect silicon crystal without surfaces, the coefficients $\alpha_{m,\mathbf{k}_0}$, reflecting contribution from each valley, can be found considering symmetry properties only and applying methods

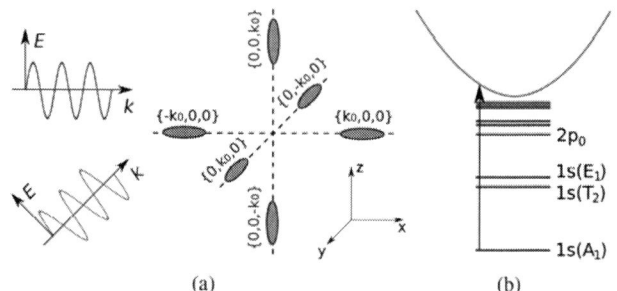

Fig. 1. Energy structure in the vicinity of the conduction band minima of doped silicon: a) six conduction band valleys are placed along [001], [010] and [100] axes in reciprocal space, b) the sketch of the energy spectrum consisting of discrete and continuum parts.

of group theory. As a result, one gets following values of coefficients for the 1s bound states of the phosphorous donor atom [9]:

$$\begin{aligned}
\alpha_{1s(A_1),\mathbf{k}_0} &= 1/\sqrt{6}\{1,1,1,1,1,1\} \\
\alpha_{1s(T_2),\mathbf{k}_0} &= 1/\sqrt{2}\{1,-1,0,0,0,0\} \\
\alpha_{1s(T_2),\mathbf{k}_0} &= 1/\sqrt{2}\{0,0,1,-1,0,0\} \\
\alpha_{1s(T_2),\mathbf{k}_0} &= 1/\sqrt{2}\{0,0,0,0,1,-1\} \\
\alpha_{1s(E_1),\mathbf{k}_0} &= 1/2\{1,1,-1,-1,0,0\} \\
\alpha_{1s(E_1),\mathbf{k}_0} &= 1/2\{1,1,0,0,-1,-1\}
\end{aligned} \quad (2)$$

The envelope functions for each valley satisfy the single-valley Schrödinger equation in the framework of the hydrogenic model for the donor atom. Besides the potential energy of the impurity, the potential energy term in this Schrödinger equation may also include external electrostatic fields, confinement potential and image charges simulating the effect of dielectric screening caused by the interfaces and surfaces in the semiconductor structure.

According to the derivation presented in Ref. [10], in the single-band approximation, the dipole matrix element over the wave functions for the radiation polarized along axes x reads:

$$\begin{aligned}
\langle m|q\hat{x}|l\rangle = q \sum_{\mathbf{k}_0,\mathbf{k}_0'} \alpha_{m,\mathbf{k}_0}^* \alpha_{l,\mathbf{k}_0'} \times \\
\times \int_{L^3} d\mathbf{r} x f_{m,\mathbf{k}_0}^*(\mathbf{r}) f_{j,\mathbf{k}_0'}(\mathbf{r}) e^{i(\mathbf{k}_0'-\mathbf{k}_0)\mathbf{r}}
\end{aligned} \quad (3)$$

where q is the elementary charge.

978-1-4799-0019-0/13 $31.00 © 2013 IEEE

Note that Eq. (3) does not contain the periodic Bloch functions. As has been stated in Ref. [10], this feature of the dipole matrix element holds when the wave functions $\langle r|m\rangle$ and $\langle r|l\rangle$ are constructed from the same set of periodic Bloch functions belonging to the same band. In other words, Eq. (3) has been obtained in the single band approximation.

In order to compute the dipole matrix element for the photoionization process, we substitute the slow-varying plane wave $e^{i\mathbf{kr}}$, which is the solution of the Schrödinger equation for a free propagating particle to the envelope function $f_{l,\mathbf{k}_0'}$. That implies that the total wave function of the electron final state after photoionization is $\langle r|f\rangle = u_{\mathbf{k}_0'}(\mathbf{r})e^{i\mathbf{k}_0'\mathbf{r}}e^{i\mathbf{kr}}$, where $u_{\mathbf{k}_0'}(\mathbf{r})$ is the periodic Bloch function that is utilized in the expansion (1) and corresponds to one of the conduction band minima. The quantum number l in this case designates the set of vectors $\mathbf{k}_0'$ and $\mathbf{k}$. The resulting equation for the dipole matrix element reads:

$$\langle m|q\hat{x}|l\rangle = q\sum_{\mathbf{k}_0}\alpha_{m,\mathbf{k}_0}^*\int_{L^3}d\mathbf{r}x f_{m,\mathbf{k}_0}^*(\mathbf{r})e^{i(\mathbf{k}+\mathbf{k}_0'-\mathbf{k}_0)\mathbf{r}} \quad (4)$$

In this work we consider the photoionization process of 1s states of the impurity atom only. In this case, the envelope function $f_{m,\mathbf{k}_0}(\mathbf{r})$ is an even function and only the sine part of the exponential factor gives a non-zero contribution:

$$\langle m|q\hat{x}|l\rangle = iq\sum_{\mathbf{k}_0}\alpha_{m,\mathbf{k}_0}^*\int_{L^3}d\mathbf{r}x f_{m,\mathbf{k}_0}^*(\mathbf{r})\sin((\mathbf{k}+\mathbf{k}_0'-\mathbf{k}_0)\mathbf{r})$$
$$(5)$$

Let us assume that the photoionization is induced by low-energy photons just above the bottom of the continuum spectrum in the vicinity of the conduction band minima (see Fig. 1b). In order to satisfy the energy conservation laws, the photoelectron's wave vector $\mathbf{k}_0'+\mathbf{k}$ should be close to the wave vector of one of the conduction band minima. Also, the matrix element is negligibly small if the sine factor is a highly-oscillating function. Considering these issues, we get following requirements: $\mathbf{k}_0 = \mathbf{k}_0'$ and $\mathbf{k} << \mathbf{k}_0$ that leads to:

$$\langle m|q\hat{x}|l\rangle = iq\sum_{\mathbf{k}_0}\alpha_{m,\mathbf{k}_0}^*\int_{L^3}d\mathbf{r}\delta_{\mathbf{k}_0,\mathbf{k}_0'}x f_{m,\mathbf{k}_0}^*(\mathbf{r})\sin(\mathbf{kr}) \quad (6)$$

The matrix element in Eq. (6) depends on the wave number and propagation direction of the final outgoing plane wave. To get the angular dependence of the matrix element explicitly, we transform Eq. (6) using spherical coordinates:

$$\langle m|q\hat{x}|l\rangle = \frac{4\pi iq}{3}|\mathbf{k}|\sin(\theta)\cos(\phi)\times$$
$$\sum_{\mathbf{k}_0}\delta_{\mathbf{k}_0,\mathbf{k}_0'}\alpha_{m,\mathbf{k}_0}^*\int_0^\infty dr r^4 f_{m,\mathbf{k}_0}^*(r) \quad (7)$$

where θ and ϕ are angular coordinates of the vector $\mathbf{k}$.

The observables computed via the Fermi's Golden Rule are proportional to the squared modulus of the dipole matrix

element which reads:

$$|\langle m|q\hat{x}|l\rangle|^2 = \frac{16\pi^2q^2}{9}I_{m,\mathbf{k}_0}^2\alpha_{m,\mathbf{k}_0}^2|\mathbf{k}|^2\sin^2(\theta)\cos^2(\phi) \quad (8)$$

where:

$$I_{m,\mathbf{k}_0} = \int_0^\infty dr r^4 f_{m,\mathbf{k}_0}^*(r)$$

It is useful to derive the explicit dependence of Eq. (8) on the kinetic energy of photoelectrons which is also related to the energy of absorbed photon. This can be done using the parabolic dispersion law for electrons that is a good approximation of the band structure in the vicinity of the conduction band minima:

$$E = \frac{\hbar^2}{2m_0}\left(\frac{k_x^2}{m_{xx}(\mathbf{k}_0)} + \frac{k_y^2}{m_{yy}(\mathbf{k}_0)} + \frac{k_z^2}{m_{zz}(\mathbf{k}_0)}\right) \quad (9)$$

where $m_{xx}(\mathbf{k}_0)$ is a component of the relative effective mass tensor.

All components of the effective mass tensor can be expressed in terms of two masses $m_\perp = 0.192$ and $m_\parallel = 0.916$. The direct correspondence between these two masses and tensor components is determined by the orientation of the energetic ellipsoids of the valley in k-space that causes the dependence of the effective masses $m_{zz}(\mathbf{k}_0)$ on the wave vector $\mathbf{k}_0$. In spherical coordinates the expression (9) takes the form:

$$E = \frac{\hbar^2|\mathbf{k}|^2}{2m_0}\xi^{-1}(\mathbf{k}_0,\phi,\theta) \quad (10)$$

where:

$$\xi^{-1}(\mathbf{k}_0,\phi,\theta) = \frac{\cos^2(\phi)\sin^2(\theta)}{m_{xx}(\mathbf{k}_0)} + \frac{\sin^2(\phi)\sin^2(\theta)}{m_{yy}(\mathbf{k}_0)} + \frac{\cos^2(\theta)}{m_{zz}(\mathbf{k}_0)}$$
$$(11)$$

Taking into account the band structure given by Eq. (10), the final expression for the dipole matrix element can be represented as:

$$|\langle m|q\hat{x}|l\rangle|^2 = \frac{32\pi^2m_0q^2}{9h^2}EI_m^2\times$$
$$\alpha_{m,\mathbf{k}_0}^2\xi(\mathbf{k}_0,\phi,\theta)\sin^2(\theta)\cos^2(\phi) \quad (12)$$

One of the most important feature of Eq. (12) is that the square modulus of the dipole matrix element is linearly dependent on the kinetic energy of photoelectrons and, consequently, on the energy of the absorbed photons. However, such a dependence follows from the parabolic band approximation and, therefore, it is an good approximation for small kinetic energies only.

The integral I_m entering the expression (12) contains the envelope function for the bound state of the donor atom and the larger is spreading of the envelope function, the larger it is. Therefore, a shallower confinement potential of the donor atom leads to larger probabilities of photoionization.

To get the total dipole matrix element, Eq. (12) must be summed up over all possible orientations of the wave vector

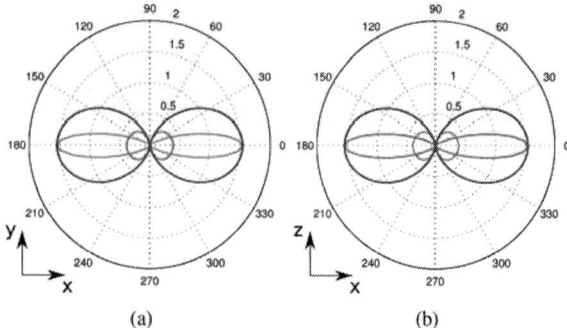
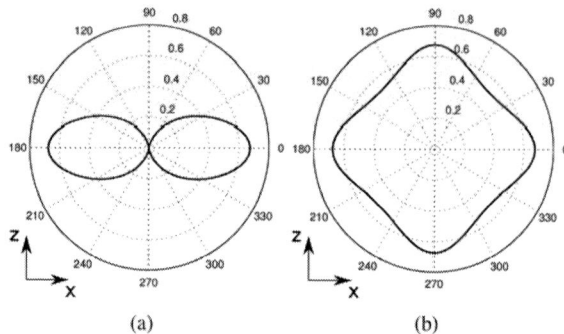

Fig. 2. The angular distribution of photoelectrons a) in the plane $x-y$ and b) in the plane $x-z$. Each curve on the diagrams corresponds to the contribution from the single valley specified by the wave vector $\mathbf{k}_0$. The photoelectrons are generated by the linearly-polarized optical pulse with the photon energy of $E = 50$ meV and the axis of polarization oriented along the crystallographic axis [100].

Fig. 3. The angular distribution of photoelectrons generated by the linearly-polarized optical pulse with the photon energy of $E = 50$ meV and the axis of polarization oriented along a) the crystallographic axis [100] and b) the crystallographic axis [101]

$\mathbf{k}_0$. However, it is interesting to estimate the contribution to the angular dependence of the dipole matrix element (see Fig. 2) from each valley separately. The presented results have been obtained for the polarization directed along axis x.

The observed non-equal contributions from the different valleys are due to the anisotropy of the effective mass and to the different orientations of the isoenergetic ellipsoids in the vicinity of the conduction band minima with respect to the polarization axis.

In Fig. 3, we present results for the angular dependence of the dipole matrix element computed for an optical pulse propagated along the axis [100] and the axis [101]. In both cases, the photoionization is considered at a photon energy of $E = 50$ meV that corresponds to the transition energy between the ground state $1s(A_1)$ and the continuum spectrum in close vicinity of the conduction band minima. In the first case (see Fig. 3 a), the outgoing photoelectrons are very focused and their paths lie very close to the polarization axis. In other case, when the polarization axis is directed along the [101] axis, the photoelectrons can take any direction with almost equal probability.

To summarize, the photoionization process of the single donor atom in silicon is very sensitive to the polarization direction of the incident radiation. An optical radiation linearly-polarized along crystallographic axes [100], [010] and [001] leads to extremely narrow angular distribution of photoelectrons, while all other directions are characterized by approximately uniform angular distributions. This feature can be used in the single-atom nano-devices in order to provide the electron transport between specified donor atoms or leads in certain directions. This can be considered as an alternative to electron focusing by applied magnetic fields which is encountered in the single-electron quantum dot devices [11], [12].

This work is supported by the proactive collaborative project TOLOP (318397) of the Seventh Framework Programme of European Commission. FR acknowledges support from Fonds National de la Recherche Scientifique, Belgium.

REFERENCES

[1] M. Martin Fuechsle, J. A. Miwa, M. Mahapatra, H. Ryu, S. Lee, O. Warschkow, L. C. L. Hollenberg, G. Klimeck, and M. Simmons, "A single-atom transistor," *Nature Nanotechnology*, vol. 7, p. 242246, 2012.

[2] G. P. Lansbergen, R. Rahman, C. J. Wellard, I. Woo, J. Carol, N. Collaert, S. Biesemans, G. Klimeck, L. C. L. Hollenberg, and S. Rogge, "Gate-induced quantum-confinement transition of a single dopant atom in a silicon finfet," *Nature Physics*, vol. 4, pp. 656 – 661, June 2008.

[3] R. Rahman, G. P. Lansbergen, S. H. Park, J. Verduijn, G. Klimeck, S. Rogge, and L. C. L. Hollenberg, "Orbital stark effect and quantum confinement transition of donors in silicon," *Phys. Rev. B*, vol. 80, p. 165314, Oct 2009. [Online]. Available: http://link.aps.org/doi/10.1103/PhysRevB.80.165314

[4] J. A. Mol, J. Verduijn, R. D. Levine, F. Remacle, and S. Rogge, "Integrated logic circuits using single-atom transistors," *Proceedings of the National Academy of Sciences*, vol. 108, no. 34, pp. 13 969–13 972, 2011. [Online]. Available: http://www.pnas.org/content/108/34/13969.abstract

[5] Y. Yan, J. A. Mol, J. Verduijn, S. Rogge, R. D. Levine, and F. Remacle, "Electrically addressing a molecule-like donor pair in silicon: An atomic scale cyclable full adder logic," *The J. Phys. Chem. C*, vol. 114, no. 48, pp. 20 380–20 386, 2010. [Online]. Available: http://pubs.acs.org/doi/abs/10.1021/jp103524d

[6] P. Yu and M. Cardona, *Fundamentals of Semiconductors: Physics and Materials Properties*, 4th ed. Springer, 2010.

[7] A. Baldereschi, "Valley-orbit interaction in semiconductors," *Phys. Rev. B*, vol. 1, pp. 4673–4677, Jun 1970. [Online]. Available: http://link.aps.org/doi/10.1103/PhysRevB.1.4673

[8] T. H. Ning and C. T. Sah, "Multivalley effective-mass approximation for donor states in silicon. i. shallow-level group-v impurities," *Phys. Rev. B*, vol. 4, pp. 3468–3481, Nov 1971. [Online]. Available: http://link.aps.org/doi/10.1103/PhysRevB.4.3468

[9] H. Fritzsche, "Effect of stress on the donor wave functions in germanium," *Phys. Rev.*, vol. 125, pp. 1560–1567, Mar 1962. [Online]. Available: http://link.aps.org/doi/10.1103/PhysRev.125.1560

[10] J. M. Gilliland, "A theoretical investigation of optical absorbtion by donor impurities in silicon," 1961.

[11] G. Usaj and C. A. Balseiro, "Transverse electron focusing in systems with spin-orbit coupling," *Phys. Rev. B*, vol. 70, p. 041301, Jul 2004. [Online]. Available: http://link.aps.org/doi/10.1103/PhysRevB.70.041301

[12] M. V. Klymenko, I. M. Safonov, O. V. Shulika, and I. A. Sukhoivanov, "Ballistic transport in semiconductor superlattices with non-zero in-plane wave vector," *physica status solidi (b)*, vol. 245, no. 8, pp. 1598–1603, 2008. [Online]. Available: http://dx.doi.org/10.1002/pssb.200844086

Resonant enhancement of electromagnetic wave in the structure with refractive index gradient

A.A.Abramov[1,2], A.A.Abramova[3]

[1]Kuang-Chi Institute of Advanced Technology, Shenzhen, China
[2]A.A. Galkin Donetsk Institute for Physics and Engineering of National Academy of Sciences, Donetsk, Ukraine
[3]Kharkov National University of Radio Electronics, Kharkov, Ukraine

Abstract: Propagation of electromagnetic wave (EW) through graded refractive index material is studied theoretically. Green's function technique used to obtain the solution of Maxwell equation in the gradient region for the nonzero angle of incidence wave. Results of our calculations show a resonant increase in the amplitude of EW waves for a certain frequency. We studied effects of the size and profile of gradation region on the position (frequencies) of the resonances.

Propagation of EW in GRIN media studied in [1, 2]. The choosing of hyperbolic tangent dependence for permittivity and permeability allowed analytical results presented in [7]. Resonant enhancement of EW propagating at oblique incidence in metamaterials, with dielectric permittivity and magnetic permeability linearly changing from positive to negative values, has been predicted and theoretically studied in [2]. In [3] optical properties of one dimensional photonic crystals containing graded materials investigated. Effect of gradation profiles on photonic band gap engineering was shown.

In this paper we theoretically demonstrate resonant enhancement of EW propagating at oblique incidence in non magnetic right-hand materials with linear graded index profile. Existence of resonant enhancement is possible due to the presence of the gradation region and a nonzero angle of incidence. Resonant position (frequencies) corresponds to pole of electromagnetic wave amplitude.

Let us consider the propagation of EM waves in a medium which is electrically inhomogeneous but isotropic. We consider nonmagnetic material assume we put everywhere for dielectric permeability $\mu=1$) and assume that the material optical properties can be described by its effective dielectric permittivity ε, which vary only in one direction (x-axis): $\varepsilon=\varepsilon(x)$. In a case of TE (or H-) polarized wave there are nonzero H_z, E_x, and E_y components of electromagnetic field. Taking into account the chosen coordinate system we can write for electric and magnetic field vectors

$$\vec{H}(\vec{r}) = -H(x,y)\vec{z}_0$$

$$\vec{E}(\vec{r}) = E(x,y)\cos(\theta)\vec{y}_0 - E(x,y)\sin(\theta)\vec{x}_0 \quad (1)$$

And then from the Maxwell equations we obtain for field H

$$\frac{d^2H}{dx^2} + \frac{d^2H}{dy^2} - \frac{1}{\varepsilon}\frac{d\varepsilon}{dx}\frac{dH}{dx} + \omega^2\mu\varepsilon H = 0 \quad (2)$$

Assuming that the medium is homogeneous in the y direction the magnetic field strength can be written as $H=H(x)exp(ik_y y)$, where k_y – wave vector in a y direction. Then we have from (2)

$$\frac{d^2H(x)}{dx^2} - \frac{1}{\varepsilon}\frac{d\varepsilon}{dx}\frac{dH(x)}{dx} + \left(\frac{\omega^2\varepsilon}{c^2} - k_y^2\right)H(x) = 0 \quad (3)$$

Introducing dimensionless length $x \rightarrow x/L$ (L – length of gradient region) and using substitution $H(x)=\sqrt{\varepsilon(x)}F(x)$ we have

$$\frac{d^2F(x)}{dx^2} + \left[k^2\varepsilon + \frac{1}{2\varepsilon}\frac{d^2\varepsilon}{dx^2} - \frac{3}{4}\left(\frac{1}{\varepsilon}\frac{d\varepsilon}{dx}\right)^2\right]F(x) = k_y^2F(x) \quad (4)$$

where $k=\omega L/c$ - dimensionless wave vector, and then $k_y=ksin(\theta)$, where θ – angle of incidence.

To construct the solution of (4) we introduce the Green's function G(x,x') via the equation

$$\frac{d^2G(x,x')}{dx^2} + \left[k^2\varepsilon + \frac{1}{2\varepsilon}\frac{d^2\varepsilon}{dx^2} - \frac{3}{4}\left(\frac{1}{\varepsilon}\frac{d\varepsilon}{dx}\right)^2\right]G(x,x') = \delta(x-x')$$

Then the expression for F(x) can be written as

$$F(x) = F_0(x) + \int_0^1 G(x,x')k_y^2F(x')dx' \quad (5)$$

where $F_0(x)$ is a solution of homogeneous equation (4), limits of the integral corresponds to the dimensionless coordinates of gradient region. For linear dependence of $\varepsilon(x)$ the expression for G(x,x') is given by

$$G(x,x') = \frac{\pi}{3b}\left[g(x,x')\theta(x-x') + g(x',x)\theta(x'-x)\right]$$

with

$$g(x,x') = \sqrt{\varepsilon_1 + bx}\sqrt{\varepsilon_1 + bx'}J_{2/3}(t(x))Y_{2/3}(t(x')),$$

$$t(x) = \frac{2k\sqrt{b}}{3}\left(\frac{\varepsilon_1 + bx}{b}\right)^{3/2}, \quad J_{2/3}(t(x)) \text{ and } Y_{2/3}(t(x)) \text{ -}$$

Bessel functions of the first and second kind respectively.

The way to solve the equation (5) numerically described in [5]. Dependences of the magnetic field amplitude versus frequency calculated in the middle of gradient region is shown on Fig.1. In the calculations we used $\sin(\theta)=0.8$ for incidence angle θ, the number of points on the dimensionless length of the gradient region N=150. Changing of N from 100 to 150 led to a change in the position of the resonance in less than 0.5 %. We chose expression $\sqrt{\varepsilon_1 + bx}\,J_{2/3}(t(x))$ for a homogeneous solution F_0 of equation (4). It should be noted, that an expression for function F_0 does not affect the existence and position of the resonance.

The results in Figs.1 are presented for the value of b=20. Position of resonance shifts with increase of b, and we can explain it as follows. The most effective scattering in the

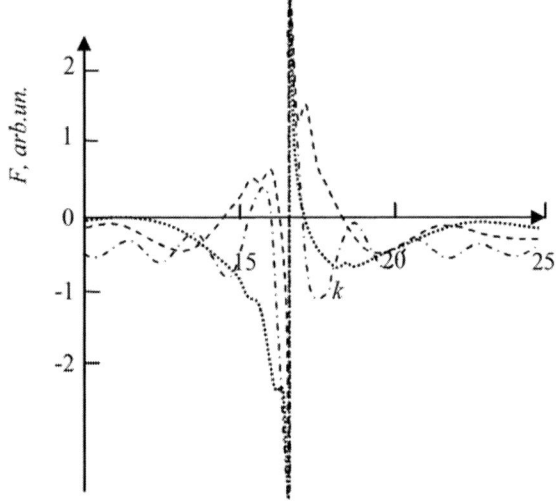

Fig.1 Frequency dependence of EM wave amplitude for three coordinates x in the GRIN region, $\Delta\varepsilon$=20: dashed – x=0.2; dash-dot – x=0.4; dot – x=0.8

gradient region will experience fields with wavelength comparable with the change of ε per unit length (i.e. $\lambda\sim1/(d\varepsilon/dx)$ or $k\sim d\varepsilon/dx$). Therefore, the increase of the gradient would shift of the resonance position to a larger k.

The position of resonance correspond to the values of the dimensionless wave vector $k=\omega L/c$ of 16.5. The increase (decrease) in the size L of gradient region led to increase (decrease) of resonant frequency.

The origin of resonances is obliged a negative part of potential $V_{neg} = -\dfrac{3}{4}\left(\dfrac{1}{\varepsilon}\dfrac{d\varepsilon}{dx}\right)^2 - k_y^{\,2}$ (see Eq.(4)), which has a tendency of binding electromagnetic fields inside the gradient region. Considering equation (4) for this potential (instead of the expression in square brackets) it is easy to verify that its solution can be described by function $exp(-k_{loc}x)$ with some value k_{loc} and have maximum (localization) at x=0. On the other hand we have a solution in the form of a plane wave putting V_{neg}=0. Thus we can assume that the general solution of equation (4) can be expressed as a superposition of plane and localized waves. Indeed, we find such a solution in Fig.2,

where the numerical calculation of (4) for k=5 presented as instance. The initial part actually show a sum of plane waves and flowing from x=0 functions. Then, by analogy with the situation for electrons (when there is a superposition of the

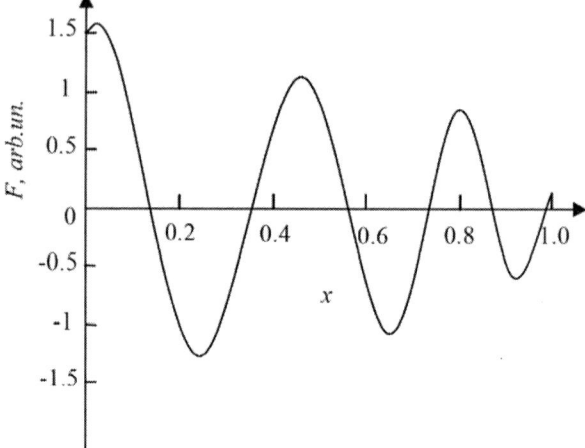

Fig.2 Amplitude of magnetic field in the GRIN region

localized and the free states [4]), we can expect the formation of a resonant state in our case.

It is clear that the positions of the resonances do not depend on the coordinates, although the frequency dependence for each point of the gradient region will, of course, different. The resonances in Fig.1 actually corresponding to singularities of the calculated functions, but in reality the absorption which must be present in the medium means that the field attain large but not infinite values. This work was supported in part by the Russian-Ukrainian project grant 06-02-12 (U).

REFERENCES

[1] M. Dalarsson, Z. Jakšić, P. Tassin, "Exact analytical solution for oblique incidence on a graded index interface between a right-handed and a left-handed material", J. Optoelectr. and BioMaterials, no.1, pp. 345 – 352, 2009

[2] N.M. Litchinitser, A.I. Maimistov, I.R. Gabitov, R.Z. Sagdeev, V.M. Shalaev, Metamaterials: electromagnetic enhancement at zero-index transition, "Optics Lett.", vol.33, no.20, pp. 2350-2352, 2008

[3] Zhu-Fang Sang, Zhen-Ya Li, "Optical properties of one-dimensional photonic crystals containing graded materials", Opt. Comm, no. 259, pp. 174-178, 2006

[4] A.A.Abramov, C.-H.Lin, C.W.Liu, "Fano interference in the quantum well–quantum dot system", Int. J. Nanosc., no. 7, pp. 181-186, 2008

[5] A.A.Abramov, Mod. Phys. Lett. B., vol. 25, pp. 89–96, 2011 .

Nanosized structures with the field localization: new opportunities for precise control of photonic and electronic devices

A. N. Yakunin[1], G. G. Akchurin[1,2], N. P. Aban'shin[3], B. I. Gorfinkel'[3]

[1]Institute of Precise Mechanics and Control of RAS, Saratov, Russian Federation
[2]Saratov State University, Saratov, Russian Federation
[3]Volga-Svet Co.Ltd, Saratov, Russian Federation

Abstract: This paper describes the results of research in the tunnel photoelectric effect in microdiode with the localization of an electrostatic field at the emitter of carbon nanosized structure. It is experimentally found that irradiation of carbon emitter by photon beams with low energy (less than the work function) within the spectral range 380 - 1550 nm and with micro- and milliwatt optical power may initiate tunneling of photocurrent by controlling the field intensity in the gap "anode - emitter". The peculiarities of the design and application of photonic and electronic devices using an photoelectric effect with "red" threshold are discussed. The estimation of investigated photoemitter broadbandness is made, an opportunity to work in a range of wavelengths from ultraviolet up to far infrared is predicted. A method for controlling a tunneling current of emission by changing the optical power density is suggest. In this case, the voltage-current characteristic becomes a linearly dependent on the level of optical power density, which ensures stability of controlling.

Researches [1] electron emission of granular films of gold and silver activated by cesium and oxygen have shown that the photoeffect in such structures is determined by the probability of tunneling nonequilibrium photoelectrons through the potential barrier formed by the active layer. In this case, an exponential decrease of the intensity of the long-wavelength part of the spectrum of photoelectron emission with increasing wavelength and the unexpected absence of a clearly defined photoelectric threshold are observed. Increasing the spectral sensitivity in the band width of about 100 nm and non-monotonic character of its dependence on the frequency was related by the authors [2] to the excitation of surface plasmons. This assumption is confirmed by the results of theoretical analysis, obtained in [3].

The phenomena of surface carrier trapping metallic nanoparticles and the search for effective tunneling conditions of undoubted practical interest in connection with the prospects for its use at creation of new high-speed photocathodes with a femtosecond performance [4].

However, in [5] covered issues which, in our opinion, were beyond the scope of research studies [1-4]. It is a very important regularities of influence on the parameters the tunnel PE level of intense electrostatic fields with high strength, which are located on nanoscale inhomogeneities of the cathode surfaces. It was theoretically predicted and observed experimentally in the field emission carbon nanosized planar-edge structures with the localization of field [6-8] the existence of photosensitivity phenomenon in relation to radiation fluxes of low energy photons in the wavelength range corresponding to a significant (up to 6 times) times the excess of "red" threshold of classical photoeffect.

Fig. shows the results describing the possibility of precise control of both the frequency dependence of the sensitivity of the photoelectric effect, and the level of the photocurrent by

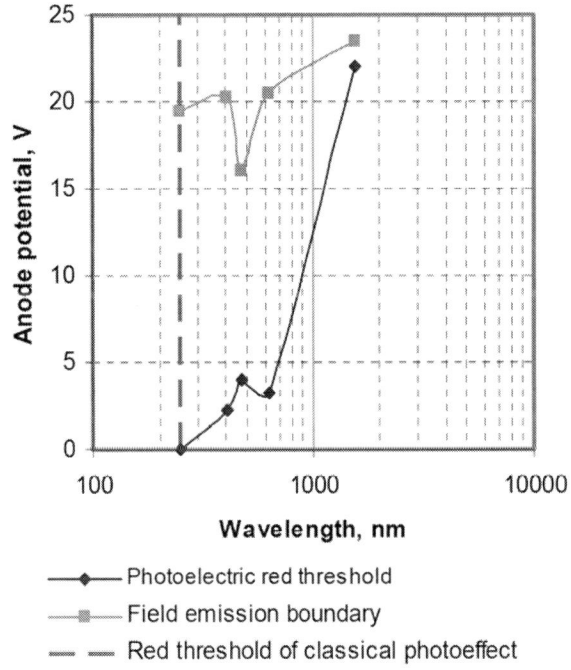

Fig. The frequency dependence of the "red" threshold of tunnel photoeffect in carbon nanostructures in the range of UV to IR vs. the anode potential

978-1-4799-0019-0/13 $31.00 © 2013 IEEE

small changes in the voltage limits "emitter-anode". The dotted line denotes the "red" threshold of the classical photoelectric effect.

On opposed to the classical photoelectric effect, which occurs only when the photon energy greater than the work function of the material (the tunneling probability is equal to 1), the tunnel photocurrent in a strong (10^7-10^8 V/cm) field occurs at a lower photon energy (the possibility of reducing by more than 6 times has been experimentally proved, see Fig.). This ensures the control of "red" threshold of tunnel photoeffect in the wavelength range from the UV to the IR.

Potentially, such a detector can detect the radiation of λ up to 50 μm, where the photon energy is comparable to kT at room temperature. Since the process of non-equilibrium electron tunneling occurs in the thin film thickness of 20 nm, the performance of the device is comparable to the speed of conventional thin-film metal photodetectors. It was shown experimentally that the observed effect is the one-photon, there is a considerable range of working potentials when the dark current is extremely small (prethreshold regime of field emission).

It is well known that the reliability, durability and operation stability of devices based on field emission is largely confined to difficulties of precision control of the emission level. The main difficulty arises from accuracy of choice of operating point on the exponential curve of the voltage-current characteristics of the device, when a small change in the accelerating voltage corresponds to a significant increase of emission current.

With the rise of emission current density the probability of occurrence of an unstable regime increases. With a voltage corresponding to the quantum probability of electron tunneling through a potential energy barrier of the "metal-vacuum", tending to 1, the potential barrier height is close to the Fermi level, and the width becomes comparable with the de Broglie wavelength of an electron. Under such conditions electrical breakdown occurs.

The physical mechanism of field emission instability associated [9, 10] with a local heating due to the effect of Nottingham on the area small fragment of the emitter, from which the maximum density tunneling current is emitted. Accordingly, there is a localization of thermionic emission, increasing the current density, and this positive feedback is increasing exponentially causes the so-called local field-emission explosive instability.

To provide dynamic precise control of the nanostructured emitter field emission it is proposed to irradiate emitter surface by laser or LED beam with a wavelength selected from the visible or ultraviolet range. The actual control of tunneling emission current is carried out by varying optical power density. In this case, the voltage-current characteristic becomes linearly dependent on the level of optical power density [5], which ensures the stability of control for each chosen value of accelerating voltage.

The slope of the voltage-current characteristics is determined by selecting of a fixed value of accelerating voltage, but now there exists an additional control parameter - the density of optical power. The speed of control by emission current is determined by the rate of change of optical power.

Acknowledgments. This study was supported by the Russian Foundation for Basic Research, project no. 12-07-12066-ofi_m.

REFERENCES

[1] Nolle E.L. "Tunneling mechanism of photoeffect in metallic nanoparticles", *Uspekhi Fizicheskikh Nauk.* 2007. Vol. 177. No 10. P. 1133-1137. (in Russian).

[2] Nolle E.L., Schelev M.Ya. "Photoelectron emission from granular gold films activated by cesium and oxygen", *Journal of Applied Physics.* 2005. Vol. 75. No 11. P. 136-138. (in Russian).

[3] Protsenko I.E., Uskov A.V. "Photoemission from metal nanoparticles", *Uspekhi Fizicheskikh Nauk.* 2012. Vol. 182. No 5. P. 543-554. (in Russian).

[4] Schelev M.Ya. "Pico-femto-attosecond photoelectronics (looking through half a century of "loupe of time")", *Uspekhi Fizicheskikh Nauk.* 2012. Vol. 182. No 6. P. 649-656. (in Russian).

[5] Akchurin G.G., Yakunin A.N., Aban'shin N.P., Gorfinkel' B.I., Akchurin G.G., Jr "Controlling the Red Boundary of the Tunneling Photoeffect in Nanodimensional Carbon Structures in a Broad (UV–IR) Wavelength Range", *Technical Physics Letters.* 2013. Vol. 39. No. 6. P. 544–547.

[6] Gorfinkel' B.I., Aban'Shin N.P., Yakunin A.N. "The cell with a field emission and method for manufacturing", RU Patent for an invention No 2446506. Reg. 27.03.2012.

[7] Aban'Shin N.P., Gorfinkel' B.I. Yakunin A.N. "Development of Durable Carbon Field Emission Nanostructures for Vacuum Electronics", 25th International Vacuum Nanoelectronics Conference IVNC 2012, Jeju, Korea July 9th -13th, 2012. P. 1-2. http://dx.doi.org/10.1109/IVNC.2012.6316901

[8] Morev S.P., Aban'shin N.P., Gorfinkel' B.I., Darmaev A.N., Komarov D.A., Makeev A.E., Yakunin A.N. "Electron-optical systems with planar field-emission cathode matrices for high-power microwave devices", *Journal of Communications Technology and Electronics.* 2013. Vol. 58. Iss. 4. P. 357-365.

[9] Aban'shin N.P., Gorfinkel B.I., Yakunin A.N. "Mechanism of ion loading of point emitters in planar edge field emission structures", *Technical Physics Letters.* 2006. V. 32. № 10. P. 892-895.

[10] Petrin A.B. "On the thermofield emission of electrons from metallic points", *Plasma Physics.* 2010. Vol. 36. No. 7. P. 671-679. (in Russian)

[11] Petrin A.B. Thermofield electron emission from metal and explosive electron emission from microperforations // *Journal of Experimental and Theoretical Physics.* 2009. Vol. 136. Iss. 2(8). P. 369-376. (in Russian).

Custom types of waves spatially periodic structure of microwires

V. A. Boiko[1], V. I. Ponomarenko[1]

[1]Taurida National V.I. Vernadsky University, Simferopol, Ukraine

Abstract: To date, developed many different radar absorbing structures located above the metal surface. Structure of microwires creates special conditions for the absorption of electromagnetic waves related to the fact, that the metal is formed near the junction of the electric field and magnetic field antinode. We considered absorbers disposed on a flat metallic mirror.

Consider an infinite periodic structure of parallel microwires located above a perfectly conducting plane at a distance h between ideal metal mirror and the structure[1], [2], the distance between the wires - a. This structure is equivalent of a quarter-wave layer.

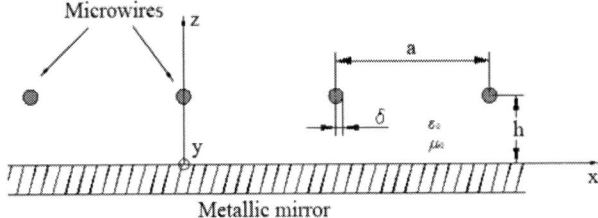

Fig. 1. Equivalent of a quarter-wave layer

Suppose that on the structure of the electromagnetic plane wave falling in the axial direction of z axis, electric vector is $e^{ik_0 z}$ and directed along the axis y (time factor is $e^{-i\omega t}$) [3]. Obviously, in the space field and in the structure is periodic along x, y axes. What is also apparent, for reasons of symmetry, the fact, that [4]:

In the plane yz have $E_x = 0$, $\dfrac{\partial E_y}{\partial x} = 0$,

In the plane xz have $H_y = 0$, $\dfrac{\partial H_x}{\partial y} = 0$. $\qquad$ (1)

So $H_z = 0$ in plane yz. Similarly $E_z = 0$ in the plane xz. It follows that we can assume that the planes xz are ideally magnetic (infinitely thin) wall, and in the planes - ideally conducting wall [5].

This allows us to reformulate the problem of finding the reflected field from periodic structure, namely, to consider now the rectangular waveguide in which two electric wall, and two others - magnetic, and in the waveguide is only one microwire [6].

From Maxwell's equations, using (1), we obtain the solution for the components of the electric and magnetic fields:

$$
\begin{cases}
\vec{e}_n = A_n \cos\dfrac{2n\pi}{a} x \cdot \vec{j}, \\[2mm]
\vec{h}_n = \dfrac{-1}{W_n} A_n \cos\dfrac{2n\pi}{a} x \cdot \vec{i}, \\[2mm]
W_n = \dfrac{\omega\mu}{\Gamma_n}, A_n = \sqrt{\dfrac{2}{a}} |W_n|, \Gamma_n = \sqrt{\omega^2 \varepsilon\mu - \dfrac{4n^2\pi^2}{a^2}} \\[2mm]
\quad n = 0,1,2,...
\end{cases}
\qquad (2)
$$

Where $\vec{e}_n, \vec{h}_n$ - vectors of electric and magnetic component of electromagnetic field in rectangular waveguide; W_n - propagation constant, ω - cyclic frequency of electromagnetic wave; a - width of rectangular waveguide, n - whole number.

Using the basic formula for the normalization of the electromagnetic field in the waveguide, we obtain the expression for finding the current wave excited H_{10} mode, we have equation for microwave current flowing through the microwire:

$$
I = \frac{-c_0 B_1 \cos\left(\dfrac{2\pi}{a} x_0\right) e^{-i\Gamma_1\delta}}{Z + \dfrac{1}{2a} \displaystyle\sum_{n=1}^{\infty} W_n \cos^2\left(\dfrac{2n\pi}{a} x_0\right) \cos\left(\Gamma_n \delta\right)}
\qquad (3)
$$

Where x_0 - location of the microwire in waveguide, Z - impedance of microwire.

Using equation (3) and expression for the impedance of the microwire we can find the value for the reflection coefficient of the structure [7].

Varying distance between the microwires and the distance between the structure of microwires and the metal mirror we can vary reflectance and bandwidth for this structure.

REFERENCES

[1] Chung B.-K. Modeling of RF absorber for application in the design of anechoic chamber / B.-K. Chung, H.-T. Chuah // Progress In Electromagnetics Research, PIER. – 2003. – Volume 43. – P. 273–285.

[2] W. L. Stutzman and G. A. Thiele, Antenna Theory and Design, 2nd ed., Wiley, New York, 1998.

[3] S. Silver, ed., Microwave Antenna Theory and Design, Peter Peregrinus, Ltd, London, 1984.

[4] C. G. Someda, Electromagnetic Waves, Chapman and Hall, London, 1998.

[5] L. B. Felsen and N, Marcuvitz, Radiation and Scattering of Waves, IEEE Press, New York, 1994.

[6] J. A. Kong, Electromagnetic Wave Theory, 2nd ed., Wiley, New York, 1990.

[7] H. J. Riblet, "General Synthesis of Quarter-Wave Impedance Transformers," IRE Trans. Microwave Theory Tech., MTT-5, 36 (1957).

Hybrid Plasmon Resonances in the Scattering and Absorption of Light by a Circular Silver Nanotube

Elena A. Velichko, *Student Member, IEEE*, Alexander I. Nosich, *Fellow, IEEE*

Institute of Radio-Physics and Electronics NASU, Kharkiv 61085, Ukraine

Abstract: We study the scattering and absorption of an H-polarized plane electromagnetic wave by a silver nanotube, in the visible range of wavelengths. The tube is assumed to have circular cross-section. The analytical solution of the wave-scattering problem is obtained in classical manner, using the separation of variables in the polar coordinates. In computations, we use the Johnson and Christy data for the silver refractive index. The computed spectra of the total scattering cross-section and the absorption cross-section display two main surface-plasmon resonances, of the dipole and quadrupole type. This data can be useful for the design of nanotube-based sensors of the changes in refractive index of host medium.

In today's nanotechnologies, gold and silver nanowires are widely used as building blocks of biosensors, lasers, and photovoltaic devices. This is because of the effects of resonant scattering and absorption mediated by the localized surface plasmons (plasmons, for brevity). The resonance wavelengths are fixed and determined by the wire cross-sectional shape [1]. The simplest case is a secular cross-section, which simplifies the analysis greatly. The metal nanotubes are even more attractive for applications because their plasmon resonances can be tuned by the variation in tube thickness [2-4].

To analyze the nanotube resonances in more detail, we consider an H-polarized plane wave normally incident on a circular silver tube of the inner radius a and the thickness h placed into free space - see the inset in Fig. 1(b). The electromagnetic field in the presence of such a scatterer must satisfy the Helmholtz equation, the tangential components continuity conditions, and the condition of the local power finiteness; its scattered-filed part must also satisfy the radiation condition at infinity. Using the separation of variables [5], the field is found analytically in the form of infinite Fourier series with known coefficients.

In Fig. 1, we present the visible-range spectra of the total scattering cross-section (TSCS) and absorption cross-section (ACS) of the nanotube with $a = 30$ nm and several values of thickness h. The tube inner material and the host medium are free space. As one can see, if the tube thickness approaches 100 nm, there is only one well visible plasmon-resonance peak at the same wavelength (≈ 350 nm) as for a solid nanowire. However thinner tubes show two resonances significantly red-sifted from the circular-wire resonance wavelength.

For a nanosize tube, approximate wavelengths of plasmon resonances of the azimuthal order m are found from (10) of [4],

$$\varepsilon_{Ag}(\lambda_{Pm}^{(\mp)}) \approx \sim -\varepsilon_{host} \pm 2[(\varepsilon_{host} + h/a)^m \pm 1]^{-1}, \quad m = 1,2,\dots \quad (1)$$

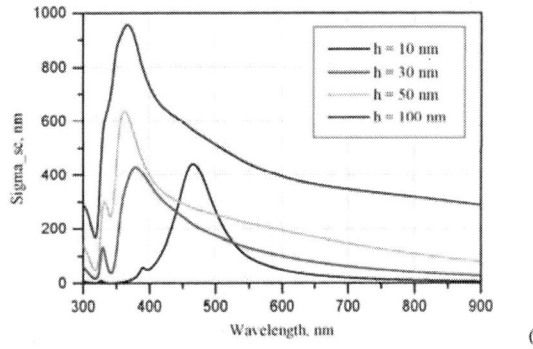

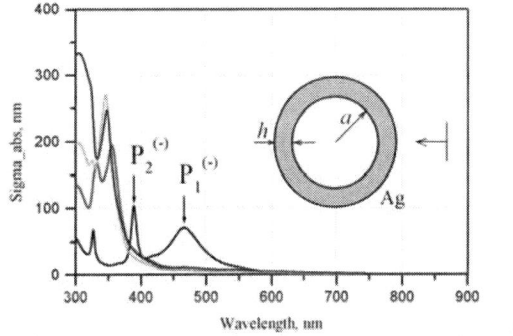

Fig. 1 TSCSs (a) and ACSs (b) versus the wavelength, for an Ag tube of the inner radius $a_1 = 30$ nm and the thickness h varying from 10 to 100 nm.

where ε_{Ag} and ε_{host} are the dielectric permittivities of the metal and the host medium ($\mathrm{Re}\,\varepsilon_{host} > 0$), respectively. For each m Eq. (1) has two roots $\lambda_{Pm}^{(\mp)}$ corresponding to two plasmons; the wavelength of each is specific for every metal and depends on the parameters, ε_{host} and h/a. One of them sits at a larger wavelength than for a circular silver nanowire, and the other at a smaller wavelength. The use of (1) enables us to determine that two resonances visible in Fig. 1 on the red side of the solid-wire resonance wavelength are the dipole ($m = 1$) and the quadrupole ($m = 2$) plasmons corresponding to the sign (-).

These plasmons have the modal fields localized on the inner and outer boundaries of the nanotube – see the magnetic field patterns (maps of $|H_z|$) in Fig. 2. They demonstrate $2m$ bright spots around the tube where m is the plasmon azimuthal index.

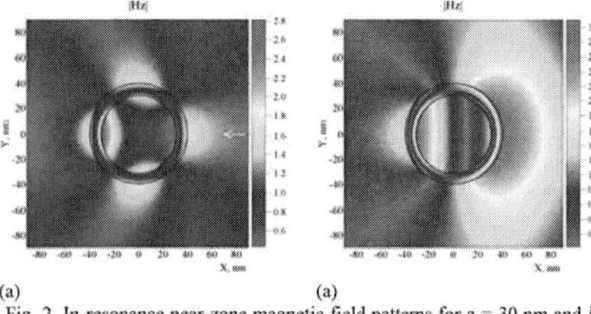

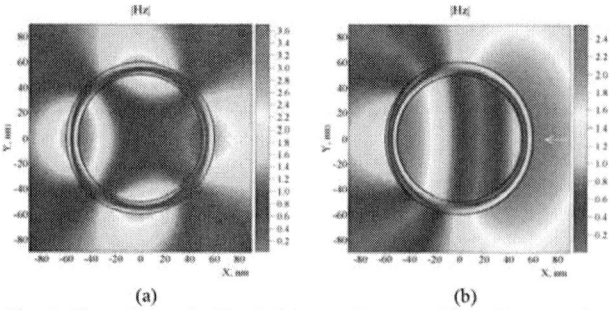

Fig. 2. In-resonance near-zone magnetic field patterns for a = 30 nm and *h* = 10 nm at the wavelengths of (a) $\lambda = 389$ nm and (b) $\lambda = 466$ nm

Fig. 4. The same as in Fig. 2 for *a* = 50 nm and *h* = 10 nm at the wavelengths of (a) $\lambda = 437$ nm and (b) $\lambda = 545$ nm

Note that the quadrupole plasmon has larger Q-factor than the dipole one, apparently because of a larger number of the electric-field minima in the tube (at the locations of maxima of the magnetic field). A sensor built on this resonance promises a better sensitivity to small changes in the host medium.

In Fig. 3, presented are the spectra of TSCS and ACS for the silver nanotube with the inner radius of 50 nm. The same two plasmons are visible, shifted even more to the red side of spectrum. The separation between them is considerably large than in Fig. 1, for the same values of the tube thickness. Note that the absorption can serve as a finer instrument of detecting the plasmon resonances than the scattering, especially in the violet part of the spectrum. The in-resonance field patterns are demonstrated in Fig. 4 and reveal more intensive maxima. Interestingly, in the quadrupole resonance, the largest outer field maximum is located at the shadow side of the nanotube.

Inside the tube, the largest field maxima are found near the shadow part of the inner wall both in the dipole plasmon resonance and in the quadrupole one. The peak values of the inner field are almost the same as the peak values of the outer field near the illuminated part of the outer wall.

Closer study reveals that each of the mentioned above plasmons at $\lambda_{Pm}^{(-)}$ is a supermode (see analogous effect for the modes of circularly-layered micro and nanocavities in [6,7] built on optically coupled anti-phase plasmon modes of the index *m* of the outer boundary and the inner (void) boundary. Their sister supermodes (in-phase ones) can be observed at the wavelengths $\lambda_{Pm}^{(+)}$ below the plasmon position on a solid wire at 350 nm. They have generally higher Q-factors because their electric fields have minima on the tube median line.

Note that the $\varepsilon_{Ag}(\lambda)$ values were taken from paper [8].

This work has been partially supported by the National Academy of Sciences of Ukraine via the State Target Program "Nanotechnologies and Nanomaterials."

REFERENCES

[1] O. J. F. Martin, "Plasmon resonances in nanowires with a non-regular cross-section," in J. Tominaga and D. P. Tsai (Eds.), *Optical Nanotechnologies,* Topics Appl. Phys., Berlin: Springer, vol. 88, pp. 183-210, 2003.

[2] U. Schroster, A. Dereus, "Surface plasmon-polaritons on metal cylinders with dielectric core," *Phys. Rev. B,* vol. 64, pp. 125420 /10, 2001.

[4] J. Zhu, "Theoretical study of the light scattering from gold nanotubes: effect of wall thickness," *Materials Sci. Eng.,* vol. 454-455, pp. 685-689, 2007.

[3] H.-Y. She, L.-W. Li, O. J. F. Martin, J. R. Mosig, "Surface polaritons of small coated cylinders illuminated by normal TM and TE plane waves," *Opt. Exp.,* vol. 16, pp. 1007-1019, 2008.

[5] J. W. Strutt (Lord Rayleigh after 1902), "On the electromagnetic theory of light," *Philos. Mag.,* vol. 12, pp. 81-101, 1881.

[6] E. I. Smotrova, J. Ctyroky, T. M. Benson, *et al.,* "Lasing frequencies and thresholds of the dipole-type supermodes in an active microdisk concentrically coupled with a passive microring," *J. Opt. Soc. Am. A*, vol. 25, pp. 2884-2892, 2008.

[7] E. A. Velichko, D. M. Natarov, A. I. Nosich, "Plasmon-assisted scattering of light by a circular silver nanowire with concentric dielectric coating," in *Proc. Int. Conf. Transparent Optical Networks (ICTON-2013)*, Cartagena, 2013.

[8] P. B. Johnson, R. W. Christy, "Optical constants of the noble metals," *Phys. Rev. B,* vol. 6, no 12, pp. 4370-4378, 1972.

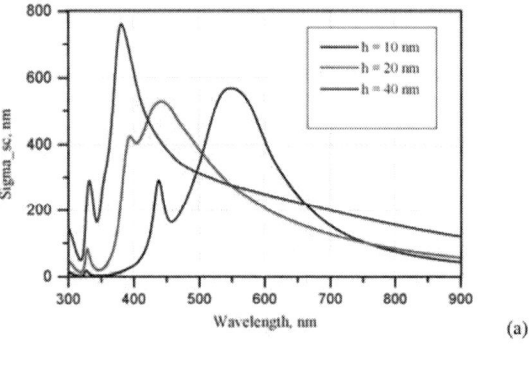

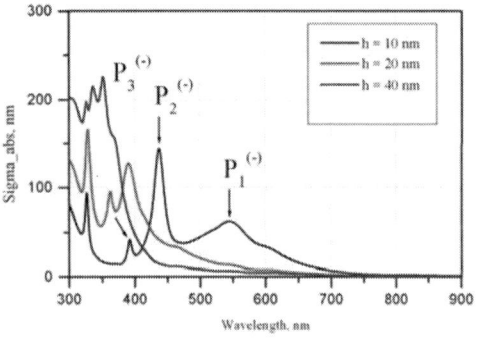

Fig. 3. The same as in Fig. 1 however for an Ag tube of the inner radius $a_1 = 50$ nm and the thickness *h* varying from 10 to 40 nm.

The diffraction of the laser radiation on the two-band axial microelement

Savelyev D.A., Khonina S.N.
Image Processing Systems Institute of the RAS, Samara, Russia

Abstract: Numerically shown, that an optical microelement consisting just two annular zones is useful for tight focusing of laser radiation if the radius of the central zone is half wavelength. The numerical calculations executed in approach of a thin element by means of the plane wave expansion method show good accordance with finite-difference time-domain method (in view three-dimensional structure of an element). Characteristics and features of diffraction of Gaussian beam with linearly polarization on a considered element are investigated.

Using diffraction microelements in a variety of optical systems provides a smaller size, weight and cost in a huge variety of functions performed. One of the effective applications diffraction microelements - focusing of the laser radiation in the near-field diffraction [1–3].

Note that the structure of the periphery of the zone plate with a short focus approaches the axicon type, i.e. rings are of equal width. In fact, opposed axicon with high numerical aperture and a zone plate with a short focus is determined only by the central part.

Thus, the influence of the size of the central part of the microelement is very important when focusing in the near field.

In this paper we investigate the possibility of focusing the laser radiation with the microelement, which consists of only two central zones, in which the phase difference is π radians.

Numerical calculations are performed in the approximation of a thin element using the method of expansion in plane waves [4], and for the volume element based on the finite-difference time method (FDTD), implemented in software product MEEP [5]. The numerical calculation in the approximation of a thin element The numerical calculation in the approximation of a thin element

1. The numerical calculation in the approximation of a thin element

Numerical calculations were performed axial intensity of a plane wave, transmitted through the optical element of the form:

$$\tau(r) = \begin{cases} \exp(i\pi), & r \le r_1, \\ 1, & r_1 < r \le R, \\ 0, & r > R. \end{cases} \qquad (1)$$

In the calculations we used the fast algorithm developed in [4]. This algorithm is based on the method of expansion in plane waves and takes into account the radial symmetry of the problem, and the Fresnel transmission coefficients. In the simulation following parameters were used: wavelength $\lambda = 0,532$ μm, the refractive index of the optical element $n = 2$, which correspond to the optical glasses of the super heavy flint.

Fig. 1 shows the simulation results when falling a plane wave with a linear y-polarized at the element (1) with $r_1 = 5\lambda$. For comparison, the full radius of an element selected from the eligibility conditions of the maxima:

$$R^2 = 2r_1^2 + \lambda^2/2, \qquad (2)$$

$R = 7,11\lambda$ (the solid line in Fig. 1) and several large $R = 10\lambda$ (the dashed line in Fig. 1).

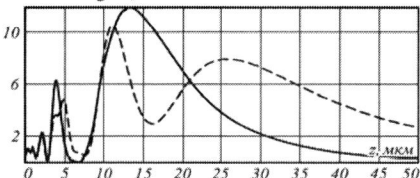

Fig. 1. Axial intensity distribution at $r_1 = 5\lambda$ for agreed with (2) of radius $R = 7,11\lambda$ (solid line) and a larger radius $R = 10\lambda$ (dashed line)

The results of numerical simulations show that increasing the maximum value is achieved when agreeing the inner and outer radii of the zones of the element (1).

The distance over which should form the greatest interference peak is equal to $z_{mx} = 24,75\lambda = 13,17$ μm. In accordance with the expression

$$I_{\text{lux}}(0,z) = \left(1 + \frac{2z}{\sqrt{z^2 + r_1^2}} + \frac{z}{\sqrt{z^2 + R^2}}\right)^2 \qquad (3)$$

and light transmittance, intensity $I_{mx} = 12,14$. These analytical estimates, as can be seen from Fig. 1, are very similar to the numerical results.

In the Tab. 1 shows the simulation results for the various parameters of the element (1). Focal spot size on the level of halftime maximum intensity in different directions denoted by $S(|)$ and $S(-)$. As follows from the results of simulation reduction of the inner zone radius reduces the size of the focal spot, which is accompanied by a loss of energy in the focus and offset to the plane of element.

Theoretically, reducing the size of the central zone has meaning only to $r_1 = \lambda/2$, when the maximum is formed at the boundary with the element $z_{mx} = 0$.

2. The numerical calculation for the volume element

In the case of subwavelength dimensions of the zones of relief optical element, the three-dimensional structure plays an important role, so in this section, calculations using a more accurate model.

Calculations for the axial volume element performed based on the finite-difference time method (FDTD), implemented in software product MEEP [5]. The radiation source is selected Gaussian beam with radius $\sigma = 0{,}6\lambda$ and a linear y-polarized. The radiation source located within the substrate at a distance of $0{,}1\ \lambda$ from the top plane of the substrate (see Fig. 2).

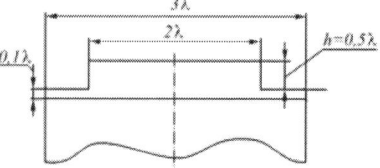

Fig.2. The calculation scheme of the volume microelement

In [6] were compared calculations using the integral method based on the expansion of plane waves and the FDTD method. Comparison shown qualitative agreement the results with quantitative differences, caused by real light optical characteristics such as the thickness of the substrate, material of the element, position and shape of the radiation source.

In the calculations we used the following simulation parameters: wavelength λ – $0{,}532\ \mu m$, the radius of the substrate – $1{,}5\lambda$. The refractive index of the element $n = 2$, the height of the relief, corresponding phase jump π radian $h = 0{,}5\lambda$.

The construction of computational grid: a spatial discretization is 100 counts per one micron, time discretization – according to the Courant condition. The absorbing layer around the perimeter of the estimated volume is $1{,}5\lambda$. The layers in the calculation are superimposed according to the model of infinite thickness of the substrate. The radius of the central zone microelement ranged from $0{,}1\lambda$ to λ. In this section, the outer radius was not chosen from the condition (2), and was fixed as radius of the substrate, to investigate the effect on the diffraction pattern of only one parameter - the size of the central zone.

Consider in more detail the effect size of the central zone when fixed incident beam and the outer radius of the optical element. Table. 2 show the results of this simulation.

The results qualitatively confirm the simulation in the previous section, but considered a different type of incident beam. Qualitative coincidence consists in the fact that by using a two-zone axial element can carry focus of the incident radiation, with the most tight focus (outside of the optical element) is achieved with a radius of the central zone $r_1 = \lambda/2$.

Note that we are considering a simple optical element provides a degree of focus is not worse than the diffractive axicon with a period close to the wavelength. In particular, in [7] was shown that 5-zone binary axicon with a numerical aperture $NA = 0{,}95$, provides the focusing of linearly polarized radiation in an elongated spot with a minimum size $S(-) = 0{,}44\lambda$. The radius of the central zone of the axicon was significantly subwavelength – $0{,}26\lambda$.

In [3] considered the zone plate with a focus equal to the length of the wave (in this case, the numerical aperture has a limiting importance for the free space). The central zone of the element $r_1 = 1{,}12\ \lambda$, but the rest 12 of the peripheral zone is a ring width of less than $\lambda/2$. The minimum spot size was $S(-) = 0{,}42\lambda$.

Dual-zone axial element with $r_1 = \lambda/2$ provides a minimum spot size $S(-) = 0{,}386\lambda$.

Reducing the radius of the central zone less critical (see lines in tab. 2 corresponding to $r_1 = 0{,}3\lambda$, $r_1 = 0{,}2\lambda$) leads to the formation of the focus inside the optical element. At the boundary between the glass and the air arise damped waves. The dimensions of the light spot just beyond relief are approximately the same as in the glass.

In this paper, analytically and numerically shown, that the optical microelement, consisting of just two coaxial annular zones, can be used for tight focusing of the laser radiation. And the greatest degree of focusing is achieved when the radius of the central zone $\lambda/2$.

Numerical calculations in the approximation of a thin element, using the expansion in plane waves, show qualitative agreement with the analytical estimates for the quantitative errors in the position of maximum intensity.

The application a more accurate method of calculation based on the finite-difference time-domain shows a good qualitative and quantitative agreement with the analytical estimates. Moreover, it was shown, that when the radius of the central zone of $0{,}1\lambda < r_1 < 0{,}5\lambda$, focus of optical element is formed inside and outside the border gets only the energy of evanescent waves.

It is shown that the linear polarization of the incident beam dual-zone axial element with a radius of the central zone $r_1 = \lambda/2$ provides a near element (at $z = 0{,}02\lambda$) focusing in the elongated light spot, the minimum amount of which is equal to the level of half-width at half maximum intensity $0{,}386\lambda$. Thus, the result is better than using binary axicon with radius of the central zone $0{,}26\lambda$.

REFERENCES

[1] P.-K. Wei, H.-L. Chou and Y.-C. Chen, "Subwavelength focusing in the near field in mesoscale air-dielectric structures", Opt. Lett., 2004. vol. 29. No 5. P. 433-435.

[2] V.V. Kotlyar, S.S. Stafeev, L. O'Faolain and V.A. Soifer, "Tight focusing with a binary microaxicon", Opt. Lett., 2011. vol. 36. No 16. P. 3100-3102.

[3] V.V. Kotlyar, A.A. Kovalev and S.S. Stafeev, "Intensity and power flow symmetry of subwavelength focal spot", Computer Optics, 2012. vol. 36. No 2. P. 190-198. [In Russian].

[4] S.N. Khonina, A.V. Ustinov, S.G. Volotovsky and A.A. Kovalev, "Calculation of diffraction of the linearly-polarized limited beam with uniform intensity on high-aperture binary micro-axicons in a near zone", Computer Optics, 2010. vol. 34. No 4. P. 443-460. [In Russian].

[5] A.F. Oskooi, D. Roundy, M. Ibanescu, P. Bermel, J.D. Joannopoulos and S.G. Johnson, "Meep: A

flexible free-software package for electromagnetic simulations by the FDTD method", Comp. Phys. Comm., 2010. vol. 181. P. 687-702.

[6] D.A. Savelyev, "Comparison of simulation diffraction linearly polarized Gaussian beam by a binary axicon with a high numeric aperture integral and difference method", News of Samara Scientific Center of RAS, 2012. vol. 14. No 4. P. 38-46. [In Russian]

[7] D.A. Savelyev, S.N. Khonina, "Maximising the longitudinal electric component at diffraction on a binary axicon linearly polarized radiation" Computer Optics, 2012. vol. 36. No 4. P. 511-517. [In Russian]

Table. 1. The intensity distribution in the approximation of a thin element (1) for a linear y-polarized

Parametres of element (1)	Longitudinal intensity in plane YZ $y\,z$ (size [$4\lambda\times10\lambda$])	The intensity in the plane of maximum $y\,x$ (size [$4\lambda\times4\lambda$])	The distribution parameters
$r_1 = \lambda$, $R = 1{,}58\lambda$			$z_{max} = 1{,}1\lambda, (z_{mx} = 0{,}75\lambda),$ $I_{max} = 3{,}5,$ $S(-) = 0{,}69\lambda$, $S(|) = 0{,}81\lambda$
$r_1 = \lambda/2$, $R = \lambda$			$z_{max} = 0{,}47\lambda, (z_{mx} = 0),$ $I_{max} = 0{,}48,$ $S(-) = 0{,}46\lambda$, $S(|) = 0{,}89\lambda$
$r_1 = \lambda/4$, $R = 0{,}79\lambda$			$z_{max} = 0{,}63\lambda, (z_{mx} = -0{,}19\lambda),$ $I_{max} = 0{,}72,$ $S(-) = 0{,}65\lambda$, $S(|) = 0{,}89\lambda$

Table. 2. Diffraction of Gaussian beam at the element with different size of the central zone (FDTD).

r_1	Distribution in the plane [$4{,}9\lambda \times 3{,}8\lambda$]		Transverse section [$3\lambda \times 3\lambda$]
	$x\,z$	$y\,z$	$y\,x$
$1{,}0\lambda$			$z_{max} = 0{,}669\lambda,$ $(z_{mx} = 0{,}75\lambda),$ S(−)=0,991λ, S(|)=1,065λ
$0{,}5\lambda$			$z_{max} = 0{,}021\lambda,$ $(z_{mx} = 0),$ S(−)=0,386λ, S(|)=0,802λ
$0{,}3\lambda$			$z_{max} = -0{,}021\lambda$ (inside the element) S(−)=0,336λ, S(|)=0,395λ
$0{,}2\lambda$			$z_{max} = -0{,}007\lambda$ (inside the element) S(−)=0,35λ, S(|)=0,492λ

Subwavelength elliptical focal spot generated by a binary zone plate

M. I. Kotlyar[1], S. S. Stafeev[1]

[1]Image Processing Systems Institute of the Russian Academy of Sciences, Samara, Russia

Abstract: Using a near-field scanning optical microscope (NSOM) with a small-aperture metal tip, we show that a glass zone plate having a focal length of one wavelength focuses a linearly polarized Gaussian beam into a weak ellipse with the Cartesian axis diameters $FWHM_x = (0.44 \pm 0.02)\lambda$ and $FWHM_y = (0.52 \pm 0.02)\lambda$ and the depth of focus $DOF = (0.75 \pm 0.02)\lambda$, where λ is the incident wavelength. The comparison of the experimental and simulation results suggests that NSOM with a hollow pyramidal aluminum-coated tip (with 70° apex and 100-nm-diameter aperture) measures the transverse intensity, rather than the power flux or the total intensity. The conclusion that the small-aperture metal tip measures the transverse intensity can be inferred from the Bethe-Bouwkamp theory.

I. INTRODUCTION

In recent years, significant advances have been reported in the subwavelength focusing. For example subwavelength focus can be achieved using planar plasmonic structures [1], plasmonic lenses [2], or conventional optical elements like a microaxicon [3], a zone plate [4], a microlens [5], or a solid immersion lens (SIL) [6]

Note, however, that the above-cited articles have not been concerned with processes whereby the near-field radiation is registered using a NSOM small-aperture tip. The following questions have remained unanswered so far. Which radiation component – the power density or the power flux – is being registered by the near-field microscope?

Using a NSOM with a low NA metal tip, we experimentally show that a glass binary zone plate with a wavelength focal length focuses the linearly polarized Gaussian beam into a weakly elliptical focal spot with the Cartesian axis diameters $FWHM_x = (0.44 \pm 0.02)\lambda$ and $FWHM_y = (0.52 \pm 0.02)\lambda$ and the depth of focus $DOF = (0.75 \pm 0.02)\lambda$, where λ is the incident wavelength. The comparison of the experimental results with the FDTD-based numerical simulation suggests an unambiguous conclusion that the NSOM tip measures the transverse intensity (power density) rather than the power flux or the total intensity. The fact that the low-NA metal tip measures the transverse intensity is in compliance with the Bethe-Bouwkamp theory.

II. MANUFACTURING AND EXPERIMENT

The high-quality ZP was fabricated by first spinning a glass substrate with hydrogen silsesquioxane (HSQ), a spin-on dielectric, which was then hardbaked at 400°C. The pattern was defined in ZEP520A resist using electron beam lithography and then transferred into HSQ hardbaked hydrogen silsesquioxane using Reactive Ion Etching. The etch rates of HSQ and the glass substrate differ giving good depth control.

Fig. 1 shows AFM (Fig. 1a) and SEM (Fig. 1b) images of the ZP: groove depth – 510 nm, diameter – 14 μm, and the peripheral zone – $0.5\lambda = 266$ nm. The ZP has 12 rings and a central disk. The numerical aperture is equal NA = 0.997.

The ZP radii were derived from the well-known formula $r_m = (m\lambda f + m^2\lambda^2/4)^{1/2}$, where $f = \lambda = 532$ nm is the ZP focal length and m is the radius number.

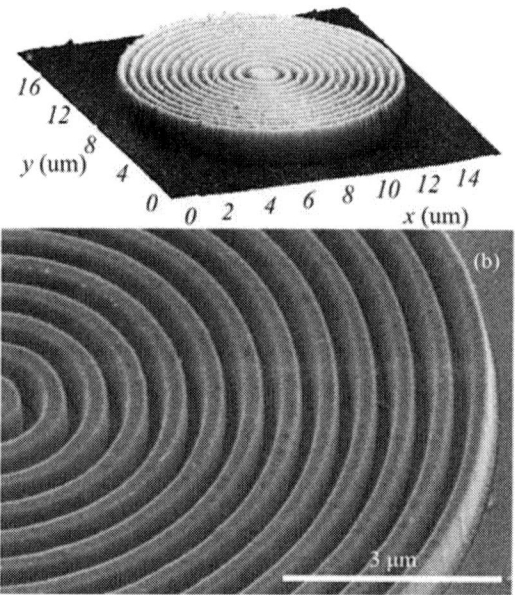

Fig. 1. AFM (a) and SEM (b) images of the ZP.

The propagation of a linearly polarized Gaussian beam of wavelength $\lambda = 532$ nm through the ZP of focus $f = \lambda$ was experimentally studied using NSOM Ntegra Spectra (NT-MDT). The NSOM arrangement is shown in Fig. 2. A linearly polarized light beam from a 532-nm laser L was focused with a lens L1 onto the substrate bottom. Following the diffraction by the ZP, the transverse intensity distributions in the planes parallel to the ZP were measured at different distances (at ~100-nm intervals) using a hollow pyramidal aluminum-coated tip C with 70° apex and 100-nm-diameter aperture (Fig. 3). The tip-coupled portion of light was then focused with lens L2 and transmitted through spectrometer S to filter out the irrelevant radiation, before being registered by the CCD-camera. Fig. 4

shows an example of the focal spot intensity pattern obtained directly on the microscope.

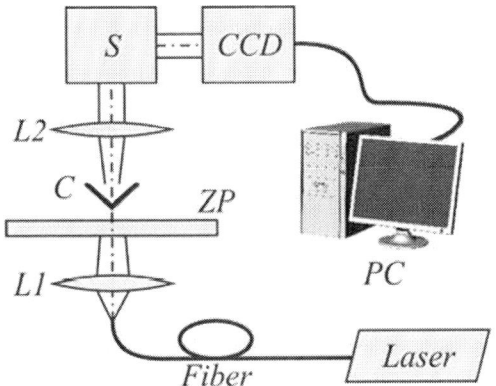

Fig. 2. The NSOM-aided experimental arrangement.

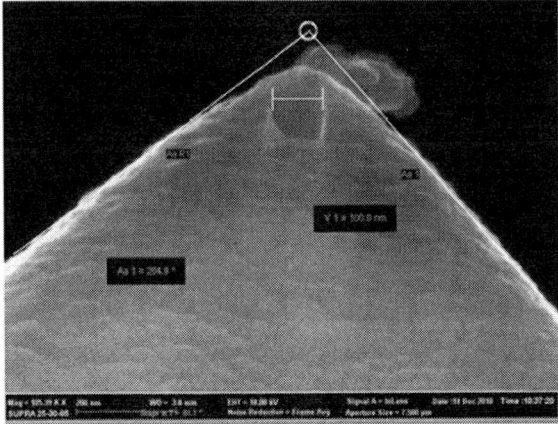

Fig. 3. SEM image of the hollow pyramid-shaped metallic cantilever tip with a 100-nm aperture and 70° tip apex of the NSOM.

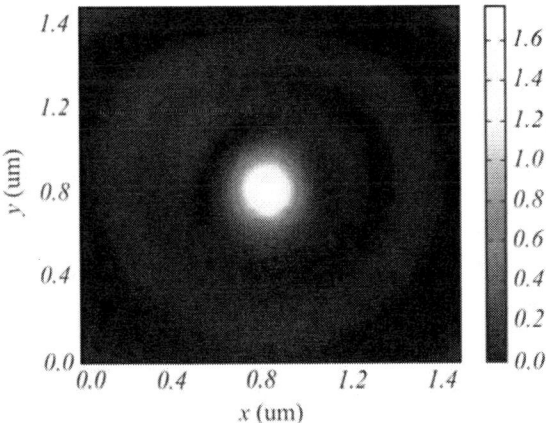

Fig. 4. The focal spot cross-section at the focal length $f = \lambda = 532$ nm (the vertical axis is in the polarization plane).

III. COMPARISON WITH NUMERICAL MODELING

The simulation was performed using the BOR-FDTD method, the grid quantization was $\lambda/50$ in space and $\lambda/100c$ in time, where c is the speed of light in free space.

Fig. 5 depicts the profiles on the x-axis of simulated intensity (curve 1) and power flux (curve 3), and the experimentally measured distribution obtained on the NSOM (curve 2). Fig. 6 shows the focal spot profiles on the y-axis, which is parallel to the polarization plane, namely, the calculated distribution of power flux on the z-axis (curve 3), the experimental intensity distribution (curve 2), and the calculated intensity distribution (curve 1) taken as a superposition of (Fig. 6a) all components and (Fig. 6b) only transverse components.

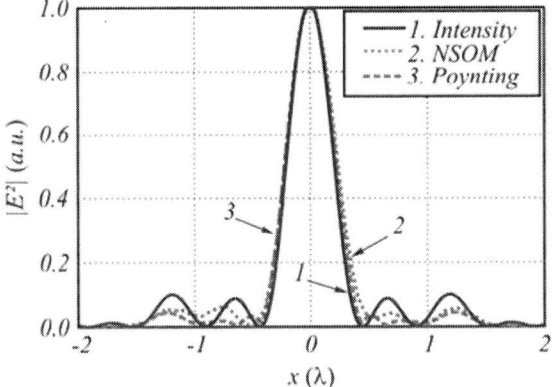

Fig. 5. Comparison of the experimental and calculated distribution in the focal spot on the x-axis: calculated intensity profile (curve 1), experimental intensity profile (curve 2), and the calculated distribution of the Poynting vector's absolute value onto the z-axis (curve 3).

Fig. 6a suggests that the longitudinal intensity component is not measured in the course of the experiment, because the total intensity peak (FWHM = 0.84λ) is wider than the experimental peak (FWHM = 0.52λ) by a value larger than the measurement error $\pm 0.02\lambda$. In the meantime, the experimental peak is wider than the calculated peak for the power flux (FWHM = 0.45λ) by a value larger than the measurement error, thus posing the question – what exactly is measured in the experiment? In Fig. 6b, the comparison of the experimental intensity (curve 2) and the transverse intensity $|E_x|^2 + |E_y|^2$ (curve 1) shows the peak widths to be the same: FWHM = 0.52λ. Therefore, we can unambiguously inferr from Fig. 6 that the NSOM with a hollow metal pyramidal tip having a 100-nm aperture and 70° apex (Fig. 3) measures the energy density in the form of transverse intensity $|E_x|^2 + |E_y|^2$ rather than the power flux or the total intensity $|E_x|^2 + |E_y|^2 + |E_z|^2$.

The propagation of the electromagnetic field through the small hole in a metal screen is described within the Bethe-Bouwkamp theory [7]. The theory states that a linearly polarized plane tilted wave incident onto the metal screen with a small hole of diameter $a \ll \lambda$ induces an electric dipole found in the perpendicular plane to the slit and a magnetic dipole found in the hole plane. Therefore, when a tilted plane wave **E** is incident on the small hole, the far-field relationship is described by the electric dipole **P** and magnetic dipole **M**:

$$\mathbf{P} = -\frac{4}{3}\varepsilon_0 a^3 \left(\mathbf{E}\mathbf{n}_z \right)\mathbf{n}_z,$$
$$\mathbf{M} = -\frac{8}{3}a^3 \left[\mathbf{n}_z \times \left[\mathbf{E} \times \mathbf{n}_z \right] \right], \tag{1}$$

where $\mathbf{n}_z$ is the unit vector of the optical axis that is perpendicular to the hole plane.

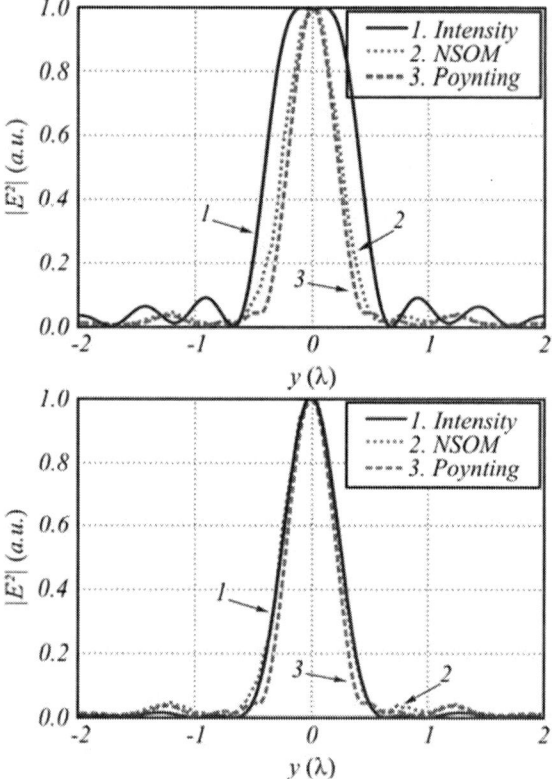

Fig. 6. Comparison of the experimental and calculated distribution in the focal spot on the y-axis, which is parallel with the polarization plane: calculated distribution of Poynting's vector absolute value onto the z-axis (curve 3), the experimental intensity distribution (curve 2), and the calculated intensity distribution (curve 1) taken as a superposition of (a) all field components and (b) only transverse field components.

From Eq. (1), the electric dipole is seen to be formed only by the longitudinal component of the electric field **E**. Note, however, that a dipole oriented along the optical axis radiates in the transverse direction, not radiating along the optical axis itself. On the contrary, the magnetic dipole in Eq. (1) is formed only by the transverse electric field components, because the internal vector product on the right-hand side of Eq. (1) equals zero for the longitudinal electric field component. Thus, the longitudinal electric field component is not registered by a photoreceiver put on the optical axis at a distance from the small hole in the metal screen.

IV. CONCLUSION

The experiment conducted using the NSOM with a hollow, aluminum-coated pyramidal tip with a 100-nm-aperture and a 70° tip apex has shown that when illuminated by a linearly polarized Gaussian beam, the binary zone plate of one-wavelength focal length generates a focal spot in the form of a weak ellipse with the Cartesian axis diameters $FWHM_x = (0.44 \pm 0.02)\lambda$ and $FWHM_y = (0.52 \pm 0.02)\lambda$ and the DOF = $(0.75 \pm 0.02)\lambda$. The comparison of the experimental and simulation results suggests that NSOM measures the energy density in the form of transverse intensity, rather than the power flux or the total intensity. The conclusion that the metal tip with a small aperture measures the transverse intensity $|E_x|^2 + |E_y|^2$ can be inferred from the Bethe-Bouwkamp theory.

ACKNOWLEDGMENT

The work was financially supported by the Russian Federal program "Research and Academic Cadres of Innovative Russia" (state contract # 14.740.11.0016), Russian Federation Presidential grants for support of the leading scientific schools (NSh-4128.2012.9), the Young Candidate of Science grant (MK-3912.2012.2), and the Russian Foundation Basic Research grants (12-07-00269, 12-07-31115, 12-07-31117, 13-07-97008).

REFERENCES

[1] K.R. Chen, W.H. Chu, H.C. Fang, C.P. Liu, C.H. Huang, H.C. Chui, C.H. Chuang, Y.L. Lo, C.Y. Lin, H.H. Hwung, A.Y.-G. Fuh, "Beyond-limit light focusing in the intermediate zone," *Opt. Lett.*, vol. 36, pp. 4497-4499, 2011.

[2] Y. Yu, H. Zappe, "Effect of lens size on the focusing performance of plasmonic lenses and suggestions for the design," *Opt. Express*, vol. 19, pp. 9434-9444, 2011.

[3] V.V. Kotlyar, S.S. Stafeev, L. O'Faolain, V.A. Soifer, "Tight focusing with a binary microaxicon," *Opt. Lett.*, vol. 36, pp. 3100-3102, 2011.

[4] R.G. Mote, S.F. Yu, A. Kumar, W. Zhou, X.F. Li, "Experimental demonstration of near-field focusing of a phase micro-Fresnel zone plate (FZP) under linearly polarized illumination," *Appl. Phys. B*, vol. 102, pp. 95-100, 2011.

[5] J.-S. Ye, G.-A. Mei, X.-H. Zheng, Y. Zhang, "Long-focal-depth cylindrical microlens with flat axial intensity distributions," *J. Mod. Opt.*, vol. 59, pp. 90-94, 2012.

[6] K. Huang, Y. Li, "Realization of a subwavelength focused spot without a longitudinal field component in a solid immersion lens-based system," *Opt. Lett.*, vol. 36, pp. 3536-3538, 2011.

[7] L. Novotny, B. Hecht, *Principles of Nano-optics*, Cambridge University Press, 2006

Tight light localization in a hyperbolic secant planar slit lens

A.G. Nalimov, V.V. Kotlyar

Image Processing Systems Institute of the RAS, Samara, Russia

Abstract: We show by FDTD simulation that a photonic-crystal hyperbolic secant binary microlens in silicon (with the refractive index n=3.47) of size 2 μm x 5 μm with a slit 50 nm wide and 250 nm long can focus a plane TM-wave into a near-surface focal spot of size λ/23 with a 44% diffraction efficiency. The focus intensity is 60 times higher than the incident wave.

The tight focusing of light near the interface surface has been performed with the aid of nanoslits [1, 2]. In Ref. [1] a focal spot of size ~ λ/2 was generated using a "subwavelength generator" consisting of two 80-nm slits. Lenses with nanoscale slits made in gold were reported in Ref. [2] in which the incident light having passed through 20-nm slits was focused into a focal spot of about half-wavelength in width. In Ref. [3], light confinement in metamaterial was discussed, with the waves passing through two or more slits. Having passed through 20-nm slits in a screen, the light waves underwent diffraction in the metamaterial, forming a focal spot of the full width at half-maximum intensity FWHM = λ/17. However, the authors have failed to explain in which way, being so sharp, the focal spot could be output from the metamaterial. A nanoslit tens of nanometers wide can be used for confining and guiding light waves, similarly to a waveguide [4]. The nanoslits have been used in this way in near-field microscopy [5]. The light scattered by the surface of the specimen under study passes through a slit in the cantilever and arrives to a high NA objective lens.

On the other hand, it is possible to perform the tight focusing of light using graded-index microlenses, in particular, planar photonic crystal lenses [5, 6]. In Ref. [7] the focusing into a spot of size about λ/4 was performed with the aid of an array of holes in a taper. The diffraction of light from a nanoslit in metamaterial was analyzed in Refs. [8, 9]. Obtaining of focal spots of size FWHM=λ/10 [8] and FWHM= λ/5 [9] was reported. The lens utilized in [8, 9] contained dozens of 5-nm thick silver layers on top of 10-nm thick SiC. Note, however, that such a lens is difficult to realize in practice.

In this work, we combine the advantages of using a several-dozen nanometers wide slit for light confinement [4] and a gradient lens for the sharp focusing of light [10]. The FDTD simulation has shown that a planar binary microlens in silicon with a 50-nm wide slit can generate a near-surface focal spot of size FWHM=λ/23 with a 44% energy efficiency. This value is smaller than those reported in the above-quoted works. With the focus occurring near the lens surface, it may find use in various nanophotonics applications. The nanoslit placed near

the lens output surface in the focus region serves the dual function

Figure 1 depicts a scheme of a planar graded-index slit lens. We consider a hyperbolic secant (HS) lens whose refractive index depends on the transverse coordinate x as follows [10]:

$n(x) = n_0 ch^{-1}[\pi x/(2H)]$, where n_0 is the refractive index on the axis and H is the lens length.

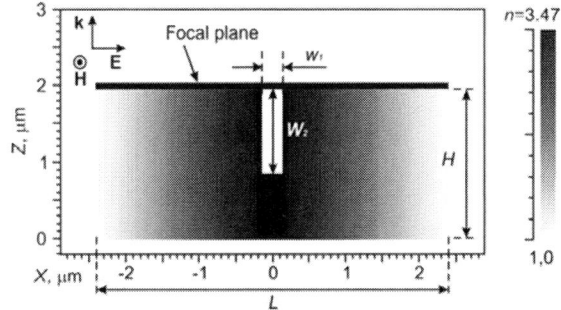

Fig. 1. (Gray-level) refractive index distribution of a HS slit lens: **k** is the wave vector, **E** and **H** are the electric and magnetic field strength.

The slit in the planar HS lens of width W_1 is located on the optical axis, stretching as far as the lens output plane. The slit can be extended through the entire lens ($W_2 = H$) or found in the lens' rear section ($W_2 < H$). The light was propagated through the lens by the FDTD simulation implemented using the commercial software FullWave (by RSoft) and the software MEEP. The simulation parameters were as follows: the computation domain- 8 μm x 4 μm, incidence wavelength - 1.55 μm, computation domain step on the X- and Y-axes- λ/500.

From Fig. 2a the focal spot width FWHM is seen to be a linear function of the slit width W_1. The focal spot is slightly wider than the slit. Figure 2b shows the intensity profile in the lens focus at W_1 = 50 nm.

The light confinement in the nanoslit can be explained in a similar way to generating a fundamental TM-mode in a planar slitted waveguide [4]. In this case, the field within the slit is described by the relation

$$E_x(x,z) = \exp(i\beta x) ch\left[\left(\beta^2 - k^2\right)^{1/2}|x|\right], |x| < a,$$

where a is the nanoslit boundary coordinate, $2a=W_1$, β is the propagation constant of the fundamental TM-mode, and k is the wavenumber in vacuum.

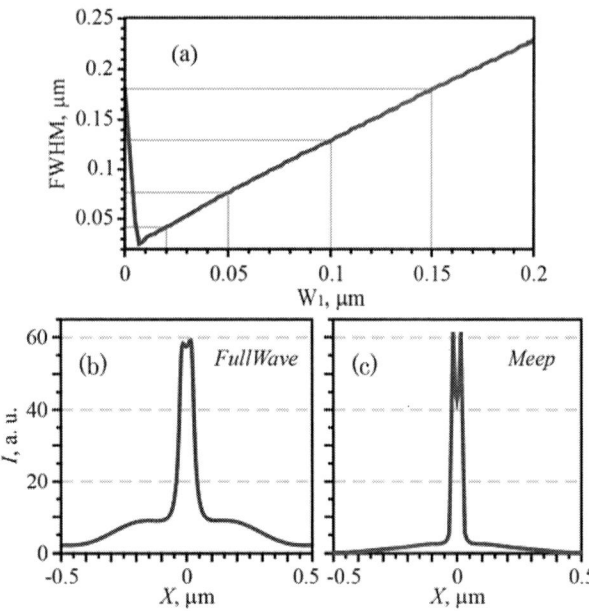

Fig. 2. The focal spot width FWHM as a function of the slit width W_1 (a) and the intensity profile $I = |E_x|^2 + |E_z|^2$ in the focal plane (10 nm after the lens) at W_1 =50 nm, simulated using the FullWave (b) and MEEP software (c).

Figure 2b shows that there is an intensity dip in the focal spot center defined by a hyperbolic cosine. The field amplitude $E_x(x,z)$ undergoes a breakdown at the slit boundary $|x| = a$, so that near the slit at $|x| > a$ the amplitude is n^2 times decreased. Considering that Fig. 2b shows the intensity profile calculated at distance 10 nm after the lens, the focal spot intensity peak is wider if compared to that produced by the lens. Besides, the intensity peak in the focus center is smaller compared to that observed in the slit.

The diffraction efficiency (DE) of focusing η_D depends on the slit width W_1. The DE is maximal at $W_1 \approx 40$ nm. With increasing slit width, the DE η_D is decreasing (Fig. 3a) because the light intensity in the focus is decreasing with increasing W_1 (Fig. 3b). The DE η_D was calculated as the ratio of the energy contained in the central lobe of the focal diffraction pattern (on the interval -75 nm $< x <$ 75 nm) to the entire energy reaching the output plane of width L. It can also be seen from Fig. 3a that for $W_1 <$ 5 nm the focal spot size starts to grow. The minimal focal spot size FWHM is achieved at W_1 =5 nm, being equal to FWHM = λ/119 at distance 10 nm

after the lens. For comparison, the focal spot without the slit has FWHM=λ/8 and the DE η_D=60%.

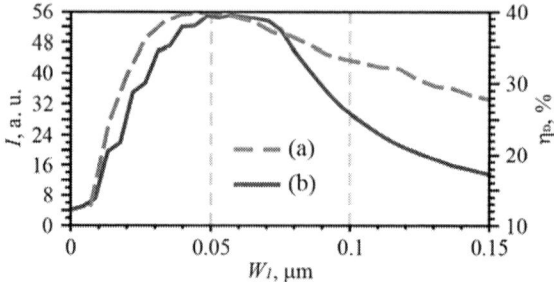

Fig. 3. (a) The DE η_D and (b) intensity at the lens focus (10 nm apart from the lens) as a function of the slit width W_1, at $W_2 = H$.

From Fig. 3 it is seen that the maximal DE is η_D=39.9%, whereas the focal spot size is FWHM = λ/28 (Fig. 2a). In Ref. [4] it was demonstrated that in the silicon waveguide less than 30% of the waveguide mode optical energy was confined in nanoslits of an arbitrary size (ranging from 10 nm to 150 nm).

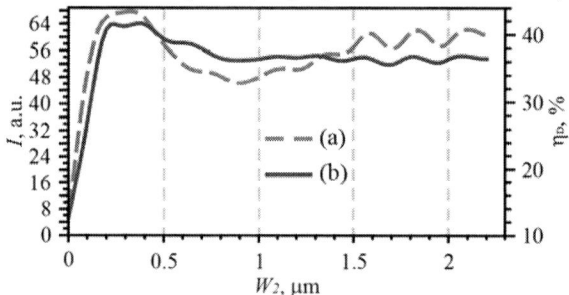

Fig. 4. (a) The DE η_D and (b) intensity I in the lens focus as a function of the slit length W_2, at W_1 =50 nm.

Figure 4 depicts the DE η_D (a) and the focal intensity I (b) as a function of the slit length W_2, assuming the slit width W_1 =50 nm. In calculating η_D, the focal spot size was again considered up to the nearest side-lobes, -75 nm $< x <$ 75 nm. From Fig. 4 it is seen that the maximal values of the DE η_D=43.4% and intensity I are observed when the slit length is chosen so as to provide a λ/2 phase delay, i.e. at $W_2 = \lambda \left[2 (n_0 - 1) \right]^{-1} = 0{,}314$ µm. We note that the optical intensity in the lens focus is about 20% larger than for W_2 =H.

Considering that the fabrication of a graded-index lens using the state-of-the-art nanolithography techniques presents a challenge, the above-described sharp focusing of the TM-wave can be implemented using a photonic crystal lens with a slit on the optical axis, which would be analogous to the graded-index lens in terms of the average refractive index distribution. Figure 5a depicts the refractive index distribution in the XZ-plane of a photonic crystal lens similar to the graded-index lens of Fig. 1.

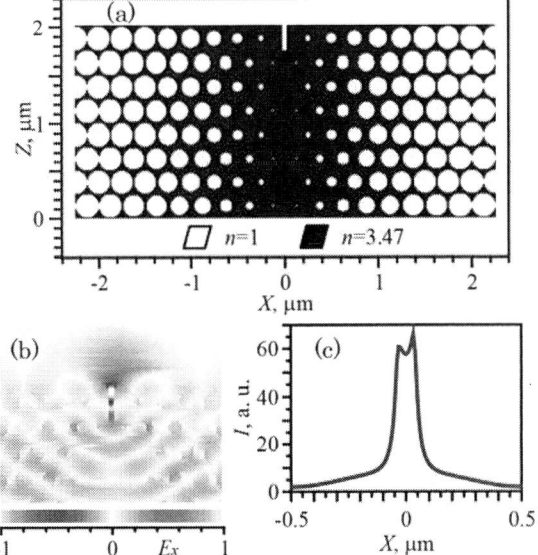

Fig. 5. (a) Refractive index distribution in a photonic crystal lens with a slit, (b) the instantaneous distribution of the field E_x in such a lens at instance cT=32 µm and (c) intensity profile in the transverse focal plane 10 nm after the lens.

The lens in Fig. 5a consists of 8 rows of holes on the Z-axis and 20 rows of holes on the X-axis arranged in a staggered order. The holes' centers are arranged with a period of 266 nm on the X-axis and 250 nm on the Y-axis. The minimal hole diameter is 30 nm, maximal is 250 nm, the lens length is 2 µm, width is 4.8 µm, the lens material refractive index is n=3.47, W_1=50 nm, and W_2=0.25 µm. For such a lens, the value of η_D depends on the slit length W_2 at W_1=50 nm in a similar way to the graded-index lens, as seen from Fig. 5b. The maximal DE is η_D=44.3% at W_2=250 nm. In this case, the focal spot size is FWHM=λ/23, with the focal spot intensity being 60 times higher than the incident wave intensity. If such a lens is illuminated by a Gaussian beam with a waist radius of σ=2.4 µm, the focal spot shape is preserved but the focal spot intensity is 1.7-times decreased.

Summing up, using the 2D FDTD simulation, we have shown that if a planar graded-index hyperbolic secant microlens of size 2 µm x 5 µm in silicon that contains on the optical axis a nanoslit of width 50 nm and length ~300nm is illuminated by a plane TM-wave, a focal spot of width FWHM = λ/28c is generated at the lens output with the DE η_D=43%.

The work was partially funded by the RF Ministry of Education and Science under Federal program "Research and Academic Cadres of Innovative Russia" (Agreement # 8027), RF President's grants for Support of Leading Scientific Schools (NSh-4128.2012.9), a Young Candidate of Science grant (MK-3912.2012.2), and RFBR grants (12-07-00269, 12-07-31117, 13-07-97008).

REFERENCES

[1] Chen, K.R. Focusing of light beyond the diffraction limit of half the wavelength / K.R. Chen // Opt. Letters, 2010. - V. 35, No. 22. - P. 3763-3765.

[2] Ishii, S. Gold Nanoslit Lenses / S. Ishii, A.V. Kildishev, V.M. Shalaev, K.-P. Chen, V.P. Drachev // Proceedings of Quantum Electronics and Laser Science Conference, CLEO'2011, Baltimore, Maryland.

[3] Ren, G. Off-axis characteristic of subwavelength focusing in anisotropic metamaterials / G. Ren, C. Wang, Z. Zhao, X. Tao, X. Luo // J. Opt. Soc. Am. B, 2012. - V. 29, No. 11. - P. 3103-3108.

[4] Almeida, V.R. Guiding and confining light in void nanostructure / V.R. Almeida, Q. Xu; C.A. Barrios, M. Lipson // Opt. Letters. – 2004. – Vol. 29, No. 11. – P. 1209-1211.

[5] Chien, H.T. Focusing of electromagnetic waves by periodic arrays of air holes with gradually varying radii / H.T. Chien and C.C. Chen // Opt. Exp. - 2006. - V. 14. – P. 10759.

[6] Kurt, H. Graded index photonic crystals / H. Kurt, D.S. Citrin / Optics Express, 2007. – V. 15. – P.1240-1252.

[7] Cheng, Z. Focusing subwavelength grating coupler for mid-infrared suspended membrane waveguide / Z. Cheng, X. Chen, C.Y. Wong, K. Xu, C.K.Y. Fung, Y.M. Chen, H.K. Tsang // Opt. Letters, 2012. - V. 37, N.7. - P. 1217-1219.

[8] Ren, G. Subwavelength focusing of light in the planar anisotropic metamaterials with zone plates / G. Ren, Z. Lai, C. Wang, Q. Feng, L. Liu, K. Liu, X. Luo // Opt. Express, 2010. - V. 18, No. 17. - P. 18151-18157.

[9] Li, G. Subwavelength focusing using a hyperbolic medium with a single slit / G. Li, J. Li, K.W. Cheah // Appl. Optics, 2011. - V. 50, N. 31. - P. G27-G30.

[10] Kotlyar, V.V. High resolution through graded-index microoptics / V.V. Kotlyar, A.A. Kovalev, A.G. Nalimov, S.S. Stafeev // Advances in Optical Technologies, 2012. - V. 2012. - P. 1-9.

A solution of the wave equation for planar gradient waveguides in a frequency domain

V.M. Fitio[1], V.V. Romakh[1], Y.V. Bobitski[1,2]

[1]Department of Photonics, Lviv Polytechnic National University, Lviv, Ukraine

[2]Institute of Technology, University of Rzeszów, Rzeszów, Poland

Abstract- **Possibilities of a numerical method for solving the wave equation of gradient planar waveguides are studied. The method is based on both the Fourier transform application and the wave equation solution in a frequency domain. Finally, a task to find propagation constants and field Fourier transforms in a discrete form is led to the eigenvalue/eigenvector problem. The new method provides high accuracy subject to the conditions of the Whittaker-Shannon sampling theorem, and it is characterized by high numerical stability.**

I. INTRODUCTION

To define propagation constants of localized modes of gradient planar waveguides, a number of approximate methods are used [1], which for the first time have been developed for the analysis of quantum mechanics problems. In Fig. 1, a planar symmetric waveguide is shown where the permittivity of layers takes certain values.

If, in this waveguide $\varepsilon_1 > \varepsilon_o$, propagation of the waveguide mode with the propagation constant β is possible. But even in this simplest case a search for the propagation constant is led to the transcendental algebraic equation solution. A problem is significantly complicated if the permittivity is changed by a complex function along the axis x. Moreover,

$$\varepsilon(x) = \varepsilon(-x). \qquad (1)$$

To solve the wave equation for gradient waveguides, the efficient numerical method in a frequency domain is proposed, which is characterized by a high accuracy of the propagation constant search [2]. A current state of computer technology and sophistication of standard software allows applying numerical methods to the analysis of gradient planar waveguides.

Therefore, the aim of this work is to study a possibility of the new numerical method. For simplicity in this study only symmetric waveguides are given while the waveguide TE polarization mode is propagating.

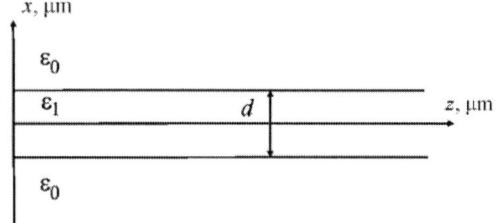

Fig. 1. An image of a planar symmetric waveguide with two fixed values of the permittivity.

II. ONE-DIMENSIONAL WAVE EQUATIONS AND THEIR FOURIER TRANSFORMS

If, in the waveguide mode, an electric field is perpendicular to the plane xz (TE polarization), the appropriate wave equation will look like:

$$\frac{d^2 E(x)}{dx^2} + \left(\frac{2\pi}{\lambda}\right)^2 \varepsilon(x) E(x) = \beta^2 E(x). \qquad (2)$$

If, in a waveguide, the TM polarization wave is propagating, the appropriate wave equation with regard to a magnetic field can be written as:

$$\frac{d^2 H(x)}{dx^2} - \frac{\frac{d\varepsilon(x)}{dx}}{\varepsilon(x)}\frac{dH}{dx} + \left(\frac{2\pi}{\lambda}\right)^2 \varepsilon(x) H(x) = \beta^2 H(x). \qquad (3)$$

Functions $E(x)$, $H(x)$ that describe fields in waveguide localized modes and their first derivatives tend towards zero if $x \rightarrow \pm\infty$. That's why for these functions, their first and second derivatives the Fourier transform exists. One can write the appropriate equations for $E(x)$. So Fourier transforms for $E(x)$, first and second derivatives of $E(x)$ are:

$$E(u) = \int_{-\infty}^{\infty} E(x)\exp(-i2\pi ux)\,dx, \qquad (4)$$

$$i2\pi u E(u) = \int_{-\infty}^{\infty} \frac{dE(x)}{dx}\exp(-i2\pi ux)\,dx, \qquad (5)$$

$$-(2\pi u)^2 E(u) = \int_{-\infty}^{\infty} \frac{d^2 E(x)}{dx^2}\exp(-i2\pi ux)\,dx. \qquad (6)$$

Besides, for functions for which Fourier transforms exist, i.e., $F\{G(x)\} = G(u)$, $F\{H(x)\} = H(u)$, the next equation is yet right:

$$F\{G(x)H(x)\} = \int_{-\infty}^{\infty} G(u-v)H(v)\,dv, \qquad (7)$$

where $F\{...\}$ is the Fourier transform. Equation (7) is named the convolution theorem.

Taking Fourier transforms of left and right parts of (2, 3), as a result we obtain:

$$-4\pi^2 u^2 E(u) + \left(\frac{2\pi}{\lambda}\right)^2 \int_{-\infty}^{\infty} \varepsilon(u-v)E(v)\,dv = \beta^2 E(u), \qquad (8)$$

$$-4\pi^2 u^2 H(u) - 2i\pi u \int_{-\infty}^{\infty} F\left\{\frac{d\ln\varepsilon(x)}{dx}\right\}\bigg|_{u \to u-v} vH(v)\,dv +$$

$$+\left(\frac{2\pi}{\lambda}\right)^2 \int_{-\infty}^{\infty} \varepsilon(u-v)H(v)dv = \beta^2 H(u), \qquad (9)$$

where $\varepsilon(u) = \int_{-\infty}^{\infty} \varepsilon(x)\exp(-i2\pi ux)dx$.

So we have moved from the differential equations (2, 3) on eigenfunctions and eigenvalues to the integral ones (8, 9). In these last equations we can replace the integral on the sum. For example, if we take (8), resulting in the replacement of continuous values u and v on discrete ones we obtain:

$$-4\pi^2\left(u_s\right)^2 E(u_s) + \left(\frac{2\pi}{\lambda}\right)^2 \sum_{k=-(N-1)/2}^{(N-1)/2} \varepsilon(u_s-v_k)E(u_v)\Delta = \beta^2 E(u_s),$$

where $\Delta = u_{max}/N, u_s = s\Delta, v_k = k\Delta,$
$$-(N-1)/2 \le s,k \le (N-1)/2.$$

N should be taken large enough and unpaired.

One can write the last equation for a set of discrete frequencies $u_s = s\Delta$. Moreover, s is changing from $-(N-1)/2$ to $(N-1)/2$. Then a set of these equations will be written as the matrix equation while β^2 is common to all s:

$$(\mathbf{P}+\mathbf{U})\mathbf{E} = \beta^2\mathbf{E},$$

where $\mathbf{P}$ – diagonal matrix with elements $-4(\pi s\Delta)^2$, $\mathbf{U}$ – square symmetric matrix with elements $\left(\frac{2\pi}{\lambda}\right)^2 \varepsilon(s\Delta-k\Delta)\Delta$,

$\mathbf{E}$ – vector-column with elements $E(s\Delta)$.

So, the problem was led to the eigenvalue (the square propagation constant) problem and the eigenvector (the discrete Fourier transform $E(x)$) problem which corresponds to the preset value β. We can have few eigenvalues and appropriate eigenvectors which are orthogonal. By carrying out the inverse discrete Fourier transform of the eigenvector we obtain the field distribution $E(x)$.

III. THE EXAMPLES OF NUMERICAL MODELING

Practical use of the new method will be demonstrated on three planar waveguides whose permittivity is described by the following expressions:

$$\varepsilon(x) = \varepsilon_0 + (\varepsilon_1 - \varepsilon_0)\mathrm{rect}\left(\frac{x}{d}\right), \qquad (10)$$

where $\mathrm{rect}(x/d) = \begin{cases} 1, & if\ |x| \le d/2, \\ 0, & if\ |x| > d/2. \end{cases}$

Two next waveguides are gradient:

$$\varepsilon(x) = \varepsilon_0 + (\varepsilon_1 - \varepsilon_0)\left[1 - \left(\frac{2x}{d}\right)^2\right], \qquad (11)$$

$$\varepsilon(x) = \varepsilon_0 + (\varepsilon_1 - \varepsilon_0)\exp\left[-\pi\left(\frac{2x}{d}\right)^2\right]. \qquad (12)$$

The numerical analysis for waveguides with parameters $\varepsilon_0 = 2.25$, $\varepsilon_1 = 2.89$, $d = 20\,\mu m$, $\lambda = 1\,\mu m$ was carried out at $N = 3001$. Herewith the dependences of propagation constants on u_{max} were removed to identify areas of change of u_{max} in

which waveguide propagation constants were practically not changed.

For the waveguide described by (10) for one calculating cycle 32 propagation constants were found, among which $\beta_0 = 10.68030459\,\mu m^{-1}$, and $\beta_{31} = 9.49500057\,\mu m^{-1}$. The same propagation constants found by solving the transcendental equation are: $\beta_0 = 10.680304594\,\mu m^{-1}$, $\beta_{31} = 9.495000602\,\mu m^{-1}$. The value $\beta_0 = 10.68030459\,\mu m^{-1}$ found by the method proposed is remained constant in the frequency range from $u_{max} = 15\,\mu m^{-1}$ to $u_{max} = 135\,\mu m^{-1}$, and $\beta_{31} = 9.49500057\,\mu m^{-1}$ is constant in the frequency range from $u_{max} = 80\,\mu m^{-1}$ to $u_{max} = 90\,\mu m^{-1}$. It follows that β_{31} was received with a greater mistake than β_0 but with a very high accuracy.

For the waveguide with the permittivity described by (11) 25 propagation constants were found. Moreover, $\beta_0 = 10.65785964\,\mu m^{-1}$ is constant at the frequency change from $u_{max} = 5\,\mu m^{-1}$ to $u_{max} = 150\,\mu m^{-1}$, and $\beta_{24} = 9.46280873\,\mu m^{-1}$ is constant for the frequency range from $u_{max} = 10\,\mu m^{-1}$ to $u_{max} = 80\,\mu m^{-1}$.

For the waveguide with the permittivity described by (12) 23 propagation constants were found. In particular, $\beta_0 = 10.64018290\,\mu m^{-1}$ is constant at the frequency change from $u_{max} = 1.5\,\mu m^{-1}$ to $u_{max} = 135\,\mu m^{-1}$, and $\beta_{22} = 9.42517619\,\mu m^{-1}$ is constant for the frequency range from $u_{max} = 2.3\,\mu m^{-1}$ to $u_{max} = 18\,\mu m^{-1}$.

Fig. 2 shows the dependency of propagation constants on the mode number for three waveguides. For the waveguides described by (10, 11) the approximating curves (linear and quadratic) are well consistent with the dependency of β_m on the number M. This result is similar to such problems of quantum mechanics as a rectangular well with infinity depth and a harmonic oscillator. Suitable discrete energy levels are imposed exactly on the quadratic and linear dependencies.

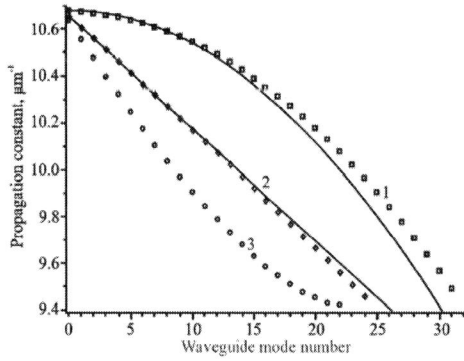

Fig. 2. The dependency of propagation constants on the waveguide mode number M: 1 – the dependency of $\varepsilon(x)$ according to (10), 2 – the dependency of $\varepsilon(x)$ according to (11), 3 – the dependency of $\varepsilon(x)$ according to (12). Continuous curves are approximating ones: 1 – the parabolic dependency, 2 – the linear dependency.

In Fig. 3, the field distributions of $E(x)$ for the modes with numbers 0, 1 and 22 are shown. It is seen that modes with numbers 0 and 1 occupy much less space in length than the mode with the number 22. For the last mode, the propagation constant is very close to the value $2\pi n_0 / \lambda = 9.42477960 \ \mu m^{-1}$, that is why $E(x)$ of this mode occupies much greater space than 50 μm in the coordinate domain according to Fig. 3.

IV. Conslusions

For the waveguides given above, we obtain sufficiently accurate values of propagation constants if u_{max} is on a certain range, and this range is decreasing while the propagation mode number is increasing. Also, this range depends on the waveguide type, i.e., on the second term of (10, 11, 12). It should also be noted that $\lim_{u \to \pm\infty} E(u) = 0$, i.e., our results must be analyzed in terms of Whittaker-Shannon sampling theorem. Therefore, u_{max} must have such a value so that $E(\pm u_{max}/2)$ can be virtually zero. On the other hand, the Fourier transform in the discrete form to get $E(x)$ must be carried out in the interval $[-0.5/\Delta u, 0.5/\Delta u]$, where $\Delta u = u_{max} / N$. Obviously, at the ends of this interval $E(\pm 0.5/\Delta u)$ should be very close to zero.

For the waveguides presented, the admissible interval in some way is defined by a behavior of the function which is related with Fourier transforms of the second terms of (10, 11 and 12). The second term of these equations can be expressed as $\Delta\varepsilon(x)$. For example, let us consider a function of this type:

$$I(u_{max}) = \int_{-0.5u_{max}}^{0.5u_{max}} \left| F\left\{ \Delta\varepsilon(x) \right\} \right|^2 du. \qquad (13)$$

The last integrals $I(\infty)$ for the waveguides described by (10, 11 and 12) are accordingly equal to 20, 10.66666667, and 7.07106781. Fig. 4 shows the standardized appropriate

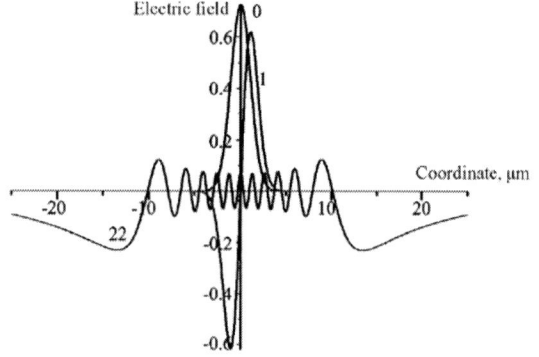

Fig. 3. The distribution of $E(x)$ for the modes with numbers 0, 1 and 22 of the gradient waveguide with the permittivity described by (12).

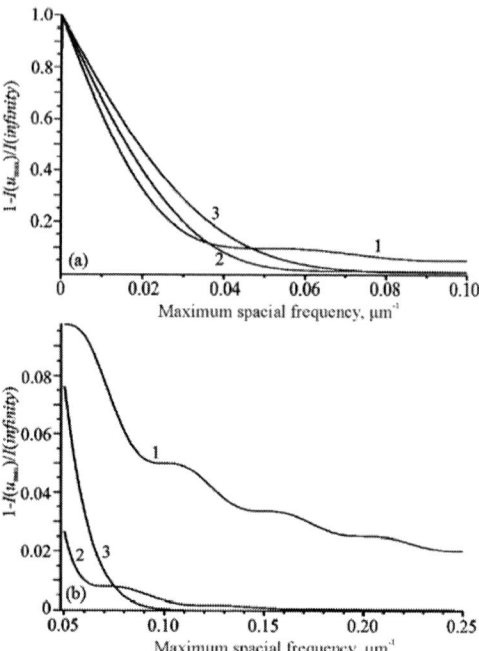

Fig. 4. The dependency of $1 - I(u_{max})/I(\infty)$ on the maximum frequency. Curve 1 corresponds to (10), curve 2 – to (11), and curve 3 – to (12).

dependencies. The integral (13) for (12) (Fig. 4b, curve 3) is most rapidly convergent. That's why for this waveguide at $u_{max} = 2.3 \ \mu m^{-1}$ we obtain all propagation constants of high accuracy. But the top admissible boundary of $u_{max} = 18 \ \mu m^{-1}$ is too small. This is because of a great interval in the coordinate space that the appropriate mode $E_{22}(x)$ occupies.

The integral (13) for the waveguide described by (10) is most slowly convergent, therefore, we find all propagation constants of high accuracy at $u_{max} = (80...90) \ \mu m^{-1}$. To increase the interval u_{max}, N should be increased. Thus, this can be argued that the bottom boundary of admissible values u_{max} is determined by how rapidly the integral (13) tends towards a constant value, and the top boundary is determined by the interval, $E(x)$ occupies, in the coordinate region.

References

[1] A.W. Snyder, J.D. Love, *Optical Waveguide Theory*, Chapman and Hall, London. New York, 1983.

[2] V. Fitio, "Localized modes of the gradient planar waveguides. Analogies in the quantum mechanics", *Scientific Journal of Ivan Franko National University of Lviv "Electronics and information technologies"*, issue 1, 2011, pp.134-141. (in Ukrainian)

Plasmon resonance, periodical structures and absorption spectra induced by laser beam in composite waveguide AgCl-Ag films

L. A. Ageev, V. K. Miloslavsky, V. M. Reznikova, E. D. Makovetsky

Karazin Kharkiv National University, Physics Department, Kharkiv, Ukraine

Abstract: Linearly polarized laser beam has been demonstrated to give rise to photoinduced transformations in AgCl-Ag composition, consisting of thin waveguide AgCl film on the glass and covered by Ag nanoparticle coating. Optical density spectra are investigated to treat physical processes involved.

Metal-dielectric compositions, wherein metal forms two-dimensional ordered nanostructures on dielectric planar waveguide surfaces [1, 2], have interesting optical properties. Localized plasmons may be excited by light in metal [3], and plasmon energy may be transmitted to waveguide modes. Being sensitive to laser radiation, thin film composition of silver chloride and silver (AgCl-Ag) has similar features. DPSS laser linear polarized beam at the wavelength $\lambda = 532$ nm is applied to irradiate samples in this study. It has been shown that periodical structures (PS) are formed in the film due to waveguide mode excitation. Absorption spectra are studied before and after irradiation.

AgCl film of $h \approx 25$ nm thickness deposited on glass and covered by granular Ag layer (of about 8 nm thickness) was researched. The sample scheme is shown in Fig. 1a. Light absorption occurs in Ag nanoparticles and defines AgCl-Ag photosensitivity to the intense light [4]. Plasmons are excited by light in nanoparticles, and it results in photoeffect with electron capture in AgCl and appearance of Ag^+ ions. Ag particles are destroyed by these processes. New particles are created in light field interference minima as Ag^+ ions move to the minima and recombine with electrons captured by traps. Traps concentration is maximal at the film-substrate boundary, and Ag^+ ions are transferred mainly to the substrate (Fig. 1b).

AgCl film on glass substrate is an asymmetrical planar waveguide. Waveguide modes are excited by laser light in AgCl-Ag composition due to Rayleigh scattering and the incident wave interferes with these modes. As Ag is transported to interference minima, the mode is amplified by positive feedback mechanism. Ag accumulation in the minima results in coincidence of effective refractive index $n_{ef} = n \sin\theta$ of mode (n is film refractive index, θ is mode forming zigzag rays incidence angle on film boundaries) with one of pure AgCl film.

The interference leads to periodic silver distribution. Polarization being linear with $\mathbf{E_0}$ vector, TE-modes scattering perpendicular to $\mathbf{E_0}$ prevails and PS with preferred vectors $\mathbf{K} \perp \mathbf{E_0}$ are formed (Fig. 1b).

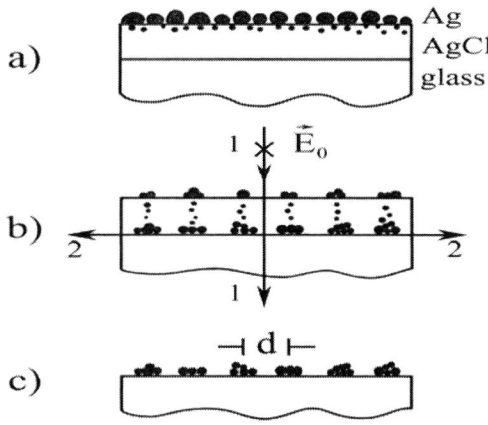

Fig. 1. Scheme of structure transformations in AgCl-Ag: a) the film before irradiation; b) during laser irradiation process (1 - beam direction, × - direction of laser beam polarization $\mathbf{E_0}$, 2 - beams diffracted by PS); c) the sample after fixer treatment (d is PS period).

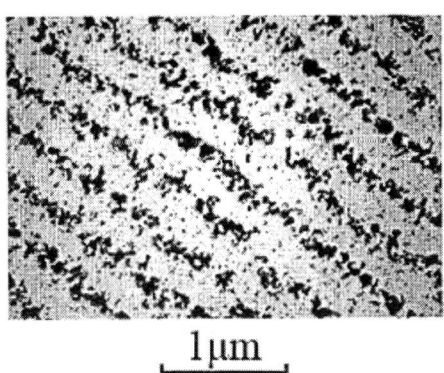

Fig. 2. PS on the substrate surface, that corresponds to the scheme in Fig. 1c.

After laser exposure AgCl has been dissolved by photographic fixer treatment. Relatively small Ag fraction is washed away during the fixing procedure. Electron micrograph of PS after fixing the sample is shown in Fig. 2.

Continuous DPSS laser beam power was $P\sim10$ mW. Irradiated spot dimensions were 6×8 mm on the irradiated film. Exposure time was $t\sim30$ minutes.

Laser beam was normally incident. In doing so, for each of the excited modes PS period has to be equal to $d = \lambda / n_{cf}$. Varying film thickness h makes it possible to vary n_{ef} value within the range determined by the values of substrate n_s and film n refractive indices: $n_s \leq n_{ef} \leq n$ at $n \geq n_s$. We have chosen the case of PS growth on TE_0-mode at $n_{ef} = n_s$ that is realized at $h < h_0$, here h_0 is TE_0-mode cut-off thickness:

$$h_0 = \frac{\lambda}{2 \cdot \pi \cdot \sqrt{n^2 - n_s^2}} \cdot \arctan \sqrt{\frac{n_s^2 - 1}{n^2 - n_s^2}} \ .$$

Except for TE_0-mode, waveguide modes can not be excited because their cut-off thicknesses are substantially larger. At $\lambda = 532$ nm and reference values of $n = 2.08$ (AgCl) and $n_s = 1.52$ (K8 glass) the calculation results in $h_0 = 40$ nm > $> h \approx 25$ nm (AgCl thickness). At that, calculated period value is $d = 350$ nm, and diffraction measurements gave corresponding to calculated PS period value $d = (350\pm0.3)$ nm.

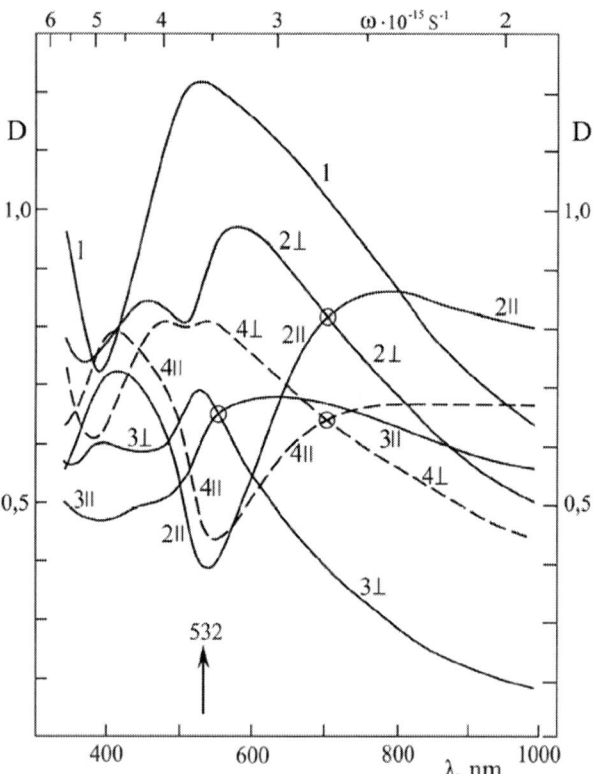

Fig. 3. Optical density spectra: 1 - unirradiated sample; 2 - the sample after irradiation; 3 - after fixer treatment; 4 - after fixer treatment and reposition of AgCl film of about initial thickness (~25 nm). Symbols $\perp$ and $\parallel$ correspond to the measuring beam polarization $E \perp E_0$ and $E \parallel E_0$ respectively.

Sample transmission spectra in the wavelength range $340 \div 1000$ nm were measured (Fig. 3). The unirradiated sample has an absorption band **1** with maximum at $\lambda = 525$ nm ($\omega_{max} = 3.6\cdot10^{15}$ s^{-1}). It is close to the value $\omega_F \approx 3.7\cdot10^{15}$ s^{-1} obtained through well-known Frohlich formula for small Ag particles. That is why band **1** may be related to plasmon excitation in Ag nanoparticles of unirradiated sample.

Polarized spectra **2** $\perp$, $\parallel$ were measured after laser irradiation of the film and PS had been formed in it. Spectra **3** $\perp$, $\parallel$ were measured after the fixing of the same film. The measurements were performed using two polarizations: $E \perp E_0$ and $E \parallel E_0$, here E is measuring beam polarization. Polarization spectra difference indicates dichroism related to PS anisotropy. Polarized spectra have intersection points (marked by circles), wherein the dichroism equals to zero. The dichroism $\Delta D = D_\perp - D_\parallel$ has different signs at the left and the right of intersection points. After the fixer treatment, intersection point has substantially shifted to the short-wave side, positive dichroism was reduced dramatically and negative one was increased slightly.

Spectral dip at $\lambda \approx 532$ nm is observed in $E \parallel E_0$ spectrum **2**$\parallel$ after the laser irradiation but before the fixer treatment. It is connected with plasmon excitation and their energy transfer to waveguide modes [1]. There is no spectral dip in spectra **3** after fixer treatment, but, as seen on **4**, polarized spectral dip and dispersion are restored after AgCl film of about initial thickness redeposited after fixer treatment.

It has been shown that, due to plasmon excitation in Ag nanoparticles, AgCl-Ag composition is sensitive to polarized laser beam exposure. As a result, PS are formed consistent with waveguide modes. Absorption spectra give dichroism connected with PS anisotropy and indicate the existence of plasmon-polariton binding between plasmons and waveguide modes.

REFERENCES

[1] V.V. Klimov, "Nanoplazmonika", *Usp. Fiz. Nauk*, vol. 178, no.8, pp. 875-880, 2008. (in Russian)

[2] S. Linden, J. Kuhl, H. Giessen, "Controlling the interaction between light and gold nanoparticles: Selective suppression of extinction", *Phys. Rev. Lett.*, vol. 86, no. 20, pp. 4688-4691, 2001.

[3] N.A. Gippius, S.G. Tikhodeyev, A. Krist, J. Kuhl, H. Giessen, "Plazmon-volnovodnie polaritoni v metallodielektricheskih fotonno-kristallicheskih sloyah", *Fiz. Tverd. Tel.*, vol. 47, no. 1, pp. 139-143, 2005. (in Russian)

[4] S.G. Tikhodeyev, N.A. Gippius, "Plazmon-polaritonniye effekti v nanostrukturirovannih metallodielektricheskih sloyah", *Usp. Fiz. Nauk*, vol. 179, no. 9, pp. 1003-1007, 2009. (in Russian)

[5] L.A. Ageev, V.K. Miloslavsky, "Photoinduced effects in light-sensitive films", *Opt. Eng.*, vol. 34, no. 4, pp. 960-972, 1995.

Shift of surface plasmon-polaritons resonances via acoustic waves in hybrid metal-semiconductor structures

N.E. Khokhlov[1], V.I. Belotelov[1], B.A. Glavin[2]

[1]Lomonosov Moscow State University, Moscow, Russia
[2]Institute of Semiconductor Physics, National Academy of Sciences, Kiev, Ukraine

Abstract: We discuss theoretical results proving feasibility of an approach capable to perform high-frequency (up to terahertz band) acoustic modulation of the plasmonic structures. The numerical calculations performed for the case of gold grating on the top of the GaAs/AlAs-based acoustic cavity suggest that the relative modulation of the light reflection can be as high as few percents for the strain amplitude of the acoustic mode about 10^{-4}, the value which can be easily achieved with the use of picosecond acoustics technique.

Nowadays, plasmonics has attracted more attention of researchers. This interest is due to the opportunities it offers: new ways of processing information in nanocircuits [1, 2], an increase of light absorption in solar cells [3], the creation of bio-sensors [4], etc. Surface plasmon polaritons (SPP) are the main objects of plasmonics study. SPP are the coupled oscillations of the electromagnetic field and the electrons plasma in metals [5]. SPP wave is localized near the metal-dielectric interface and may extend to tens and even hundreds of microns along the interface. Modern communication systems require the creation of "active" materials for plasmonics, which change their optical properties under external influence [6]. Since 2004, when the term "active plasmonics" was introduced first time [7], the number of proposed methods for SPP propagation control via external influence is on the rise [8-15]. The main idea of such control is based on the changing of the SPP's wavevector k_{SPP} [12]:

$$k_{spp} = k_0 \sqrt{\frac{\varepsilon_m \varepsilon_d}{\varepsilon_m + \varepsilon_d}}, \qquad (1)$$

where k_0 is the vacuum wavevector of the light; ε_m and ε_d are the metal and dielectric permittivities respectively. Reversible changes in the permittivities may be caused by external influences of different nature – optical excitation [12], electric [13] and magnetic fields [14] etc. which entail changes in the SPP wave vector.

Also controllable changes of the SPP's wavevector in the periodic structure can be produced by the variation of the light to SPP coupling condition [12]:

$$k_{spp} = k_0 \sin(\theta) + m \frac{2\pi}{d_{gr}}, \qquad (2)$$

where θ the incident angle of the light, d_{gr} is the period of the structure, m is an integer. Equations (1) and (2) is correct

approximately for the real metal grating case; k_{SPP} depends on the geometrical parameters of the metal layer such as thickness l, widths of the air slits w_{air} and metal stripes w_m (fig. 1).

Modulation of the geometrical parameters of the plasmonic structure can be achieved by dynamical strain caused by the acoustic wave propagation [15]. In conventional acousto-optics the maximal frequency at which electromagnetic wave can be efficiently modulated is controlled by the energy and momentum conservation under photon-phonon scattering, being restricted commonly by the value about few tens of GHz. Recent experiments demonstrate acoustic modulation of the plasmonic structures in the frequency range up to 10 GHz [16]. The frequency of modulation can be increased in the case of acousto-plasmonic coupling of high-order diffracted acoustic waves [17].

In this work we discuss theoretical results proving feasibility of an alternative approach capable to perform high-frequency (up to terahertz band) acoustic modulation of the plasmonic structures. The coupling at such a high frequency is due to integration of vertical acoustic cavity and lateral plasmonic grating with the period d_{gr} and thickness l (fig. 1). The layered grating's substrate consists of the layer of material "1" of thickness L (cavity), followed by the superlattice (SL) formed by a periodic sequence of layers of materials "2" and "1" of thicknesses d_2 and d_1 respectively, SL's period is $d=d_1+d_2$. The elastic mismatch of the layers gives rise to narrow acoustic stop-band formation and, consequently, to confined acoustic

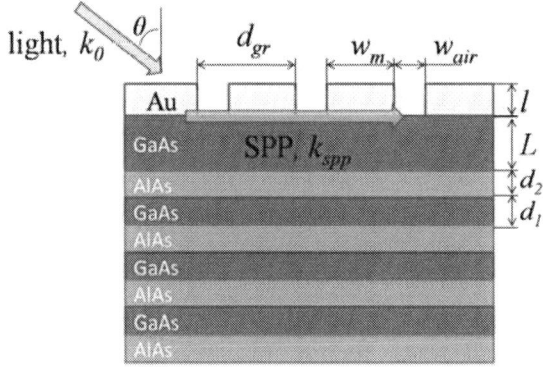

Fig. 1. Sketch of the considered acousto-plasmonics structure and scheme of the coupling light to the SPP in the metal grating deposited on the substrate. Symbols description is within the text.

mode under proper selection of the near-surface cap layer parameters. Thereby such combination of the SL and cavity opens up the way for acoustic wave's localization near the palsmonic layer increasing the efficiency of phonon-SPP coupling up to terahertz band frequencies.

The coupling of the acoustic wave within the cavity and electromagnetic radiation is due to two effects:

- photoelastic perturbation in the cavity and SL. Two components of dielectric permittivity of the materials "1" and "2" have perturbation in diagonal components: $\delta\varepsilon_{xx}$ and $\delta\varepsilon_{zz}$. Both perturbation components depend on the strain due to acoustic vibrations and spatial coordinates;
- perturbation of the metal layer thickness δl.

These factors act mainly on the near-field zone of the electromagnetic wave lifting modulation frequency restriction inherent for conventional light waves.

For the most effective phonon-SPP coupling the parameters of the SL and metal film were calculated for observation the same acoustic mode structures within the cavity with and without metal film. This corresponds to the condition

$$\sin\frac{2\pi v_1 l}{d v_f} = 0,$$ where v_1 and v_f are sound velocities in material

"1" and the film, respectively. For $\cos\frac{2\pi v_1 l}{d v_f} = 1$ there is no

perturbation of the metal film thickness ($\delta l = 0$), and for

$\cos\frac{2\pi v_1 l}{d v_f} = -1$ we have $\delta l = \dfrac{\alpha U_{zz} d}{\pi \sin\frac{2\pi L}{d}}$, where α depends on

the elastic mismatch in the SL layers and d_1 and d_2, U_{zz} is the characteristic magnitude of strain. The last case ($\delta l \neq 0$) is more interesting for observation relative reflectance changes $\Delta R/R$ because both perturbations are involved in changing SPP excitation conditions.

For the research gold was considered as a metal and GaAs and AlAs as the materials "1' and "2", respectively. Geometrical parameters of the SL and cavity are $d_1 = 12.5$ nm, $d_2 = 45$ nm, $L = 25$ nm. Accordingly the case of nonzero gold film perturbation ($\delta l \neq 0$) corresponds, in particular, to $l \approx 118$ nm. The characteristic magnitude of strain $U_{zz} = 10^{-4}$.

The geometrical parameters of the gold grating (d_{gr} and w_{air}) were calculated in the way for SPP resonance condition fulfillment for both Au/Air and Au/GaAs interfaces at normal incidence of the light. Equations (1) and (2) were used as a first approximation. However, gold film contains air slits, Au and GaAs media are not semi-infinite numerical calculations by the RCWA method [18] were performed.

Numerical calculation showed the highest relative modulation of the reflectivity $\Delta R/R$ is observed near R minima (fig. 2a) at wavelength of both SPP resonances (clearly seen at field distribution map, fig. 2b). For grating parameters $d_{gr} = 730$ nm and $w_{air} = 400$ nm value $\Delta R/R$ achieves at wavelength 815 nm the maxima in 3.2% which is two orders of magnitude greater than the corresponding values in other researches [16]. Such a high value is achieved by combination acoustic cavity and SL.

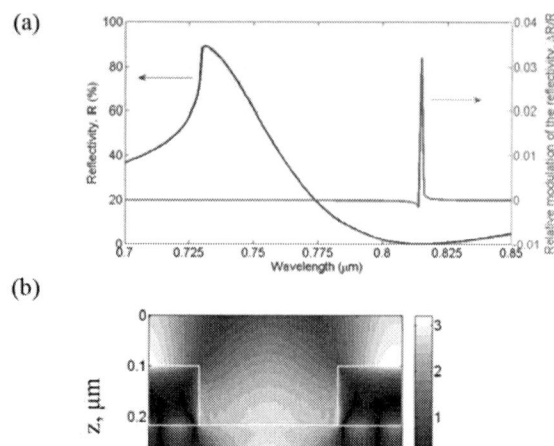

Fig. 2. (a) Spectrum of reflectivity R and relative modulation of the reflectivity $\Delta R/R$; (b) spatial distribution of H_y field component. White stripes show Au and air boundaries; wavelenght of the light is 815 nm (peak of $\Delta R/R$ value). One period of the grating is shown. Grating parameters: $d_{gr} = 730$ nm, $w_{air} = 400$ nm

The peak of the $\Delta R/R$ is very narrow due to coincidence two SPP resonances. In the case of only one resonance's excitation maximum $\Delta R/R$ value reduces by two orders of magnitude, but the peak is broadened (fig. 3a) what it is more preferable for practical purposes.

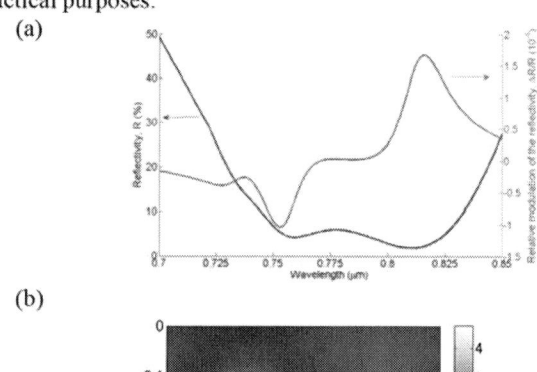

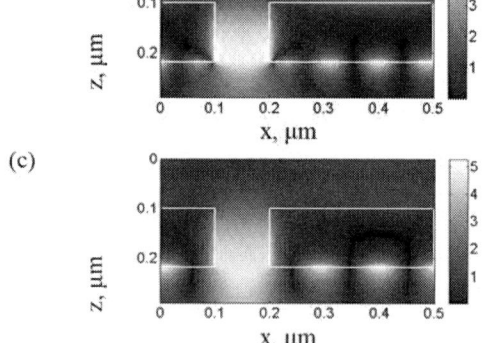

Fig. 3. (a) Spectrum of reflectivity R and relative modulation of the reflectivity $\Delta R/R$; (b), (c) spatial distribution of H_y field component. White stripes show Au and air boundaries; wavelenghts are 755 nm (b) and 815 nm (c). One period of the grating is shown. Grating parameters: $d_{gr} = 500$ nm, $w_{air} = 100$ nm

REFERENCES

[1] M.L. Brongersma, V.M. Shalaev, "The Case for Plasmonics," *Science*, vol. 328. pp. 440-441, 2010.

[2] J.A. Schuller, E.S. Barnard, W. Cai, Y.C. Jun, J.S. White, M.L. Brongersma, "Plasmonics for extreme light concentration and manipulation," *Nat. Mater.*, vol. 9, pp. 193-204, 2010.

[3] H.A. Atwater, A. Polman, "Plasmonics for improved photovoltaic devices," *Nat. Mater.*, vol. 9, p. 205213 2010.

[4] A.V. Kabashin, P. Evans, S. Pastkovsky, W. Hendren, G.A. Wurtz, R. Atkinson, R. Pollard, V.A. Podolskiy, A.V. Zayats, "Plasmonic nanorod metamaterials for biosensing," *Nat. Mater.*, vol. 8. pp. 867-871, 2009.

[5] S.A. Maier, *Plasmonics - Fundamentals and applications*. New York: Springer, 2007.

[6] K.F. MacDonald, Z.L. Sámson, N.I. Stockman, N.I. Zheludev, "Ultrafast active plasmonics" *Nature Photon.*, vol. 3, pp. 55-58 2009.

[7] A.V. Krasavin, N.I. Zheludev, "Active plasmonics: Controlling signals in Au/Ga waveguide using nanoscale structural transformations," *Appl. Phys. Lett.*, vol. 84, pp. 1416–1418, 2004.

[8] T. Nikolajsen, K. Leosson and S.I. Bozhevolnyi, "Surface plasmon polariton based modulators and switches operating at telecom wavelengths" *Appl. Phys. Lett.*, Vol. 85, pp. 5833–5835, 2004.

[9] A.L. Lereu, A. Passian, J.P. Goudonnet, T. Thundat and T.L. Ferrell, *Appl. Phys. Lett.*, vol. 86, p. 154101, 2005.

[10] R.A. Pala, K.T. Shimizu, N.A. Melosh and M.L. Brongersma, *Nano Lett.*, vol. 8, pp. 1506–1510, 2008.

[11] D. Pacici, H.J. Lezec, H. A. Atwater, *Nature Photon.*, vol. 1, pp. 402–406, 2007.

[12] J.N. Caspers, N. Rotenberg, and H.M. van Driel, Opt. Express vol. 18, p. 19761, 2010.

[13] N.M. Hassan, V.V. Mkhitaryan and E.G. Mishchenko, Phys. Rev. B, vol. 85, p. 125411, 2012.

[14] V.V. Temnov, G. Armelles, U. Woggon, D. Guzatov, A. Cebollada, et al., Nature Photon. vol. 4, p. 107, 2010.

[15] Y.-L. Chiang, C.-W. Chen, C.-H. Wang, C.-Y. Hsieh, Y.-T. Chen, et al., Appl. Phys. Lett. vol. 96, p. 041904, 2010.

[16] H.-P. Chen, Y.-C. Wen, Y.-H. Chen, C.-H. Tsai, K.-L. Lee, et al., Appl. Phys. Lett. vol. 97, p. 201102, 2010.

[17] C. Brueggemann, A. V. Akimov, B. A. Glavin, V. I. Belotelov, I. A. Akimov, et al., "Modulation of a surface plasmon-polariton resonance by subterahertz diffracted coherent phonons," *Phys. Rev. B*, vol. 86, p. 121401(R), 2012.

[18] L. Li, "Fourier modal method for crossed anisotropic gratings with arbitrary permittivity and permeability tensors," *J. Opt. A: Pure Appl. Opt.*, vol. 5, pp. 345-355, 2003.

Effective coupling between THz electromagnetic radiation and 2D plasmons

Vadym V. Korotyeyev and Yurii M. Lyaschuk

V. Lashkaryov Institute of Semiconductor Physics, National Academy of Sciences of Ukraine, 03028 Kyiv, Ukraine

Abstract: Theory of the plasmonic structure "grating - 2DEG" that consists of subwavelength metallic grating and a layer of the 2D-electron gas under the grating is presented. It is shown that frequency dependencies of the transmission, reflection and losses coefficients in THz frequency range have resonance behavior which relates to the 2D-plasmon excitation. The influence of geometrical and electrical parameters of the system on plasmonic resonance is studied in detail. The structure of the electromagnetic field in the near-field zone is analyzed. The spatial dependencies of the electric field components, density of the energy and the polarization of electromagnetic wave are found. It is shown the essential increasing of the local magnitude of the electric field in near-field zone at the plasmonic resonance.

Introduction

Nowadays, the terahertz range of the electromagnetic radiation remains poorly studied. In spite of the numerous potential applications the modern THz optoelectronic and microelectronic devices have a number of disadvantages that restrict their wide usage in practice. For instance, they have small output power, or do not allow easy frequency tuning, sometimes they require deep-cryogenic temperatures, or can not satisfy the consumer in terms of compactness, etc . The novel trend of the THz technology is associated with the using of plasmonic semiconductor structures. In such structures, the effective manipulation of THz radiations (detection, rectification, modulation, emission) can be achieved using the various mechanism of the 2D-plasmons excitation.

The effective coupling of plasmons waves (short wavelength) and electromagnetic radiation (long wavelength) is possible only in the case of spatially inhomogeneous conditions. For example, it can be structure that consists of the subwavelength metallic grating placed near the plane of the 2D gas (see Fig.1). It was shown that similar plasmonic structures are the perspective for detection [2] and amplification [3] of the

terahertz radiation.

Spectral characteristics of plasmonic structure

The electrodynamic theory of these structures was developed and described in detail in papers [3, 4]. In the frame of this theory we investigated the spectra of transmission T_ω, reflection R_ω, and losses L_ω coefficients of terahertz electromagnetic waves as well the spatial distribution of the electromagnetic fields in the near-field zone. All calculations are carried out for 2 DEG with parameters of GaN. The III-nitride compounds have small electron relaxation time and, consequently, large electron response up to the THz frequency range[5] that is favorable for THz applications. The analysis of THz properties of nitride materials, including hot-electron effects and magnetic field effects can be found in ref.[6-7].

The result of the calculations of T_ω, R_ω, L_ω is presented in Fig. 2. As can be seen, these quantities have resonance –like

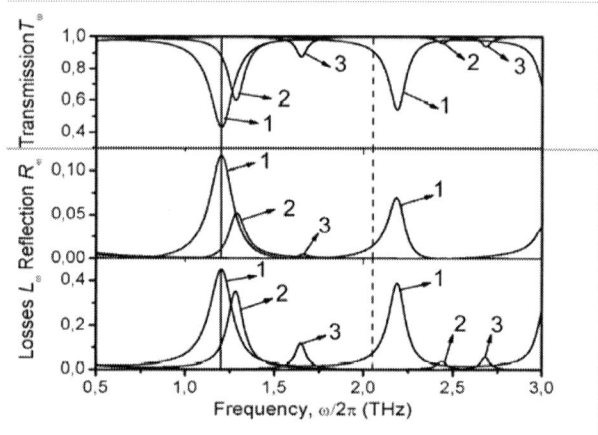

Fig. 2. The frequency dependences of the coefficients T_ω, R_ω, L_ω. The curves 1, 2, 3, correspond to the different filling factors of the grating, $b/a = 0.9, 0.6, 0.3$ respectively. The conductivity of the metal strips, $\sigma_g = 4\times10^{15}\,s^{-1}$, thickness of the strips, $d_g = 0.02\,\mu m$, period of grating, $a = 3\,\mu m$, distance between grating and 2DEG, $D = 0.1\,\mu m$, electron concentration of the 2D gas, $n^{2D} = 10^{12}\,cm^{-2}$ and relaxation time in 2DEG, $\tau = 3\,ps$.

behavior that associated with excitation of plasmons in 2D gas. The frequencies of these resonances are located between frequencies of gated and ungated plasmons which corresponds

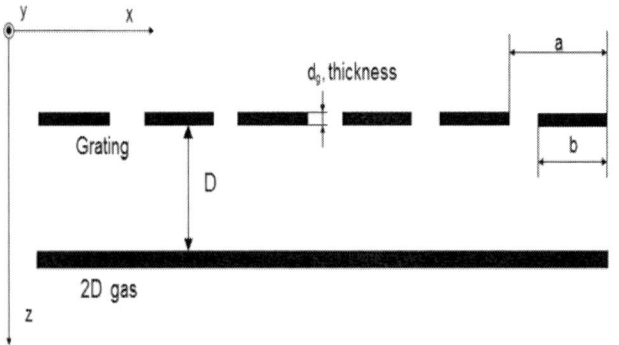

Fig. 1. Illustration of the geometry of the considered plasmonic structure.

978-1-4799-0019-0/13 $31.00 © 2013 IEEE   248

to wave vectors, $q_m = 2\pi m / a, m = 1, 2 \ldots$, where a is grating period. The curves 1-3 demonstrate the variation of the frequency and intensity of plasmon resonances in dependence of the geometry of grating.

The vertical solid and dashed lines denote the frequencies of the gated and ungated plasmons, respectively for the $m = 1$. With an increasing of the grating filling factor b/a the resonance frequencies shift to the frequency of the gated plasmons an it is accompanied by increasing of the intensity of the resonances.

Near-field properties

In the near field-zone the metallic grating produces a complicate vector structure of electromagnetic field (it has two components: lateral or x-component and transversal or z-component) with strong spatial redistribution of the energy.

The spatial mapping of the time-averaged density of the electric energy in the near-field zone of grating structure is shown in Fig. 3 for three selected frequencies: below plasmon resonance, at resonance and beyond.

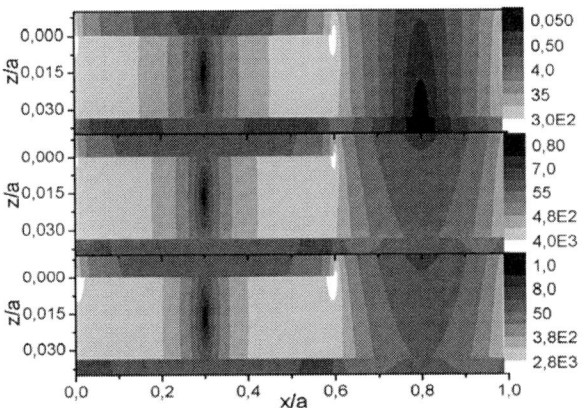

Fig. 3. The spatial distribution of the normalized time-averaged density of the electric energy, $W(x,z) = \left(\left| E_x(x,z) \right|^2 + \left| E_z(x,z) \right|^2 \right) / \left| E_0 \right|^2$, E_0 is the amplitude of incident plane wave. The upper, middle and bottom mappings correspond to the frequencies 1.0 THz, 1.25 THz (resonant frequency) and 1.4 THz, respectively. The other parameters are taken the same as for curve 2 in Fig. 2.

The significant local concentration of the electric energy is observed at all frequencies. Especially, it is well-pronounced in the region under the metallic strip. Here, the strong concentration of energy is related to the great increasing of the z-component of electric field. This component appears only in the near-field zone, but it is absent in the far-field zone. The maximal local concentration of electromagnetic energy is achieved at the plasmons resonance (see middle panel in Fig. 3).

Polarization characteristics

The electromagnetic wave in the near-field zone have local elliptic polarization as result of the existing of x and z-components of electric field. Moreover, their amplitudes and relative phase shift between oscillations strongly depend on spatial coordinates. Therefore the structure of near field can be characterized by the set of polarization ellipses (see insert in Fig.4) with parameters δ (ellipticity), and α (the orientation angle). The parameters of polarization ellipse are determined by the ratio $r = \left| E_z \right| / \left| E_X \right|$ and the relative phase shift $\Delta\phi = \phi_{EX} - \phi_{EZ}$, as follows:

$$\delta = \left[1 - \sqrt{1 - \beta^2} \right] / \left[1 + \sqrt{1 + \beta^2} \right]$$

$$\tan\alpha = \left[2r^2 - \left(1 + r^2 \right)\left(1 - \sqrt{1 - \beta^2} \right) \right] / \left[2r\cos\Delta\phi \right]$$

where, $\beta = 2r\sin\Delta\phi / \left(1 + r^2 \right)$. The spatial dependences of the parameters δ and α are depicted in Fig. 4.

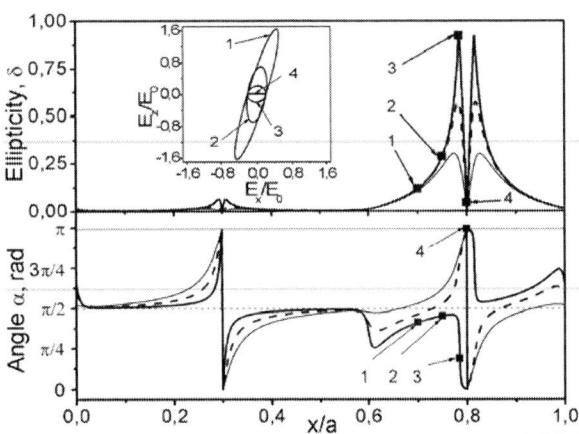

Fig. 4. The dependences of the ellipticity δ and the orientation angle α of the polarization ellipses on spatial coordinate x at the fixed coordinate z. The thick solid line correspond to $z/D = 0.2$, the dashed line correspond to $z/D = 0.5$, and thin line correspond to $z/D = 0.9$. Frequency, $\omega/2\pi = 1.0\,THz$. All other parameters are taken the same as for curve 2 in Fig. 2.

As can seen, under the metallic strip the electromagnetic wave is practically linear-polarized that is stipulated by prevalence of the z-component over the x-component. Under the window, x- and z- components have amplitudes of same order and, as a result, the electromagnetic wave has an essential ellipticity. Due to the symmetry, there are two isolated points, corresponding to the middle of strip and middle of window. In these points, z- component tends to zero as a result the ellipticity is zero and orientation angle has discontinuity.

Summary

In summary, the plasmonic structure with subwavelength grating ensures the effective coupling between THz electromagnetic radiation and 2D plasmons and allows to effectively control the concentration of electromagnetic field and its polarization. Such properties can be applied for both the study of THz properties of the low-dimensional carries and the manipulations of quantum states of individual molecules and atoms.

REFERENCES

[1] R M. Tonouchi, "Cutting-edge terahertz technology," *Nat. photonics*, vol. 1, pp. 97-105 2007.

[2] V. V. Popov, "Plasmon Excitation and Plasmonic Detection of Terahertz Radiation in the Grating-Gate Field-Effect-Transistor Structures" *J. Infrared Milli Terahertz Waves*, vol. 32, pp. 1178-1191, 2011.

[3] S. A. Mikhailov, "Tunable solid-state far-infrared sources: New ideas and prospects," *Recent Res Devel. Applied Phys.*, vol. 2, pp. 65-108 1999; S. A. Mikhailov, "Plasma instability and amplification of electromagnetic waves in low-dimensional electron systems," *Phys. Rev. B*, vol. 58, no. 2, pp.1517-1532, 1998.

[4] 4) Yu. M. Lyaschuk, V. V. Korotyeyev, "Interaction of terahertz electromagnetic field with a metallic grating: Near-field zone," *Ukr. J. Phys. Opt.*, vol. 13, pp. 142-150, 2012.

[5] V. N. Sokolov, K. W. Kim, V. A. Kochelap and D. L.Woolard, "High-frequency small-signal conductivity of hot electrons in nitride semiconductors," *Appl. Phys. Lett.* vol. **84**, no. 18, pp. 3630-3632, 2004

[6] G.I. Syngayivska and V.V. Korotyeyev "Electrical and high-frequency properties of compensated GaN under electron streaming conditions," *Ukrainian Journal of Physics* vol. **58**, no.1, pp. 40-55 2013.

[7] G I Syngayivska, V V Korotyeyev and V A Kochelap "High-frequency response of GaN in moderate electric and magnetic fields: interplay between cyclotron and optical phonon transient time resonances," *Semicond. Sci. Technol.* vol. **28** 035007, 2013.

Theory of near field management via magnetoplasmon tunneling

A. N. Kalish[1], V. I. Belotelov[1], S. N. Andreev[2], V. P. Tarakanov[2], A. K. Zvezdin[2]

[1]Lomonosov Moscow State University, Moscow, Russia

[2]Prokhorov General Physics Institute of Russian Academy of Science, Moscow, Russia

Abstract: We consider the effect of plasmon tunneling through a metallic film when the latter is surrounded by magnetic dielectrics. We develop a theory based on coupled oscillators approach. The application of magnetic field allows manipulation of tunneling coefficient and tunneling length.

I. INTRODUCTION

Recently plasmonics has become one of the principal areas of photonics due to numerous applications [1]. One of the distinguishing features of plasmons is their great affection on electromagnetic near field. The properties of plasmons can be affected by magnetic field. It was shown recently that plasmon excitation leads to enhancement of magnetooptical effects so magnetoplasmonics is very intriguing [2]. In particular, if the plasmonic structure contains a magnetic dielectric magnetized transversally, its dispersion is linearly dependent on magnetization. The magneto-optical properties of a medium are defined by gyration vector $\mathbf{g}$ that is usually directly proportional to magnetization.

If a plasmonic structure contains two interfaces supporting surface plasmon propagation, then at plasmon excitation at only one of them and its further propagation the energy periodically transfers to the other one fully or partially. This effect is plasmon tunneling through the metal film. If the system contains magnetic dielectrics then magnetic field allows management of tunneling, i.e. the tunneling length and the tunneling coefficient that is the transferred energy fraction.

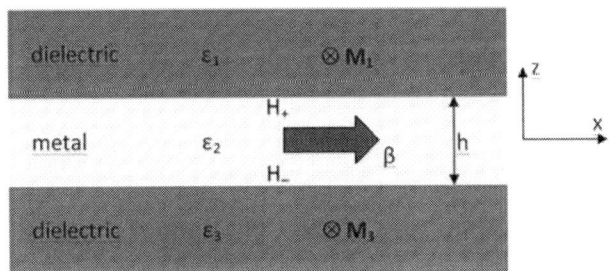

Fig. 1. Metallic film surrounded by transversally magnetized dielectrics. The blue arrow denotes plasmon propagation. H_+ и H_- are the amplitudes of H_y at the interfaces.

II. MAGNETOPLASMONS IN METALLIC FILMS

We consider a metallic film of thickness h and dielectric constant ε_2 surrounded by semi-infinite transversally magnetized dielectrics with dielectric constants ε_1 и ε_3 and gyration coefficients g_1 и g_3 (Fig. 1). The dispersion equation for magnetoplasmons is the following:

$$\tanh(\gamma_2 h) = -\frac{\alpha_2(\alpha_1 + \alpha_3 + \nu_1 - \nu_3)}{\alpha_2^2 + (\alpha_1 + \nu_1)(\alpha_3 - \nu_3)}, \quad (1)$$

where $\alpha_i = \gamma_i/\varepsilon_i$, $\nu_i = g_i\beta/\varepsilon_i^2$, $\gamma_{1,2} = \left(\beta^2 - k_0^2\varepsilon_{1,2}\right)^{1/2}$ are localization coefficients, β is the propagation constant, k_0 is the vacuum wavenumber. ν_i are small parameters. Eq. (1) has two solutions corresponding to plasmonic modes with different propagation constants and different electromagnetic field configuration.

If $\varepsilon_1 = \varepsilon_3$ and $g_1 = g_3 \neq 0$, then Eq. (1) doesn't contain terms linear in g. The quadratic in g contribution becomes significant only for rather thick films ($\gamma_2 h \gg 1$). For thin films this constribution can be neglected, and Eq. (1) splits into two separate equations for the symmetric and the antisymmetric modes:

$$\tanh\left(\frac{\gamma_2 h}{2}\right) = -\frac{\alpha_1}{\alpha_2}, \quad (2)$$

$$\tanh\left(\frac{\gamma_2 h}{2}\right) = -\frac{\alpha_2}{\alpha_1}, \quad (3)$$

respectively. So the dispersion of magnetoplasmons isn't affected by magnetization. However, as the system is no longer symmetrical upon transformation $z \to -z$ the electromagnetic field configuration changes:

$$H_+ = \pm H_-\left(1 + 2\frac{\alpha_1\nu_1}{\alpha_2^2 - \alpha_1^2}\right), \quad (4)$$

where H_+ and H_- are the amplitudes of H_y component at the two interfaces of the film, and the signs "+" and "−" at the right side correspond to symmetric and antisymmetric modes respectively. With the increase of thickness Eq. (2) becomes invalid, and the modes become localized at one of the interfaces.

If $g_1 = -g_3 \neq 0$, then the dispersion equations acquire linear in g terms:

$$\tanh\left(\frac{\gamma_2 h}{2}\right) = -\frac{\alpha_1 + \nu_1}{\alpha_2}, \qquad (5)$$

$$\tanh\left(\frac{\gamma_2 h}{2}\right) = -\frac{\alpha_2}{\alpha_1 + \nu_1}, \qquad (6)$$

for the symmetric and antisymmetric modes, respectively.

At this the field amplitudes at the film interfaces remain equal, as the system is symmetric upon transformation $z \to -z$:

$$H_+ = \pm H_-. \qquad (7)$$

III. Magnetoplasmon Tunneling in Thick Films

If the metallic film is thick enough the plasmons at the two interfaces propagate almost independently of each other. The finiteness of the film thickness leads to the coupling between them, and the thinner is the film the stronger is the coupling.

Electromagnetic field is the superposition of the modes described in Sec. 1, so the waves propagating along x direction have the form:

$$H_+(x) = \kappa_1 A_1 \exp(i\beta_1 x) + \kappa_2 A_2 \exp(i\beta_2 x)$$
$$H_-(x) = \kappa_1 B_1 \exp(i\beta_1 x) + \kappa_2 B_2 \exp(i\beta_2 x), \qquad (8)$$

where $\beta_{1,2}$ are the propagation constants defined from Eqs. (1-3) or (5-6), $A_{1,2}$ and $B_{1,2}$ are normalized field amplitudes and $\kappa_{1,2}$ are defined by initial conditions at $x = 0$.

It follows from Eq. (8) that the field amplitudes satisfy the following equations:

$$\frac{\partial^2 H_+}{\partial x^2} + \beta_+^2 H_+ = \eta_+ H_-$$
$$\frac{\partial^2 H_-}{\partial x^2} + \beta_-^2 H_- = \eta_- H_+ \qquad , \qquad (9)$$

where

$$\beta_+^2 = \frac{\beta_1^2 - \mu_1 \mu_2 \beta_2^2}{1 - \mu_1 \mu_2}, \quad \beta_-^2 = \frac{\beta_2^2 - \mu_1 \mu_2 \beta_1^2}{1 - \mu_1 \mu_2},$$
$$\eta_+ = \frac{\mu_2 \left(\beta_1^2 - \beta_2^2\right)}{1 - \mu_1 \mu_2}, \quad \eta_- = \frac{\mu_1 \left(\beta_2^2 - \beta_1^2\right)}{1 - \mu_1 \mu_2}, \qquad (10)$$
$$\mu_1 = \frac{B_1}{A_1}, \quad \mu_2 = \frac{A_2}{B_2}$$

Eq. (9) has the form of two coupled oscillators equations with time substituted by coordinate x, β_+ and β_- are the 'partial frequencies', and η_+ and η_- are the coupling coefficients [3].

The solution of Eq. (9) has the form of beats when the energy fully or partially periodically transfers from one oscillators to the other. In our case it is equivalent to tunneling of plasmon energy through the metallic film. Let's assume that $H_- = 0$ at $x = 0$. Then the tunneling coefficient is

$$K = \frac{4\eta^2}{\left(\beta_+^2 - \beta_-^2\right)^2 + 4\eta_+ \eta_-}, \qquad (11)$$

and the tunneling length is

$$l_{tun} = \frac{\pi}{|\beta_1 - \beta_2|}. \qquad (12)$$

For the further analysis the two cases are to be considered separately.

1. $\beta_+ = \beta_-$ (further they are denoted by β). This corresponds to the case when the system possesses $z \to -z$ transformation symmetry. In particular, it is the case of $\varepsilon_1 = \varepsilon_3$ and $g_1 = -g_3$. β_1 и β_2 correspond to symmetric and antisymmetric modes respectively, so according to Eqs. (7) and (10)

$$\mu_1 = -1, \quad \mu_2 = 1. \qquad (13)$$

β_1 and β_2 can be estimated analytically from Eq. (1), taking into account that $\tanh(\gamma_2 h) \approx 1 - 2\exp(-2\gamma_2 h)$ at $h \to \infty$. The coupling coefficients are

$$\eta_+ = \eta_- = \frac{4\alpha_1}{\dfrac{1}{\gamma_1^2} - \dfrac{1}{\gamma_2^2} + \dfrac{\nu_1 \left(\beta^2 - \gamma_1^2\right)}{\alpha_2 \beta^2 \gamma_1^2}} \exp(-\gamma_2 h). \qquad (14)$$

According to Eq. (11) $K = 1$. The propagation constants satisfy the relation $\beta_{1,2} = \beta \pm \dfrac{\eta_\pm}{2\beta}$, so $l_{tun} = \dfrac{\pi\beta}{\eta_\pm} \sim \exp(\gamma_2 h)$. The tunneling length increases with the thickness. As both β and $\eta_\pm$ contain linear in magnetization contributions, so does the tunneling length:

$$l_{tun} = l_{tun0} + ag. \qquad (15)$$

So the tunneling coefficient is maximal and doesn't depend on magnetization, and the tunneling length depends on it linearly.

2. $\beta_+ \neq \beta_-$. It is valid, in particular, if $\varepsilon_1 = \varepsilon_3$ and $g_1 = g_3$. β_1 and β_2 correspond to modes localized at the high and low interfaces respectively. For μ_1 and μ_2 the following approximate equations can be obtained:

$$\mu_1 \approx \left(\left|\frac{\alpha_2}{\nu_1}\right| \exp(-\gamma_2 h)\right)\Bigg|_{\alpha_1 + \alpha_2 + \nu_1 = 0}$$
$$\mu_2 \approx -\left(\left|\frac{\alpha_2}{\nu_1}\right| \exp(-\gamma_2 h)\right)\Bigg|_{\alpha_1 + \alpha_2 - \nu_1 = 0} \qquad . \qquad (16)$$

The thick film approximation can be applied only if μ_1 and μ_2 given by Eq. (16) are small enough. From Eqs. (10) and (16) we come to the following:

$$\beta_+ \approx \beta_1, \quad \beta_- \approx \beta_2$$
$$\eta_\pm \approx \frac{\alpha_2}{\nu_1}\left(\beta_2^2 - \beta_1^2\right)\exp(-\gamma_2 h). \qquad (17)$$

For the tunneling coefficient and the tunneling length the following formulas are obtained:

$$K \approx 4\left(\frac{\alpha_2}{\nu_1}\right)^2 \exp(-2\gamma_2 h), \qquad (18)$$

$$l_{tun} = \frac{\pi}{|\beta_1 - \beta_2|} \approx \frac{\pi}{|\beta_+ - \beta_-|} \propto \frac{1}{g_1} \qquad (19)$$

Therefore, both the tunneling coefficient and the tunneling length are inversely proportional to magnetization. With the decrease of magnetization this relation becomes invalid because the thick film approximation becomes inapplicable.

IV. MAGNETOPLASMON TUNNELING IN THIN FILMS

The problem is analyzed without the involvement of the coupled oscillations theory. In thin metallic film surrounded by similar magnetic dielectrics the field is the superposition of the field of symmetric and antisymmetric modes:

$$H_+(x) = \kappa_s H_{s+} \exp(i\beta_s x) + \kappa_a H_{a+} \exp(i\beta_a x)$$
$$H_-(x) = \kappa_s H_{s-} \exp(i\beta_s x) + \kappa_a H_{a-} \exp(i\beta_a x) \tag{20}$$

The propagation constants $\beta_{s,a}$ are defined by Eqs. (2-3) or (5-6), and $\kappa_{s,a}$ are defined by initial conditions at $x = 0$.

Let's fix the initial conditions such that $H_-(0) = 0$, i.e.

$$\kappa_1 H_{s-} = -\kappa_2 H_{a-} = H_0. \tag{21}$$

Then

$$|H_-(x)| = 2H_0 \left| \sin\left(\frac{\beta_a - \beta_s}{2} x \right) \right|, \tag{22}$$

Therefore the tunneling length is

$$l_{tun} = \frac{\pi}{\beta_a - \beta_s}. \tag{23}$$

The denominator $(\beta_a - \beta_s)$ increases at the thickness decrease, so the tunneling length decreases.

If the magnetizations of the dielectrics are directed oppositely, then $H_{s+} = H_{s-}$ and $H_{a+} = -H_{a-}$. Therefore

$$|H_+(x)| = 2H_0 \left| \cos\left(\frac{\beta_a - \beta_s}{2} x \right) \right|, \tag{24}$$

and the energy is fully transferred at the tunneling.

As the dispersion linearly depends on magnetization according to Eqs. (5-6), the tunneling length also depends on magnetization. Eq. (15) is valid in the linear approximation.

For the co-directed magnetizations

$$H_+(x) = H_0 \left\{ \left(1 + 2\frac{\alpha_{1s} v_{1s}}{\alpha_{2s}^2 - \alpha_{1s}^2} \right) \exp(i\beta_s x) + \left(1 + 2\frac{\alpha_{1a} v_{1a}}{\alpha_{2a}^2 - \alpha_{1a}^2} \right) \exp(i\beta_a x) \right\}, \tag{25}$$

and

$$|H_+(x = l_{tun})| = 2H_0 \left| \frac{\alpha_{1s} v_{1s}}{\alpha_{2s}^2 - \alpha_{1s}^2} - \frac{\alpha_{1a} v_{1a}}{\alpha_{2a}^2 - \alpha_{1a}^2} \right|. \tag{26}$$

Hence the energy transfers partially. According to Eqs. (22) and (26) the tunneling coefficient has the form:

$$K = 1 - \bar{a} g_1^2. \tag{27}$$

The tunneling length doesn't depend on magnetization, as Eqs. (3-4) don't contain magnetization contributions.

V. GENERAL PROPERTIES OF MAGNETOPLASMON TUNNELING

To conclude, changes of magnetization configuration of the system lead to change of tunneling coefficient and tunneling length. The general properties of magnetoplasmon tunneling are presented in Table 1.

Table 1. Dependence of magnetoplasmon tunneling characteristics on magnetization.

Magnetization configuration	Thin film	Thick film
$M_1 = M_3$	$K = 1 - \bar{a} g^2$ $l_{tun} = l_{tun0}$	$K \sim g^{-1}$ $l_{tun} \sim g^{-1}$
$M_1 = -M_3$	$K = 1$ $l_{tun} = l_{tun0} + ag$	

The thick film approximation can be applied if the dispersion of plasmonic modes are close to those for singular interfaces. The critical value of the film thickness for this approximation decreases at the magnetization increase. Without the magnetization this approximation is not valid.

The numerical calculations demonstrate that the effect in metallic film is slight. At the thickness of gold film of 80 nm the effect is noticeable only at gyration of about $0.1 - 0.5$. At the oppositely directed magnetizations the tunneling length is about 6 µm, and at the zero magnetization it is 5 µm. At the codirected magnetizations the tunneling coefficient reduces to 10% and the tunneling length is 1.6 µm.

The similar effect takes place also in more complex structures, for example, in periodic metallic grating. At this the slit width should be rather small (about 10% of the period) to decrease scattering losses. According to the empty lattice approximation for the magnetoplasmon dispersion the plasmon properties in grating are close to those of metallic film. In this case the results presented here for the film remail valid for the grating as well.

ACKNOWLEDGMENT

The work is supported by Russian Foundation for Basic Research (13-02-01122, 13-02-91334, 12-02-33100, 11-02-00681, 12-02-31298, 12-02-31396) and Federal Targeted Program "Scientific and Scientific-Pedagogical Personnel of the Innovative Russia".

REFERENCES

[1] S. A. Maier, "Plasmonics: fundamentals and applications", Springer, New York, 2006.

[2] M. Inoue, M. Levy, A.V. Baryshev (ed.), "Magnetophotonics: From Theory to Applications", Springer, 2013.

[3] K. Y. Bliokh, Y. P. Bliokh, V. Freilikher, S. Savel'ev, F. Nori "Colloquium: unusual resonators: plasmonics, metamaterials, and random media" Rev. Mod. Phys. Vol. 80. No. 4. Pp. 1201-1213, 2008.

Gaussian beam in linear gradient-index medium

A.A. Kovalev, V.V. Kotlyar

Image Processing Systems Institute of the Russian Academy of Sciences, Samara, Russia

Abstract: **We have obtained an integral transform describing paraxial propagation of a light beam in gradient-index medium with linear dependence of dielectric permittivity on transverse Cartesian coordinates. We have shown that propagation of light in such medium is equivalent to passing through the prism, propagating in homogeneous medium and again passing through the same prism. We have also shown that for the Gaussian beam, propagating in such medium, its center is being shifted along a parabola, its radius is coinciding with radius of the Gaussian beam in homogeneous medium.**

To describe the propagation of light fields in homogeneous medium and various optical systems the integral transforms are often used, such as the Rayleigh-Sommerfeld transform[1], Kirchhoff, Fresnel and Fourier transforms [2]. If light propagates through the paraxial optical system, described by ABCD-matrix, then distributions of complex amplitudes in the input and output planes are related with each other by the ABCD-transform [2]:

$$
E(x,z) = \sqrt{\frac{-ik}{2\pi B}} \times
$$
$$
\times \int_{-\infty}^{+\infty} E(\xi,0) \exp\left[\frac{ik}{2B}\left(A\xi^2 - 2x\xi + Dx^2\right)\right] d\xi.
\tag{1}
$$

where ξ and x are transverse Cartesian coordinates in input and output planes, $k = 2\pi/\lambda$ is the wave number and λ is wavelength.

Later it occurred that the ABCD-transform describes the propagation of light in a gradient-index waveguide with a parabolic dependence of the refractive index on the transverse coordinates: $n(x) = n_0[1 - x^2/(2a^2)]$ [3, 4], where n_0 is the refractive index at the axis of the waveguide ($x = 0$), a is the rate of decreasing of the refractive index from the waveguide center to its edge, z is the propagation distance from the input to the output plane.

In Ref. [5] a solution of the paraxial propagation equation has been obtained for a planar gradient-index medium, inhomogeneous along optical axis and with linear transverse profile. This solution was found in a form of integral transform from the unknown spatial Fourier-spectrum of the initial field.

In this paper, based on [5], we show that within paraxial approximation the quadratic-phase integral transform describes also propagation of light in a 2D linearly gradient-profile medium with dielectric permittivity linearly dependent on the transverse coordinate

$$
n^2(x) = n_0^2(1 - \alpha x),
\tag{2}
$$

where x is the Cartesian coordinate in the plane transverse to the optical axis z, n_0 is the refractive index at the optical axis, α is the rate of changing of the dielectric permittivity along axis x. Properties of the Gaussian beam and the Airy beam, propagating in such a gradient medium, has also been studied.

Let us consider the gradient-index medium (2). Paraxial Helmholtz equation for the complex amplitude U of TE-polarized beam, propagating in such medium, reads as

$$
2ik\frac{\partial U}{\partial z} + \frac{\partial^2 U}{\partial x^2} - k^2\alpha x U = 0,
\tag{3}
$$

where $k = k_0 n_0$ is the wave number at the optical axis ($x = 0$), $k_0 = 2\pi/\lambda_0$ is the free space wave number and λ_0 is the free space wavelength. We will seek the solution of (3) in the following form:

$$
U(x,z) = A \times
$$
$$
\times \int_{-\infty}^{+\infty} S(u)\exp\left[i\left(Bx^2 + Cu^2 + Dux + Ex + Fu\right)\right] du,
\tag{4}
$$

where A, B, C, D, E, F are functions of z and S is an arbitrary function.

Substituting (4) into (3), we obtain a system of six first-order ordinary differential equations. Solving this system, we obtain the integral transform relating the complex amplitudes of light in the two planes transverse to the optical axis:

$$
U(x,z) = \sqrt{\frac{-ik}{2\pi z}}\exp\left(-\frac{ik\alpha^2 z^3}{96}\right) \times
$$
$$
\times \int_{-\infty}^{+\infty} U(\xi,0)\exp\left[\frac{ik}{2z}(\xi-x)^2 - \frac{ik\alpha z}{4}(x+\xi)\right] d\xi.
\tag{5}
$$

It is easy to see that at $\alpha = 0$ the integral transform (5) becomes the well-known Fresnel transform.

It is seen from (5) that the propagation of light along the distance z in the medium (2) is equivalent to passing through the prism with its slope being proportional to z, subsequent propagation in a homogeneous medium with a refractive index n_0, second passage through said prism, and an additional constant phase shift, which depends cubically on the propagation distance z.

For example, let us consider the distribution of the 2D Gaussian beam with waist radius of w_0:

$$
U(\xi,0) = \exp\left(-\frac{\xi^2}{w_0^2}\right).
\tag{6}
$$

At a distance of z from its waist the beam will have the following complex amplitude:

978-1-4799-0019-0/13 $31.00 © 2013 IEEE

$$U(x,z) = \left\{ \frac{w_0}{w(z)} \exp\left[i\zeta(z)\right] \right\}^{\frac{1}{2}} \times$$
$$\times \exp\left\{ -\frac{\left[x - x_0(z)\right]^2}{w^2(z)} + \frac{ik\left[x - x_1(z)\right]^2}{2R(z)} + i\Phi(z) \right\}. \tag{7}$$

where $z_R = kw_0^2/2$ is the Rayleigh range, $w(z)$ is the beam radius at distance z:

$$w(z) = w_0\sqrt{1 + \frac{z^2}{z_R^2}}, \tag{8}$$

$x_0(z)$ is the beam center (intensity maximum) at distance z:

$$x_0(z) = -\frac{\alpha z^2}{4}, \tag{9}$$

$x_1(z)$ is the beam center of curvature:

$$x_1(z) = -\left[x_0(z) - \frac{\alpha z_R^2}{2}\right], \tag{10}$$

$\zeta(z)$ is the Gouy phase shift:

$$\zeta(z) = -\arctan\left(\frac{z}{z_R}\right), \tag{11}$$

$R(z)$ is the radius of curvature of the beam wavefront:

$$R(z) = z\left[1 + \left(\frac{z_R}{z}\right)^2\right], \tag{12}$$

and $\Phi(z)$ is the additional phase shift:

$$\Phi(z) = \frac{k\alpha^2 z^2}{24}\left[2z - 3R(z)\right]. \tag{13}$$

It is seen in (7)–(13) that in the linearly graded-index profile medium (2) the center of the paraxial Gaussian beam shifts along a parabola, proportionally z^2, and its radius coincides with the radius of the Gaussian beam, propagating in a homogeneous medium with a refractive index n_0.

Numerical simulation of propagation of the Gaussian beam in medium (2) has been conducted with use of the finite-difference time-domain FDTD-method of solving Maxwell's equations. We considered a gradient-index medium with the dielectric permittivity linearly changing from $\varepsilon_a = 1$ (air) to $\varepsilon_g = 2.25$ (glass). Other simulation parameters were the following: free space wavelength $\lambda = 633$ nm, simulation area $-20\lambda \leq x \leq 20\lambda$, $0 \leq z \leq 50\lambda$, propagation time $0 \leq t \leq 100\lambda/c$ (c is the speed of light in vacuum), sampling step along both coordinates $\lambda/16$, time step $\lambda/32$. Dielectric permittivity at the optical axis was $\varepsilon(x = 0) = (\varepsilon_a + \varepsilon_g)/2 = 1.625$ (i.e. $n_0 = 1.27$). The α-parameter has been chosen such that $\varepsilon(x = -20\lambda) = \varepsilon_g$, $\varepsilon(x = +20\lambda) = \varepsilon_a$, therefore $\alpha = 1/(52\lambda) \approx 0.03$. The Gaussian beam waist radius was $w_0 = 2\lambda$, polarization – TE, i.e. $\mathbf{E} \equiv (0, E_y, 0)$.

Shown in Fig. 1 is the time-averaged intensity in the plane Oxz. Bright dots indicate centers of the Gaussian beam at different distances z, calculated with use of (9). It is seen that despite paraxial nature of (9) it rather accurately describes center of the Gaussian beam simulated by nonparaxial FDTD-method.

Now let us consider the Airy beam [6] in medium (2). We suppose that in the initial plane the following complex amplitude is given:

$$U(x, z = 0) = \mathrm{Ai}\left(\frac{x}{x_0}\right), \tag{14}$$

where x_0 is scaling factor. Substituting (14) into the integral transform (5), we find that at a distance of z the amplitude reads as:

$$U(x,z) = \exp\left[i\left(\frac{\alpha}{2} - \frac{k^2 x_0^3 \alpha^2}{6} - \frac{1}{3k^2 x_0^3}\right)\frac{z^3}{4kx_0^3}\right] \times$$
$$\times \exp\left[i\left(\frac{x}{x_0} - k^2\alpha x_0^2 x\right)\frac{z}{2kx_0^2}\right] \times \tag{15}$$
$$\times \mathrm{Ai}\left[\frac{x}{x_0} + \left(\alpha - \frac{1}{k^2 x_0^3}\right)\frac{z^2}{4x_0}\right].$$

In general, a light beam with an amplitude (15) propagates along the parabolic trajectory as in a homogeneous medium. However, if parameters x_0 and α chosen so that $\alpha = 1/(k^2 x_0^3)$, the argument of the Airy function becomes independent on z and the amplitude (15) becomes the following:

$$U(x,z) = \mathrm{Ai}\left(\frac{x}{x_0}\right). \tag{16}$$

The amplitude (16) describes the modal solution of (3) as described in [7]. It is seen in (16) that the Airy beam with its scale being matched to the properties of medium propagates in a linearly gradient-profile medium (2) along a straight line (Fig. 2). Fig. 2 is obtained by the FDTD-simulation, simulation parameters are the same as in Fig. 1, however, the region of simulation has been expanded along the x axis: $-40\lambda < x < 40\lambda$, therefore α-parameter was $1/(104\lambda)$, and x_0 was equal to $1/(k^2\alpha)^{1/3} = (26/\pi^2)^{1/3}\lambda \approx 0.87$ μm. Small bend of the beam in Fig. 2 is due to finiteness of the beam at the input plane.

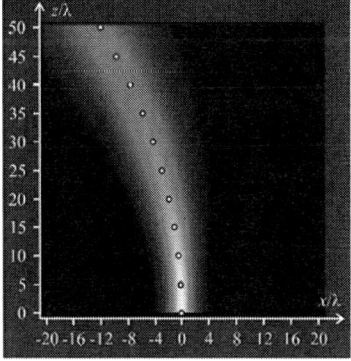

Fig. 1. Time-averaged intensity in the plane Oxz of the Gaussian beam propagating in medium (2).

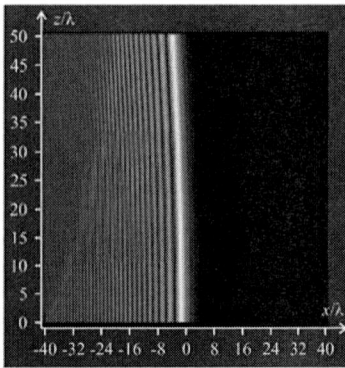

Fig. 2. Time-averaged intensity in the plane Oxz of the straight Airy beam propagating in medium (2).

So, the following results have been obtained:

– An integral ABCD-type transform has been obtained which describes the paraxial propagation of a light beam in a medium with linear dependence of the dielectric permittivity on the transverse coordinate. It is shown that the propagation of light along the distance z in such a medium is equivalent to passing through the prism with its slope being proportional to z, subsequent propagating in a homogeneous medium with a refractive index n_0, second passage through the said prism, and an additional constant phase shift, which depends cubically on the propagation distance z.

– It is shown that at the propagation of a Gaussian beam in a gradient medium (2) its center shifts along a parabola, proportionally to z^2, and its width equals to the width of the Gaussian beam propagating in a homogeneous medium with a refractive index on the optical axis of the medium (2).

– Depending on the relationship between the parameters of the medium (2) and the initial scale-factor of the Airy beam (14), its path can be either a parabola, curved in a negative or positive direction of x, or a straight line along the axis $x = 0$.

The work was supported by The Ministry of education and science of Russian Federation (project 8027), the RF Presidential grants for leading scientific schools (NSh-4128.2012.9), young doctors and candidates of sciences (MD-1929.2013.2, MK-3912.2012.2) and RFBR grants (#12-07-00269, #12-07-31117, #13-07-97008).

REFERENCES

[1] R.K. Luneburg, *Mathematical Theory of Optics*, University of California Press, Berkeley, 1966.

[2] A.E. Siegman, *Lasers*, University Science, 1986.

[3] S.N. Khonina, A.S. Striletz, A.A. Kovalev, and V.V. Kotlyar, "Propagation of Laser Vortex Beams in a Parabolic Optical Fiber," *Proc. SPIE*, vol. 7523, pp. 7523B, 2009.

[4] M. Bandres and J. Gutiérrez-Vega, "Airy-Gauss Beams and Their Transformation by Paraxial Optical Systems," *Opt. Express*, vol. 15, pp. 16719-16728, 2007.

[5] N.K. Efremidis, "Ary trajectory engineering in dynamic lin-ear index potentials," *Opt. Lett.*, vol. 36. pp. 3006-3008, 2011.

[6] M.V. Berry and N.L. Balazs, "Nonspreading Wave Packets," *Am. J. Phys.*, vol. 47, pp. 264–267, 1979.

[7] T. Touam and F. Yergeau, "Analytical Solution for a Linearly Graded-Index-Profile Planar Waveguide," *Appl. Opt.*, vol. 32, pp. 309-312, 1993.

Nature of extraordinary transmission through a metal grating at frequencies closed to the Rayleigh-Wood anomaly

M. I. Panov, A. A. Shmat'ko

Karazin Kharkiv National University, Kharkiv, Ukraine

Abstract: The simple two-dimensional theoretical model of a conducting screen with a periodic array of rectangular holes is developed. It is discovered that the narrow extraordinary transmission peak which occurs at frequencies closed to the first Rayleigh-Wood anomaly is caused by the Fabry-Perot resonance of the fundamental eigenmode propagated in the holes.

Since 1998, when optical extraordinary transmission (ET) through subwavelength hole arrays was reported, this physical phenomenon has attracted a great amount of scientists' interest. Initially, terms of Plasmonics [1] was employed for its explanation. According to this theory, the main physical reason of ET is that a real metal acting as a lossy dielectric supports electromagnetic field modes localized in the immediate vicinity of the air-metal interfaces. These confined modes were called surface-plasmon-polaritons (SPPs).

Later, the well-known fact [2] was rediscovered that metal gratings exhibit the similar effect of resonant full transmission at microwaves and terahertz waves. At these spectrum regions real metals are successfully treated as perfect conductors, and there is only negligible field penetration into a metal volume. Thus, dielectric properties of real metals do not reveal the nature of the effect and the conception of SPPs makes no physical sense at low frequencies. Therefore, the meaningful articles (examining the diffraction gratings as the typical frequency-selective surfaces) are mainly focused on the conventional methods of diffraction [3] and circuit theory [4].

Firstly, because of its simplicity, theoretical analysis considered the two-dimensional (2D) geometries: individual slits and slit arrays perforated in a metal screens. So far it is known that an empty slit (or the slit filled with a double positive magnetodielectric) perforated in a metal screen acts as the Fabry-Perot (FP) cavity. Its fundamental mode, TEM-mode, experiences a periodic sequence of resonances with increasing of a slit depth. They emerge due to constructive interference between the forward and backward waves propagated in a slit. It should be noted that the slit fundamental mode has not a cut-off frequency; therefore it is propagating at all frequency ranges. It is also discovered that FP resonances can red-shift from their expected positions [2, 5].

Subsequently, researchers embarked on the study of the 3D hole gratings. The frequency dependence of the zeroth order transmission efficiency through a periodic rectangular hole grating is shown in Fig. 1. The presented dependence obtained

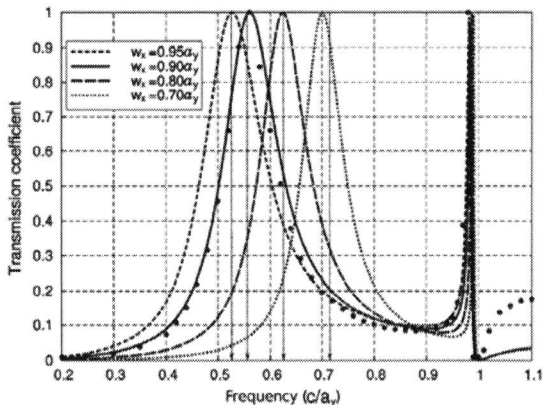

Fig. 1. Transmission spectrum of a perforated thick conducting plate for several lengths w_x of the rectangular holes. The graph is taken from [4]. Here, frequency is normalized by the first Rayleigh-Wood minimum defined by the E-plane grating period a_y; $a_x = a_y$ is the H-plane period, $w_y = 0.2a_y$ is a hole width, $h = 0.2a_y$ is a screen thickness.

in [4] is also typical for periodic arrays of holes with another form of the cross section [3]. As shown in Fig. 1, two transmission resonances (known as "extraordinary") emerge. While the first, low-frequency ET resonance is attributed to the transverse waveguide resonance, the physical nature of the second, narrow ET peak occurring at frequencies closed to the first Rayleigh-Wood anomaly is not very clear.

In order to simplify the theoretical analysis, we propose to approximate the screen structuring in the H-plane of incident wave using the continuous plasma-like material with the same dispersive properties as shown in Fig. 2. Possibility of plasma simulation by parallel-plate media is well-known [6]. Note that the effective dielectric permittivity ε_{eff} can be both positive and negative. Last situation corresponds to the below-cutoff regime of holes. Such simplification of the 3D diffraction problem leads us to the 2D slit geometry.

Fig. 2 (b, c) shows schematically the structure under study. Periodic slits of width w_y are perforated in a perfectly

978-1-4799-0019-0/13 $31.00 © 2013 IEEE

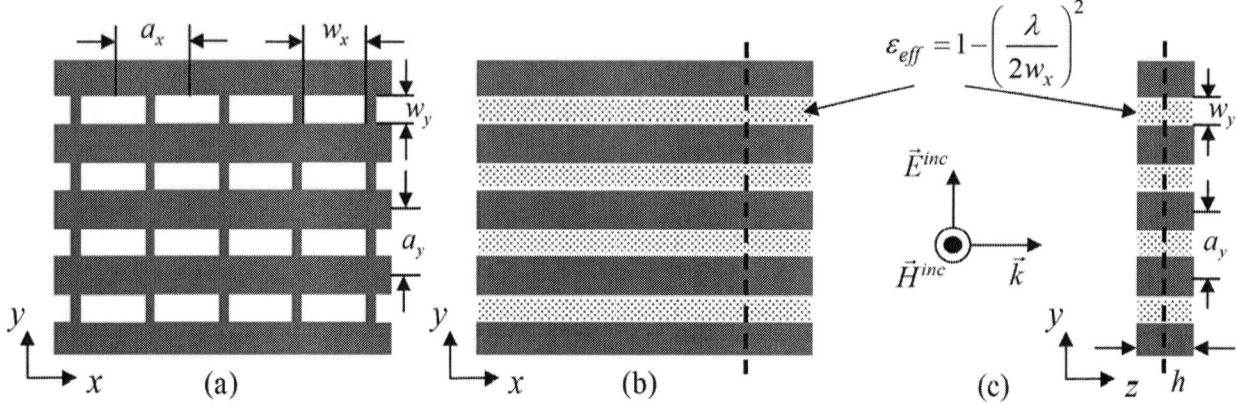

Fig. 2. The problem geometry: (a) the metal screen with the 2D periodic array of rectangular holes and (b, c) the equivalent metal screen with the 1D periodic array of slits filled with the plasma-like medium.

conducting screen of a thickness h. A grating period is a_y. The system is illuminated by a normally incident TM-polarized plane wave with wavelength λ. The slits are filled with a medium with material parameters ε_{eff} and μ_{eff} (μ_{eff} is involved to provide some generality of the problem formulation, $\mu_{eff} = 1$). Scattered fields inside and outside the slits are expanded in Fourier series. By matching the appropriate boundary conditions, the reflection and transmission coefficients for the zeroth diffraction order (in the singlemode approximation of the slit field) can be expressed as

$$R = -1 + \frac{2i\xi Z \sin\varphi \left(\cos\varphi - igZ\sin\varphi\right)}{1 - \left(\cos\varphi - igZ\sin\varphi\right)^2},$$

$$T = \frac{2i\xi Z \sin\varphi}{1 - \left(\cos\varphi - igZ\sin\varphi\right)^2}.$$

Here, $\xi = w_y/a_y$ is the grating fill factor, $Z = \sqrt{\mu_{eff}}/\sqrt{\varepsilon_{eff}}$ is the wave resistance of the slit filling, $\varphi = kh\sqrt{\varepsilon_{eff}\mu_{eff}}$ is the phase shift along the slit depth, $g = \xi(1 - i\sigma)$ is the reactivity of entrance and exit sides of the slits, $\sigma = 2\sum_{m=1}^{\infty} \mathrm{sinc}^2 m\pi\xi / \sqrt{m^2 \left(\lambda/a_y\right)^2 - 1}$.

Fig. 3 shows the frequency dependence of the energy transmittance $|T|^2$ at the same geometry parameters that were used in [4] for drawing Fig. 1. The out-of-y0z-plane size w_x is included in ε_{eff}. As can be seen, the obtained expressions provide qualitative agreement with the exact solution. Thus, the proposed model captures the underlying physics of the effect. Moreover, because of its simplicity, such idealization is more preferred for analyzing of the 3D hole geometry.

Analysis of the expressions for reflection and transmission efficiency clearly reveals the nature of the second ET peak. It occurs due to the first FP resonance of the hole fundamental

mode. There is one field variation along the hole depth at this resonance. Note that the narrow ET peak stands at frequencies closed to the Rayleigh-Wood anomaly only for the case of the not-very-thick screen. Fig. 4 illustrates the peak relocation with increasing of the hole depth h. It should be pointed out that this resonance experiences the significant red-shift which can hide its real origin when the screen thickness is small enough.

Fig. 5 shows the transmission spectrum in the case when h is so large that the second FP resonance is excited at frequencies below the Rayleigh-Wood minimum. The dependence of $|T|^2$ upon h is depicted in Fig. 6. As can be seen, the hole fundamental mode experiences a periodic sequence of resonances with increasing of h. Note that labels 1 (as well as labels 2) in different figures correspond to the same set of the problem parameters.

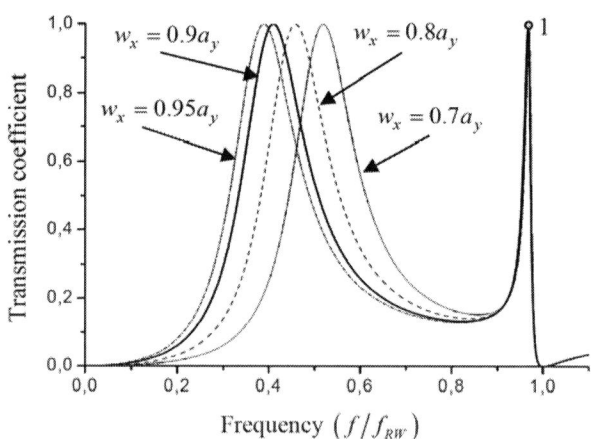

Fig. 3. Transmittance versus frequency normalized by the first Rayleigh-Wood minimum for the different hole lengths w_x ($a_x = a_y$, $w_y = 0.2a_y$, $h = 0.2a_y$, $\lambda_{RW} = a_y$).

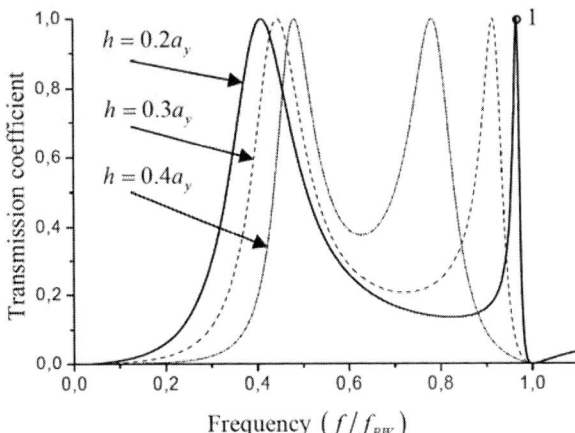
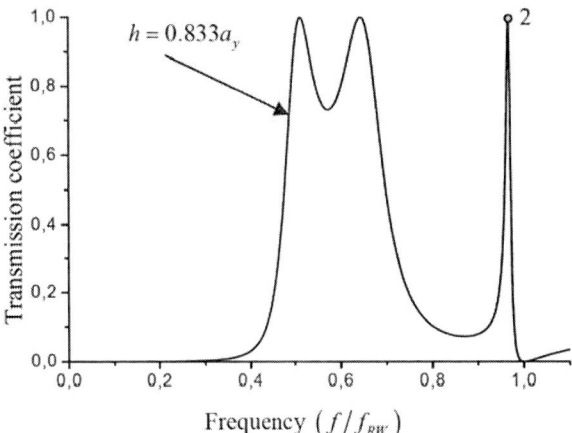

Fig. 4. Transmittance versus normalized frequency for the different screen thickness h ($w_x = 0.9a_y$, $a_x = a_y$, $w_y = 0.2a_y$, $\lambda_{RW} = a_y$).

Fig. 5. Transmittance versus normalized frequency ($w_x = 0.9a_y$, $a_x = a_y$, $w_y = 0.2a_y$, $\lambda_{RW} = a_y$).

By using the resonance condition $\mathrm{Im}\,T = 0$, one can derive the analytical expression for the resonant hole depth:

$$h_n = \frac{n\lambda}{2\sqrt{\varepsilon_{\textit{eff}}\mu_{\textit{eff}}}} - \frac{\lambda}{2\pi\sqrt{\varepsilon_{\textit{eff}}\mu_{\textit{eff}}}} \arctan \frac{2\alpha\sigma}{1+\alpha^2(1-\sigma^2)},$$

where n is the order of the FP resonances, $\alpha = \xi Z$. It is evident that the second term in the formula above leads to the shortening of the resonant hole depth. This is the reason why the FP resonances shift from their expected positions towards longer wavelengths.

It should be pointed out that the grating periodicity is not crucial for full transmission through a perforated metal screen. It is not surprising because it is known that some waveguide-diaphragm structures without periodicity can exhibit ET [4].

Thus, nature of the studied phenomenon consists in the coupling of fields at the entrance and exit air-metal interface due to the excitation of the Fabry-Perot resonances of the fundamental mode propagated in the screen holes. This physical conclusion remains the same not only for rectangular holes but also for another form of the hole cross section. Features of the strong red-shift of the Fabry-Perot resonances will be the subject of our subsequent research.

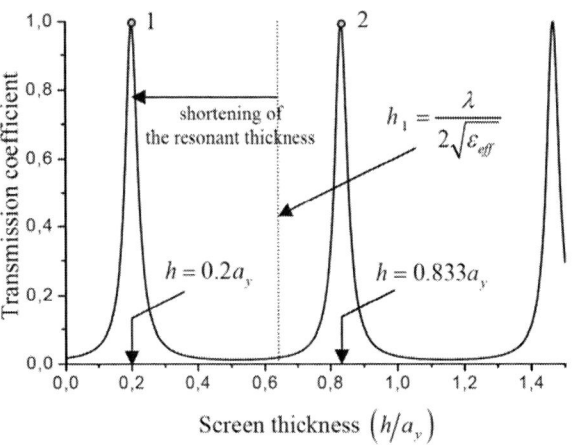

Fig. 6. Transmittance versus normalized screen thickness at frequency $f = 0.966f_{RW}$ ($w_x = 0.9a_y$, $a_x = a_y$, $w_y = 0.2a_y$).

REFERENCES

[1] S. A. Maier. Plasmonics: Fundamentals and Applications. – New York: *Springer*, 2007. – 201 p.

[2] V. P. Shestopalov, L. N. Litvinenko, S. A. Masalov, V. G. Sologub. Difraktsiya voln na reshetkax. (in Russian). – Kharkov: Kharkov University, 1973. – 288 p.

[3] A. A. Kirilenko, A. O. Perov. On the common nature of the enhanced and resonance transmission through the periodical set of holes. *IEEE Trans. on Antennas and Propagation*, 2008, vol. 56, no. 10, p. 3210–3216.

[4] F. Medina, F. Mesa, R. Marques. Extraordinary transmission through arrays of electrically small holes from a circuit theory perspective. *IEEE Trans. on Microwave Theory and Techniques*, 2008, vol. 56, no. 12, pp. 3108–3120.

[5] Y. Takakura. Optical resonance in a narrow slit in a thick metallic screen. *Phys. Rev. Lett.*, 2001, no. 86, pp. 560–5603.

[6] W. Rotman. Plasma simulation by artificial dielectrics and parallel-plate media. *IRE Trans. Antennas Propag.*, vol. 10, issue 1, 1962, pp. 82–95.

Gaussian beam tunneling through a gyrotropic–nihility finely stratified structure

Vladimir R. Tuz[1,2] and Volodymyr I. Fesenko[1]

[1]Institute of Radio Astronomy of National Academy of Sciences of Ukraine, 4, Krasnoznamennaya st., Kharkiv 61002, Ukraine
[2]School of Radio Physics, Karazin Kharkiv National University, 4, Svobody Square, Kharkiv 61077, Ukraine.

Abstract— The three–dimensional Gaussian beam transmission through a ferrite–semiconductor multilayer structure in the Faraday geometry is considered. The beam field is represented by an angular continuous spectrum of plane waves. In the long-wavelength limit, the studied structure is described as a gyrotropic medium defined by the effective permittivity and effective permeability tensors. The investigations are carried out in the frequency band where the real parts of the diagonal elements of both effective permittivity and effective permeability tensors are close to zero. In this frequency band the studied structure is referred to a gyrotropic–nihility medium. It is found out that a Gaussian beam can pass through such a system practically without any distortion.

I. INTRODUCTION

The conception of *nihility* was firstly introduced in the paper [1] for a medium, whose material parameters are close to zero ($\varepsilon \approx 0$, $\mu \approx 0$). Further this conception was extended for an isotropic chiral medium [2]. In this case the permittivity and permeability of the medium remain to be close to zero while the chirality parameter is maintained at a finite value. It was found out that in such an isotropic *chiral–nihility* medium there are two eigenwaves with right and left circularly polarized states, and one of these eigenwaves experiences the backward propagation. In particular, it results in some exotic characteristics in the waves transmission through a single layer and multilayer system consisted of chiral-nihility media [3].

The circularly polarized eigenwaves are also inherent to magneto-optic (gyrotropic) media (e.g. ferrites or semiconductors) in the presence of an external static magnetic field biased in the longitudinal geometry to the direction of wave propagation (Faraday configuration) [4]. Such media are characterized by the permeability or permittivity tensor with non-zero diagonal elements (gyrotropic parameters). Combining gyromagnetic (ferrite) and gyroelectric (semiconductor) materials into a certain unified structure, it is possible to reach *gyrotropic–nihility* condition within a narrow frequency range where the real parts of diagonal elements of both permeability and permittivity tensors simultaneously acquire zero while the gyrotropic parameters are far from zero. In particular, in a finely stratified ferrite–semiconductor structure such a condition is valid nearly the frequencies of ferromagnetic and plasma resonances [5]. This structure has many interesting features, such as complete transmission, impedance matching, backward propagation which are of special interest in the transformation optics.

The goal of this report is to show the ability of such a gyrotropic–nihility finely stratified structure to transmit a Gaussian beam practically without any distortion.

II. PROBLEM STATEMENT AND SOLUTION

A stack of N identical double–layer slabs (unit cells) which are arranged periodically along the z axis is investigated (Fig. 1). Each unit cell is composed of ferrite (with constitutive parameters ε_1, $\hat{\mu}_1$) and semiconductor (with constitutive parameters $\hat{\varepsilon}_2$, μ_2) layers with thicknesses d_1 and d_2, respectively. The structure's period is $L = d_1 + d_2$, and in the x and y directions the system is infinite. An external static magnetic field $\vec{M}$ is directed along the z axis (Faraday configuration). The input $z \leq 0$ and output $z \geq NL$ half-spaces are homogeneous, isotropic and have constitutive parameters ε_0 and μ_0.

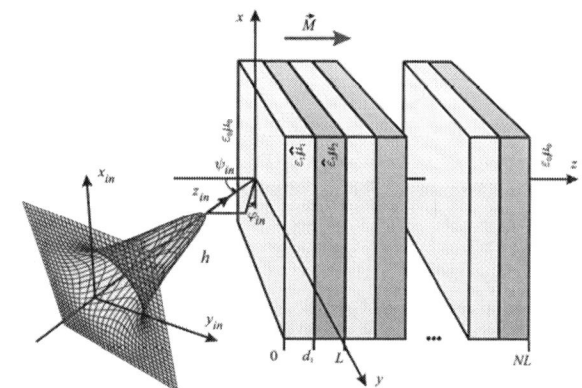

Fig. 1. A periodic stack of one–dimensional double–layer ferrite–semiconductor structure in the Faraday configuration under Gaussian beam illumination.

The auxiliary coordinate system x_{in}, y_{in}, z_{in} is introduced to describe the incident beam field [6], [7]. In it, the incident field $\vec{\Psi}_{in} = \vec{E}_{in}, \vec{H}_{in}$ is written as a continued sum of the partial plane waves with the spectral parameter $\vec{\kappa}_{in}$ (it has a sense of the transverse wave vector of the partial plane wave):

$$\vec{\Psi}_{in} =$$
$$\vec{v} \iint_{-\infty}^{\infty} U(\vec{\kappa}_{in}) \exp[i\vec{\kappa}_{in}(\vec{r}_{in} + \vec{a}_{in}) + i\gamma_{in}(z_{in} + a_3)]d\vec{\kappa}_{in}.$$

$$(1)$$

In Eq. (1) the vector $\vec{v}$ is related to E ($\vec{v} = \vec{e}_{in}$) or H ($\vec{v} = \vec{h}_{in}$) field, respectively, $\vec{e}_{in} = \vec{P}V_p - \vec{b}_{in} \times \vec{P}V_s$, $\vec{h}_{in} = \vec{P}V_s + \vec{b}_{in} \times \vec{P}V_p$, where the vector $\vec{P} = \vec{z}_0 \times \vec{n}$ describes the field polarization. In the structure coordinates x, y, z, the vector $\vec{n}$ is characterized via the components $(\cos\theta_{in}\cos\varphi_{in}, \ \cos\theta_{in}\sin\varphi_{in}, \ 0)$, $\theta_{in} = 90° - \psi_{in}$, $\vec{z}_0$ is the basis vector of z-axis, and the vector $\vec{b}_{in}$ describes the direction of the incident beam propagation $\vec{b}_{in} = (\cos\theta_{in}\cos\varphi_{in}, \ \cos\theta_{in}\sin\varphi_{in}, \ -\sqrt{\varepsilon_0\mu_0 - \cos^2\theta_{in}})$, $U(\vec{\kappa}_{in})$ is the spectral density of the beam in the plane $z_{in} = 0$, $\gamma_{in} = \sqrt{k_0^2 - \vec{\kappa}_{in} \cdot \vec{\kappa}_{in}}$, $0 < \arg(\sqrt{k_0^2 - \vec{\kappa}_{in} \cdot \vec{\kappa}_{in}}) < \pi$, and $\vec{a}_{in} = (a_1, \ a_2)$.

The transmitted field is written as

$$\vec{\Psi}_{tr} = \vec{U}_{tr}^{vv} + \vec{U}_{tr}^{v'v} =$$
$$\vec{P}V_v \iint_{-\infty}^{\infty} U(\vec{\kappa}_{in})\tau^{vv}(\vec{\kappa}) \exp[i\vec{\kappa}\cdot\vec{r} + i\gamma(z - NL)]d\vec{\kappa}_{in} \mp$$
$$\vec{b}_{tr} \times$$
$$\vec{P}V_{v'} \iint_{-\infty}^{\infty} U(\vec{\kappa}_{in})\tau^{v'v}(\vec{\kappa}) \exp[i\vec{\kappa}\cdot\vec{r} + i\gamma(z - NL)]d\vec{\kappa}_{in},$$
$$(2)$$

where $\gamma = \sqrt{k_0^2 - \vec{\kappa}\cdot\vec{\kappa}}$, $0 < \arg(\sqrt{k_0^2 - \vec{\kappa}\cdot\vec{\kappa}}) < \pi$, and τ^{vv} and $\tau^{v'v}$ are the complex transmission coefficients $(v, v' = s, p)$ of the partial plane electromagnetic waves. They depend on the frequency of the incident field, angles $(\psi_{in}, \varphi_{in})$ and other electromagnetic and geometric parameters of the structure. The coefficients with coincident indexes (vv) describe the transformation of the incident wave of the perpendicular $(v = s)$ or the parallel $(v = p)$ polarization into the co-polarized wave, and the coefficients with distinct indexes $(v'v)$ describe the transformation of the incident wave into the cross-polarized wave at the structure output. The left and right indexes correspond to the polarization states of the incident and transmitted waves, respectively. The corresponding transmission coefficients are determined using some long-wave approximation which allows us to transform the rigorous solution of the Cauchy problem [4] related to the tangential field components.

In the long-wavelength limit, when the characteristic dimensions of the structure (d_1, d_2, L) are significantly smaller than the wavelength in the corresponding layer $(d_1 \ll \lambda, d_2 \ll \lambda, L \ll \lambda)$, the interaction of electromagnetic waves with the studied periodic gyromagnetic-gyroelectric structure is described analytically using the effective medium theory [5]. From its viewpoint, the periodic structure is presented approximately as an anisotropic (gyrotropic) uniform medium whose optical axis is directed along the structure periodicity, and this medium is described with some effective permittivity $\hat{\varepsilon}_e$ and permeability $\hat{\mu}_e$ tensors (Fig. 2). By this means, the investigation of the wave interaction with an inhomogeneous periodic structure is reduced to the solution of the boundary-value problem of conjugations of an equivalent homogeneous anisotropic layer with surrounding spaces.

It is revealed [5] that there is a frequency $f_{gn} \approx 4.94$ GHz where μ_e^T and ε_e^T simultaneously reach zero. It is significant

that, by special adjusting ferrite and semiconductor type, external static magnetic field strength and thicknesses of layers, it is possible to obtain the condition when μ_e^T and ε_e^T acquire zero at the same frequency. Exactly this gyrotropic–nihility condition is marked in the insets of Fig. 2 with circles. Note that at this frequency, the gyrotropic parameters α_e and β_e are far from zero and the medium losses are small.

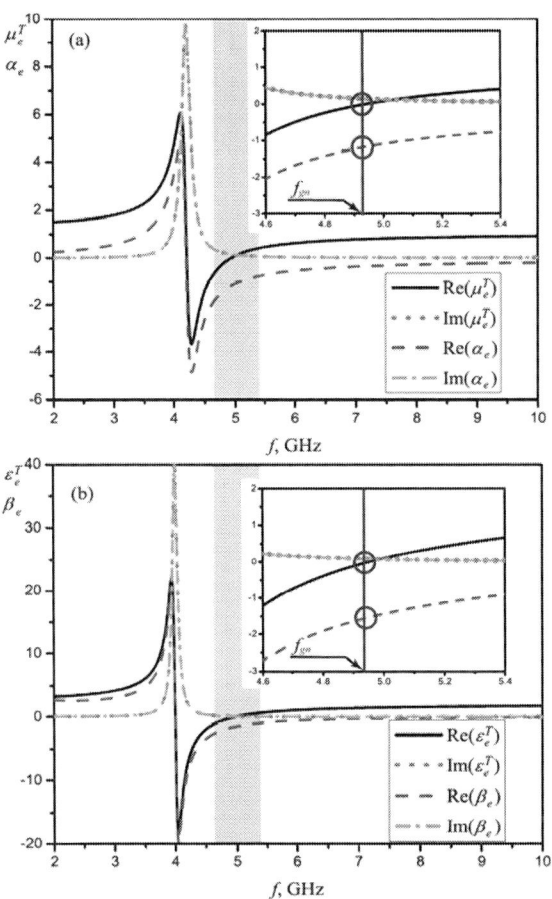

Fig. 2. Frequency dependences of the effective permeability (a) and effective permittivity (b) of the homogenized ferrite–semiconductor medium. We use typical parameters for ferrite and semiconductor in the microwave region. For the ferrite layers, under saturation magnetization of 2000 G, parameters are: $\omega_0/2\pi = 4.2$ GHz, $\omega_m/2\pi = 8.2$ GHz, $b = 0.02$, $\varepsilon_f = 5.5$. For the semiconductor layers, parameters are: $\omega_p/2\pi = 4.5$ GHz, $\omega_c/2\pi = 4.0$ GHz, $\nu/2\pi = 0.05$ GHz, $\varepsilon_0 = 1.0$, $\mu_s = 1.0$. Other structure parameters are: $d_1 = 0.05$ mm, $d_2 = 0.2$ mm.

It is anticipated that if the frequency of the electromagnetic wave which incidents on a finite layer of such composite medium is chosen to be nearly the frequency of the gyrotropic-nihility condition f_{gn}, the transmitted field will acquire some unusual properties. In order to demonstrate this, in the long-wavelength limit, the angular dependences of the transmission coefficient of the co-polarized and cross-polarized partial plane monochromatic waves at two different frequencies are calculated (Fig. 3). The first frequency is chosen to be far from the frequencies of the ferromagnetic and plasma resonances and the second one is selected to be at the gyrotropic-nihility frequency. One can see that in the first case, at the frequency of $f = 20.0$ GHz, the curves of the magnitude of the transmission coefficient have typical form where the total transmission at a certain angle of incidence occurs only for the wave of parallel polarization. Note that at this frequency, the cross-polarized transmission is small. On the other hand, at the gyrotropic-nihility frequency ($f = 4.94$ GHz), the curves of the transmission coefficient magnitude are different drastically from that ones in the first case. Thus, the level of the transmission remains to be invariable almost down to the glancing angles. The cross-polarized transmission is considerable, and the conditions $|\tau^{pp}| \approx |\tau^{ss}|$ and $|\tau^{ps}| \approx |\tau^{sp}|$ are valid in all range of angles.

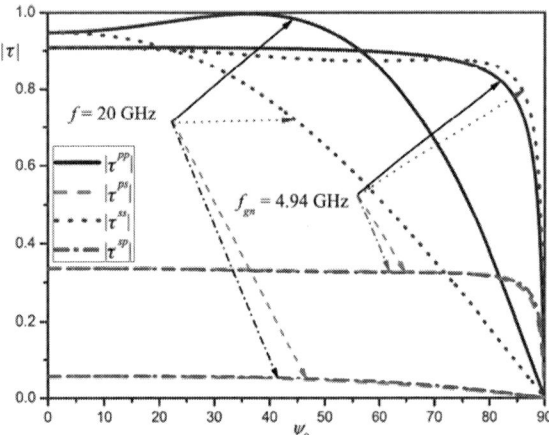

Fig. 3. The angular dependences of the magnitude of the transmission coefficient of the equivalent gyrotropic layer with finite thickness. Parameters of the ferrite and semiconductor layers are the same as in Fig. 2. $d_1 = 0.05$ mm, $d_2 = 0.2$ mm, $NL = 2.5$ mm.

Since the beam field is represented by an angular continuous spectrum of plane waves, the transmitted beam distribution depends on the angular characteristic of the transmission coefficient of spatial plane monochromatic waves at a particular frequency. We consider an incident Gaussian beam with the spectral density assigned due to the law $U(\vec{\kappa}_{in}) = \exp(-(\vec{w} \cdot \vec{\kappa}_{in})^2/16) H_m(k_{xin} w_x/\sqrt{2}) H_n(k_{yin} w_y/\sqrt{2})$, where $\vec{w} = \{w_x, w_y\}$, w_x and w_y are the beam widths along x_{in} and y_{in} axis, respectively, $H_v(\cdot)$ is the Hermit polynomial of v-th order ($v = m, n$). In this report we restrict ourselves to the case of the zeroth-order ($m = n = 0$) beam. The final distribution of the transmitted beam is given in Fig. 4 where the results are presented for two distinct frequencies and orthogonal polarizations. One can see that at the frequency of the gyrotropic–nihility condition the transmitted beam distribution is without any significant distortion and practically the same for s- and p-polarized beams even under the oblique incidence of the primary beam.

ACKNOWLEDGMENT

This work was partially supported (V. R. Tuz) by Ministry of Education and Science of Ukraine under the Program "Elec-

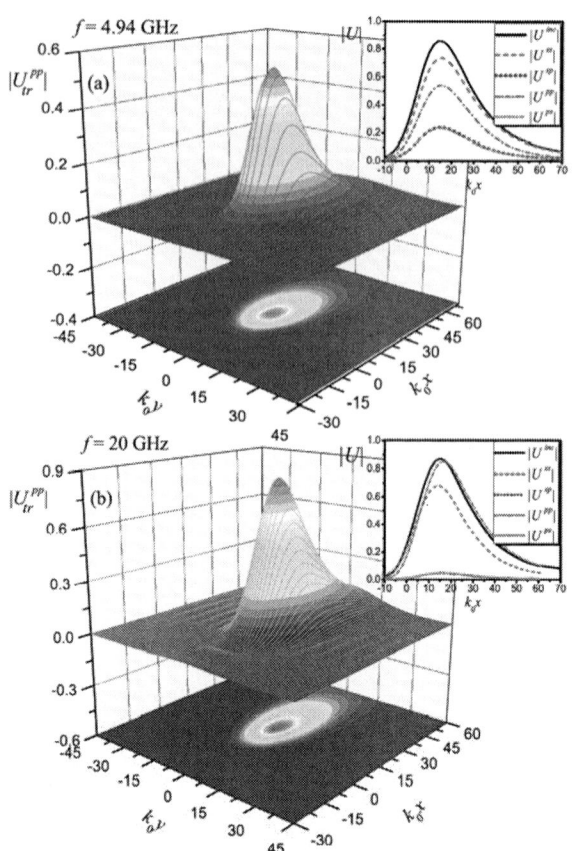

Fig. 4. The distribution of the absolute value of the field $|U_{tr}|$ of the transmitted beam. The field distribution is normalized to the maximum value of the normally incident beam. Parameters of the ferrite and semiconductor layers are the same as in Fig. 2. The incident beam parameters are: $k_0 w_x = k_0 w_y = k_0 h = 10$, $\psi_{in} = 60°$, $\varphi_{in} = 0°$. Other structure parameters are: $d_1 = 0.05$ mm, $d_2 = 0.2$ mm, $NL = 2.5$ mm.

trodynamics of layered composites with chiral properties and multifunctional planar systems", Project No. 0112 U 000561.

REFERENCES

[1] A. Lakhtakia, *An electromagnetic trinity from "negative permittivity" and "negative permeability"*, Int J Infrared Millimeter Waves, **22**, 1731–1734, 2001.

[2] S. Tretyakov, I. Nefedov, A. H. Sihvola, S. Maslovki, and C. Simovski, *Waves and energy in chiral nihility*, J. Electromagn. Waves Appl., **17**, 695–706, 2003.

[3] V. Tuz and C.–W. Qiu, *Semi-infinite chiral nihility photonics: Parametric dependence, wave tunelling and rejection*, Prog. Electromagn. Res., **103**, 139–152, 2010.

[4] V. R. Tuz, M. Y. Vidil, and S. L. Prosvirnin, *Polarization transformations by a magneto–photonic layered structure in the vicinity of a ferromagnetic resonance*, J. Opt., **12**, 095102, 2010.

[5] V. R. Tuz, O. D. Batrakov, and Y. Zheng, *Gyrotropic–nihility in ferrite–semiconductor composite in Faraday geometry*, Prog. Electromagn. Res. B, **41**, 397–417, 2012.

[6] S. N. Shulga, *Two-dimensional wave beam scattering on an anisotropic half-space with anisotropic inclusion*, Optics and Spectroscopy, **87**, 503–509, 1999 (in Russian).

[7] V. Tuz, *Three-dimensional Gaussian beam scattering from a periodic sequence of bi-isotropic and material layers*, Prog. Electromagn. Res. B, **7**, 53–73, 2008.

Spiral beams: New Results and Application

(Invited Paper)

V. Volostnikov, S. Kishkin, S. Kotova

Samara Branch of the Lebedev Physical Intitute, Russian Academy of Sciences
Novo-Sadovaya 221, 443011 Samara, Russian Federation

Abstract: The new results concerning spiral beams and their applications are considered.

1. INTRODUCTION

It is well known that the propagation of a light field is a wave phenomenon and as any oscillatory process is characterized by a complex-valued amplitude. When the distribution of a complex field amplitude is defined in some plane, the subsequent field evolution in the course of its propagation is described by some differential equation. Hence, it follows that generally speaking the light field undergoes quantitative and qualitative changes.

However, with the discovery of lasers and advance of coherent optics describing the propagation of laser beams, it has been theoretically and experimentally shown that lasers can radiate self-consistent light beams i.e. those maintaining their structure during propagation and focusing, up to scale. Such beams are the eigen modes of laser resonators, have a strictly defined form and are described by two families of special functions with different types of symmetry: Hermite - Gauss and Laguerre - Gauss beams.

On the other hand, as have been shown earlier, there exist so called spiral beams or light fields which retain the form of their intensity during propagation up to scale and rotation. Moreover, the spiral beams can acquire the form of any curve $\zeta(t)$, $t \in [0,T]$. The complex amplitudes of these beams are the following:

$$S\left(z,\overline{z}\,|\,\zeta(t),\ t \in [0,T]\right) = \exp\left(-\frac{z\overline{z}}{\rho^2}\right) \times$$

$$\times \int_0^T \exp\left[-\frac{\zeta(t)\overline{\zeta}(t)}{\rho^2} + \frac{2z\overline{\zeta}(t)}{\rho^2}\right] \times$$

$$\times \exp\left[\frac{1}{\rho^2}\int_0^t \left(\overline{\zeta}(\tau)\zeta'(\tau) - \zeta(\tau)\overline{\zeta}'(\tau)\right)d\tau\right]|\zeta'(t)|dt$$

It should be noted that the intensity distribution of the beam in the form of a closed curve is only observed for curves with the surface S satisfying the quantization condition:

$$\frac{1}{i\rho^2}\int_0^T \left(\overline{\zeta}(\tau)\zeta'(\tau) - \zeta(\tau)\overline{\zeta}'(\tau)\right)d\tau = \frac{4S}{\rho^2} = 2\pi n,\ n \in \mathbb{N}_0$$

It's easily seen that the Laguerre - Gauss modes $LG_{0,n}(x,y) = exp(-x^2 - y^2)(x+iy)$ are a special case of quantized spiral beams when a circle is selected as the generating curve.

The entire family of Laguerre-Gauss modes can be obtained in terms of generating curves by means of the birth operator (see fig.1).

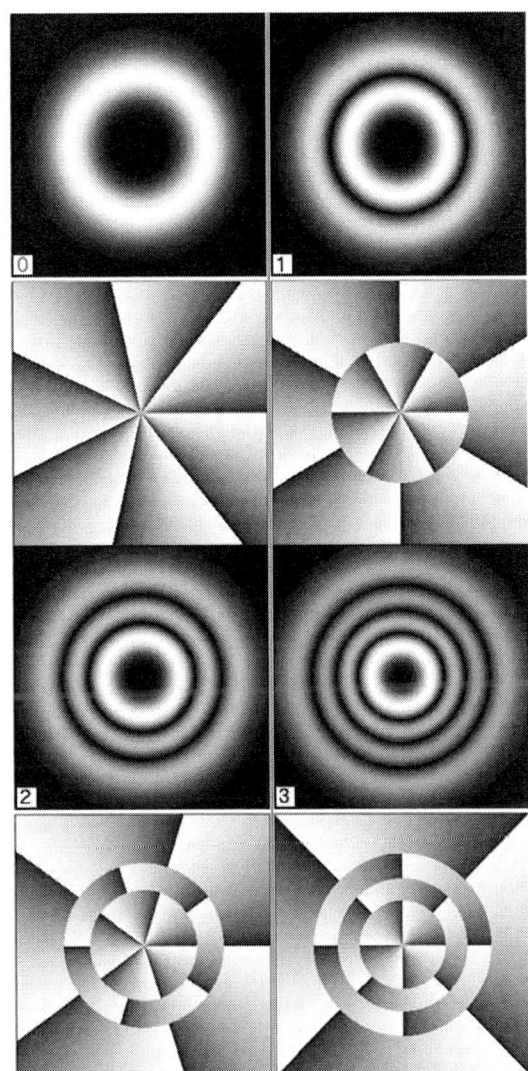

Fig.1. Intensity and phase of $LG_{n,7-n}(x,y)$ under $n = 0,1,2,3$.

For the spiral beams the appropriate formula is the follows:

$$S_m(z,\bar{z}) = e^{z\bar{z}}\frac{\partial^m}{\partial Z^m}(e^{-2z\bar{z}}f(Z)) = \left(\frac{\partial}{\partial Z} - \bar{Z}\right)^m S(z,\bar{z}),$$

here $Z = (x+iy)/\rho$.

The spiral beams in the form of closed curves can be regarded as a generalization of Laguerre-Gauss beams $LG_{m,n}(x,y)$. Hence, we can further develop this analogy and construct, for every generating curve, a family of spiral beams corresponding to the complete family of Laguerre - Gauss beams (see fig.2). The properties of such beams constitute the first part of this report. The other part of the report is dedicated to the application of spiral beam optics to the contour analysis.

The results of numerical simulations are presented.

2. N-DERIVED BEAMS

At first we show that the result of the birth operator action is the follows:

$$B(Z) = \left(\frac{\partial}{\partial Z} - \bar{Z}\right)^N$$

Secondly we obtain the quantization condition for Laguerre-Gauss beams:

$$\frac{1}{i\rho^2}\int_0^T \left(\bar{\zeta}(\tau)\zeta'(\tau) - \zeta(\tau)\bar{\zeta}'(\tau)\right)d\tau = \frac{4S}{\rho^2} = 2\pi n,$$

where $n \neq N_1$ (here N_1- is the number of zeroes inside the domain enclosed by the generating curve. At the end of this part of our report we obtain the formula for the optical angular momentum for these beams:

$$L(S_N) = \frac{\sum_n \frac{\pi}{2}\frac{n!}{2^n}2^n N!|C_n|^2 - \sum_n \frac{\pi}{2}\frac{n!}{2^n}2^n N!|C_n|^2 \cdot N}{\sum \frac{\pi}{2}\frac{n!}{2^n}2^n N!|C_n|^2} =$$

$$= L(S_0) - N,$$

where $L(S_0)$ is the orbital angular momentum of the initial beam.

3. CONTOUR ANALYSIS

The quantization condition is an addition to the condition when the beam intensity is independent of the initial point of integration (see formula (2)). This property is very useful for the contour analysis because the initial point on a contour when we give the finite part of its decomposition effects this decomposition. And the dependence on the initial point presents serious difficulties for contour analysis as a rule [2].

We show that this difficulty can be overcome when the spiral beams are analyzed but not the contour. The results of numerical simulations are offered.

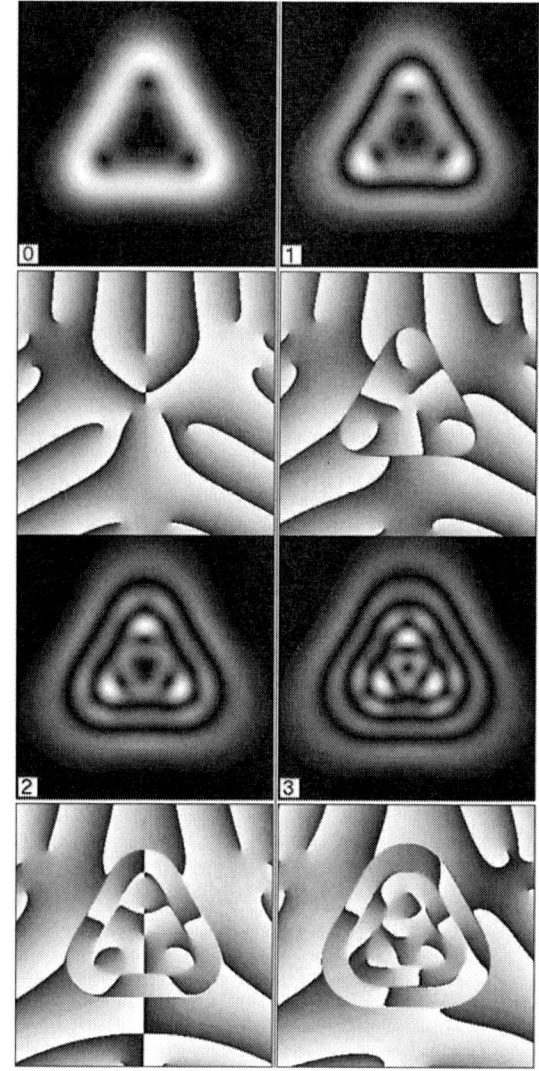

Fig.2. Intensity and phase of the spiral beams $S_n(z,\bar{z})$ for the generating curve in the form of triangular hypocycloid under $n = 0, 1, 2, 3$.

The work is supported by Program of Fundamental Researches of the RAS PSD "Fundamental aspects of physics and technology of semiconductor lasers as the main elements photonics and quantum electronics".

REFERENCES

[1] E.Abramochkin, V.Volostnikov. "Spiral-type beams: optical and quantum aspects", *Optics Commun.* vol.125, N 4-6, pp.302- 323, 1996.

[2] Yu.V. Vizilter, S.Yu. Gheltov, A.V. Bondarenko, M.V. Ososkov, A.V. Morghin. *Image processing and analysis in machine vision,* M., Fizmatkniga, 2010, 672 p.

Broadband similariton: applications to ultrafast optics and photonics

(Invited Paper)

**Levon Kh. Mouradian[1], Aram S. Zeytunyan[1], Garegin L. Yesayan[1], Frédéric Louradour[2],
Alain Barthélémy[2], Ruben Zadoyan[3]**

[1]Ultrafast Optics Laboratory, Faculty of Physics, Yerevan State University, Armenia
[2]XLIM-UMR 6172 Université de Limoges/CNRS, France
[3]Technology & Applications Center, Newport Corporation, USA

Abstract: Applications of broadband nonlinear-dispersive similariton to pulse temporal and spectral compression, femtosecond signal characterization, and CARS spectroscopy are presented, based on our experimental and numerical studies and supported by the concept of similariton-induced temporal lens.

Urgent problems of ultrafast optics and photonics stimulate the interest in similariton due to its prospective applications [1], particularly, to femtosecond signal generation, manipulation, delivery and characterization. Generally, the studies are related to the parabolic similaritons, pulses with the parabolic temporal, spectral, and phase profiles, generated in active fibers, such as rare-earth-doped fiber amplifiers and Raman fiber amplifiers, or in passive dispersion-decreasing fibers, as well as in the laser resonator. Another type of similariton was generated recently, in a conventional uniform and passive fiber (without gain), under the combined impacts of Kerr nonlinearity and dispersion [2]. This nonlinear-dispersive similariton is of spectronic nature, it has parabolic phase only, but maintains its temporal shape during the propagation, as well. The spectral interferometric studies of this type of pulses showed the linearity of their chirp, with a slope given only by the fiber dispersion. This property leads to the spectrotemporal similarity and self-spectrotemporal imaging of nonlinear-dispersive similariton, with accuracy given both by spectral broadening and pulse stretching. Both the parabolic similariton of active fiber and nonlinear-dispersive similariton of passive fiber demonstrate prospective features in view of the signal analysis and synthesis in ultrafast photonics, demanding, however, the generation and study of broadband similaritons. The applications of broadband similariton to the urgent problems of nonlinear ultrafast photonics, based on our experimental and numerical studies and supported by the concept of similariton-induced temporal lens are the subjects of this review.

- *Generation of broadband similariton.* The signal analysis and synthesis problems on the femtosecond time scale demand the application of a broadband similariton. We generate broadband nonlinear-dispersive similaritons of ~50 THz-bandwidth (>100 nm at 800 nm wavelength), coupling the 100-fs pulse radiation of a standard commercial laser (~76 MHz repetition rate) of a few 100-

mW average power into a short piece of standard single-mode fiber (~1 m).

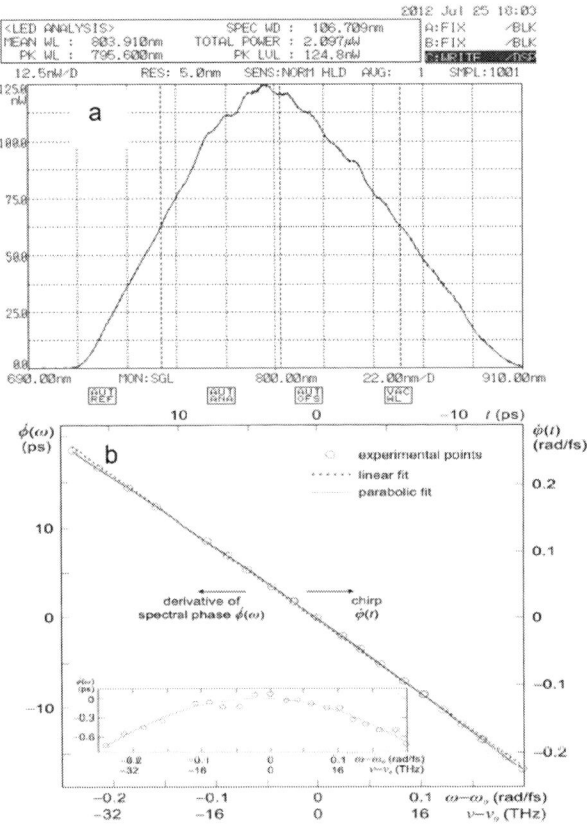

Fig. 1. Spectrum (a) and chirp (b) of broadband nonlinear-dispersive similariton.

- *Similariton pulse compression.* Applications of similariton essentially improve the techniques of pulse compression and shaping, leading to accurate, aberration-free methods, since the chirp of similariton is linear (the phase is parabolic) and its spectrotemporal profile is smooth and bell-shaped. We generate broadband nonlinear-dispersive similariton and compress it in a conventional prism compressor down to 15-20 fs for the average powers of

300-500 mW, comparable with the parameters of commercial 10-fs lasers. The use of a hybrid prism-grating compressor or grism-line, free of third-order dispersion, instead of the prism compressor, will provide pulse compression down to a few femtoseconds for the broadband similariton of 50 THz bandwidth.

- *Spectral focusing in similariton-induced spectrotemporal lens.* Similariton-induced parabolic temporal lens provides an effective aberration-free spectral compression, in the analogy of the beam collimation in the space domain. This method is based on the dispersive stretching of the pulse and afterwards cancellation of the dispersion-induced phase in a quadratic nonlinear process by adding a reference parabolic phase instead of the Kerr lensing of traditional spectral compression. The method can be implemented in the frequency mixing processes, such as sum- or difference-frequency generation, CARS, etc, using the nonlinear-dispersive similariton as a reference pulse generated from the part of signal. Experimentally, we achieve an effective (up to 22 times) aberration-free spectral compression through sum-frequency generation. This type spectral focusing is of special interest in CARS spectroscopy in view of its resolution improvement.

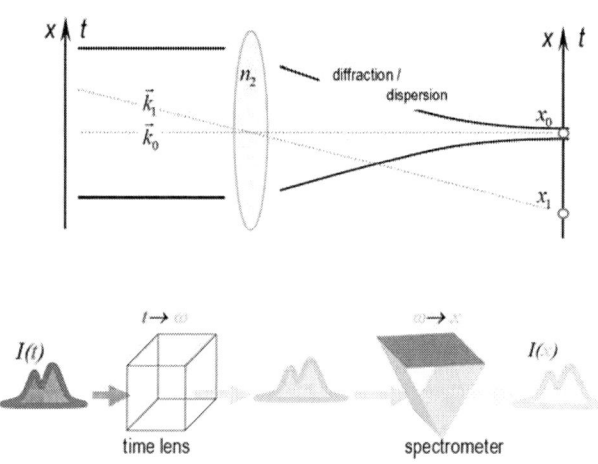

Fig. 2. Illustration of the principles of the time lens (a), and spectrotemporal imaging (a). The use of similariton makes the time lens parabolic and aberration-free

- *Noise suppression-filtering in similariton-induced temporal lens.* Similariton regulates the radiation parameters, and thus, also the parameters of the compressed pulse. The noise filtering of radiation is prospective also through spectral focusing in the similariton-induced temporal lens.

- *Spectral control of signal in the similariton-induced temporal lens.* The temporal delay between the interacting similariton and dispersively stretched signal pulses leads to the frequency shift of the spectrally compressed radiation. This method of spectral focusing and fine frequency tuning of the signal in the similariton-induced parabolic temporal

lens can serve for resonant spectroscopy and optical communication. This technique allows also measuring the similariton chirp and the dispersion of material, where the similariton is generated.

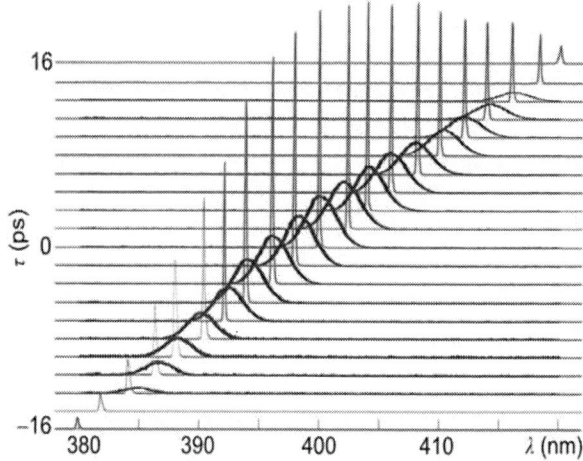

Fig. 3. 3D frequency tuning pattern of similariton obtained in the result of spectral compression.

- *Similariton-based chirped CARS spectroscopy and microscopy.* The use of broadband femtosecond laser pulses in CARS microscopy with high spectral resolution was demonstrated by the use of pulse chirping and spectral focusing, providing both spectral tuning and high contrast imaging [6]. Another approach of special interest is based on the periodical amplitude-modulation of broadband radiation with high-repetition rate, in resonance with the Raman oscillations of medium [7]. The application of broadband similariton, with the mentioned two approaches, is prospective-promising for CARS microscopy in view of the exploitation of a single laser only, and a substantially simplified setup.

- *Pulse spectrotemporal imaging in similariton-induced temporal lens*, i.e. conversion of the temporal information to the spectral domain for both the intensity and phase: Fourier transformation (FT). The temporal lens serves as an optical processor, which performs the mathematical operation of FT. Direct, real-time, high-resolution temporal measurements are carried out through the spectral imaging of temporal pulse in the parabolic, aberration-free similariton-induced temporal lens, leading to the development of femtosecond optical oscilloscope [3]. The resolution of measurements is given by the bandwidth of similariton, and it is at the level of ~ 5 fs for 50-THz bandwidth similariton.

- *Similariton-based self-referencing spectral interferometry* (SI) for femtosecond pulse complete characterization. The classic SI is based on the interference of the signal and reference beams spectrally dispersed in a spectrometer, with the spectral fringe pattern caused by the difference of spectral phases. The SI measurement is accurate as any interferometric one, but its application range is restricted by the bandwidth of the reference. We improve the method by generating the similariton from the part of signal and using it as a reference. Thus, the method of similariton-based SI combines the simplicity of the principle and configuration of the classic SI with the self-referencing performance [4]. Our comparative experiments of similariton-based SI and spectrotemporal imaging, carried out together with autocorrelation measurements, evidence the quantitative accordance and high precision of both the similariton-referencing methods for accurate femtosecond-scale temporal measurements.

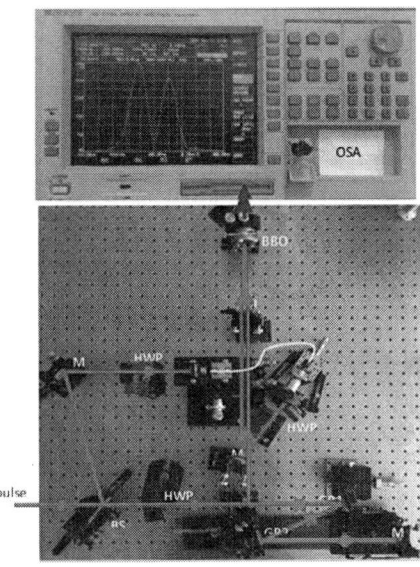

Fig. 4. Prototype of femtosecond optical oscilloscope based on the spectrotemporal imaging of signal in the similariton induced time lens.

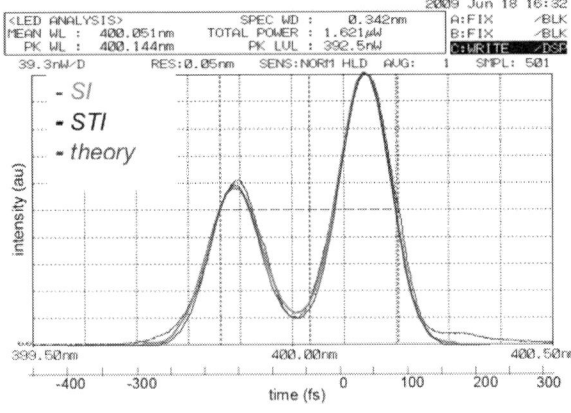

Fig. 5. Measured spectrotemporal image and SI-retrieved profile of a double-peak femtosecond pulse.

- *Reverse problem of nonlinear-dispersive similariton generation* in view of femtosecond pulse characterization. The nonlinear-dispersive similariton asymptotically has a linear chirp, independent of the pulse initial parameters; practically, only the fiber dispersion determines the chirp slope. Thus, the information on the initial pulse parameters is transferring to the intensity profile, which can be measured by a spectrometer due to the self-spectrotemporal imaging of similariton. The solution of the reverse problem of the generation of such a similariton is helpful for femtosecond pulse delivery [8]. It, together with the measured spectrum and known phase, provides complete information about the initial signal, serving for signal characterization. This way a short piece of fiber can serve as an alternative to the FROG device [5].

The results of our studies lead to the development of new techniques and tools for ultrafast photonics.

REFERENCES

[1] C. Finot, J. M. Dudley, B. Kibler, D. J. Richardson, G. Millot, "Optical parabolic pulse generation and applications," IEEE J. Quantum Electron. **45**, 1482 (2009).

[2] A. Zeytunyan, G. Yesayan, L. Mouradian, P. Kockaert, P. Emplit, F. Louradour, A. Barthélémy, "Nonlinear-dispersive similariton of passive fiber," J. Europ. Opt. Soc. Rap. Public. **4**, 09009 (2009).

[3] T. Mansuryan, A. Zeytunyan, M. Kalashyan, G. Yesayan, L. Mouradian, F. Louradour, A. Barthélémy, "Parabolic temporal lensing and spectrotemporal imaging: a femtosecond optical oscilloscope," J. Opt. Soc. Am. B **25**, A101 (2008).

[4] A. Zeytunyan, A. Muradyan, G. Yesayan, L. Mouradian, F. Louradour, A. Barthélémy, "Generation of broadband similaritons for complete characterization of femtosecond pulses," Opt. Commun. **284**, 3742 (2011).

[5] R. Trebino, *Frequency-Resolved Optical Gating. The Measurement of Ultrashort Laser Pulses* (Kluwer Academic Publishers, 2002).

[6] A. F. Pegoraro, A. Ridsdale, D. J. Moffatt, Y. Jia, J. P. Pezacki, A. Stolow, "Optimally chirped multimodal CARS microscopy based on a single Ti:sapphire oscillator," Opt. Express **17**, 2984 (2009).

[7] E. Gershgoren, R.A. Bartels, J. T. Fourkas, R. Tobey, M. M. Murnane, H. C. Kapteyn, "Simplified setup for high-resolution spectroscopy that uses ultrashort pulses," Opt. Lett. **28**, 361 (2003).

[8] M.Kalashyan, C.Lefort, L.Martínez-León, T.Mansuryan, L. Mouradian, F. Louradour, "Ultrashort pulse fiber delivery with optimized dispersion control by reflection grisms at 800 nm", Opt. Express **20**, 25624 (2012).

Transient processes and steady state regimes in dynamic WGM microcavities with time dependent material parameters

(Invited Paper)

N. K. Sakhnenko, *Senior Member, IEEE*

Kharkov National University of Radio Electronics & Institute of Radio Electronics NASU, Kharkov, Ukraine

Abstract: Accurate analysis of electromagnetic fields evolution in whispering gallery mode (WGM) resonators with time dependent material parameters is presented. Wavelength conversion and other possibilities of light manipulation are discussed.

Electromagnetic wave propagation in time-varying media gives rise to new physical phenomena and possibility of novel applications. Dynamic resonators and photonic systems in which material parameters can be varied by external forces have great opportunities for their use in all-optical switchers and tunable filters [1] and represent a powerful approach for all-optical control of light. Permittivity modulation within microcavity systems can be exploited to stop, store and time-reverse of light pulse [2]. Modulation of permittivity in photonic crystals leads to changes of light colour [3, 4] and reversed Doppler shifts [5]. Dynamic control of photonic bandgap can be used to coherently convert a propagating light pulse into stationary excitation which is effectively trapped in the medium [6].

Effective wavelength conversion plays a significant role in many telecom applications. Efficiency of the conversion based on nonlinear effects depends on the input light intensity and pulse travelling distance. However, using linear time-varying structure one can obtain nearly 100% efficient conversion for weak light in a tiny volume [4].

In practice, the temporal switching of the material refractive index can be realized by varying the input signal in a nonlinear structure [7]; by voltage control [8]; by a focused laser beam as a local heat source [9] or else by a free carrier plasma injection [1]. Typically the value of change in refractive index attainable with present day technology is of the order of 10^{-4} [10-12].

Dielectric resonators with WGM have been the subject of significant interest in recent years as they exhibit properties useful for a wide range of applications. However, much of the theoretical work focused upon resonators has concentrated upon prediction of their frequency domain properties, although accurate time domain modelling is essential for active devices and circuit components design.

The main goal of this paper is to demonstrate possibilities of light manipulation in WGM resonators with time varying properties.

Theoretical studies of modern WGM microresonator devices require highly accurate simulations that are complicated by both the open nature and very high Q factors of these

structures. The most widely used today numerical approach is finite difference time domain (FDTD) method that is flexible but demands large computer memory. Moreover, conventional FDTD codes have problems with visualization of the high Q resonances [13].

In this paper the rigorous mathematical method that allows analyzing problems both in the frequency domain and in the time domain has been used. An analytical solution of the problem in the Laplace transform domain has been derived. The time domain electromagnetic field has been recovered by virtue of the computation of the inverse Laplace transform via the residue evaluation at singular points associated with eigenvalues of the structure. This approach guarantees accurate back transformation with controllable accuracy and allows extracting and interpreting physical phenomena easily. This method has been already successfully applied to a variety of time domain problems with different geometries [14-17].

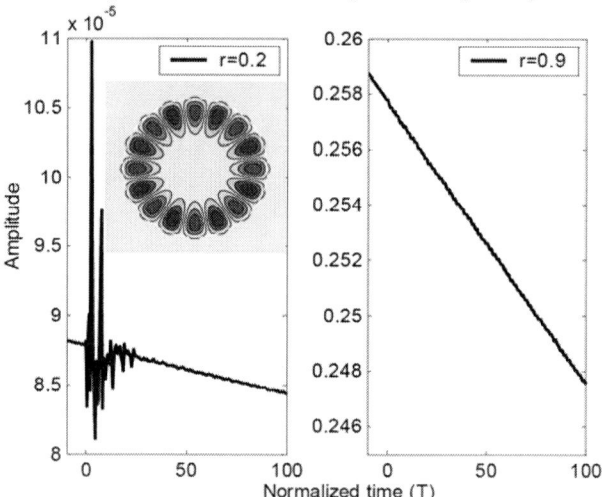

Fig. 1. Electromagnetic field evolution in WGM resonator
($n_1 = 2.631$, $n_1 = 2.63$).

First, we study transient response of WGM that exited in a single disk resonator on time variation of the refractive index. For thin disks 3D problem can be replaced by its 2D model within the effective refractive index approximation. The material is considered to be linear and non-magnetic. The formulation of the problem in the above manner permits construction of an analytical solution for abrupt time change of the refractive index. Arbitrary time dependence of the refractive index can be approximated by stair-case function. Here, dimensionless values are introduced: $T = tc/a$ is the normalized time, $r = \rho/a$ is the normalized distance, a is radius of the cavity, ρ is polar distance, c is the speed of light in a vacuum.

$WGH_{8,1}$ mode is considered as initial field (see inset in Fig. 1), its $ka = 4.5418 + 0.000399i$, refractive index of the material is $n_1 = 2.631$. The notation of the mode $WGH_{k,m}$ stands for WGM of TE polarization; the indices k and m determine the number of the angular field variation and number of field variations along the radius, respectively. Assume, that at zero moment of time, the refractive index changes to the value $n_2 = 2.63$.

Fig. 1 presents the time domain behaviour of the total field versus the normalized time near the centre of the resonator (r=0.2) and near the rim (r=0.9). Changing the refractive index leads to the excitation of all modes (leaky and whispering gallery ones) with the same angular dependence as initial one. However amplitudes of leaky modes are negligibly small (see Fig. 1, left panel). With growing of r transient process becomes smooth (Fig. 1, right panel). Consequently, abrupt change of refractive index transforms WGM into WGM with the same field distribution. Radiation losses that described by leaky modes are negligibly small.

Change of the refractive index leads to the frequency shift of mode from the initial value $\omega_0 = \omega_0' + i\omega_0''$ to the transformed value $\omega_1 = \omega_1' + i\omega_1''$.

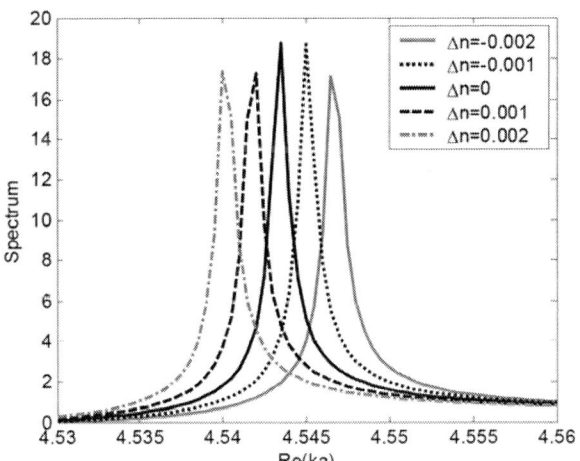

Fig. 2. Spectra inside WGM resonator with and without index turning ($\Delta = n_2 - n_1$).

Fig. 2 shows spectra of electromagnetic field with and without refractive index turning. These spectra obtained from analytical solution of the problem if we consider abrupt time change of the material (zero value of the turning time). If the change of the refractive index occurs over a certain time

interval Δt, the problem can be solved semianalytically. In this case 'slow' variation is approximated by step-like function. It should be noted that we will have the same spectrum for different tuning times. Approximate value in frequency shift $\Delta \omega' = \omega_1' - \omega_0'$ is a little smaller than $\Delta n/n \cdot \omega_0'$. It asymptotically tends to this value with growing of mode Q.

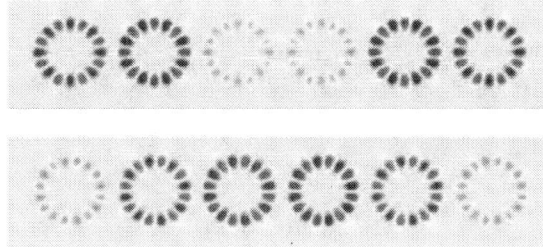

Fig. 3. Near field patterns of antibonding all-even (upper panel) and all-odd (low panel) $WGH_{8,1}$ coupled modes in the linear chain.

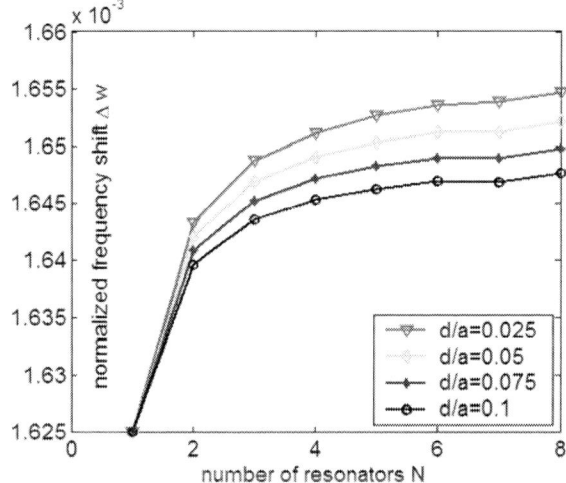

Fig. 4. The normalized frequency shift versus number of the resonators in the chain.

Enhancement in frequency shift can be achieved in coupled resonators. Four families of coupled modes with different types of symmetry can be excited in the linear chain of resonators. Coupled modes can be viewed as symmetric and antisymmetric combinations of single resonator modes. For distant cavities eigenfrequencies are very close to eigenfrequencies of single resonator. When the cavities are brought together splitting of the resonant frequencies occurs. Fig. 3 represent near field distributions of two coupled antibonding $WGH_{8,1}$ modes with all-even and all-odd symmetry along the horizontal and vertical axes in a chain of 6 resonators. If the material is a subject of time variation of the refractive index resonant frequency shift occurs. However in this case it depends on the mode field distribution. The more the area of overlap between the initial field with transient region the more the observed frequency shift. Therefore the greatest frequency shift is observable for coupled antibonding antisymmetrical modes with fields the most localized within cavities. Fig. 4 represents the normalized frequency shift versus number of the resonators in the chain for such modes. If the number of resonators in the chain with small air-gaps gets larger, this leads to increase in the frequency shift for these coupled modes.

During presentation it would be shown that in WGM resonator with the step-periodic time-varying medium it is possible to control the amplitude of the whispering gallery mode that speeds up or slows down the process of the radiation of modes. This effect increases with increasing of the modulation depth.

For the case of refractive index switching in circular coaxial region or in a ring region near the rim a dependence of the frequency shift on the degree of overlap between the initial field and the transient region is observable [15].

It would be discussed the response of the electromagnetic field in a circular WGM resonator to temporal variations of the material properties of a small circular inclusion contained within it.

Due to the double-degeneracy splitting when the inclusion occurs, a redistribution of energy between even and odd excited modes is possible. Misalignment of the initial mode symmetry axes and the inclusion axes leads to a rotation of the transformed field pattern. The maximum rotation occurs for inclusions whose diameter is approximately equal to half of the radius of the cavity [18].

The optical energy transport [19] between two WGMs micro-resonators and sensitivity of such a transport to the time material variation of individual resonator will be discussed.

REFERENCES

[1] K. Djordjev, S. – J. Choi, S. – J. Choi, R. Dapkus, "Microdisk tunable resonant filters and switches," IEEE Phot. Technol. Lett., 14 (6), pp. 823-830, 2002.

[2] M. Yanic, S Fan, "Time reversal of light with linear optics and modulators," Phys. Rev. Lett., 93 (17), 173903, 2004.

[3] E. J. Reed, M. Soljacic, J. D. Joannopoulos, "Color of shock waves in photonic crystals," Phys. Rev. Lett., 90 (20), 203904, 2003.

[4] M. Notomi, S. Mitsugi, "Wavelength conversion via dynamic refractive index tuning of a cavity," Physical Rev. A 73, 051803(R), 2006.

[5] E. J. Reed, M. Soljacic, J. D. Joannopoulos, "Reversed Doppler effect in photonic crystals," Phys. Rev. Lett., 91 (13), 133901, 2003.

[6] A. Andre, M. D. Lukin, "Manipulating light pulses via dynamically controlled photonic bandgap," Phys. Rev. Letters, 89 (14), 143602, 2002.

[7] F. Blom, D. Dijk, H. Hoekstra, M. Driessen, A. Popma, "Experimental study of integrated-optics microcavity resonators," Appl. Phys. Lett., 71 (6), pp. 747-749, 1997.

[8] A. A. Savchenkov, V. S. Ilchenko, A. B. Matsko, L.

Maleki, "High-order tunable filters on a chain of coupled crystalline whispering-gallery-mode resonators," IEEE Photonics Techn. Lett., 17 (1), pp. 136-138, 2005.

[9] M. Benyoucef, S. Kiravittaya, Y. Mei, A. Rastelli, O. Schmidt, "Strongly coupled semiconductor microcavities: A route to couple artificial atoms over micrometric distances," Physical Rev. B 77, 035108, 2008.

[10] B. R. Bennett, R. A. Soref, J. A. Del Alamo, "Carrier-Induced Change in Refractive Index of InP, GaAs, and InGaAsP," IEEE J. Quantum Electron., 26, pp. 113-122, 1990.

[11] A. Liu, R. Jones, L. Liao, D. Samara-Rubio, D. Rubin, O. Cohen, R. Nicolaescu, M. Paniccia, "A High-Speed Silicon Optical Modulator Based on a Metal-Oxide-Semiconductor Capacitor," Nature, 427, pp. 615-618, 2004.

[12] V. R. Almeida, C. A. Barrios, R. R. Panepucci, M. Lipson, M. A. Foster, D. G. Ouzounov, A. L. Gaeta , "All-optical switching on a silicon chip," Opt. Lett., 29 (24), pp. 2867-2869, 2004.

[13] A.V. Boriskin, S. V. Boriskina, A. Rolland, R. Sauleau, A. I. Nosich, "Test of the FDTD accuracy in the analysis of the scattering resonances associated with high-Q whispering-gallery modes of a circular cylinder," J. Opt. Soc. Am. A, 25 (5), pp. 1169-1173, 2008.

[14] E.V. Bekker, A. Vukovic, P. Sewell, T. M. Benson, N. K. Sakhnenko, A. G. Nerukh, "An assessment of coherent coupling through radiation fields in time varying slab waveguides," Opt. and Quantum Electron., 39 (7), pp. 533-551, 2007.

[15] N. K. Sakhnenko, A. G. Nerukh, T. M. Benson, P. Sewell, "Whispering Gallery Mode transformation in a switched micro-cavity with concentric ring geometry," Opt. and Quantum Electron., 40 (11-12), pp. 818-820, 2008.

[16] N. K. Sakhnenko, A. G. Nerukh, T. Benson, P. Sewell, "Near field pattern images in 2D circular resonator with time varying plasma," IEEE Trans. on Plasma Science, 36 (4), pp. 1222-1223, 2008.

[17] A. Nerukh, T. Remaeva, N. Sakhnenko, "Frequency change of partial spherical waves indused by time change of medium permittivity," Opt. Quant. Electron, 41, pp. 327-335, 2009.

[18] N. Sakhnenko, A. Nerukh, T. Benson, and P. Sewell, "Frequency conversion and field pattern rotation in WGM resonator with transient inclusion", Optical and Quantum Electronics, 39 (9), pp. 761 – 771, 2007.

[19] A. Nerukh, N. Sakhnenko, T. Benson, and P. Sewell, "Non-stationary electromagnetics", Singapore: Pan Stanford Publishing, 596 pp., 2012.

Interpretation of the ultrafast optical measurements of time delay in the ionization of Coulomb systems

Vladimir L. Derbov, Vladislav V. Serov, and Tatyana A. Sergeeva,
Saratov State University, Saratov, Russia

(Invited paper)

Abstract—We consider the time delay of electron detachment from a Coulomb center in the process of ionization. It is shown that the attosecond streaking, an ultrafast optical method most commonly used for ionization time delay measurement, can be formally described by placing a virtual detector of the arrival time delay at a certain distance from the center of the system. This approach allows derivation of a simple formula for Coulomb-laser coupling that perfectly agrees with the results of numerical solution of the appropriate time-dependent Schrödinger equation.

Recently the appearance of laser systems that can generate super-intense pulses as short as a few hundreds of attoseconds gave rise to new capabilities in studying electron dynamics in atoms and molecules, which were not available earlier. Among these of particular interest are the methods of measuring the delay of electron ejection from the atom subject to photoionization [1]. At present the measurements of ionization delay are already performed in noble gas atoms using the methods of attosecond streaking [2] and RABITT (reconstruction of attosecond beating by interference of two-photon transition) [3]. The time delays in photoionization of molecules are a subject of growing theoretical interest (see [4] and references therein). However, the very concept of inoization time delay, as well as the interpretation of indirect ultrafast optical measurements of this delay, need further clarification. In particular, is is necessary for correct understanding of large negative time delays numerically found in diatomic targets [4] that seemingly contradict the causality principle. In the present paper the physical meaning of Wigner time delay t_W [5], initially introduced for short-range potentials, is consistently reconsidered for Coulomb systems. The general question of choosing the origin for the ionization time delay in the case of Coulomb field is considered. We demonstrate that the known technique of attosecond streaking is equivalent to placing a detector of the electron arrival delay at a certain distance from the center, and that the Coulomb-laser coupling [6], [7], arising in the theory of the attosecond streaking, is caused by the Coulomb advance of the electron arrival at this virtual device. We use the atomic units of measurement, unless noted otherwise. For details see [8].

Consider the wave function of a system, ionized by an external laser field. When the external action is terminated,

it can be presented in the form

$$\psi(\mathbf{r},t) = \int f(\mathbf{k})\varphi_{\mathbf{k}}^{(-)}(\mathbf{r})e^{-i\frac{k^2}{2}t}d\mathbf{k}, \qquad (1)$$

where $f(\mathbf{k})$ is the ionization probability amplitude, $\varphi_{\mathbf{k}}^{(-)}(\mathbf{r})$ is the wave function of the continuous spectrum, describing the free particle with the momentum $\mathbf{k}$ at the infinitely large distance. If the potential coupling the particle with the center is short-range, then for $r \to \infty$ the wave function can be expressed as

$$\psi(\mathbf{r},t) \sim \int_0^\infty |f(k\mathbf{n})|e^{iS_+}dk - \int_0^\infty |f(-k\mathbf{n})|e^{iS_-}dk. \quad (2)$$

Here the phases of the integrands $S_\pm = \pm kr + \delta(\pm k\mathbf{n}) - k^2/2t$ include the phase $\delta(\mathbf{k}) = \arg f(\mathbf{k})$ of the ionization complex amplitude. For $t \to \infty$ the major contribution comes from the vicinity of the stationary points $k_0 = k_0(\mathbf{r},t)$, in which the derivative of the phase of the integrand is equal to zero, since otherwise the integrands are fast-oscillating. For $k \geq 0$ only S_+ possesses an extremum, so the wave function is expressed as $\psi(\mathbf{r},t) \sim f(k_0\mathbf{n})e^{ik_0 r - i\frac{k_0^2}{2}t}$. This fact can be interpreted as follows. At the point $\mathbf{r} = r\mathbf{n}$ at the moment of time $t = r/k + (1/k)\partial\delta(k)/\partial k$ one can detect the particle having the momentum $\mathbf{k} = k\mathbf{n}$ with the maximal probability. Since the ratio r/k is the time, required for the arrival at the point $\mathbf{r}$ of the particle that left the center $r = 0$ at the time $t = 0$ and moved with the uniform velocity k, the expression

$$t_W = \frac{1}{k}\frac{\partial\delta(\mathbf{k})}{\partial k} = \frac{\partial\delta}{\partial E}(\mathbf{k}). \qquad (3)$$

has the physical meaning of the time delay of the particle arrival at the distance r from the center with respect to that in the case of uniform rectilinear motion. The interpretation of the energy derivative of the phase as the time delay was first proposed by Wigner [5], hence, below we will refer the energy derivative of the phase, t_W, as *Wigner time delay*.

If the external impact is weak enough to use the first-order perturbation theory, then for a centrosymmetric system the ionization amplitude can be expanded in terms of partial spherical waves with definite orbital quantum number ℓ. For each of them the Wigner time delay depends only on the corresponding partial phase

$$t_W = \frac{d\delta_\ell}{dE}. \qquad (4)$$

This work was supported by the RFBR grant No. 11-01-00523a.

In other words, if in an one-center system the transition occurs into a state with fixed ℓ, then the Wigner time delay appears to be completely determined by the energy and the quantum number ℓ, and is independent of the initial state and the particular form of the perturbation potential.

If the potential is not short-range and at large distance tends to Coulomb one, Z/r, then the asymptotic form of the wave function becomes essentially different, since the phase in Eq.(2) acquires the term logarithmic in r. The calculation of the stationary point from yields the time of arrival of the electron, having the momentum k, at the point at the distance r from the Coulomb center with the charge Z

$$t(r) = \frac{r}{k} - \frac{Z}{k^3} \ln 2kr + \frac{Z}{k^3} + t_W. \qquad (5)$$

To understand the physical meaning of Wigner time delay in the case of Coulomb field, the comparison should be made with the classical motion in Coulomb field rather than with the free-particle motion, used for short-range potentials. A reasonable choice is the motion of a particle with zero angular momentum, starting from the center $r = 0$ at the moment of time $t = 0$. Using the classical law of one-dimensional motion

$$t_C(r) = \int_0^r \frac{dr}{p(r)}, \qquad (6)$$

where $p(r) = \sqrt{k^2 + 2Z/r}$ is the momentum of the particle at the distance r from the center, we arrive at the asymptotic formula

$$t_C(r \to \infty) = \frac{r}{k} - \frac{Z}{k^3} \ln \frac{2k^2 r}{Z} + \frac{Z}{k^3}. \qquad (7)$$

Let us define the delay t_0 of a quantum particle moving in the asymptotic Coulomb field with respect to a classical particle with the angular momentum $\ell = 0$, ejected from the center at the time moment $t = 0$ as $t_0 = \lim_{r \to \infty}[t(r) - t_C(r)]$. From Eqs. (5) and (7) it is clear that the ejection delay is related to to Wigner time delay as

$$t_0 = t_W + \frac{Z}{k^3} \ln \frac{k}{Z}. \qquad (8)$$

Fig. 1a shows the Wigner time delays for the continuum of a hydrogen atom at different values of the angular momentum ℓ, mutiplied by k^3 for clearness. It is seen that at small energies $t_W \to -\frac{Z}{k^3} \ln \frac{k}{Z}$ for all values of ℓ. In Fig.1b the delays t_0 are shown. At $k \to 0$ they tend to infinity, but not as $-\frac{Z}{k^3} \ln \frac{k}{Z}$, but much slower, as $1/k^2$. Besides, $t_0 > 0$ for all values E and grows with increasing ℓ. This can be interpreted as follows. The centrifugal potential can be considered as a short-range repulsive one, and the particle motion through it results in the delay with respect to the motion in the pure Coulomb field.

Let us also consider the time delay for the particle, arriving at the detector located at the distance r from the center, in comparison with the case of uniform and rectilinear motion

$$t_D(r) = t_0 + t_C(r) - \frac{r}{k} \simeq t_0 - \frac{Z}{k^3} \ln \frac{2k^2 r}{Z} + \frac{Z}{k^3}. \qquad (9)$$

Obviously, at large r, when the logarithmic term, describing the Coulomb advance

$$t_{CA}(r) = t_C(r) - \frac{r}{k} \simeq -\frac{Z}{k^3} \left[\ln \frac{2k^2 r}{Z} - 1 \right], \qquad (10)$$

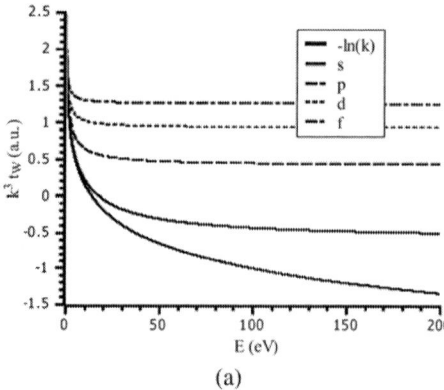

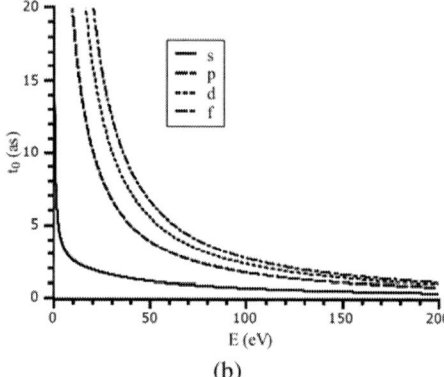

Fig. 1. Dependence of the time delay upon the energy of the ejected electron for a hydrogen atom: a) Wigner time delay; b) the ejection delay, Eq. (8) .

becomes dominant, the time delay is negative.

Direct detection of time delay discussed above is a gedanken experiment rather than a real one. The methods of attosecond streaking [2] and reconstruction of attosecond beating by interference of two-photon transition (RABITT) [3] provide indirect measurements. Here we demonstrate that the correction to the attosecond streaking time delay measurement, referred to as Coulomb-laser coupling [6], in fact is nothing but the Coulomb advance $t_{CA}(r)$, accumulated by the moment of essential variation of the laser field strength. This actually means that the attosecond streaking technique is equivalent to placing a virtual detector measuring the time delay (9) of the particle arrival at a fixed distance from the center.

In the method of attosecond streaking [10] an atom is simultaneously affected by the extreme ultraviolet (XUV) ionizing pulse with the duration of about 100 attoseconds and the auxiliary femtosecond IR laser pulse. The latter changes the momentum of the ejected electron, depending upon the time of ejection, which allows measurement of the ejection delay. Typically the electrons are detected in the direction of the auxiliary field polarization vector of the probing and the electron has a time to get far enough from the center compared to the atomic radius during the action of the IR field. Then the angular motion can be neglected and the interaction potential in which the radial motion occurs takes the form $V(r,t) = \mathcal{E}(t)r$. For simplicity, we first consider the constant

IR field, suddenly switched off at the time T, i.e., $\mathcal{E}(t) = \mathcal{E}_0$ if $t < T$ and $\mathcal{E}(t) = 0$ if $t > T$. Note, that although in real experiments the IR field is strong enough to provide detectable momentum change, the field strength $\mathcal{E}_0$ does not affect the result of temporal measurements, so it can be assumed small to simplify the theoretical consideration.

At the time τ an ultrashort pulse of XUV radiation acts on the atom and an electron is ejected having the energy E, equal to the difference between the mean energy of the XUV photon ω_0 and the ionization potential I. Until the change of $\mathcal{E}(t)$ the energy is obviously conserved, so that $E = p^2(t)/2 + \mathcal{E}_0 r(t)$. One can express the additional momentum, acquired before the field $\mathcal{E}(t)$ is switched off, as $\Delta p = -\mathcal{E}_0 r(T)/p(T)$. If in this formula we assume $r(T) \simeq r^{(0)}(T - \tau)$ and $p(T) \simeq p^{(0)}(T - \tau)$, i.e., instead of the exact position and momentum we use their values for a particle that appeared in the center at the time τ and moves in the absence of the auxiliary IR field, then only the second-order terms in $\mathcal{E}_0$ will be changed. In the first-order approximation with respect to $\mathcal{E}_0$ we obtain

$$\Delta p = -\mathcal{E}_0 \frac{r^{(0)}(T - \tau)}{p^{(0)}(T - \tau)}. \qquad (11)$$

Both for short-range and Coulomb potentials $p^{(0)}(T - \tau) \approx k$ at sufficiently large $T - \tau$. For systems with short-range potential $r^{(0)}(T - \tau) \approx k(T - \tau - t_W)$ and, correspondingly,

$$\Delta p = -\mathcal{E}_0(T - \tau - t_W). \qquad (12)$$

Thus, the time of ejection manifests itself as a horizontal shift of the plot $\Delta p(\tau)$ with respect to $-\mathcal{E}_0(T - \tau)$.

In systems with Coulomb potential $r^{(0)}(T - \tau) \approx k\{T - \tau - t_D[r^{(0)}(T - \tau)]\}$. Since t_D (see Eq.(9)) logarithmically depends upon the distance, one can use $t_D[r^{(0)}(T - \tau)] \approx t_D[k(T - \tau)]\}$, which yields

$$\Delta p = -\mathcal{E}_0\{T - \tau - t_D[k(T - \tau)]\}. \qquad (13)$$

Thus, the attosecond streaking works as a detector, located at the distance $r_{\text{eff}} = k(T - \tau)$ from the point of free particle ejection.

It should be emphasized that although the particle acceleration starts immediately after the transition to the unbound state, the energy conservation relates the velocity change to the remote particle motion, so that the unknown details of the particle behavior near the center become unimportant.

If the field strength $\mathcal{E}(t)$ is a smooth function of time, then the particle energy increment ΔE obeys the first-order equation $\frac{d(\Delta E)}{dt} \approx \frac{d\mathcal{E}}{dt}(t)r^{(0)}(t - \tau)$. If the natural condition $\mathcal{E}(t \to \infty) = 0$ holds, then $\Delta E(t \to \infty) = k\Delta p$ and the momentum increment is expressed as

$$\Delta p = \int_\tau^\infty \frac{d\mathcal{E}}{dt} \frac{r^{(0)}(t - \tau)}{k} dt. \qquad (14)$$

For short-range potential this expression approximately yields [8]

$$\Delta p \simeq -\mathcal{A}(\tau + t_W), \qquad (15)$$

where $\mathcal{E}(\tau) = -\frac{d\mathcal{A}}{dt}(\tau)$ is the vector potential, $A(\tau) = \int_\tau^\infty \mathcal{E}(t)dt$. Thus we arrived at the fundamental expression in the conventional streaking formalism [10], i.e. the momentum

of the electron plotted versus time differs from the vector potential plotted versus time by the *horizontal shift, equal to the Wigner time delay*.

For Coulomb potential similar considerations yield

$$\Delta p \simeq -\mathcal{A}(\tau + t_0) + \Delta p_{CL}. \qquad (16)$$

Here the notation

$$\Delta p_{CL} = \int_0^\infty \mathcal{E}(t + \tau)\left[\frac{1}{p^{(0)}(kt)} - \frac{1}{k}\right]kdt. \qquad (17)$$

is introduced. The change of the momentum Δp_{CL}, arising under the joint action of Coulomb and laser probing field, is usually referred to as Coulomb-laser coupling [6], [7], [9]. It is easily seen that if one substitutes the step function for the field strength in Eq. (17), then $\Delta p_{CL} = \mathcal{E}_0 t_{CA}[k(T - \tau)]$, and, therefore, Eq.(16) turns into Eq.(13). Hence, the momentum correction Δp_{CL} is associated with the Coulomb advance $t_{CA}(r_{\text{eff}})$ of the particle arrival at the point, separated by the distance $r_{\text{eff}} = k(T - \tau)$ from the center, which, in turn, represents the distance, that the particle can reach until the streaking laser field changes significantly. For a periodic field with the frequency ω the characteristic time of the streaking field variation is $\sim 1/\omega$, that yields an estimate $r_{\text{eff}} \sim k/\omega$.

Let the streaking pulse have the form $\mathcal{E}(t) = \mathcal{E}_0(t)\sin(\omega t)$ with the envelope $\mathcal{E}_0(t)$. Assuming the decrease of the integrand in Eq.(17) to be much faster, than the variation of $\mathcal{E}_0(t)$, so that $\mathcal{E}_0(\tau + t) \approx \mathcal{E}_0(\tau)$, we get:

$$\Delta p_{CL} = \frac{\mathcal{E}_0(\tau)}{\omega}\left(\sin(\omega\tau)I_{\cos}(a) + \cos(\omega\tau)I_{\sin}(a)\right). \qquad (18)$$

Here we use the notation

$$I_{\cos}(a) = \int_0^\infty \cos(t)\left[\frac{1}{\sqrt{1 + 2a/t}} - 1\right]dt; \qquad (19)$$

$$I_{\sin}(a) = \int_0^\infty \sin(t)\left[\frac{1}{\sqrt{1 + 2a/t}} - 1\right]dt, \qquad (20)$$

where the parameter a is defined as $a = Z\omega/k^3$.

Under the condition $a \ll 1$ the integrals have simple expressions

$$I_{\cos}(a) \approx -a\left[\ln\frac{2}{a} - 1 - \gamma\right] - \frac{3\pi}{4}a^2; \qquad (21)$$

$$I_{\sin}(a) \approx -\frac{\pi}{2}a + \frac{3}{2}a^2\left[\ln\frac{2}{a} - \frac{1}{6} - \gamma\right], \qquad (22)$$

where $\gamma = 0.57721\ldots$ is the Euler constant. If the variation of the envelope function $\mathcal{E}_0(t)$ during the period $T_{IR} = 2\pi/\omega$ is small and $\mathcal{E}_0(t \to \infty) = 0$, then Eq.(18) can be rewritten as

$$\Delta p_{CL} = -\frac{d\mathcal{A}}{dt}t_{CA}(r_{\text{eff}}) - \mathcal{A}(\tau)\alpha, \qquad (23)$$

where

$$r_{\text{eff}} = \frac{k}{\omega}\exp\left[-\gamma + \frac{3\pi}{4}a\right], \qquad (24)$$

is the effective radius and

$$\alpha = -I_{\sin}(a) = \frac{\pi}{2}a - \frac{3}{2}a^2\left[\ln\frac{2}{a} - \frac{1}{6} - \gamma\right]. \qquad (25)$$

978-1-4799-0019-0/13 $31.00 © 2013 IEEE

Finally, the increment of the momentum is expressed as

$$\Delta p = -(1+\alpha)\mathcal{A}(\tau + t_S), \qquad (26)$$

where

$$t_S = \frac{t_D(r_{\text{eff}})}{1+\alpha} \qquad (27)$$

is the time shift. Thus, in analogy with the above model example of a stepwise field, the time delay $t_D(r_{\text{eff}})$ is recorded at the moment of the field switch-off, and r_{eff} is the distance that the particle reaches by this moment.

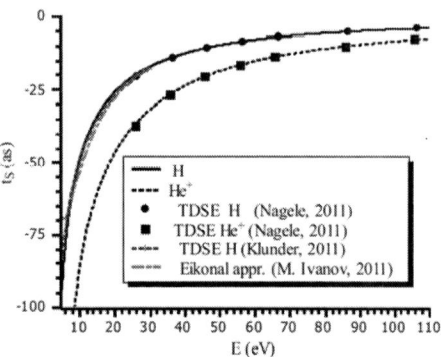

Fig. 2. Dependence of t_S upon the energy of the ejected electron with the angular momentum $\ell = 1$ for the ionization of a hydrogen atom (solid line) and a helium ion He^+ (dashed line) for the frequency of the IR field $\omega = 1.55$ eV. The results are calculated analytically using Eqs. (27), (24), (25), and () (lines); for comparison the results of solving the time-dependent Schrödinger equation for an atom in the joint XUV and IR fields [11] (points) and [3] (dotted line) are shown, as well as the result of the calculation based on the eikonal approximation [7].

Figure 2 demonstrates that the simple analytical expression (27) provides complete agreement with the results [11] and [3], obtained from the solution of time-dependent Schrödinger equation for an electron affected by the ion field and the field of an IR laser pulse.

Let us explicitly express the time shift in terms of the Wigner time

$$t_S = \frac{t_W}{1 + \alpha(\omega Z/k^3)} + t_{IR}, \qquad (28)$$

where the additive part of the contribution from the Coulomb-laser coupling to the observed time shift is

$$t_{IR} = -\frac{1}{1+\alpha(\omega Z/k^3)}\frac{Z}{k^3}\left[\ln\frac{2k^2}{\omega} - 1 - \gamma + \frac{3\pi}{4}\frac{\omega Z}{k^3}\right]. \qquad (29)$$

Using the eikonal approximation, the authors of [7] also derived an analytical expression for t_{IR}. Although the dominant term in this expression, logarithmically depending upon $1/\omega$, is similar to that of Eq. (29), the minor terms are not similar, and even the power of the momentum that enters them is different. At $k \to \infty$ the formula from [7] differs from our one by $2\gamma/k^3$. From Fig. 3 it is apparent that our Eq. (29) and the formula from [7] yield close but not identical results. For $E = 10$ eV the difference between the results amounts to 7 attoseconds, which is detectable by the present-time experimental facilities. In Fig. 2 one can see that our results are apparently closer to those obtained by means of the TDSE

numerical solution than the result of [7]. Yet more important is the difference in the approach itself and the interpretation of the results. In [7] t_{IR} was obtained by evaluation of the initial electron position r_0. Naturally, this makes it hard to generalize the results of [7] over potentials, strongly different from the Coulomb one near the center, which may be of particular importance for ionization of multi-center systems. We were basing on the interpretation of attosecond streaking method as equivalent to placing a virtual detector at the large distance r_{eff} from the center, which allows considering only the classical behavior of the electron remote from the nucleus. The Coulomb-laser coupling in this interpretation is a manifestation of the Coulomb advance of the particle arrival at the point r_{eff}.

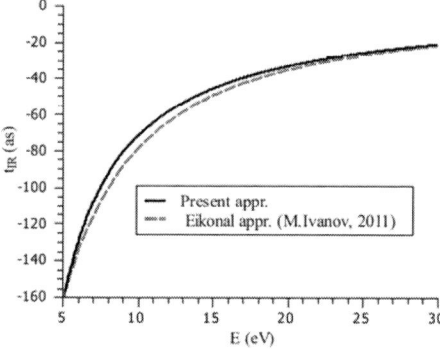

Fig. 3. Dependence of the Coulomb-laser coupling contribution t_{IR} to the time shift upon the energy of the electron: our formula (solid line) and the formula, derived using the eikonal approximation [7] (dashed line).

REFERENCES

[1] A. Scrinzi, M.Yu. Ivanov, R. Kienberger, and D.M. Villeneuve, "Attosecond physics," J. Phys. B: At. Mol. Opt. Phys., vol. 39, p. R1–R37, 2009.

[2] A. Scrinzi et al., "Delay in photoemission," Science, vol. 328, p. 1658-1662, 2010.

[3] K. Klünder et al., "Probing single-photon ionization on the attosecond time scale," Phys. Rev. Lett., vol. 106, p. 143002, 2011.

[4] I.A. Ivanov, A.S. Kheifets, V.V. Serov, "Attosecond time-delay spectroscopy of the hydrogen molecule," Physical Review A, vol. 86, p. 063422, 2012.

[5] E.P. Wigner, "Lower limit for the energy derivative of the scattering phase shift," Phys. Rev., vol. 98, p. 145–147, 1955.

[6] O. Smirnova, A.S. Mouritzen, S. Patchkovskii, M.Yu. Ivanov, "Coulomb-laser coupling in laser-assisted photoionization and molecular tomography," J. Phys. B: At. Mol. Opt. Phys., vol. 40, p. F197F206, 2007.

[7] O. Smirnova, M.Yu. Ivanov, "How accurate is the attosecond streak camera?" Phys. Rev. Lett., vol. 107, p. 213605, 2011.

[8] V.V. Serov, V.L. Derbov, T.A. Sergeeva, "Interpretation of the time delay in the ionization of Coulomb and two-center systems," http://arxiv.org/abs/1304.2686

[9] C.-H. Zhang and U. Thumm, "Electron-ion interaction effects in attosecond time-resolved photoelectron spectra," Phys. Rev. A, vol. 82, p. 043405, 2010.

[10] J. Itatani, F. Quere, G.L. Yudin, M.Yu. Ivanov, F. Krausz, and P.B. Corkum, "Attosecond streak camera," Phys. Rev. Lett., vol. 88, p. 173903, 2002.

[11] S. Nagele, R. Pazourek, J. Feist, K. Doblhoff-Dier, C. Lemell, K. Tökési, J. Burgdörfer, "Time-resolved photoemission by attosecond streaking: extraction of time information," J. Phys. B: At. Mol. Opt. Phys., vol. 44, p. 081001, 2011.

Impact of electric field on dissipative micro-patterns in cholesteric-nematic mixtures (COC-5CB) doped by multiwalled carbon nanotubes

M.S. Soskin[a], V.V. Ponevchinsky[a], L.N. Lisetski[b], O. Deriabina[c] A.I. Goncharuk[c], N.I. Lebovka[c]

[a]Institute of Physics, National Academy of Sciences of Ukraine, Kiev, Ukraine
[b]Institute for Scintillation Materials of NAS of Ukraine, Kharkiv, Ukraine
[c]Institute of Biocolloidal Chemistry of NAS of Ukraine, Kyiv, Ukraine, lebovka@gmail.com

Abstract: Microstructure and electrooptical properties of multiwalled carbon nanotubes (NTs), dispersed in the cholesteric liquid crystal (cholesteryl oleyl carbonate, COC), nematic 5CB and their mixtures, were studied at the fixed temperature of 298 K. The relative concentration $X=COC/(COC+5CB)$ was varied within 0.0-1.0, the concentration C of NTs was varied within 0.01-1% wt. The value of X affected agglomeration and stability of NTs inside COC+5CB mixtures. High-quality dispersion, exfoliation, and stabilization of the NTs were observed in COC solvent ("good" solvent). From the other side, the aggregation of NTs was very pronounced in nematic 5CB solvent ("bad" solvent). The dispersing quality of solvent influenced the percolation concentration, corresponding to transition between the low conductive and high conductive states. Moreover, external electric field influenced the structure of dissipative micro-patterns and orientation of NTs in dependence of the value of X. The mixtures of COC and 5CB were found to be promising for application as functional media with controllable useful chiral and electrophysical properties.

Introduction

The doping of a liquid crystal (LC) by carbon nanotubes (NTs) allows improving of many valuable characteristics of LC cells. E.g., the effects of reduction of the response time and driving voltage, suppressing of the parasitic back flow and image sticking, the electro-optical memory effects, ultra-low percolation thresholds were recently discovered[1]. Particularly interesting are LC composites based on cholesteric (chiral) liquid crystals (CLC). These materials exhibit selective reflection and giant optical activity that can be regulated by electric field and temperature. Moreover, previous experiments have demonstrated the impact of NTs on the selective reflection spectra[2]. The functional ability of such composites is determined by the nature of integration of NT networks into the LC structure. Poor dispersability of NTs in many solvents and their tendency for formation of the bundles or large aggregates presents a serious obstacle to their good functionality and realization of attractive electrical, mechanical and thermal properties. It was shown recently that the dispersing quality of NTs can finely regulated using dispersed nano-particles, e.g., platelets of Laponite[3], as an additive and in a mixtures of good and bad solvents[4]. The spatial arrangement of NTs in different media can be also regulated by application of external electric fields.

However, the effects of NT integration inside the mixtures of different LC substances are still unexplored and little is known about the impact of electric field on LC mixtures doped by carbon NTs.

This work is devoted to the study of the impact of electric field on microstructure and topological dynamics of dissipative micro patterns in suspensions of multiwalled carbon nanotubes in a cholesterol oleil carbonate (COC), nematic pentyl-4'-cyanobiphenyl (5CB), and their mixtures.

Materials and methods

Pure COC exhibits the isotropic (I) $\rightarrow$ cholesteric (Ch) transition at $T_{ICh}\approx309$ K, the cholesteric (Ch) $\rightarrow$ smectic A (SmA) transition at $T_{ChSm}\approx295$ K and the smectic A (SmA) $\rightarrow$ solid(C) transition at $T_{SmC}<273$ K. Pure 5CB demonstrates the nematic-isotropic transition at $T_{NI} \sim 308 - 309$ K, and the crystal- nematic transition (melting) at $T_{CN} = 295.5$ K. Note that both COC and 5CB have similar nematic phase diapasons. It is reasonable to expect that nematic phase diapasons of COC+5CB mixtures are also similar and are in vicinity of the room temperature. It was really justified by direct DCS investigations.

The multiwalled carbon nanotubes (NTs) were prepared from ethylene using the chemical vapour deposition method (TMS pets mash Ltd., Ukraine) with FeAlMo as a catalyst. The typical outer diameter of NTs was 25-30 nm, while their length ranged from 5 to 10 μm. The specific surface area of the NT powder, determined from N_2 adsorption, was 130 ± 5 m^2/g.

The COC+5CB+NTs suspensions were prepared by addition of the relevant quantities of NTs to the COC, 5CB, or COC+5CB mixtures in isotropic (T=320 K) phase with subsequent 20-30 min sonication at the frequency of 22 kHz and the output power of 150 W using an ultrasonic disperser UZDN-2T (Ukrrospribor, Sumy, Ukraine). After sonication, the suspensions were rapidly cooled to the temperature ~350K.

The relative concentration $X=COC/(COC+5CB)$ was varied within 0.0-1.0, the concentration C of NTs was varied within 0.01-1% wt.

The effects of electric field on microstructure of NT suspensions in LC composites was studied using a home made electro-optical cell consisting of two parallel cylindrical electrodes with diameter of 70 μm and distance between them of 240 μm.

The samples were not oriented and the voltage (50 Hz) was changed between 0 and 150 V.

The optical microscopy images were obtained using sandwich-type LC cell of 20 μm thickness. The glass plates were covered by conducting layer and polyimide, which guaranteed planar orientation of LC matrix.

Fine topological evolution of nanocomposites was investigated be the modified polarization microscope, equipped with a light diode emitting 10 nm band around 620 nm, thus providing precise stokes polarimetry of the studied nanocomposites and changes, related to impact of the applied electric field.

The temperature was fixed at T=298 K in all experiments.

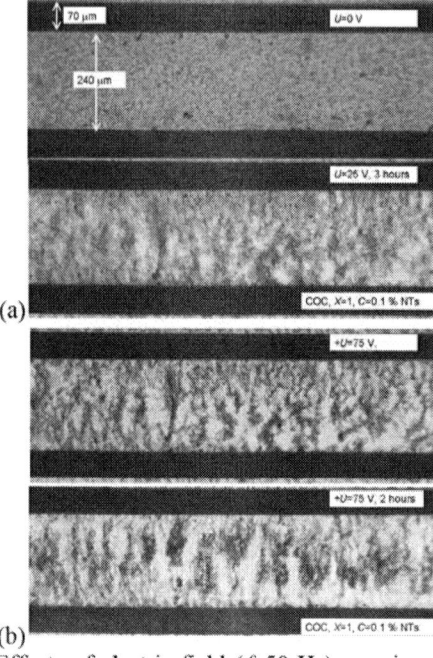

(a)

(b)

Fig. 1. Effects of electric field (f=50 Hz) on microstructure of 0.1% suspension of NTs in pure COC (X=1). The voltage was 25 V (a) and 75 V (b). The layer was un-oriented, T=298 K.

Results and discussion

Preliminary studies of microstructure and phase transitions of NTs, dispersed in the COC+5CB mixtures, have shown that level of X=COC/(COC+ 5CB) affected agglomeration and stability of NTs inside COC + 5CB. High-quality dispersion, exfoliation, and stabilization of the NTs were observed in COC solvent ("good" solvent). From the other side, aggregation of NTs was very pronounced in nematic 5CB solvent ("bad" solvent).

The dispersing quality of solvent influenced the percolation concentration C_p, corresponding to transition between the low conductive and high conductive states: e.g., percolation was observed at C_p≈1% and C_p≈0.1% for pure COC and 5CB, respectively.

A rather high percolation concentration of NTs in pure COC can reflect two effects:

- high exfoliation and stabilization of the NTs in COC, and suppressing of aggregation ability of NTs.

- formation of isolating solvate layers of COC molecules on the surface of individual NTs. These layers can inhibit the hopping transport of charge carriers between different NTs at small concentrations of NTs.

Figure 1 presents the effects of electric field (f=50 Hz) on microstructure of 0.1% suspension of NTs in the pure COC (X=1). The NTs were almost homogeneously spatially distributed in the sample in absence of electric field. The application of ac voltage between two electrodes produced the alignment of NTs above some threshold value (≈20 V).

Fig. 2. Square grid patters and folded oily streaks in a cholesteric-nematic COC+5CB mixture, X=0.05, X=COC /(COC +5CB)) .Voltage was U=10 V and frequency was f=1 kHz. The 20 μm thick layer had planar orientation, T=298 K.

The microscopic observations evidence the presence of *unidirectional* orientation and end-to-end contact between NTs. So, using of the COC and 5CB mixtures provided the excellent tool for regulation of aggregation of NTs in LC+NTs composites. Changing of X=COC/(COC+ 5CB) allowed also regulation of the value of helical pitch, e.g., at T=306.5 K:

$$\lambda = \lambda_m / X \qquad (1)$$

where λ_m=0.355 μm is a helical pitch for pure COC at X=1.

Finally, formation of long-range and electrically conducting clusters that span the entire gap between the electrodes was observed at large voltage (75 V) and long time curing (Fig.1b). The similar effects of electric fields on microstructure of NT composites were recently observed in different systems [5].

Figure 2 presents the effects of electric field (f=1 kHz) on microstructure of cholesteric-nematic COC+5CB mixture, X=0.05, X=COC/ (COC+ 5CB). In this mixture, COC acted as a chiral dopant to the nematic 5CB. The estimated value of helical pitch for this mixture λ ≈7.1 μm was comparable with the thickness of the cell (20 μm). The characteristic square grid patterns (at $U \approx 8$ V) and folded oily streaks (at $U \approx 10$ V) were observed in this mixture,. Note that previous studies revealed a large variety of patterns and structural deformations, induced by electric field, with different spatial uniformity[6].

Note that previously it was shown that a cholesteric to nematic transition can be induced by a relatively weak electric field. The square grid patterns, the double spirals of bright lines, zigzag instability, abnormal roll instability and more complex instabilities, depending on electric field strength and frequency, were observed in the conduction regime (at small frequencies)[6–8].

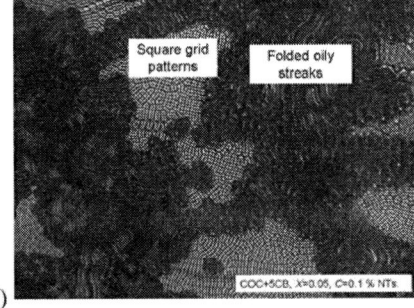

Fig. 3. Square grid patters, oily streaks and folded oily streaks of 0.1% suspension of NTs in a cholesteric-nematic COC+5CB mixture. $X=CC/(CC+5CB))$. $U=5.4$ V (a) and $U=10.4$ V (b), $f=1$ kHz.

In NT-loaded suspensions on the basis of COC+5CB mixture ($X=0.05$), the electric field initially provokes formation of linear oily streaks between aggregates of CNTs and folded oily streaks that cover the surface of aggregates (Fig3a, at $U \approx 5.4$ V). The birefringent structure of composites, observed between the crossed polarizer and analyzer, show that the structure of the interfacial layer, as well as its birefringence, drastically changes when the applied field strength is above the Freedericksz transition threshold. The mean thickness of interfacial layers increases when structure of LC undergoes the Freedericksz transition from planar to homeotropic orientation under the applied transverse electric field.

Moreover, the folded oily streak layers around the NT clusters were becoming thicker and square grid patterns appeared in the unperturbed spaces between NT clusters at larger voltage (Fig3b, at $U \approx 10.4$ V).

Summary

The observed effects evidence that a nematic-cholesteric COC+5CB mixture with highly oriented NTs can be prepared by application of an ac electric field. The degree of orientation in an electric field was dependent on dispersion ability of NTs in LC solvent that was related with the content of COC+5CB mixtures. We can conclude that application of ac electric field to NTs in a COC+5CB mixture is an efficient tool for manipulation by NTs and their alignment. Moreover, the NT aggregates initiate formation of long-ranged folded oily streaks instabilities in a cholesteric liquid crystal with long pitch (i.e. at small X),.

REFERENCES

[1] L. Dolgov, O. Kovalchuk, N. Lebovka, S. Tomylko, and O. Yaroshchuk, "Liquid crystal dispersions of carbon nanotubes: Dielectric, electro-optical and structural peculiarities," in *Carbon nanotubes*, J. M. Marulanda, Ed. InTech, 2010, pp. 451–484.

[2] S. Schymura and J. Lagerwall, "Carbon nanoparticles in cholesteric liquid crystals," in *37. Arbeitstagung Flussigkristalle, Deutschen Bunsen-Gesellschaft e.V.*, 2009, pp. P33(1–4).

[3] M. Loginov, N. Lebovka, and E. Vorobiev, "Laponite assisted dispersion of carbon nanotubes in water," *Journal of Colloid and Interface Science*, vol. 365, no. 1, pp. 127–136, 2012.

[4] O. Deriabina, N. Lebovka, L. Bulavin, and A. Goncharuk, "Regulation of dispersion of carbon nanotubes in a mixture of good and bad solvents," *In Press*, vol. xx, p. xx, 2013.

[5] M. Monti, M. Natali, L. Torre, and J. M. Kenny, "The alignment of single walled carbon nanotubes in an epoxy resin by applying a DC electric field," *Carbon*, vol. 50, no. 7, pp. 2453–2464, 2012.

[6] N. Miessen, J. Strauss, A. Hoischen, K. Kurschner, and H.-S. Kitzerow, "Durable Micropatterns Obtained from Dissipative Structures in Liquid Crystals," *ChemPhysChem*, vol. 2, no. 11, pp. 691–694, 2001.

[7] J. J. Wysocki, J. Adams, and W. Haas, "Electric-Field-Induced Phase Change in Cholesteric Liquid Crystals," *Phys. Rev. Lett.*, vol. 20, no. 19, pp. 1024–1025, 1968.

[8] O. D. Lavrentovich and D.-K. Yang, "Cholesteric cellular patterns with electric-field-controlled line tension," *Physical Review E - Statistical Physics, Plasmas, Fluids, and Related Interdisciplinary Topics*, vol. 57, no. 6, pp. R6269–R6272, 1998.

Optimal spectral decomposition of Raman gain profile using time response test

M.Y. Dyriv, G.S. Felinskyi, P.A. Korotkov

Taras Shevchenko National University of Kyiv,

Kyiv, Ukraine

Abstract: Simulation of Raman response function is proposed using confined part of Raman gain efficiency spectrum. Comparison of our response-function model with other theoretical models is presented. Received response-function curve based on inverse Fourier transform of complex third-order susceptibility shows that we may use only 10 Gauss spectral components instead of 12 Gauss components calculated before. Moreover, proposed principle of Gauss spectral decomposition method is applicable for other types of silica fibers.

Raman gain efficiency (RGE) as a function of frequency shift between pump wavelength ω_p and Stokes wavelength ω_s is one of the main optical fiber parameters. It is widely adopted in nonlinear optics, especially for silica Raman amplifier's synthesis with using of quasi-monochromatic pulses. In the case of ultra-short high-intensity optical pulses the coefficient RGE can be conveniently described by a time-dependent Raman response function (RRF) [1]. In general analytical form RRF contains two terms: the first of them is responsible for medium instantaneous electronic response, the second one – for vibrational with time delay [2-4]. In previous years, many scientific papers have been devoted to fundamental research of Raman gain profile both in frequency and in time domains. Stolen et al [1] were pioneers in obtaining of Raman response function from experimental Raman gain profile (RGP) in single-mode pure silica fiber. At that time RGP simulation was realized using single-Lorentzian model [2, 3]. However, Lorentzian approximation is not widely used because of inaccuracy of RGP reproduction on small (under 10 THz) and large (above 15 THz) frequency shifts [2]. The intermediate-broadening model (IB model) was proposed for Raman response function approximation on the basis of spectral components superposition, which is the convolution of 13 Gaussians and 13 Lorentzians [5]. The RRF shape is fitted to a superposition of six phase-shifted under-damped functions in order to use it in the numerical simulation of ultra-short pulse propagation [6]. Though, the parameters, which are used for simulation, don't have any physical interpretation and mismatch with individual peaks of RGP; they have only qualitative character.

In view of practical interest to properties of Raman amplification, we proposed response function model based on composition of only Gaussian spectral components. It is shown [5, 7] that spectral overlapping of Gaussian functions approximating response function is unsatisfactory. However, we proved that with 12 Gaussian components [4] one can obtain even better approximation accuracy instead of 13 spectral components [5] in presented IB model (see Fig. 1). In this way we can present another argument in favor of the Gaussian spectral decomposition for accurate reproduction of RGPs and thus the impulse response function.

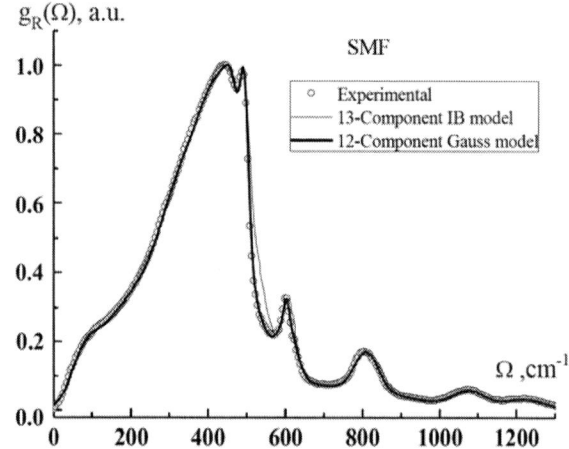

Fig. 1. Raman gain profile in single-mode silica fiber: white circles – experimental data, fine line – 13-component intermediate broadening model [5], coarse line – 12-component Gauss model.

One can obtain an explicit form of Raman response function $h_R(t)$ directly either from imaginary or from real parts of 3-rd order nonlinear susceptibility $N_2(\Omega)$ utilizing of inverse Fourier transform. Here $\Omega = \omega_p - \omega_s$ is frequency difference between pump wave and, respectively, Stokes wave, which take part in process of stimulated Raman scattering. In this case, we obtain the next expression for response function, assuming that all waves are linearly polarized with parallel polarizations [1] and using RGP (in direct proportion to imaginary part of $N_2(\Omega)$) due to Gauss decomposition method:

$$h_R(t) = B \cdot \int_0^\infty g_R(\Omega) \cdot Sin(2\pi c\Omega \cdot t) d\Omega, \qquad (1)$$

where $B = \dfrac{c}{\pi \cdot N_{2R} \cdot \omega_p}$, c – electromagnetic constant, $N_{2R} = \alpha \cdot N_{20} = \alpha \cdot N_2(0)$ – nonlinear susceptibility, which responds to vibrational contributions and gives time response function delay, $\alpha \approx 0.18$ – the fractional contribution of

$h_R(t)$), $g_R(\Omega)$ – the Raman gain efficiency coefficient, which has such expression:

$$g_R(\Omega) = \sum_{i=1}^{m} A_i e^{-\frac{(\Omega - \omega_{v,i})^2}{\Gamma_i^2}}, \qquad (2)$$

where $m = 12$ – spectral components number, A_i – the i-th Gauss component amplitude, normalized to its maximal value, $\omega_{v,i}$ – the i-th Gauss component peak frequency in cm^{-1} units, Γ_i – the half width of Gauss component in cm^{-1} units. Various models simulating pulse response functions with maximal correct fitting are compared and shown in Fig. 2.

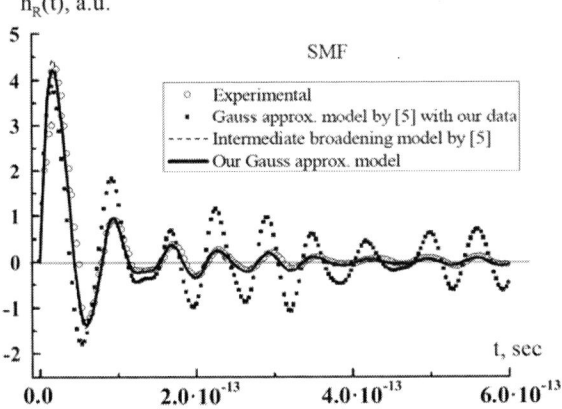

Fig. 2. Raman response function in single-mode fiber (SMF): white circles – calculated by [1] from experimental data, black circles – Gauss approximation model, presented in [5] with our input data, fine line – 13-component intermediate broadening model [5], coarse line – our 12-component Gauss approx. model.

Accuracy evaluation criterion is a minimal value of mean-square error (MSE) (by least-squares method). In our model MSE is 0.111 whereas in IB model MSE is 0.152. In the case of Gauss fitting model from [5] the approximation is inadequate: MSE = 0.483. It is confirmed by plots in Fig. 2. Also from results (see Fig. 1 and Fig. 2) one can see that the better is the analytical model reproducing experimental RGP of investigated fiber (in our research – SMF) the better is Raman response function fitting. Output parameters of Gauss decomposition of Raman amplification spectrum used in our calculations are taken from [4].

Analyzing obtained curves, it can be argued that time response, which is almost identical with the response function calculated on the basis of experimental profile SRS amplification, may be reproduced limiting the gain spectrum by the optimal number of spectral components. If part of RGE above 900 cm^{-1} may be neglected due to insignificance of its value in comparison with RGE on central frequency shifts (Fig. 1) it is enough only 10 Gauss components instead of 12 for plotting of sufficient RRF approximation. Moreover, fitting accuracy improves roughly on 7 %. According to model results, further neglecting of Raman gain profile components, especially close to 805 cm^{-1}, leads to accuracy degradation. That's why it's inadmissible to use less than 10 spectral

components. Spectral parameters of 10-component Gauss decomposition are presented in Table 1 and comparative plots of $h_R(t)$ are shown in Fig. 3.

TABLE 1
Initial parameters of 10-components Gauss decomposition
for RGP and RRF approximations

№	A_i	$\omega_{v,i}$, cm^{-1}	Γ_i, cm^{-1}
1	0.105	78.89	49.18
2	0.175	158.09	84.98
3	0.759	372.17	151.05
4	0.258	432.30	54.63
5	0.247	469.61	32.19
6	0.356	495.25	13.92
7	0.071	602.69	9.77
8	0.136	606.07	31.72
9	0.068	733.95	239.08
10	0.110	810.08	42.30

Simulation results indicate that one can limit the number of spectral Gauss components by optimal ten for the graphical construction of Raman response function in pure quartz fiber. However, the number of components is less than 10 don't provide sufficient accuracy of approximation. The cause of it lies in correspondence between components number and spectral singularities of Raman gain efficiency profile. While local extrema of Raman gain spectrum are neglected in relation to intensity of main part of Raman gain efficiency, one can leave them out the calculation. As shown, approximation accuracy becomes even better than when we use 12-component fitting. In particular, we received minimal mean-square error, which is tolerable judging from obtained graphics. These results are usable for other types of silica fibers too and may be applied for synthesis wideband Raman amplifiers in the future.

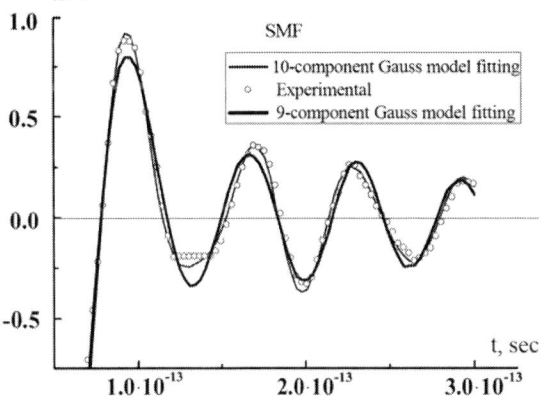

Fig. 3. Comparison of 10- and 9-component Gauss approximations of impulse Raman response function with calculated from experimental RGP by [1].

REFERENCES

[1] R.H. Stolen, J.P. Gordon, W.J. Tomlinson, and H.A. Haus, "Raman response function of silica-core

fibers", J. Opt. Soc. Am. B, vol. 6, No. 6, pp. 1159–1166, 1989.

[2] Q. Lin, G.P. Agraval, "Raman response function for silica fibers", Opt. Lett., vol. 31, No. 21, pp. 3086–3088, 2006.

[3] K.J. Blow and D. Wood, "Theoretical description of transient stimulated Raman scattering in optical fibers", IEEE J. of Quant. Electr., vol. 25, No. 12, pp. 2665–2673, 1989.

[4] P.A. Korotkov, G.S. Felinskyi, "Forced-Raman-scattering-based amplification of light in one-mode quartz fibers", Rev. Ukr. J. Phys. vol. 5, No. 2, pp. 103-169, 2009.

[5] D. Hollenbeck, C.D. Cantrell, "Multiple-vibrational-mode model for fiber-optic Raman gain spectrum and response function", J. Opt. Soc. Am. B, vol. 19, No. 12, pp. 2886–2892, 2002.

[6] G. Salceda-Delgado, A. Martinez-Rios, B. Ilan, D. Monzon-Hernandez, "Raman response function and Raman fraction of phosphosilicate fibers", Opt. Quant. Electron., 2012.

[7] D. Hollenbeck, "Dynamics of a fiber-optic Raman amplifier", diss. for the degree of Ph.D., Univ. of Texas, 439 pages, 2000.

Submicron structures for nonlinear photonics in the mid-infrared spectral range

E.A. Romanova, A.I. Konyukhov, E.V. Borisov, D.S. Zhivotkov

Saratov State University, Saratov, Russia

Abstract: Chalcogenide glass based submicron photonic structures have been considered as prospective nonlinear devices operating in the mid-infrared spectral range. Tailoring of the group velocity dispersion and ultra-fast pulse duration enables to achieve the efficient spectrum broadening in the highly nonlinear glass.

The on-chip optical signal processing is based on integrated optical devices of the submicron dimensions. Group velocity dispersion (GVD) engineering is an important task in the problem of design and fabrication of the highly nonlinear photonic devices. Various geometries of the photonic structures that can be used for the GVD tailoring are shown in Fig.1.

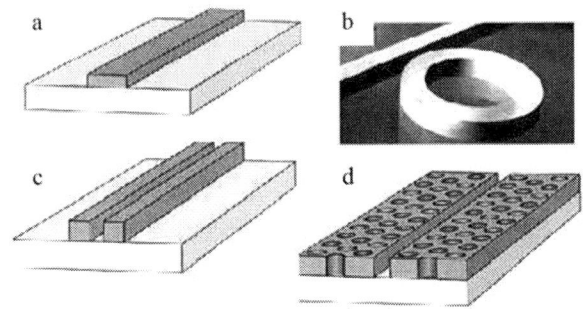

Fig.1. Basic integrated optical structures: rib waveguide (a), microresonator (b), slot waveguide (c), photonic-crystal waveguide (d).

Using of the wavelength range within 1-2 μm in optical communications is dictated by the transparency windows of silica optical fibres. The fibres and all supporting technology are already mature in the form of compact sources, detectors, connectors, gain media, which can route and exploit near-infrared light. Exploration of the mid-infrared (mid-IR) spectral range potentially provides very sensitive spectroscopic tools because many inorganic and organic molecular substances exhibit here strong fundamental absorption bands. Intelligent and sensitive high-resolution detection of individual molecular gas species requires bright broadband radiation.

Chalcogenide glasses are prospective optical materials for the creation of broadband sources of coherent radiation in the mid-IR [1,2]. The range of transparency of chalcogenide glasses is extending from 0.5 to 15-20 μm. Transmittance of the glass samples of some compositions near the fundamental absorption band edge is shown in Fig.2. Chalcogenide glasses

have the greatest magnitudes of linear refractive index n and Kerr constant n_2 among the inorganic-compound glasses. The values of n and n_2 evaluated far from the bandgap wavelength λ_g are shown in Table 1.

Microstructuring of chalcogenide glasses is of a particular importance because dispersion of a high-contrast waveguide or microresonator can be used to compensate large material dispersion of chalcogenide glasses that have zeros of material dispersion in the spectral range 5-8 microns (Table 1). For the group velocity dispersion (GVD) tailoring, a combination of different compositions of chalcogenide glasses can be used.

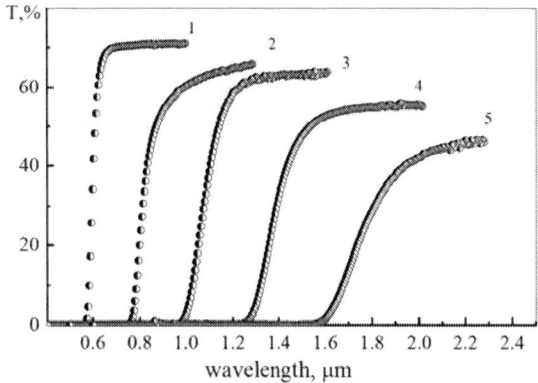

Fig. 2. Transmittance of the chalcogenide glass samples: 1- As_2S_3; 2- As_2Se_3; 3- $As_{40}Se_{30}Te_{10}$; 4- $As_{40}Se_{20}Te_{40}$; 5- $As_{30}Se_{10}Te_{60}$. The samples thickness was about 1 mm.

Table 1

Glass composition	Zero dispersion wavelength, μm	Refractive index, n $\lambda \gg \lambda_g$	Kerr constant, n_2, cm^2/W $\lambda \gg \lambda_g$
As_2S_3	4.9	2.4	$1.2 \cdot 10^{-14}$
As_2Se_3	7.2	2.8	$6.6 \cdot 10^{-14}$
$As_{30}Se_{50}Te_{20}$	8.2	3.0	$1.9 \cdot 10^{-13}$
$As_{40}Se_{20}Te_{40}$	not available	3.3	$3.4 \cdot 10^{-13}$
$As_{30}Se_{10}Te_{60}$	not available	3.8	$1.1 \cdot 10^{-12}$

In this paper, we present a comparative analysis of the basic integrated optical structures that can be potentially used for the nonlinear frequency conversion in mid-IR. The 2D analogs of the structures shown in Fig.1 have been considered.

The GVD coefficients of the guided modes of planar multilayered waveguides (Fig.3) and slot waveguides (Fig.4)

$$D = -\frac{2\pi c}{\lambda^2}\left(\frac{\partial^2 \beta}{\partial \omega^2}\right) \qquad (1)$$

and the GVD coefficients [4] of 2D resonators (Fig.5)

$$D = \frac{\omega_l^2}{2\pi Rc}\frac{\Delta(\Delta\omega_l)}{(\Delta\omega_l)^3} \qquad (2)$$

have been evaluated. Longitudinal propagation constants $\beta(\omega)$ of the planar multilayered waveguides and slot waveguides were calculated by using the Fast Fourier Transform BPM and the Transfer Matrix Method [3]. The resonance frequencies of a resonator were defined by [4]:

$$\omega_l = \frac{l \cdot c}{R \cdot n(\omega)} \qquad (3)$$

In (1)-(3), c is the speed of light in vacuum, R is the resonator radius, $n(\omega)$ is refractive index of a chalcogenide glass, l is integer mode number. In the structures design, As-S-Se and As-Se-Te glass systems have been used. Optical parameters of the glass compositions have been evaluated by a semi-empirical approach based on spectroscopic measurements of the transmittance and reflectance of the glass samples [5].

Waveguide dispersion of bandgap modes of an all-chalcogenide structure with periodic cladding can compensate the large normal dispersion of chalcogenide glass and shift the zero dispersion wavelength from mid-IR to near-IR spectral range (Fig.3). This is not possible by using a guided mode of the all-chalcogenide step-index waveguide with uniform cladding [6].

The GVD tailoring in the mid-IR by using guided modes is possible if a chalcogenide structure has one-side air cladding. Flattened near-zero dispersion curves of the TM modes of non-symmetric slot waveguides composed of various chalcogenide glasses (Fig.4) enable soliton formation and efficient supercontinuum generation in this spectral range. However in order to achieve $D > 0$ in near-IR, another side cladding should be composed of air or fused silica [7].

Modes of a 2D circular microresonator are equidistant if the refractive index is frequency independent. If $n=n(\omega)$, the mode spacing $\Delta\omega_l$ is not equidistant, and variation of the mode spacing $\Delta(\Delta\omega_l)$ is not equal to zero (as can be directly seen from (3)). The GVD coefficient (2) has been evaluated for the circular resonators made of the As$_2$S$_3$ and As$_2$Se$_3$ glasses (Fig.5). In the resonators, D changes its sign at the longer wavelengths than in bulk glass samples.

Propagation of the ultra-short optical pulses in the guiding structures with the embedded Kerr nonlinearity has been numerically simulated as follows. Variation of each spectral

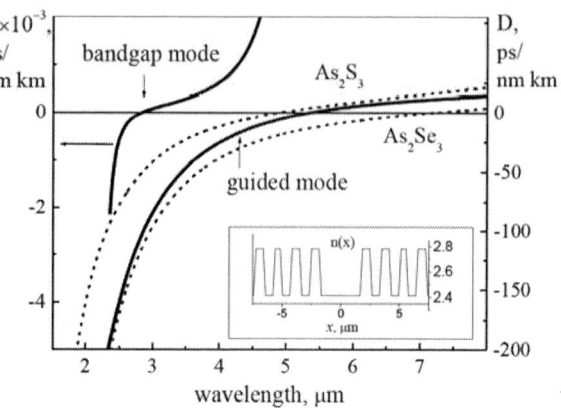

Fig.3. The GVD coefficient of a TE bandgap mode of the planar waveguide with a periodic cladding made of As$_2$Se$_3$ and As$_2$S$_3$ glass (refractive index profile is shown on the inset).

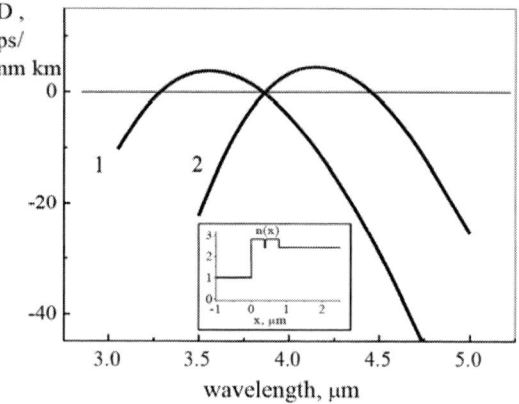

Fig.4. The GVD coefficient of guided TM modes of the non-symmetric slot structures (with one-side air cladding) made of As$_2$Se$_3$ and As$_2$S$_3$ (line 1) and As$_{30}$Se$_{50}$Te$_{20}$ and As$_2$S$_3$ (line 2).

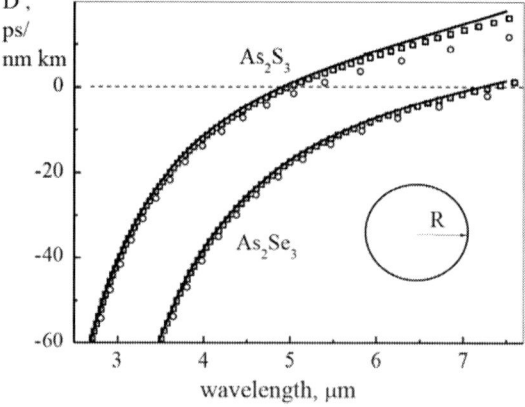

Fig.5. The GVD coefficient of the circular microresonator: R = 5 µm (circles), 20 µm (rectangles).

component $F(x,z,\omega)$ of the electric field in space was described by the equation:

$$2i\beta(\omega)\frac{\partial F}{\partial z} + \frac{\partial^2 F}{\partial x^2} + \left(\omega^2 n^2(x,\omega)/c^2 - \beta(\omega)^2\right)F(x,z,\omega) = \mathbf{H}_\omega \quad (4)$$

where $\mathbf{H}_\omega$ is an operator corresponding to the nonlinear response at the frequency ω. It generally depends on the field amplitudes at other frequencies. In the numerical model, a split-step approach was used. At the linear step ($\mathbf{H}_\omega = 0$), Eq.(1) was solved by the Finite-Difference BPM [3]. The nonlinear phase modulation was taken into account in the time domain:

$$i\beta_0\frac{\partial A(x,z,t)}{\partial z} + \frac{kn_0^2}{2Z_0}n_2\mid A(x,z,t)\mid^2 A(x,z,t) = 0 \quad (5)$$

where $A(x,z,t)$ is a Fourier transform of $F(x,z,\omega)$, Z_0 is the free-space impedance, n_0 and β_0 are defined at the pulse central frequency ω_0, $k=2\pi/\lambda$.

In Fig.6, the results of modeling the pulse propagation in a waveguide with infinitive periodic cladding (the refractive index profile shown in Fig.3, inset) in the soliton regime ($D > 0$) are presented. The number of solitons can be defined as $N = (L_d/L_{nl})^{1/2}$, where $L_d = \tau_0^2/\beta_0$ is the dispersion length and $L_{nl} = (n_2 I^0 k)^{1/2}$ is the nonlinear length (I^0 is the peak intensity). For the curved shown in Fig.6, $N = 1.52$ (1), 7.6 (2), 1.59 (3), 6.38 (4). In the soliton regime (curves 2-4), the spectrum broadening was symmetric at $\tau_0 = 500$ fs. At the shorter pulse durations, the pulse shape and spectrum were not symmetric. Typical modulation of the 50 fs pulse spectrum is shown on the inset. This effect can be explained by evaluating the impact of the third-order dispersion having the characteristic length $L_d^{(3)} = 2\tau_0^3/\beta^{(3)}$. In the case of $\tau_0=500$ fs, $L_d^{(3)} \gg L_d$, contrary to $\tau_0=50$ fs when $L_d^{(3)} \approx L_d$.

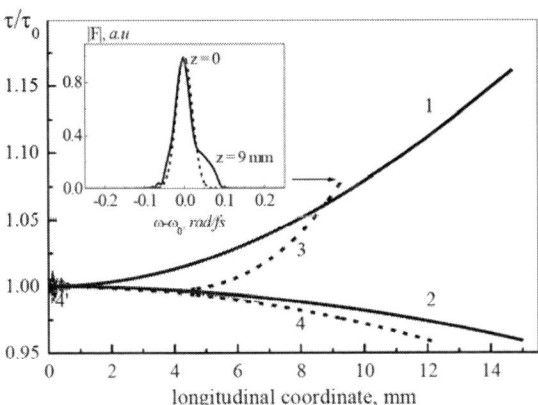

Fig.6. Variation of the pulse duration in the waveguide with infinite periodic cladding. The peak wavelength is 3.2 μm (solid lines), 2.94 (dashed lines). Initial pulse duration is 100 fs (1), 500 fs (2), 50 fs (3) and 200 fs (4).

In summary, photonic structures and optical materials suitable for exploration of the mid-IR spectral range are currently available. The micro- and nano- structuring is of a particular importance for compensation of the large material dispersion of chalcogenide glasses. However using of the guided modes of all-chalcogenide structures cannot significantly manage the GVD. For significant GVD tailoring, chalcogenide glasses are to be used in combination with air or fused silica. All-chalcogenide periodic structures can provide a blue-shift of the GVD zeros in the bandgap modes, but large third-order dispersion violates the femtosecond pulse compression in the soliton regime (D>0). Picosecond lasers in the spectral range 1.5-3.5 μm can provide an efficient pulse spectrum broadening in waveguides with uniform cladding (D<0) due to the phase self-modulation effect. Creation of the ultra-fast lasers in the spectral range above 4 μm will make it possible to use the femtosecond pulses to enhance the effect of frequency conversion in chip-scale structures.

REFERENCES

[1] A. Zakery, S. R. Elliott, "Optical nonlinearities in chalcogenide glasses and their applications", Springer-Verlag, Berlin, Heidelberg, New York, 2007.

[2] J. H. V. Price, T. M. Monro, H. Edendorff-Heidepriem, F. Poletti et. el, "Mid-IR supercontinuum generation from nonsilica microstructured optical fibers", IEEE J. of Selected Topics in Quantum Electronics, vol 13, no. 3, p. 738-749, 2007.

[3] Methods for modeling and simulation of guided-wave optoelectronic devices: waves and interactions. - W.P.Huang, Ed., PIERS 11, EMW Publishing, Cambridge, Massachusets, USA, 1995. – 410p.

[4] P. Del'Haye, "Optical frequency comb generation in monolithic microresonators", Dissertation an der Fakultat fur Physik der Ludwig-Maximilians-Universitat, Munchen, 2011.

[5] Yu. Kuzutkina, A. Melnikov, E. Romanova, V. Kochubey, N. Abdel-Moneim, D. Furniss, A. Seddon, "Dispersion of linear and nonlinear refractive index in chalcogenide glass", Proc. of: 14th International Conference on Transparent Optical Networks (Coventry, UK) July 2-5, 2012.

[6] A. I. Konyukhov, E. A. Romanova, and V. S. Shiryaev, "Chalcogenide glasses as a medium for controlling ultrashort IR pulses: Part I", Optics and Spectroscopy, V. 110, No. 3, 2011, pp. 442–448.

[7] E. Romanova, A.Konyukhov, Yu. Kuzutkina, A. Melnikov, "Dispersion tailoring in chalcogenide slot waveguides, slot arrays and bandgap structures", Proc. of: 14th International Conference on Transparent Optical Networks Networks (Coventry, UK), July 2-5, 2012.

978-1-4799-0019-0/13 $31.00 © 2013 IEEE

Pulse shape transformation upon reflection from Bragg resonators with asymmetric feedback

V. F. Borulko, *Senior Member, IEEE*, O.O. Drobakhin, *Senior Member, IEEE*, and D.V. Sidorov
Oles Honchar Dnipropetrovsk National University, Dnipropetrovsk, Ukraine

Abstract: The transformation of the shape of pulses reflected by Bragg resonators with step-up and step-down perturbation types of period contrast under conditions than carrier frequency is in the vicinity of the resonance Bragg frequency have been considered. Integral and differential estimates of the delay time have been compared. The coefficients of skewness and kurtosis for the reflected pulses versus carrier frequency have been calculated within Bragg reflection band. The conditions for the appearance of negative mass center delay of the reflected pulses have been determined. The conditions of the anomalous values of the delay time have been generalized for the case of asymmetric feedback in Bragg resonators.

The phenomenon of constructive interference is one of the basic principles for matching, filtering, dispersion compensation in transmission lines, optical devices and microwave feeders [1]. For specific frequency range (for the structures in the open space in the range of incidence angles as well) strong coupling of waves (modes) propagating in opposite directions is achieved by periodic changes in system parameters (refractive indices, wave impedance) along the direction of wave propagation [2, 3].

Distortion of the periodicity of the structure parameters allows obtaining the desired reflection and transmission characteristics. Step disturbance of the periodic phase parameters leads to appearing eigenmode with high Q-factor in the Bragg reflection band. Smooth periodic perturbation of the amplitude parameters can provide relative decrease of Q-factor of the side resonance frequencies (outside the band of Bragg reflection frequencies) [3]. The structure with a spatially modulated optical (electrical) thickness of "period" (chirped mirrors) are well known and widely used in sources of ultrashort laser pulses to compensation of group delay dispersion [4].

In many applications pulse propagation should be considered instead of monochromatic wave propagation. Pulse propagation in a medium, reflection and transmission through the structure are always accompanied by distortion due to dispersion, which can be ignored only at the specified frequency range at specific times and distances of propagation. The dispersion phenomenon is associated with the commensurability of media or structures parameters with the parameters of the electromagnetic wave. In case of media such parameters are the distances between the energy levels of atoms or molecules, in the structures such parameter is the value of the period.

If the center frequency of pulse's spectrum is near one of the eigenfrequencies of structures or the medium, the analysis of propagation, reflection and transmission of these pulses can not be carried out only in terms describing the pulse as a whole (speed, delay time) [5]. It is necessary to consider the widening, the deformation of the pulse form, and to specify the limits within such physical characteristics as delay time have sense.

In present work integral and differential estimates of the delay time will be compared. The coefficients of skewness and kurtosis for the reflected and transmitted pulses with π-sinusoidal envelop will be calculated versus carrier frequency within Bragg reflection band. The values of group delay for layered structures with a small chirp variation of optical (electrical) thickness of the "period" along longitudinal coordinates have been obtained.

According to the formalism of the transmission matrix method of light propagation through multilayer structures [2], any homogeneous layer associates with 2 by 2 matrix that relates the total electric and magnetic fields between the boundaries of one layer.

In the future we shall use the normalized frequency f / f_0, where f_0 has sense of the first Bragg resonance frequency. The consideration will be carried out for dielectric material ($\mu_1 = \mu_2 = 1$), thus the frequency f_0 is defined in following way $f_0 = c / \left(2 \left(h_1 \sqrt{\varepsilon_1} + h_2 \sqrt{\varepsilon_2} \right) \right)$, and the values $h_1 \sqrt{\varepsilon_1}$ and $h_2 \sqrt{\varepsilon_2}$ are the electric thickness of layers of the structure's period, c is the speed of light in vacuum.

Traditionally, if the pulse spectrum is concentrated near the own central frequency, the pulse can be considered as one quasi-monochromatic waves group. For overcoming the discontinuity of formally determined phase, the reflected pulse group delay time (GD) is convenient to calculate through derivatives of the real and imaginary parts of the reflection coefficient (RC), viz.

$$\Delta \tau_g = \frac{\mathrm{Im}(R(\omega))}{|R(\omega)|^2} \frac{\mathrm{d\,Re}(R(\omega))}{\mathrm{d}\omega} - \frac{\mathrm{Re}(R(\omega))}{|R(\omega)|^2} \frac{\mathrm{d\,Im}(R(\omega))}{\mathrm{d}\omega}. \quad (2)$$

The analysis of the transformation of the pulse shape will be carried out using its representation as Fourier transform. This approach requires that every incident, reflected and transmitted waves is presented as superposition of harmonic waves (at the same angle of incidence with varying amplitude) [5].

978-1-4799-0019-0/13 $31.00 © 2013 IEEE

If the pulse propagates with distortion, we define the delay time for each part of the pulse: for maximum, pulse rise and fall, etc. We will consider the delay of the power center of the pulse. The center mass delay (CMD) is calculated as the initial mathematical moment of first order [4, 5]:

$$\Delta\tau_e = \int_{-\infty}^{+\infty} \tau |s_N(\tau)|^2 d\tau , \quad |s_N(\tau)|^2 = |s(\tau)|^2 \Big/ \int_{-\infty}^{+\infty} |s(\tau)|^2 d\tau . \quad (3)$$

here $s(\tau)$ is the envelope of the pulse. We have defined the pulse width as the central moment of the second order for quantitative estimate of reflected and transmitted pulses expansion:

$$\sigma^2 = \int_{-\infty}^{+\infty} (\tau - \Delta\tau_e)^2 |s_N(\tau)|^2 d\tau . \quad (4)$$

Similarly, we have defined the coefficient of skewness and kurtosis as the central moments of the third and fourth order, respectively:

$$\eta = \frac{\int_{-\infty}^{+\infty} (\tau - \Delta\tau_e)^3 |s_N(\tau)|^2 d\tau}{\sigma^3} , \quad \xi = \frac{\int_{-\infty}^{+\infty} (\tau - \Delta\tau_e)^4 |s_N(\tau)|^2 d\tau}{\sigma^4} . \quad (5)$$

It is clear that the method of determining the parameters of the deformation of the pulse shape by mathematical moments of various orders have sense only if the corresponding integrals (3) - (5) converge. For finite time pulses, this condition is automatically satisfied.

Earlier in Ref [6] the transformation of the pulse shape was considered for the pulse falling on the Bragg structure. It was also revealed the effect of negative group delay and the center mass delay. This effect was due to the asymmetry of the resonance layer location. It occurs when the resonant layer is disposed closer to the end of the structure.

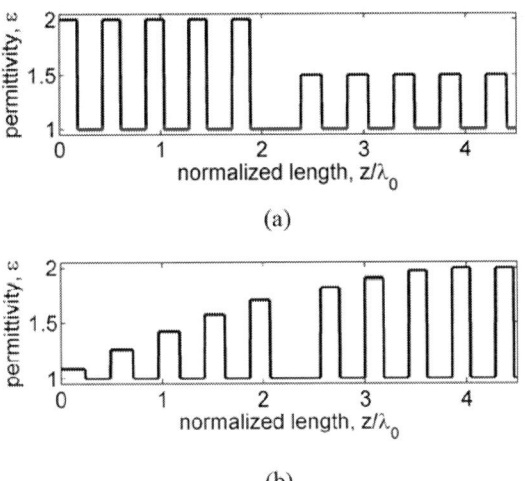

(a)

(b)

Fig. 1. Permittivity profiles. (a) The permittivity profile of the Bragg resonator with step-down perturbation of period contrast and (b) the permittivity profile of Bragg resonator with monotonically increasing perturbation of period contrast.

In this paper, we show that the negative effect of the delay is due not only to the symmetry of the resonance layer location, but it can be observed in the more general case. It depends on the symmetry or asymmetry of feedback (Figure 1), which is provided by the resonator mirrors. The asymmetric feedback can be obtained in several ways: either by a different number of layers of the resonator mirrors, or by changing the average value of the permittivity on the period thickness (the contrast of the period). It is also possible to use the combinations of these two methods.

Let us consider the reflection of limited pulses with π-sinusoidal envelop from Bragg resonators with asymmetric perturbation of the period contrast. The perturbation of permittivity contrast may be monotonous (profile shown in Figure 1b), and step-wise (Figure 1a). However, no significant differences in the behavior of the group delay for these types of perturbations when the step-wise and monotonous perturbations simultaneously increase or decrease. Therefore, in what follows, we will compare the characteristics of the reflected pulse transformation for structures with step-up and step-down perturbation of period contrast.

When the period contrast has step-down perturbation type the reflectivity of the back resonator mirror is smaller and group delay near the Bragg frequency takes negative value (Figure 2b, line1). Conversely, when the reflectivity of front mirror of the resonator is smaller, the group delay takes large positive value (Figure 2b, line2). Figure 2a shows the relative positions on the time axis of the incident pulse (Figure 2a, line 2) as well as pulse reflected from the structure with step-down perturbation of period contrast (Figure 2a, line 3) and step-up perturbation (Figure 2a, line 1).

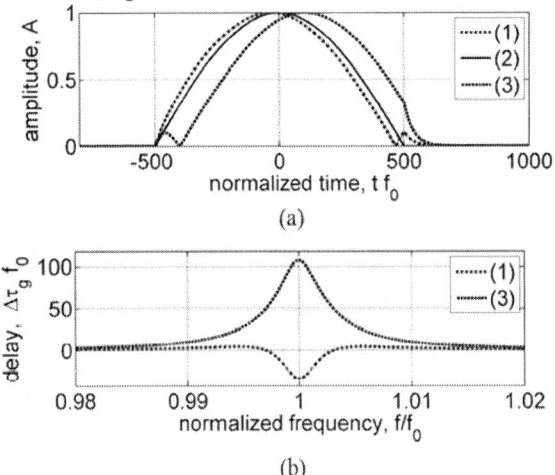

(a)

(b)

Fig. 2. Pulse reflection from Bragg resonator. (a) The time dependences of pulse reflected from Bragg resonator with step-down perturbation (line 1), resonator with step-up perturbation (line 3), and incident pulse (line 2). (b) The group delay frequency dependences for Bragg resonator with step-down perturbation (line 1), resonator with step-up perturbation (line 3).

978-1-4799-0019-0/13 $31.00 © 2013 IEEE

In numerical simulation the incident pulse had an absolute duration $T f_0 = 200$ and $N = 19$ is the total number of structure layers, $h_j \sqrt{\varepsilon_j} = \lambda_0 / 2$ is the electrical thickness of the resonance layer, $j = 10$ is the number of resonance layer. Figure 3 and Figure 4 show the integral estimates of the delay time and deformation parameters of the reflected pulses for the Bragg resonator with step-down perturbation type of period contrast in the Bragg reflection band. Similar to group delay, the center mass delay of sufficiently extended pulses with carrier frequency near the Bragg reflection frequency can take negative values (Figure 3a, line 2). This holds true for resonators with step-down perturbation type of period contrast.

In this case, in vicinity of Bragg reflection frequency the widening has strongly nonmonotonic behavior, and on the Bragg frequency the pulse becomes narrower (Figure 3b, line 2), right-skewness (Figure 4a, line 2), and more peakedness (Figure 4b, line 2). In contrast, the pulse reflected from the resonator with step-up perturbation type of period contrast has positive mass center delay (Figure 3a, line 1) and a left-skewness (Figure 4a, line 1).

Negative values of group delay have occurred for the Bragg layered structures under the condition when the resonator has step-wise or monotonically decreasing perturbation of period contrast. For small deformations of the pulse, group delay estimation and center mass delay have given similar values. The anomalous value of center mass delay is caused by pulse widening, changing of it asymmetry and kurtosis.

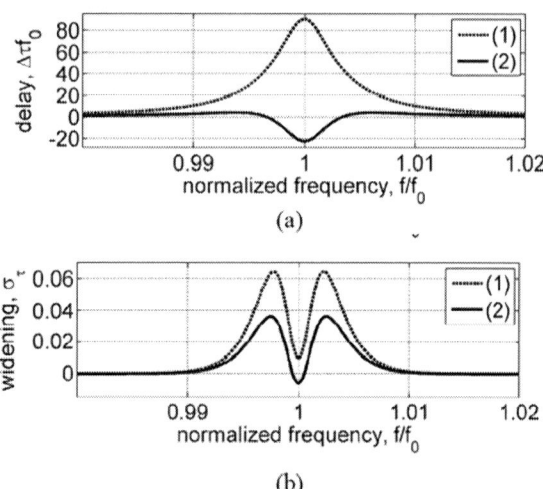

(a)

(b)

Fig. 3. Integral estimates. (a) The CMD of reflected pulse for Bragg resonator with step-up perturbation (line 1), and step-down perturbation (line 2). (b) The widening of pulse reflected from Bragg resonator with step-up (line 1), and step-down (line 2) perturbation of period contrast.

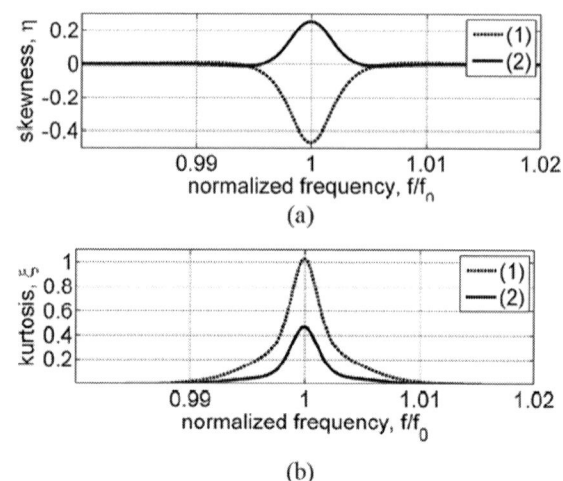

(a)

(b)

Fig. 4. Integral estimates. (a) The skewness of reflected pulse for Bragg resonator with step-up perturbation (line 1), and step-down perturbation of period contrast (line 2). (b) The kurtosis of pulse reflected from Bragg resonator with step-up perturbation (line 1), and step-down perturbation of period contrast (line 2).

The finite pulse with π-sinusoidal envelope reflected by the Bragg resonator has negligibly small deformation if the pulse duration is large enough and the carrier frequency located away from one of the eigenfrequencies of the resonator. If carrier frequency of the pulse lies near the main resonant frequency of the Bragg structures as well in the band of reflection, the group delay or the center mass delay can not be used as the single estimate of pulse propagation time.

REFERENCES

[1] C. Elachi "Waves in active and passive periodic structures: A review," *in Proc. of the IEEE*, vol. 64, no. 12, pp. 1666-1698, 1976.

[2] M. Born and E. Wolf *Principles of Optics*, Oxford: Pergamon Press, 1975.

[3] V. Borulko, O. Drobakhin, and D. Sidorov, "Eigenfrequencies of Periodic and Quasiperiodic Apodized Bragg Structures," *Telecommunications and Radio Engineering*, vol. 71, no. 16, pp. 1433-1445, 2012.

[4] S.O. Yakushev, I.A. Sukhoivanov, O.V. Shulika, V.V. Lysak, S.I. Petrov, "Simulation of interaction of the femtosecond laser pulse with chirped mirror", *International Conference on Numerical Simulation of Semiconductor Optoelectronic Devices, NUSOD '06*, China, pp. 99 – 100, 2006.

[5] V. L. Ginzburg *Propagation of electromagnetic waves in plasma*, Oxford, Pergamon Press, 1970.

[6] V. F. Borulko "Superluminal tunnelling and negative group delay in layered Bragg structures," *in Proc. of the DIPED*, Tbilisi, pp. 71-73, 2006.

Optical properties of layered organic-inorganic perovskite $(CH_3NH_3)_2PbBr_4$

S. Ahmadi-Kandjani[1], H. Ghanbari[1], M. S. Zakerhamidi[1]

[1]Photonics group, Research Institute for Applied Physics and Astronomy, University of Tabriz, Tabriz,

Abstract: The studies of layered organic-inorganic hybrids materials are very interesting due to their unique electronic structure, electrical and optical properties. The optical properties of layered organic-inorganic perovskite properties such as Photoluminescence (PL) spectra can be change with temperatures. In this study the Photoluminescence properties characterize at various temperature ranging from 12K to 285K. At lower temperatures than 140K, the PL spectra of $(CH_3NH_3)_2PbBr_4$ show two overlapped lines (peaks). The temperature dependences of these two PL lines are discussed by using the exciton phonon coupling and polaron model.

Layered organic-inorganic hybrids materials get lots of attention due to their unique electronic structure, electrical and optical properties. This kind of materials provides considerable opportunities in optoelectronic devices manufacturing because of the combining properties of inorganic (high mobility, electrical pumping, band engineering…) and organic materials (low cost technology, high luminescence quantum yield at room temperature …) [1-6].

Organic-inorganic provskite hybrids class of materials, generally expressed as $(R-NH3)2AX4$ or $(NH3-R-NH3)AX4$ (R=organic group, A=Ge, Sn, Pb and X=Cl, Br, I), Includes of extended network of corner-sharing metal halide octahedrons, alternating with a bilayer or monolayer organic moiety. R-NH3+ is a primary amine cation with appropriate profile to fit the inorganic framework. The organic/inorganic perovskite hybrids can be obtained in solution by solvent evaporation or cooling the saturated solution slowly [7-9]. The obtained organic-inorganic provskite hybrids have the layered inorganic sheet with interconnect organic amine salt due to the ammonium heads hydrogen bonding to the inorganic sheet. The spectroscopic parameter of these materials such as photoluminescence spectra can be easily controlled by changing the organic chain of amine as well as by changing the metal and halogen [10-11].

In this work, the spectroscopic properties of layered organic-inorganic perovskite, ([bi(methylamunium) tetra bromo plumbate]) properties such as Photoluminescence (PL) spectra at various temperatures ranging from 12K to 285K and photoluminescence excitation spectra measured at room temperature. The $(CH_3NH_3)_2PbBr_4$, absorbs light in UV wavelength and emitted in green region of wavelength. At lower temperatures than 140K, the PL spectra of $(CH3NH3)2PbBr4$ show two overlapped lines (peaks). The temperature dependences of these two PL lines are discussed by using the exciton phonon coupling and polaron

The layered organic-inorganic self assembled hybrid, $(CH_3NH_3)_2PbBr_4$, was synthesized and purified according to the common procedure in our laboratory and used as sample [12]. Fig. 1 represents the X-ray scattered pattern from the synthesized sample in powder from. The sharp lines in diffraction pattern prove the high crystallinity of the synthesized hybrid and the fine periodicity of the layers.

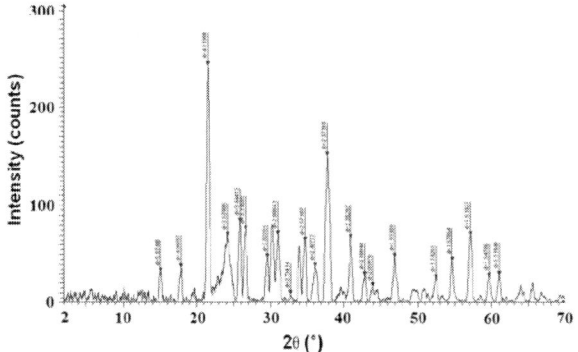

Fig. 1. XRD pattern of $(CH3NH3)2PbBr4$ powder

Double beam Shimadzu UV-2450 Scan UV–visible spectrophotometer was used to record the absorption spectra over a wavelength range 200–800 nm, which combined with a cell temperature controller (22 °C). Fluorescence of the sample in room temperature was studied with JASCO FP-6200. To perform the PL measurements in various temperatures,10-285 K, we used mercury UV lamp with optical filter in front of it as excitation source in 364nm and spectrometer model USB-4000 product of Ocean Optics used for spectra detection. To cool down the hybrid and control it's temperature, we used CTI cryogenic model M22 made by Helix Technology Co. which can control the temperature of the sample placed on it's cold finger from 9K to room temperature by helium closed cycle.

The absorption and fluorescence spectra of the PbBr2 and synthesized hybrid were recorded at room temperature in solid form (Fig 3).The synthesized $(CH3NH3)2PbBr4$ absorbance spectra show the broad shift as compared PbBr2 , in other words, the added organic CH3-NH3+ cation change the PbBr2 molecular structure and structural configuration in synthesized hybrid. As it can be seen (Fig2), the absorption spectra of the $(CH3NH3)2PbBr4$ red shifted as compared to PbBr2. Larger red shifts indicate relatively strong interactions between organic cation and PbBr2 in the hybrid $(CH3NH3)2PbBr4$

978-1-4799-0019-0/13 $31.00 © 2013 IEEE

.These interaction modify the electronic structure of (CH3NH3)2PbBr4 and then this sample shows the intense fluorescence as compared of PbBr2 (PbBr2 sample do not show fluorescence).Fig (2) represents the fluorescence spectra of (CH3NH3)2PbBr4 . There is two significant peaks in it, the intense band occur in 376nm and the lower one can be seen in 532nm (P1 and P2 respectively). We attribute the P1 peak to transitions in which electrons is excited to conduction band from valance band and then form the excitons and the P2 peak resulted from direct formations of excitons. P1 peak has higher intensity because the valance and conduction bands are much wider than excitonic energy levels, so electrons transitions probability/possibility in transition of valance and conduction bands is more than excitonic transitions.

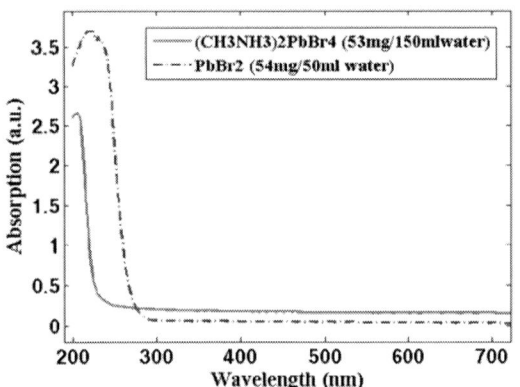

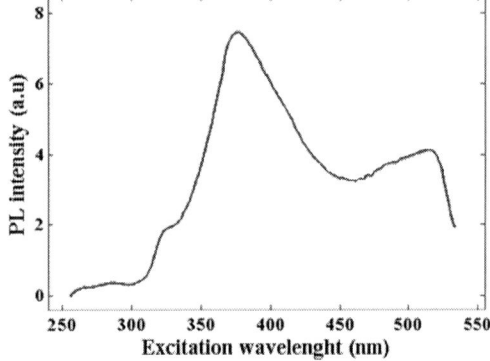

Fig. 2. Up: Absorption spectra of initial substance, PbBr2, (dashed curve) and synthesized hybrid (solid curve). Down: spectra at room temperature (emission at 550 nm)

Fig (3) shows the PL spectra peak excited in 364nm, measured at various temperatures ranging from 12K to 295K. The PL peak wavelength changes with temperature and show blue shift for (CH3NH3)2PbBr4 sample. In other words, the temperature dependence of the PL peak position shows that the exciton binding energy depended on temperature. It should be noted exciton binding energy defined mainly by the dielectric confinement in DQWs [13]. So, changing in the system

dielectric constant $(\varepsilon_{ow} - \varepsilon_B)$ increasing or decreasing exciton binding energy with temperature. Furthermore, the system dielectric constant depended on the composition of the barrier [14] and temperature change the barrier properties. If $(\varepsilon_{ow} - \varepsilon_B)$ increase with temperature, exciton binding energy show decreases.

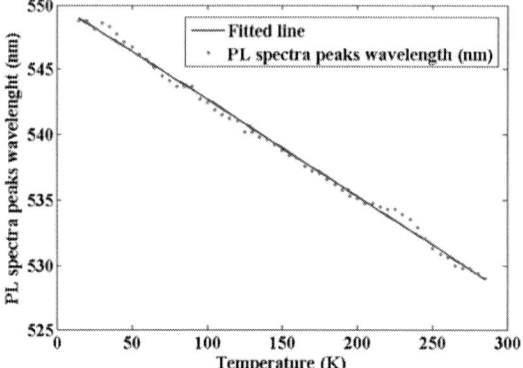

Fig. 3. PL peak wavelength vs. temperature

In fig (3) PL spectra peak wavelength as a function of temperature is linear and has negative slope. We can deduce the exciton binding energy increases with temperature so the system (barrier and well) effective dielectric constant decreases with temperature in sample.

The full width at half maximum (FWHM) of S1 PL line in temperatures range from 12K to 285K illustrated in Fig (7). We imputed this widening to exciton interactions with phonons and used the phenomenological model to justify it, which usually used for inorganic quantum well structures [15, 16].

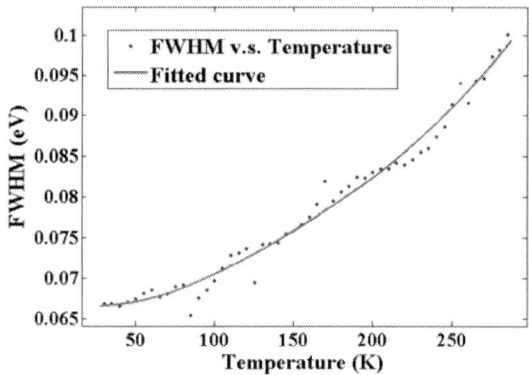

Fig. 3. FWHM as a function of temperature. The red line is the calculated FWHM following the model explained in the text

The layered organic-inorganic self assembled hybrid, (CH3NH3)2PbBr4, was synthesized and recorded the absorption and fluorescence spectra. The absorbance spectra show the broad peak and shift as compared precursor matters. The fluorescence spectra of (CH3NH3)2PbBr4 show two significant peaks the intense band occur in 376nm and the lower one can be seen in 532nm (P1 and P2 respectively). We attribute the P1 peak to transitions in which electrons is excited to conduction band from valance band and then form the excitons and the P2 peak resulted from direct formations of excitons. Because of existing two overlapped lines in PL spectra.

REFERENCES

[1] Y. Takeoka, M. Fukasawa, T. Matsui, K. Kikuchi, M. Rikukawa and K. Sanui, "Intercalated formation of two-dimensional and multi-layered perovskites in organic thin films", *Chem. Commun.*, pp. 378–380, 2005.

[2] D.B. Mitzi, K. Chondroudis and C.R. Kagan, "Organic-Inorganic electronics", *IBM J. Res. Dev.*, vol. 45, no. 1, pp. 29-45, 2001.

[3] D. B. Mitzi, C.A. Field, W. T.A.Harrison and A.M.Guloy, "Conducting Tin Halides with a Layered Organic-Based Perovskite Structure", *Nature*, vol. 369, pp. 467-469, 1994.

[4] N. Kitazawa, "Optical Absorption and Photoluminescence Properties of Pb(I, Br)-Based Two-Dimensional Layered Perovskite", *Jpn. J. Appl. Phys.*, vol.36, pp. 2272-2276,1997.

[5] Z. Y. Cheng, H. F. Wang, Z. W. Quan, C. K. Lin, J. Lin and Y. C. Han, "Layered organic–inorganic perovskite-type hybrid materials fabricated by spray pyrolysis route", *J. Cryst. Growth*, vol. 285, pp. 352-357, 2005.

[6] Y. Li, G. Zheng and J. Lin, "Synthesis, Structure, and Optical Properties of a Contorted <110>-Oriented Layered Hybrid Perovskite: $C_3H_{11}SN_3PbBr_4$" *Eur. J. Inorg. Chem.*, vol. 2008, pp. 1689-1692, 2008.

[7] N.L. Calabrese, R.L. Jones, R.L. Harlow, N. Herron, D.L. Thorn and Y. Wang, "Preparation and characterization of layered lead halide compounds", *J. Am. Chem. Soc.*, vol. 113,no. 6, pp. 2328-2330, 1991.

[8] C.-Q. Xu, T. Kondo, H. Sakakura, K. Kumata, Y. Takahashi, R. Ito, "Optical third-harmonic generation in layered perovskite-type material $(C_{10}H_{21}NH_3)_2PbI_4$", *Solid State Commun.*, vol. 79, no. 3, pp. 245-248, 1991.

[9] A.M. Guloy, Z. Tang, P.B. Miranda and V.I. Srdanov, "A New Luminescent Organic–Inorganic Hybrid Compound with Large Optical Nonlinearity",*Adv. Mater.*, vol. 13, no. 11, pp. 833-837, 2001.

[10] D. B. Mitzi, Synthesis, "Crystal Structure, and Optical and Thermal Properties of $(C_4H_9NH_3)_2MI_4$ (M = Ge, Sn, Pb)", *Chem. Mater.*, vol. 8, no.3, pp. 791-800,1996.

[11] G. C. Papavassiliou, G. A. Mousdis and I. B. Koutselas, "Some new organic–inorganic hybrid semiconductors based on metal halide units: structural, optical and related properties", Adv. Mater. Opt. Electron., vol. 9, pp. 265-271, no. 6, 1999.

[12] Z. Cheng and J. Lin, "Layered organic–inorganic hybrid perovskites: structure, optical properties, film preparation, patterning and templating engineering", *CrystEngComm*, vol. 12, pp. 2646-2662,2010.

[13] M. Kumagai and T. Takagahara, "Excitonic and nonlinear-optical properties of dielectric quantum-well structures", *Physical Review B*, vol. 40, no.18, pp. 12359-12381, 1989.

[14] S. Zhang, G. Lanty, J. S. Lauret, E. Deleporte, P. Aude-bert and L. Galmiche, "Synthesis and optical properties of novel organic-inorganic hybrid nanolayer structure semiconductors", *Acta Mater.*, , vol. 57, no.11, pp. 3301-3309, 2009.

[15] J. Lee, E. S. Koteles and M. O. Vassel, "Luminescence linewidths of excitons in GaAs quantum wells below 150K", Phys. Rev. B, vol. 33, no. 8, pp. 5512-5516, 1986.

[16] X. B. Zhang, T. Taliercio, S. Kolliakos and P. Lefebvre, "Influence of electron-phonon interaction on the optical properties of III nitride semiconductors", *J. Phys.: Condens. Matter*, 2001, vol. 13, no. 32, 7053-7074, 2001.

Femtosecond Parabolic Pulse Formation in All-Normal Dispersion Photonic Crystal Fiber

I. A. Sukhoivanov[1] S. O. Iakushev[2], J. A. Andrade Lucio[1], A. García Pérez[1], O. Ibarra Manzano[1]
[1]University of Guanajuato, Mexico
[2]Kharkov National University of Radio Electronics, Kharkov, Ukraine

Abstract: We have shown numerically that all-normal dispersion fiber with a flat top at 800 nm can be used for producing of parabolic pulses with femtosecond pulse duration by means of passive nonlinear reshaping in the fiber. Ultrashort pulses delivered by typical Ti:Sapphire lasers or fiber lasers can be used as a pump source.

In the field of all-optical signal processing and high speed optical communications applications of different optical pulse waveforms are highly important [1]. As compared to the traditional electronics in the optical frequency range bulky and complex devices are used usually to shape optical pulses. Therefore, the compact fiber-based approaches for pulse shaping are desirable in order to fulfill telecommunication system requirements. An important optical pulse shape is a parabolic one due to its unique properties [2]. Parabolic pulses are required in numerous applications such as pulse amplification, compression, regeneration of optical signals [1-3]. Several approaches have been proposed for producing of parabolic pulses in fiber systems such as fiber Bragg gratings [4], arrayed waveguide gratings [5], dispersion decreasing fibers [6]. Then it was found by Finot at al. that parabolic pulses can be produced in conventional optical fibers as a result of interaction between nonlinear effects such as self-phase modulation (SPM) and normal group velocity dispersion (GVD) [7]. However, this approach implies that parabolic pulses are formed as a transient state of pulse evolution in the fiber, i.e. parabolic pulses shape is not preserved when pulse propagates in the fiber ahead. However, recently we have shown that parabolic pulses can be formed in the fiber in the steady-state regime, when optical pulse remains parabolic waveform during pulse propagation far ahead [8]. Mentioned above approaches however, preliminary are intended for formation of picosecond parabolic pulses, whereas femtosecond parabolic pulses would be more attractive in many applications. Here we investigate numerically the possibility of femtosecond parabolic pulses generation via nonlinear reshaping in all-normal dispersive photonic crystal fiber (ANDi PCF) applying both approaches.

For producing the parabolic pulses one can use any type of optical fiber with normal dispersion at the pump wavelength [9], but here we investigate parabolic pulse formation in all-normal dispersive photonic crystal fibers with a flat-top located at the pump wavelength. Such fibers are very attractive for femtosecond parabolic pulse reshaping because of the minimal impact of third-order dispersion if pumping at the flat-top of

dispersion curve. Another attractive feature is a flexibility of the dispersion tailoring, which allows for example to reduce a second order dispersion at the pump wavelength providing smaller dispersion broadening of the pulse. Recently we have designed ANDi PCF for supercontinuum generation at 800 nm [10]. Here we show that this fiber can be also applied for pulse reshaping at 800 nm. It has a solid core and hexagonal lattice of air holes, with the geometrical design parameter pitch $\Lambda = 1\,\mu m$ and a relative hole size $d/\Lambda = 0.5$. The dispersion profile and mode field diameter (MFD) of the PCF were generated with the analytical method described in [10]. Fig. 1 shows dispersion parameter D and MFD of the designed PCF. Fiber's parameters at 800 nm used in simulations are following: $\beta_2 = 1.36 \times 10^1\ ps^2/km$, $\beta_3 = -8.65 \times 10^{-3}\ ps^3/km$, $\beta_4 = 1.67 \times 10^{-4}\ ps^4/km$, $\gamma = 113\ 1/(W \cdot km)$.

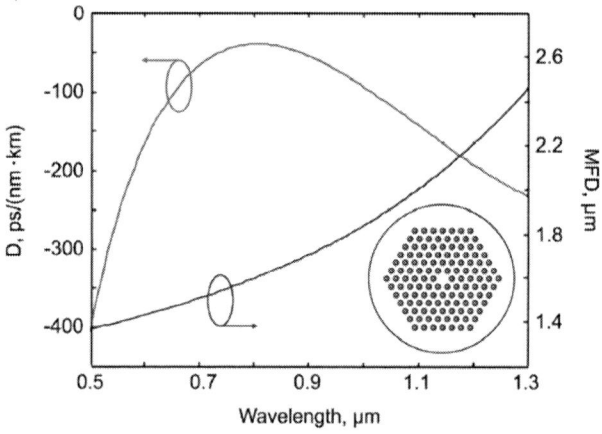

Fig. 1. Calculated dispersion profile and MFD of the designed ANDi PCF.

The evolution of an ultrashort pulse during its propagation in a normal-dispersion fiber with Kerr nonlinearity was investigated by solving generalized nonlinear Schrödinger equation with Runge–Kutta in the interaction picture method. Initial pulse parameters used in simulations are chosen to be similar to that one produced by Ti:Sapphire lasers or Erbium-doped fiber lasers. In simulations we used following initial pulse parameters: pulse shape – Gaussian and secant, pulse duration ~100 fs, pulse energy – up to 100 pJ. Pulse reshaping towards parabolic waveform during its propagation in the fiber is

978-1-4799-0019-0/13 $31.00 © 2013 IEEE

analyzed using a misfit parameter M, which shows the deviation between the pulse temporal intensity profile and a parabolic fit of the same energy. Fig. 2 and Fig. 3 shows results of simulations. We analyze femtosecond parabolic pulse formation using misfit parameter maps $M(E_0, z)$ and contour curves of pulse duration (FWHM) $\tau(E_0, z)$ superimposed above, where E_0 is initial pulse energy and z is propagation distance in the fiber. This allows finding the areas of parameters where parabolic pulses are formed dark areas ($M < 0.04$) and simultaneously analyzing the duration of parabolic pulses. Two approaches for pulse reshaping in normal dispersive fibers are analyzed and compared. The first one implies parabolic pulse formation as a transient state of pulse evolution in the fiber, when z is smaller than dispersion length L_D [7]. The second approach proposed in our group implies parabolic pulse formation in the steady-state regime, when $z > L_D$ [8], in this case pulse remains parabolic shape futher with increasing of propagation distance in the fiber. Fig. 2 shows results of simulation in the transient regime for Gaussiant and secant intial pulses.

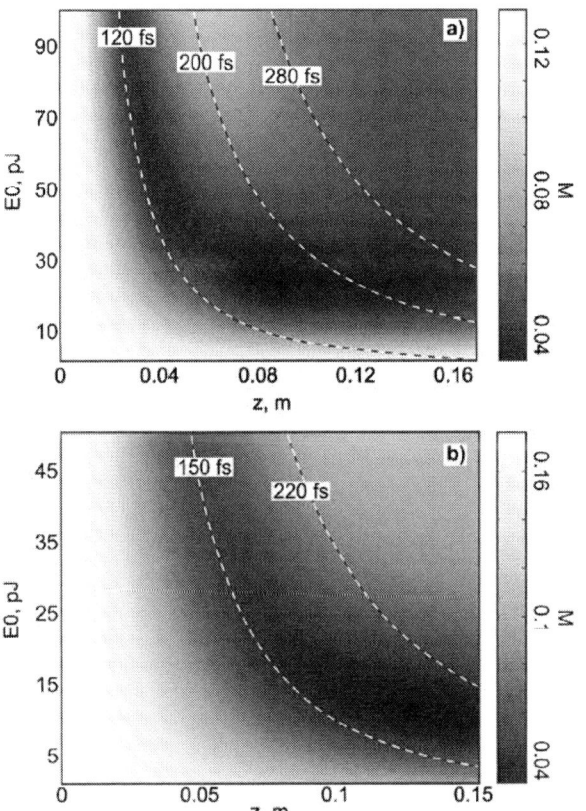

Fig. 2. Misfit parameter maps $M(E_0, z)$ and contour curves of pulse duration $\tau(E_0, z)$ (dashed lines) for the case $z \leq L_D$. a) – initial unchirped Gaussian pulse $\tau_0 = 80\,\text{fs}$; b) – initial unchirped secant pulse $\tau_0 = 80\,\text{fs}$.

From Fig. 2 a) we can see that from initial unchirped Gaussian pulse one can obtain quite short parabolic pulses (140-260 fs) for initial pulse energy range 15-70 pJ. Fiber length has to be quite small 4-16 cm. For the same secant pulse (Fig. 2 b)) the dark area becomes smaller and it is shifted towards longer fiber lengths and lower energies, because the secant shape varies stronger from the parabolic one, and thus, pulse needs longer distance to be transformed. However, in this case it is also possible to obtain 160-220 fs parabolic pulses.

Fig. 3 shows results of simulation in the steady-state regime for Gaussiant and secant intial pulses. Only chirped initial pulses are considered here, because evolution of unchirped Gaussian and secant pulses does not tend to the parabolic waveform in the steady-state regime [8].

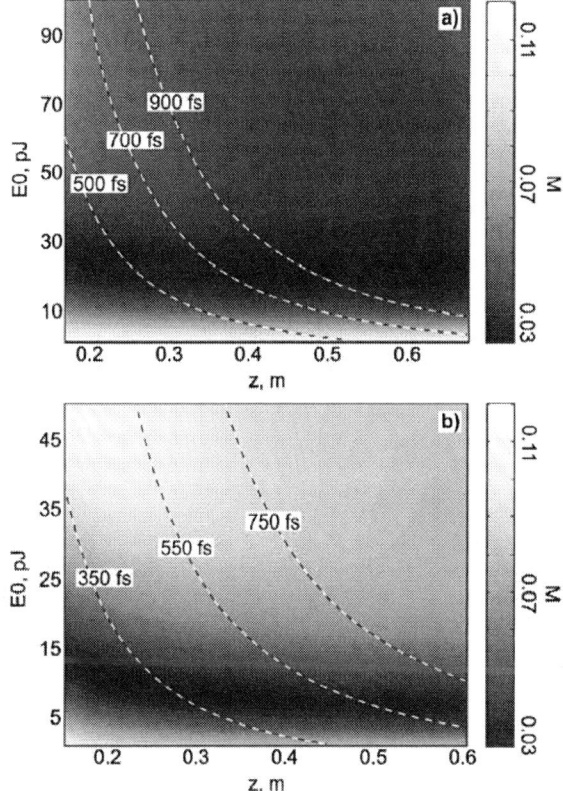

Fig. 3. Misfit parameter maps $M(E_0, z)$ and contour curves of pulse duration $\tau(E_0, z)$ (dashed lines) for the case $L_D < z < 4L_D$. a) – initial Gaussian pulse $\tau_0 = 80\,\text{fs}$ with chirp $C = 1.24$; b) – initial secant pulse $\tau_0 = 80\,\text{fs}$ with chirp $C = 0.24$.

From Fig. 3 a) we can see that from initial Gaussian pulse one can obtain parabolic pulses with duration as short as 500 fs for initial pulse energy 10-50 nJ and fiber 20-30 cm. From secant initial pulse (Fig. 3 b)) one can obtain shorter parabolic pulse 320-350 fs in the steady-state regime at the similar fiber length (20-30 cm), but smaller initial pulse energy has to be used ~10 pJ and the dark area is smaller here.

Fig. 4 shows examples of parabolic pulses generated via passive nonlinear reshaping in ANDi PCF in the transient regime and in the steady-state regime. We can see that both pulses are perfectly parabolic. However, at the short propagation distance (10 cm) pulse width (FWHM) is 167 fs; in the steady-state regime (30 cm) it is larger, 513 fs, due to the longer propagation distance in the fiber required for appearance of the steady-state regime.

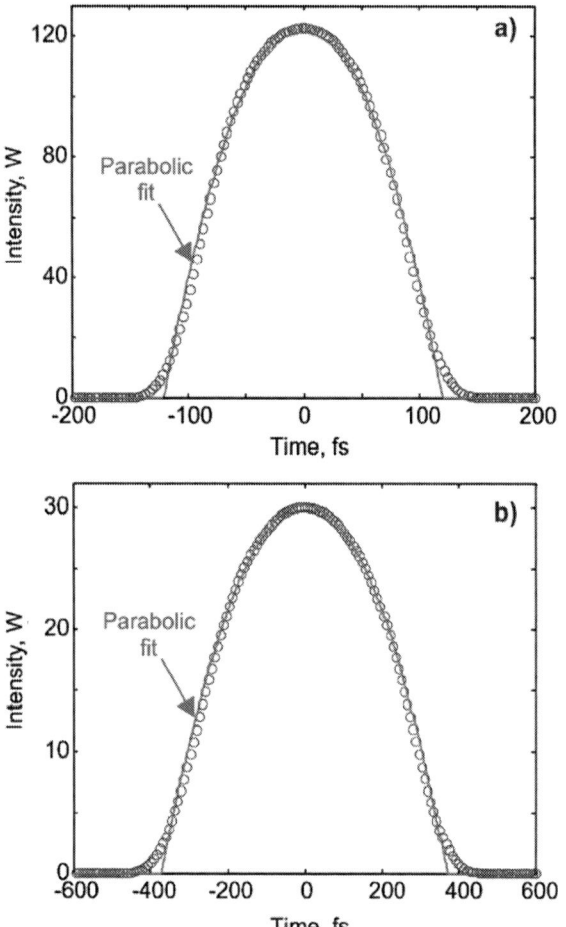

Fig. 4. Parabolic pulses generated in the ANDi PCF (circles) and appropriate parabolic fitting curves. a) – parabolic pulse generated from initial unchirped Gaussian pulse in the transient regime ($\tau_0 = 80\,\mathrm{fs}$, $E_0 = 20\,\mathrm{pJ}$, $z = 10\,\mathrm{cm}$), misfit parameter is $M = 0.035$. b) – parabolic pulse generated from initial chirped Gaussian pulse in the steady-state regime ($\tau_0 = 80\,\mathrm{fs}$, $E_0 = 15\,\mathrm{pJ}$, $C = 1.24$, $z = 30\,\mathrm{cm}$), misfit parameter is $M = 0.033$.

To conclude we have investigated numerically formation of femtosecond parabolic pulses by means of passive nonlinear reshaping in ANDi PCF at 800 nm. In the transient regime we can obtain the shortest parabolic pulses from a Gaussian pulse ~167 fs applying 10 cm fiber. In the steady-state regime we can obtain longer parabolic pulses as short as ~513 fs applying 30 cm fiber. A common feature is that for obtaining shorter pulses one has to use shorter initial pulses and shorter fiber's pieces. Another important result is that parabolic pulses both at the short propagation distance and in the steady-state regime are obtained within a limited range of pulse energy (tens of pJ). For initial Gaussian pulses the suitable range of pulse energy is larger as compared to the secant one pulses.

REFERENCES

[1] S. Boscolo and C. Finot, "Nonlinear pulse shaping in fibres for pulse generation and optical processing" *Int. J. Optics*, vol. 2012, p. 159057, 2012.

[2] J. M. Dudley, C. Finot, D. J. Richardson, G. Millot, "Self-similarity in ultrafast nonlinear optics", *Nature Physics*, vol. 3, pp. 597-603, 2007.

[3] C. Finot, J.M. Dudley, B. Kibler, D.J. Richardson, G. Millot, "Optical parabolic pulse generation and applications", *IEEE J. Quantum Electron.*, vol. 45, pp. 1482-1489, 2009.

[4] D. Krcmarik, R. Slavik, Y. Park, and J. Azana, "Nonlinear pulse compression of picosecond parabolic-like pulses synthesized with a long period fiber grating filter," *Opt. Express,* vol. 17, pp. 7074–7087, 2009.

[5] T. Hirooka, M. Nakazawa, and K. Okamoto, "Bright and dark 40 GHz parabolic pulse generation using a picosecond optical pulse train and an arrayed waveguide grating," *Opt. Lett.*, vol. 33, pp. 1102–1104, 2008.

[6] T. Hirooka and M. Nakazawa, "Parabolic pulse generation by use of a dispersion-decreasing fiber with normal group-velocity dispersion," *Opt. Lett.*, vol. 29, pp. 498-500, 2004.

[7] C. Finot, L. Provost, P. Petropoulos, and D. J. Richardson, "Parabolic pulse generation through passive nonlinear pulse reshaping in a normally dispersive two segment fiber device," *Opt. Express*, vol. 15, pp.852-864, 2007.

[8] S. O. Iakushev, O. V. Shulika, I.A. Sukhoivanov, "Passive nonlinear reshaping towards parabolic pulses in the steady-state regime in optical fibers", *Opt. Comm.*, vol. 285, pp. 4493–4499, 2012.

[9] I. A. Sukhoivanov, S. O. Iakushev, O. V. Sulika, A. Diez, M. Andrés, "Femtosecond parabolic pulse shaping in normally dispersive optical fibers," *Opt. Express*, vol. 21, pp. 17769-17785, 2013.

[10] S. O. Iakushev, O. V. Shulika, and I. A. Sukhoivanov, "Sub-10-fs Pulses Produced From Compression of Supercontinuum Generated in All-Normal Dispersion Photonic Crystal Fiber," in *Frontiers in Optics Conference, OSA Technical Digest (online) (Optical Society of America, 2012)*, paper FW3A.41.

Similariton pulse compression using hybrid grating-prism compressor

Aram S. Zeytunyan, Garegin L. Yesayan, Levon Kh. Mouradian

Ultrafast Optics Laboratory, Faculty of Physics, Yerevan State University, Yerevan, Armenia

Abstract: We generate broadband nonlinear-dispersive similaritons of up to 75 THz FWHM bandwidth (160 nm at 800 nm central wavelength) in LMA-5 single-mode photonic crystal fiber, and compress them to 15 fs using a hybrid grating-prism dispersive delay line. We numerically and experimentally show the large tunability of third-order dispersion of such a hybrid dispersive delay line.

The processes of AM-FM / FM-AM type conversions in optics, and particularly the technique of fiber-optic pulse compression, are strongly supporting the advancement of contemporary ultrafast optics and laser technology. In the systems of pulse compression, the positive chirp of a pulse spectrally broadened in a normally dispersive single-mode fiber is compensated by the negative one in a dispersive delay line (DDL) [1]. On the femtosecond time scale, high-order nonlinear and dispersive effects are substantially limiting the pulse compression ratio. To compress a pulse to the transform limit, one should not only compensate the quadratic component of spectral phase induced due to the group-velocity dispersion (GVD), but also its cubic, quartic and higher order terms. The impact of residual third-order dispersion (TOD) at the system output is the most substantial among all high-order effects. Various dispersive devices, including grating [2] and prism pairs [3], have been used as a DDL for dispersion compensation. The combination of gratings and prisms was used for the TOD balance and pulse compression to 6 fs, using the opposite TOD signs of gratings and prisms [4]. The drawback of this approach is that the grating and prism sequences are used separately, thus one has to use a large prism sequence. Mixing the prism and grating sequences was proposed in [5]. In this setup, Brewster-angle prisms are used in between the two gratings, allowing for a compact setup. This hybrid grating-prism compressor provides TOD tunability in a wide range, and has been applied for parabolic pulse compression [6] and for ten-cycle pulse generation from a fiber laser [7]. The modified version of this hybrid compressor – diffraction gratings coupled to prisms – was also demonstrated and called "grisms" [8].

Our new approach to the pulse compression problem is based on the generation of nonlinear-dispersive similariton in a conventional passive uniform fiber [9] and its further compression [10]. In general, pulse compression is carried out in the regime of "rectangular" pulse shaping [11], which can be considered as an earlier step of similariton generation. The

time-bandwidth product in the regime of "rectangular" pulses is larger than that for bell-shaped pulses, thus the compressed pulse will be longer for the same spectral bandwidth. In the similariton regime, the spectral broadening ratio exceeds that for "rectangular" pulse, and the chirp becomes more linear and given by the fiber length only [9]. Thus, in the similariton regime it is possible to compress pulses to the transform-limit with the spectral broadening ratio and with higher peak intensity. Therefore, the application of similariton can essentially improve the techniques of pulse compression and shaping, leading to accurate, aberration-free methods, since the chirp of similariton is linear (the phase is parabolic) and its spectrotemporal profile is smooth and bell-shaped.

The signal analysis-synthesis problems on a few-femtosecond time scale demand the generation and study of broadband similaritons [12,13], when the impact of higher-order dispersion is significant. In this case, the spectral phase of similariton is also given only by the fiber dispersion [12]. In [10], the similariton pulse compression was demonstrated by generating broadband nonlinear-dispersive similariton in fiber and its chirp cancellation in a DDL consisting of a pair of 6-m separated fused-silica dispersive prisms. The phase compensation was carried out using the fact that the signs of TOD in fiber and prism DDL are opposite. Approximately transform-limited pulses of down to 25 fs autocorrelation duration, i.e. ~18 fs pulse duration at FWHM for a Gaussian pulse, were synthesized from a similariton of 77-nm bandwidth. However, such phase compensation limits the parameters of fiber and DDL, and thus, the compression ratio. This means that there is a need for a compact compressor device that can compensate for the high orders of dispersion in a wide range. In this paper we propose and study the application of hybrid grating-prism compressor for similariton pulse compression.

In our experiment, we use the Coherent Verdi V10 + Mira 900F femtosecond laser system with the following parameters of radiation: 100 fs FWHM pulse duration, 76 MHz repetition rate, 1.6 W average power, 775-800 nm central wavelength. We couple up to ~1 W of the laser radiation (corresponding to pulse energy of up to ~13 nJ) into the normally-dispersive 30-cm long LMA-5 photonic crystal fiber by means of an aspheric lens (8 mm focal length) and generate a broadband nonlinear-dispersive similariton with FWHM bandwidths of up to 160 nm. Figure 1 shows the similariton spectrum (a) measured by optical spectrum analyzer Ando 6315, and corresponding calculated transform-limited (zero-phase) pulse (b). The similariton pulse can be compressed to ~6 fs, which shows the potential of the technique of similariton pulse compression for obtaining few-cycle optical pulses.

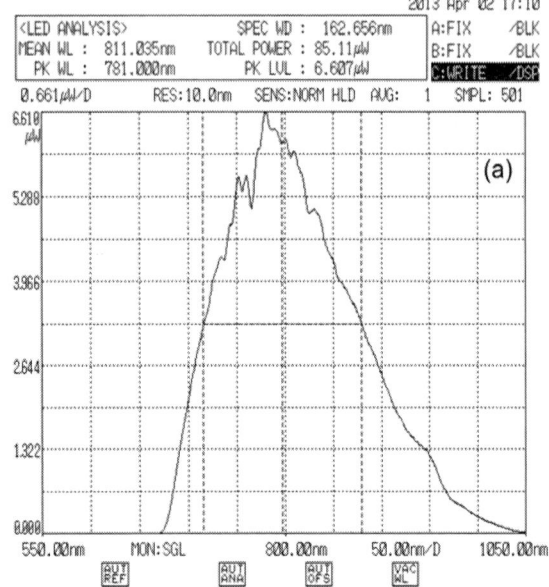

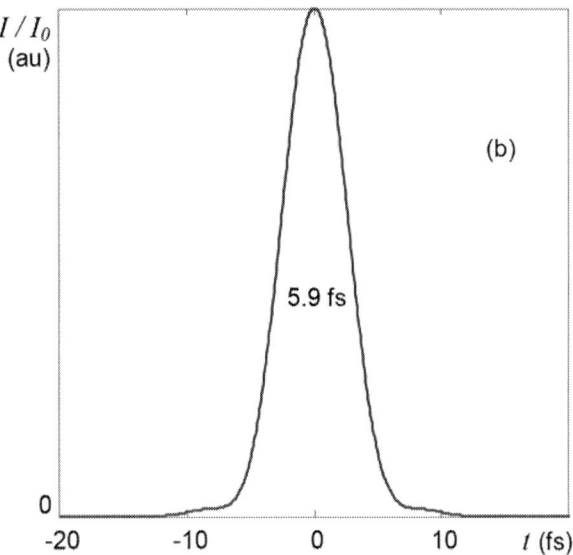

Fig. 1. Measured spectrum of broadband similariton (a), and corresponding calculated transform-limited pulse (b).

Afterwards we direct the spectrally broadened and positively chirped pulse from the fiber to the hybrid grating-prism DDL for pulse compression. The reflection gratings and Brewster-angle prisms are located in uncrossed configuration, allowing the first prism to enhance the beam divergence from the first grating. We simulated the dispersion of different combinations of gratings and prisms (up to the fourth order) using the Sheriff's method of dispersion calculation in arbitrary prism

sequences [14] to find the tuning range of the TOD-to-GVD ratio β_3 / β_2. Figure 2 shows the dependence β_3 / β_2 vs. the grating incident angle γ for the combinations of 300 mm^{-1} gratings and SF11 prisms (blue), 600 mm^{-1} gratings and SF11 prisms (red), 300 mm^{-1} gratings and fused silica (FS) prisms (green), and 600 mm^{-1} gratings and fused silica prisms (black). Using these combinations, it is possible to tune the TOD-to-GVD ratio from −0.5 fs to 2 fs. Since the fiber TOD coefficient β_3 was unknown, we tried to dechirp the similariton pulses from the fiber with different DDL configurations. We obtained the shortest pulse in case of 300 mm^{-1} gratings and SF11 prisms for the prism apex distance $L = 10$ cm and grating incident angle $\gamma = 50°$. Figure 3 illustrates the results of this best realization. The shortest pulse did not correspond to the broadest spectrum, which evidently had some uncontrollable phase. The spectrum of the dechirped pulse after DDL is shown in Fig. 3(a) by the black curve, along with the similariton spectrum after fiber (gray). The calculated autocorrelation track of the transform-limited pulse corresponding to the dechirped spectrum (blue), and the measured autocorrelation track of dechirped pulse (red) are shown in Fig. 3(b). The shortest measured autocorrelation has 22 fs FWHM duration, corresponding to ~15 fs duration for a Gaussian pulse, and the average power at the system output is 80 mW (corresponding to ~70 kW pulse peak power). We attribute the bottom pedestal of the measured autocorrelation track and the difference between the measured and calculated autocorrelation tracks to uncompensated fourth-order dispersion, since the β_4 coefficients of all components of our pulse compression system have the same sign.

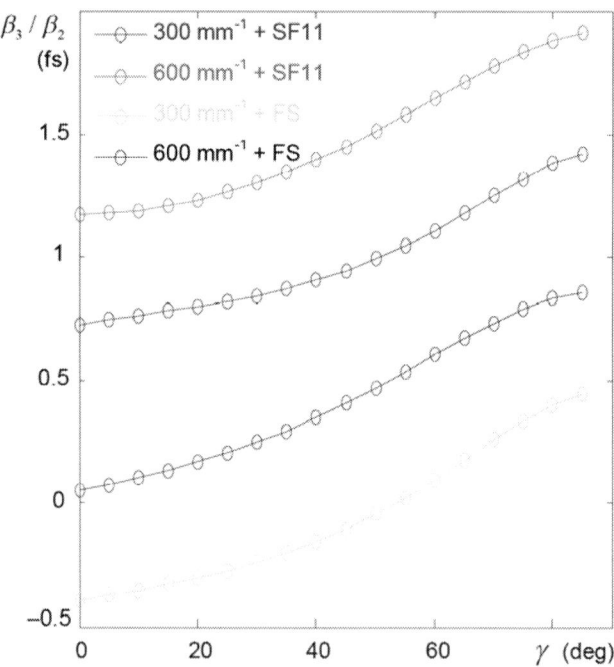

Fig. 2. TOD-to-GVD ratio vs. grating incident angle for different combinations of gratings and prisms.

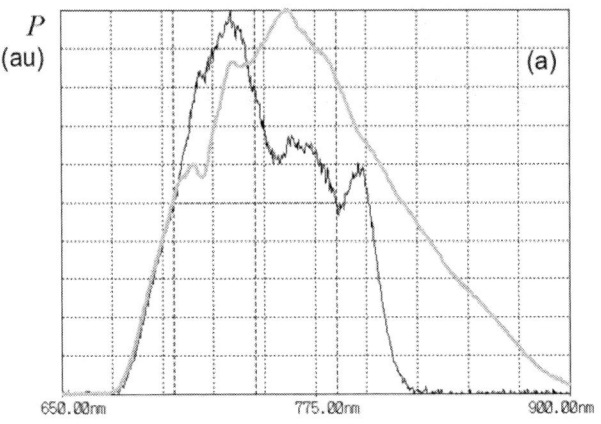

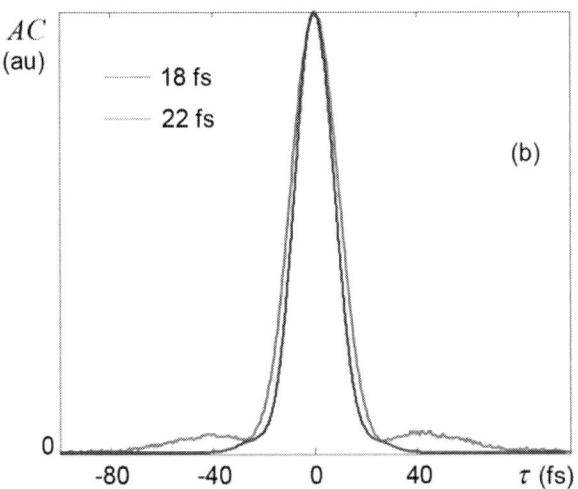

Fig. 3. (a) Spectra of similariton (gray) and dechirped (black) pulses. (b) Calculated autocorrelation track of transform-limited pulse (blue), and measured autocorrelation track of dechirped pulse (red).

Concluding, we generated broadband nonlinear-dispersive similaritons of up to 160-nm FWHM bandwidth in a normally dispersive photonic crystal fiber, and compressed these similaritons using a hybrid grating-prism DDL. We showed that it is possible to tune the TOD-to-GVD ratio of such a DDL from –0.5 fs to 2 fs using 300 mm^{-1} and 600 mm^{-1} gratings and SF11 and fused-silica prisms. We obtained 10 times pulse compression (22 fs autocorrelation duration or ~15 fs FWHM pulse duration). The generation of nonlinear-dispersive similariton in a passive fiber makes the pulse compression technique more effective due to the induced phase defined only by the fiber dispersion. Further efforts will address the technical optimization of the system to obtain shorter pulses with less energetic losses, comparable with the output parameters of commercial 10-fs lasers.

The authors acknowledge Hui Liu and Frank Wise, Cornell University, for providing the LMA-5 fiber and for helpful discussions.

REFERENCES

[1] W.J. Tomlinson, R.H. Stolen, and C.V. Shank, "Compression of optical pulses chirped by self-phase modulation in fibers," *J. Opt. Soc. Am. B*, vol. 1, pp. 139-149, 1984.

[2] E.B. Treacy, "Optical pulse compression with diffraction gratings," *IEEE J. Quantum Electron.*, vol. QE-5, pp. 454-458, 1969.

[3] R.L. Fork, O.E. Martinez, and J.P. Gordon, "Negative dispersion using pairs of prisms," *Opt. Lett.*, vol. 9, pp. 150-152, 1984.

[4] R.L. Fork, C.H. Brito Cruz, P.C. Becker, and C.V. Shank, "Compression of optical pulses to six femtoseconds by using cubic phase compensation," *Opt. Lett.*, vol. 12, pp. 483-485, 1987.

[5] S. Kane, J. Squier, J.V. Rudd, and G. Mourou, "Hybrid grating-prism stretcher-compressor system with cubic phase and wavelength tunability and decreased alignment sensitivity," *Opt. Lett.*, vol. 19, pp. 1876-1878, 1994.

[6] Y. Zaouter, D.N. Papadopoulos, M. Hanna, F. Druon, E. Cormier, and P. Georges, "Third-order spectral phase compensation in parabolic pulse compression," *Opt. Express*, vol. 15, pp. 9372-9377, 2007.

[7] J.R. Buckley, S.W. Clark, and F.W. Wise, "Generation of ten-cycle pulses from an ytterbium fiber laser with cubic phase compensation," *Opt. Lett.*, vol. 31, pp. 1340-1342, 2006.

[8] S. Kane and J. Squier, "Grism-pair stretcher-compressor system for simultaneous second- and third-order dispersion compensation in chirped-pulse amplification," *J. Opt. Soc. Am. B* 14, 661-665, 1997.

[9] A. Zeytunyan, G. Yesayan, L. Mouradian, P. Kockaert, P. Emplit, F. Louradour, and A. Barthélémy, "Nonlinear-dispersive similariton of passive fiber," *J. Europ. Opt. Soc. Rap. Public.*, vol. 4, 09009, 2009.

[10] K. Palanjyan, A. Muradyan, A. Zeytunyan, G. Yesayan, and L. Mouradian, "Pulse compression down to 17 femtoseconds by generating broadband similariton," *Proc. SPIE*, vol. 7998, 79980N, 2010.

[11] G.P. Agrawal, *Nonlinear Fiber Optics*, third ed., Academic, 2001.

[12] A. Zeytunyan, A. Muradyan, G. Yesayan, L. Mouradian, F. Louradour, and A. Barthélémy, "Generation of broadband similaritons for complete characterization of femtosecond pulses," *Opt. Commun.*, vol. 284, pp. 3742-3747, 2011.

[13] A. Chong, H. Liu, B. Nie, B.G. Bale, S. Wabnitz, W.H. Renninger, M. Dantus, and F.W. Wise, "Pulse generation without gain-bandwidth limitation in a laser with self-similar evolution," *Opt. Express*, vol. 20, pp. 14213-14220, 2012.

[14] R.E. Sherriff, "Analytic expressions for group-delay dispersion and cubic dispersion in arbitrary prism sequences," *J. Opt. Soc. Am. B*, vol. **15**, pp. 1224-1230, 1998.

2D wave structure induced by femtosecond laser pulse in semiconductor

Vyacheslav A. Trofimov[1], Maria M. Loginova[1] and Vladimir A. Egorenkov[1]
[1] Lomonosov Moscow State University, Moscow, Russia

Abstract. We simulate interaction of a femtosecond laser pulse with semiconductor. As it is well known, under certain conditions the optical bistability (OB) takes place due to a nonlinear absorption of the optical energy, for example. We investigate the 2D wave process, occurring in such conditions and appearing because of instability of one from the bistable element states. Due to laser pulse propagating and electrons diffusion, the instability of the states of the optical radiation-semiconductor system causes the 2D wave propagation of free-electrons and ionized donors in semiconductor. This process can essentially change properties of the bistable element.

STATEMENT OF 2D PROBLEM

We consider propagation of the femtosecond laser pulse in a semiconductor in the framework of 2D case and this process is described by the following set of dimensionless equations [1,2]:

$$\frac{\partial^2 \varphi}{\partial x^2} + \frac{\partial^2 \varphi}{\partial y^2} = \gamma(n - N), \quad \frac{\partial N}{\partial t} = G(n, N, \varphi) - R(n, N), \quad (1)$$

$$\frac{\partial n}{\partial t} = D_x \frac{\partial}{\partial x}(\frac{\partial n}{\partial x} - \mu_x n \frac{\partial \varphi}{\partial x}) + D_y \frac{\partial}{\partial y}(\frac{\partial n}{\partial y} - \mu_y n \frac{\partial \varphi}{\partial y}) +$$

$$+ G(n, N, \varphi) - R(n, N),$$

$$\frac{\partial I}{\partial y} + \delta_0 \delta(N, n, \varphi)I = 0, 0 < x < L_x, \ 0 < y < L_y, \ t > 0.$$

The boundary and initial conditions correspond to semiconductor placed in the external uniform electric field, directed along x –axis and they are:

$$\frac{\partial \varphi}{\partial x}\Big|_{x=0, L_x} = -E_{ex}, \frac{\partial \varphi}{\partial y}\Big|_{y=0, L_y} = 0, \quad (2)$$

$$D_x(\frac{\partial n}{\partial x} - \mu_x n \frac{\partial \varphi}{\partial x}) = D_y(\frac{\partial n}{\partial y} - \mu_y n \frac{\partial \varphi}{\partial y}) = 0,$$

$$I\big|_{y=0} = \exp\left(-\left(\frac{x - 0.5L_x}{0.1L_x}\right)\right)(1 - \exp(-10t)),$$

$$n\big|_{t=0} = N\big|_{t=0} = n_0 e^{\mu_x \varphi}\big|_{t=0}, \ \varphi\big|_{t=0} = -E_{ex}x, \ I_{t=0} = 0 .$$

The functions G and R, describing generation and recombination of free electrons and ionized donors in the semiconductor, are given by the formulas:

$$R = \frac{nN - n_0^2}{\tau_p}, G = q_0 I \delta(n, N, \varphi), \ q_0 > 0 . \quad (3)$$

In the set of equations (1) – (3) the following variables are introduced: x, y- dimensionless spatial coordinates, t – dimensionless time, n - concentration of free electrons in the conductivity zone of semiconductor, N – concentration of ionized donors. Function φ - dimensionless an electric field potential. Function I denotes intensity of a laser pulse propagating along the y coordinate. Coefficients of electrons diffusion, denoted as D_x, D_y, and coefficients of electrons mobility, denoted as μ_x, μ_y, are positive constants.

Light energy absorption $\delta(n, N, \varphi)$ can be approximated by various functions in dependence on physical mechanism of the absorption. Below we use the following its approximation

$$\delta(N, n) = (1 - N) \ e^{-\psi(1-\xi n)}, \ \psi, \xi > 0, \quad (4)$$

which is close to one of the experimental dependencies corresponding to concentration optical bistability (OB) [3,4]. It takes into account the Burstein-Moss effect: depletion of donor level and dynamic levels saturation in the conduction band of semiconductor. Parameter δ_0 is a maximal value of the absorption.

Parameter n_0 describes an equilibrium value of the free electrons concentration and ionized donors one. Parameter τ_p characterizes recombination time. Parameter q_0 denotes a maximal value of the laser intensity incident on the semiconductor.

It should be noticed, that similar systems with increasing absorption have been widely studied early [3-8] for the laser pulse interaction with various media. In [9, 10] the OB, based on bandgap changing if action of a strong electric field takes place, was predicted in the framework of 1D case. This dependence can appear if a femtosecond pulse act on the semiconductor. Nevertheless, for practice the 2D case is more of interest. Therefore, in [11, 12] we have developed the finite-difference schemes for such kind of the problem by using various iterative processes. Below we show an opportunity of wave process developing if laser pulse with its duration from hundred femtoseconds to hundred picoseconds.

COMPUTER SIMULATION RESULTS FOR 2D PROBLEM

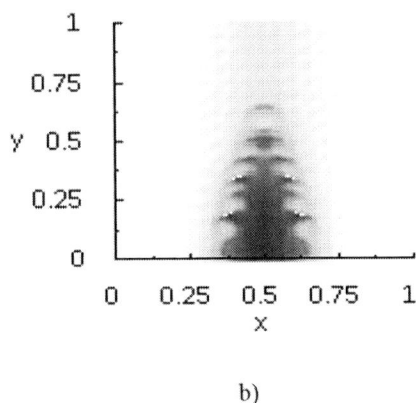

a)

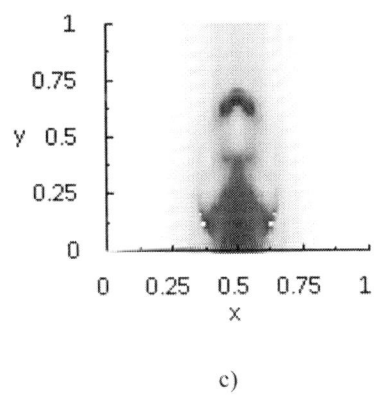

b)

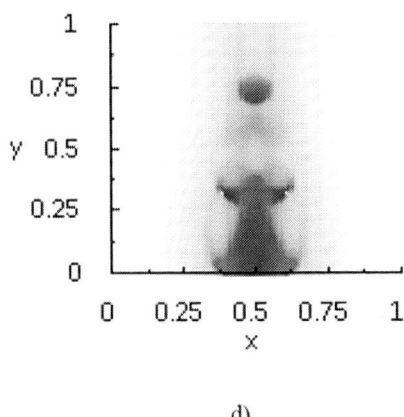

d)

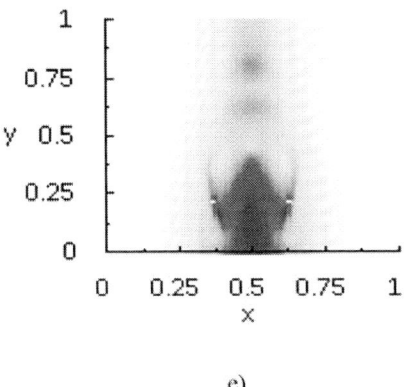

e)

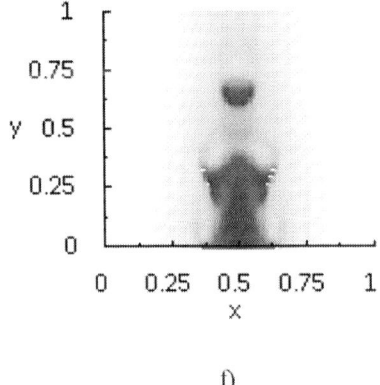

c)

f)

Fig 1. Spatial distributions of free electrons concentration occurring at time moments t=20(a), 210(b), 400(c), 510(d), 560(e), 620(f).

Computer simulation shows some different modes of laser pulse interaction with the semiconductor. Among them, we emphasize the optical bistable mode, of course. Other mode

corresponds to appearance of transverse wave of free charges. The last mode is 2D wave process of free electrons propagation and ionized donors one. This process is the more interesting for practice. As example, in Fig. 1 we show spatial distributions of free electrons concentration for parameters $D_x = D_y = 10^{-5}, \mu_x = \mu_y = 1, \gamma = 10^3, q_0 = 1,$

$n_0 = 0.01$, $\psi = 2.553$, $\xi = 3$, $\tau_p = 1$, $E_{ex} = 0$, which corresponds to absence of the external electric field. We can distinguish a few stages of laser pulse interaction with the semiconductor. First stage spreads until 30 dimensionless units in time. During this time interval the domain with high absorption appears. Then irregular waves of free electrons concentration forms and propagates during a few hundred units in time. After that, since time about 500 units the periodic wave process occurs. Its period is equal to 110 units (let us compare the Fig. 1 d and Fig. 1 f). We see very complicated spatial structure developing in time. For its computation, it is necessary to use high effective finite-difference scheme.

It should be stressed, that the spatial distribution of free electrons concentration strongly depends on the interaction parameters. Their changing leads to inducing of more complicated charged particles distribution in comparison with the distribution depicted in the Fig. 1.

Physical reasons of an appearance of the wave process are an existence of the optical bistability; instability of the states of the bistable system (laser radiation-semiconductor); diffusion of free electron concentration.

CONCLUSION

We showed that the interaction of femtosecond pulse with semiconductor, having an absorption increasing with growth of the free electron concentration, can result in developing of complicated wave process which change dramatically the properties of bistable system. In particular, optical pulse intensities corresponding to switching from state to another state of bistable system may be increased because of the appearance of additional domains with high concentration of charged particles. In problems of THz radiation generation such complicated spatial distribution of free electrons can cause distortions of the generated pulse shape.

ACKNOWLEDGEMENT

This paper was partly financially supported by Russian Foundation for Basic Research (grant number 12-01-31496 mol_a).

REFERENCES

[1] R.A. Smith, *Semiconductors*, Cambridge University Press, Cambridge, 1959.

[2] N.B. Delone, V.P. Kraynov, *Nonlinear ionization of atoms by laser radiation*, Moscow: Phismatlit, 2001 (in Russian).

[3] H. Gibbs, *Optical Bistability: Controlling Light with Light*, Academic Press, New York, 1985.

[4] N.N. Rozanov, *Optical Bistability and Hysteresis in Distributed Nonlinear Systems*, Nauka, Moscow, 1997 (in Russian).

[5] H.M. Gibbs, G.R. Olbright, N. Peyghambarian, et. al., "Kinks: Longitudinal excitation discontinuities in increasing-absorption optical bistability," *Phys. Rev. A.* vol.32, no.1, pp.692-694, 1985.

[6] S.W. Koch, H.E. Schmidt, H. Haug, "Optical bistability due to induced absorption: Propagation dynamics of excitation profiles," *Appl. Phys. Lett.* vol.45, no.9, pp. 932-934, 1984.

[7] O.A. Gunaze, V.A. Trofimov, "Formation of an "inverse" kink in optically bistable systems based on increasing absorption," *Technical Physics Letters* 23 (11), pp. 846 – 847, 1997.

[8] Yu.N. Karamzin, S.V. Polyakov, V.A. Trofimov, I.G. Zakharova, "Numerical simulation of some optical bistability problems in semiconductor system," *Proceedings of SPIE* 1840, pp. 113-129, 1992.

[9] V.A. Trofimov, M.M. Loginov, "On the possibility of transverse size oscillations in a domain with a high free electron concentration under the action of a short light pulse on a semiconductor," *Technical Physics* vol.49, no.11, pp. 1517–1520, 2004.

[10] V.A. Trofimov, M.M. Loginova, "Anomalous influence of electrons diffusion on absorption optical bistability realization," *In "Nonlinear Optics Applications" Ed. Karpierz M.M., Boardman A.D., Stegeman G.I. Proceedings of SPIE* 5949, 59491J-1 – 59491J-7, 2005.

[11] V.A. Trofimov, M.M. Loginova, "Conservative Finite-Difference Scheme for a Two-Dimensional Problem of Plasma Generation in a Semiconductor under a Laser Pulse Action," *Proceedings of "10th International Conference on Laser&Fiber-Optical Networks Modeling*, Ukraine, pp.186-188, 2010.

[12] V.A. Trofimov, M.M. Loginova, "Finite-difference schemes for a two-dimensional problem of femtosecond pulse interaction with semiconductor," *Proceedings of the International Conference CMMSE-2011*, Spain, vol. IV. pp. 1641-1651, 2011.

Nonlinear interference effects in different types of three-level quantum systems

A. A. Orudzhev[1], I. L. Plastun[1], V.L. Derbov[2]
[1]Saratov State Technical University, Russia
[2]Saratov State University, Russia

Abstract: Coherent population trapping (CPT) resonance formation is modeled numerically in a three-level Λ- and Ξ-systems with one of the near-resonance fields being frequency-modulated. The model is based on density matrix equations in RW approximation with atomic relaxation properly taken into account. Slow modulation is shown to be equivalent to CW excitation with the frequency changed point by point. As the modulation period approaches the relaxation times, the delayed response of the system is shown to cause the CPT resonance shift and reshaping, as well as the appearance of transient oscillations of the energy level populations.

Coherent population trapping (CPT) [1-4] is a technique in ultrahigh-resolution spectroscopy, developed in recent years. It relies on the nonlinear coherent interaction between an atomic system and incident electromagnetic radiation and is followed by an analysis of the fine structure of the response to this interaction, which contains information about the spectral characteristics of the quantum system.

To observe a CPT resonance it is common to fix the frequency of one laser wave and to vary that of the other. Here we consider this variation as a real-time process, i.e., the second laser is assumed to be frequency-modulated and the upper energy level population is researched versus the time.

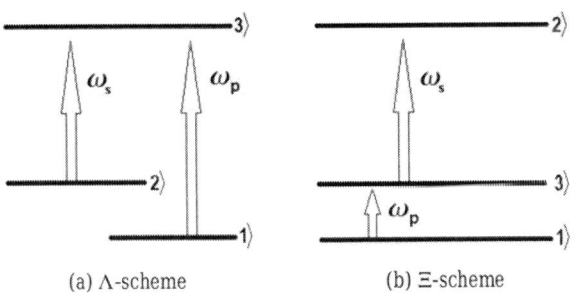

(a) Λ-scheme   (b) Ξ-scheme

Fig. 1. Three-level atom interacting with the two coherent fields ω_s and ω_p.

Numerical scheme based on the system of six equations for density matrix.

For Λ-system:

$$\dot{\tilde{\rho}}_{21} = \frac{1}{i}[\omega_{21} - (\omega_p - \omega_s) - i\gamma_{12}]\tilde{\rho}_{21} + \frac{1}{i}(\tilde{\rho}_{31}(V_s^-)_{23} - \tilde{\rho}_{32}^*(V_p^+)_{31}]$$

$$\dot{\tilde{\rho}}_{31} = \frac{1}{i}[\omega_{31} - \omega_p - i\gamma_{13})\tilde{\rho}_{31} + \frac{1}{i}((\rho_{11} - \rho_{33})(V_p^+)_{21} + \tilde{\rho}_{21}(V_s^+)_{32}]$$

$$\dot{\tilde{\rho}}_{32} = \frac{1}{i}[\omega_{32} - \omega_s - i\gamma_{32})\tilde{\rho}_{32} + \frac{1}{i}((\rho_{22} - \rho_{33})(V_s^+)_{32} + \tilde{\rho}_{21}^*(V_p^+)_{31}]$$

$$\dot{\rho}_{11} = \frac{1}{i}[(V_p^-)_{13}\tilde{\rho}_{31} - (V_p^-)_{13}^*\tilde{\rho}_{31}^*] + W_{21}\rho_{22} + W_{31}\rho_{33}$$

$$\dot{\rho}_{22} = \frac{1}{i}[\tilde{\rho}_{32}(V_s^-)_{23} - c.c.] + W_{32}\rho_{33} - W_{21}\rho_{22}$$

$$\dot{\rho}_{33} = \frac{1}{i}[-\tilde{\rho}_{31}(V_p^-)_{13} + c.c. - \tilde{\rho}_{32}(V_s^-)_{23} + c.c.] - W_{32}\rho_{33} - W_{31}\rho_{33}$$

For Ξ-system:

$$\dot{\tilde{\rho}}_{31} = \frac{1}{i}[\omega_{31} - \omega_p - i\gamma_{31} + (V_p^+)_{31}(\rho_{11} - \rho_{33}) + (V_s^-)_{32}\tilde{\rho}_{21}]$$

$$\dot{\tilde{\rho}}_{23} = \frac{1}{i}[\omega_{32} - \omega_s - i\gamma_{32} + (V_s^+)_{23}(\rho_{33} - \rho_{22}) - (V_p^-)_{12}\tilde{\rho}_{21}]$$

$$\dot{\tilde{\rho}}_{21} = \frac{1}{i}[\omega_{21} - \omega_p - \omega_s - i\gamma_{21} + (V_s^+)_{23}\tilde{\rho}_{31} - (V_p^+)_{31}\tilde{\rho}_{23}]$$

$$\dot{\rho}_{11} = \frac{1}{i}[(V_p^-)_{13}\tilde{\rho}_{31} - c.c.] + W_{31}\rho_{33} + W_{21}\rho_{22}$$

$$\dot{\rho}_{22} = \frac{1}{i}[(V_s^+)_{23}\tilde{\rho}_{23}^* - c.c.] - W_{21}\rho_{22} - W_{23}\rho_{22}$$

$$\dot{\rho}_{33} = \frac{1}{i}[(V_p^+)_{31}\tilde{\rho}_{31}^* - c.c. + (V_s^-)_{32}\tilde{\rho}_{23} - c.c.] + W_{23}\rho_{22} - W_{31}\rho_{33}$$

Here the γ_{ij} are the transition line widths, W_{ij} are the relaxation transition rates, c.c. stands for complex conjugate. The slow envelopes of the interaction operators are expressed via the transition dipole matrix elements d_{ij} and the complex amplitudes of the fields:

$$(V_{p,s}^+)_{ij} = -\frac{1}{h}d_{ij}E_{p,s}^0 \quad (V_{p,s}^-)_{ij} = -\frac{1}{h}d_{ij}E_{p,s}^{0*}$$

Modulation frequency varies according to the following formula:

$$E = E_0 \exp(-\frac{r^2}{2a^2})\exp(\frac{i\omega_1}{\Omega}\cos\Omega t)$$

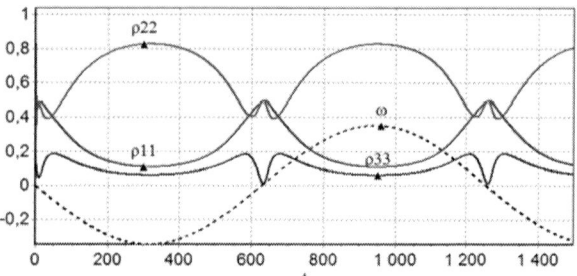

Fig. 2. Slow modulation of frequency produces dips in the level 3 population at the time of crossing the point where the two frequency detuning's coincide.

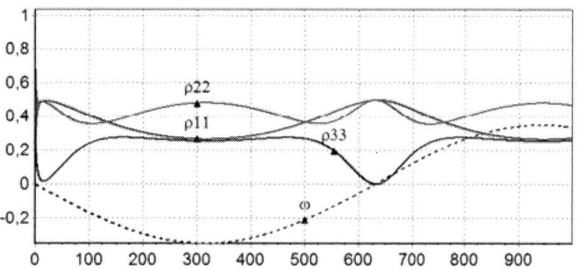

Fig. 3. Disappearance of the effect of CPT with increasing frequency modulation.

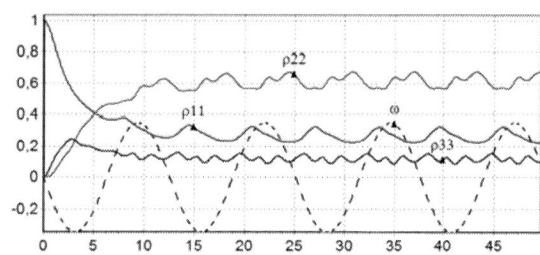

Fig. 4. The spreading of the resonance CPT.

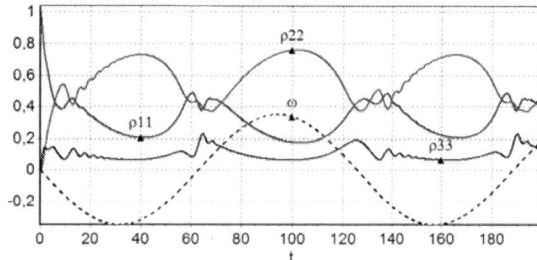

Fig. 5. Increasing the modulation frequency results in delayed CPT dips and transient oscillations.

Based on numerical solutions of equations for the density matrix describes the known problem of the interaction-level system excited by two quasiresonance fields in the scheme,

under conditions of the coherent population trapping (CPT). Unlike well-known works is the frequency modulation of one of the fields in real time. In the slow modulation reproduced the known effects of formation of CPT resonances. With the growth of the modulation frequency demonstrated transient effects associated with delayed response of the environment, to vary the instantaneous frequency modulated field. Constructed model, implemented in the developed program allows you to quickly play specific situations that arise when trying to observe CPT in atomic systems with various parameters like the atomic system, and acting on it fields, in particular, examine the various options of analytical approximations used by other authors and assess their applicability by means of direct numerical simulation.

REFERENCES

[1]   Altshuler G.B. Nonlinear lenses and their applications /G.B.Altshuler, M.V.Inochkin //Uspekhi Fizicheskih Nauk 1993.T.163, №7. -C.65-84 (in Russian)

[2]   Dutton Z. Analysis and optimization of channelization architecture for wideband slow light in atomic vapors / Z. Dutton, M. Bashkansky, M. Steiner, J. Reintjes // Optics Express. 2006 Vol. 14, № 12 -P.4978-4991

[3]   Y.R. Shen The principles of nonlinear optics /A Wiley-Interscience Publication, John Wiley&Sons, Inc. N.Y., 1984

[4]   Plastun I.L. Investigation of the nonstationary coherent effects and resonant self-action influence on the characteristics of a frequency-modulated laser beam / I.L.Plastun, V.L.Derbov //Computer Optics 2009. V.33, №3, P.233-239 – ISSN 0134-2452 (in Russian)

A dynamically optical fiber loop memory using a nonlinear regeneration element

I.A. Malevich, A.V. Polyakov, S.I. Chubarov
Belarussian State University, Minsk, Republic of Belarus

Abstract: We report a multi- wavelength optical fiber loop memory with in-loop waveguide grating router and a method for increasing the information capacity and enhancing the reliability of information storage. Modified nonlinear amplifying mirror for phase-preserving 2R regeneration of wavelength division multiplexed return-to-zero signals is proposed.

Optical computing systems have attracted wide attention of researchers in connection with the necessary of entry and processing large arrays of optical information in a number of fundamental areas of modern science, including the problem of space and oceans exploration. Information processing in the optical calculator may be in the process of transferring the optical information field through the optical computing environment, and in transformation of optical information by optoelectronic elements. Nowadays research and developments laid the physical basis of the analysis of high–speed optical processes nanosecond and subnanosecond ranges in real time.

In this paper considered the structure of active type optoelectronic processor (OEP) based on the method of regenerative storage and optical information recording in manystable data storing element with a fiber–optical delay line. In the simplest case, the laser manystable memory element with a regenerative principle of information storage is an optical oscillation system that is sensitive to the flow of information from different types of carriers that are in the process of controllable optical storage may be transformed by the specified calculations program. When the return ratio is greater than 1, in this optical memory element is a stationary mode, which allows to realize the full cycle of information recording, storing and reading. The advantage of regenerative OEP is the ability to record and store information in digital and analog form both. In this case the data burn speed depends on the response time of the injection laser and can range from a nanosecond to the picosecond length. If necessary to write in OEP memory digital information into a linear code or a random sequence of optical pulses is possible, using an optical frequency synthesizer, to generate digital data as a predetermined sequence of discrete moments of switching of the injection laser, thus, is an optoelectronic processor programming. If necessary OEP analog recording information such as time intervals between the optical pulses in the memory circuit of optoelectronic processor formed the timelines, the dimension of which corresponds to the information attributes of analog signals.

The main way to increase the capacity of optical fiber informational channels is the technology of dense wavelength (frequency) division multiplexing (seals) channel using wavelength division, called DWDM-technology (dense wavelength division multiplexing). Economy of DWDM-systems with a large total data rate largely depends on efficiency of using the working spectrum for transmission of information, from increasing so-called spectral efficiency. We have developed an architecture of the fiber–optical dynamic storage device (FODSD) with wavelength division multiplexing of informational channels that can be used as a high-speed buffer memory. This structure has the following distinctive features. The combination of standard single-mode fiber and fiber with negative dispersion allowed to increase the time of storing of information of more than one order, using a given error probability. Use as a linear amplifier directly in an optical fiber recirculation loop and power amplifier two erbium-doped fiber amplifiers excludes use electronic amplifier at the output of each photodetector and provides data signals regeneration circulating directly in the optical range, which allows working with gigahertz speed recording optical information flow.

Implementation of the technology WDM / DWDM, which leads to a significant increase in the optical fiber input power as well as increasing rates of up to 10 Gbit/s and above, requires the consideration of nonlinear effects in fiber-optic line in the study of fiber-optic information systems. Self-phase modulation (SPM) is due to the fact that the refractive index of the fiber contains a non-linear depending on the intensity component, which causes a phase shift proportional to the intensity of the pulse. For this reason, the various components of pulse undergo different phase bias, causing changes in the linear frequency modulation (LFM) of pulses, regardless of their shape. Changes in pulses LFM in turn leads to an increase in their length due to their dispersion. Thus, SPM modifies the effect of dispersion on the expansion of the pulse. Since this effect of LFM change is proportional to transmitted signal power, SPM more noticeable in systems using high power transmission. Therefore, the LFM change caused by SPM affects pulse widening due to the dispersion and therefore must be considered in systems with high bit rates, which also have significant limitations because of dispersion. It is shown that under the influence of phase modulation associated with the dependence of the refractive index of the optical power during recycling pulse duration first decreases and then increases. This effect increases with an increase in transmit power (i.e., increase the number of data channels) and has a significant influence on the optical–fiber memory informational parameters.

Several 2R-regenerator (reamplifying and reshaping) schemes for RZ-signals have already been proposed. Most of them use nonlinear effects in a single fiber such as self-phase modulation or four-wave-mixing (FWM) in order to regenerate the signal. These regeneration methods work only for signals comprised of a single wavelength. However, in order to use the bandwidth of a transmission fiber efficiently, in current systems wavelength division multiplexed signals are used. Trying to regenerate a WDM signal with 2R regenerators that exploit the ultra fast Kerr effect in a single fiber, common for all WDM channels, would result in amplitude and phase distortions due to cross-phase modulation (XPM) and FWM between different wavelength channels. The reason for this is that in order to regenerate a certain wavelength channel all XPM and FWM contributions from other channels must also be taken into account. Such a regenerator scheme would demand fixed relationships between all WDM channels in bit sequence, channel power and time. However, in operational systems this is generally not the case. Therefore, it is not possible to regenerate all channels at the same time by this means. We propose an adapted demultiplexing-multiplexing approach integrated within a nonlinear amplifying loop mirror (NALM) with an asymmetrical splitting ratio. A NALM is a nonlinear fiber Sagnac interferometer based on SPM with an optical bidirectional amplifier in the loop. In order to prevent XPM and FWM effects from disturbing the SPM in the loop we propose for WDM signal regeneration a modified NALM setup.

A NALM is a nonlinear fiber Sagnac interferometer based on SPM with an optical bidirectional amplifier in the loop. In order to prevent XPM and FWM effects from disturbing the SPM in the loop we propose for WDM signal regeneration a modified NALM setup. This setup uses wavelength multiplexer and demultiplexer inside the interferometer loop to split and recombine the WDM signals. Between the multiplexer and demultiplexer separate nonlinear fibers are provided for each channel of the WDM signal. This setup has the advantage that only separate nonlinear fibers are used in contrast to the common approach where for each channel a separate complete regenerator is necessary. This concept significantly decreases the complexity of the setup because now only one bidirectional EDFA and fiber coupler are needed. The operation principle of the modified NALM is as follows: The incoming WDM signal is split asymmetrically at the fiber coupler into two counter-propagating partial signals. The weaker signal propagates through the EDFA first where it is strongly amplified. Then it passes the first demultiplexer after which different wavelength channels propagate in separate nonlinear fibers. Because the signals in this propagation direction have been strongly amplified, they acquire significant phase shifts due to SPM in the nonlinear fibers. Afterwards the different channels are again multiplexed by the second multiplexer. The counter-propagating stronger signal is first split at the second multiplexer so that again the different channels propagate separately through the nonlinear fibers. However, the power of the signals in this direction is much lower compared to the

amplified, originally weak signals. Therefore, the acquired nonlinear phase shifts in this direction are almost negligible. After the propagation through the nonlinear fibers the wavelength channels are recombined by the multiplexer "B" and the resulting WDM signal is amplified in the EDFA with the same gain as the weak partial signal. At the output port of the fiber coupler the weak partial signal with a large nonlinear phase shift interferes with the strong partial signal which has an almost unchanged phase. The latter, being much stronger, mainly determines the phase of the output signal and thus ensures negligible phase distortions. At the same time a nonlinear power characteristic with a plateau region is obtained for all wavelength channels, which is necessary for amplitude noise reduction.

A mathematical model for calculating the storage time in FODSD on the basis of a generalized nonlinear Schrödinger equation (NSE) has proposed:

$$ i\frac{\partial A}{\partial z} - \frac{B_2}{2}\frac{\partial^2 A}{\partial t^2} - i\frac{B_3}{6}\frac{\partial^3 A}{\partial t^3} + \gamma \left| A^2 \right| A = -iaA , \qquad (1) $$

where A is the envelope of the pulses electric field; z is coordinate along the fiber; t is "retarded" time associated with physical time t_{phis} by ratio $t = t_{phis} - z / v_{group}$; v_{group} is the group velocity of the package; B_2 is the group velocity dispersion; B_3 is third-order dispersion; α is attenuation coefficient; $\gamma = 2\pi n_2 / \lambda A_{eff}$ is the coefficient of nonlinearity; n_2 is nonlinear refractive index.

NSE (1) with given initial conditions (the pulse shape at the entrance to the fiber) and a set of coefficients B_2, B_3, γ, α that determine the parameters of the simulated fiber, describes the dynamics of shapes (power, duration) changing of an optical pulse as it passes through an optical fiber line of a given length.

The equation was solved numerically on a uniform rectangular grid by separating the imaginary and the real part with subsequent replacement of partial derivatives to the finite difference approximation.

Thus, a new scheme for a WDM loop buffer memory with phase-preserving NALM-based 2R regenerator has been proposed that uses a multiplexer and demultiplexer inside the NALM so that each wavelength channel of a WDM signal propagates in a separate nonlinear fiber. Moreover, the nonlinear fibers used in the NALM setup had different lengths and even different signs of the group velocity dispersion thus demonstrating that the modified NALM can operate easily under a large range of conditions. Considering recent advances in highly nonlinear materials also a full integration of all components of this NALM such as multiplexers, nonlinear waveguides and amplifying waveguides seems to be possible.

978-1-4799-0019-0/13 $31.00 © 2013 IEEE

Acoustic Modes on $As_2S_3/Y^{+128}X$ LiNbO$_3$ Crystal

R. M. Taziev, *Member, IEEE*

Institute of Semiconductor Physics SB RAS, Novosibirsk, Russia, taziev@thermo.isp.nsc.ru

Abstract: By means of effective permittivity function for layered system the frequency dependence of excitation of various acoustic modes in a chalcogenide film As_2S_3 on rotated 128^0 Y-cut lithium niobate is investigated.

Waveguide structure similar to thin layer of chalcogenide As_2S_3 (or As_2Se_3) over surface of lithium niobate substrate allows to improve acoustooptic interaction characteristics in comparison with a usual waveguide on LiNbO$_3$, preparing by diffusion method [1-2]. However, the structure of the acoustic modes excited by a linear point charge source in a such layered structure, practically is not studied. The aim of the present paper is numerically to study characteristics of the acoustic modes launched by linear point charge source, located on the layer-substrate interface by effective permittivity method. For its derivation usually one uses a method of transfer matrix [3], which essence is to transfer the amplitudes of acoustic fields from one layer to another, satisfying boundary conditions on interfaces of layers.

Let us consider briefly the derivation of effective permittivity function for computation of excitation characteristics of acoustic modes in layered system: a thin film of chalcogenide As_2S_3-Y^{+128},X-cut of lithium niobate. It can be derived if we solve system of the wave equations with corresponding boundary conditions on interfaces of every section of these layered structure [3]. The acoustic fields in the medium (see Fig. 1) are described by the wave equations [4]:

$$\begin{cases} \rho\, \dfrac{\partial^2 U_i}{\partial t^2} = \dfrac{\partial T_{ij}}{\partial x_j}, \ i = 1,2,3, j = 1,3 \ ; \\[2mm] \dfrac{\partial D_i}{\partial x_i} = 0, \ i = 1,3, \end{cases} \quad (1)$$

and material constitutive relations

$$\begin{cases} T_{ij} = C_{ijkl}\, \dfrac{\partial U_k}{\partial x_l} + e_{lij}\, \dfrac{\partial \varphi}{\partial x_l}, \ i,j,k = 1,2,3 \ ; \\[2mm] D_i = e_{ikl}\, \dfrac{\partial U_k}{\partial x_l} - \varepsilon_{il}\, \dfrac{\partial \varphi}{\partial x_l}, \ l = 1,2,3, \end{cases} \quad (2)$$

where C_{ijkl}, e_{ikl} и ε_{ik} are the elastic, piezoelectric and dielectric material constants; D_i, U_i are the electrical and elastic displacements, respectively; φ, T_{ij} are the electrical potential and elastic stress tensor, respectively; ρ is the density of medium.

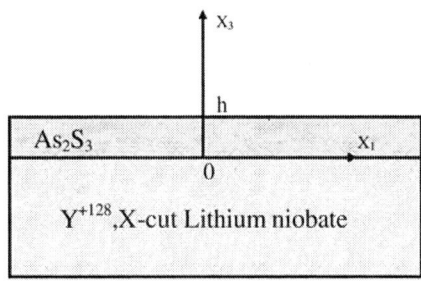

Fig.1. Chalcogenide layer As_2S_3– Y+128^0,X-cut of lithium niobate.

Elastic, piezoelectric and dielectric constants for chalcogenide and lithium niobate were taken from [5], [6] and [7], respectively.

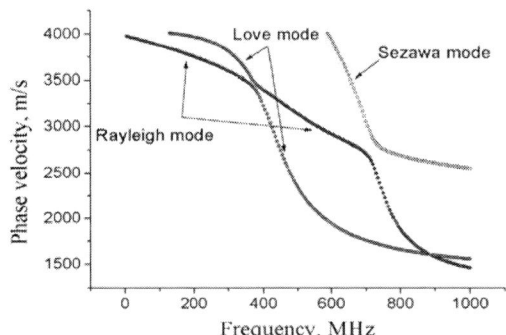

Fig.2a. Phase velocities of acoustic modes in layered system $As_2S_3/Y+128^0$,X-cut of LiNbO$_3$.

Fig.2 shows acoustic modes in the layered structure $As_2S_3/Y+128^0$,X-cut LiNbO$_3$ for thickness h=0.8 µ of chalcogenide film. Several acoustic modes are observed in the layered structure: besides the Rayleigh wave, the guided modes such as Love and Sezawa waves, which arise when the sound velocity in the chalcogenide layer is lower than one in the substrate. It should be noted, that leaky wave modes may also appear in a such structure. The main mode is a surface acoustic wave (Rayleigh wave) mode, propagating without attenuation

978-1-4799-0019-0/13 $31.00 © 2013 IEEE

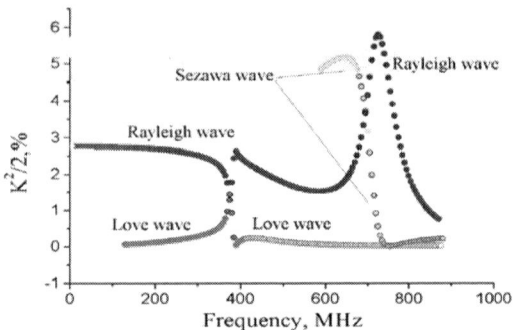

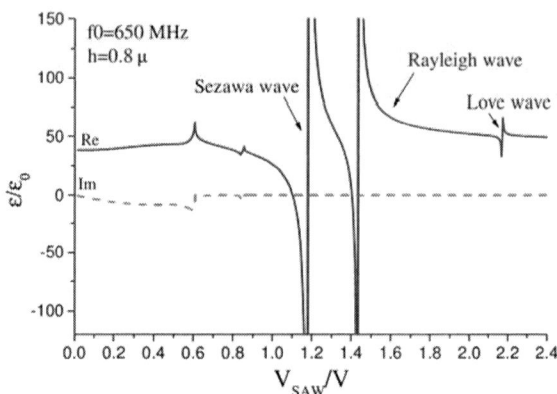

Fig.2b. Electromechanical coupling coefficients for acoustic modes in layered system $As_2S_3/Y+128^0$,X-cut of $LiNbO_3$.

in the layered structure for arbitrary As_2S_3 film thickness. Other acoustic modes propagates without attenuation only for appropriate thickness h of As_2S_3 film and frequency f0. For other values of thickness and frequency they are the leaky modes or skimming surface bulk acoustic waves with a strong or weak attenuation along propagation direction. These modes may be interpreted as Lamb waves in chalcogenide layer, perturbed by the presence of lithium niobate substrate. The first such nonsymmetric Lamb wave mode is known as a Sezawa wave [8]. One can see from Fig.2, the magnitude of electromechanical coefficient ($K^2/2$) of Rayleigh wave has maxima around the frequency of 780 MHz for chalcogenide film thickness of 0.8 μ. Sezawa wave has a lower value of $K^2/2$, than that for Rayleigh wave. At frequency of 375 MHz there is a strong interaction of Rayleigh and Love wave modes, and, as a result, they are transformed into each other without crossing their dispersion curves. Nevertheless, to observe in details the excitation properties of these modes it is necessary numerically investigate the effective permittivity function for layered structure $As_2S_3/Y+128^0$,X-cut $LiNbO_3$. Fig.3 displays the effective permittivity function as a function of slowness for different values of frequencies with its discrete step of 50 MHz. The sharp peaks (residues) on these curves belong to propagating modes without attenuation, which is described approximately by function $\gamma/(s-s_0)$, where γ is proportional to the value of electromechanical coupling coefficient of wave. The greater the value, the broader the resonanse peak in Fig.3. From Fig.3 one can see Sezawa wave emergence. For frequency less than f0=600 MHz this wave is a leaky wave, which attenuation magnitude depends on the frequency of source. The broader and smaller peak, the larger attenuation of wave. The leaky wave in the range of frequency f0=400÷550 MHz has small attenuation and may be considered as a pure propagating mode in the layered structure. With increasing the excitation frequency its strong attenuation appears again, and at frequency of 650 MHz it transforms to undamping Sezawa wave mode.

Fig.3. Effective permittivity function as a function of dimensionless slowness V_{SAW}/V on layered structure $As_2S_3/Y+128^0$,X-$LiNbO_3$. Solid and dashed lines are the real and imaginary parts of the effective permittivity function, respectively.

REFERENCES

[1] V.V.Atuchin, C.C.Ziling, D.P. Shipilova et al, "Crystallographic, ferroelectric and optical properties of TiO_2-doped $LiNbO_3$ crystals", *Ferroelectrics,* vol.100, pp.261-269, 1989.

[2] V.V.Atuchin, H. Nagata, K. Namakura et al, "Structure and refractive indeces of proton-implanted $LiNbO_3$", *Japanese Journal of Applied Physics*, part 1, vol.39, pp.2653-2656, 2000.

[3] E.L.Adler, "Matrix methods applied to acoustic waves in multilayers", *IEEE Trans. on Ultrason. Ferroelectrics and Freq. Control*, Vol.37, N6, pp.485-490, 1990.

[4] R. M. Taziev, "FEM/BEM for simulation of LSAW devices", *IEEE Trans. on Ultrason. Ferroelectrics, and Freq. Control*, 2007, Vol.54, N10, pp.2060-2069.

[5] Y. Ohmachi, "Acousto-optical light diffraction in thin films", *J. Appl. Phys.*, vol.44, pp.3928-3933, 1973.

[6] I.C.M. Litter, L.B. Fu, E.C. Magi et al, "Widely tunable, acousto-optic resonances in chalcogenide As_2Se_3 fiber", *Optic Express*, vol.14, pp.8088-8095, 2006.

[7] G. Kovacs, M. Anhorn, M.E. Engan, G. Visintini, and C.C. Ruppel, "Improved material constants for $LiNbO_3$ and $LiTaO_3$", *Proc.IEEE Ultrason. Symp.*1990, Vol.1, pp.435-438.

[8] H.F. Tiersten, "Elastic surface waves quided by thin films", *J.Appl.Phys.*, vol.40, pp.770-789, 1969.

Dispersion oscillating fibers for the fusion of optical solitons

M.A. Dorokhova[1], A.I. Konyukhov[2]
[1]Saratov State Technical University, Saratov, Russia
[2]Saratov State University, Saratov, Russia

Abstract - **In this article the results of numerical modeling of soliton's collisions in fiber with variable dispersion are presented. We use the nonlinear Schrödinger equation. Different results of the soliton's propagation will be shown. One of them, namely, the formation of the single pulse is of greatest interest.**

I. INTRODUCTION

Nonlinear effects and dispersion have a strong influence on the ultra-short laser pulses. The balance between dispersion and nonlinear effects in the medium leads to the formation of special kind of optical pulses. This pulses don't change its shape when they propagate through optical fiber. It is so-called soliton pulses (solitons). Their properties, such as saving of initial shape, localization within region, interaction with other solitons, make them suitable for data transmission over fiber.

In this article we are interested by interaction of solitons. Effects of soliton's interaction have drawn much attention in recent years. There are different evolution pictures of propagation of solitons after their collision. The one of the interesting opportunity is merging of solitons with formation one pulse. Earlier the different optical systems were used for collision of solitons and their merging. For example, in [2] formation of alone pulse was result of the merge of gap solitons in fiber Bragg grating; in [3] soliton's pair merge in mode-locked lasers.

In our article for generation alone pulse after collision of two solitons we offer to use optical fiber with changing diameter. Such fibers are interested because their dispersion change along fiber's length due to verify of diameter. Such fibers are interesting because their dispersion varies along the length of the fiber due to oscillations of the diameter. We use nonlinear Schrödinger equation and we simulate evolution of two soliton pulses, which was launched in this fiber. As it turned out, when the period of modulation of the fiber dispersion approaches the soliton period, two solitons merge in one pulse after collision.

II. COLLISION AND MERGING OF SOLITON PAIR

As mentioned above, optical solitons arise due to interplay between anomalous dispersion ($\beta_2 < 0$) and nonlinearity (Kerr self-phase modulation). To describe the evolution of solitons in optical fibers, it is convenient to use nonlinear Schrödinger (NLS) equation (1):

$$i\frac{\partial u}{\partial z} + \frac{\beta_2}{2}\frac{\partial^2 u}{\partial t^2} + \gamma |u|^2 u = 0 \tag{1}$$

where $u(z,t)$ is the amplitude envelope of the wave packet, β_2 represents dispersion of the group velocity, γ is nonlinear parameter. It is seen from equation (1), we neglect by the high-order dispersions, that is $\beta_m = 0$, (m=3, 4, 5...). But we consider only second-order dispersion (β_2). In addition, effects of self-steepening and stimulated Raman scattering are not considered.

NLS equation (1) have many solutions. And one of them is soliton solution, which interesting for us.

$$A(0,\tau) = N\sqrt{\frac{|\beta_2|}{\gamma\tau_0}}[\mathrm{sech}\left(\frac{\tau - s/2}{\tau_0}\right) + \\ + \mathrm{sech}\left(\frac{\tau + s/2}{\tau_0}\right)] \tag{2}$$

(2) is initial field, where τ_0 - initial pulse width, N is order of soliton.

N is defined as

$$N^2 = \frac{L_D}{L_{NL}} = \frac{\gamma P_0 \tau_0^2}{|\beta_2|} \tag{3}$$

Where L_D, L_{NL} are dispersion and nonlinear length, respectively; P_0 is peak power of initial pulse.

If $N=1$, soliton (2) corresponds to so-called fundamental soliton. One of the properties of the fundamental soliton: shape of this pulse does not change on propagation in the optical fiber, if dispersion is fixed (β_2 =const).

Figure 1 shows dynamics of propagation of fundamental soliton pair, when dispersion does not change, i.e. $\beta_2 = const$.

For modeling of pulse propagation we use fiber with variable dispersion along length. The second-order dispersion coefficient of this fiber varies along its length according to the following law

$$\beta_2(z) = \beta_2^{(0)}[1 + 0.2\sin(2\pi\frac{z}{z_m} + \varphi)] \qquad (4)$$

Where z_m is period of dispersion's modulation.

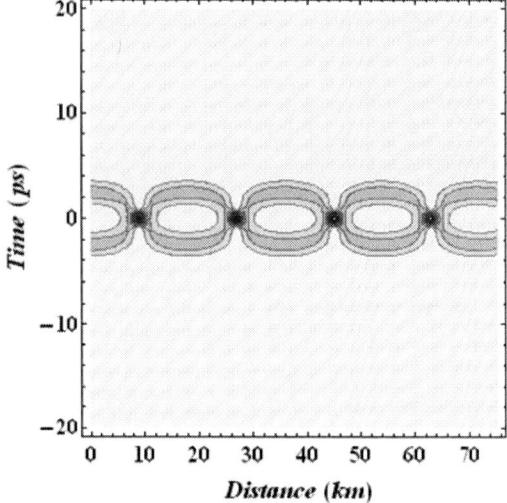

Fig. 1. Propagation of two fundamental solitons in optical fiber with fixed dispersion ($\beta_2 = const$). It is evolution over four soliton periods for the first-order solitons. $N = 1$, $\beta_2^{(0)} = 1 ps^2/km$, $\varphi = 0$, $s = 5 ps$.

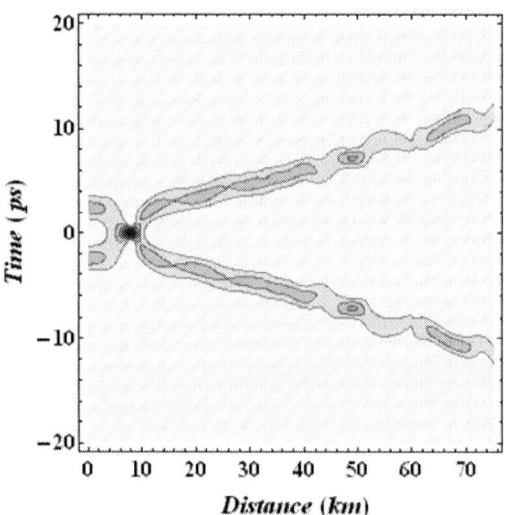

Fig. 2. Collision of two fundamental solitons in fiber with the sinusoidal modulation of the local dispersion coefficient as (4). Soliton's period does not coincide with period of modulation of fiber's dispersion. $N = 1$,

$\beta_2^{(0)} = 1 ps^2/km$, $\varphi = 0$, $z_m = 9.91 km$, $s = 5 ps$.

If we use the fiber with dispersion which varies according to the sine law (4), we don't obtain periodical picture as on figure 1. One of the possible scenario is shown in Figure 2. Two fundamental solitons are launched into the fiber with variable dispersion. Solitons collide after going some distance. Then they diverge without any further collisions.

If we consider the case of resonance, i.e. when the soliton's period coincides with a period of modulation fiber's dispersion, we obtain a more interesting picture (see fig. 3). Figure 3 shows that solitons collide and merge in alone pulse after going some distance.

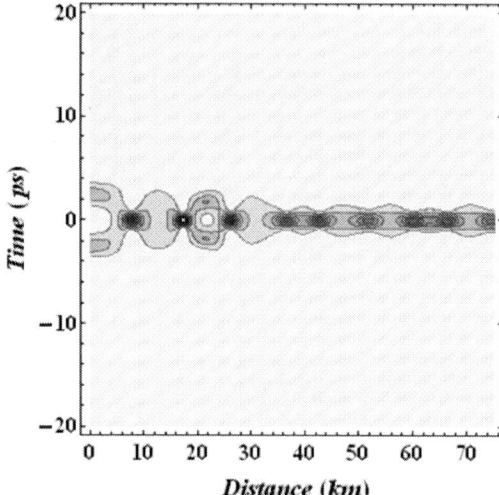

Fig. 3. Merging of two fundamental solitons in fiber with modulation of the dispersion's coefficient. Soliton's period coincide with period of modulation of fiber's dispersion. $N = 1$, $\beta_2^{(0)} = 1 ps^2/km$, $\varphi = 0$,

$z_m = 22 km$, $s = 5 ps$.

III. CONCLUSION

The fibers with varying along length dispersion are of interest for nonlinear fiber optics. In this article collision and merging of two fundamental solitons were demonstrated. Simulating of propagation of solitons shown that when two fundamental (first-order) solitons are launched in dispersion oscillating fiber, they collide and diverge in different directions. But if the period of modulation of the fiber dispersion approaches the soliton period, two solitons merge in one pulse after collision.

REFERENCES

[1] S.A. Akhmanov, V.A. Vysloukh, A.S. Chirkin, "Optics of femtosecond laser pulses," American Institute of Physics, 1992.

[2] W.C. Mak, B.A. Malomed, P.L. Chu, "Formation of a standing-light pulse through collision of gap solitons," Phys Rev E , №68, 2003.

[3] M.J. Ablowitz, T.P. Horikis, S.D. Nixon, Y. Zhu, "Asymptotic analysis of pulse dynamics in mode-locked lasers," Studies in Applied Mathematics. Volume 122, Issue 4, pages 411–425, May 2009.

[4] A.A. Sysoliatin, A.I. Konyukhov, L.A Melnikov, "Dynamics of optical pulses propagating in fibers with variable dispersion," Numerical Simulations of Physical and Engineering Processes. – InTech, 2011.

[5] R. G. Bauer and L. A. Melnikov, "Multi-soliton fission and quasiperiodicity in a fiber with a periodically modulated core diameter," Opt. Commun., vol. 115, pp. 190–195, 1995

Spatial anisotropy of electro- and nonlinear optical effects in the LiNbO₃ and KTP crystals: calculation and experiment

A. S. Andrushchak[1,2], O. V. Yurkevych[2], I. M. Solskii[3], A. Rusek[1]

[1]Faculty of Electrical Engineering, Czestochowa University of Technology,
17 Al. Armii Krajowej, Czestochowa, PL-42200, Poland
[2]Department of Telecommunications, Lviv Polytechnic National University,
12 S. Bandery Str., 79013, Lviv, Ukraine, e-mail: anat@polynet.lviv.ua
[3]Scientific Research Company "Carat", 202 Stryjska Str., 79031, Lviv, Ukraine

Abstract: We report the spatial analysis of nonlinear optical effects in potassium titanyl phosphate (KTP) and electrooptic effect in lithium niobate (LiNbO₃) crystals. Calculations of the electrooptical anisotropy in LiNbO₃ are supported by experimental measurements verifying the directional dependence of the electrooptical properties in this crystal. Employing the spatial 3D-analysis to the indicative surfaces the extreme directions characterizing by maximal magnitudes of these effects are found. A good agreement between the experiment and calculations confirms correctness of relevant approaches.

Designing and production of high quality electrooptic and nonlinear optical devices, that are applied for control or conversion of the laser radiation, usually require the geometry coupling optimization to improve the performance of the electro- or nonlinear optical cells made of crystal materials. It appears to be especially crucial for low symmetry crystals characterizing in many cases by large spatial anisotropy of their electro- and/or nonlinear optical properties [1, 2]. In relation to this problem several recent works, see e.g. [3, 4], have explored the spatial anisotropy 3D-analysis of the electrooptic effect applying the formalism of the indicative surfaces as well as their stereographic projections. In relevant calculations such analysis is based on a complete set of electro-optical tensor coefficients measured experimentally in LiNbO₃ crystals [5]. Accordingly, the spatial analysis of the indicative surfaces gives the angular parameters of the directional maxima, i.e. provides the most efficient crystal orientation which should be employed in cell production.

Actual work is aimed to present the spatial anisotropy of the nonlinear optical effects in KTP crystals and the linear electrooptic effect in LiNbO₃. Particularly, by using the magnitudes of the nonlinear optical constants for KTP crystals, as being determined in Ref. [6], we explore the spatial anisotropy of their nonlinear optical properties.

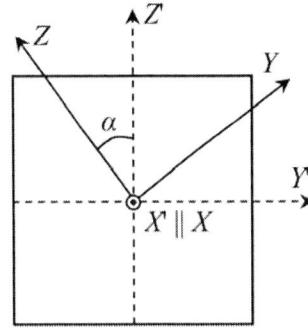

Fig. 1. Samples orientation in the electro-optical measurements

The spatial analysis of the electrooptic properties in lithium niobate crystals, on the other hand, is subjected to a careful experimental verification. For this purpose eight different rectangular crystal slabs representing α/X_1-cuts, where $\alpha = 0°$, 10°, 23.3°, 36°, 46.8°, 54°, 66.7°, 80° (see Fig. 1), have been prepared and measured. From the one hand, such choice is suggested by preliminary analysis indicating on expecting extreme directions in this angular range. On the other hand, it enables to apply the electric field or set the light polarization for 16 different directions defined by the angle θ with respect to the X_3-axis which ranges from 0° to 180° (here θ and φ are the angles of the spherical coordinates system). For each sample geometry defined in such a way it was performed a series of all possible interferometric and polarization-optical measurements to determine characteristic half-wavelength voltages for the longitudinal or transverse electrooptic effects. The measurements were carried out at room temperature using He-Ne laser with a wavelength of 633 nm.

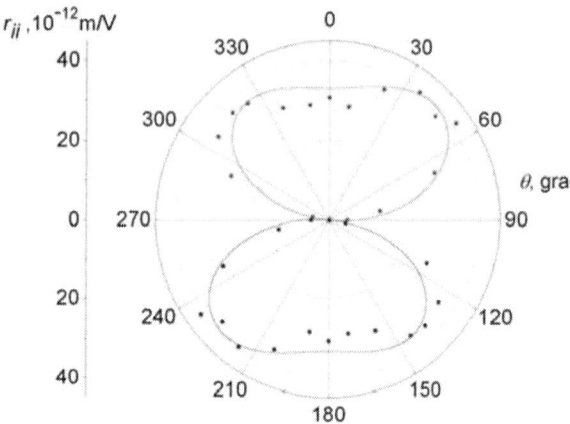

Fig. 2. Angular dependence of the effective electrooptic coefficient corresponding to the longitudinal electrooptic effect in lithium niobate crystals. Points - experiment; lines - calculation.

As example, Fig. 2 shows the angular dependence of the effective coefficient corresponding to the longitudinal electrooptic effect as determined experimentally (points) and calculated theoretically (solid line). The latter one represents a cross-section of the indicative surface $r_{ii}(\theta,\varphi)$, as calculated in Ref.[3], by the (X_2X_3)-plane. Similar dependences have been obtained also for the transverse electrooptic effect. Fig.3 explores the angular dependence of the electrically induced optical birefringence characterizing by the effective coefficient $r_{k\ell}^{\prime*(\ell)}(\theta)$. Here again the solid lines represents the cross-section of the indicative surface $r_{k\ell}^{\prime*(\ell)}(\theta),\varphi)$ by the (X_2X_3)-plane whereas the points mark the magnitudes obtained experimentally on eight crystal slabs as described above.

A good agreement between the experiment and calculations confirms a correctness of relevant approaches.

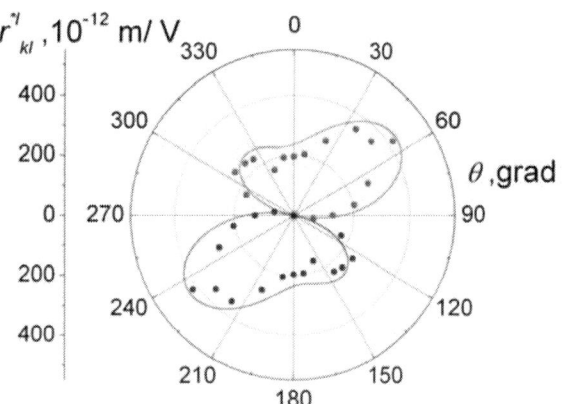

Fig. 3. Angular dependence of the electrically induced birefringence in lithium niobate crystals. Points - experiment; lines - calculation.

In addition, we explore the spatial anisotropy of the nonlinear optical effects in KTP crystals. The analysis of the indicative surfaces is focused mainly on their cross-section by a cone defining the spatial directions that fulfill a phase matching condition. Following this route we have determined the directional maximum on this surface which provides the crystal orientation with the best conversion efficiency.

Acknowledgement

The authors acknowledge numerous fruitful discussions with Professor A. V. Kityk (Czestochowa University of Technology, Poland). One of the authors (A. S. Andrushchak) acknowledges a Marie Curie International Incoming Fellowship within the 7th European Community Framework Programme (project N°272715).

REFERENCES

[1] Y. R. Shen "The Principles of Nonlinear Optics" Wiley-Interscience 2002, 576 p.

[2] A. S. Andrushchak "Spatial anisotropy of electro-, piezo- and acoustooptical interactions in crystalline materials of solid state optoelectronics": Thes. Hab. Dr. Tech. Sci. : , Lviv: Lviv Polytechnic National University, 2009, 405 p.

[3] A. S. Andrushchak, B. G. Mytsyk, N. M. Demyanyshyn, M. V. Kaidan, O. V. Yurkevych, S. S. Dumych, A. V. Kityk, W. Schranz "Spatial anisotropy of linear electro-optic effect for crystal materials : II. Indicative surfaces as efficient tool for electro-optic coupling optimization" Opt. Lasers. Eng., 2009, vol. 47, P. 24–30.

[4] O. A. Buryy, S. B. Ubizskii, A. S. Andrushchak "New method of extremal surfaces for most efficient application of crystalline materials in electro-optic devices" Proceed. of 11th Internat. conference on modern problems of radio engineering, telecommunications and computer science TCSET'2012, 21-24 February 2012, Lviv-Slavske (Ukraine). – P. 495–497.

[5] A. S. Andrushchak, B. G. Mytsyk, N. M. Demyanyshyn, M. V. Kaidan, O. V. Yurkevych, I. M. Solskii, A. V. Kityk, W. Schranz "Spatial anisotropy of linear electro-optic effect for crystal materials: I. Experimental determination of electro-optic tensor by means of interferometric technique" Opt. Lasers. Eng., 2009, vol. 47, P. 31–38.

[6] G. R. Bulka, O. Ph. Butyagin, G. A. Ermakov, N. I. Pavlova, T. N. Kharcieva "Methods of measurements and physical properties of KTP crystals" Laser technique and optoelectronics, 1992, No. 1-2, P. 69–76.

Light beams interaction in the cell with thermal optical nonlinearity at the presence of a feedback system

G.A. Knyazev, D.A. Davtyan, A.P. Sukhorukov
Lomonosov Moscow State University, Moscow, Russia

Abstract: The excitation process of oscillations of light and temperature fields in a media with thermal optical nonlinearity of the refractive index at the presence of the optoelectronic feedback was investigated. Both experimental and numerical simulation results are presented. The conditions of self-oscillation regimes for various types of the feedback were obtained theoretically.

I. INTRODUCTION

Recently in the photonics, much attention is paid to research and development of optoelectronic oscillating systems. These systems are based on such devices as electro-optical and acousto-optic modulators covered by a positive feedback circuit. Optical emission in these systems is detected by a photodetector. The signal from the photodetector is used to control the modulation index of the laser radiation. In such systems with time delay in the feedback circuit can be implemented stabilization modes of laser radiation and the oscillatory modes. In the latter case, there may be realized both regular and chaotic oscillations [1,2].

It is known that in case of the interaction of two optical beams in media with cubic nonlinearity there can be implemented total internal reflection of the signal beam from the pump beam [3-7]. Due to the fact that the interaction of light in a medium with thermal nonlinearity is nonlocal and inert [7,8] it is possible to implement a regime of self-oscillations similarly to the existing optoelectronic systems.

II. DESCRIPTION OF THE EXPERIMENT

This self-oscillating system was implemented in the experiment. In the fig. 1 the scheme of the experimental setup is shown. Liquid epoxy resin was used as the nonlinear medium, since the material has a high thermal nonlinearity and high viscosity. In the experiment the signal beam is generated by a semiconductor laser with a wavelength of 633 nm. The second harmonic of YAG-laser was used as the pump wave. Power of the pump wave could be varied from 10 mW to 310 mW.

Pump beam spreading in the nonlinear cell heated the medium and formed the inhomogeneity of the refractive index.

The work was supported by grants RFBR 12-02-90023-Bel_a, 12-02-01119-a, 11-02-00681-a.

The wavelengths of the signal beam and the pump beam were chosen so that the absorption of the signal wave in a nonlinear medium was absent. As a result the signal beam did not affect on the medium properties. Signal beam directed at a small angle to the pump was totally internally reflected from the optical inhomogeneity in case of the temperature distribution in the medium had come to stationary state.

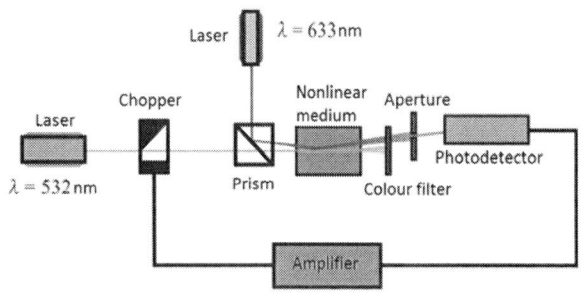

Fig. 1. The scheme of the experimental setup.

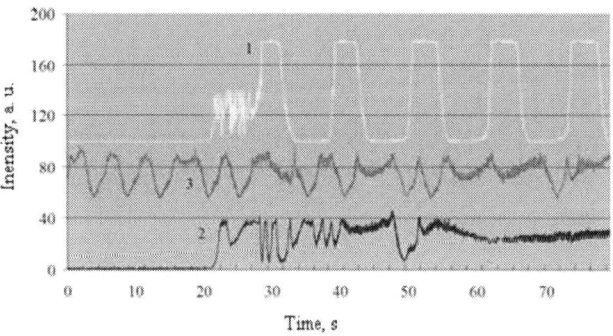

Fig. 2. The distribution of intensity on the photodiode in the experiment (1 – regular oscillations, 2 and 3 – aperiodic oscilations).

Thus while the pump beam was heating the medium the signal beam was sweeping. The angle of deflection of the beam was increasing with time. As soon as the intensity of the signal wave fell in the detector (see fig. 1) reached a certain value the chopper shut the laser aperture and the pump intensity decreased to zero. Due to the fact that the interaction of beams in a medium with thermal nonlinearity of the refractive index has a significant inertia, the deflection angle signal wave decreased gradually after turning off the pump. After the signal beam was deflected by a certain angle, the intensity of light falling to the detector decreased below the switching level of the chopper and the pump beam was directed to the nonlinear medium again.

In the fig. 2 shows the experimentally measured distribution of the intensity of the signal wave radiation falling to the photodetector. It can be seen that, depending on the position of the sensor there can be implemented regularly (curve 1) or aperiodic (lines 2 and 3) oscillations in the system. The aperiodicity is caused by the finiteness of the response chopper time.

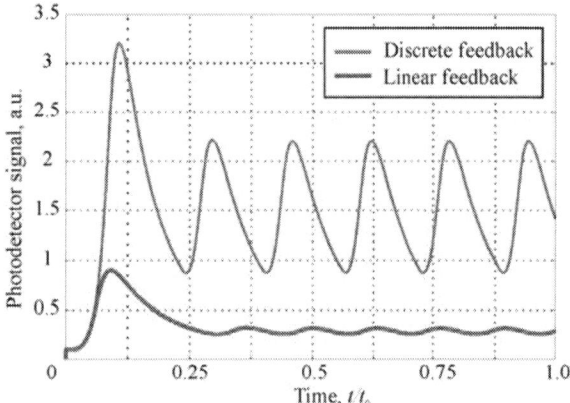

Fig. 3. Calculated distribution of signal beam intensity on the photodetector for discrete and linear regimes of the feedback.

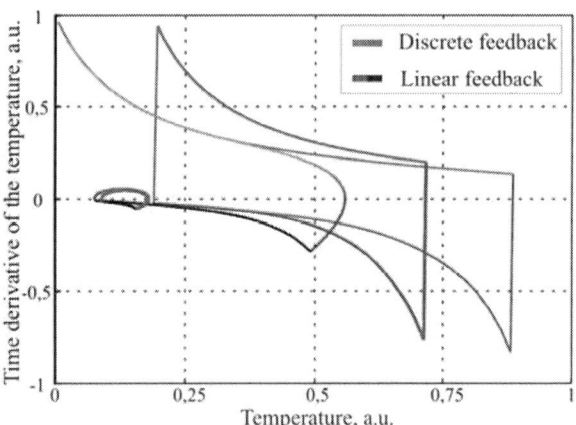

Fig. 4. Phase trajectory of oscillations in the system for discrete (blue-green) and linear (red-black) regimes of the feedback.

III. THEORETICAL INVESTIGATION

In addition to the experimental study a numerical simulation of the self-oscillating system was carried out. The simulation

was prepared on the base of the solution of the system of equations [7]:

$$\frac{\partial A_1}{\partial z} + iD_1 \frac{\partial^2 A_1}{\partial x^2} = -\alpha_1 (T - T_0) A_1 - \delta A_1, \tag{1}$$

$$\frac{\partial A_2}{\partial z} + iD_2 \frac{\partial^2 A_2}{\partial x^2} = -\alpha_2 (T - T_0) A_2, \tag{2}$$

$$\rho c_p \frac{\partial T}{\partial t} - \chi \frac{\partial^2 T}{\partial x^2} = \delta |A_1|^2, \tag{3}$$

where $A_j(x, z, t) = \sqrt{c_0 n_j} E_j(x, z, t)$ – normalized amplitude of electric field of the pump ($j = 1$) and signal ($j = 2$) optical waves, c_0 – velocity of light, n_j – refractive indexes, $T(x, z, t)$ – temperature perturbation in the media, T_0 – ambient temperature, $D_j = \frac{1}{2k_j}$ – diffraction coefficients, k_j – wave numbers, χ – heat conductivity index, $\alpha_j = \frac{k_j}{n_j} \frac{dn_j}{dT}$ – thermal nonlinearity coefficient, , x_1, x_2 – photodetector aperture coordinates, l length of the nonlinear media.

The feedback condition in the case of discrete pump beam modulation by mechanical chopper (as in the experiment) has the following form:

$$A_1(x, 0, t) = \begin{cases} 0, & \text{if } d \geq I_0 \\ A_0 e^{-(x-x_0)^2/\sigma^2}, & \text{if } d < I_0 \end{cases} \tag{4}$$

where I_0 – the switching level of the feedback system, σ – pump beam size, $d(t) = \int_{x_1}^{x_2} A_2^2(x, l, t) dx$ – amplitude of the photodetector signal. If the linear law of the pump beam modulation is used (for example due to electro-optic modulator) it is necessary to use other feedback condition:

$$A_1(x, 0, t) = \begin{cases} 0, & \text{if } d(t) \geq I_0, \\ A_0 \left(1 - \frac{d(t)}{I_0}\right) e^{-\frac{(x-x_0)^2}{\sigma^2}}, & \text{if } d(t) < I_0. \end{cases} \tag{5}$$

In the fig. 3 shows the intensity distribution curves of the signal wave on the photodetector for two types of the feedback for similar other parameters of the system. It can be seen that the amplitude of oscillations lower in the case of the linear feedback. The trend is caused by the non-oscillating component of the signal on the photodetector. This component decreases the pump intensity permanently if the feedback linear but does not affect in the discrete case. It should be noted that difference between the shape of the signal in the experiment (see fig. 2 curve 1) and the calculations (blue curve in the fig. 3) is explained by the saturation of the photodetector because of the high intensity of the pump beam. The fig. 4 demonstrates the phase trajectory of the temperature oscillation. The green and black colors corresponds to the zero value of the pump beam. In the figure the temperature at the center of the pump beam is shown. Calculations proves that in the linear regime continuous oscillations may be observed only in the region including zero intensity pump beam point.

IV. THE PHASE BALANCE CONDITION

To obtain the most fundamental principles of design the examined oscillation system the temperature equation (3) was investigated analytically. If ignore the effects of diffraction and self-defocusing of optical beams and suppose that the pump beam very narrow with respect to the sizes of the media and distance to the signal beam, the equation (3) should be transform to the following form:

$$\frac{\partial T}{\partial t} = a^2 \frac{\partial^2 T}{\partial x^2} + f(x, t),$$
$$T(0, t) = T(m, t) = 0, \qquad (6)$$
$$T(x, 0) = 0,$$

where $a^2 = \frac{\chi}{\rho c_p}$, m – width of the nonlinear media (along x axis), $f(x, t) = A_0 \, \delta(x - x_0) \theta(t)$ – pump beam, possessing delta function shape, $\theta(t)$ – unit function of the time evolution of the pump intensity.

If the feedback regime is discrete the value $\theta(t)$ periodically takes the value 1 and 0. Analysis of the equation (6) proves that in the quasi-equilibrium regime i.e. in the case of $t \gg t_0$, where t_0 – period of the oscillations the duty-cycle is equal 50%. In this case the solution of the system (6) is given by the expression:

$$\frac{\partial T}{\partial x} = \sum_{n=0}^{\infty} \frac{2 A_0}{m n^2} \sin(\frac{\pi n x_0}{m}) \sin(\frac{\pi n x}{m}) \begin{cases} 1 - \dfrac{e^{-(2n+1)^2 \delta t'}}{1 + e^{-(2n+1)^2 t'_0/2}}, \text{if } \theta(t) = 1, \\ \dfrac{e^{-(2n+1)^2 \delta t'}}{1 + e^{-(2n+1)^2 t'_0/2}}, \text{if } \theta(t) = 0. \end{cases} \quad (7)$$

In the equation (7) $\delta t'$ – the time interval after last shutting of the chopper. Time is equal $t' = N t'_0 + \delta t'$ if pump beam is turned off and $t' = (N+0.5) t'_0 + \delta t'$ if the pump is turned on. N is the number of happened oscillations. The values t', $\delta t'$ and t'_0 are normalized time by the relaxation time of the media $t_{rel} = l^2/(\pi a)^2$, i.e. $t' = t/t_{rel}$, $\delta t' = \delta t/t_{rel}$, $t'_0 = t_0/t_{rel}$. From the expression (7) may be obtained the phase balance condition. Combining the equation

$$\frac{\partial T(x_{eff}, \delta t = t_0/2)}{\partial x} - \frac{\partial T(x_{eff}, \delta t = 0)}{\partial x} = 0 \qquad (8)$$

and equation (7) it is possible to obtain the phase balance equation:

$$\frac{\partial T}{\partial x} = \sum_{n=0}^{\infty} \frac{2 A_0}{m n^2} \sin(\frac{\pi n x_0}{m}) \sin(\frac{\pi n x_{eff}}{m}) \tanh\left(\left(\frac{\pi(2n+1)a}{2l}\right)^2 t_0\right) (9)$$

In the equations (8) and (9) a parameter x_{eff} is the effective average coordinate describing the trajectory of the signal beam in the nonlinear media. The effective coordinate is the combination of the initial distance and angle between the optical beams, and position of the photodetector (x_1 and x_2).

V. CONCLUSION

The examined system may be applied for quick measurement of the time of thermal relaxation in nonlinear media. Obtained results prove to investigate the problem of design a similar oscillation system with all-optical feedback.

REFERENCES

[1] V.I. Balakshy, Y.I. Kuznetsov, "Optoelectronic systems with delayed acousto-optic feedback" *Proc. SPIE, Photonics, Devices, and Systems V*, vol. 8306, pp. 83060W–1–83060W–7, October 11, 2011,

[2] V.I. Balakshy, I.M. Sinev, "Competition of modes in an optically heterodyned acoustooptic generator", *Quantum Electron*, vol. 34, no. 3, pp. 277–282, 2004.

[3] A.P. Sukhorukov, A.K. Sukhorukova and V.E. Lobanov, "Diffraction of optical waves by nonlinearly induced cylinders". *Bull. Russ. Acad. Sci.: Phys.*, vol. 72. no. 12, pp. 1593-1596. 2008.

[4] A.A. Kalinovich, V.E. Lobanov, A.P. Sukhorukov and A. L. Tolstik. "Tunneling of Optical Beams through Inhomogeneity of a Refractive Index", *Bull. Russ. Acad. Sci.: Phys.*, vol. 74, no. 12, pp. 1718-1720, 2010.

[5] D.V. Gorbach, O.G. Romanov, A.P. Sukhorukov and A.L. Tolstik, "Nonlinear interaction and reflection of incoherent light beams". *Bull. Russ. Acad. Sci.: Phys.*, vol. 74, no. 12, pp. 1637-1641, 2010.

[6] G.A. Knyazev, A.P. Sukhorukov, "Interaction of optical beams in medium possessing thermal nonlinearity," *IEEE Xplore. Proceedings of 1st International Workshop on Nonlinear Photonics (NLP)*, pp.1,2, 6-8 Sept. Kharkov, Ukraine, 2011, doi:10.1109/NLP.2011.6102673

[7] V.E. Lobanov, A.A. Kalinovich, A.P. Sukhorukov, F. Bennet and D. Neshev, "Nonlinear reflection of optical beams in the media with a thermal nonlinearity", *Las. Phys.*, vol., 19, no. 5, pp. 1112-1116, 2009.

[8] A.N. Rubinov, I.M. Korda, E.A. Zinkevich, "Dynamics of a laser with a nonlinear TIR Q switch", *Quantum Electron*, vol. 32, no. 4, pp. 319–323, 2002.

978-1-4799-0019-0/13 $31.00 © 2013 IEEE

Vibration spectra evaluation of dyes from their stimulated Raman scattering in multiple scattering media

V. P. Yashchuk, A. A. Sukhariev

Kyiv T.Shevchenko Nat.University, Phys.dep.

Abstract: According to the properties of stimulated Raman scattering in multiple-scattering medium an algorithm for dye molecules vibration spectrum evaluation from their random lasing spectra was developed. Developed algorithm achieves vibration spectrum from the set of random lasing spectra received under different experimental conditions.

I. INTRODUCTION

Vibration spectra (VS) evaluation of dyes is complicated problem because of their strong luminescence that extinguishes Raman lines. There are some techniques for getting Raman spectra of dyes such as SE-RRS [1] and CARS [2] but they are very difficult to implement.

In a multiple scattering medium (MSM) random lasing [3] (RL) and stimulate Raman scattering (SRS) of dyes interact between them forming integrated nonlinear process with common radiation spectrum. This spectrum contains solid and linear components caused by RL and SRS respectively. As SRS in MSM occurs at all Stokes frequencies coinciding with RL spectrum [4,5] there is possibility to evaluate the corresponding section of the molecule VS. It needs decomposition of the spectrum by the solid and linear components (Fig. 1). The decomposition must be carried out very accurate because errors of solid component determination influence intensity and contour of the Raman line. The problem of decomposition is complicated because of RL spectrum sensitivity to pump intensity and sample parameters.

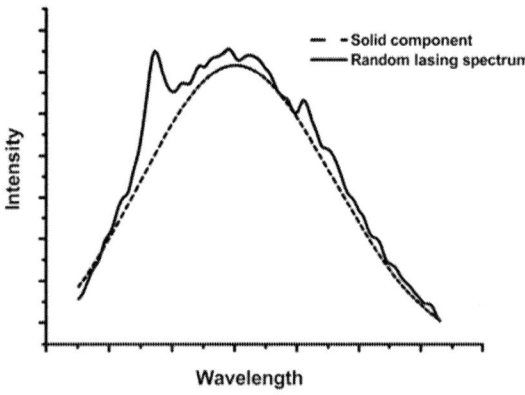

Fig. 1. Decomposed random lasing spectrum. The solid line

shows random lasing spectrum. The dashed line shows solid component of random lasing spectrum.

II. DEVELOPED TECHNIQUE

According to the properties of SRS in MSM, VS of dye molecule can be presented as a proper nonlinear susceptibility. Nonlinear susceptibility can be evaluated from random lasing spectrum itself I_{RL} and it's solid component I_B as follows [5]:

$$\chi_{srs}^{(3)}(\omega) = \frac{1}{gLI_p} \ln\left(\frac{I_{RL}(\omega)}{I_B(\omega)}\right),$$

Where I_p — pump intensity, L — emission's effective path length in MSM, g — aspect ratio (constant). $\chi_{srs}^{(3)}$ — it is a characteristic of dye molecule and it doesn't depend on experimental conditions and sample parameters. On the other hand it can be evaluated from random lasing spectrums. Basing on this idea an algorithm of accurate vibration spectrum evaluation was developed. As input it has set of random lasing spectrums of dye received under different pump intensities.

The algorithm consists of next steps:

1. Zero estimation: spectral sections containing sharp lines are removed from the RL spectrum and rest of it is approximated [6] by two Gaussians to describe its asymmetry. Sharp lines detection is based on analyzing first derivative of RL. Spectral regions near maximums with width that exceeds some threshold value are removed.

2. For each RL spectrum individual $\chi_{srs}^{(3)}$ and hence the average $\langle\chi_{srs}^{(3)}\rangle$ is calculated. From the average $\langle\chi_{srs}^{(3)}\rangle$ individual $\widetilde{I_B}$ (solid component) is calculated as follows:

$$\widetilde{I_B}(\omega) = I_{RL}(\omega)\exp\left(-LgI_p\langle\chi_{srs}^{(3)}(\omega)\rangle\right)$$

3. Each individual $\widetilde{I_B}$ is approximated by two Gaussians. On this step also residual $R_\Sigma = \sum_{i=1}^{n}\int\left(I_B(\omega) - \widetilde{I_B}(\omega)\right)^2 d\omega$ is calculated to control the effectiveness of the developed algorithm. Hence we receive solid component of RL spectra in k-th estimation.

4. Repeat steps 2-3 controlling residual.

The flowchart of the described algorithm is shown in Fig. 2.

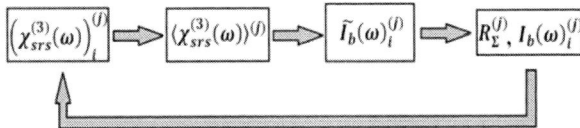

Fig. 2. The flowchart of the described algorithm. Here i - number of processed RL spectrum, j - algorithm step.

III. EXPERIMENTAL CONDITIONS AND SAMPLES

Dye p597 was investigated. Its RL spectra contain sharp SRS line under excitation by second harmonic of YAG-Nd^{3+} laser (pulse duration 15 ns, λ = 532 nm). Multiple scattering was fulfilled by placing dye particles into strongly diffusive shall of Al_2O_3. The emission spectra of the sample were single-shot registered by diffraction spectrograph (grating 1200 gr/mm) and CCD camera. Investigations were carried out under helium temperatures.

IV. RESULTS

Developed algorithm was implemented using C++ programming language and applied to dye p597. Some decomposed RL Spectra of p597 and evaluated vibration spectrum are shown in fig.3.

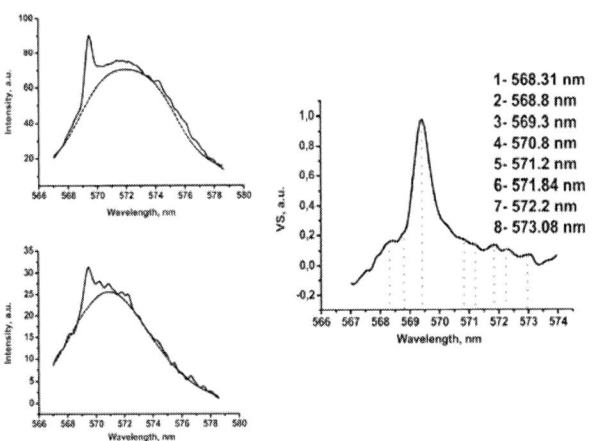

Fig. 3. Decomposed RL spectra and VS spectrum of p597. The solid line shows random lasing spectrum. The dashed line shows solid component of random lasing spectrum.

Residual that actually describes the difference between the average VS and individual ones is shown in fig. 4.

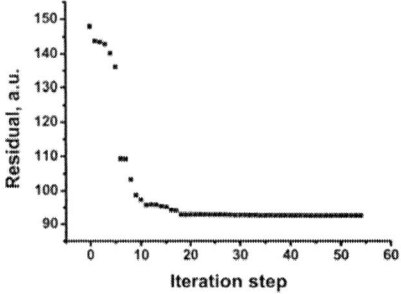

Fig. 4. Residual diminishing and stabilization (residual dependence on iteration step).

Diminishing and stabilization of residual proves the effectiveness and adequacy of the developed algorithm.

Developed algorithm gives very important results. First of all presented VS evaluation is based on properties of dye RL spectra. It uses condition of independence RL spectra on experimental factors and it's free from individual spectrum's processing subjectivity. Also it demonstrates low intensity SRS lines (lines 1, 2, 4-8 in fig. 3).

V. CONCLUSIONS

The algorithm of vibration spectra evaluation of dyes from their random lasing spectra was developed and applied to dye p597. The effectiveness and adequacy of the developed algorithm was shown. Vibration spectra of dye p597 were evaluated using developed technique.

REFERENCES

[1] Zhou Zeng-Hui, Liu Li, Wang Gui-Ying and Xu Zhi-Zhan, "Surface-enhanced resonance Raman scattering spectroscopy of single R6G molecules", Chinese Physics, 15(1), 126-131 (2006).

[2] V.P. Yashchuk, E.A. Tikhonov, A.O. Bukatar, O.A. Prigodiuk, A.P. Smaliuk, "Stimulated resonance Raman scattering from organic dyes in a multiple-scattering medium as a potential probe of their vibrational spectra", Quant. electronics, vol. 41(10), p. 875, 2011.

[3] N.M. Lawandy, R.M. Balachandran, A.S. Gomes, E. Sauvain, "Laser action in strongly scattering media", Nature Lett., vol. 436, p. 368, 1994.

[4] V.P.Yashchuk, E.A Tikhonov, O.A. Prigodiuk, "Effect of stimulated Raman scattering on the formation of the random lasing spectrum of dyes", JETP Letters, vol. 91(4), p. 174, 2010.

[5] Vasil P.Yashchuk, E.Tikhonov, O. Prygodiuk, "Stimulated Raman scattering of dyes under random lasing in polymeric vesicular films", Mol.Cryst.Liq.Cryst., vol. 535, p. 156. 2011.

[6] Гилл Ф. Практическая оптимизация / Ф. Гилл, У. Мюррей, М. Райт. - М. : Мир, 1985. – 509 с.

PROBLEM OF COHERENCE IN MODERN OPTOELECTRONICS AND RELAXED OPTICS

P.P. Trokhimchuck, I.P. Dmytruk

Lesya Ukrayinka's East European National University, Lutsk, Ukraine

Abstract The problem of coherence in modern optoelectronics (MOE) and Relaxed Optics (RO) is discussed. Classical and quantum conceptions of coherence are analyzed. Problem of coherence in RO is represented as problem of creation and change of coherent structures. Other problems of coherence breakdown, including phase coherence, and here applications in MOE and RO are discussed too.

Key Words: coherence, optoelectronics, Relaxed Optics, coherent structures, coherence breakdown.

The problem of a coherence (from Latin word coaherens – is being in bond) is one of the central problems of modern optics. This problem has three representation and applications in classic, quantum and relaxed optics and therefore in modern optoelectronics [1,2].

In physics coherence is an ideal property of waves that enables stationary (i.e. temporally and spatially constant) interference. It contains several distinct concepts, which are limit cases that never occur in reality but allow an understanding of the physics of waves, and has become a very important concept in quantum physics. More generally, coherence describes all properties of the correlation between physical quantities of a single wave, or between several waves or wave packets [3–9].

In classical optics basic principle of coherence is Rayleygh criterion [10]

$$\Delta k_x \Delta x = \Delta k_y \Delta y = \Delta k_z \Delta z = \Delta \omega \Delta t = 1. \qquad /1/$$

Where Δk_x, Δk_y, Δk_z, Δx, Δy, Δz, $\Delta \omega$, Δt – changes of proper wave numbers, coordinates, frequency and time [1,2]. This criterion is used for the separation of the spectral lines. Formally spatial part of Eq. /1/ may be used as criterion of spatial coherence and time part – as criterion of time coherence. Strictly speaking, for the separation of spectral lines sign = before 1 must be change on sign ≥, and for coherence – on sign ≤.

Basic principle of Quantum Mechanics (uncertainty principle) and basic principle of quantum theory of coherence (Quantum Optics) are receiving from Eq. /1/ after here multiplication by Planck's constant $\hbar$ according by N.Bohr [10]. Therefore we have

$$\Delta p_x \Delta x = \Delta p_y \Delta y = \Delta p_z \Delta z = \Delta E \Delta t \leq \hbar. \qquad /2a/$$

$$\Delta p_x \Delta x = \Delta p_y \Delta y = \Delta p_z \Delta z = \Delta E \Delta t \geq \hbar. \qquad /2b/$$

Where Δp_x, Δp_y, Δp_z, ΔE – changes of proper components of linear momentum and energy. Eq. /2a/ is the basic principle of Quantum theory of coherence and Eq. /2b/ is the uncertainty principle.

The spatial and time coherence can't be selected really.

Therefore in classical optics the difference between spatial and time coherence (phase coherence) $\varphi = \vec{k}\vec{r} - \omega t$ is used.

In classic optics coherence is the coordinate passage in space and times few oscillate or wave processes, which is appeared for its addition.

Oscillations are called coherent if its phase difference is stable (or change according to some law) in time and foe an addition it determined a amplitude of summary oscillation.

Harmonic oscillation may be represented in the next form:

$$V(t) = A \cos(\omega t + \varphi), \qquad /3/$$

where basic characteristics of an oscillation – amplitude A, frequency ω and phase φ are constant.

For the addition two oscillations with one frequency ω and various amplitudes A_1, A_2 and various phases φ_1, φ_2 the resulting harmonic oscillation has frequency ω too. An amplitude of its oscillation

$$A_s = \sqrt{A_1^2 + A_2^2 + 2A_1 A_2 \cos(\varphi_1 - \varphi_2)} \qquad /4/$$

may be change from $A_1 + A_2$ to $A_1 - A_2$ depending on phase difference $\varphi_1 - \varphi_2$.

For the cardinal estimation the coherency of oscillations a correlative function $R(\tau)$ was introduced, where τ is time interval of the change of phase in interval $\varphi_1 - \varphi_2 < \pi$.

An amplitude of the addition two oscillations from one source with time interval τ has next form

$$A_s = \sqrt{A_1^2 + A_2^2 + 2A_1 A_2 R(\tau) \cos \overline{\omega} \tau}, \qquad /5/$$

where $\overline{\omega}$ is average frequency of an oscillation.

Value of τ, for its $R(\tau) = 0,5$, is called the time of coherency or period of harmonic train.

For the propagation of plane electromagnetic wave in homogeneous matter phase of oscillation is constant at the time τ_0. In this time wave is propagated on distance $c\tau_0$, where c is the velocity of light. This distance $l_{coh} = c\tau_0$ is called the length of coherency or length of train.

A notion of coherency is used for the representation autooscillative oscillations with constant amplitude.

For the representation of coherent properties wave in the perpendicular direction to a direction of wave propagation the notions of coherent space and coherent length are introduced. Coherent length may be determined with the help of correlative function $R_\perp(l)$, where l is the corresponding space size. Condition $R_\perp(l) = 0,5$ is determined size or radius of coherency. All space of wave propagation may be fractured on

978-1-4799-0019-0/13 $31.00 © 2013 IEEE

the regions with constant coherency. The volume of its region (coherent volume) is determined as multuiplication the length of coherency l_{coh} on the area of figure, which is bounded of line $R_\perp(l) = 0.5R_\perp(0)$.

In general case the correlative function isn't be pure time or space. But in experiment these two cases may be discriminated: Michelson interferometer as time coherency and Yung interference as space coherency [2–4]. In mathematical sense for this case the correlation function may be represented in the next form

$$R(r,t) = R_1(r)R_2(t). \qquad /6/$$

The conception of coherency is used for the representation wave properties electrons, neutrons and other particles. In this case the coherency is called the directional coordinated flux of particles.

The Hanbary-Brown-Twiss stellar interferometer (Fig.1) allows measuring the fluctuations of intensity, which fall into detector, only. This experiment is differed from Michelson method. It is detected the mean value from multiplication from two probabilistic intensities but no one. A signal of quadratic detector [1] is proportional to value

$$\left|E^{(+)}(r_1,t)\right|^2 = |A|^2 + |B|^2 + AB^* e^{i(k-k')r_1} + A^*B e^{-i(k-k')r_1}. \qquad /7/$$

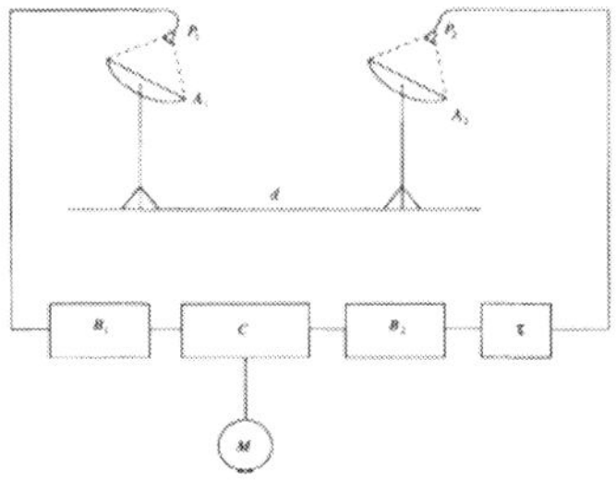

Fig.1. Schematic diagram of Hanbury Brown – Twiss stellar intensity interferometer [1]. Here P_1 and P_2 are the photodetectors, A_1 and A_2 are the mirrors, B_1 and B_2 are the amplifiers, τ is the delay time, C is a multiplier, and M is the integrator.

This signal isn't included of quick oscillation the detected wave, but mean value from this transforming signal doesn't include interference component (since mean value $\langle AB^* \rangle = 0$).

Hanbary Brown and Twiss multiplied these two transformed signals and after this they measured statistic mean value. Mean value from the multiplication of two intensities of type /7/ may be represented in the next form

$$\left\langle \left|E^{(+)}(r_1,t)\right|^2 \left|E^{(+)}(r_2,t)\right|^2 \right\rangle = \left\langle \left(|A|^2 + |B|^2\right)^2 \right\rangle +$$
$$+ 2\left\langle |A|^2|B|^2 \right\rangle \cos[(k-k')(r_1 - r_2)], \qquad /8/$$

where next conditions were taken into account: $\left\langle |A|^2 A^*B \right\rangle = 0$, $\left\langle |B|^2 AB^* \right\rangle = 0$ and other.

Though interferometric experiments are interpreted with help of universe average but in these experiments time averaging is realized too.

But time averaging is more difficult as universe average. For the time averaging interferometric measurements necessity give into account that plane waves aren't absolute monochromatic in general case. It allow to conclude that A and B are stochastic time functions $A(t)$ and $B(t)$.

Quantum theory of coherence is based on the idea of E.Schrödinger in 1927 and was developed by R.Glauber and E.C.D. Sudarshan [3–8]. Phase of wave functions in quantum mechanics has a deeper physical content than in classical physics. It includes energy and impulse characteristics of proper physical interaction. Therefore coherence is one of important notion of Quantum Mechanics. Roughly speaking wave functions of exponential form with phase term in form $\vec{p}\vec{r} - Ht$ are coherent functions, where H – Hamiltonian of system. Hamiltonian of system is the full energy of system. Therefore, the coherence of the process or phenomenon is related to the fundamental physical quantities of energy and momentum. These functions after additional transformations may be used as orthonormal basis of Quantum Mechanics. Therefore we can represent each quantum state as function from coherent states. Representation of coherent states is more physical as, for example, Fock representation [8]. But two representation is bonded with help next formula [8]

$$|\alpha\rangle = e^{-|\alpha|^2/2} \sum_{n=0}^{\infty} \frac{\alpha^n}{\sqrt{n!}} |n\rangle, \qquad /9/$$

where $|\alpha\rangle$ – coherent state, $|n\rangle$ – Fock state.

Equations /1/ and /2a/ aren't selected coherent and incoherent part of interaction. Roughly speaking it is the condition of interference two waves. But these correlations aren't allowed to select only coherent or incoherent part of this interaction. This problem was resolved in quantum theory of coherence (Quantum Optics). In this chapter of modern physics concept of harmonic oscillator is basis [8]. Therefore uncertainty principle has next form

$$\Delta p_x \Delta x = \Delta p_y \Delta y = \Delta p_z \Delta z = \Delta E \Delta t \geq \hbar/2. \qquad /10/$$

It is connected with minimal (zero) energy of harmonic oscillator $\hbar\omega/2$. This energy is corresponded the energy of noncoherent part of electromagnetic field, other words difference between full energy of field and coherent part of energy [8].

In Nonlinear Optics conditions of phase matching and phase synchronism are the representation of conditions of coherence. But condition of phase synchronism has spatial geometrical nature [6–8]. For example, for generation of second harmonic angle diapason of observation of maximal effect is equaled 10' – 30' [6, 9]. Second example is generation of laser radiation. For solid state laser angle diapason for stable resonator is equaled ~10', for unstable resonators – ~10"–20" [9]. For semiconductor lasers this value may be large [5]. For

978-1-4799-0019-0/13 $31.00 © 2013 IEEE

holography we must have large value of coherent length, this value is caused of coherence of light source (lasers) [9].

Coherence breakdown is caused a failure of proper phenomena. For our examples, it may be generation of second harmonic, lasing and holograms. Coherence breakdown is associated with symmetry breakdown.

But for the interaction light and matter we have two partners of interaction: light and matter. Symmetry and nature of light, matter and its interactions are caused the generation of proper process or phenomenon [1,2].

The concept of coherent structures (CS) is represented in RO [1,2], chapter of modern physics of irreversible interaction light and matter.

The CS (coherent structures) are called the structures with next properties:

1. these structures have the some order and symmetry;
2. it may be source for the generation and transformation of radiation, including light, (linear and nonlinear optics);
3. it may be source for the generation and transformation of irradiated matter, including amorphous and crystal materials, (the pure effects of RO);
4. it may be source for the generation and transformation of the radiation and irradiated matter, (holography, mixed phenomena of RO);
5. it may be formed with help coherent irradiation.

Practically it is the "traces" of the interaction laser irradiation and matter in matter. These structures may be classified as first-order and second-order structures. One has various energetic and time conditions of the creation and stability. The problem of the classification and modeling of these structures may be resolved with help short-range, long-range action and mixed approximations of the interaction light and solid.

The first-order range CSs have quantum character and may be represented as photochemical or crystal photochemical processes. One of the interesting problems of these phenomena is the problem of the nonadiabatic scattering light on valent and impurities bonds. This problem in the classical physics of status solid is represented as adiabatic. In radiation physics of status solid this problem is neglected. In RO it is one of the interesting problems (the oriental effect and the creation donor layers in semiconductors [1,2]).

The second-order range CS have, as rule, the wave nature (thermochemical, including annealing, plasmic, electromagnetic hydrodynamic structures, interferometrical phenomena and other [1,2]).

Quantum CS may be classified as multiphotonic $(h\nu < E_g)$, including monophotonic $(h\nu \sim E_g)$, and fractured structures $(h\nu >> E_g)$; where $h\nu$ – quantum energy, E_g – band gap of irradiated materials. The experimental research in this part of physics is almost absented.

Concept of CS allows associating radiative and nonradiative parts of interaction light and matter with point of change coherent characteristics of this interaction.

Transition from one coherent state to another may be represented as breakdown initial coherent state or structure respectively.

Concept of CSs allows associate the light-induced phase transformations in irradiated matter with change of here coherent properties [1,2]. Therefore methods of theories of phase transitions and transformations may be used for the resolutions of problems of coherence and contrarily, methods of coherence may be used for the resolutions of problems of the theory of phase transformations.

The basis of the RO is phenomenological kinetic-energy classification of the interaction of optical radiation with matter. Classification of CSs is corresponded of next physical systems and phenomena: basic phenomena of RO; polymorphoid, crystal and quasicrystal phases and other [1].

Concept of CSs allows realizing transitions from nonlinear optical processes, which can be represented as nonequilibrium phenomena, to relaxed optical processes, these processes are irreversible.

Problems of transition from nonequilibrium to irreversible processes is the problem of the aging the elements of optoelectronic systems. Therefore the search of waves the transformation irreversible processes to nonequilibrium and equilibrium is one of the central problems of creation more stable optoelectronic system. But this problem is connected with structural properties and here changes of irradiated matter, which affects the reliability of the optoelectronic systems as a whole and their components.

This approach allows flaring the application on coherent conception on all aspects of problem of interaction light and matter.

Thus basic concepts of coherence and ways of its development are analyzed. Possible applications of this concept in RO and MOE are discussed too.

References

[1] P.P. Trokhimchuck. *Nonlinear and Relaxed Optical Processes,* Lutsk: Vezha–Print, 2013.

[2] P.P. Trokhimchuck. *Foundations of Relaxed Optics,* Lutsk: Volyn' University Press Vezha, 2011.

[3] L. Mandel, E. Volf. Optical Coherence and Quantum Optics, Moscow: Fizmatlit, 2000. (In Russian).

[4] R. J. Glauber. *Quantum Theory of Optical Coherence. Selected Papers and Lectures,* New-York e.a.: Wiley@Sons, 2006.

[5] V.P. Gribkovskiy. *Semiconductor lasers,* Minsk: University Press, 1988. (In Russian)

[6] I.R. Shen. *Principles of Nonlinear Optics,* Moscow Nauka Publishers, 1989. (In Russian)

[7] H. Haken. *Laser light dynamics,* Moscow: Mir Publishers, 1988. (In Russian)

[8] J. Perina. *Coherence of light.* Moscow: Mir Publisher, 1974. (In Russian)

[9] B. Ziętek. *Optoelektronika,* Toruń Wydawnictwo universytetu Nikolaja Kopernika, 2005. (In Polish)

[10] N. Bohr. *The Quantum postulate and the Recent Development of Atomic Theory.* // Nature, Supplement, vol.121, 1928, pp.580–590.

About fluorescence excitation spectrums

(Invited Paper)

N. Kh. Gomidze[1], Z. Kh. Shashikadze[1], K. A. Makharadze[1], M. R. Khajishvili[1], O. M. Nakashidze

[1]Batumi State University, Batumi, Georgia, e-mail: gomidze@bsu.edu.ge

Abstract: Diagnostics of mineral oil breaks up on several problems, among which primary goal definition of concentration weighed and dissolved in water mineral oil. For the correct decision of this problem it is required researches mineral oil in water in various forms (dissolved, emulsive, a film), at various stages of ageing, for different mineral oil. fluorescence spectrums gives us sufficient the information on the quantitative composition of water mediums.

A series of methods of the analysis oil pollution (for example: methods infrared - spectroscopy, a nuclear magnetic resonance, on the basis of effect Shpolsky, weight methods, a liquid and gas chromatography, etc.) is developed and is applied. Among them are applied also optical method, including the laser spectroscopy method which advantage is expressivity and distancivity.

We for measuring applied a method of an interior reference point when the signal of fluorescence mineral oil in water is compared to a signal Stokes component a water or Hexane Raman effect (for Hexane extracts and solutions) [1].

By test preparation emulsive oil was poured on a surface of the distilled water then for correlation the jarring by means of the ultrasonic correlator was spent. Light dispersion on frequency excitation from the sample of an emulsion in 2-6 times exceeded the Rayleigh scattering on water molecules [2].

The shape of a spectrum of fluorescence dissolved of mineral oil depends on type mineral oil on initial stage a little and does not depend on time of finding the mineral oil in water. It is stable also intensity of fluorescence in the course of ageing and feeblly depends on type mineral oil on initial stage [3-5].

The method consists in examination of the processes natural fluctuation which are taking place in fluids (fig. 1.). The diffusion of these fluctuations is characterised by usual macroscopical kinetic coefficients. An effective method of studying of processes of a diffusion fluctuation in the transparent fluids is the molecular dispersion of light. The physical reasons of the molecular light dispersion is thermal fluctuations of an exponent light refractive. These fluctuations are in turn caused by fluctuations of the thermodynamic quantities which function is the refraction index. Generally unitary dipole dispersion of light in an intermixture of two fluids for intensity of a diffused light validly expression:

$$I = \left(\frac{\partial n}{\partial \rho}\right)_{T,X}^2 \cdot \left(\frac{\partial \rho}{\partial p}\right)_{s,X}^2 \cdot (\overline{\Delta p})_{T,X}^2 + \left(\frac{\partial n}{\partial \rho}\right)_{T,X}^2 \cdot \left(\frac{\partial \rho}{\partial s}\right)_{P,X}^2 \cdot (\overline{\Delta s})^2 + \left(\frac{\partial n}{\partial x}\right)_{P,T}^2 \cdot (\overline{\Delta x})^2$$

where n - a refraction index, ρ - density, P -pressure, T - temperature, s -entropy, x - molar concentration of a solution.

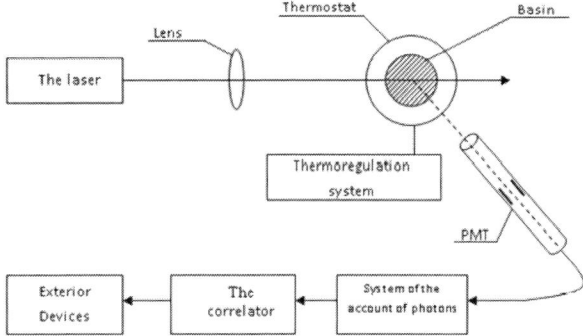

Fig. 1. The Block-scheme of a correlation spectrometer of optical bias.

From the analysis (1) follows that light dispersion is caused with three reasons: the adiabatic fluctuations of pressure (spontaneous sound waves of the thermal nature), isobaric fluctuations of entropy (heat) and concentration fluctuations. The diffused light spectrum is diagrammatically figured on fig.2. Bias rather ω_0 - frequencies of exciting light on a $\pm\Omega$ line, are termed as builders of Mandelstam - Brillion (MB). Detrusion component is caused by a Doppler effect at light dispersion on running spontaneous sound waves and, accordingly, is proportional to velocity of a sound in medium. The line broadening occurs at the expense of absorption of sound waves.

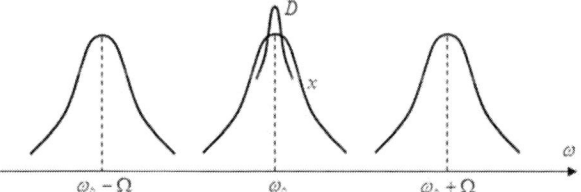

Fig. 2. Spectrum of a diffused light an intermixture of two fluids.

Central a builder located on frequency of exciting light, consists of two lines: the narrow line is caused by signal attenuation of fluctuation of concentration - diffusion processes, wide - signal attenuation of fluctuation of heat - thermal conduction processes. According to a hypothesis of Onsager's, process in a fluid can be featured a fluctuation diffusion by means of macroscopical magnetohydrodynamic equations. So, for process of a diffusion of fluctuations of concentration $\delta x(r,t)$ expression is valid:

$$\frac{\partial \delta x(r,t)}{\partial t} = D\nabla^2 \delta x(r,t), \qquad (1)$$

978-1-4799-0019-0/13 $31.00 © 2013 IEEE

from this:

$$\delta x(k,t) = \delta x(k,0) \cdot e^{-\Gamma t}, \quad \Gamma = Dk^2. \quad (2)$$

Γ - The characteristic frequency signal attenuation of fluctuation of concentration, D - a coefficient of diffusion, k - change of a wave vector of light at dispersion.

Fluctuation signal attenuation is convenient for studying using a temporary autocorrelation function:

$$G(\tau) = \frac{1}{T}\int_0^T f(t) \cdot f(t+\tau)dt = < f(t)f(t+\tau) > . \quad (3)$$

Then for the fluctuations featured by expression (2), autocorrelation function looks like:

$$G_x(\tau) = <|\delta x(k,t)|^2> e^{-\Gamma \tau}. \quad (4)$$

According to (4), the autocorrelation function of light diffused on decadenting fluctuations of concentration will register in a view (for an electric field of a light wave with frequency ω_0):

$$G_E(\tau) = E_p \cdot e^{-\Gamma \tau} \cdot e^{-i\omega_0 \tau}, \quad (5)$$

the optical spectrum can be found under Wiener-Khintchin theorem:

$$S_E(\omega) = \frac{1}{2\pi}\int_{-\infty}^{\infty} e^{i\omega\tau} \cdot G_E(\tau)d\tau . \quad (6)$$

From the solution (6) follows that the optical signal with exponential autocorrelation functions (5), has a spectrum in a view:

$$S_E(\omega) = E_p \frac{\Gamma/\pi}{\Gamma^2 + (\omega - \omega_0)^2} . \quad (7)$$

Here Γ - a half-breadth of a spectroscopic line at semiheight. Gained above expression allow to calculate, using the corresponding measured value of breadth central builders of diffused light Γ, diffusion constant D.

For Brownian particles the diffusion constant and radius of particles r_0 is related by following expression (Einstein-Stokes formula):

The characteristic frequency of process of diffusion $\Gamma \approx 10^5\,sec^{-1}$ and usual spectrometer methods do not allow to measure such narrow spectroscopic lines. Overcoming of the arisen difficulties probably at use of a method of spectroscopy of optical mixture. The essence of a method consists that bending around a spectrum from optical frequency ω_0 is transferred to fields of low frequencies where the further analysis is carried out by spectrum analyzers or correlators of signals. Spectrum transport to frequency domain - the non-linear transformation grounded on method quadratic detection of light.

The radius of an emulsion for different tests was spotted. Their values lay in limits 0,270÷0,338 micron. Any monotonous changes of radius of an emulsion in the course of "ageing" emulsify oil product were not observed. Attempts to estimate breadth of allocation of particles on radiuses have shown a system monodispersity. It was not observed also changes of spectral characteristics of an old emulsion after additional crushing on ultrasonic monochromator. Last fact

speaks also in favour of equilibrium inconvertible character of emulsions. Experimental excitation spectrums of intensity fluorescence for usual water, Hexane and oil extracts are given on fig. 3-5.

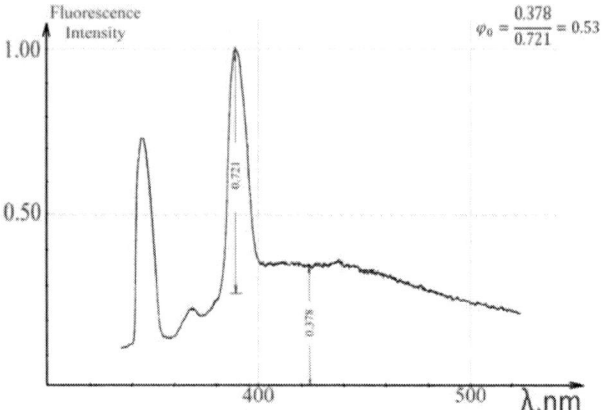

Fig. 3. Spectrum of an echo-signals of sea water an excitation by the nitrogen laser ($\lambda_0 = 337$ nm)

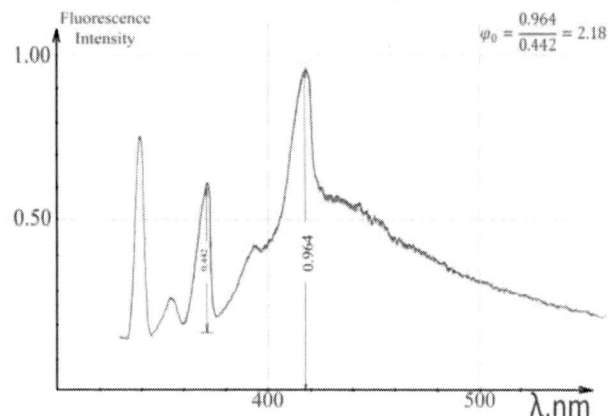

Fig. 4. Spectrum of an echo-signals of Hexane an excitation by the nitrogen laser ($\lambda_0 = 337$ nm)

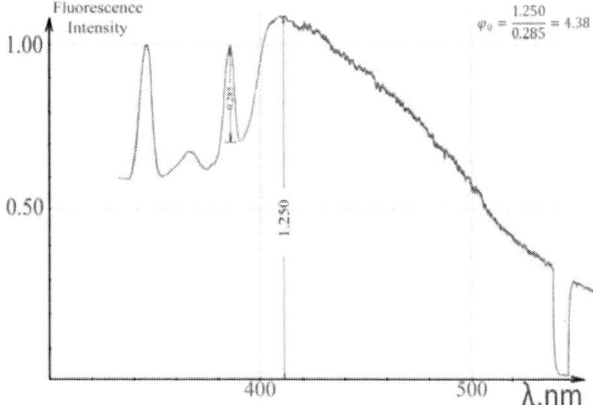

Fig. 5. Spectrum of an echo-signals of water with oil an excitation by the nitrogen laser ($\lambda_0 = 337$ nm)

REFERENCES

[1] V. V. Fadeev, V. V. Chubarov. The Quantitative definition of oil products in water laser spectroscopy methods. T.261, №2, 342-346, 1981.

[2] I. L. Fabelinski. The Molecular dispersion of light. M: the Science, 1965.

[3] Z. Shashikadze, Z. Davitadze, O. Nakashidze, N. Gomidze. Problems of diagnostics sea water with the method laser spectroscopy. *Works of RSU, series: Natural Science and Medicine.* Vol. 15, pp.282-285, Batumi, 2009.

[4] N. Gomidze, I. Jabnidze, K. Makharadze, M. Khajishvili, Z. Shashikadze, Z. Surmanidze, I. Surmanidze. Numerical Analyses of Fluorescence Characteristics of Watery Media via Laser Spectroscopy Method. *Journal of Advanced Materials Research Vol. 590 (2012), pp. 206-211.* www.scientific.net/AMR.590.201.

[5] N. Gomidze, K. Makharadze, M. Khajishvili, Z. Shashikadze. About Numerical Analyses of Sea Water with Laser Spectroscopy Method. *2011 XXXth URSI General Assembly and Scientific Symposium, 30TH 2011 (5 VOLS), pp. 1620-1624,* ISBN 978-1-4244-5117-3.

978-1-4799-0019-0/13 $31.00 © 2013 IEEE

PRECISE LASER MEASURINGS OF A MATERIAL INDEX REFRACTION ON BREWSTER ANGLE

(Invited Paper)

E. A. Tikhonov, *Senior Member IEEE & SPIE,*
A. K. Lyamets, Institute of Physics Academy of
Sci., Kiev, Ukraine

Abstract: It is studied Brewster angle refractometry with one instrumental error at availability to measuring's of materials in the form of samples of different topologies with one or two surfaces of optical quality. The nature of the reflected radiation and its polarization in small neighborhoods of Brewster angle is analyzed experimentally. It is found no infringements of Fresnel theory on Brewster angle, related to behavior of polarization and intensity in reflected light exist. New optical scheme of precise measurement is proposed.

In the report the previously offered laser method of a refraction index measuring for materials of various topology (volume, lamellar and film) on a Brewster angle is considered profoundly [1,2]. Refractometry on Brewster condition n=tan(φ_{br}), accompanied one instrumental error (IE) and at possibility to carry out measuring of various materials at availability only one (or even two) optical quality surfaces is supposed to be settled in for studying and the subsequent applications. IE source in this measuring procedure are: nonzero azimuth angle α for the p-polarized beam, a limited polarization degree p and an error of installation of the zero position of the reflected surface of a studied sample. Understanding of the errors $\alpha \neq 0$ and p$\neq$ 1,000 has allowed to offer the new optical measuring scheme with the polarization analyzer in a reflected beam shoulder, suppressing as far as possible s-polarized component, resulting due to termed errors and responsible for their effects. Until now the precise determination the reflected power minimum under Brewster conditions was yielded on integrated on s, - and p-polarization of radiation. Thereof, the s-polarized component of the radiation which are inevitably presented in measuring beam (and laser also) owing to the termed errors and also given some rise owing to the anisotropic Rayleigh scattering of the incident radiation by a rough surface, led to bias of the angle power minimum to smaller values. On Fig.1. typical angular dependences of reflected beam power of the diode injection laser 660nm, $\approx$20mW are presented. The initial degree of polarization of laser emission was equal p=99,9 %. S-, and P- polarized components of reflected beam power were registered independently at two transmission of the analyzer different by turning point on 90^0.

The initial degree of polarization of laser emission was equal p=99,9 %. S-, and P- polarized components of reflected beam power were registered independently at two transmission of the analyzer different by turning point on 90^0. On angles of incidence 36^0, 50^0, 55^0 grades the beam power of 1P-dependence was incremented in 2,9, 8,4, 2,9 times by magnification of incident power; power S-components of the reflected beam on 2S- dependence corresponds to a case crossed polarization and is spotted by an unit ellipticity of linear polarized laser emission $\approx$600:1.

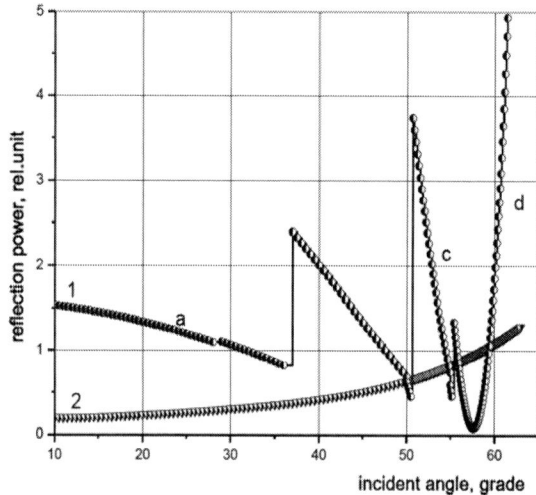

Fig.1. Reflection power from K8-surface for the case of p-polarized (1P) and s-polarized laser light 660nm of diode injection laser

It is visible that in a small neighborhood of Brewster angle (an intersection points of curves 1P and 2S) for integrated on both polarization power under one -quarter wave analyzer twist it is possible to observe the continuous polarization change: circular→linear →circular only due to the expense of the changing contribution to the total radiation of s-components power containing in an incident p-polarised laser beam. Therefore without the analyzer in the measuring shoulder the resulted minimum of the total power 1P+2S in the reflected beam inevitably moves on some minutes towards smaller values about true Brewster angle. The same time relation of powers 1P/2S on incident 10^0 and on Brewster angles was more than 3 order (4390)!

Drude' adjustments made on the basis of the similar observed polarization change in a small neighborhood of Brewster angle relates the changes in the frame of Fresnel theory with p-polarized light component transformation solely [3]. In opposite that under measu rement sensibility for low light restricted by noises of

978-1-4799-0019-0/13 $31.00 © 2013 IEEE   320

the monitoring scheme we did not find out any power springs at transition through exact Brewster angle.

Therefore the multiply reproduced behavior of 1P,- and 2S-dependences presented fig.1. does not allows us to consider presence of minimum power in reflection on Brewster angle as infringement of Fresnel theory. Correspondingly it is not necessary to consider adjustments on this emission reflection and polarization transformation by entering into consideration the supposed interface layer with thickness $<<\lambda$. Our interpretation of existence of week emission on Brewster angle is supported presence of not disappearing s-polarized component of light in a measuring beam.

Brewster angle measuring for the thin layer materials shows specific questions to a procedure. To thin layer materials we carry film on substrates, the free films and plates. General specificity for transparent thin layer material under Brewster condition procedure is recording 2-beam reflected by forward and back surfaces of a plates. Presence of appreciable absorption in layer on a measuring wave length excludes the second beam so we meet a case of reflection measuring by one surface viewed above. For transparent layer of plane-parallel surfaces there is an interference of equal slope angles of 2 beams and Brewster refractometry gets some features. For not plane-parallel surfaces it is easy to distinguish one beam for processing, but with thickness reduction plane parallelism becomes the prevailing performance. The arising interference in reflected light from a plane-parallel plate (film) is characterized by spatial period Λ, visibility, depending on a wave length λ, thickness of film T and refraction angles ψ (incident φ) and intensities of beams.

Formulas give magnitude of phase shifts in an

$$(\cos\psi_1 - \cos\psi_2) = -2\sin\psi_{ave}\sin\delta\psi = -\frac{2}{n^2}\sin\varphi_{ave}\sin\delta\varphi = \frac{\lambda}{2nT}$$

$$\delta\varphi \approx \frac{\lambda n}{4T\sin\varphi_{ave}}$$

angular position of maximums on interference pattern through refraction angles and other parameters of a plate. With their help it is possible to estimate the linear size of period Λ as product of distance R from a surface of a plate to a photo detector on $\delta\varphi$. It is evident that the thickness T and an incidence angle have solving influence on this period and its relation to the photo detector aperture $\varnothing$ determines character of a registered interference power: if $\Lambda \geq \varnothing$ the interference oscillations

are registered by a photo detector; if $\Lambda < \varnothing$ the interference oscillations of power by a photo detector are not registered.. As example we will give results of Brewster minimum measuring's for fused quartz plate with T=1030mkm and glass plate T=150mkm.

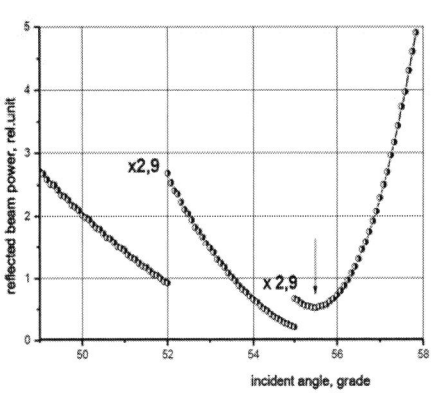

Fig.2a,b Determination of reflected power minimum of p-polarized beam with the analyzer: a- plate fused quartz d- glass plate. The spatial period Λ: a$<\varnothing$, b$\geq\varnothing$.

Conclusion from presented results follows: 2-beam interference does not limit the precision of index refraction measurement in Brewster refractometry. The same successful occurs to be Brewster index refraction measurement for the nanometric thickness films, birefringence plates and films and film with powerful absorption.

REFERENCES

1) E.A. Tikhonov, V.A. Ivashkin, Studying of refractive index measurements in reflected light, arxive Cornall univer., 2010, **http://arxiv.org/abs/1008.4256**
2) E.A.Tikhonov,V.A Ivashkin., A.K Lyamets, Reflection refractometry for nearly normal incidence and at Brewster angle, J. Appl. Spectr., 2012,79,#1, pp.148-154
3) Д.В. Сивухин, Общий курс физики, Москва, Наука, ГРФМЛ, 1980, 751сс.

A nondestructive validation of reverse impact experiment based on shape measurement using high speed photographs

D. Kohdadad[1], T. Sjöberg[2]

[1] Experimental Mechanics, Luleå University of Technology, SE-971 87, Luleå, Sweden
[2] Division of Mechanics of Solid Materials, Luleå University of Technology, SE-97187 Luleå, Sweden

Abstract: High speed photography of a reverse impact scenario was taken in order to make shape measurement. The results from the shape measurements were then compared with results from numerical simulations in order to evaluate the possibility to use noncontact shape measurement as a validation tool in future simulations.

Keywords: reverse impact experiment, nondestructive validation, shape measurement, strain.

I. Introduction

In order to improve the ability to validate results from numerical simulation with the use of experiments new reliable parameters to compare are always sought for. Usually the validating parameters need to be measured by a physical device included in the experimental set-up. The downside of this is that the measuring devices themselves are invasive and influence the controlled part meaning that they can introduce new unknown effects into the experiment. Therefore noncontact measurement is always desirable. Several different optical noninvasive technologies are already used to measure and characterize surface shape and deformation [1, 2, 3, 4]. In this work we track surface (edge) deformation and displacements in a sequence of images while keeping the method as simple as possible. It means the only thing needed is photos of the experiment as well as numerical software such as MATLAB.

In order to test the feasibility of using shape measurements as a way to find new validating parameters high speed photography were used to capture the impact of a reverse impact test.

II. Method

A. Reverse impact experiment

The experiment consisted of an instrumented slender steel rod, with a diameter of 10 mm, onto which a half spherical tip of tungsten-carbide was attached, an air-gun used to accelerate discs of sheet material towards the rod at speeds up to 70 m/s, and a cylindrical container made of PTFE to guide the specimens and make sure they hit the tip as straight as possible. The discs were made of Nickel alloy 718, had a diameter of 46 mm and a thickness of 1.6 mm. A schematic of the experiment can be seen in Fig. 1.

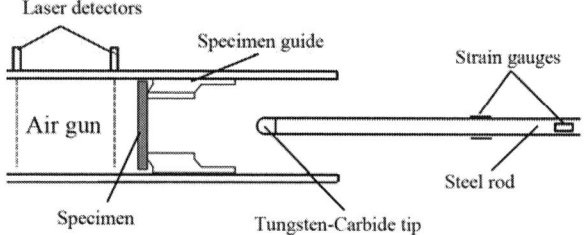

Fig. 1. Schematic of the reverse impact experiment.

The rod is instrumented with two sets of strain gauges mounted in Wheatstone half-bridges as to compensate for bending in the two principal directions perpendicular to the rod length. The strain gauges are used to capture the elastic wave formed at the impact. From this elastic wave the force of the impact can be calculated. This force is used to evaluate the plastic properties of the plates throughout the impact event. For more details on the reverse impact set-up refer to [5].

B. High speed photography

In order to capture the high velocity impact a Phantom v1610 high speed camera from Vision Research was used. This camera was able to record images of the size 80 by 512 pixels at speeds up to 280 000 images/second. This made it possible to capture around 50 images of the impact that takes around 200 µs. In order to get enough light at these short exposure times the set-up included two flashes rigged to bounce on a white screen put behind the impact position, thereby creating good contrast between the background and the discs. The flashes had a burn time of around 2 ms meaning the entire impact event could be captured.

C. Shape measurements

Initial measurement step requires the segmentation of certain structures in the captured images. We need a set of structure that can be robustly tracked in all image sequences. A surface structure can be considered simple to segment if its associated intensities are fairly homogeneous. For this we used contrast enhancement with adaptive histogram algorithm. Therefore, the best candidates for segmentation that are easily visible in all images are the contours and edges because they exhibit higher contrast and are easier to track the changes. An adaptive local thresholding algorithm based on the minimum and maximum intensity of each image combined with Canny operator used to extract the surface and edges. Fig. 2 shows the specimen and extracted edges. To measure the strain of the specimen the edges of specimen surface are repeatedly tracked in different sequences and their respective one-dimensional displacement vectors determined. A moving average filter is used to remove random noise and mutations while retaining a sharp step response, Cumulative surface edge displacement Δ_{surf} can be determined from the incremental displacements δ_{surf} between consecutive images using Eq. (1)

$$\Delta_{surf} = \sum' \delta_{surf} .$$

(1)

By calculating the difference of displacements $\Delta l = \Delta_{surf\,II} - \Delta_{surf\,I}$ and dividing it by a selectable but constant base length l_0, any strain value can be determined as:

$$\varepsilon = \frac{\Delta l}{l_0} = \frac{\Delta_{surf\,II} - \Delta_{surf\,I}}{l_0} .$$

(2)

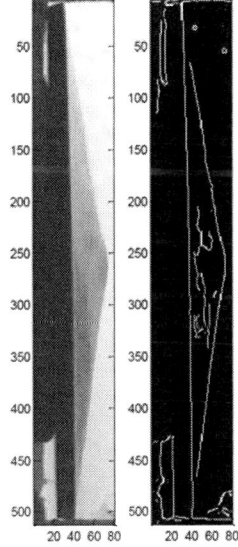

Fig. 2. An arbitrary image of specimen from the sequence of the images and corresponded extracted edge information are shown in left image and right one respectively. Tracking the edge displacement and using Eq.(2) lead to strain measurement.

D. Numerical analysis

A numerical analysis based on the reverse impact was performed using the commercial FE-code LS-Dyna. The numerical analysis modeled the plate as free-flying since the effect of the cylindrical guide could be neglected. The rod was modeled as purely elastic material since no plastic deformation in the rod could be observed after the experiments. The plates of Alloy 718 were modeled using the Johnson-Cook plasticity model [6]. The evaluation and validation of the parameters used for the Johnson-Cook model are described in [7].

III.   Results

The results obtained from the shape measurements could then be compared with results from the numerical simulations. In Fig. 3 the average velocity for displacement of points on the specimen (dashed line for simulation and solid line for results from shape measurement) are plotted throughout the entire impact event. The trend of change in velocity of displacements can then been used to give initial feeling of how force is spreading through the specimen which can be one of the parameters of validation.

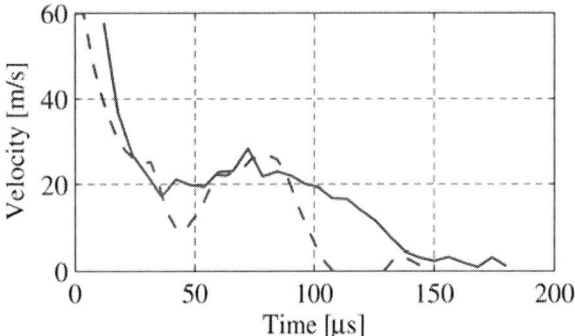

Fig. 3. Average velocity of displacement for the plate at the first 200 µs. The solid line is the results from the shape measurements while the dashed line shows results from the simulation.

Three arbitrary points on a plate are chosen to compare the results of shape measurement with FE-model. The strain evolution evaluated using shape measurement is plotted in Fig. 4 together with strain calculations from the FE-model. General increment of strain in both methods is seen along the time axis with closely same trend.

IV.   Discussion

The results in Fig 3 and 4 show agreement with the simulations which verifies that shape measurement method can be a nondestructive validation tool for the results from numerical simulation with the use of experiments. In this work both the spatial as well as the temporal resolution of the high speed camera is quite low. The low spatial resolution makes edge detection difficult and the low temporal resolution means that the small quick changes in the impact become hard to resolve. Also in some images of sequence the specimen has been rotated or moved which can cause some errors in displacement calculation and later on in image based strain measurement. Then image registration methods [8, 9] can be used in order to obtain a transformation between the images of sequence to monitor and compensate the movement or rotations between the images from different time. To make the validation more accurate and robust, recording images with pulsed digital holographic methods can be an alternative. In this case images with higher resolution including phase information are easily in access to use for shape measurement. Then the ability of tracking very small changes and displacements in holographic images will increase the accuracy of validation.

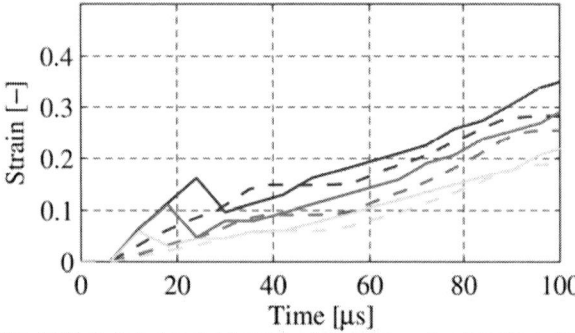

Fig. 4. The evaluated strain plotted for three points on the plate. The solid lines show results from the shape measurements and dashed lines from the simulations. The color grouping shows results taken from the same position on the plate radial axis.

ACKNOWLEDGMENT

The Authors would like to thanks Prof. Mikael Sjödahl and Dr. Per Gren from the experimental mechanics group of Luleå University of Technology for valuable discussions.

REFERENCES

[1] F. Chen, G.M. Brown, M. Song, "Overview of three-dimensional shape measurement using optical methods," Optical Engineering, vol. 39, pp. 10-22, 2000.

[2] D. Khodadad ; E. Hällstig and M. Sjödahl " Shape reconstruction using dual wavelength digital holography and speckle movements ", Proc. SPIE 8788, Optical Measurement Systems for Industrial Inspection VIII, 87880I (May 13, 2013); doi:10.1117/12.2020471; http://dx.doi.org/10.1117/12.2020471

[3] D. Khodadad, E. Hällstig, M. Sjödahl, "Dual-wavelength digital holographic shape measurement using speckle movements and phase gradients", Opt. Eng. 52(10), 101912 (Jun 28, 2013), http://dx.doi.org/10.1117/1.OE.52.10.101912

[4] I. Yamaguchi, T. Ida, M. Yokota, "Measurement of Surface Shape and Position by Phase-Shifting Digital Holography," Strain, vol. 49, pp. 349-356, 2008.

[5] T. Sjöberg, K. G. Sundin and M. Oldenburg, "Comparative investigation of parameters in the Johnson-Cook model for Alloy 718 through instrumented reverse impact experiments," Unpublished.

[6] J.R. Gordon and W.H. Cook, "A constitutive model and data for metals subjected to large strains, high strain rates and high temperatures," Proceedings of the 7th Int. Symposium on Ballistics, 21 (1983) 541-547.

[7] T. Sjöberg, K.G. Sundin and M. Oldenburg, "Calibration and validation of plastic high strain rate models for Alloy 718," unpublished.

[8] B. Zitova, and J. Flusser. "Image registration methods: a survey."Image and vision computing 21.11 (2003): 977-1000.

[9] D. Khodadad, A. Ahmadian, M. Ay, A. F. Esfahani, H. Y. Banaem, H. Zaidi. "B-spline based free form deformation thoracic non-rigid registration of CT and PET images," Proc. SPIE 8285, International Conference on Graphic and Image Processing (ICGIP 2011), 82851K (September 30, 2011), doi:10.1117/12.913422; http://dx.doi.org/10.1117/12.913422

Features of Optical Image Jitter in a Random Medium with a Finite Outer Scale

L. A. Bolbasova[1], *Member, SPIE, OSA*, P. G. Kovadlo[2], V. P. Lukin[1], *Fellow Member, OSA, SPIE*,
V. V. Nosov[1], A. V. Torgaev[1]

[1]V.E. Zuev Institute of Atmospheric Optics SB RAS, Tomsk, Russia
[2]Institute of Solar Terrestrial Physics SB RAS, Irkutsk, Russia

Abstract: Features of optical wave fluctuations while propagating through a randomly inhomogeneous turbulent medium with a finite outer scale are considered, including conditions when areas with dominating influence of one large scale coherent structure are observed in the atmosphere, for which the spectrum of atmospheric turbulence can differ significantly from the Kolmogorov model spectrum. Using an approximate model of the spectrum for coherent turbulence, described earlier in our works, the variance of jitter of an optical image is calculated (under the applicability condition for the smooth perturbation method). The comparison of these equations with known similar equations for Kolmogorov turbulence has shown that the variance of fluctuations is significantly weaker in coherent turbulence than in the Kolmogorov theory under similar conditions. This means that phase fluctuations of optical radiation decrease significantly in coherent turbulence. The importance of this conclusion is noted for interpretation of the results of optical sounding of atmospheric turbulence.

INTRODUCTION

Experimental studies show that there are atmospheric regions that deviate significantly from Kolmogorov turbulence conventionally used for the description. One of the possible reasons is the influence of the finiteness of the outer turbulence scale. In particular, conditions when one large-scale structure dominates can be implemented. Turbulence is commonly called coherent in such regions. A compact structure including a long-lived spatial structure cell (originating from a long-term action of thermodynamic gradients) and products of its discrete cascade decay is called a hydrodynamic coherent structure. In an extended consideration, a coherent structure is a solution of the hydrodynamic equations; it includes both large-scale and small-scale turbulence.

EFFECT OF THE OUTER TURBULENT SCALE ON THE VARIANCE OF IMAGE JITTER

It is known that given a random jitter of the position of the center of gravity of the image, a distant optical source that forms a plane wave-front is characterized by the position of the energy centroid which is defined by the following equation in the first approximation (when neglecting amplitude fluctuations):

$$\vec{\rho}_F^{pl} = -\frac{F}{k\Sigma}\iint_\Sigma \nabla S(\vec{\rho}_1)d^2\rho_1 \quad , \tag{1}$$

where k is the optical radiation wavenumber, $S(\vec{\rho}_1)$ is phase fluctuation for plane waves.

By analogy with the some calculations, the equation of the variance of the jitter of the image center of gravity can be written as

$$<(\vec{\rho}_F^{pl})^2> = \frac{F^2}{\Sigma^2}\int d^2\rho_1 \int_\Sigma d^2\rho_2 \int_0^X d\xi_1 \int_0^X d\xi_2 \iint(\vec{\kappa}_1\vec{\kappa}_2)\backslash \tag{2}$$
$$exp(i\vec{\kappa}_1\vec{\rho}_1)exp(-i\vec{\kappa}_2\vec{\rho}_2)<d^2n(\vec{\kappa}_1,X-\xi_1)d^2n(\vec{\kappa}_2,X-\xi_2)> .$$

Then, using the Gaussian receiving aperture, as well as will use atmospheric turbulence model, which consider finiteness of the outer turbulence scale,

$$\Phi_n(\vec{\kappa},\xi) = 0.033C_n^2(\xi)(\kappa^2+\kappa_0^2)^{-11/6}exp(-\kappa^2/\kappa_m^2) . \tag{3}$$

Now let us use a so called effective outer turbulence scale for the atmosphere in general, which can be introduced on the basis of the following equation:

$$(\kappa_0^*)^{-1} = [\int_0^\infty d\xi C_n^2(\xi)\kappa_0^{1/3} / \int_0^\infty d\xi C_n^2(\xi)]^{-3} \tag{4}$$

Using the coherence radius of atmospheric turbulence r_0 in the form

$$r_0 \approx (k^2\int_0^\infty d\xi C_n^2(\xi))^{-3/5} , \tag{5}$$

the equation for the variance of angular jitter of an image in the focal plane of the telescope can be finally derived in the form

$$<(\varphi_F^{pl})^2> \approx 3.23R^{-1/3}r_0^{-5/3}k^{-2}[1-2^{-1/6}(\kappa_0^*R)^{1/3}]. \tag{6}$$

Let us analyze the effect of the second term in the parenthesis in (6), which causes the difference in the behaviour of image jitter variance as a function of the size of the receiving aperture from the power function of the form $R^{-1/3}$.

Table 1.

$(\kappa_0^*R)^{-1}$	1000	300	100	50	30	10	5
$[1-2^{-1/6}(\kappa_0^*R)^{1/3}]$	0.91	0.87	0.80	0.75	0.70	0.57	0.42

The Table gives the calculation results of the image jitter variance for different values of the effective outer turbulence scale for the atmosphere in general versus the size of the telescope receiving aperture **R**. Analysing the table, one can see that behaviour of the image jitter variance differs significantly from the power law, even for a ratio of the outer turbulence scale to the receiving aperture size of about 10*3, i.e., the outer scale affects image jitter significantly.

978-1-4799-0019-0/13 $31.00 © 2013 IEEE

EFFECT OF COHERENT TURBULENCE

Let us compare the behaviour of image jitter for Kolmogorov and coherent turbulence. The approximate model of turbulence spectrum under conditions of manifestation of one coherent turbulent structure was derived earlier in our works:

$$\Phi_n^{\kappa o c}(\vec{\kappa},\xi) = 0.033(C_n^2(\xi))^{\kappa o c}(\kappa^2 + \kappa_0^2)^{-7/3}\exp(-\kappa^2/\kappa_m^2), \quad (7)$$

we have for coherent turbulence

$$<(\varphi_F^{nn})^2>_{\kappa o c} = \frac{16}{5}\pi^2 0.033\frac{\Gamma(1/3)}{\Gamma(7/3)}\int\limits_0^\infty d\xi C_n^2(\xi)\kappa_0^{1/3}. \quad (8)$$

The measurements were carried out at the Sayan Solar Observatory of Institute Solar-Terrestrial Physics SB RAS (Mondy, Buratiya) with the automated horizontal solar telescope. The variance of solar limb jitter was measured as a function of the size of the receiving mirror. A Brandt sensor, which is a detector of the image jitter due to an atmospheric optical source, was used as a field meter. The Brandt sensor was successfully tested during several tens of years and used earlier in similar studies. The measurements were carried out for five different diameters of the receiving mirror (aperture **2R**) of the telescope, i.e., 5, 10, 30, 50, and 80 cm (Fig.1). Simultaneously with the optical measurements, an ultrasound meteorological system monitored the meteorological conditions with certain types of turbulence (coherent or incoherent Kolmogorov turbulence) near the receiving telescope.

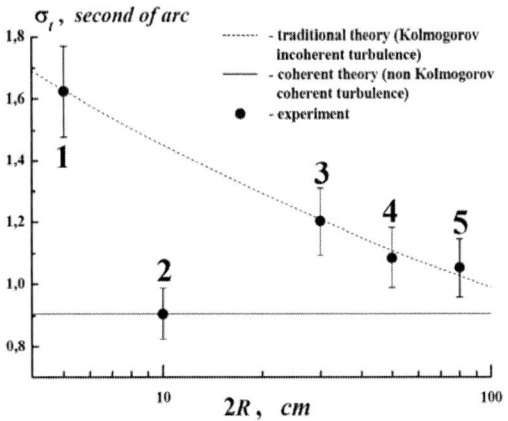

Fig. 1. Standard deviation of astronomical image jitter of the solar limb as a function of the diameter of telescope input aperture $2a_t$. Sayan Solar observatory; measurements in summer 2010; $W_T \sim f^{-8/3}$ - for the experimental point with $2R = 10$ cm and $W_T \sim f^{-5/3}$ for other points.

The effect originates in the presence of large-scale coherent structures in the atmosphere (regions with coherent turbulence) and consists in the decrease in phase and amplitude fluctuations of optical radiation as compared to Kolmogorov turbulence; it is caused by a more rapid decrease in the spectrum of coherent turbulence and, hence, a lower contribution of small-scale components. It turns out that real atmospheric turbulence is a (incoherent) mixture of different coherent structures with incommensurable frequencies of primary energy carrying vortices. Therefore, a coherent structure can be considered as a structural element forming Kolmogorov turbulence.

ACKNOWLEDGMENT

The reported study was supported by The Ministry of education and science of Russia, project 8877.

REFERENCES

[1] V.P. Lukin and V.V. Pokasov, "Phase Fluctuations of an Optical Wave which Propagates in a Turbulent Atmosphere," *Radiophys. Quant. Electr.*, vol 16, no. 11, pp. 1335–1337, 1973.

[2] V.L. Mironov, V.P. Lukin, V. . Pokasov, and S.S. Khmelevtsov, "Phase Fluctuations of Optical Waves Propagating in a Turbulent Atmosphere," *Izv.Akad. Nauk SSSR, Radiotekhn. Elektron.*, vol. 20, no. 6, pp. 1164–1170, 1975.

[3] V.P. Lukin, V.V. Pokasov, N.S. Time, and L.S. Turovtseva, "Retrieval of the Pulsation Spectrum of the Refraction Index in the Atmosphere from Optical Measurements," *Izv. Akad. Nauk SSSR, Fiz. Atmos. Okeana*, vol. 13, no. 1, pp. 90–94, 1977.

[4] V.P. Lukin and V.V. Pokasov, "Optical Wave Phase Fluctuations," *Appl. Opt.*, vol. 20, no. 1, pp. 121–135, 1981.

[5] A.S. Monin and A.M. Yaglom *Statistical Hydromechanics* (Nauka, Moscow, 1967), vol. 1.

[6] V.V. Nosov, V.M. Grigor'ev, P.G. Kovadlo, V.P. Lukin, E.V. Nosov, and A.V. Torgaev, "Astroclimate of Specialized Rooms at the Large Solar Vacuum Telescope. Part 2" *Atmos. Ocean. Opt.* vol. 21, no. 3, pp. 180–190, 2008.

[7] V.V. Nosov, O.N. Emaleev, V.P. Lukin, and E.V. Nosov, "Semiempirical Hypotheses of Turbulence Theory in the Anisotropic Boundary Layer," *Atmos. Ocean. Opt.* vol. 18, no. 10, pp. 756–772, 2005.

[8] A.S. Gurvich, A.I. Kon, V.L. Mironov, and S.S. Khmelevtsov, *Laser Radiation in Turbulent Atmosphere* (Nauka, Moscow, 1976). [in Russian].

[9] A.I. Kon, V.L. Mironov, and V.V. Nosov "Fluctuations of the Centers of Gravity of Light Beams in a Turbulent Atmosphere," *Radiophys. Quant. Electr.*, vol. 17, no. 10, pp. 1147–1155, 1974.

[10] V.L. Mironov, V.V. Nosov, and B.N. Chen, "Quivering of Optical Images of Laser Sources in a Turbulent Atmosphere," *Radiophys. Quant. Electr.*, vol. 23, no. 4, pp. 319–325, 1980.

[11] V.P. Lukin, "Intercomparison of Models of the Atmospheric Turbulence Spectrum," *Atmos. Ocean. Opt.*, vol. 6 no.9, pp. 628–631, 1993.

[12] V.P. Lukin, "Comparison of the Spectral Model of Atmospheric Turbulence," *Proc. SPIE* vol. 2222, pp. 527–535, 1994.

[13] V.P. Lukin, B.V. Fortes, and E.V. Nosov, "The Efficient Outer Scale of Atmospheric Turbulence," *Atmos.Ocean. Opt.*, vol. 10, no. 2, pp. 100–106, 1997.

[14] V.P. Lukin, S.M. Gubkin, O.N. Emaleev, N.G. Mutnitskii, and V.V. Pokasov, "Experimental Studies of Seeing Characteristics of the Vicinity of the Mt. Elbrus," *Astronom. J.*, vol.60, no. 4, pp. 789–794, 1983.

[15] V.V. Nosov, V.M. Grigor'ev, P.G. Kovadlo, V.P. Lukin, E.V. Nosov, and A.V. Torgaev, "Recommendations for the Selection of Sites for the Ground Based Astronomical Telescopes," *Opt. Atmos. Okeana*, vol. 23, no. 12, pp. 1099–1110, 2010.

Using digital camera as metering device in geometrical, spectral and intensity measurements

A.V.Kraiski, T.V.Mironova, T.T.Sultanov

P.N.Lebedev Physical Institute RAS, Moscow, Russia

Abstract: We discuss a possibility of using the standard commercial digital camera as a metering device in correlation, colorimetric and intensity measurements. It is shown that after extracting linear data from the camera matrix and carrying out an appropriate calibration procedure the camera characteristics are comparable with those of technical cameras.

In traditional methods of image registration the digital matrix practically displaced the photo film. Matrices have a whole number of advantages in comparison with films:
- instant conversion of the intensity into digital data (as compared with the procedure of film processing);
- high value of the photographic latitude and possibility of its increasing by special means;
- rigid coordinate system of pixels numbers in the image ;
- high repeatability of the geometry and energy parameters of the matrix receiving elements;
- good correspondence of the image data flow with the computer storing and processing technique;

These advantages are particularly essential for the optical measurements. The computer productivity growth and calculation technique progress led to a quick step forward of the digital image processing that is now widely used in all fields of research practically displacing pure optical analog methods.

Geometrical and correlation measurements.

A significant feature of the digital image is the fixed size of the pixel and the fixed spacing of their arrangement, which sets spatial discretization of the data. One could consider that the period of the pixel arrangement fixes the spatial resolution of the registration. However the precision of measurement of an object position in a digital picture can be considerably better if the size of the object image is larger than two pixels of the receiving matrix. An appropriate approximation of the optical system instrument function (or of the correlation peak shape, for exact determination of its maximum in the case of correlation measurements) allows one to calculate the object coordinates with the accuracy that is nearly two degrees better than the distance between the receiving matrix cells. Thus in studying digital images it is possible to detect object shifts and distortions with the scale much smaller than the distance between the receiving elements of the matrix.

Digital image correlation technique is widely used in strain analysis. The cross-correlation between two images is studied, one of the object in the loaded state and the other one in the unloaded state. Measuring velocity field and the refractive index in the flow of liquids and gases are also based on different variations of the digital calculations of the correlation function. The flow under investigation is visualized either by the Background Oriented Schlieren method or by measuring the shift of small tracer particles suspended in the flow (Particle Image Velocimetry). The Background Oriented Schlieren method allows one to obtain the gradient of the refraction index of a transparent medium (and therefore the density gradients using the Gladstone–Dale relation) from digital image processing. Two pictures are taken: one of a pure background image and the other one distorted by the optical inhomogeneity under study. Then, the pair of pictures is examined with the use of Digital Image Correlation method. Many researchers from various fields of science and technology use this method and it is almost impossible to review all the papers. As for the information content, the results obtained by this method in some applications can compete with the interference methods which are much more difficult to implement. For example in [1] for the diffusion process of two liquids, the results obtained by two experiments are compared. The first one is the holographic interferometry with a nonstationary reference wave with linear dependence of the frequency on the coordinate and time, and the second one was made as a digital image correlation in the Background Oriented Schlieren scheme. It is shown that both methods lead to similar results, but experimental implementation of the Digital Image Correlation method does not require such a precision alignment and vibroprotection of optical system as interferometry methods do. In the experiment with random background image the images for correlation development were taken by an ordinary pocket photocamera.

One can take the optical system which forms the image as the transparent medium under study which leads to optical inhomogeneities in the output image. In this case the digital image correlation method allows one to obtain the distortion of this optical system [2, 3]. This allows one to turn an ordinary camera into a metric one. A lot of papers are devoted to the distortion measurements. In most of them the distortion is measured with the use of a special calibration object consisted of contrast aligned elements. The calibration object is photographed, and, from related coordinates of the image points and those of the object the magnitude of the distortion is determined in the grid nodes. Then, the polynomial coefficients of the distortion are determined. A very simple method of the optical system distortion determination is proposed in [2, 3]. The method is based on the cross-correlation analysis of two digital images. The first one is the calibration object which is a known random pattern with small correlation length. The other one is a digital photo of this pattern that was printed on a flat surface. This image is distorted when passing through the optical system that forms the image. A projective transformation of the image which arise with the optical axis deflection from the perpendicular to the object plane can be compensated by calculating the transformation coefficients from the correlation correspondence of four pairs of points in the image and in the calibration object. This allows one to perform shooting in a simple way, without any fine camera adjustment.

This method gives a possibility of calculating at any point of the image not only the distortion value but also the value of mismatch of the color channels connected with chromatic aberration of the optical system. The accuracy of distortion measurement is about 0.01%, and the accuracy of the chromatic mismatch measurement is about 0.001%.

Linear data extraction from the matrix of a commercial camera.

The digital camera can be used not only for geometrical measurements in a variety of fields. It can be used also for measuring the light intensity distribution and even for a sort of spectral measurements with the use of colorimetric ratio of the values of the matrix signals in the case when the light radiation is narrow-band.

In digital cameras high-quality solid-state photo sensors are used. That allows using inexpensive cameras as a measuring device. The main problem is that the information about the recorded radiation essentially changes in the course of conversion of the recorded data into the image of any graphic format (e.g., BMP, TIFF, JPEG). Manufacturers try to make the intensity proportion and the color reproduction such that the digital image would possess maximal similarity with the picture, perceived by the human eye. For example, the relative brightness of the image points is varied to match the effective light sensitivity of the camera to the logarithmic sensitivity of the eye. Thus, the linearity of the signal and dynamic range are irreversibly corrupted, as well as the ratio of signals in the color channels. For the purpose of measurements the unprocessed RAW (read-after-write) format is more suitable. Without conversion it is not intended for visual representation. Besides the technical descriptions of shooting conditions and data, concerning the camera, the RAW file contains an array of digitized data from each pixel of the light-sensitive matrix of the camera. After saving the image in the RAW-format the dcraw-converter [4] should be employed in the so-called 'document' regime for obtaining the linear data. It gives a 16-bit TIF file which can be later either analyzed by means of the graphical analyzer NIP2 (as in [5]) or used for calculations as a usual numerical matrix (with MATLAB or by self-written software as in [6]). Dealing with this matrix one has to take into account that each element contains data on the light flux passed through the corresponding element of the RGGB Bayer color filter array.

Measurement of radiation spatial intensity distribution.

The question of the laser radiation intensity measurement is considered in detail in [5, 7]. In [5] the possibilities of a commercial camera in the regime of linear data extraction are demonstrated. It is shown that the camera can be used for measuring the laser radiation spatial intensity distribution and the estimations of the measurement accuracy are given for digital camera Canon EOS 400D. In the regime of linear data extraction all the basic measurement characteristics of the camera are obtained: the radiometric function, deviation from linearity, dynamic range, temporal and spatial noises (both dark and those depending on the signal value). The parameters obtained correspond to those of technical measuring cameras. The linear dynamic range of camera is 58 dB with the maximal deviation from linearity 2.7 %; the full dynamic range is 59 dB. The dark temporal noise is 1.6 digital units (the saturated signal for the camera is 3470 digital units). The spatial dark noise is 0.01 % from the maximal signal. The photo response nonuniformity is 0.5 %. These characteristics are comparable with those of technical cameras.

If the registered radiation is monochromatic, the dynamic range of the digital camera can be considerably widened. The main idea is to use data from color filters on the photosensor to estimate the true value of oversaturated pixels. First, camera's response function to the desirable light source is obtained. During this calibration process, camera's response function to the light source is calculated as well as the correction coefficients for the image linearization. Then, an oversaturated image in the main measuring channel (for example, red for the He-Ne laser) is analyzed and the saturated pixels are replaced using information from the neighbour green and blue pixels. Finally, the constructed image is linearized using the correction coefficients, which were obtained at the calibration step.

A similar approach is used in several techniques of increasing the camera dynamic range such as the Spatially Varying pixel Exposures [8] (SVE) and Assorted Pixels [9]. In [10] it is shown that the reconstructed High Dynamic Range signals are linear as the normal signals are. The increase of dynamic range of signal's registration from 58 dB up to 73 dB is obtained.

Colorimetric and spectral measurements.

Spectral sensitivity of the camera receiving matrix differs from that of the human eye. As a result of this mismatch, the data obtained from the camera can not lead to restoring the correct colors. Thus, different color perceptions from the human point of view can correspond to the save values obtained from the camera sensors. And vice versa.

Colorimetric measurements with the use of the digital camera are very hard to implement not only for this reason but also because, as it was mentioned, the build-in camera computer processes the data recorded on the matrix to obtain the image close to that perceiving by the human eye. Visualizing the RAW images transforms the color coverage of the camera into the sRGB system, which is accompanied with certain side effects. For example, when using the camera for recording

978-1-4799-0019-0/13 $31.00 © 2013 IEEE

monochromatic radiation with different wavelengths the Hue of the digital image remains the same within the wavelength range of 540 – 570 nm. Moreover, in the far red and violet regions of the spectrum the relative response of color channels changes drastically. Still it is possible to make not only colorimetric, but also spectral measurements with the use of the usual digital camera. In papers [5, 11, 12] we describe the colorimetric method for determination of the spatial two-dimensional distribution of the mean wavelength of narrow-band radiation. The method we propose allows one to fulfil this with photographic resolution accuracy with no using any spectral device. Similarly to measuring the intensity distribution, one has to use unprocessed data from the matrix and perform a preliminary calibrating. The calibrating procedure is as follows. One makes a picture of the spectrum (continuous with basic lines imposed) in the RAW format and gets from the receiving matrix of the camera the unprocessed signal and divides it into the color channels taking into account locations of the Bayer cells with corresponding filters. Then, one associates with each wavelength some function of intensities in the three channels. As this function, it is convenient to choose a counterpart of the standard Hue function used in RGB calculations, but taken as a function not of R, G and B, but of the signal values in the corresponding matrix cells. When calibrating the camera, the Hue of the spectrum varies from 0 to 240, then the wavelength is determined in the range from 455 to 625 nm. In principle, a similar calibrating procedure can be also done for the camera without option of giving out data in the RAW format. In this case, the working range in wavelengths consists of two parts: those in the blue-green and red-yellow parts of the spectrum [11, 12]. Besides, the same wavelength depending on the light intensity corresponds to different values of the Hue function, even for the weak intensities.

REFERENCES

[1]   A. V. Kraisky and T. V. Mironova. Comparison of the results of digital image correlation method and the data obtained in holographic interferometry experiment with nonstationary reference wave for refractometry measurement of diffusion process. To be published in *Quantum Electronics*.

[2]   A. V. Kraisky and T. V. Mironova. Способ калибровки оптической системы. Патент на изобретение №2381474, 2010.

[3]   A. V. Kraisky and T. V. Mironova. Optical system calibration by the correlation method. Bulletin of the Lebedev Physics Institute, 2008, Volume 35, Number 8, Pages 231-237

[4]   http://www.cybercom.net/~dcoffin/dcraw.

[5]   M.V. Konnik, E.A. Manykin, S.N. Starikov. Extension of the possibilities of a commercial digital camera in detecting spatial intensity distribution of laser radiation, Quantum Electronics **40** (4) 314 – 320 (2010).

[6]   A.V. Kraiskii, T.V. Mironova, T.T. Sultanov. Narrow-band radiation wavelength measurement by processing digital photographs in RAW format. *Quantum Electronics* **42** (12) 1137 – 1139 (2012).

[7]   M.V. Konnik, E.A. Manykin, S.N. Starikov. Increasing linear dynamic range of commercial digital photocamera used in imaging systems with optical coding. arXiv:0805.2690v1 [cs.CV] 17 May 2008.

[8]   Nayar, S.K. and Mitsunaga, T., High Dynamic Range Imaging: Spatially Varying Pixel Exposures, IEEE Conference on Computer Vision and Pattern Recognition (CVPR), 2000, V1, pp.472-479.

[9]   Nayar, S.K. and Branzoi, V., Adaptive Dynamic Range Imaging: Optical Control of Pixel Exposures Over Space and Time, , IEEE Conference on Computer Vision, 2003, V2, pp1168-1175.

[10]   M.V.Konnik, E.A.Manykin, and S.N.Starikov. Optical-Digital Correlator with Increased Dynamic Range Using Spatially Varying Pixels Exposure Technique. Optical Memory and Neural Networks (Information Optics), 2009, V18, N2, pp.61-71.

[11]   А.В.Крайский, Т.В.Миронова, Т.Т.Султанов, В.А.Постников, В.И.Сергиенко, В.Е.Тихонов. Способ измерения длины волны узкополосного светового излучения колориметрическим способом. Патент РФ № 2390738, 2010.

[12]   A V Kraiskii, T V Mironova and T T Sultanov. Measurement of the surface wavelength distribution of narrow-band radiation by a colorimetric method. *Quantum Electronics* **40** (7) 652 – 658 (2010).

Scanning devices for the beam profile measurement of laser irradiation

N.G. Kokodij [1,2], B.V. Safronov [1], I.A. Priz [1], V.P. Balkashin [1], M.P. Perepechaj [1]

[1] V.N. Karazin Kharkov National University, Kharkov, Ukraine
[2] Kharkov National University of Pharmacy, Kharkov, Ukraine

Abstract: The paper describes a scanning bolometer receiver for measuring the parameters of optical radiation beam, which operates in visible and infrared regions of the spectrum. By means of the device the beam profile can be reconstructed, which is essential for specialists who work with laser equipment.

INTRODUCTION

Beam profile or intensity distribution function is an important characteristic of the laser radiation [1]. On the one hand in manufactured lasers it must be close to Gaussian distribution which minimizes the size of the focal spot, on the other hand beam profile of lasers used in interferometers, must be close to flat. It is also necessary to know the beam profile when calculating the parameters specified by ISO11146 and ISO13694 Standards [2].

To measure the beam profile in the visible and near infrared regions of the spectrum matrices made of photodiodes are used (CCD-matrices). These matrices are able to measure the laser beam power from 100 µW up to 10 W (and even up to 5 kW in case when light attenuators are used) and have the input window of a size of 10 to 50 mm. For measurements in the middle infrared range matrices made of pyroelectric elements are used.

There are devices in which the hollow metal needles intersect the beam in different directions. Radiation propagates inside the cavity optical fiber towards the receiver, which is a photodiode or pyroelectric element. Signals from the receiver are input into a computer. In other devices instead of hollow metal needles scanning knives are used. Depending on a modification such devices have the following characteristics: an aperture is of 3 to 9 mm; resolving power is 1 µm and limited by diffraction phenomena of radiation; spectral range comes up to 190 to 1800 nm; power range is 10 mW to 1 W (in case when light attenuators are used).

To measure the characteristics of laser radiation of power up to 10 kW and large beam diameter grid bolometer devices are designed. They consist of several grids made of thin metal wire-bolometers intersecting the laser beam. Bolometers heated by radiation change their resistance [1]. Having measured the resistance of each bolometer it can be possible to restore the radiation beam profile, incident on the device.

In the direct radiation beam only scanning and grid devices can operate. The latter are designed for measurements within wide laser beams (with the cross-sectional dimensions of a few centimeters or more). For narrow laser beams scanning devices are designed. In the following part of our paper we will describe one of such devices - scanning bolometer receiver.

SCANNING BOLOMETER RECEIVER

The receiver is designed to measure the intensity distribution in narrow radiation beams (a few millimeters in diameter) with power of 10 W to 100 W.

The receiver circuit is shown in Fig. 1. Bolometers 1 - 5 are arranged in a slit with length of 100 mm and width of 20 mm. Bolometer 3 is vertical, bolometers 2 and 4 are at angle of 30° to the vertical, bolometers 1 and 5 are at 60° to the vertical. Length of bolometer 3 is 20 mm, bolometers 2 and 4 - 24 mm, bolometers 1 and 5 - 40 mm.

Bottom edges of bolometers are connected together; dc voltage is applied to the top edges of bolometers via $R_1 - R_5$ resistors. Signals from the bolometers are input in a computer via ADC.

The receiver is designed to measure the intensity distribution in narrow radiation beams (a few millimeters in diameter) with power of 10 W to 100 W.

The receiver circuit is shown in Fig. 1. Bolometers 1 - 5 are arranged in a slit with length of 100 mm and width of 20 mm. Bolometer 3 is vertical, bolometers 2 and 4 are at angle of 30° to the vertical, bolometers 1 and 5 are at 60° to the vertical. Length of bolometer 3 is 20 mm, bolometers 2 and 4 - 24 mm, bolometers 1 and 5 - 40 mm.

Bottom edges of bolometers are connected together; dc voltage is applied to the top edges of bolometers via $R_1 - R_5$ resistors. Signals from the bolometers are input in a computer via ADC.

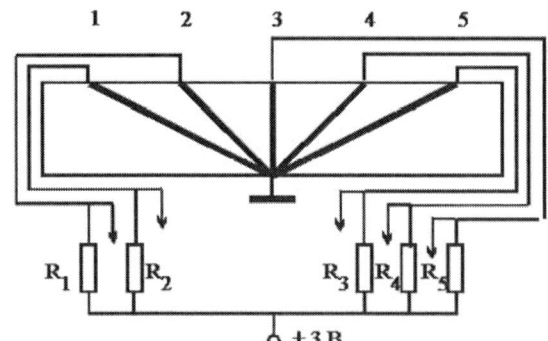

Fig. 1. The circuit of scanning bolometer receiver.

Two receivers have been developed. In the first receiver platinum wires of a diameter of 20 µm serve as bolometers. It is designed to measure the radiation power of 50 W or more.

In the second receiver strips with a cross-section of 100 x 10 mm made of platinum-silver alloy serve as bolometers. The receiver is designed for measurements the lower radiation output power starting from 10 W.

In front of the board with bolometers the screen is placed. The screen is a metal plate with a slit of a size of 90 x 10 mm. It prevents the board bolometers from heating by scattered radiation.

The receiver plane is perpendicular to the direction of radiation propagation. The receiver can move at a speed of 3.2 mm/s at a distance of 104 mm. Bolometers alternately cross the radiation beam and the signals from them are recorded by means of a computer.

Fig. 2 shows the plot of five signals from the receiver, following one after another. Horizontal axis represents time in seconds.

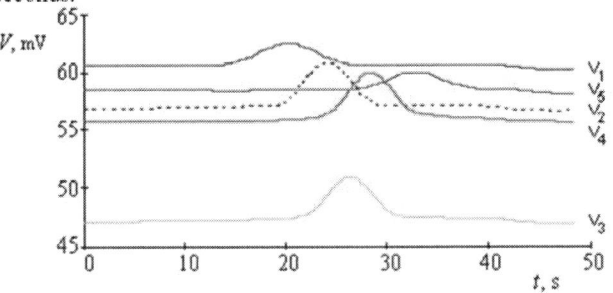

Fig. 2. Signals from the receiver

For signal processing it is necessary to know the radiation power P absorbed by each bolometer. From the theory of heat conduction it is known that

$$P = \alpha_p \cdot L \cdot \Delta T , \qquad (1)$$

where ΔT is the bolometer temperature increase, L is bolometer length, α_p is linear heat transfer factor. For platinum bolometer of a diameter of 20 μm factor $\alpha_p = 0.03$ W/(m·deg).

The temperature difference ΔT has been determined by the change in bolometer resistance:

$$\Delta T = \frac{1}{\alpha_r} \cdot \frac{\Delta R}{R_0} , \qquad (2)$$

where α_r is temperature coefficient of resistance, ΔR is change in resistance, R_0 is the initial resistance of the bolometer. For platinum and platinum-silver alloy $\alpha_r = 0.004$ deg[-1].

In the described bolometer circuit the output signal is proportional to the resistance of bolometers, so that

$$\Delta T = \frac{1}{\alpha_r} \cdot \frac{\Delta U}{U_0} , \qquad (3)$$

where ΔU is change in signal, U_0 is signal initial value.

Substituting (3) into the expression (1), we get:

$$P = \frac{\alpha_p \, L}{\alpha_r} \cdot \frac{\Delta U}{U_0} . \qquad (4)$$

Signal processing has been carried out by the program based on the integral Radon transform.

The plot of signals from bolometers is shown in Fig. 3.

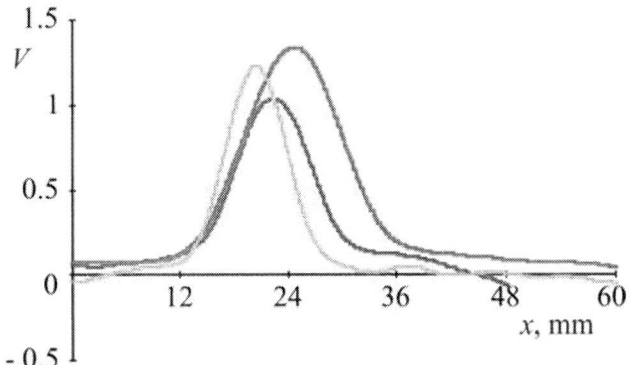

Fig. 3. Signals from the receiver.

Fig. 4 shows the beam profile of optical radiation reconstructed by means of above mentioned program. The size of the rectangle in Fig. 4a is 40 x 30 mm. It is divided into smaller squares with a side of 10 mm. Fig. 4b shows the boundaries of the light spot, created by the light source in the receiver location. As you can see it is an ellipse with a horizontal axis of 17 mm and vertical axis of 10 mm. Fig. 4a also highlights the rectangle of the size of 10 x 17 mm, which covers the brightest part of the beam profile.

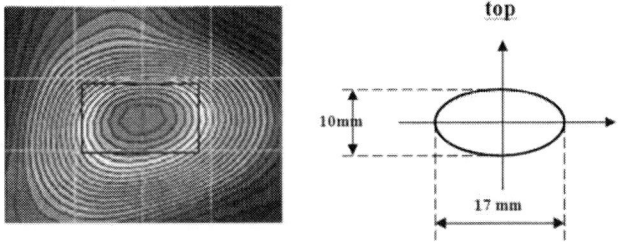

Fig. 4. The profile of optical radiation beam.

CONCLUSION

We have described scanning bolometer for measuring the parameters the of optical radiation beam, specified by ISO11146 and ISO13694 Standards. The device operates in visible and infrared regions of the spectrum with beams of a diameter of 1 to 10 mm and of power of 10 to 500 W.

REFERENCES

[1]   Kuzmichev V.M., Priz I.A., Safronov B.V., Balkashin V.P., Pogorelov S.V., Kohns P. "Measuring of the factor of absorption efficiency of thin-wire bolometer," Proc. of CAOL-2005 – Vol.2. – P.313-315.

[2]   ISO 13694:2000. "Optics and optical instruments - Lasers and laser-related equipment - Test methods for laser beam power (energy) density distribution," www.iso.org

978-1-4799-0019-0/13 $31.00 © 2013 IEEE

Gap in pagination due to unavailable paper.

Pages 332-333

Absolute Calibration of Profile Thin-Wire Bolometric Gauge of Laser Pulse Energy

S.V. Pogorelov

National Pharmaceutical University, Department of Pharmaceutical informatics,
53, str. Pushkinskaja, Kharkiv, 61002, Ukraine, ph. +380 57 706 21 89 e-mail: svpog@yahoo.co.uk

Abstract: the technique of absolute calibration of profile thin-wire bolometer of laser pulse energy which does not require additional absolute gauges is offered

Thin-wire bolometers in the form of lattices from thin metal wires are used for continuous power, pulse energy [1], state of polarization [2], generalized square of a beam, coordinates of energy center, profile distribution along the fixed directions and another parameters of intensive and wide-aperture laser radiation measurements. High levels of radiation intensity lead to essential heating of bolometer elements, and temperature dependencies of their basic physical parameters which determine bolometer transformation coefficient. This heating leads to appearance of nonlinearity of transformation characteristic which become stronger with non-uniformity of radiation distribution along the bolometer, and systematic errors of measurements of optic radiation parameters are risen that require bolometer calibration.

The main disadvantages of calibration with absolute gauge are, firstly, one have to use an additional absolute gauge that makes the calibration significantly complex, leads to additional systematic errors and gauges with wide aperture square are not industrially produced as well; secondly, at the measurements within wide dynamic range it is required to take into account the nonuniformity of radiation distribution that leads to high errors of calibrated parameters.

The object of the work is making of absolute calibration technique of the thin-wire bolometer for measurements of wide-aperture intense energy of laser radiation within long wave IR range. The absolute calibration of thin-wire bolometer one can perform with usage of suggested dependence of relative resistance increase on the linear specific energy of laser pulse and determination of coefficients of this dependence. It allows us to produce bolometers with ready calibration characteristics.

The output signal of profile thin-wire bolometer at measurements of laser radiation pulse energy is a sum of every bolometric elements signals of two lattices with mutually perpendicular elements. It is proportional to maximal relative increase of the whole bolometer and a measuring equation may be in the form

$$U = \eta_0 \left(1 + K_D\right) F\left(\frac{\delta \overline{E}}{m}\right) \frac{\overline{E}}{m}, \qquad (1)$$

where $U = \dfrac{\Delta R}{R_0}$ – is output signal of bolometer; ΔR and R_0 – are the bolometer resistance increase and its initial value; $\eta_0 = \dfrac{\alpha_0 \, q_0^E}{c_0}$ – is the transformation coefficient of the bolometer in the linear mode; α_0, q_0^F, c_0 – are the temperature coefficient of resistance, factor of absorption efficiency for E-polarized radiation, and specific heat capacity of the bolometer at an environment temperature T_0, respectively; K_D – is coefficient of dichroism of the bolometer that is equal to ratio of factors of absorption efficiency for H- and E-polarization of radiation; $F\left(\dfrac{\delta \overline{E}}{m}\right)$ – is a transformation coefficient of the bolometer normalized on η_0 that determines the nonlinearity of the bolometer transformation characteristic and it may be represented with high accuracy as polynomial of the second order on the effective energy of radiation $\dfrac{\delta \overline{E}}{m}$ in the form of mathematical expression

$$F\left(\frac{\delta \overline{E}}{m}\right) = 1 + a\frac{k_1^2 + k_2^2}{1 + K_D}\frac{\delta \overline{E}}{m} + b\frac{k_1^3 + k_2^3}{1 + K_D}\left(\frac{\delta \overline{E}}{m}\right)^2, \quad (2)$$

where k_1 and k_2 – are polarized coefficients of interaction of radiation with bolometric elements of the first and second lattices and may be presented as

$$k_{1,2} = \cos^2\left(\varphi - \psi_{1,2}\right) + K_D \sin^2\left(\varphi - \psi_{1,2}\right), \qquad (3)$$

where $\psi_{1,2}$ – are angles directions of bolometric elements of the first and second lattices relative to chosen coordinate axis (in the case the axis coincides with direction of bolometric elements of the first lattice then $\psi_1 = 0^o$, $\psi_2 = 90^o$); φ – is an angle between a direction of electric vector of radiation and direction of bolometric elements of the first lattice.

Values k_1, k_2 and δ are measured by profile gauge with a high accuracy on the method described in the work [1], and the value K_D has to be previously measured or theoretically calculated. Then the measuring equation (1) with consideration of the expression (2) is a cubic one relative to linear specific energy $\frac{\overline{E}}{m}$. In order to obtain the solution of the equation (1) in the absolute form, i.e. in J/g, one has to determine the value of η_0 in absolute form g/J and values a and b in dimensions g/J and g²/J² with minimal relative standard deviations (RSD).

The absolute calibration of the thin-wire profile bolometer of laser radiation pulse energy one performs with usage of tabulated temperature dependencies of specific resistance, specific heat-capacity, complex index of refraction within longwave approximation of classical optic, and a solution of heat balance equation averaged along the total bolometer length. The each of the mentioned parameter is well-known for specific bolometer material.

The temperature dependence of temperature coefficient of resistance $\alpha(T)$ one should determine at first. For that the tabulated temperature dependence of specific resistance of bolometer material [3] should be approximated with polynomial

$$\alpha(T) = \alpha_0 + \alpha_1 T .$$

The total relative increase of resistance one can determine with integration along the whole bolometer length in view of nonuniform distribution of temperature along the bolometer $T(x)$, where x – is a coordinate of point along the total bolometer length l

$$\frac{\Delta R}{R_0} = \left(\alpha_0 + \alpha_1 \, \delta \overline{T}\right)\overline{T} \qquad (4)$$

where $\overline{T} = \frac{1}{l}\int_0^l T(x)dx$ – is the averaged value of bolometer temperature.

The complex index of refraction may be presented in view $m = m' - i\,m''$ within longwave radiation range, and its refraction index and absorption index are equal to one another

$$m' = m'' = 9{,}487 \cdot 10^5 \sqrt{\frac{\mu \sigma}{f}} , \qquad (5)$$

where μ – is relative magnetic conductivity that is equal to unit for metals; σ – is specific conductivity; f – is radiation frequency. Dependencies of factors of absorption efficiency (FAE) on bolometer diameter, radiation polarization and temperature are calculated with usage of expressions cited in [4] which have been obtained at the exact solution of flat electromagnetic wave diffraction on the circle cylinder with absorption. The dependence of FAE on temperature one can approximates with the second degree polynomial

$$q(T) = q_0' + q_1' \, T + q_2 \, T^2 . \qquad (6)$$

Coefficients q_0', q_1', q_2 and their relative RSD $\sigma_{q_0'}$, $\sigma_{q_1'}$ and σ_{q_2} are determined with least-squared method under environment temperature $T_0 = 0°C$.

The tabulated dependence of specific heat capacity of platinum [5] is approximated with the second degree polynomial as

$$c(T) = c_0' + c_1' T + c_2 T^2 , \qquad (7)$$

where c_0' and c_1' correspond to environment temperature $T_0 = 0°C$.

Coefficients a and b in the expression (2) don't depend on the direction of radiation polarization and so one can determine them in the case when the first lattice is irradiated with E-polarized radiation and $k_l=1$.

The averaged along the bolometer length equation of heat balance is

$$\frac{c_0 + c_1 \delta \overline{T}}{q_0 + q_1 \overline{T} + q_2 \left(\delta \overline{T}\right)^2} d\left(\delta \overline{T}\right) = d\left(\frac{\delta \overline{E}}{m}\right), \qquad (8)$$

where $c_0 = c_0' + c_1 T_0$; $c_1 = c_1' + 2c_2 T_0$; $q_0 = q_0'^E + q_1' T_0 + q_2 T_0^2$; $q_1 = q_1' + 2q_2 T_0$, and T_0 – is environment temperature during the measuring. The solution of the equation (8) with initial condition $\delta \overline{E}/m = 0$ at effective temperature $\delta \overline{T} = 0$ is

$$\frac{\delta \overline{E}}{m} = \frac{1}{\sqrt{-\Delta}}\left(c_0 - c_1 \frac{q_1}{2q_2}\right) *$$
$$* \ln\left[\frac{\left(2q_2 \, \delta \overline{T} + q_1 - \sqrt{-\Delta}\right)\left(q_1 + \sqrt{-\Delta}\right)}{\left(2q_2 \, \delta \overline{T} + q_1 + \sqrt{-\Delta}\right)\left(q_1 - \sqrt{-\Delta}\right)}\right] + , \qquad (9)$$
$$+ \frac{c_1}{2q_2}\ln\left[\frac{q_0 + q_1 \delta \overline{T} + q_2 \left(\delta \overline{T}\right)^2}{q_0}\right]$$

where $\Delta = 4q_2 q_0 - q_1^2 < 0$. The obtained solution determines the functional dependence between $\delta \overline{E}/m$ and $\delta \overline{T}$. In this case the relative resistance increase will be

$$\frac{\Delta R}{R_0} = \left(\alpha_0 + \alpha_1 \delta \overline{T}\right)\overline{T} =$$
$$= \frac{\alpha_0 q_0}{c_0} \cdot \frac{c_0}{q_0}\left(1 + \frac{\alpha_1}{\alpha_0}\delta \overline{T}\right)\frac{\delta \overline{T}}{\delta \overline{E}/m} \frac{\overline{E}}{m} , \qquad (10)$$

where $\frac{\alpha_0 q_0}{c_0} = \eta_0$ – is bolometer transformation coefficient in linear mode and normalized transformation coefficient is

$$F\left(\frac{\delta \overline{E}}{m}\right) = \frac{c_0}{q_0}\left(1 + \frac{\alpha_1}{\alpha_0}\delta \overline{T}\right)\frac{\delta \overline{T}}{\delta \overline{E}/m} . \qquad (11)$$

The transformation coefficient (11) should be approximated with the second degree polynomial

$$F\left(\frac{\delta \overline{E}}{m}\right) = 1 + a\frac{\delta \overline{E}}{m} + b\left(\frac{\delta \overline{E}}{m}\right)^2 \qquad (12)$$

and coefficients a and b are determined.

Polarized coefficients of radiation interaction with bolometric elements of the first and the second lattices k_1 and k_2 as well as coefficient of dichroism K_D one can find out with accordance to [1] for specific laser radiation. The measuring equation (1) with usage of expression (2) and determined coefficients η_0, a and b is a cubic one relative to linear specific energy

$$\frac{\Delta R}{R_0} = \eta_0 \cdot (1 + K_D)*$$

$$*\left(1 + a\frac{k_1^2 + k_2^2}{1 + K_D}\frac{\delta \overline{E}}{m} + b\frac{k_1^3 + k_2^3}{1 + K_D}\left(\frac{\delta \overline{E}}{m}\right)^2\right)\frac{\overline{E}}{m} \qquad (13)$$

and has the only solution.

The invention is performed in the following way. One determines values α_0, q_0 and c_0 for given environment temperature and calculates the coefficient $\eta_0 = \dfrac{\alpha_0 q_0}{c_0}$. Values $\delta \overline{E}/m$ are determined for concrete bolometer with diameter d with the expression (9) for fixed values $\delta \overline{T} = 0, 100, 200, \ldots 1500°C$, and obtained tabulated dependence is approximated with the second degree polynomial (12) according to (11). Its coefficients a and b as well their relative RSD are determined with the least square method.

The cubic equation (13) with known coefficients has the only solution relative to averaged specific incident optic energy $\overline{E}/m$ with elimination of dominant systematic error of nonlinearity of bolometer transformation characteristic by measuring results of calibrated bolometer. The total energy of laser pulse may be found with expression

$$E = \frac{m}{d}\left(\frac{\overline{E}}{m}\right)S, \qquad (14)$$

where S – is an area of lattice entrance aperture, which has not any limitation on maximal size.

Thus, the technique of absolute calibration of profile thin-wire bolometer of laser pulse energy which does not require additional absolute gauges, is easier than previous ones, and reduces bolometer production costs due to refusal from additional calibration is offered.

REFERENCES

[1] Kuzmichov V.M., Solov'yev V.A., Lapko A.V. Measuring of energy parameters of intense laser radiation with profile thin-wire bolometer // Radiophysics and radioastronomy – 1999. – Vol.4, N3 – P.286-295.

[2] Kuzmichov V.M., Kuzmichova E.V. Measuring of elliptic polarization of intense laser radiation with nonlinear thin-wire bolometers // Measuring technique – 1998 – N6 – P.19-22.

[3] Tables of physical values. Reference-book / edited by Kikoin I.K. – Moscow, Atomizdat, 1976. – 1006 p.

[4] Wan de Hulst G. Light dispersion with small particles. Moscow, foreign literature publish house, 1961. – 536 p.

[5] Zinov'ev V.E. Heat-physical properties of metals under high temperatures – Moscow, Metallurgy, 1989. – 384 p.

Features of the wavefront sensor based on the Talbot effect

Dmytro V. Podanchuk, Andrey A. Goloborodko, Myhailo M. Kotov
Taras Shevchenko National University of Kyiv, Ukraine.

Abstract: The influence of aperture size on the accuracy of wavefront reconstruction with the sensor based on the Talbot effect is experimentally investigated. The analysis of set of spherical and astigmatic lenses shows that the aperture effects grow with the magnitude of the wavefront aberrations.

The problem of noncontact testing of spatially heterogeneous objects is essential today. Wavefront sensing is the one of the optical diagnostic techniques. It provides the measurements and analysis of the wave front of the probe laser beam passing through the object or being reflected from its surface. The Shack-Hartmann sensor is widely used for this purpose [1, 2]. It consists of the lenslet array and CCD-photodetector located at the focal plane of the lenslets. The lenslets produce the point images on the photodetector when the spatial coherent light beam falls on lenslet array. The shift of each image from respective optical axis is directly proportional to the local slope of the wave front in the corresponding lenslet subaperture. The reconstruction of the wavefront shape from the local slopes is a typically inverse problem because a local slope is a partial derivative of a wavefront phase. For instance, it can be solved by the modal method, in which the wavefront phase is represented as the expansion in basis functions. Generally, the basis of Zernike polynomials is used, since they describe standard aberrations.

Unfortunately, the usage of the Shack-Hartmann sensors is complicated by the contradiction between their angular sensitivity and spatial resolution. Usually, the spatial resolution is determined by the lenslet size. Reducing the diameter of the lenslets leads, in fact, to reducing its focal length that affects the accuracy of the wavefront reconstruction.

The possible solution of this problem is the wavefront sensor based of the Talbot effect. The effect principle is the self-reconstruction of the image of a periodical object at the regular distances (called Talbot distance or Talbot length) behind the object plane without any optical system while the object is illuminated by monochromatic plane beam. The image in the Talbot plane is distorted if the incident wave has aberrations. The degree of the image distortion is defined by the wavefront shape. The measurement principle and data processing with the sensor based on the Talbot effect (or Talbot sensor, for simplicity) are similar to the Shack-Hartman sensor working if a two-dimensional diffractive grating is used as the periodical object. At the same time the Talbot sensor has a higher spatial resolution and better producibility because grating period can be quite smaller in comparison with the lenslet size and grating making is simpler than the making of the lenslet array [3, 4].

But perfect self-imaging occurs only in case of infinite periodic grating [5]. In practice, the grating aperture has several periods only, so we need to take into account this fact in real experiments. This work concerns the influence of grating extent on the accuracy of the Talbot sensor.

Let's consider the light diffraction by a periodic structure under the Huygens-Fresnel approximation. Let U_0 is a plane wave incident normally on a periodic structure. Its amplitude transmission function $T(\xi,\eta)$ can be represented as the superposition of $T(\xi)$ i $T(\eta)$:

$$T(\xi,\eta) = T(\xi)T(\eta) \quad , \qquad (1)$$

or:

$$T(\xi,\eta) = \sum_{n=-\infty}^{\infty} C_n \exp\left(j\frac{2\pi n}{\Delta}\xi\right) \sum_{m=-\infty}^{\infty} C_m \exp\left(j\frac{2\pi m}{\Delta}\eta\right) , \quad (2)$$

where Δ is a period, C_k are coefficients of the expansion:

$$C_k = \frac{1}{\Delta} \int_{-\Delta/2}^{\Delta/2} T(\xi)\exp\left(j\frac{2\pi k}{\Delta}\xi\right)\mathrm{d}\xi . \qquad (3)$$

The field in the fare region can be written with the diffraction integral [5]:

$$U(x,y) = \int_{-\infty}^{+\infty}\int_{-\infty}^{+\infty} U'(\xi,\eta) h(x-\xi, y-\eta) d\xi d\eta, \qquad (4)$$

where $U'(\xi,\eta)$ is the field distribution behind the diffractive structure, $h(x-\xi, y-\eta)$ is the Point Spread Function (PSF). Since the distribution $U(x,y)$ can be represented by superposition $U(x)U(y)$, we can consider below one-dimensional projection of the distribution. Then, using (1) and (3), the amplitude distribution in the Talbot plane can be written as

$$U(x) = \exp\left(j\frac{kx^2}{2z}\right)\sqrt{\frac{U_0\exp(jkz)}{j\lambda z}}\sum_{n=-\infty}^{\infty} C_n \delta\left(\frac{2\pi n}{\Delta} - \frac{kx}{z}\right), \quad (5)$$

where λ is a wavelength, $k=2\pi/\lambda$ is a wave vector, U_0 is amplitude of the plane wave, z is a distance, $\{x, y\}$ are coordinates in the Talbot plane, $\{\xi, \eta\}$ are coordinates in the grating plane ($z=0$). One can easy see that this distribution is periodical.

978-1-4799-0019-0/13 $31.00 © 2013 IEEE

To take into account aperture effects, let us limit the one-dimensional series in (2) by using additional multipliers:

$$T_a(\xi) = \text{rect}\left(\frac{\xi}{a}\right) \sum_{n=-\infty}^{\infty} C_n \exp\left(j\frac{2\pi n}{\Delta}\xi\right), \qquad (6)$$

where a is illuminated aperture of the grating. Then the distribution (5) can be rewritten in form:

$$U(x) = \exp\left(j\frac{kx^2}{2z}\right)\sqrt{\frac{U_0 \exp(jkz)}{j\lambda z}} \times$$

$$\times a \sum_{n=-\infty}^{\infty} C_n \text{sinc}\left(\frac{\pi na}{\Delta} - \frac{kxa}{2z}\right) \qquad . \qquad (7)$$

Thus it is evident that the influence of the aperture is determined by the localization of the basis functions sinc(...). And for the binary grating, the distribution takes the form:

$$U(x) = \frac{\sin\left(\dfrac{2\pi x(2N-1)\Delta}{\lambda z}\right)}{\sin\left(\dfrac{2\pi x\Delta}{\lambda z}\right)} \text{sinc}\frac{\pi x\Delta}{(N-1)\lambda z}, \qquad (8)$$

where N is the number of periods that get into in the grating aperture. From (8) it follows that image of the aperture is defined by function sinc($\pi x\Delta/(N-1)\lambda Z_T$). Thus, the grating image is not formally limited by aperture, but because of interference, determined by many factors in the sum (7). Obviously, the actual image area is limited by the area, which is comparable to the real aperture grating. It should be mentioned, that image of transparent areas is blurred and deformed due to the stronger influence of diffraction on the edge of the grating aperture. Thus one must take into account this fact when using the Talbot effect in wavefront sensors.

The capability of the wavefront sensor based on Talbot effect is assessed on the experimental setup depicted on Fig.1.

He-Ne laser (λ=0.63 μm) is used as a coherent light source. The collimated laser beam falls on the tested lens L. Then the beam is divided into two parts by the beamsplitter BS. The beam, passing through BS, gets in plane of the two-dimensional diffractive grating DG. The square diaphragm D, placed in front of the grating, allows illuminating only the part of the grating with a required number of periods. Photodetector CCD_1 records the grating image at the Talbot distance Z_T. The beam reflected from BS is analyzed by the Shack-Hartmann sensor which consists of the lenslet array LA and photodetector CCD_2. The lenslets of the array have a diameter of 0.4 mm and focal length of 24 mm. Thus, we can compare the results of wavefront measurements obtained by the Talbot sensor and the Shack-Hartmann sensor.

The two-dimensional binary amplitude grating with 80 μm square holes is used in experiments. It has a period d of 160 μm and the Talbot distance $Z_T=2d^2/\lambda\approx80$ mm. The grating was produced on photoplate PFG-1 projecting the reduced images of necessary structures. The full size of the grating is 8×8 mm or 49×49 periods.

Several spherical and astigmatic lenses are used as the testing objects. Spherical waves are formed by the set of lens which are placed at a distance of ~0.4 m before the measurement planes. The lenses have a power from −2.0 D to +0.5 D.

The analysis of the wavefront has been carried out by the Talbot sensor with different size of the grating aperture. The aperture size has been reduced from 49×49 to 9×9 periods by putting the corresponding diaphragm D in front of the grating. Wave fronts have been reconstructed from the array of 7×7 spots on the grating image. Fig.2 shows the deviations of the Zernike defocus coefficients obtained with the Talbot sensor (average values over five realizations) from the coefficients obtained with the Shack-Hartmann sensor. Values of the coefficients are normalized to unit circle. It should be mentioned that the measurement error, related to diaphragm misalignment, does not exceed 1% of the average value. In case of the full aperture, the experimental results are equal for all tested lenses and are practically identical with the reference Shack-Hartmann data. In the range of aperture size from 25 to 13 grating periods the deviation is less than 0.1λ.

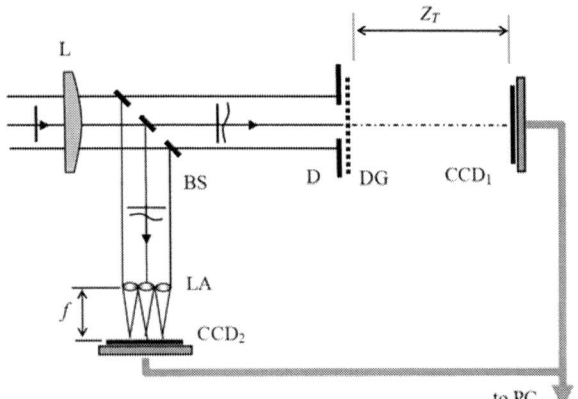

Fig. 1. Experimental setup for testing the sensor based on the Talbot effect.

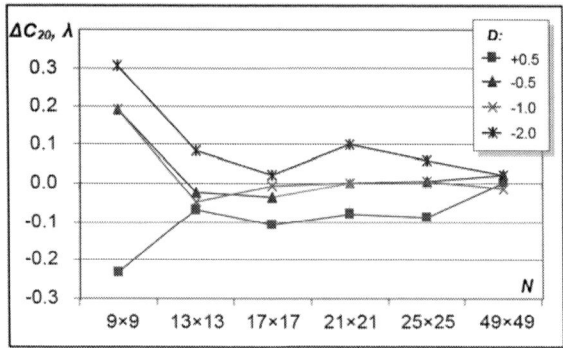

Fig. 2. The deviations of defocus coefficient from the reference values obtained for different sizes of the grating aperture in the Talbot sensor.

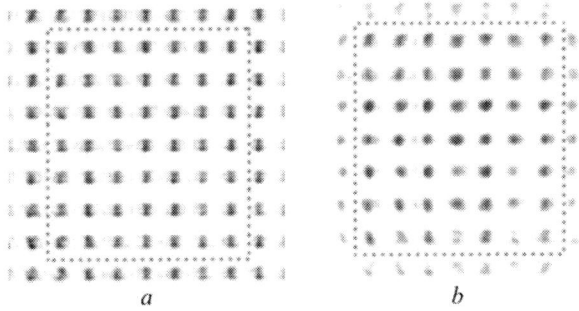

Fig. 3. Fragments of the grating image in the Talbot plane: 49×49 (a) and 9×9 (b) periods of the grating are illuminated. Dotted frames show the sub-area of wavefront reconstruction.

The deviation of the defocus coefficients from the reference values spikes for $N = 9$. The grating image is getting worse when the illuminating aperture has been reduced closely to the reconstruction area (Fig.3b). The spots on the image change the form that leads to the error of the centroid measurement and invalid aberration estimation of the tested wave front.

In the next experiment the influence of the aperture effects on the measurements of astigmatic wave fronts by the Talbot sensor has been investigated. We use two astigmatic lenses. The first lens has a sphere of +0.75 D and a cylinder of −0.25 D, the second lens has a sphere of +2.5 D and a cylinder of −0.5 D. The lenses are placed at the distance of ~0.25 m before the measurements plane. The excessive sphericity of the second lens has been compensated by placing the additional thin divergent lens with a power of −3.0 D before it. The Zernike coefficients of defocus ($C_{2;0}$) and astigmatism ($C_{2;2}$ and $C_{2;-2}$) obtained with the Shack-Hartmann sensor are considered as the reference values. In the Talbot sensor wavefront reconstruction has been carried out on the grid of 7×7 periods. The illuminated area of the diffractive grating has been changed from 49×49 to 9×9 periods (Fig.4).

The reference values of the coefficients obtained by the Shack-Hartmann sensor are represented on the diagrams by the solid lines. The experimental results show that if the aberration value is less than 0.5 λ (the astigmatism coefficient $C_{2;-2}$ on the Fig.4a and defocus and astigmatism coefficients $C_{2;0}$ and $C_{2;-2}$ on the Fig.4b) then coefficients measured with the Talbot sensor are equal within the limits of error for all used apertures. For stronger aberrations ($C_{2;0}$, $C_{2;2}$ on the Fig.4a and $C_{2;2}$ on the Fig.4b) there is severe departure of the data (up to 0.5 λ) from the reference values when illuminated aperture has been reduced to 9×9 periods. It is also concerned with the aperture effect influence on the grating self-imaging intensity distribution in the Talbot plane.

The analysis of the aperture effect influence on the quality of wavefront reconstruction by the Talbot sensor shows that in case of full-aperture illuminating wavefront reconstruction on small dimension array gives accurate results agreeing with the Shack-Hartmann sensor measurements. The grating image is getting worse when the illuminating aperture has been reduced

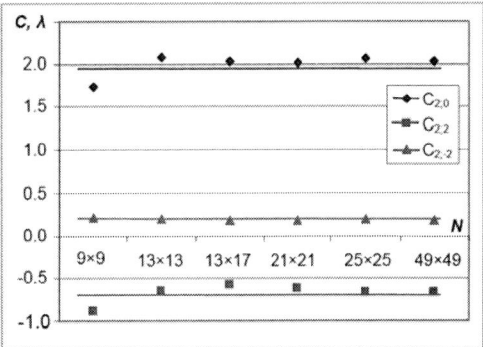

a

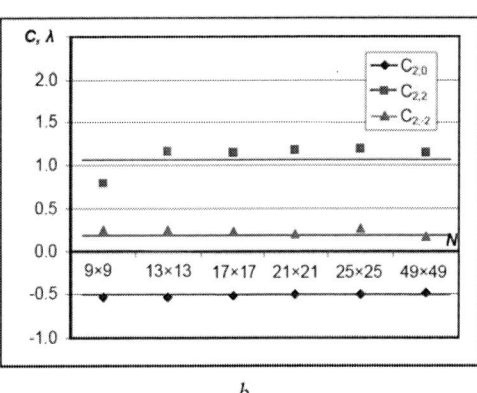

b

Fig. 4. Zernike coefficients obtained by the Talbot sensor with different sizes of the grating for two tested astigmatic wave. The lines show the reference values obtained by the Shack-Hartmann sensor.

closely to the reconstruction area. The image blurring leads to the reduction of the aberration estimation accuracy. It is shown that the aperture effect influence becomes stronger when aberrations of the tested wave front grow.

REFERENCES

[1] R. Tyson, *Principles of Adaptive Optics*, 3rd ed, CRC Press, 2010.

[2] A. Goloborodko, V. Grygoruk, V. Kurashov, D. Podanchuk, N. Sutyagina, "Determination of local surface defects using a Shack-Hartmann wavefront sensor," *Ukr. J. Phys.*, vol. 53, pp. 946-951, 2008.

[3] D. Podanchuk, V. Kurashov, V. Dan'ko, M. Kotov, O. Parkhomenko, "Wavefront sensor based on the Talbot effect," *Bulletin of the University of Kiev. Series: Radiophysics and electronics*, no. 16, pp.43-46, 2011.

[4] D. Podanchuk, V. Kurashov, A. Goloborodko, V. Dan'ko, M. Kotov, N. Goloborodko, "Wavefront sensor based on the Talbot effect with the precorrected holographic grating," *Applied Optics*, vol., pp. C125-C132, 2012.

[5] J.F. Barrera, R. Henao, Z. Jaroszewicz, A. Kolodziejczyk, "Talbot effect for periodical objects limited by finite apertures: a new interpretation", *Optik*, vol. 116, pp. 144-148, 2005.

Modern approach for estimating uncertainty of a precision optoelectronic phase noise measurement

P. Salzenstein[1], E. Pavlyuchenko[1,2]

[1]Centre National de la Recherche Scientifique (CNRS), Franche Comté Eléctronique Mécanique Thermique, Optique Sciences et Technologies, Besançon, France
[2]Donetsk University, Donestk, Ukraine

Abstract: Modern approach according to recent standards is used to determine uncertainty on phase noise for an optoelectronic measurement system. Deduced global uncertainty on the spectral density of phase noise is 1.6 dB.

Applications in metrology, fundamental physics or telecommunication require that the optoelectronic oscillators (OEO) deliver an ultra stable signal [1–4]. Unfortunately, the resonance frequency of these kind of OEO is not predictable as it depends inherently on geometry or physics of the resonator. The measurement of phase noise is then not obvious to implement. A new category of instruments specially dedicated to the measurement of phase noise has been developed to determine the phase noise for any delivered signal in X-band (8.2–12.4 GHz). Starting with a brief presentation of the main principle, we then detail how the uncertainty is determined by a modern method.

Oscillator frequency fluctuation is converted to phase frequency fluctuation through the delay line. Short-term instabilities of signal are characterized in terms of single sideband noise spectral density $S\varphi(f)$ in dB.rad²/Hz or rather £ (f) expressed in dBc/Hz. The transfer function of optical delay lineis $\left| H\varphi(jf) \right|^2 = 4.\sin^2(\pi f\tau)$. The offset frequency is noted f. τ is the delay related to optic fibers [5]. The phase noise in dBc/Hz of the OEO to be characterized is defined by Eq. (1) :

$$\text{£}(f) = [V^2 out(f)] / [2K^2\varphi. \left| H\varphi(jf) \right|^2 G^2 DC\ B] \qquad (1)$$

where Vout is the amplitude of the output signal, $\varphi(t)$ is the phase fluctuation, GDC is the gain of DC amplifier, $K\varphi$ depends on the mixer, B is the bandwidth.

Equation (1) shows that the sensitivity of the system depends directly on $K^2\varphi$ and $\left| H\varphi(jf) \right|$. In practice, we need a Fast Fourier Transform (FFT) analyzer to measure the spectral density of noise amplitude $V^2 out(f)/B$.

The background phase noise of the bench shown on Figure 1 is determined after averaging with cross-correlation method, when removing the optical delay lines. In this case, phase noise of the X-band synthesizer is rejected. The noise floor without optical transfer function is respectively then better than -150 and -170 dBc/Hz at 101 and 104 Hz from the 10 GHz carrier. When optical fiber is introduced noise floor of such a system is up to -90 and -170 dBc/Hz at 101 and 104 Hz from the 10 GHz carrier.

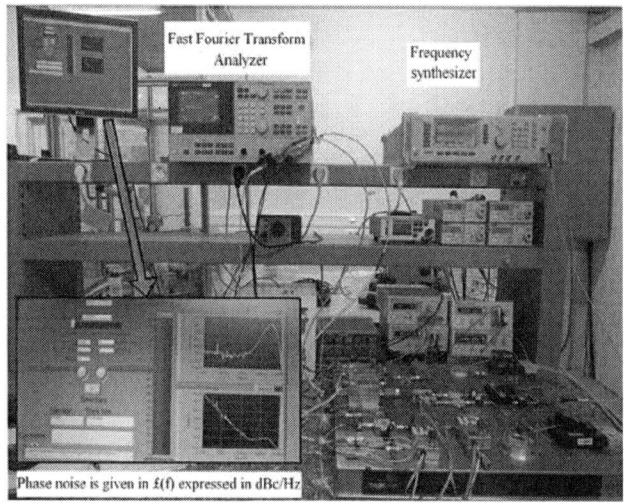

Fig. 1. High precision optoelectronic phase noise measurement system characterization of the phase noise delivered by OEOs.

In order to estimate the uncertainty, the use of the main guideline [6] delivered by the Bureau International des Poids et Mesures (BIPM) is in the guide "Evaluation of measurement data – Guide to the expression of uncertainty in measurement (GUM)". Ideas of error and uncertainty were mixed up until the GUM clarified their meanings [7]. Thanks to our experience on similar issues, we must take into account all contributions and assess their weight in the determination of the uncertainty [8,9]. However, taking into account certain characteristics due to optics, the result is significantly different from the conventional method [10,11].

According to the guideline, the uncertainty in the result of a measurement generally consists of several components which may be grouped into two categories according to the way in which their numerical value is estimated.

The first category is called "type A", is those which are evaluated by statistical methods such as reproducibility, repeatability, special consideration about Fast Fourier Transform analysis, and the experimental standard deviation. The components in category A are characterized by the estimated variances. Second family of uncertainties contributions is for those which are evaluated by other mean. They are called "type B" and due to various components and temperature control. Experience with or general knowledge of the behavior and properties of relevant materials and instruments, manufacturer's specifications, data provided in calibration and other certificates, uncertainties assigned to reference data taken from handbooks. The components in category B should be characterized by quantities which may be considered as approximations to the corresponding variances.

By taking into account the contributions mentioned above, the accuracy of determining the 1.6 dB overall uncertainty is improved.

REFERENCES

[1] Eliyahu D. et al., "Phase noise of a high performance OEO and an ultra low noise floor cross-correlation microwave photonic homodyne system," IEEE Freq. Contr. Symp., Honolulu, may 2008, pp. 811–814.

[2] Volyanskiy K., Salzenstein P., Tavernier H., Pogurmirskiy M., Chembo Y. K. and Larger L., "Compact Optoelectronic Microwave Oscillators using Ultra-High Q Whispering Gallery Mode Disk-Resonators and Phase Modulation," *Optics Express*, vol. 18, no. 21, pp. 22358–22363, 2010.

[3] Salzenstein P., Voloshinov V. B. and Trushin A. S., "Investigation in acousto-optic laser stabilization for crystal resonator based optoelectronic oscillators," *Optical Engineering*, vol. 52, no. 2, pp. 024603, 2013.

[4] Coillet A., Henriet H., Salzenstein P., Phan Huy K., Larger L. and Chembo Y. K., "Time-domain Dynamics and Stability Analysis of Optoelectronic Oscillators based on Whispering-Gallery Mode Resonators," *IEEE Journal of Selected Topics in Quantum Electronics*, vol. 19, no. 5, pp. 6000112, 2013.

[5] P. Salzenstein, J. Cussey, X. Jouvenceau, H. Tavernier, L. Larger, E. Rubiola, G. Sauvage, "Realization of a Phase Noise Measurement Bench Using Cross Correlation and Double Optical Delay Line," *Acta Physica Polonica A*, vol. 112, no. 5, pp. 1107-1111, 2007.

[6] GUM: Guide to the Expression of Uncertainty in Measurement, fundamental reference document (2008). http://www.bipm.org/en/publications/guides/gum.html

[7] R. Kacker, K. D. Sommer and R. Kessel, "Evolution of modern approaches to express uncertainty in measurement," *Metrologia*, vol. 44, no. 6, pp. 513–529, 2007.

[8] Salzenstein P., Kuna A., Sojdr L., Sthal F., Cholley N. and Lefebvre F., "Frequency stability measurements of ultra-stable BVA resonators and oscillators," *Electronics Letters*, vol. 46, no. 10, pp. 686–688, 2010.

[9] Salzenstein P., Cholley N., Kuna A., Abbé P., Lardet-Vieudrin F., Sojdr L. and Chauvin J., "Distributed amplified ultra-stable signal quartz oscillator based," *Measurement*, vol. 45, no. 7, pp. 1937–1939, 2012.

[10] Salzenstein P., Pavlyuchenko E., Hmima A., Cholley N., Zarubin M., Galliou S., Chembo Y. K. and Larger L., "Estimation of the uncertainty for a phase noise optoelectronic metrology system," *Physica Scripta*, vol. 2012, no. T149, pp. 014025, 2012.

[11] Won-Kyu Lee, Dai-Hyuk Yu, Chang Yong Park and Jongchul Mun, "The uncertainty associated with the weighted mean frequency of a phase-stabilized signal with white phase noise," *Metrologia*, vol. 47, no. 1, pp. 24–32, 2010.

978-1-4799-0019-0/13 $31.00 © 2013 IEEE

Pulsed laser/ion beam treatment of Ge/Si and Ge/Al$_2$O$_3$ thin film structures

R.I. Batalov[1*], R.M. Bayazitov[1], H.A. Novikov[1], V.A. Shustov[1], I.A. Faizrakhmanov[1], N.M. Lyadov[1],
K.N. Galkin[2], P.I. Gaiduk[3], G.D. Ivlev[3], S.L. Prakopyeu[3]

[1]Kazan Physical-Technical Institute of RAS, Kazan, Russia
*E-mail: batalov@kfti.knc.ru
[2]Institute of Automation and Control Processes of RAS, Vladivostok, Russia
[3]Belarusian State University, Minsk, Belarus

Abstract-Vacuum deposition of Ge thin films onto Si and Al$_2$O$_3$ substrates by magnetron and ion-beam assisted sputtering was studied. During deposition sputtering time and substrate temperature were varied. Nanosecond pulsed annealing of deposited Ge films by powerful laser or ion beams was performed. The dependence of structural and optical properties of Ge/Si and Ge/Al$_2$O$_3$ films on parameters of pulsed treatments was investigated. Optimum parameters for deposition and pulsed treatments resulted in light emitting layers are determined.

I. INTRODUCTION

For the last 20 years there has been significant interest to thin film Si-based structures emitting light in visible and near infrared region. Such interest is due to the possibility to create LEDs and lasers by means of highly developed planar CMOS technology and their integration with micro- and optoelectronic devices and fiber optic lines. Since the wavelength range of 1.2-1.6 μm is of the most interest for optoelectronics (transparency region for Si and SiO$_2$), currently there are four basic directions in creation of Si-based light emitting structures: (1) doping of Si by rare earth Er ions ($^4I_{13/2} \rightarrow {}^4I_{15/2}$ optical transition at λ=1.54 μm) [1], (2) synthesis of direct bandgap silicides such as β-FeSi$_2$ (E_g = 0.85 eV) [2], (3) defect engineering including introduction of point (W- and G-centers, λ=1.2-1.3 μm) and extended (dislocations, D1 line, λ=1.53 μm) defects [3,4] and (4) growth of SiGe alloys, Ge quantum dots in Si and continuous Ge films on Si [5-7]. The most significant progress in the creation of effective light emitting structures has been reached for Ge/Si layers within the last method mentioned above where laser generation under optical and electrical pumping was obtained recently at room temperature [7]. The main methods for growth of such monocrystalline Ge films are UHV molecular-beam epitaxy and chemical-vapor deposition. In our work we have used alternative deposition methods that is magnetron sputtering and ion-beam assisted deposition in combination with subsequent pulsed annealing by powerful nanosecond laser or ion beams. Besides Si we have used also monocrystalline sapphire (Al$_2$O$_3$) substrates in order to exclude Si-Ge intermixing.

II. EXPERIMENTAL

Double-side polished Cz-Si wafers (15x15x0.4 mm) with (100) and (111) orientation and resistivity of 5-10 Ohm cm and also C-plane Al$_2$O$_3$ wafers (20x1 mm) were used as substrates for Ge deposition. Vacuum deposition of Ge films onto Si and Al$_2$O$_3$ substrates was performed by unbalanced DC magnetron sputtering or ion-beam assisted deposition. Monocrystalline p-Ge wafer (40-mm in dia.) was used as a sputtered target. The distance between Ge target and Si substrate was 30 mm. The base pressure of the chamber was evacuated to be less than 5×10^{-5} Torr with diffusion pump. During deposition the substrate temperature was 20-200 ^{0}C and deposition time was 3-20 min. The thickness of deposited Ge films was 90-800 nm. Pulsed annealing of deposited Ge layers was carried out by ruby laser pulses (λ = 690 nm, τ = 80 ns) in open air (Pulsed laser annealing - PLA) or by high current (up to 150 A/cm^2) nanosecond ion beams (C$^+$, H$^+$, E = 300 keV, τ = 50 ns) in vacuum chamber of ion accelerator TEMP (Pulsed ion-beam treatment - PIBT). During PLA and PIBT we varied pulse energy density (W = 0.3-1.5 J/cm^2) and number of pulses (N = 1-10). The dose of carbon and hydrogen ions implanted into Ge films during PIBT did not exceed $\Phi = 3 \times 10^{13}$ cm^{-2} per pulse. The separate samples were thermally annealed in quartz furnace (T = 400-800 ^{0}C/30 min) in nitrogen ambient.

The structural measurements of deposited Ge films before and after pulsed annealing were carried out by several methods such as atomic-force microscopy (AFM), scanning and transmission electron microscopy (SEM/TEM), Rutherford backscattering (RBS) and X-ray diffraction (XRD) in grazing incidence. Optical properties of irradiated Ge/Si layers were studied by optical spectroscopy (transmission/reflection) at 300 K and photoluminescence (PL) at 77 K.

III. RESULTS AND DISCUSSION

AFM measurements of surface morphology for as deposited Ge films on Si showed that the sizes of Ge islands varied in the range from 40 to 100 nm and root-mean square (RMS) roughness was less than 1.6 nm. XRD measurements showed an amorphous film structure. PLA of Ge/Si films with increasing pulse energy density (W = 0.3-1.5 J/cm^2) resulted in reduction of Ge amorphous fraction, increase of both Ge islands (up to 0.5 μm) and RMS value (up to 25 nm). RBS data showed that Ge films lost significant part of its thickness (up to 2/3) as a result of PLA. This effect can be related to the deposition of maximum energy of laser radiation near the film surface and its oxidation to GeO and evaporation during the heating in open air.

In contrast to PLA annealing by powerful ion beams (PIBT) allows to perform recrystallization of more thick films (up to 1 μm) and to reduce Ge loss due to deposition of maximum ion

978-1-4799-0019-0/13 $31.00 © 2013 IEEE

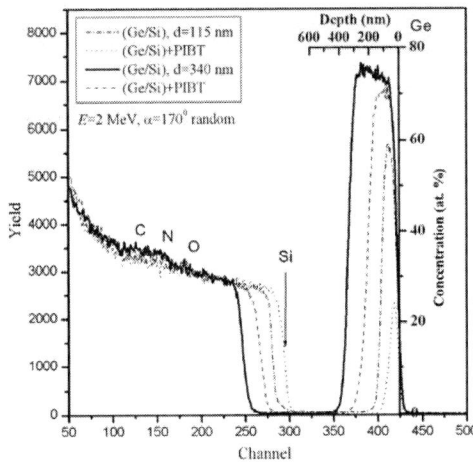

Fig. 1. Random RBS spectra for two Ge films on Si before and after PIBT ($W = 1.0$ J/cm^2).

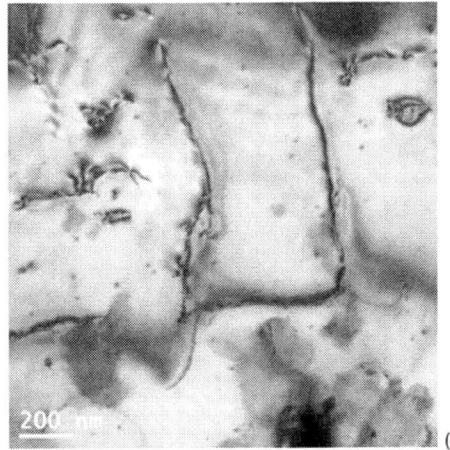

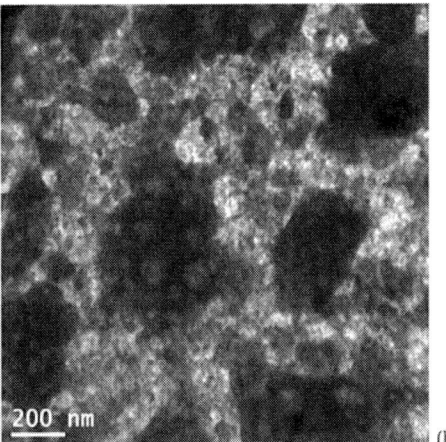

Fig. 2. Plan-view TEM images for two Ge films after PIBT ($W = 1.0$ J/cm^2) as described in Fig.1.

energy in deeper layers (ion projection range $R_p \sim 0.7$ μm) and film irradiation in oxygen-free ambient (vacuum). The results of RBS measurements for two Ge films of different initial thickness on Si (Fig. 1) showed smaller erosion degree after PIBT. Moreover one can see that the composition of the thick Ge film ($Si_{25}Ge_{75}$) changes slightly after PIBT ($Si_{30}Ge_{70}$) compared to the thin one which composition decreases approx. 3 times ($Si_{75}Ge_{25}$). XRD measurements demonstrated the efficient film recrystallization leading to polycrystalline or epitaxial layers. The formation of monocrystalline $Si_{75}Ge_{25}$ alloy is due to full melting of the thin Ge film and underlying Si layers and their mixing in the liquid state. In case of thick Ge layers incomplete melting takes place, liquid/solid interface does not reach Si substrate, no mixing occurs and the Ge film becomes polycrystalline.

Plan-view TEM measurements of as-deposited Ge films demonstrated the formation of Ge particles with sizes of 30-50 nm in amorphous Ge matrix (amorphous halo in microdiffraction). PIBT resulted in creation of monocrystalline SiGe layers with threading dislocations (Fig.2a) for thin films or polycrystalline dislocation-free Ge layers with large Ge particles of 0.4-0.7 μm (Fig. 2b).

Optical measurements (300 K) of deposited Ge/Si films in transmission mode in the wavelength range of 900-2500 nm showed the decrease of transmission level and red shift of absorption edge from 1000 to 1100 nm with increase of Ge film thickness from 100 to 800 nm. Transmission measurements of Ge thin films on transparent Al_2O_3 wafers showed the real absorption edge of Ge which is moved from 800 nm to 600 nm as a result of thermal annealing (450 ^{0}C/30min).

PL spectra measured at 77 K for thin Ge/Si films after PIBT and additional thermal annealing (800 ^{0}C/20 min) demonstrated light emission in the range of 1200-1700 nm with maximum at 1550 nm. Asymmetrical line shape and film microstructure (Fig. 2a) indicate the radiative recombination of carriers in SiGe alloy and at dislocations. The absence of PL signal from thick Ge layers is probably due to its polycrystal-

line structure and increased nonradiative recombination at grain boundaries.

ACKNOWLEDGEMENT

This work was supported by RFBR grant No 13-02-00348.

REFERENCES

[1] A.J. Kenyon, "Erbium in silicon," *Semicond. Sci. Technol.* vol. 20, pp. R65–R84, 2005.

[2] M. Suzuno, T. Koizumi, T. Suemasu, "p-Si/β-FeSi$_2$/n-Si double-heterostructure light-emitting diodes achieving 1.6 μm electroluminescence at room temperature," *Appl. Phys. Lett.* vol. 94, 213509 (3 pp), 2009.

[3] J. Bao, M. Tabbal, T. Kim, et al., "Point defect engineered Si sub-bandgap light-emitting diode," *Optics Express* vol. 15, pp. 6727-6733, 2007.

[4] L. Xiang, D. Li, L. Jin, D. Yang, "Dislocation-related electroluminescence of silicon after electron radiation," *Solid State Commun.* vol. 152, pp.1956-1959, 2012.

[5] L. Vescan, T. Stoica, "Room-temperature SiGe light-emitting diodes," *J. Luminesc.* vol. 80, pp. 485-489, 1999.

[6] Z.F. Krasilnik, A.V. Novikov, D.N. Lobanov et al., "SiGe nanostrucrures with self-assembled islands for Si-based optoelectronics," *Semicond. Sci. Technol.* vol. 26, 014029 (5pp), 2011.

[7] J. Liu, L.C. Kimerling, J. Michel, "Monolithic Ge-on-Si lasers for large-scale electronic-photonic integration," *Semicond. Sci. Technol.* vol. 27, 094006 (13pp), 2012

Some features of estimation of the diffusion length of minority carriers in cathodoluminescence microscopy

Yu.E. Gagarin[1], N.N. Mikheev[2], N.A. Nikiforova[3], M.A. Stepovich[3]

[1]Bauman Moscow State Technical University, Kaluga Branch, Kaluga, Russia
[2]Research Center for Space Materials Science, Shubnikov Institute of Crystallography, Russian Academy of Sciences, Kaluga, Russia
[3]Tsiolkovsky Kaluga State University, Kaluga, Russia

Abstract: Some possibilities for cathodoluminescence identification of electrophysical parameters of homogeneous direct-gap semiconductor materials are examined by mathematical modeling methods. Mathematical model of dependences of the intensity of monochromatic cathodoluminescence on the electron beam energy due to both linear and quadratic recombination of minority charge carriers (MCC) proposed by our group was used. It is shown how the proposed method allows for interval estimation of the diffusion length of MCC.

Earlier studies [1, 2] have shown that the analysis of cathodoluminescence (CL) emission allows us to study the recombination processes in semiconductors and, basing on the analysis of the dependence of monochromatic CL intensity I on the electron beam energy E_0 of scanning electron microscope, evaluate the electrophysical parameters of studying structures.

The dependence of the intensity of monochromatic CL I on the electron beam energy E_0 to determine the electrophysical parameters of direct-gap semiconductors is proposed to use in classic works [3, 4]. For the case of linear recombination of nonequilibrium minority carriers (NMC), capabilities of this technique were previously studied in details experimentally and theoretically (see, for example, [1, 2, 5]), while the case of quadratic recombination of NMC and the general case (the presence of both linear and quadratic recombination) is not studied well.

We have considered the possibility and some features of the method of confluence analysis (MCA) to obtain estimates of the diffusion length of the MCC L using the dependence the CL intensity on the energy of the electron beam due to both linear [6] and quadratic [7] recombination of the MCC that has not previously been done.

Mathematical modeling of the dependence of the CL intensity on the energy of a wide beam of electrons was carried out for the target parameters which are characteristic for direct-gap semiconductor optoelectronic materials, for which the best agreement between theory and experiment was observed while using the linear and quadratic recombination simultaneously.

Estimation model of the diffusion length of semiconductor materials based on the CL intensity of the electron beam energy in the MCA was as follows:

$$\begin{cases} E_{0i} = E^{(i)}{}_0 + \varepsilon_i, \\ I_i = I(E^{(i)}{}_0, L) + \delta_i. \end{cases}$$

Here, $E^{(i)}{}_0$, $I(E^{(i)}{}_0, L)$ are unknown true values of primary electron beam energy and intensity of the CL, n is the number of measurements, ε_i and δ_i are the errors in magnitude of E_{0i}, I_i, normally distributed random variables, of zero means, variances $\sigma(E_{0i})$, $\sigma(I_i)$, and zero correlation coefficient, L is the diffusion length of MCC.

Required estimates of diffusion length of MCC L for the functional dependence of CL intensity of electron beam energy $I(E_0^{(i)}, L)$ were found by minimizing the functional of following form [2]:

$$F = \frac{1}{2} \sum_{i=1}^{n} \left[\frac{\left(E_{0i} - E_0^{(i)}\right)^2}{\sigma^2(E_{0i})} + \frac{\left(I_i - I\left(E_0^{(i)}, L\right)\right)^2}{\sigma^2(I_i)} \right].$$

Parameter estimation for L was found by minimizing the functional F [2]:

$$\frac{\partial F}{\partial L} = 0, \qquad \frac{\partial F}{\partial E_0^{(i)}} = 0, \qquad i = \overline{1, n}.$$

Here, F is the functional to be minimized.

Using of considered method allowed mathematical modeling of estimation of the diffusion length of the MCC L for some well-studied semiconductor materials [1, 2, 7].

In case of a wide electron beam and a low level of generation, functional dependence linking the CL intensity I with energy E_0 had following form [1, 2]:

978-1-4799-0019-0/13 $31.00 © 2013 IEEE

$$I(E_0) \approx \left\{1 + 0.155\left[1 - \exp(-\frac{z_c}{L})\right]\right\} \int\limits_0^\infty dz_0 \int\limits_{l_s}^\infty \Delta\rho(z, z_0) \exp[-\alpha z] dz.$$

Here, z is a coordinate measured from the surface into a semiconductor, z_c is a value of the barycenter of the region of energy scatter of the electron beam, $\Delta\rho(z, z_0)$ is the distribution over the depth z of minority carriers generated in the semiconductor after their diffusion from a planar source located at depth z_0, l_s is the depth of the surface region with depletion of majority carriers, α is the absorption coefficient of radiation.

The distribution function of MCC $\Delta\rho(z, z_0)$ for the model was defined by equation from the paper [8].

For quantitative description of monochromatic CL intensity for wide electron beam case and quadratic recombination only following equation, which was proposed at the paper [7], was used:

$$I(E_0) \approx \left\{\frac{G\tau}{\sigma L}\left[1 - \frac{S-1}{S+1}\exp(-\frac{2z_c}{L})\right]\right\}^2 \Phi(z_c, L).$$

Here, G is the number of MCC of the same polarity, excited by an electron beam per unit time, excluding the surface recombination, S is the rate of surface recombination, τ is the lifetime of MCC, L is the diffusion length of MCC, σ is square of a flat MCC source, which is located parallel to the surface at depth z_0, z_c is a value of coordinate of MCC generation area barycenter, $\Phi(z_c, L)$ is a correction function takes into account the decrease of CL intensity due to the reduction of the spectral density of radiation in the generation area of MCC [9],

$$z_c = \frac{1}{G_0}\int\limits_0^\infty z\rho(z)dz, \quad G_0 = \int\limits_0^\infty \rho(z)dz,$$

$$\rho(z) = \int\limits_{-\infty}^\infty \int\limits_{-\infty}^\infty \rho(x, y, z)dxdy.$$

Distribution $\rho(x, y, z)$ were defined according to as well [7].

Analysis of the results showed that using MCA to determine the diffusion length of MCC in CL microscopy allows to obtain more accurate estimates of the electrophysical parameters.

Investigations were carried out with partial financial support of Russian Ministry of Science and Education (Project № 1.6107.2011), and the Russian Foundation for Basic Research and the Government of the Kaluga region (project № 12-02-97519).

REFERENCES

[1] N.N. Mikheev, A.N. Polyakov, M.A. Stepovich., "On applicability of confluence analysis in cathodoluminescence microscopy for interval estimation of diffusion lengths of minority carriers and depths of subsurface regions with depletion of majority carriers," *Journal of Surface Investigation. X-ray, Synchrotron and Neutron Techniques*, vol. 3, no. 5, pp. 820-825, 2009.

[2] Yu.E. Gagarin, M.A. Stepovich, "Some possibilities of the use of confluence analysis for an interval parameter estimation of semiconductors in a cathodoluminescent microscopy," *Proc. SPIE*, vol. 5398, pp. 179-185, 2004.

[3] D.B. Wittry, D.F. Kyser, "Measurement of diffusion lengths in direct-gap semiconductors by electron-beam excitation," *J. Appl. Phys.*, vol. 38, no. 1, pp. 375-382, 1967.

[4] D.F. Kyser, D.B. Wittry, "Spatial distribution of excess carriers in electron-beam excited semiconductors," *Proc. IEEE.*, vol. 55, no. 5, pp. 733-734, 1967.

[5] A.N. Polyakov, M. Noltemeyer, T. Hempel, J. Christen, M.A. Stepovich, "Experimental cathodoluminescence studies of exciton transport in gallium nitride," *Bulletin of the Russian Academy of Sciences. Physics*, vol. 76, no. 9, pp. 970-973, 2012.

[6] N.N. Mikheev, V.I. Petrov, M.A. Stepovich, "Independent source modeling of excess carriers in calculation of cathodoluminescence in semiconductors," *Bulletin of the Russian Academy of Sciences. Physics*, vol. 56, no. 3, pp. 426-430, 1992.

[7] N.N. Mikheev, I.M. Nikonorov, V.I. Petrov, M.A. Stepovich, V. Al Shaer, "Cathodoluminescence of semiconductors in conditions of a high excitation level," *Bulletin of the Academy of Sciences of the USSR. Physical Series*, vol. 54, no. 2, pp. 75-78, 1990.

[8] N.N. Mikheev, M.A. Stepovich, E.V. Shirokova, "Allowing for matrix effects in local electron_probe analysis using a new model of the distribution function for the depth of characteristic X-ray radiation," *Bulletin of the Russian Academy of Sciences. Physics*, vol. 76, no. 9, pp. 974-977, 2012.

[9] Yu.E. Gagarin, N.N. Mikheev, A.N. Polyakov, M.A. Stepovich, "On the choice of initial approximation in the problem of parameter identification of direct-gap semiconductors by cathodoluminescence microscopy," *Journal of Surface Investigation. X-ray, Synchrotron and Neutron Techniques*, vol. 2, no. 5, pp. 709-715, 2008.

The possibilities of the colorimetric method for measuring of wavelength's distribution of light radiation in standard and the RAW formats

A.V.Kraiski[1], T.V.Mironova[1], T.T.Sultanov[1], V.A.Postnikov[2]

[1]P.N.Lebedev Physical Institute RAS, Moscow, Russia

[2]Institute of Physico-chemical medicine, Moscow, Russia

Abstract: We compare results of the colorimetric method use for measurements of the wavelength distribution from digital images in standard formats (BMP, JPEG, TIFF) and in RAW-format. It is shown that with the use of the last one the operating range is several times larger (up to 455-625 nm).

When one uses the holographic sensors, the space inhomogeneity of processes leads to the problem of the registration of the reflected wavelength distribution along the sensor surface. The sensor is a multilayer periodic structure [1-3]. Therefore, its reflection line is of narrow spectral width. Basing on this, for such problems we proposed and developed a colorimetric method of measuring the wavelength distribution of the radiation reflected by the sensor from its digital image [4-5]. In the experimental implementation of the method, at the first stage [5] we focused on common simple cameras and used standard digital image formats (BMP, JPEG, TIFF). It appeared that in all cameras (we tested about 10) not the whole spectral range is suitable for measuring the wavelength. The point is that for this method to operate, it necessarily requires a monotonic dependence of the hue on the registered wavelength [5]. In digital images registered in standard formats, for a significant wavelength range in the green part of the spectrum the hue does not change with the change of the wavelength. It is shown in Fig. 1c for the camera Canon EOS 10D. This is due to the lack of overlapping of the spectral sensitivities of the red and blue sensors in these formats (Fig.1b). As a result, the spectral range is divided into two separated areas: blue-green and yellow-red. In [5] the experiments were done in the yellow-red part of the spectrum in the range of 550 - 605 nm.

The spectral sensitivities of the human eye are shown in Fig.2 [6]. Those of the camera sensors in the RAW-format (for the same camera Canon EOS 10D) are given in Fig.1a. It can be seen that the overlap of the spectral sensitivities of the red and blue sensors exist although it is smaller than for the eye. In this case the dependence of the hue on the wavelength becomes monotonic. The operating range is extended to almost the entire range of the matrix sensitivity and is from 455 to 625 nm (Fig.1c) [7].

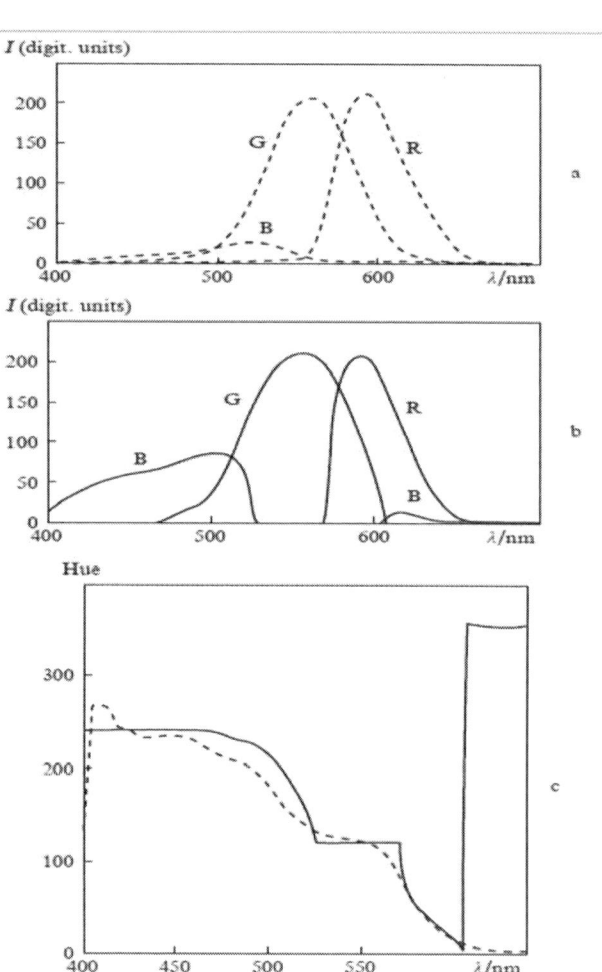

Fig. 1. Color components of the image in the continuous spectrum (R – red, G – green, B – blue) without processing by the camera processor or visualization software (a) and after the camera processor (b), as well as (c) the relation between the wavelength and the hue function (Hue) for processed (solid line) and unprocessed (dashed line) images. Hue = 0 corresponds to the red, Hue = 120 to the green, and Hue = 240 to the blue colors.

978-1-4799-0019-0/13 $31.00 © 2013 IEEE

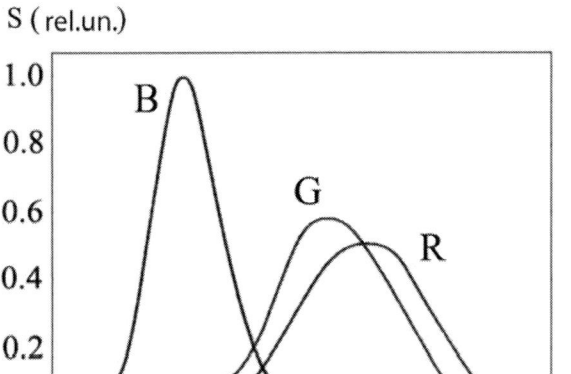

Fig. 2. The spectral sensitivity of the human eye: signals of red (R), green (G), and blue (B) receptors [6].

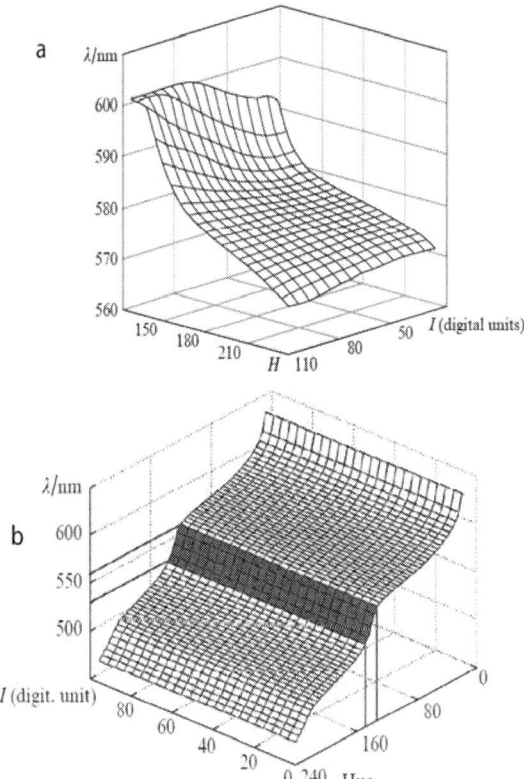

Fig. 3. Characteristic surface for standard format for the Sony F717 (a) and for RAW-format for the Cannon EOS 10D (b).

The camera Sony F717 was used when working with standard formats and Cannon EOS 10D with the RAW-format. The spectral sensitivities of these cameras in the standard formats are almost the same, only the operating range borders are slightly different [5].

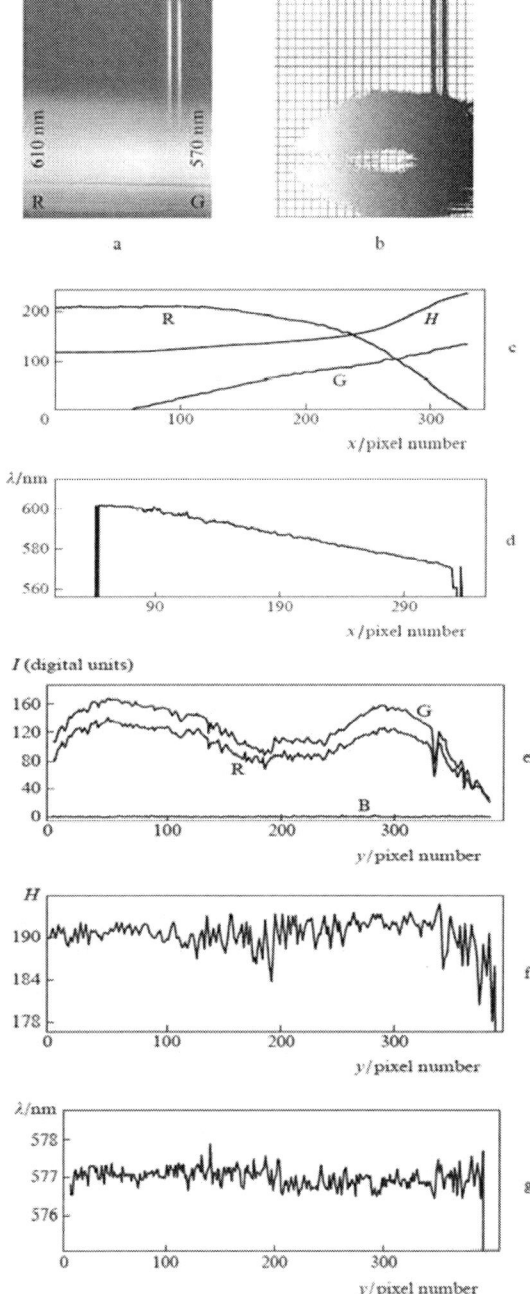

Fig.4. Illustration for an operating test of the method. Shot of spectrum fragment (a); the map of wavelength distribution over the image (b); the distribution of signals in red (R) and green (G) channels and colour hue (H) in a horizontal cross section of the shot (c); the horizontal cross section of the wavelength map (d); the distribution of signal in colour channels (e), colour hue (f), and calculated wavelengths (g,) in a vertical cross section of the shot along the mercury doublet lines 577 nm (e, f, g). A rectangular hatch on the wavelength map marks the domains in which the signals are beyond the working range and, hence, are excluded from calculation

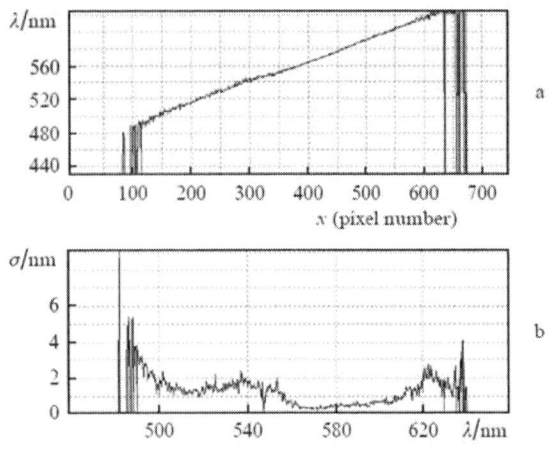

Fig. 5. Calculated distribution of wavelengths over the frame section in the direction of dispersion (a) and the corresponding dependence of the mean square deviation of the wavelength (b).

To perform the measurements, first of all one has to calibrate the camera in order to obtain the possibility of determining the wavelength of the incident radiation from the color of the pixel [5]. It appeared that in the standard formats the hue depends not only on the wavelength but also on the intensity of the incident light. In the RAW-format the hue practically does not depend on the intensity. When calibrating the camera, a characteristic function was built, which uniquely determines the radiation wavelength from the intensity and the hue data. For calibrating we used a set of images of incandescent lamp spectrum with reference lines from the spectrum of mercury taken in a wide range of exposure. The characteristic functions obtained are depicted in Fig.3a for the standard formats [5] and in Fig.3b for the RAW-format [7].

In Fig.4, there are: the image of a part of spectrum of the incandescent lamp with the yellow mercury doublet of 577 and 579 nm, the map of wavelengths for this image, the horizontal and vertical (along 577 nm line) sections of sensor signals, the hue functions calculated basing on these signals and the corresponding sections of the wavelength map. The mean square deviation of the wavelength is up to 0.16 nm.

In Fig.5a, there are depicted: the horizontal section of the wavelength map in the incandescent lamp spectrum obtained from the images in the RAW-format and the mean square deviation of the wavelength as a function of the wavelength [7].

The mean square deviation of the wavelength amounts to 0.3 – 3 nm. The value of the error depends mainly on two factors. The first one is the brightness of the image fragment: in case of low brightness of the image the noise strongly grows and the accuracy of the wavelength determination becomes worse. The second factor is the wavelength and the slope of the calibration surface at the appropriate wavelength. In the green region (530 – 560 nm) a small

change of hue (15 units) corresponds to the difference of 30 nm in the wavelength scale (Fig. 3). The mean square deviation in this range even at sufficient intensity of radiation amounts to 1.0 – 2.0 nm. In the yellow spectral region the calibration surface is sloping milder, and in the region 570 – 580 nm the hue changes by almost 40 units. In this region of the spectrum the wavelength can be determined with higher accuracy, and the mean square deviation amounts to 0.3 – 0.5 nm. We used the described method for measuring the distribution of the mean wavelength of reflected radiation over the surface of a hologram. Thus one can monitor different parameters of holographic sensors: the degree of swelling, homogeneity of the hydrogel emulsion, rate of reaction on the change of the solution composition, etc.

Acknowledgements. The work was partially supported by a grant within the program of fundamental research 'Fundamental Sciences to Medicine'.

REFERENCES

[1] Lowe Ch.R., Millington R.B., Blyth J., Mayes J.E., Patent USA №5989923, data publ. 1999-11-23.

[2] Lowe Ch.R. Holographic sensors. Handbook of biosensors and biochips (Eds R.S.Marks, D.C.Cullen, I.Karube, C.R.Lowe & H.Weetall) Wiley interscience, Sussex, UK, pp.587-596.

[3] A.V. Kraiskii, V.A. Postnikov, T.T. Sultanov, A.V. Khamidulin. Holographic sensors for diagnostics of solution components., Quantum Electronics 40 (2) 178 - 182 (2010)

[4] Kraiskii A.V., Mironova T.V. et al., *Sposob izmereniya dliny volny...*, Pat. of RF No. 2390738, prior. 21.05.2008.

[5] A.V. Kraiskii, T.V. Mironova, T.T. Sultanov. Measurement of the surface wavelength distribution of narrow-band radiation by a colorimetric method. *Quantum Electronics* 40 (7) 652 – 658 (2010)

[6] Judd D.B., Wyszecki G. *Color in Business, Science and Industry* (New York: Wiley, 1975; Moscow: Mir, 1978).

[7] A.V. Kraiskii, T.V. Mironova, T.T. Sultanov. Narrow-band radiation wavelength measurement by processing digital photographs in RAW format. *Quantum Electronics* 42 (12), 1137-1139, (2012)

Peculiarities of interferometric studies of surfaces with phase fluctuations of various statistical characteristics

O.V. Gnatovskyi, L.A.Derzhypolska, A.M. Negriiko

Institute of Physics of National Academy of Sciences of Ukraine, 46 Prospect Nauki, 03028 Kiev, Ukraine

Abstract. Considered are the problems of transfer of optical information within holographic interferometry context under random phase distortions in an optical tract. Predicted is a shift of interference fringes in holographic interferometry experiment, which is connected solely to statistical peculiarities of roughness of object surface.

I. INTRODUCTION

Within the new branch of interferometry of randomly inhomogeneous media investigated are the interconnections between the contrast of interference fringes of mean intensity observed in scattered light and the statistical characteristics of inhomogeneities of scattering medium. The two optical signals formed after scattering from the surface of the object in the two comparable states subject to phase changes passing through some medium. These changes described by statistical distribution over the cross-section of the beam cause the decorrelation of phase of the signals. This leads to formation of impaired interference pattern under interference comparison of the signals.

In this area many times posed was the problem of establishing the connection between the contrast of interference fringes observed in scattered field and the characteristics of scattering object or (e.i. surface roughness) and developing corresponding measurement techniques. This practically important task had not been solved yet. Suggested model predicts the appearance of non-interferometric statistical shift in interference pattern in holographic interferometry experiments.

II. THEORETICAL DESCRIPTION OF STATISTICAL PHASE SHIFT IN HOLOGRAPHIC INTERFEROMETRY EXPERIMENTS

In the works [1,2] considered in details is the scheme of holographic interferometer with focused image with unpredictable phase noise in the process of transmission of spatial optical information. Within this consideration obtained is the expression of intensity of interference pattern formed under phase fluctuations in optical tract:

$$I \sim [1 + cos(\alpha x' - \psi)] = 1 + I_\psi \qquad (1)$$

where $\psi(x', y')$ is phase fluctuation in the point x', y', $\alpha x'$ is the interferometric phase shift caused by test

deformation of the investigated object (tilt by small angle α with respect to y' axis.

As soon as ψ is a stochastic variable hence $\cos(\alpha x' - \psi)$ gets different value for the same $\alpha x'$ in different points of interference pattern. To determine the contrast of interference pattern it is necessary to define a maximum and minimum values of intensity, for this we should average I_ψ over all the ψ values occurred in all the points $\alpha x'$. Then the expression for average intensity will be the following:

$$\tilde{I}_\psi = M \cos(\alpha x' - \psi) = \text{Re} \int_{-\infty}^{+\infty} \rho(\psi) \exp[i(\alpha x' - \psi)] d\psi =$$

$$= \text{Re} \left[\exp[i\alpha x] \int_{-\infty}^{+\infty} \rho(\psi) \exp[i\psi] d\psi \right] =$$

$$= \text{Re} \left[(\cos(\alpha x') + i\sin(\alpha x')) \left(\int_{-\infty}^{+\infty} \rho(\psi) \cos(\psi) d\psi + i \int_{-\infty}^{+\infty} \rho(\psi) \sin(\psi) d\psi \right) \right] =$$

$$= \cos(\alpha x') \int_{-\infty}^{+\infty} \rho(\psi) \cos(\psi) d\psi - \sin(\alpha x') \int_{-\infty}^{+\infty} \rho(\psi) \sin(\psi) d\psi$$

$$(2)$$

Where taken is [3] that

$$A \cos(x) + B \sin(x) = \sqrt{A^2 + B^2} \cos(x + \varphi_0) \text{ and } \varphi_0 \text{ is:}$$

$$\varphi_0 = arctg \frac{\int_{-\infty}^{+\infty} \rho(\psi) \sin(\psi) d\psi}{\int_{-\infty}^{+\infty} \rho(\psi) \cos(\psi) d\psi} \qquad (3)$$

Here one should note that to the interferometric phase $\alpha x'$ added is the pure statistical phase shift φ_0 and hence the location of interference maxima and minima are shifted relative to the location defined by interferometric phase $\alpha x'$. As seen from expression (3) φ_0 is defined by the statistics of phase noise thereby detailed consideration of this phenomenon is possible with known particular distribution $\rho(\psi)$.

978-1-4799-0019-0/13 $31.00 © 2013 IEEE

III. PECULIARITIES OF INTERFEROMETRIC STUDIES OF SURFACES WITH PHASE FLUCTUATION OF VARIOUS STATISTICAL CHARACTERISTICS

Further analyzed are the particular cases of statistical distribution. Performed was the simulation of interferometric experimets with the two comparable states of the surface of virtual objects. For each point of the surface given is the phase value including interferomertic phase and statistical phase with particular distribution representing the microrelief of the surface. Obtained in such a way arrays of numbers are further processed in computational environment.

Gaussian (normal) distribution. Normal distribution is the most popular in natural phenomena. Distribution density is defined as:

$$\rho(\psi) = \frac{1}{\sigma\sqrt{2\pi}} \exp\left[-\frac{\psi^2}{2\sigma^2}\right] \tag{4}$$

where σ is square dispersion.

Taking expressions (3) and (4) obtain:

$$\varphi_0 = \arccos\left(sign\left(\int_{-\infty}^{+\infty} \rho(\psi)\cos(\psi)\,d\psi\right)\right) = \tag{5}$$

$$= \arccos\left(sign\left(\exp\left[-\frac{\sigma^2}{2}\right]\right)\right) = 0$$

From here it appears that for normal distribution there is no statistical phase shift as soon as $\varphi_0 = 0$ and hence the statistical shift in interference pattern never occurs. This conclusion explains why this phenomenon have never been noticed in earlier investigations. Because mostly used distribution for description of scattering media is Gaussian [4, 5].

Uniform distribution. Distribution density for uniform distribution is:

$$\rho(\psi) = \begin{cases} \psi > a \Rightarrow \rho(\psi) = 0 \\ \psi < a \Rightarrow \rho(\psi) = \frac{1}{2a} \end{cases} \tag{6}$$

where a is halfwidth of distribution.

Taking expressions (3) and (6) obtain:

$$\varphi_0 = \arccos\left(sign\left(\int_{-\infty}^{+\infty} \rho(\psi)\cos(\psi)\,d\psi\right)\right) = \tag{7}$$

$$= \arccos(sign(\sin c(a))) = \begin{cases} \sin c(a) < 0 \Rightarrow \varphi_0 = 0 \\ \sin c(a) > 0 \Rightarrow \varphi_0 = \pi \end{cases}$$

From here appears that for uniform distribution statistical phase shift gets two possible values: $\varphi_0 = 0$ for $\sin c(a) > 0$ and $\varphi_0 = \pi$ for $\sin c(a) < 0$.

These results have to be taken into account when measuring deformation. As shown on Fig.1 it may occur for some object that has a statistical phase changes described by uniform distribution but with different dispersion in different parts of surface.

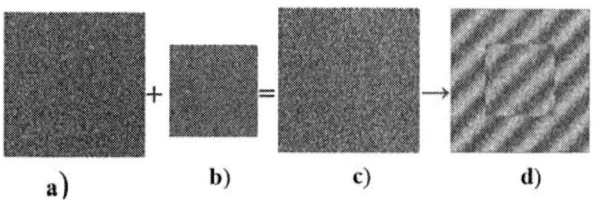

Fig.1. Diffusors obtained with uniform distribution with the following parameters: **a)** a=0,8π **b)** a=3π/2, **c)** resultant diffuser and **d)** interferogram obtained from simulation (at the edge a=0,8π, in the center a=3π/2)

This may yield a hop of interference fringes. Considering a graph of visibility [2] Fig.2 ($V = |\text{sinc}(a)|$) it is understood that the in the region of second maximum (a=1,5π) sinc(a)<0 and $\varphi_0 = \pi$. Which is counterphase to the region of the first maximum, where a=0,8π belongs and sinc(a)>0 and $\varphi_0 = 0$.

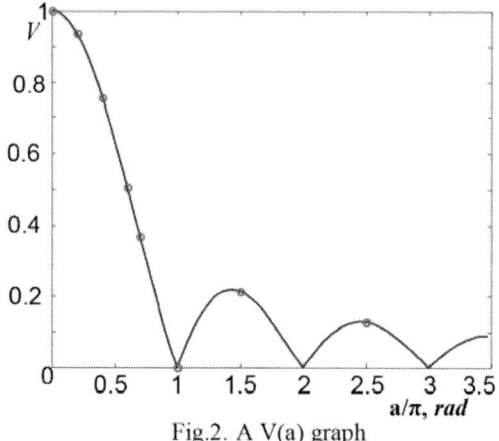

Fig.2. A V(a) graph

Thus, having the same type of statistical distribution of phase distortion but with different dispersion in different areas of the medium, a shift or a hop of interference fringes may be observed. On Fig.3 shown are some other situations. If parameter **a** gets a value when interference contrast drops to zero then interference fringes are not observed at all giving false information about absence of any deformation of the object.

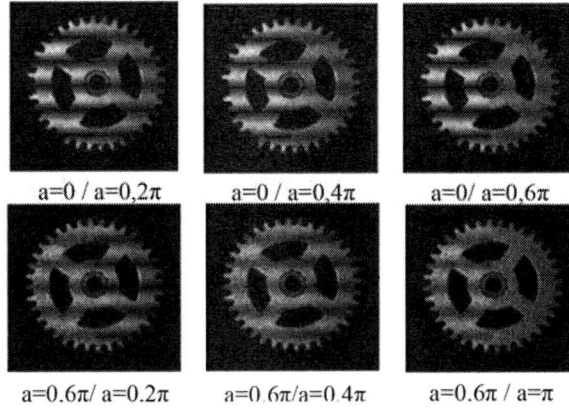

a=0 / a=0,2π   a=0 / a=0,4π   a=0/ a=0,6π

a=0.6π/ a=0.2π   a=0.6π/a=0.4π   a=0.6π / a=π

Fig. 3. Simulation of interferograms for uniform distribution (the first indicated a value is for left part of the image, the second one is for right part)

Analyzing the whole given description of phenomenon of shift/hop of interference fringes an important conclusion may be carried out that such a shift is not connected with interferometric information but is an appearance of statistical parameters of the scattering medium. This in turns allows alternative interpretation of known concept of scattered light. This concept assumes that part the light passed through scattering medium converts to diffuse component. In the same time non-scattered part is assumed to remain unchanged. From the given results it appears, on the contrary, that light beam after scattering medium in some cases contains no memory about the primary beam. The new created beam consists of the diffuse component and the plane wave. This, from the other hand, follows from the Huygens-Fresnel principle. But, at this point, the phase of the plane wave varies from 0 to π depending on the statistical properties of the scattering medium. This in turn means that interferometric study of diffuse surfaces may have an indetermination up to half of period of interference pattern.

IV.   CONCLUSION

Demonstrated and explained is the shift of interference fringes in holographic interferometry experiment. The shift is not connected with the interferometric information but is an appearance of statistical properties of scattering medium. It is an artifact in interferometric study but, as shown in the work, it's different for different statistical distributions. Analysis of this shift gives additional information about statistical properties of scattering medium.

REFERENCES

[1]   Derzhypolska L.A. Holographic interferometry under phase distortions / L.A. Derzhypolska, O.V. Gnatovskij.   //   Semiconductor   Physics,   Quantum Electronics and Optoelectronics: -2006.- № 3. – P. 56-59.
[2]   Derzhpolska L. Investigation of operation of holographic interferometer under phase distortions in probe beam / L. Derzhypolska, N. Medvid, L. Priadko // Proceedings of the SPIE – 2008 – Volume 7008 – pp. 70081U-70081U-8.
[3]   Korn G. Reference book on mathematics. / G.Korn, T.Korn – M., "Nauka", 1978 – 832ст.
[4]   V. P. Ryabukho, V. V. Tuchin. Diffraction of interference fields on random phase objects in: Handbook of Coherent Domain Optical Methods. Biomedical Diagnostics. // Environmental and Material Science. – Kluwer Academic Publishers, Boston, – 2004. – Vol. 1, – P. 235-318 .
[5]   O.V.Angeleky, I.I.Magun, and P.P.Maksimyak, Optical   correlation   methods   in   the   studies   of inhomogeneous   phase   samples,   Proc.   IIId Internal.Symp.Mod.Opt., vol.2, Budapest. 1968, c.337.

AGITATED REACTOR WITH IN SITU NANOPARTICLE SIZE CONTROL BY LIGHT SCATTERING PHOTON CORRELATION SPECTROSCOPY

A. G. Lazarenko[1], A. N. Andreev[1], M. Ben Amar[2], K. Chhor[2], A. V. Kanaev[2]

[1]National Technical University KhPI, Kharkiv, Ukraine

[2]LSMP - CNRS, Université Paris 13, Villetaneuse, France

Abstract: We propose a simple design of the precipitation reactor vessels permitting measurements of the mean size of nanoparticles by light-scattering photon correlation spectroscopy. The measurements are conducted in connecting transparent capillary array in conditions free of turbulence, which otherwise perturb such measurements. A satisfactory quality of the autocorrelation function has been obtained in test measurements of 100 nm latex nanoparticles.

Essential parts of a reactor for nanoparticles fabrication are mixer and staff for particles size and polydispersity control.

The mixing quality of the reacting chemicals is a key factor in optimization of the precipitation processes [1]. Ander conditions that the characteristic physical mixing time (τ_{Phys}) becomes smaller than that of the chemical transformations (τ_{Chem}), the particles precipitation process attains its optimum conditions, correspondent to regimes of small Damköhler numbers: $Da=\tau_{phys}/\tau_{chem}\leq 1$. It corresponds to an enhanced homogeneity of the reactive media and results in most narrow polydispersity of particles. A powerful mixer is than an essential part of a fluid reactor for fabrication of nanoparticles, since reactivity of the small units is generally much higher compared to macroscopic powders.

For fabrication of nanoparticles of a given size and their growth process control, it is desirable to monitor in situ their mean size. A common procedure consists in applying the light-scattering photon-correlation spectroscopy (LSPCS) method to size measurements in a reactor, which main parts are a bath with the active medium in which nano-particles grow and a device for the chemicals mixing and thermal stabilization (hereinafter, we shall remember the location of the heater and cooler only, because the shapes of and connections between other components of thermal stabilization such as temperature sensors, controllers, power supplier and so on are not fundamental in our device). The fiber-optic probe is then immersed in the reactor solution and connected to the coherent light source and to sensor communicated withthe photon correlator and computer [2-3].

An intense mixing required for nanoparticles fabrication creates turbulence in the reacting fluids [4-5] producing strong fluctuations of the optical density and modifying the diffusion coefficient by a local convection. These fluctuations registered by the correlator distort the acquired data (autocorrelation function - ACF) and perturb measurements of the real particles size. Consequently, the apparent size issued from LSPCS measurements is related to the local specific power input and reduces with its increase. Therefore, the size measurements need a renormalization procedure [2]. Moreover, these measurements become impossible still at relatively moderate input powers far below those needed for the reactive fluids homogenization. Therefore, in order to accurately measure the true particles size in an agitated reactor, one has to stop the mixing device and wait for the energy dissipation (thereby breaking up a technological process of particles preparation).

One possible solution is presented by ACOS (Automatic Continuous Online Sizing) [6]. In this device thanks to the solution flow from the reactor bath by special pump through a thin optical cell, which permits the particles size measured by LSPCS method, the formation of turbulence in the area of measurement is prevented. After passing through the optical cell, the solution is returned to the reactor bath, thereby forming a communicating vessels connected in two places. This system is hybrid one composed of two parts of largely different dimensions: the reactor bath (large, wide, in which turbulence occurs) and volume for LSPCS measurements with connecting tubes (small, narrow, in which turbulence is supressed). Thus, by separation of the particle elaboration volume from that of optical control by slow pumping a small part of the reactive solution through a closed optical bench circuit, one can provide a quality particle size control in real time. The above solution has a drawback of a need of an additional motive element - fluid propulsion device. A contact of the chemically active media with mechanical components of this device may be an additional source of the solution contamination (e.g. by humidity), reduction of the process reproducibility and excessive energy consumption.

In the present communication we describe a simple and robust solution [7] for the particles size monitoring by the LSPCS method in agitated precipitation reactors.

This device can be free of all the above mentioned defects if at least one of thermal stabilization means is connected only to one of the reactor vessels and/or if both heater and cooler are attached to different vessels, forming a closed loop for solution circulation like a gravity heating machine. In addition, mixer has to be conceived and placed in

978-1-4799-0019-0/13 $31.00 © 2013 IEEE

an anisotropic geometry, e.g. a propeller with blades bent in the direction of the solution circulation caused by the influence of the said thermal stabilization flows and the bath itself to be imparted a form that would support the solution circulation in the same direction.

More concretely, the reactor bath is performed in a form of two communicating vessels, connected in two or more places, of which at least one possesses a transparent envelope permitted exit of photons scattered by the fluid towards an external LSPCS detector. The particles in a fluid, circulating in the connecting capillary staff, will provide non-distorted ACF function, since the turbulence is suppressed at low Reynolds's numbers. Taking into account a small amount of fluid in the measurements loop and it's low speed, the thermal stabilization and mixing equipment (if appropriately formed and placed) can serve as a fluid pump device in ACOS.

The reactor includes a bath in a form of two communicating vessels 1 and 2 filled with a solution. The vessels are connected by tubing and/or capillary (6). Figures 1 and 2 show only the reactor parts located directly in the bath: propeller 3 of mixing device, heater 4 and cooler 5 of thermal stabilization device. The sensors, control units and power supply are not shown, since their locations are not essential for the reactor performance. Rings 6 mark the places where the narrow parts of connected vessels (e.g. transparent capillary), suitable for the particle size monitoring, can be placed. Cooler 5 and heater 4 are placed at the opposite extremities of the vessels, as shown in Figure 1. The dashed arrows in Figure 1 indicate clockwise direction of the fluid flow, which is opposite to the direction of the pressure gradient in the containers.

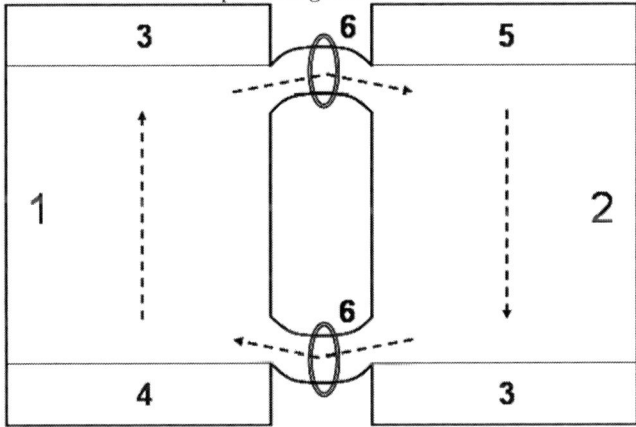

Fig. 1. Principal scheme of reactor.

The principle of implementation and deployment of the circulation units and respective designs of the reactor bath are depicted in Figure 2 in two perpendicular frames: horizontal XZ (a) and vertical YZ (b). The stirrer 3, if chosen in form of a propeller with blades that are bent clockwise, can afford clockwise circulation of the reactor fluid, pushing it through the connecting tubing and/or capillary in the XZ plane (Figure 2a). Accordingly, stirrer 3 and/or thermostatic units, heater 4 and cooler 5 can be placed in a way to afford a desirable circulation of the reactor fluid in the YZ plane (Figure 2b). The rings 6 in Figure 2 mark possible placements of LSPCS detectors.

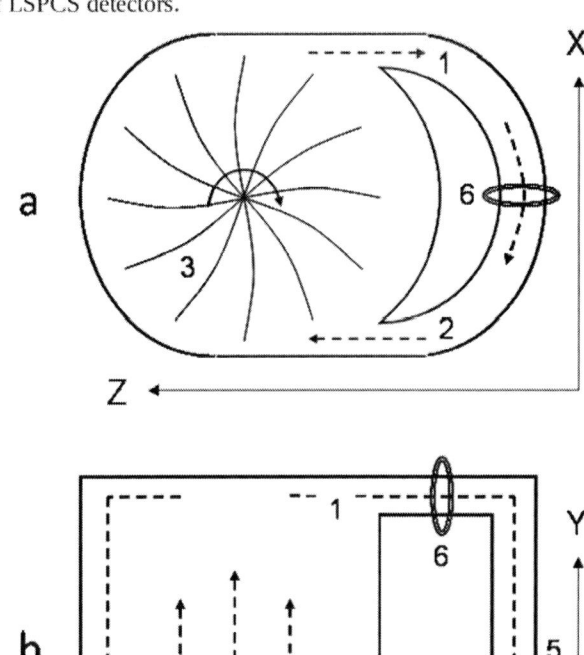

Fig. 2. Principle of implementation and deployment of the fluid circulation units in the reactor bath.

The validation of LSPCS measurements in a capillary system has been performed. A radiation from 640nm/40mW CUBE laser (Coherent) was introduced into the monomode quartz optical fibre (5-μm core) by using an optical fibre adapter (Newport). The core was than inset into the capillary with a slowly circulating fluid containing latex nanoparticles of R=50±2 nm (Sigma).

The 150 μm ($\varnothing$) capillary TEFLON AF 2400 (Biomed Inc., CA) was tested for a rapid sampling of the reactor media. The turbulence is strongly suppressed inside such capillary because of low Reynolds's numbers (Re∝$\varnothing$), which enables direct particle size measurements. Because of a low refraction index of the capillary walls (n_c=1.29), the confined fluid with the refraction coefficient n≥n_c forms a liquid waveguide for the laser radiation, which is axially introduced by the monomode optical fibre.

The capillary walls do not absorb laser radiation, however strongly scatter it. The light scattered by the particles inside the capillary is passed through the capillary walls and collected by another monomode optical fibre placed perpendicularly to the capillary surface; this fibre captures and conducts the scattered light to the photo-multiplier and photon correlator. Despite of a lower coupling efficiency with the laser beam compared to the multimode fibres (20% against 80% efficiency), the use of a monomode optical fibre results in a higher contrast of the autocorrelation function (ACF) and shorter accumulation time, which provide satisfactory precision of measurements [8].

The measurements were made in the homodyne technique of photon-correlation spectroscopy by using 32 bits 288 channels digital correlator Photocor-FC with software Dynals. An image of the liquid waveguide inside the capillary filled with scattered light nanoparticles is shown in Figure 3.

Fig. 3. Image of optical fiber with propagated He-Ne laser beam along the liquid core ($\varnothing$=150 μm) with nanoparticles.

A typical measured data and fitted ACF curve are shown in Figure 4. Taking into account the 2-propanol viscosity 20°C and refraction index, the mean particle radius of R=49.3±0.6 nm has been calculated from Stokes-Einstein formula. The obtained value is in a good agreement with the

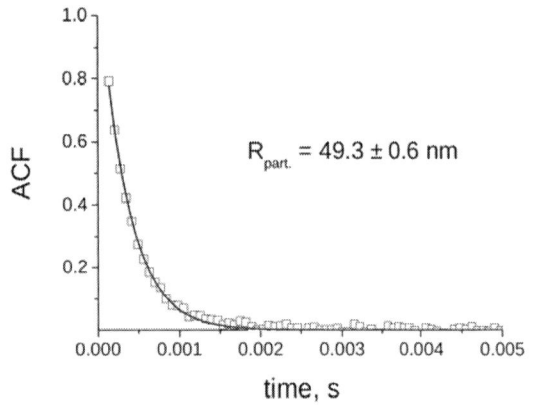

Fig. 4. Autocorrelation function of 50-nm latex nanoparticles measured in capillary.

true radius of the latex spheres. We notice that the suspensions of TiO$_2$ nanoparticles of $\varnothing$=18 nm in 2-propanol prepared in the laboratory, were also tested and provided satisfactory results.

In conclusion, the method is proposed for in-situ measurements of the means size of nanoparticles in agitated reactor by light-scattering photon-correlation spectroscopy. A very simple alignment of the optical system and turbulence free conditions permit the in situ measurements without normalization in large gamme of operation conditions.

REFERENCES

[1] J. Bałdyga and R. Pohorecki, "Turbulent micromixing in chemical reactors - a review", *The Chem. Eng. J.*, vol. 58, pp. 183-195, 1995.

[2] M. Rivallin, M. Benmami, A. Kanaev and A. Gaunand, "Sol-gel reactor with rapid micromixing: modelling and measurements of titanium oxide nano-particles growth", *Chem. Eng. Res. Design*, vol. 83, pp. 67-74, 2005.

[3] R. Azouani, A. Michau, K. Hassouni, K. Chhor, J.-F. Bocquet, J.-L. Vignes and A. Kanaev, "Elaboration of pure and doped TiO$_2$ nanoparticles in sol-gel reactor with turbulent micromixing: application to nanocoatings and photocatalysis", *Chem. Eng. Res. Design*, vol. 88 pp. 1123-1130, 2010.

[4] D. L. Marchisio, L. Rivautella and A. A. Barresi, "Design and Scale-Up of Chemical Reactors for Nanoparticle Precipitation", *AIChE J.*, vol. 52, pp. 1877-1887, 2006

[5] H. C. Schwarzer and W. Peukert, "Combined Experimental/Numerical Study on the Precipitation of Nanoparticles", *AIChE J.*, vol. 50, pp. 3234-3247, 2004.

[6] "Automatic Continuous Online Sizing", http://www.brookhaveninstruments.com/products/particle_sizing/p_PS_ACOS.html

[7] A. Lazarenko, A. V. Kanaev, K. Chhor and A. O. Mikolayovich, Ukrainian patent application, a 2011 13325, Nov. 24, 2011.

[8] R. G. W. Brown, "Homodyne Optical Fiber Dynamic Light Scattering", *Appl. Opt.*, vol. 40, pp. 4004-4010, 2001.

Development of Algorithms for Image Forward Using Finite Element Method with Florescent Molecular Tomography in Homogenous Media

Sima Saleh[1,*]

1.Master student of Medical Physics, Medical Physics and Bioengineering, Tehran University of medical Sciences, Tehran, Iran

Abstract

Fluorescent Molecular Tomography (FMT) is a non-invasive method which can be used in cellular and molecular levels. The reconstruction algorithm of FMT has two main step includes Forward and Inverse problem. For solving the forward problem diffusion equation should be solved, so it can be done by using Finite Elements Method (FEM).

Keywords: Fluorescent Molecular Tomography, diffusion equation, finite element method, forward problem.

Introduction

Optical imaging is established as one of the modalities applied to molecular imaging studies. Molecular imaging can be used to visualization of molecular events in the cellular or sub cellular level. This modality is an imaging technique that uses quantum dotes or Fluorescent. The cell function can be shown by using nanoparticles (such as Quantum Dots) or fluorescent and suitable wavelength of optical radiation. Among the different methods of optical imaging Fluorescent Molecular Tomography (FMT) is a non-invasive method for imaging the biological tissue at cellular level. The image reconstruction of FMT system has two main steps, Forward problem and invers problem. Forward problem simulate the light source distribution on surface of objects and the goal of Invers problem is the recovery of optical properties using measurement intensity at the surface of abject The goal of this study was developed of a fast algorithm for forward problem based on finite element method for 2-D geometry [1,2].

Methods

Image reconstruction of optical tomography like FMT system include two steps which is described as forward and invers problem. The aim of present study was determination the fast algorithm for forward problem that is used in image reconstruction of FMT system.

For this purpose diffusion equation was solved by using finite element method which is the fast, accurate and flexible technique. The air-tissue boundary was presented by Robin boundary condition, which constrains combination of the photon density and current at boundary [3]. A mismatch between the refractive indices n within Ω and $\acute{n}$ in the surrounding medium describe in the following equation while $\acute{n}$ assumes equal 1:

$$\Phi(\xi) + 2\kappa\beta\hat{n}.\nabla\Phi(\xi) = 0$$

Where ζ is a point on the boundary and $\hat{n}$ is the outward pointing normal, and β can be derived from Fresnel's law and can be expressed in the following equation:

$$\beta = \frac{1 + R}{1 - R}$$

In order to generate simulated data, a 2-D circular mesh with linear triangular elements was used. The radius of the model was 43 mm with background optical properties of absorbing coefficient equal to. 0.02 mm^{-1} and reduce scattering coefficient equal to 0.5 mm^{-1}. For good balance between accuracy and computation time the optimum mesh size should be used. So the mesh was used here consisting 5923 nodes corresponding 11574 elements, Fig (1) shows the 2D circular mesh.

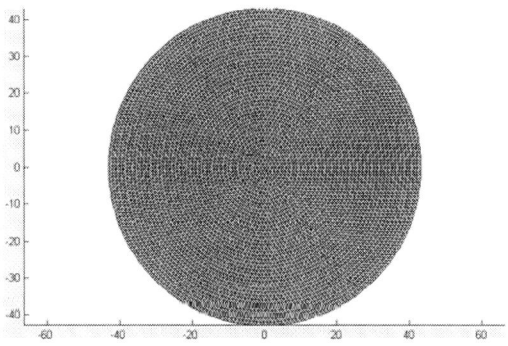

Fig 1:A 2D circular mesh with radius 43 mm

After making the mesh, one source was located at the one transport scattering distance on the boundary. The fluorescent target placed at center of mesh. Fluorescent consider as point.

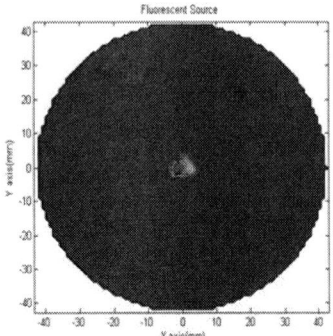

Fig 2: Fluorescent target at center of mesh.

The detectors were placed opposite the light source at boundary pointes. For tomography imaging we should image every 60 degrees.

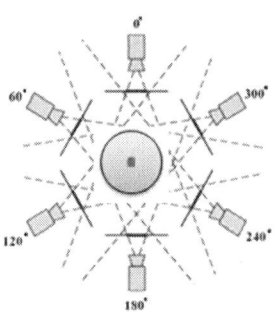

Fig 4: Geometry used for study

The algorithm based on FEM to solve the diffusion approximation was developed and written in the MATLAB programming to measure the intensity on the nodal boundary pointes. The diffusion equation in the FEM framework can be expressed as a system of linear [4]

$$[K(\kappa) + C(\mu_a) + \zeta A]\phi = q_0 \quad (2)$$

Where the matrices $K(\kappa), C(\mu_a)$ and A given by:

$$K_{i,j}^e = \int_{\Omega} \kappa(r)\, \nabla u_i(r).\, \nabla u_j(r)d\Omega,$$

$$C_{i,j}^e = \int_{\Omega} \mu_a(r)u_i(r)d, \quad (3)$$

$$A_{i,j}^e = \int_{\partial\Omega} u_i(r)u_j(r)d\partial\Omega$$

And q_0 is the source vector and give by:

$$q_{0,i} = \int_{\Omega} u_i(r)q_0(r)d\Omega.$$

ζ is the constant to apply the RBC and equal to $\frac{1}{2\beta}$.

This algorithm was evaluated by the open access software package that developed at the Darthmouth for NIR imaging [5].

Results

The intensity of nodal boundary pointes was measured at every 60^0 for tomography imaging. The FEM algorithm of diffusion approximation which was developed at this study was compared with NIRFAST. These results show the good agreement between the FEMDA codes with NIRFAST. At every angel we observed the same intensity at the boundary nodal points, by increasing in photons scattering angle, the intensity of light is decreased. Figure 5 shows the comparison of light intensity obtained from FEM forward algorithm and NIRFAST forward algorithm on the surface of objects.

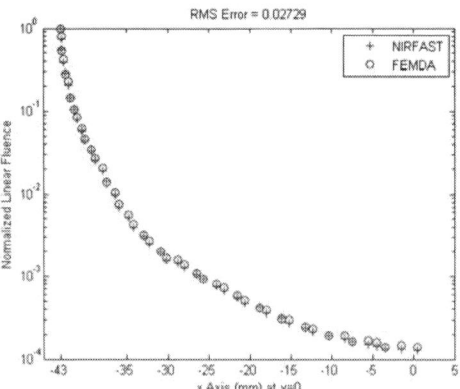

Fig 5. FEM forward algorithm and NIRFAST forward algorithm

Although the error was low but the most of error is located near the source term. The results showed the significant correlational coefficients (R >0.95) which demonstrated the high accuracy

978-1-4799-0019-0/13 $31.00 © 2013 IEEE

of the algorithm. As pervious part we put our forward algorithm and NIRFAS forward algorithm as input of reconstruction step in NIRFAST software and were observed same result for both suggested forward algorithm and NIRFAST forward algorithm.

a)

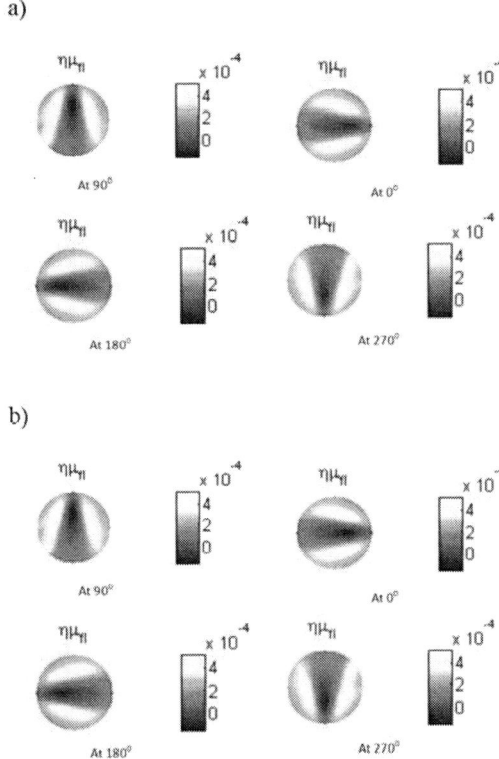

b)

Fig 6. The simulated light intensity of a fluorescent source by FEM (a) and NIRFAT (b) algorithms

For comparison, correlation test and t-test was used. There was no significant difference between FEM algorithm and the NIRFAST algorithm (p_value=0.164).

Discussion

The FEDA algorithm was shown that FEDA yields similar results to NIRFAST for the 2-D geometry. The results showed the significant correlation coefficients (R >0.95), which demonstrated the high accuracy of the algorithm. The algorithm which is designed and developed in this study is more fast, flexible and accurate than analytical method that has good agreement to NIRFAST results. The computation time was lower than analytical theory and useful for heterogonous medium with complex geometry.

This algorithm gives the user the flexibility to modify every aspect of the model, including element shape, basis functions, boundary conditions, and diffusion approximation equation. The suggested algorithm can be used for image reconstruction of FMT system and can be modified to 3-D geometry.

References

1) Diafa Wang and et-all "A Novel Finite Element Based Algorithm for Fluorescence Molecular Tomography of Heterogenous",IEEE TRANSACTION INFORMATION TECHNOLOGHY IN BIOMEDICINE(2009)

2) D. Khodadad, E.J. Hällstig, M. Sjödahl, "Shape reconstruction using dual wavelength digital holography and speckle movements," in Proc. SPIE 8788, Optical Measurement Systems for Industrial Inspection VIII, 87880I (May 13, 2013); doi:10.1117/12.2020471.

3) Arridge S, Schweiger M, Hiraoka M, Delpy D. "A finite element approach for modeling photon transport in tissue". MEDICAL PHYSICS-LANCASTER PA- ,(1993).

4) R. Aronson. Boundary conditions for diffusion of light. Journal of the Optical Society of America A,(May 1995), 12(11):2531-2538

5) .Hamide Dehghani,et all "Near infrared optical tomography using NIRFAST: Algorithm for numerical model and image reconstruction" Commun Numer Methods Eng. (2008)

Laser-microwave spectroscopy of singlet Mg I atoms in S,P,D,F,G Rydberg state

S.F. Dyubko[1], V.A. Efremov[1], A.S. Kutsenko[2], N.L. Pogrebnyak[2], K.B. MacAdam[3]

[1]Karazin Kharkov National University, Kharkov, Ukraine

[2]Institute of Radio Astronomy, NAS-Ukraine, Kharkov, Ukraine

[3]University of Kentucky, Lexington KY 40506-0055 USA

Abstract: For the first time the energy level positions of Mg I atoms have been measured with high accuracy by means of laser-microwave spectroscopy in the 79-234 GHz frequency range. The new measurements cover the range of principal quantum number n=25-40 in L=0-4 Rydberg states. Using our and optical measured frequencies the quantum defect constants for these states are obtained. The description of our spectrometer, measurement technique and results are presented.

During the years 2011-2012 we were investigating microwave absorption spectrum of Mg I atoms in S, P, D, F and G Rydberg states. The results are published in Ref. [1].

The experimental details that are specific for the investigation of Mg I are given below.

Laser excitation of Mg atoms to Rydberg states was performed by different two-step schemes.

$$3s^{2}\,^{1}S_{0}\xrightarrow[431nm]{2\hbar\omega}3s3d\,^{1}D_{2}\xrightarrow[660-680nm]{\hbar\omega}3snf\,^{1}F_{3},3s(n+1)p\,^{1}P_{1}$$

$$3s^{2}\,^{1}S_{0}\xrightarrow[285nm]{\hbar\omega}3s3p\,^{1}P_{1}\xrightarrow[740-790nm]{2\hbar\omega}3snf\,^{1}F_{3},3s(n+1)p\,^{1}P_{1}$$

$$3s^{2}\,^{1}S_{0}\xrightarrow[285nm]{\hbar\omega}3s3p\,^{1}P_{1}\xrightarrow[370-385nm]{\hbar\omega}3snd\,^{1}D_{2}$$

In the scheme of double-photon excitation of $3s3d\,^{1}D_{2}$ state a coumarin-540 dye laser working at a fixed wavelength of 431 nm was used. To excite to Rydberg $3snf\,^{1}F_{3}$ states a tunable (660-680 nm) DCM dye laser was applied. Both lasers were pumped by Xe-Cl UV laser (308 nm) that worked at the pulse repetition rate of 10 Hz. The laser pulse energy was 15 mJ and the pulse duration was 40 ns.

The UV radiation for the first step of the next scheme was obtained by doubling the R6G dye laser frequency in a KDP crystal. For a two-photon transition to Rydberg $3snf\,^{1}F_{3}$, $3s(n+1)p\,^{1}P_{1}$ states (second step) the infrared LDS 751 dye laser was used. The same laser was used in the second step of the last scheme but its frequency was doubled in a KDP crystal.

Both dye lasers were pumped by the second harmonic of a Nd:YAG laser. This laser operated at the pulse repetition rate of 12.5 Hz. Its pulse duration was 25 ns and the energy of green light pulse was about 10-15 mJ. All used in the experiments dye lasers were with perpendicular pumping and with gratings working in grazing-incidence mode.

Laser beams entered the vacuum chamber through a quartz window and passed between two electrodes perpendicular to an atomic beam. The electrode plates had diameter of 40 mm and spacing of 6.5 mm. A channel electron multiplier (CEM) was positioned behind one of the electrode plates opposite to the interaction region. From the opposite side a microwave radiation was fed to the interaction region. A computer-controlled backward wave oscillator based synthesizer with several PLL stages of frequency multiplication was used as a source of microwave radiation covering 59-117 GHz frequency range. A solid-state multiplier was used to exceed the frequency range up to 234 GHz.

Knudsen oven in a form of stainless steel cylinder heated by a tungsten coil was used as an atomic beam source. Mg grains placed in the oven were heated up to 900 K. Atomic beam passed through a 2 mm slit and reached the excitation region between the ionization plates.

Rydberg atoms were detected using a method of pulse field ionization. The ionization pulse was switched on in 1.5-2 μs after a laser pulse. The amplitude of the negative ionization pulse with pulse leading-edge time not more than 50 ns was precisely varied in the range 50-1000 V. The output signal from CEM passed to an analog-to-digital converter and after digitizing to PC. For signal-to-noise ratio enhancement multiple scans in the desired microwave frequency range were performed.

In the 79 to 234 GHz frequency range the frequencies of 68 resonances that belong to $n\,^{1}P_{1}\rightarrow(n+1)\,^{1}S_{0}$, $n\,^{1}P_{1}\rightarrow(n+1)\,^{1}P_{1}$, $n\,^{1}P_{1}\rightarrow n\,^{1}D_{2}$, $n\,^{1}D_{2}\rightarrow(n+2)\,^{1}S_{0}$, $n\,^{1}D_{2}\rightarrow n\,^{1}P_{1}$, $n\,^{1}D_{2}\rightarrow(n+1)\,^{1}P_{1}$, $n\,^{1}D_{2}\rightarrow(n+2)\,^{1}P_{1}$, $n\,^{1}D_{2}\rightarrow(n+1)\,^{1}D_{2}$, $n\,^{1}D_{2}\rightarrow(n-1)\,^{1}F_{3}$, $n\,^{1}F_{3}\rightarrow(n+2)\,^{1}P_{1}$, $n\,^{1}F_{3}\rightarrow(n+1)\,^{1}D_{2}$, $n\,^{1}F_{3}\rightarrow n\,^{1}D_{2}$, $n\,^{1}F_{3}\rightarrow(n+1)\,^{1}F_{3}$, and $n\,^{1}F_{3}\rightarrow(n+1)\,^{1}G_{4}$ type transitions in the range of principal quantum number n =25 to 40 have been assigned and measured for the first time. The results of the experiment are presented in a summary Table I.

The energy levels of singlet Mg corresponding to quantum number n, l, j may be expressed by the Rydberg-Ritz formula

$$E(n,l,j)=\frac{-R_{Mg}}{n^{*2}}=\frac{-R_{Mg}}{\left(n-\delta_{n,l,j}\right)^{2}}\ ,$$

where $\delta_{n,l,j}=E_{l,j}+\frac{A_{l,j}}{n^{*2}}+\frac{B_{l,j}}{n^{*4}}+\frac{C_{l,j}}{n^{*6}}\cdots$.

We have made an attempt of joint fitting of the whole dataset with independent variation of quantum defect parameters for S, P, D, F и G states. In this joint fit in addition to our measurements we have included also four single photon microwave transitions of $n^1F_3 \to n^1G_4$ type with $n = 19 \div 21$ falling in the frequency range $26 \div 36$ GHz [6] and 21 single photon optical transitions of $n^1S_0 \to n'^1P_1$, $n^1P_1 \to n'^1S_0$, $n^1P_1 \to n'^1D_2$, $n^1D_2 \to n'^1P_1$ and $n^1D_2 \to n'^1F_3$ type with n, n'=$3 \div 9$ [2-5,7] (third column of Table I). The observed - calculated differences are given in the fourth columns of Table I. The root mean square deviation for the joint fit of all measured frequencies was 0.66 MHz. Since we were not able to resolve

the absorption lines from different Mg isotopes we used the Rydberg constant for Mg I of $R_{Mg} = 109734.83884$ cm^{-1} that corresponds to the averaged atomic weight 24,3050 u. The quantum - defect constants $E_{l,j}$, $A_{l,j}$, $B_{l,j}$ and $C_{l,j}$ for S-G terms are given in Table II. For comparison in Table II we give also corresponding constants for S, P, D and F states determined in the work [1] basing on the results of microwave measurements.

TABLE I
Observed resonance line centers of Mg I

№	Type of transition	Frequency (*cm^{-1}, MHz)	(Obs.-cal.) (*cm^{-1}, MHz)	Ref.
1	$5^1S_0 \to 7^1P_1$	6024.054*	-0.010*	[2]
2	$5^1S_0 \to 6^1P_1$	4658.807*	0.017*	[2]
3	$5^1S_0 \to 5^1P_1$	2150.353*	0.002*	[2]
4	$4^1S_0 \to 4^1P_1$	5843.407*	0.000*	[2]
5	$4^1P_1 \to 8^1S_0$	9706.75*	0.01*	[3]
6	$4^1P_1 \to 7^1S_0$	8662.635*	0.007*	[2]
7	$4^1P_1 \to 6^1S_0$	6840.126*	-0.132*	[2]
8	$4^1P_1 \to 5^1S_0$	3209.447*	-0.004*	[2]
9	$5^1P_1 \to 8^1S_0$	4346.954*	0.013*	[2]
10	$5^1P_1 \to 7^1S_0$	3302.816*	-0.010*	[2]
11	$6^1P_1 \to 7^1S_0$	794.380*	-0.007*	[4,5]
12	$4^1P_1 \to 4^1D_2$	3787.878*	-0.007*	[2]
13	$3^1D_2 \to 5^1P_1$	8303.498*	-0.002*	[2]
14	$3^1D_2 \to 4^1P_1$	2943.701*	0.002*	[2]
15	$4^1D_2 \to 8^1P_1$	6268.517*	-0.045*	[2]
16	$4^1D_2 \to 7^1P_1$	5445.621*	-0.009*	[2]
17	$4^1D_2 \to 6^1P_1$	4080.374*	0.018*	[2]
18	$3^1D_2 \to 4^1F_3$	8273.382*	0.000*	[2]
19	$4^1D_2 \to 9^1F_3$	7166.61*	-0.17*	[2]
20	$4^1D_2 \to 8^1F_3$	6800.738*	-0.101*	[2]
21	$4^1D_2 \to 7^1F_3$	6266.13*	0.00*	[2]
22	$34^1P_1 \to 35^1S_0$	94421.6	-0.15	[1]
23	$33^1P_1 \to 34^1S_0$	103487.7	-0.50	[1]
24	$32^1P_1 \to 33^1S_0$	113753.5	0.07	[1]
25	$28^1P_1 \to 29^1S_0$	171623.8	2.30	[1]
26	$27^1P_1 \to 28^1S_0$	192014.2	2.54	[1]
27	$38^1P_1 \to 39^1P_1$	2×62661.4	1.23	[1]
28	$37^1P_1 \to 38^1P_1$	2×67962.3	0.24	[1]
29	$36^1P_1 \to 37^1P_1$	2×73879.0	0.20	[1]
30	$33^1P_1 \to 34^1P_1$	2×96329.0	-0.24	[1]
31	$32^1P_1 \to 33^1P_1$	2×105816.2	-0.15	[1]
32	$31^1P_1 \to 32^1P_1$	2×116592.0	1.26	[1]
33	$34^1P_1 \to 34^1D_2$	79639.5	-0.47	[1]
34	$33^1P_1 \to 33^1D_2$	87330.1	-0.20	[1]
35	$32^1P_1 \to 32^1D_2$	96044.1	-0.42	[1]
36	$31^1P_1 \to 31^1D_2$	105958.9	-0.12	[1]
37	$38^1D_2 \to 40^1S_0$	2×65394.0	-3.17	[1]
38	$37^1D_2 \to 39^1S_0$	2×70844.6	-2.59	[1]
39	$35^1D_2 \to 37^1S_0$	2×83705.3	-1.19	[1]
40	$34^1D_2 \to 36^1S_0$	2×91315.0	-0.26	[1]
41	$33^1D_2 \to 35^1S_0$	2×99875.0	0.31	[1]
42	$32^1D_2 \to 34^1S_0$	2×109538.0	-0.23	[1]
43	$27^1D_2 \to 27^1P_1$	-162703.5	-4.13	[1]
44	$26^1D_2 \to 26^1P_1$	-183004.0	-4.65	[1]
45	$36^1D_2 \to 37^1P_1$	80986.2	-0.95	[1]
46	$35^1D_2 \to 36^1P_1$	88178.3	-0.12	[1]

№	Type of transition	Frequency (MHz)	(Obs.-cal.) (MHz)	Ref.
47	$34^1D_2 \to 35^1P_1$	96246.0	-0.16	[1]
48	$33^1D_2 \to 34^1P_1$	105327.5	-0.43	[1]
50	$31^1D_2 \to 32^1P_1$	127224.2	0.49	[1]
51	$30^1D_2 \to 31^1P_1$	140473.5	0.33	[1]
52	$29^1D_2 \to 30^1P_1$	155626.1	0.32	[1]
53	$28^1D_2 \to 29^1P_1$	173035.8	0.38	[1]
54	$27^1D_2 \to 28^1P_1$	193138.2	0.41	[1]
55	$26^1D_2 \to 27^1P_1$	216472.8	-0.61	[1]
56	$38^1D_2 \to 40^1P_1$	184582.0	-0.26	[1]
57	$37^1D_2 \to 39^1P_1$	199878.2	0.07	[1]
58	$36^1D_2 \to 38^1P_1$	216911.6	0.09	[1]
59	$38^1D_2 \to 39^1D_2$	2×60491.0	0.25	[1]
60	$37^1D_2 \to 38^1D_2$	2×65544.7	0.02	[1]
61	$36^1D_2 \to 37^1D_2$	2×71177.7	0.46	[1]
62	$35^1D_2 \to 36^1D_2$	2×77474.6	0.13	[1]
63	$34^1D_2 \to 35^1D_2$	2×84537.0	0.48	[1]
64	$33^1D_2 \to 34^1D_2$	2×92484.0	0.10	[1]
65	$32^1D_2 \to 33^1D_2$	2×101459.3	0.26	[1]
66	$31^1D_2 \to 32^1D_2$	2×111634.5	0.77	[1]
67	$34^1D_2 \to 33^1F_3$	-79816.3	-3.27	[1]
68	$33^1D_2 \to 32^1F_3$	-87525.0	-3.48	[1]
69	$33^1F_3 \to 35^1P_1$	2×88029.8	0.41	[1]
70	$31^1F_3 \to 33^1P_1$	2×105923.0	1.40	[1]
71	$30^1F_3 \to 32^1P_1$	2×116709.0	-0.80	[1]
72	$33^1F_3 \to 34^1D_2$	79813.5	0.47	[1]
73	$32^1F_3 \to 33^1D_2$	87522.0	0.48	[1]
74	$31^1F_3 \to 32^1D_2$	96256.5	-0.07	[1]
75	$30^1F_3 \to 31^1D_2$	106194.0	-1.09	[1]
76	$26^1F_3 \to 27^1D_2$	163078.2	-0.69	[1]
77	$25^1F_3 \to 26^1D_2$	183433.6	1.07	[1]
78	$27^1F_3 \to 27^1D_2$	-192805.0	-1.69	[1]
79	$37^1F_3 \to 38^1F_3$	2×62664.6	-2.13	[1]
80	$36^1F_3 \to 37^1F_3$	2×67966.8	-1.72	[1]
81	$35^1F_3 \to 36^1F_3$	2×73884.2	-1.77	[1]
82	$34^1F_3 \to 35^1F_3$	2×80509.2	-1.38	[1]
83	$33^1F_3 \to 34^1F_3$	2×87950.3	-1.44	[1]
84	$32^1F_3 \to 33^1F_3$	2×96337.7	-0.99	[1]
85	$31^1F_3 \to 32^1F_3$	2×105826.5	-0.38	[1]
86	$30^1F_3 \to 31^1F_3$	2×116603.0	-0.75	[1]
87	$38^1F_3 \to 39^1G_4$	119955.0	-0.49	[1]
88	$37^1F_3 \to 38^1G_4$	129822.0	-0.20	[1]
89	$36^1F_3 \to 37^1G_4$	140802.0	0.58	[1]
90	$21^1F_3 \to 21^1G_4$	26707	1.4	[6]
91	$20^1F_3 \to 20^1G_4$	30898	-2.0	[6]
92	$19^1F_3 \to 19^1G_4$	36007	-1.1	[6]
93	$19^1F_3 \to 19^1G_4$	36010	1.9	[6]

TABLE II

Quantum-defect Ridberg- Ritz expansion coefficients for Mg I $^1S_0$, $^1P_1$, $^1D_2$, $^1F_3$ and $^1G_4$ terms. Standard deviations are given in parentheses in terms of the last digit shown.

Level	E_l	A_l	B_l	C_l	Range n	Reference
$^1S_0$	1.525367(11)	-0.0310(8)	1.364(12)	-3.37(5)	$5 \leq n < 40$	This work
$^1S_0$	1.525234(29)	-0.055(17)	—	—	$28 \leq n \leq 40$	[1]
$^1P_1$	1.051333(10)	-0.3679(9)	0.874(18)	-3.51(10)	$4 < n < 40$	This work
$^1P_1$	1.05119(3)	-0.272(15)	—	—	$26 \leq n \leq 40$	[1]
$^1D_2$	0.612110(11)	-3.147(3)	8.25(6)	-5.54(28)	$3 < n < 39$	This work
$^1D_2$	0.61197(3)	-3.044(14)	—		$26 \leq n \leq 39$	[1]
$^1F_3$	0.052167 (11)	-0.253(5)	2.64(30)	-29(4)	$4 < n < 38$	This work
$^1F_3$	0.052156 (42)	-0.36(5)	74(18)	—	$19 \leq n \leq 38$	[1]
$^1G_4$	0.014971(15)	-0.540(11)	87(3)	—	$19 \leq n \leq 39$	This work
$^1G_4$	0.014951(44)	-0.65(5)	158(18)	—	$19 \leq n \leq 39$	[1]

REFERENCES

[1] K. B. MacAdam, S. F. Dyubko, V. A. Efremov, A. S. Kutsenko, N. L. Pogrebnyak, "Microwave spectroscopy of singlet Mg I in *L*=0-4 Rydberg states", *J. Phys. B: At. Mol. Opt. Phys.*, vol. 45, 215002, 2012.

[2] E. Biemont, J. W. Brault, "The infrared spectrum of magnesium (1800< σ <9000cm⁻¹) and an extension of the term systems of Mg I and Mg II", *Phys. Scr.*, vol. 34, pp. 751-758, 1986.

[3] J. W. Swensson, G. Risberg, "Mg I Lines in the Solar Spectrum", *Ark. Fys.*, vol. 31, pp. 237-254, 1966.

[4] J. Brault, R. Noyes "Solar emission lines near 12 microns", *Astrophys. J.*, vol. 269, pp. L61-L66, 1983.

[5] E. S. Chang, "Solar emission lines revisited: extended study of magnesium", *Phys. Scr.*, vol. 35, pp. 792-797, 1987.

[6] B.J. Lyons, T.F. Gallagher, "Mg *3snf-3sng-3snh-3sni* intervals and the Mg⁺ dipole polarizability", *Phys. Rev. A*, vol. 57, no. 4, pp. 2426-2429, 1998.

[7] V. Kaufman, W. C. Martin, "Tables of wavelengths and energy level classifications MgI", *J. Phys. Chem. Ref. Date*, vol. 20, no. 1, pp. 86-96, 1991.

Optical properties of composite thick films based on silver nanoparticles in the acrilic polymer matrices

N. M. Ushakov[1,2], *Member, OSA*, I. D. Kosobudsky[1,2], P. A. Muzalev[1], V. Ya. Podvigalkin[1]

[1]Saratov Branch Kotel'nikov IRE of RAS, Saratov, Russia
[2]Saratov State Technical University, Saratov, Russia

Abstract: The polymer-substrate structures based on (1-5 wt.%) Ag-PMMA (polymethylmethacrylate) and Ag-PHEMA (poly-2-hydroxyethyl methacrylate) nanostructures on the glass substrates have been fabricated by the methods of chemical and photochemical reduction of metals from the solutions of their own salts in a polymer medium. The researches of optical properties of the coatings based on obtained nanomaterials allow making the conclusion about their possible use as an antireflective medium for the glass substrates.

Nanoparticles (NPs) of noble metal (eg, silver) are increasingly receiving attention as important starting points for the generation of micro and nanostructures. These NPs are under active research because they posses interesting optical properties differing considerably from that of the bulk phase. It comes from small sizes and high surface/volume ratio.
Silver NPs exhibit new optical properties, which are not observed neither in molecules nor in bulk metals [1], and can be synthesized using various methods: chemical, electron beam evaporation, vapour phase co-deposition in high vacuum etc. We used methods of chemical and photochemical reduction of metals own solutions of salts in the environment of the polymer [2]. Depending on the polymer basis can be obtain materials with specific properties. Call of acrylic polymers PMMA (polymethylmethacrylate) and PHEMA (poly-2-hydroxyethyl methacrylate) with amorphous structure have been used as matrix-stabilizing NPs of metals and their compounds. These owns can be characterized by rigid relations between the basis segments of the polymer chain and short-range order of the arrangement of links of macromolecules. Such polymeric materials are more high mechanical strength, resistance to sunlight and to many chemical reagents. Due to their structure, such polymers practically do not absorb light in the visible and near- infrared (IR) region, and therefore represent great interest as a material for optical coatings for glasses and semiconductors.

In this research, Ag NPs were synthesized in the PMMA & PHEMA matrices, and basic optical properties of the nanostructures in the form of thick films on glass substrates have been investigated.
We used methods of chemical and photochemical reduction of metals own solutions of salts in the environment of the polymer. Different approaches to the preparation of nanosized particles were used due to the features structure of these polymers.
In the first case, to a solution of PMMA in 1; 2-dichloro ethane, a metal-containing compound (MCC), namely, a silver

oxide ammonia solution (with the concentration of 10^2- 10^4 M), and an aqueous solution of the reducing agent, namely, hydrated hydrazine (2×10^2), were added. The resulting mixture was heated until evaporation of the solvent and dried.

In the second case, silver particles were obtained upon photo irradiation ($\lambda = 320$ nm) of a monomer (2-HEMA) - MCC mixture. Here, the formation of silver particles proceeded in parallel with the photo polymerization of HEMA. As a result, a polymeric nanocomposite with a cross-linked structure was obtained.
The physicochemical and optical properties of the composites produced were investigated by a number of optical spectroscopy methods, electron microscopy, and X-ray diffraction analysis (XRD).
Optical spectral measurements were performed on the basis modernized complex KSVU-2 (LOMO, Russia) and optical spectrometer LAMBDA-950 (USA) in the range of optical wavelengths ($\lambda = 300 \div 1000$ nm).
Fig. 1 shows the relative spectral transmission function of the polymer thick film composite nanostructures Ag (3 wt.%) + PMMA on glass and transmittance of a polymeric film of pure PMMA on glass. The thicknesses of the films were equal to 50 microns. The relative spectral function is the ratio of the signals of the photodetector, recorded after the passage of light through the sample. The interference pattern of the spectral function associated with different values and spectral location of the extrema in the spectra of the nanocomposite film and thick film of pure PMMA. The value of the relative spectral features throughout the measured spectral range is greater than one, indicating an increase in the transmission of the film of polymethylmethacrylate with the introduction into it of silver nanoparticles with 3 wt. %. Range of increase transparency polymeric composite thick film based on silver nanoparticles in a matrix of polymethylmethacrylate is from 15% to 40%.
Fig. 2 shows the transmission spectra of the sample in a thick polymer film of the composite nanomaterial on the polymethylmethacrylate with 5 wt.% silver nanoparticles deposited on the glass substrate (curve 2). For comparison, the spectrum of a separate glass substrate is shown (curve 1). From these relationships, it follows that the transparency of the sample with the deposited composite film starts to rise since the wavelength of 500 nm.
A similar phenomenon was observed in the Ag (3 wt.%)- PHEMA nanostructure, but it was weaker.

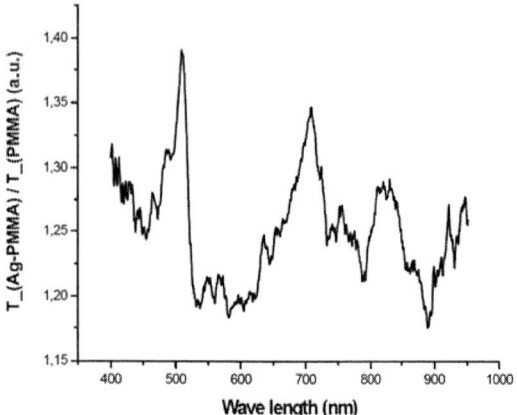

Fig. 1. Relative spectral transmission function of the polymer thick film composite nanostructures Ag (3 wt.%) + PMMA on glass and transmittance of a polymeric film of pure PMMA on glass. Film thickness of 50 microns.

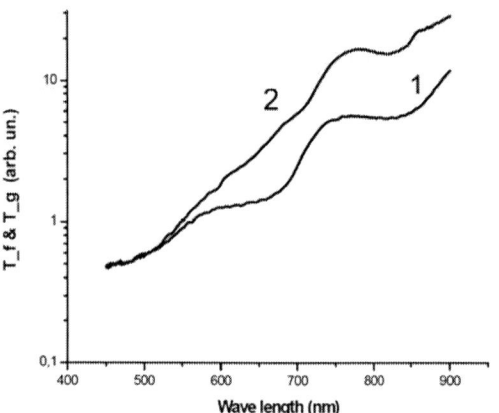

Fig. 2 The optical transparency of the glass thickness of 1 mm (curve 1) and the structure of the polymer nanocomposite films Ag (5wt.%) + PMMA thickness of 140 microns on glass (curve 2)

Strengthening of the light passing through the sample thick film polymer composite material based on polymethylmethacrylate with silver nanoparticles with a mass fraction of 5% is well illustrates the spectral dependence of the reduced transmittance of the sample $\Delta T\_f = (T\_f - T\_g)/T\_g$ (curve 1) shown in Fig. 3, wherein T_f - the transmission spectrum function of the sample with the nanocomposite film on glass, T_g - the transmission spectrum of the glass substrate. Light amplification rapidly increases after 500 nm and saturates near 1000 nm. The measured spectrum of the optical absorption in

thick-film sample (film thickness of 140 microns) (curve 2) has a maximum at 375 nm.

The observed phenomenon may be explained that the silver nanoparticles in the polymer is an unordered array emitting nanoantennas with high gain and an effective directional diagram. The peak of the optical absorption in polymer structures with silver nanoparticles (Fig.3, curve 2) associated with the surface plasmon resonance is observed at the wavelength of 375 nm.

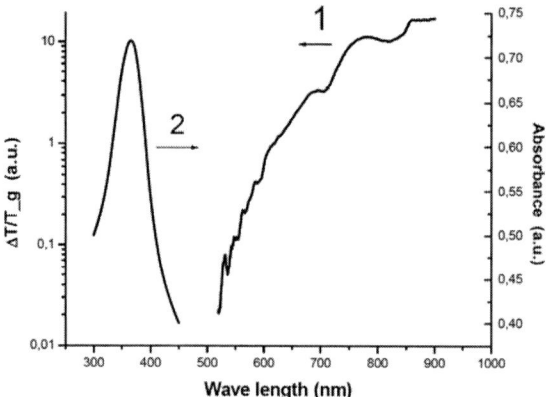

Fig. 3. Optical transmission spectra (curve 1) and absorption (curve 2) nanocomposite polymer thick film Ag (5 wt.%) + PMMA on a glass substrate. The differential transmission spectrum of the nanocomposite film is shown relative to the spectrum of pure glass substrate. Film thickness 140 um.

High light transmission polymeric nanostructures on glass substrates were found at wavelengths above 550 nm in the visible and near IR range. This new phenomenon of light amplification in polymer nanostructures can be useful for not only for anti-reflective coatings, but also for the development of a new laser media.

This study was supported in part by the Russian Foundation for Basic Research (project no. 11-08-00351-a)

REFERENCES

[1] N.M. Ushakov, O.N. Gadomsky, I.D. Kosobudskii, V.Ya. Podvigalkin, P.A. Muzalev, and D.M. Kulbatskii. "A New Optical Polymer Nanocomposite Antireflection medium for Optical Devices," Proc. 10th Intern. Conf. on Laser and Fiber-Optical networks Modeling. LENM'2010 Sevastopol, Ukraine, 12-14 September 2010. P.69

[2] P.A. Muzalev, I.D.Kosobudskii, N.M.Ushakov, and L.G.Panova. "Metal Nanoparticles in Acrylic Polymer Matrices," Inorganic Materials: Appl. Reserch, 2011. vol. 2. No 5. P. 528-530.

Dynamic holographic grating
in liquid crystalline polymer

E.O. Berezhniy, M.M. Burykin, S.G. Ilchenko, A.P. Ostroukh, R.A. Lymarenko
International Center "Institute of Applied Optics" NASU, Kyiv, Ukraine

Abstract: We study optical properties of synthesized liquid crystalline polymers containing azobenzene dye moieties. The grating recording by two linear polarized beams is investigated at wavelength 441 and 532 nm. Holographic grating recording in such system show capability of a dynamic media.

Liquid crystalline polymers [1,2] containing azobenzene dye moieties [3-5] are widely investigated as holography media. The actuality of this research derives from the necessity of effective rewritable photorecording media for experimental studying of diffractive phase structures. Synthesized azodye polymer structure has specific photooptical response on polarized light [6]. Therefore, it is possible to use them for rewritable photorecording media applying to optical data processing or other information technologies.

Liquid crystalline polymer is synthesized in cooperation of Ukrorgsyntez Ltd. (UORSY). Synthesis scheme of the monomers and polymers based on modified Mitsunobu reaction [7] (Fig. 1).

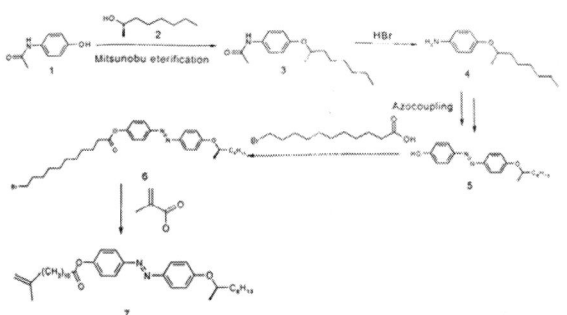

Fig. 1 Synthesis scheme of liquid crystalline polymer.

We prepare photochrome sample for investigation by pressing the medium between optical glass plates; estimated thickness of sample is about 50 μm. We do not align or twist the molecules of liquid crystalline polymer. The image of sample and its chemical structure are shown on Fig. 2.

The azobenzene unit is chemically stable at moderate temperatures [1]. Cis-trans isomerization does not change its absorption spectrum. The optically induced isomerization as well as the complete thermal relaxation back to the trans state are found to be similar to those for amorphous polymers containing azobenzene-type groups.

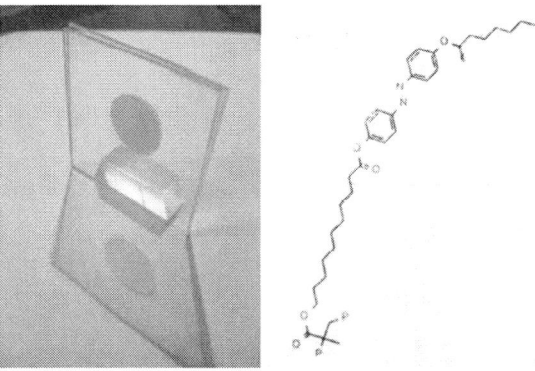

Fig. 2. The sample and its chemical structure.

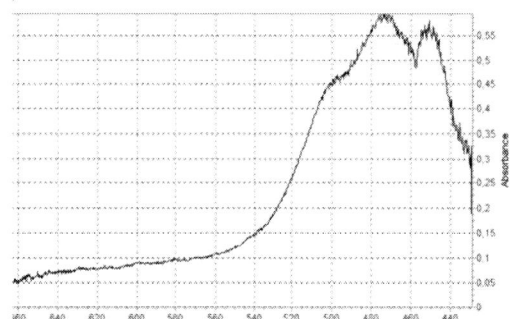

Fig. 3. Absorption spectrum of sample.

Fig. 3 shows the absorption spectrum of sample. The sample has large absorption on a wavelength of the recording laser (405 nm). The data recording on "slope" of the diagram i.e. at point of laser irradiation on a wavelength (532 nm) are of interest too. There is also presumption that diffraction structure will be recording more effectively on thinner polymer layer.

Scheme of experimental setup used for diffraction gratings recording on the samples have shown on Fig. 4. The laser irradiation pass through the objective (O) to beam-splitting cubic (BS), which divides it on two beams of equal intensity and then mirrors (M) guide them onto the sample. The readout of diffraction orders is performed by He-Ne laser on wavelength 633 nm. The distance between the sample and mirror is L = 1.2 m, the distance between mirrors is D = 10 cm. The period of recorded grating is about 15 μm.

978-1-4799-0019-0/13 $31.00 © 2013 IEEE   363

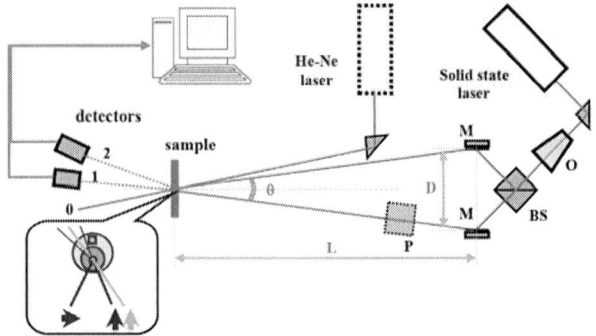

Fig. 4. Experimental setup: recording SSDP SLM lasers at wavelength 405 and 532 nm, laser power about 30 mW, readout He-Ne laser at wavelength 633 nm, distance L = 1.2 m, D = 10 cm, O – objective, P – polarizer, BS – beam splitter, M – mirror.

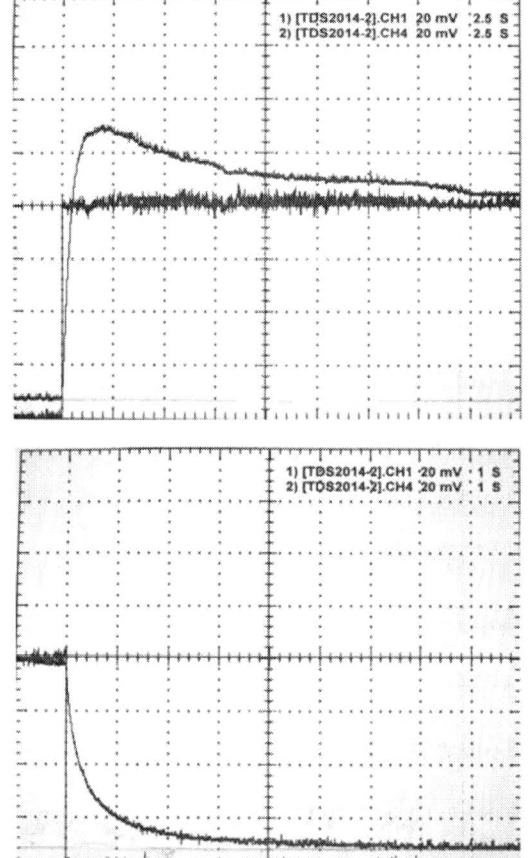

Fig.5. Evolution of diffraction efficiency at recording wavelength 532 nm: (a) recording; (b) relaxing.

We do not use the pumping beam wavelength exactly at the maximum of the n-π* absorption band, λ = 441 nm, which has to be the most efficient for polarization gratings recording.

The sample recording by polarized light is faster than on reference sample 8N10PM [7] reaching stationary level of 1-3% diffraction efficiency at the first maximum. Weak second and third diffraction orders are also observed.

We also measure the dynamic behavior of diffraction efficiency. The sample orientation influences on amplitude of grating diffraction orders. The recording time is shorter (Fig. 5) compared to previously investigated sample of homopolymer 8*NN-10PM (S-isomer) which has time of recording about minutes [7]. The behavior of grating recording on wavelength 405 nm is similar.

Holographic grating recording by two beams with mutually parallel linear polarization in the liquid crystalline polymer and evolution of the diffraction order efficiency is studied. The characteristics of this grating obtain at wavelengths 405 and 532 nm with the readout beam at 633 nm. Behavior of recording is similar to bacteriorhodopsin films [8].

Holographic grating recording in such system show capability of a dynamic response. Among those reversible materials, chiral liquid crystalline polymers are of peculiar interest for experimental studies of photonic periodic structures, plasmonic holography etc.

REFERENCES

[1] Zouheir Sekkat,Wolfgang Knoll (Ed.) Photoreactive Organic Thin Films, Academic Press, 2002.

[2] C.B.McArdle (Ed.), Applied Photochromic Polymer Systems, N.Y.: Blackie & Son Ltd., 1992.

[3] N. Platee, "Liquid crystalline polymers", *Chemistry. Nonlinear optics of liquid crystal*, 1988.

[4] M.V. Kozlovsky, V.P. Shibaev, A.I. Stakhanov, Th. Weyrauch, W. Haase, *Liquid Crystals*, vol. 24, 759, 1998.

[5] R.Rosenhauer, M.V.Kozlovsky, J.Stumpe, *J. Phys. Chem. A*, 107, 1441, 2003.

[6] O. Mitsunobu, Synthesis, pp. 1–28, 1981.

[7] Mikhail Kozlovsky, Ruslan Lymarenko, Lu Wang, Wolfgang Haase, "Chiral photochromic liquid crystalline polymers for holography applications," *Proc. SPIE Organic Holographic Materials and Applications II)* vol. 5521, pp. 85-95, 2004.

[8] Korchemskaya, E., D. Stepanchikov and N. Burykin Potentials of dynamic holography on bacteriorhodopsin films for real-time optical processing. *In Bioelectronic Applications of Photochromic Pigments (Edited by A. Der and L. Keszthelyi)*, pp. 74-89. IOS Press, Amsterdam, 2001.

Differential method of doppler laser anemometry of the objects with retroreflecting sheet

G.N. Dolya[1], A.M. Kryukov[2], V.G. Mudrik[2]
[1]V.N. Karazin National University, Kharkiv, Ukraine
[2]Academy of the Interior Troops of MIA of Ukraine

Abstract: The theoretical basis of the method of differential Doppler laser anemometry of the objects with retroreflecting sheet has been considered. The experimental research results of the method, its advantages, disadvantages and limitations caused by the spatial characteristics of the scattered radiation were presented.

Today, the method of laser Doppler anemometry (LDA) /1-3/ is widely used for noncontact fluid and gas flows remote control, the atmospheric turbulence condition, etc. Of special interest are differential laser Doppler anemometer circuits based on measurement of the difference between the Doppler frequency shift of the interacting beams, that are propagated in space in different directions from the moving object. However, the application of these circuits is limited by a low power of radiation received, which is a consequence of the low signal-to-noise ratio at the output of the transmitter. To overcome these difficulties, it may be very helpful to use the retroreflecting sheet (RRS), consisting, for example, of the microscopic glass beads, the characteristic sizes of which are several tens of micrometers (Fig. 1) /4/.

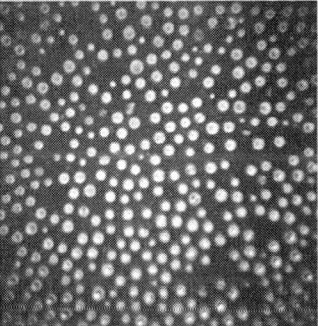

Figure 1. Structure of the RRS

The diagram of coherent light radiation scattering based on a large number of "glittering spots" in the far-field diffraction represents the chaotic combination of local maximums, the statistical regularities of which are determined as the correlation function of observed scattered field. If you introduce the field correlation function for the point source placed on the surface of the object at the point $\vec{r} = 0$, and denote this correlation function by $B(\vec{\rho}_1, \vec{\rho}_2)$, the field correlation function in the observation plane, characterized by the vector $\vec{\rho}$ can be expressed as

$$\left\langle E(\vec{\rho}_1)E^*(\vec{\rho}_2)\right\rangle = B(\vec{\rho}_1, \vec{\rho}_2)\int_\sigma I_\text{н}(\vec{r})\, exp\left[-ik\frac{\vec{r}(\vec{\rho}_1 - \vec{\rho}_2)}{L}\right]d\sigma \quad (1)$$

where $I_\text{н}(\vec{r})$ - a normalized intensity distribution on an object surface [5].

Equation (1) shows that increasing the size of the illuminated part of the object (as long as the wavelength and its range remain fixed) leads to the decrease of the area of field coherence (the speckle size) on the receiving aperture. Since the point source produces the spatial coherent field at the aperture, in the case of an coherence decrease is determined by the modulus of the integral in the right-hand member (1). The following function

$$F(\Delta\vec{\rho}) = \left|\int_\sigma I(\vec{r})\, exp\left(-ik\frac{\vec{r}\Delta\vec{\rho}}{L}\right)d\sigma\right| \quad (2)$$

defines the size of coherence area d_κ. For example, if the illuminated area represents evenly illuminated circle of radius r_0, it is easy to show that the function $F(\Delta\vec{\rho})$ vanishes range $\Delta\vec{\rho} = 1{,}22L\lambda / 2r_0$.

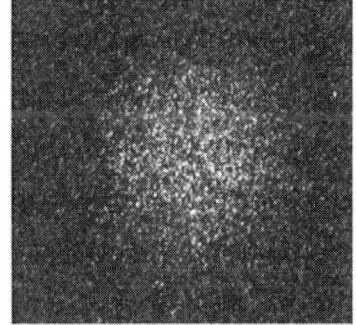

Figure 2. Speckle distribution.

From the practical application viewpoint this means that the core power of scattered radiation is contained within a few degrees sized spot of angular measure θ, in case if the average measure of the illuminated part of a microscopic glass beads several tens of micrometers when the surface is illuminated by the helium-neon laser. If the remote control distance is a few meters, then the average size of speckle is equal to several millimeters when the measure of illuminated area equals to several millimeters.

In the case of object motion, even at normal incidence of flat monochromatic wave with cyclic frequency f on its surface provided that traffic route along this surface equals to velocity V the beams within the directional pattern scattered in directions θ, that are different from the normal, will have frequency shifts dependent on the direction of scattering in accordance with the Doppler effect.

This fact is the basis for the implementation of a differential circuit laser anemometry.

Let probing beams A and B fall onto the surface element of RRS moving in space at a velocity of V, in a plane containing the beams A and B as it is shown in Figure 4. The electromagnetic wave propagates towards A, and the electric component of this wave has a complex value $\underline{E}_A$, and the phase changes in time according to the harmonic law with cyclic frequency f. When reflected from the moving surface element the wave frequency changes in accordance with the (Doppler law) [3, 8]:

$$f_A = f\frac{1 - \dfrac{V}{c}cos\alpha}{1 - \dfrac{v}{c}cos\varphi} = f\left(1 - \frac{V}{c}cos\alpha\right)\left(1 + \frac{V}{c}cos\varphi\right) =$$

$$= f\left[1 - \frac{V}{c}\left(cos\alpha - cos\varphi\right)\right] \quad (3)$$

where the angle of incidence α and the scattering angle φ are determined in accordance with Figure 4, and the velocity V is supposed to be much lower than the velocity of light c.

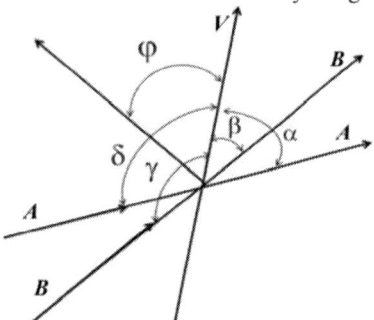

Figure 3. Circuit of probing beams path.

Similarly, the equation for the wave frequency $\underline{E}_B$ after its reflection from the surface can be written:

$$f_B = f\frac{1 - \dfrac{V}{c}cos\beta}{1 - \dfrac{v}{c}cos\varphi} = f\left(1 - \frac{V}{c}cos\beta\right)\left(1 + \frac{V}{c}cos\varphi\right) =$$

$$= f\left[1 - \frac{V}{c}\left(cos\beta - cos\varphi\right)\right] \quad (4)$$

When detected, these two waves scattered onto direction φ generate an electrical current the power of which is proportional to the resulting field intensity:

$$J \propto \left|\underline{E}_\Sigma\right|^2 = \left|\underline{E}_A + \underline{E}_B\right|^2 = \left|\underline{E}_A\right|^2 + \left|\underline{E}_A\right|^2 +$$

$$+ \underline{E}_A\,\underline{E}_B^* + \underline{E}_A^*\,\underline{E}_B \quad (5)$$

In the right-hand member of this equation the first two terms create a constant average <u>amperage</u> value when detected in time, that is much more than the light oscillations period $T \gg 1/f$.

Oscillation frequency of the third term is defined by the ratio

$$\Delta f = f_A - f_B = \frac{fV}{c}\left(cos\alpha - cos\beta\right) \quad (6)$$

and the beat frequency, that is defined by the fourth term, is negative, so it will not be considered below (note that its module is also defined by equation (6)).

Equation (6) shows the well-known fact that the difference frequency Δf of the Doppler shifts of interacting waves can be used to define the component of object velocity, which lies in the plane of interacting waves [6]. Also this value depends on the conditions of the spatial arrangement of the interacting waves in relation to the velocity vector.

The practical implementation of this method was carried out on the experimental system schematics shown in Figure 4

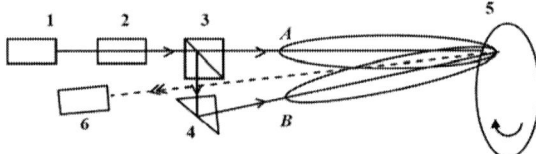

Figure 4. Implementation scheme for the differential method.

Helium-neon laser 1 radiation through collimator 2 was directed towards beam splitting cube 3 which with the help of prism 4 created beams A and B. These beams were then directed to the rotating disk 5, the surface of which was coated with RRS based on a microscopic glass beads (see Figure 1). Angular misalignment of rays A and B was selected in such a way that it was smaller than the angular value of zero diffraction order of laser radiation on the RSS, i.e. $\alpha - \beta < \theta$.

The diaphragm aperture was chosen equal to the average diameter of the speckle (a few millimeters), which provided the best signal-to-noise ratio.

Thus, the total radiation received due to the reflection of rays A and B fell on the photodetector 6 forming its output signal whose frequency was determined in the equation (6). This signal was then fed to the line input of the computer sound card and was analyzed with the program *Cool Edit*.

The typical spectrograms of photocurrent are shown in Figures 5 - 6. The most important feature of this method is the high signal-to-noise ratio in the received signal (about 30 dB), which makes this method very attractive for solving several problems, that allow coating RRS on the object under study. The position of the maximums in the spectrograms corresponds to the frequencies of difference Doppler shifts (see (6)) and can be used to determine the object velocity V. Thus, the maximum in the spectrogram for the initial position of the beam convergence point (curve 1 in Figure 5) is observed in the 4 kHz range. When the beam convergence point was moved along the disk surface in the radial direction from the center at a constant angular velocity, the linear velocity V increased, resulting in the adequate shift of the spectrogram maximum to 8 kHz range (curve 2).

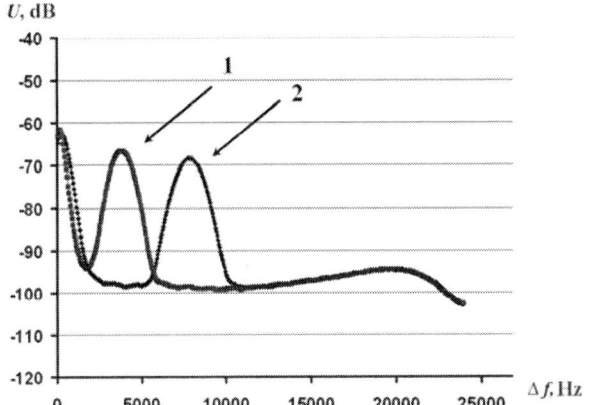

Figure 5. Spectrograms of the photocurrent for various values of object linear velocity.

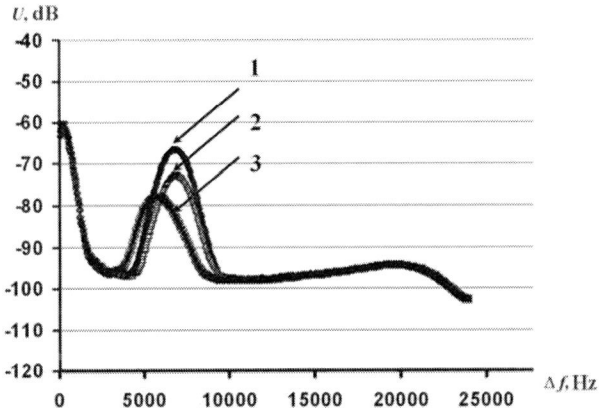

Figure 6. Spectrograms of the photocurrent for various degrees of spatial misalignment of beams.

In case of exact beam convergence on RRS, the signal-to-noise ratio was about 30 dB (curve 1, Figure 6). However, in case of spatial divergence of beams along the velocity vector at a distance of about the radius of a light spot on the surface of RRS, the maximum value decreased (curve 2). In case of spatial divergence of beams across the velocity vector, there was both the change of maximum value on the spectrogram and its shift along the frequency axis (curve 3). The decrease of the amplitude of photocurrent oscillations cannot be explained by the decreased intensity of interacting waves, and is a consequence of Siegman antenna theorem [7], and the frequency shift is caused by the change of velocity values and angle values α , β appearing in (6).

Another feature of this method implementation lies in the necessity of the exact selection of the direction of the total resultant scattered radiation reception. As noted above, it is necessary to ensure the fulfillment of the condition $\alpha - \beta < \theta$ in order to make a successful beat recording in the spectrogram. Otherwise (when the angle φ is arbitrary chosen) the amplitude $\underline{E}_A$ and $\underline{E}_B$ values in the third term of equation (5) become slight due to the characteristic features of retroreflectioin, and the definition of the difference in the Doppler frequency in the spectrogram become impossible due to low signal-to-noise ratio.

CONCLUSION

The main feature of the scattered laser radiation on the retroreflecting sheet is a significant concentration of its power in the direction within several degrees of the probing source. This paper shows that under the two-beam probing of moving objects containing RRS, the beats are recorded with a frequency proportional to the velocity, and the signal-to-noise ratio is about 30 dB. At the same time, the features of retroreflection impose limits on object probing angles and the receive direction.

REFERENCES

[1] V.P.Klochkov, L.F.Kozlov, I.V Potykevich, M.S.Soskin, "Laser anemometry, remote spectroscopy and interferometry", Handbook, Kiev, pp. 6-35,1985.

[2] B.S.Rinkevichyus, V.A.Fabrikant, "Laser diagnostics of streams", Moscow, pp. 4-18, 1990.

[3] G.N.Dolya, A.N.Katunin, K.V.Sadovy, S.I. Shmatkov, "Speckle interferometry method of the anemometry of turbulent streams", Proceedings of "Visnuk of the V.N. Karazin National University, Kharkiv", № 756, Series of "Radio Physics and Electronics", issue № 11, Kharkiv, pp.118-122, 2007.

[4] John Lloyd "A brief history of retroreflective sign face sheet materials. The principles of retroreflection", Electronic resource: http://www.rema.org.uk/pdf/history-retroreflective-materials.pdf.

[5] N.D. Ustinov, I.N. Matveev, V.V. Protopopov, "Methods of processing of optical fields in laser location", Moscow, Main edition physical and mathematical literature, 1983.

[6] R.Ditchburn, "Physical Optics", trans. from eng. ed. I.A.Yakovleva, Moscow, Main edition physical and mathematical literature, , 1965.

[7] V.V.Protopopov, N.D. Ustinov, "Laser heterodyning", ed. N.D. Ustinov, Moscow,1985.

Microstructure and chemical composition of heterogeneous crystal GaSe:AgGaS$_2$

V. V. Atuchin[1], Yu. M. Andreev[2], N. F. Beisel[3,4], A. R. Tsygankova[3], T. A. Gavrilova[5], L. D. Pokrovsky[1], A. I. Saprykin[3]

[1]Laboratory of Optical Materials and Structures, A.V.Rzhanov Institute of Semiconductor Physics, SB RAS, Novosibirsk 630090, Russia

[2]Laboratory of Ecological Devices Making, Institute of Monitoring of Climatic and Ecological Systems, Tomsk 634055, Russia

[3]Analytical Laboratory, Nikolaev Institute of Inorganic Chemistry, SB RAS, Novosibirsk 630090, Russia

[4]Department of Natural Sciences, Novosibirsk State University, 2, Pirogova St., Novosibirsk 90, 630090, Russia

[5]Laboratory of Nanodiagnostics and Nanolithography, A.V.Rzhanov Institute of Semiconductor Physics, SB RAS, Novosibirsk 630090, Russia

Abstract: The GaSe crystal doped with AgGaS$_2$ has been grown and evaluated by SEM and TEM. The chemical composition analysis has been produced by atomic spectrometry methods. Micromorphology and structural parameters of GaSe:AgGaS$_2$ crystal are very similar to those of pure GaSe. The inhomogeneous distribution of Ag over crystal probes has been detected by atomic-absorption spectrometry (AAS). Generally, the GaSe:AgGaS$_2$ crystal has been classified as a GaSe:S solid solution with Ag precipitation.

Gallium selenide, -GaSe, space group 6- 2, is among most promising materials for optical frequency conversion in infrared (IR) and THz spectral ranges [1-4]. Because of layered crystal structure, however, pure GaSe possesses high cleavage that is a key problem in optical device fabrication. From other hand, GaSe crystal lattice is capable to incorporate different doping elements at high content and with noticeable modification of structural, mechanical, electronic and optical properties [1-9]. This method of crystal property modification is well known and widely used for tuning linear and nonlinear optical properties responsible for frequency conversion or waveguide formation in different oxide or chalcogenide materials [1,2,4,6,8,10-22]. Commonly, with only few exceptions, element agents are used for the substitution at cation or anion sites in parent crystal lattice. As to -GaSe, wide range solid solutions were found for Ga substitution by In and Se substitution by S or Te. Besides doping with element admixtures, such exotic complex doping agent as AgGaSe$_2$ was proposed [23]. The GaSe crystal doped with AgGaSe$_2$ compound, 10.4 mass , was characterized by increased nonlinear coefficient of 75 pm V yielding a figure of merit six times larger than that of AgGaSe$_2$ and twice than that of ZnGeP$_2$. To develop this doping idea, the present study is aimed to crystal growth of $(Ga_2Se_2)_{1-x}(AgGaS_2)_x$, $x = 0.1$, solid solution and an observation of structural and optical properties of grown crystal with SEM, TEM and optical spectroscopy.

Doped GaSe crystal was grown by conventional Bridgman technique in evacuated quartz ampoule of 18 mm in diameter. The starting materials for GaSe synthesis were 6N Ga and 6N

Se. Stoichiometric AgGaS$_2$ was added into the charge used for crystal growth. The samples were cleaved from grown ingot parallel to the c-plane and were used for measurements without any additional treatment or polishing. Micromorphology was evaluated by scanning electron microscopy (SEM) using LEO 1430 device. Element composition was estimated with electron probe microanalysis (EPMA). Structural properties of doped crystals were observed by transmission electron microscopy (TEM) under electron energy 100 kV by BS-513A device.

The quantitative composition determination was provided with the methods of atomic spectrometry. Sample weights were dissolved in pure nitric acid and diluted with deionised water. Sulphur was determined by inductively coupled plasma optical emission spectrometry: sample solution containing less than 1 mg ml of GaSe (for eliminating matrix interferences) was introduced into argon plasma of spectrometer iCAP-6500, "Thermo Scientific", and spectral line intensity was measured at $\lambda = 180.7$ nm Sc was used as internal standard. Silver concentration was determined by atomic-absorption

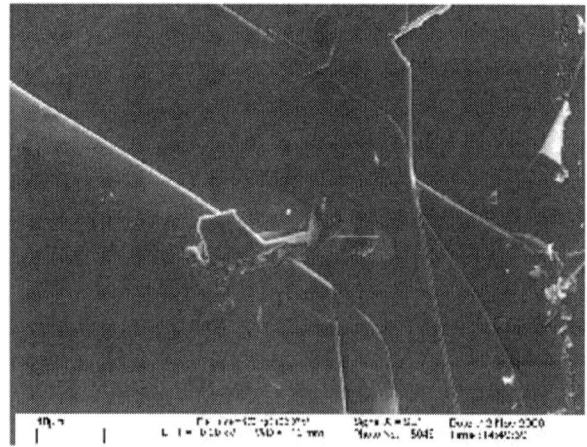

Fig. 1. SEM image of (001) cleaved surface.

spectrometry using AA spectrometer with Zeeman background correction Z8000, "Hitachi", air-acetylene flame silver atomic absorption was measured at λ = 328.1 nm. Matrix interferences up to GaSe concentration of 8-10 mg ml were not detected.

As it is shown in Fig.1, sample surface was formed by flat terraces with sharp edges that is typical for the GaSe surface. The EPMA analysis, besides gallium and selenium, reveals clear signal of sulphur and no signal related to silver. Determination of chemical composition of $GaSe:AgGaS_2$ solid solution shows stable sulphur content at the level of 2.19 0.05 wt . Contrary to that, the silver content fluctuates drastically from sample to sample within the range 0.06-1.6 wt . Generally, it can be concluded that sulphur is homogeneously distributed over the crystal bulk. As to silver, this admixture seems to be concentrated into local precipitates. Some precipitate-type structures were detected by SEM analysis. Only the GaSe-type crystalline phase was detected by TEM observation. Representative SAED patterns are shown in Figures 2 and 3. As it may be supposed, silver-bearing precipitates are either amorphous or these were not opened by

the sample preparation. As it is found by complex observation, silver atoms were not incorporated into GaSe lattice and were segregated with the formation of amorphous inclusions. Generally, the grown $GaSe:AgGaS_2$ crystal should be classified as a GaSe:S solid solution.

As it is shown by crystal growth experiment, the optical quality $GaSe:AgGaS_2$ crystals can be grown. The micromorphology of cleaved surface and SAED patterns obtained from grown crystal are close to those of pure GaSe. Chemical composition analysis, however, reveals the drastic inhomogenety of Ag distribution that indicates very low or even zero solubility of Ag in GaSe. Supposedly, similar result was achieved in Ref. 23, when $GaSe:AgGaSe_2$ crystal was grown and used in frequency conversion experiment.

Ac   ts

This study was partly supported by SB RAS under Integration Pro ect No. 46.2013 and Ministry of Education and Science of Russian Federation (state contract No. 16.518.11.7091).

Fig. 2. SAED patterns recorded from dominant high-quality crystal part at electron beam along [001].

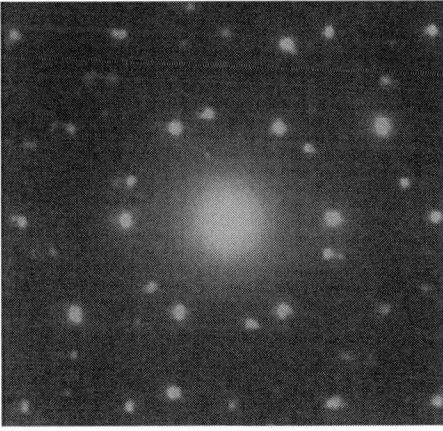

Fig. 3. SAED patterns recorded from most defective parts of GaSe:AgGaS_2 sample at electron beam along [001].

REFERENCES

[1] K. R. Allakhverdiev, R. I. Guliev, E. Y. Salaev, and V. V. Smirnov, "An investigation of linear and non-linear optical properties of GaS_xSe_{1-x} crystals," e , vol. 9, pp. 1483-1485, 1982.

[2] N. C. Fernelius, "Properties of gallium selenide single crystal," e ., vol. 28, pp. 275-353, 1994.

[3] K. C. Mandal, Sung Hoon Kang, M. Choi, Jian Chen, i-Cheng Zhang, J. M. Schleicher, C. A. Schmuttenmaer, and N. C. Fernelius, "III-VI chalcogenide semiconductor crystals for broadband tunable THz sources and sensors," $e e$ e t, vol. 14, pp. 284-288, 2008.

[4] S. A. Ku, W. C. Chu, C. W. Luo, Y. M. Andreev, G. Lanskii, A. Shaiduko, T. Azaak, V. Svetlichnyi, K. H. Wu, and T. Kobayashi, "Optimal Te-doping in GaSe for non-linear applications," x e , vol. 20, pp. 5029-5037, 2012.

[5] Yu. M. Andreev, V. V. Atuchin, G. V. Lanskii, A. N. Morozov, L. D. Pokrovsky, S. Yu. Sarkisov, and O. V. Voevodina, "Growth, real structure and applications of $GaSe_{1-x}S_x$ crystals," e , vol. 128, pp. 205-210, 2006.

[6] S. Das, C. Ghosh, O. G. Voevodina, Yu. M. Andreev, and S. Yu. Sarkisov, "Modified GaSe crystal as a parametric frequency converter," , vol. 82, pp. 43-46, 2006.

[7] Zhi-Shu Feng, Zhi-Hui Kang, Feng-Guang Wu, Jin-Yue Gao, Yun Jiang, Hong-Zhi Zhang, Yury M. Andreev, Grigory V. Lanskii, Victor V. Atuchin, Tatyana A. Gavrilova, "SHG in doped GaSe:In crystals," x e , vol. 16, pp. 9978-9985, 2008.

[8] S. Yu. Sarkisov, V. V. Atuchin, T. A. Gavrilova, V. N. Kruchinin, S. A. Bereznaya, Z. V. Korotchenko, O. P. Tolbanov, and A. I. Chernychev, "Growth and optical parameters of GaSe:Te crystals," ., vol. 53, pp. 346-352, 2010.

[9] V. V. Atuchin, N. F. Beisel, K. A. Kokh, V. N. Kruchinin, I. V. Korolkov, L. D. Pokrovsky, A. R. Tsygankova, and A. E. Kokh, "Growth and microstructure of heterogeneous crystal GaSe:InS," , vol. 15, pp. 1365-1369, 2013.

[10] V. V. Atuchin, C. C. Ziling, D. P. Shipilova, N. F. Beizel, "Crystallographic, ferroelectric and optical properties of TiO_2-doped $LiNbO_3$ crystals," *e e e* , vol. 100, pp. 261-269, 1989.

[11] V. V. Atuchin, C. C. Ziling, I. Savatinova, M. N. Armenise, and V. M. N. Passaro, "Waveguide formation mechanism generated by double doping in ferroelectric crystals," ., vol. 78, pp. 6936-6939, 1995.

[12] I. Savatinova, I. Savova, E. Liarokapis, C. C. Ziling, V. V. Atuchin, M. N. Armenise, and V. M. N. Passaro, "A comparative analysis of Rb:KTP and Cs:KTP optical waveguides," , vol. 31, pp. 1667-1672, 1998.

[13] J. Z. Huang, D. M. Ren, . Y. Hu, Y. C. u, Y. Andreev, P. Geiko, V. Badikov, and G. Lanskii, "Nonlinear optical properties of mixed $Cd_{0.35}Hg_{0.65}Ga_2S_4$ crystal," , vol. 53, pp. 3761-3765, 2004.

[14] Yu. M. Andreev, V. V. Atuchin, G. V. Lanskii, N. V. Pervukhina, V. V. Popov, and N. C. Trocenco, "Linear optical properties of $LiIn(S_{1-x}Se_x)_2$ crystals and tuning of phase matching conditions," *d e* ., vol. 7, pp. 1188-1193, 2005.

[15] J.-J. Huang, V. V. Atuchin, Yu. M. Andreev, G. V. Lanskii, and N. V. Pervukhina, "Potentials of $LiGa(S_{1-x}Se_x)_2$ mixed crystals for optical frequency conversion," , vol. 292, pp. 500-504, 2006.

[16] Jin Jer Huang, Wei Gao, Tao Shen, Bei Li Mao, Yu. M. Andreev, A. V. Shaiduko, G.V. Lanskii, Udit Chatter ee, and V. V. Atuchin, "Influence of composition ratio variations on optical frequency conversion in mixed crystals. II. Random variation of composition ratio," ., vol. 24, pp. 3081-3090, 2007.

[17] V. Badikov, K. Mitin, F. Noack, V. Panyutin, V. Petrov, A. Seryogin, and G. Shevyrdyaeva, "Orthorhombic nonlinear crystals $Ag_xGa_xGe_{1-x}Se_2$ for the mid-infrared spectral range," *e* ., vol. 31, pp. 590-597, 2009.

[18] O. V. Parasyuk, V. V. Atuchin, Y. E. Romanyuk, L. P. Marushko, L. V. Piskach, I. D. Olekseyuk, S. V. Volkov, and V. I. Pekhnyo, "The $CuGaSe_2$-$CuInSe_2$-$2CdS$ system and single crystal growth of the -phase," , vol. 318, pp. 332-336, 2011.

[19] M. V. Shevchuk, V. V. Atuchin, A. V. Kityk, A. O. Fedorchuk, Y. E. Romanyuk, S. Całus, O. M. Yurchenko, O. V. Parasyuk, "Single crystal preparation and properties of the $AgGaGeS_4$-$AgGaGe_3Se_8$ solid solution," , vol. 318, pp. 708-712, 2011.

[20] Z.-H. Kang, J. Guo, Z.-S. Feng, J.-Y. Gao, J.-J. ie, L. M. Zhang, V. Atuchin, Y. Andreev, G. Lanskii, and A. Shaiduko, "Tellurium and sulfur doped GaSe for mid-IR applications," , vol. 108, pp. 5251-5255, 2012.

[21] V. V. Atuchin, S. V. Adichtchev, B. G. Bazarov, Zh. G. Bazarova, T. A. Gavrilova, V. G. Grossman, V. G. Kesler, G. S. Meng, Z. S. Lin, and N. V. Surovtsev, "Electronic structure and vibrational properties of $KRbAl_2B_2O_7$," *e e* ., vol. 48, pp. 929-934, 2013.

[22] Jin Guo, Ji iang ie, Laiming Zhang, Ke Jiang, Fei Chen, Yu. M. Andreev, V. V. Atuchin, K. A. Kokh, G. V. Lanskii, V. F. Losev, A. V. Shaiduko, "Interaction of high intensity optical pulses with modified nonlinear GaSe crystals," , vol. 8796, pp. 87962D, 2013.

[23] N. B. Singh, D. R. Suhre, W. Rosch, R. Meyer, M. Marable, N. C. Fernelius, F. K. Hopkins, D. E. Zelmon, and R. Narayanan, "Modified GaSe crystals for mid-IR applications," , vol. 198 199, pp. 588-592, 1999.

Pyroelectric effect in X-cut LiNbO₃ optical modulators

R. S. Ponomarev[1], A.B. Volyntsev[1], I.S. Azanova[2], E. D. Voblikov[2]
[1]Perm State University, Perm, Russia
[2]Perm State Technical University, Perm, Russia

Abstract: We investigate the pyroelectric effect in lithium niobate crystal and its influence on the integrated optical Mach-Zehnder interferometer. It was shown that pyroelectric charges provoke high operating point drift and waveguide properties degradation with sample heating.

Lithium niobate (LN) optical modulators are successfully used in telecommunications, fiber-optic gyroscopes and electric field measurements [1-3]. These devices have temperature instabilities connected with pyroelectric properties of LN crystal. This effect was studied earlier for Z-cut LN intensity modulators [4, 5] but not for X-cut modulators because of simplicity of pyroelectric charge rejection using crystal faces electrical closing. Theoretical first view on X-cut LN devices shows that pyroelectric effect cause equal phase shift in optical waveguides of Y-splitter or Mach-Zehnder interferometer because of uniform electric field and waveguide proximity and optical output intensity need to be stable. But results of our experiments demonstrate very strong influence of pyroelectric effect on X-cut LN intensity modulator based on Mach-Zehnder interferometer. This paper is devoted to explanation of pyroelectric influence to X-cut LN devices.

To investigate pyroelectric effect in LN devices we have made three specimens of integrated optical Mach-Zehnder interferometers. The symmetrical waveguide topology was created by proton exchange with annealing (APE) procedure on congruent X-cut LN crystal. Along Y-axes was placed 6 μm single mode waveguides for λ=1.55 μm. APE parameters was chosen for best matching of numerical aperture and mode field diameter. Integrated optical chip was connected with polarization maintaining fibers using optical adhesive with low thermal expansion coefficient. Crystal +z and −z faces were carefully cleaned to decrease surface conductivity. The size of chip was 50×3.2×1 mm. For these chip parameters the estimated pyroelectric voltage was about 560 V/°C without pyroelectric relaxation and parallel-plate capacitor approximation.

Thermal cycles was made using Espec MC-711 thermal chamber. Experimental specimens were connected with 1.55 μm fiber laser and photodetector. We used General Photonics ERM-102 with PC connection to collect optical power data. Fiber laser and photodetector was out of chamber during thermal cycles. The temperature change from -60 to +60 °C with different (0.1 to 1.6 °C/min) rates and constant sections. Measurement time was about 60 hours. Fig 1. shows the experimental setup.

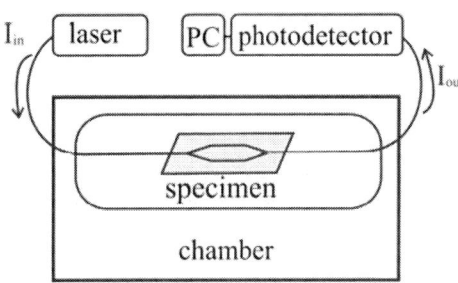

Fig. 1. Experimental setup with specimen.

All specimens show similar optical signal behavior with temperature changing. It was collected a lot of data and we show only representative parts of it. Typical curve of I_{out} (t) is shown in Fig.2 and Fig.3.

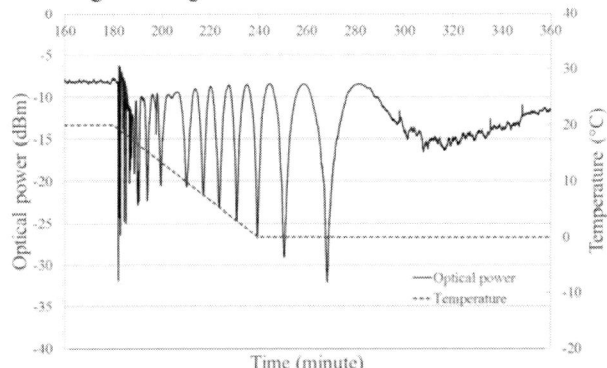

Fig. 2. Optical power with specimen cooling.

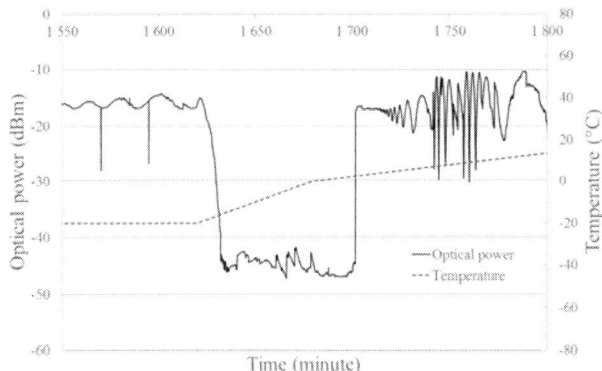

Fig. 3. Optical power with specimen heating.

978-1-4799-0019-0/13 $31.00 © 2013 IEEE

As it is shown in Fig. 2 optical power was changed immediately after start of specimen heating (180 min). It is observed periodical signal change corresponding to interferometer operating point movement along cosine transfer function in logarithmic scale. The velocity of this movement decreased after temperature stabilization and relaxation time was about 2 hours (240 to 360 min). The magnitude of this operating point drift is about 10π with ΔT=20 °C what is equal to electric field strength about 3.5 kV/cm. Using typical value of V_π=3 V and electrode gap value 10 μm this electric field is not enough to move operating point on 10π. Some effect increase the influence of pyroelectic charges to the channel waveguides.

Points of power minimums demonstrate very uncommon behavior. The envelope curve of minimum points have maximum near 205 min in Fig.2. After reach of this maximum operating point start move backward monotonically. It is equal to applying about 20 Volts to modulator electrodes. The modulation depth is changing from 11 dB to 24 dB. But usually it is constant for optical amplitude modulators with DC modulation.

Optical power with specimen heating is described by Fig.3. It was observed very fast degradation and restoration of waveguide properties, at 1625 min and 1700 min respectively. This effect was observed many times for different specimens only with specimen heating and it was not related with chip to fiber connection.

Therefore we have a few experimental facts:
- Pyroelectric effect cause to increase of phase difference and refractive index difference between two symmetrical waveguides.
- Initially the refractive index difference was increased and after certain critical point it was go back.
- The modulation depth of optical modulation is changed for account of minimal optical power changing.
- Specimen heating may cause to fast waveguide degradation or waveguide «turning-off» and «turning-on».
- The magnitude of pyroeffect is not enough to keep observed signal changing.

In our opinion the explanation of these facts is connected with structure of LN crystal and waveguide creation process. All of these facts are connected with waveguide refractive index change related to movement of charged defects in LN crystal lattice. These charged defects need to conserve near a waveguide and need to be mobile for a long time. Before proton exchange procedure LN crystal has very high electrical resistance [6] and there is no mobile charge in the LN lattice. After proton exchange procedure about 80 % of Li^+ ions are replaced to H^+ ions in regular and interstitial positions [7, 8].

The crystal lattice of waveguide region is increased simultaneously with ion exchange [9] and two lattices (initial LN and exchanged LN) need to match using transitional region. This region needs to include some oxide plane dislocations and other crystal imperfection which are negatively charged. These imperfections attract positive charged H^+ ions and conserve their mobility. Therefore we offer the waveguide model showed in Fig. 4.

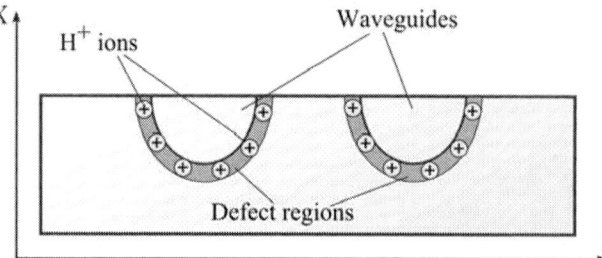

Fig. 4. Mobile charges near waveguide (model)

The pyroelectric field cause to moving of mobile charges and redistribution of charges in electrode system. This complex charge system can increase the effective electric field in the waveguide. During the heating the pyroelectric field cause decreasing of waveguide contrast Δn and tranition to the two-mode operating mode. All of these effects need to be investigated to create integrated optical devices which are able to operate in high temperature range.

REFERENCES

[1] Wooten, E., Kissa, K., & Yi-Yan, A. (2000). A review of lithium niobate modulators for fiber-optic communications systems. Selected Topics in Quantum Electronics, 6(1), 69–82.

[2] Lefevre, H. C. (1993). The Fiber-Optic Gyroscope (p. 332). Artech.

[3] Zeng, R., Wang, B., Niu, B., & Yu, Z. (2012). Development and application of integrated optical sensors for intense E-field measurement. Sensors (Basel, Switzerland), 12(8), 11406–34.

[4] Skeath, P., Bulmer, C. H., Hiser, S. C., & Burns, W. K. (1986). Novel electrostatic mechanism in the thermal instability of z-cut LiNbO3 interferometers. Applied Physics Letters, 49(19), 1221.

[5] Bulmer, C. H., Burns, W. K., & Hiser, S. C. (1986). Pyroelectric effects in LiNbO3 channel-waveguide devices. Applied Physics Letters, 48(16), 1036.

[6] Wong, K. K. (2002). Propeties of lithium niobate. (K. K. Wong, Ed.) (p. 432). London: The Institution of Engineering and Technology.

[7] Nekvindová, P., Špirková, J., Červená, J., Budnar, M., Razpet, A., Zorko, B., & Pelicon, P. (2002). Annealed proton exchanged optical waveguides in lithium niobate: differences between the X- and Z-cuts. Optical Materials, 19(2), 245–253.

[8] Korkishko, Y. N., & Fedorov, V. A. (1996). Structural phase diagram of Hx/Li1-xNbO3 waveguides: The correlation between optical and structural properties. IEEE Journal of Selected Topics in Quantum Electronics, 2(2), 187–196.

[9] Azanova, I. S., Shevtsov, D. I., Zhundrikov, A. V. et al. (2008). Chemical etching technique for investigations of a structure of annealed and unannealed proton exchange channel LiNbO₃ waveguides. Ferroelectrics, 374(1), 110–121.

Fabrication of Controllable Holographic Gratings to Manage the Energy Transfer

S. Bugaychuk[1], L. Pryadko[1], I. Pryadko[1], O. Kolesnyk[1,2], R. Conte[3], V. Gnatovskiy[1], A. Negriyko[1]

[1] Institute of Physics of National Academy of Sciences of Ukraine, 46 Prospect Nauki, 03028 Kiev, Ukraine
[2] Taras Shevchenko National University, Radio-physical Department , 4 Prospect Glushkova, 03127 Kiev, Ukraine
[3] LRC MESO. École normale supérieure de Cachan (CMLA) et CEA-DAM 61, Avenue du Président Wilson, F-94235 Cachan Cedex, France

Abstract – **The peculiarities of managements of optical beams on controlled diffraction gratings are investigating both theoretically and experimentally. The gratings are distinguished by non-uniform phase contrast along a thickness of a sample in the direction of wave propagation. The manipulation of a space shift between the light grating and the refractive-index grating is taken into account as well. The self-consistent problem of the self-diffraction of waves from shifter gratings gives that the intensity profile gets a soliton-like envelope along the thickness of the sample. We show that energy transfer is increased for a weak beam due to diffraction on a grating, which has a profile of a phase contrast in a form of a sech-function obtained from a self-consistent task.**

I. INTRODUCTION

Finding different methods of light-control-light continues to be relevant in connection technological development of integral optical circuits. Diffraction gratings are most popular because they easily can be modified to implement various, sometimes contradictory, needs, for example, such as significant amplification of individual diffraction orders, spectral division of optical beams, spectral selectivity, etc. [1]. The modern technologies permit to embed the diffraction gratings onto integrated optical circuits.

The energy transfer among different diffraction orders is one of a remarkable feature that is realized by the gratings. Many useful applications based on this effect known, such as optical phase conjugation, amplification and steering of beams, spectral filters for telecommunication, adaptive interferometry, novelty filters, different tipes of holographic storage (permanent, dynamical, multiplexing, multibeam recording), etc. The mechanisms of recording of shifted dynamic gratings as well as the various applications of the dynamical energy transfer between interacting laser beams were widely studied for the photorefractive crystals [2, 3]. In the dynamical regime, the usual way is the investigation of the self-diffraction of waves just in the case of volume gratings (for the Bragg conditions), when the only diffraction orders coinciding with the propagation direction of interacting waves exist. The energy transfer of a probe beam into high diffraction orders in dependence of the beam polarization was studied for liquid crystal cells recently, where a thin refractive index grating was recorded [4].

Study of the self-consistent problem of the self-diffraction of waves under the Bragg conditions with shifted gratings show that the intensity pattern has a non-uniform profile in a volume of a dynamical medium along the direction of the wave propagation. This problem includes that a photorefractive dynamical medium possesses both a property of the amplification being proportional to the light intensity and an exponential relaxation. In this case the intensity pattern gets a soliton-like profile (in the form either of the sech-function for the transmission geometry or of the tanh-function in the reflection geometry) in the z-direction of the volume of the medium. The refractive-index pattern, which is proportional to the light intensity distribution, will have the similar soliton-like envelope. The behavior of the wave-mixing in such dynamical medium is reduced to the complex Ginzburg-Landau equation, by which the whole problem is described that includes both the wave-mixing and the time evolution of the medium [5, 6].

In this paper we investigate both theoretically and experimentally the two-wave mixing on gratings under two key circumstances: (i) the grating is shifted relative to the light interference pattern; (ii) a refractive-index grating has non-uniform distribution of its amplitude along the direction of wave propagation. We present theoretical and experimental results, which prove a new principle to control the energy transfer and the diffraction efficiency, namely to construct a phase grating with soliton-like profile. In the experiments, we use given thin refractive-index gratings, which have non-uniform phase contrast distribution in thickness (along the z-axis of the wave propagation). The method of the preparation of the gratings is described in the Section III: Fabrication of gratings with changing phase contrast. Our experimental study of the two-wave mixing proves that the diffraction of waves from the shifted gratings with altering phase contrast along the thickness gives new possibilities of the beam steering, which is impossible to realize in the case of a standard gratings.

978-1-4799-0019-0/13 $31.00 © 2013 IEEE

II. MANAGE THE ENERGY TRANSFER DURING TWO-WAVE MIXING ON THE GRATING WITH SOLITON-TYPE PHASE PROFILE

The effect of the energy transfer can be always archived when the grating is shifted relative to the light interference pattern. Amplification of weak beams due to the energy transfer has been highly investigated for photorefractive crystals where the shifted gratings are formed owing to the diffusion mechanism of the grating recording [2, 3]. The theoretical research predicts complete energy transfer from a strong beam to a very week (noisy) beam. That means the amplification of a noisy beam which has the intensity of the order $10^{-1} \div 10^{-9}$ up to the intensity of the pump beam of the order of 1 [7]. In this case the dynamical grating gets the specific distribution of the grating amplitude, i.e. it takes a localized form of a soliton-like shape [5, 6].

From the Maxwell's equations, the two-wave mixing on a given phase grating is described by the coupled wave equations:

$$\partial A_1 / \partial z = -i\Delta n \cdot A_2 ; \qquad \partial \overline{A}_2 / \partial z = i\Delta n \cdot \overline{A}_1 \qquad (1)$$

where $A_1(z)$ and $A_2(z)$ are the complex slow variable amplitude of the two interacting waves, Δn is the grating amplitude, $z = z'k_0^2 / 2k_z$ is a dimensionless longitudinal coordinate along the thickness of a nonlinear medium, k_0 is the wave-vector amplitude if the free space, k_z is the z-component of the wave-vector in the nonlinear medium, z' is the spatial coordinate, and bar marks the complex conjugation. When the grating is shifted on the quarter grating period relative to the light pattern, $\Delta n = iN$, the wave A_1 will be amplified on the cost of the energy transfer from the wave A_2. In the case of the self-diffraction the input interacting waves form a fringe interference pattern inside a nonlinear medium that modulates the refractive index, i.e. creates a dynamical grating. Then all physical values will be functions of the time and space: $A_1 = A_1(t,z)$, $A_2 = A_2(t,z)$, $N = N(t,z)$. The self-consistent task will include an evolution equation described the gratings recording, which in the simplest case includes the amplification of the grating proportional to the light intensity (in a Kerr-like medium) and a dielectric relaxation of the grating:

$$\partial N / \partial t = N \cdot \gamma I_m / I_0 - N / \tau \qquad (2)$$

where $I_m = A_1 \overline{A}_2$ is the intensity in a given point z and time t, $I_0 = A_1 \overline{A}_1 + A_2 \overline{A}_2$ is the total intensity, γ is the amplification coefficient of the medium, τ is the relaxation constant. In a nonlocal medium the coefficient γ is complex one, which means that the dynamical grating can be shifted relative to the light interference pattern due to specific nonlinear mechanism of the grating recording: $\gamma = |\gamma| \exp(ig)$. We have shown that the self-consistent task (1)-(2) is reduced to the complex Ginzburg-Landau equation by applying the reductive perturbation technique [5]:

$$i\frac{\partial \Psi}{\partial T} + \frac{2\kappa_0}{q^3}\frac{\partial^2 \Psi}{\partial Z^2} + \frac{4\exp(-2t/\tau - 2\Im(q)z)}{q}|\Psi|^2\Psi = -i\frac{2\Im(q)}{|q|^2}\frac{\kappa_2}{\kappa_0}\Psi \quad (3)$$

$$N = -i|\gamma|(\sin g)\tau \cdot I_m \qquad (4)$$

were the functions and the variables are expanded in series of a small parameter ε: $N = \varepsilon\varphi_0 + \varepsilon^2\varphi_1 + \varepsilon^3\varphi_2 + ...,$ $\varepsilon^k z = \varsigma_k$, $\varepsilon^k t = \eta_k$. We designate $Z = \varepsilon^2 z$, $T = \varepsilon^2 t$, $\varphi_0 = \Psi(T,Z)\exp[i(qz - wt)]$. The dispersion relation is $qw + iq/\tau + i\gamma = 0$, from where it is following that if w is real then q is complex value. The coefficients $\kappa_0 = conts$ and $\kappa_2 = \kappa_2(T)$ are determined by the input intensity ratio. The expression (4) shows that the intensity profile I_m and the grating amplitude profile N have the same form, but the imaginary unit i indicates that they are shifted on phase by $\pi/2$.

The equation (3) has a steady state solution in a soliton-like form for the distribution of the grating amplitude along the thickness z:

$$E = \frac{\gamma_N C}{\cosh(2\gamma_N C z - p)} \qquad (6)$$

$$tg(u) = \exp(2\gamma_N C z - p) \qquad (7)$$

where $\gamma_N = |\gamma| \sin g$, the constants C and p are defined by the input intensity ratio I_{10} / I_{20}. The diffraction efficiency and the energy transfer are determined by the square under the profile of the grating amplitude u_d:

$$\eta(d) = \sin^2 \left| \gamma_N u_d + arctg(\sqrt{I_{10}/I_{20}}) \right| \qquad (8)$$

where $u_d = \int_0^d N(z')dz'$.

In Fig. 1 we show the energy transfer and amplification of a weak beam during the two-wave mixing by a given phase grating, which has a soliton-like profile. This profile was calculated from the self-consistent task for $\gamma_N d = 10$ and the input intensity ratio $I_{10} : I_{20} = 0.1 : 0.9$. One can see that complete energy transfer takes place in a thin layer inside the medium, where the grating amplitude is maximum; in this case a weak beam I_1 takes the value of an initially strong beam I_2.

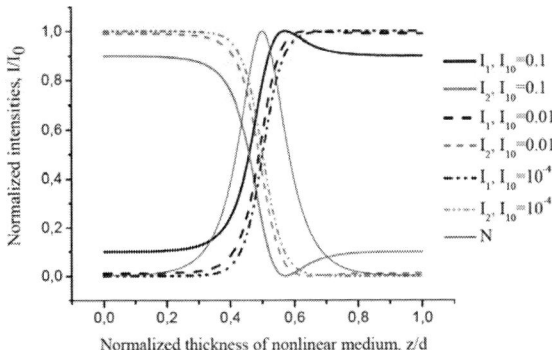

Fig. 1. The energy transfer during two-wave mixing by a phase grating, which has a soliton-like profile, for different input intensity ratios. N is the normalized soliton-like profile of the grating amplitude, which is located in the center of the nonlinear medium. d is the thickness of the medium. The total intensity is $I_0 = I_1 + I_2 = 1 = const$.

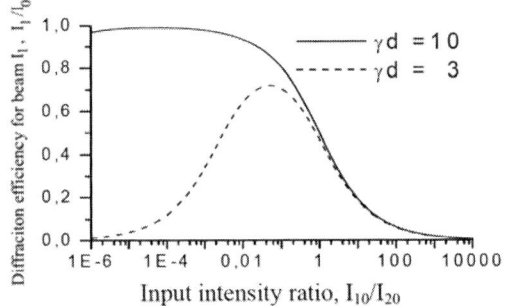

Fig. 2. The diffraction efficiency of the two-wave mixing in a nonlocal dynamical medium in dependence on the input intensity ratio. γd is the coupling constant of the nonlinear medium for the grating shifted on $\Lambda/4$, Λ is the grating period, (on the phase by $\pi/2$) relative to the light interference pattern. $I_0 = I_1 + I_2 = 1 = const$.

The typical dependence of the diffraction efficiency versus input intensity ratio for the self-consistent task of the two-wave mixing is shown in Fig.2 (for a small coupling constant ($\gamma d=3$) and the high one ($\gamma d=10$)). In all graphs the crystal occupies the distance $z/d=0\div1$; the total intensity $I_0=1$. One can see from the Fig.2 that a weak input beam I_{10} can completely get energy from a strong beam I_{20} in a medium with high nonlinearity.

III. FABRICATION OF PHASE HOLOGRAPHIC GRATINGS WITH CHANGING PHASE CONTRAST. EXPERIMENTAL RESEARCH

We have developed an experimental set-up, which is greatly allowed to implement the theoretical assumptions of the two-wave mixing of the given grating with soliton-like profile in z-direction. The complex refractive index profile has been approximated by overlaying discrete planar structures. Transparent slides were served as the basic phase plane-structures that were obtained by photographing a mask, which had either binary or sinusoidal distribution of the amplitude, with the subsequent bleaching of frames. Fine-grained (holographic) film is used, where the emulsion and the substrate have the thicknesses 10 and 100 microns respectively. By this way we may create various types of "sandwiches" to simulate the total distribution of the refractive index, for example, 100-10-10-100 μm, 10-100-10-100 μm, etc.

A holographic interferometer was mounted on the base He-Ne laser. Based on the thickness of the obtained "sandwiches", the optical scheme of the interferometer provides the conditions in which they can be considered as either "flat" or "bulk". It was achieved by precise variation of the angle of convergence of the beams, which form the interference field. A splitting system of the interferometer has provided spacing between beams from 5 mm up to 40 mm, and the base of the interferometer was varied from 1.5 m up to 5 m. Adjustment system is used to insure that the period of the interference field (a light grating) is coincided with the period of the phase grating of the "sandwich"-transporant. The intensities of input waves could be changed with the help of filters.

The "sandwich-grating" was placed in the interference field and can be moved in a transverse direction to create a shift between the light grating and the refractive index grating. An angular spectrum of the diffraction field was observed on the focal plane of an objective. The intensities in different diffraction orders after amplification with the help of a system of prisms were directed to the photodiodes. The obtained oscillograms were recorded and processed on a computer.

We compare the intensities of the main diffraction orders in dependence on intensity ratio of input beams for different types of "sandwich-gratings". We have found a specific structure of this "bulk grating" where the maximal amplification of a weak beam occurs. Our results qualitatively confirm theoretical calculations, which were made for the two-wave mixing on a given grating with the soliton-like profile.

REFERENCES

[1] M. Born and E. Wolf, *Principles of Optics*, Pergamon Press, 1970.

[2] *Photorefractive materials and their applications I*, P. Gunter and J.-P, Huignard, Eds. Springer-Verlag, Berlin, 1989.

[3] *Photorefractive materials and their applications II*, P. Gunter and J.-P, Huignard, Eds. Springer-Verlag, Berlin, 1988.

[4] P. Korneychuk, O. Tereshchenko, Yu. Reznikov, V. Reshetnyak, K. Singer, Hidden surface photorefractive gratings in a nematic liquid crystal cell in the absence of a deposited alignment layer, *J. Opt. Soc. Am. B*, vol. 23 (6), pp. 1007-1011, 2006.

[5] S. Bugaychuk and R. Conte, "Ginzburg-Landau equation for dynamical four-wave mixing in gain nonlinear media with relaxation", *Phys. Rev. E*, vol. 80, pp. 066603-1-7, 2009.

[6] S. Bugaychuk, R. Conte, "Nonlinear amplification of coherent waves in media with soliton-type refractive index pattern", *Phys. Rev. E*, vol. 86, pp. 026603-1-8, 2012.

[7] B. Sturman, E. Podivilov, M. Gorkunov, "Regimes of feedback-controlled beam coupling", *Phys. Rev. E*, vol. 72, pp. 016621-1-11, 2005.

Some optical properties of $Zn_{1-x}Mn_xTe$ semimagnetic films

O. Klymov, D. Kurbatov, O. Levchenko

Sumy State University, Rimsky-Korsakov str. 2, UA-40007, Sumy, Ukraine

e-mail: kurd@ukr.net; klymov_olexiy@mail.ru; *tel/fax:* +38 0542 334108

Abstract: Some optical properties of $Zn_{1-x}Mn_xTe$ films, obtained by close-spaced vacuum sublimation technique at different conditions are investigated. Measuring of optical characteristics layers was carried out by spectrophotometric analysis. This research allowed to obtained spectral distributions of transmittance, reflectance and absorption coefficients of films and to expect band-gap energy of material.

Nanostructured semimagnetic semiconductors have revolutionary applications in non-linear optics, fast optical switching and memory devices [1]. At present, the high-quality nanostructured II-VI films and their solid solutions, such as $Zn_{1-x}Mn_xTe$, are widely used in various fields such as light emitting diodes, photodetectors, laser diodes, gas sensors, field emission and solar cells [2].

$Zn_{1-x}Mn_xTe$ films were deposited by the method of CSVS on cleaned glass substrates from $Zn_{90}Mn_{10}Te$ powder under following conditions:

- evaporation temperature: $T_e = 873$ K;
- substrate temperature: $T_S = 423 - 823$ K;
- deposition time: 10 minutes;
- pressure: 5×10^{-3} Pa.

The experimental arrangement of the CSVS technique has been reported elsewhere [3]. The evaporant ($Zn_{90}Mn_{10}Te$ powder) was placed in a tungsten (molybdenum) boat, which was used as an evaporation source.

Optical properties of the films were investigated by "Lazany" LI-722 spectrophotometer under room temperature in the range of wavelengths $\lambda=400-1000$ nm. Further, spectral distributions of the absorption coefficient $\alpha(\lambda)$ were calculated from the $T(\lambda)$ and $R(\lambda)$ spectrums in the range of strong light absorption using the expression [4]:

$$\alpha = -\frac{1}{d}\ln\left(\frac{1}{R^2}\left(-\frac{(1-R)^2}{2T}+\sqrt{\frac{(1-R)^4}{4T^2}+R^2}\right)\right), \quad (1)$$

where d – thickness of the film.

To determine the optical band gap E_g of zinc sulfide we have applied the common expression valid for direct-gap and indirect-gap semiconductors [5]:

$$\alpha h v = A\big(h v - E_g\big)^m, \quad (2)$$

where A is a constant depending on the effective mass of the charge carriers in the material, $h v$ is incident photon energy, α is an absorption coefficient, m is determined by the mechanism of photon absorption in the semiconductors; for direct-gap materials $m=1/2$. Therefore, the extrapolation of linear part $(\alpha h v)^2 - h v$-dependence down to interception with the energy axis enables to determine the band gap of the semiconductor.

Transmission spectra $T(\lambda)$ of $Zn_{1-x}Mn_xTe$ films obtained under different growth temperature are shown in the Fig. 1.

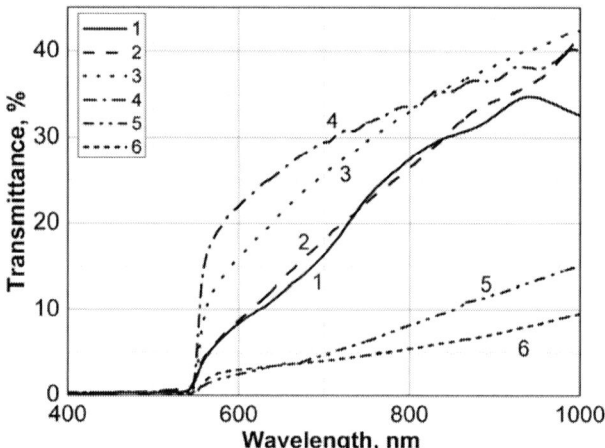

Fig. 1. Transmission spectra of $Zn_{1-x}Mn_xTe$ films obtained under various substrate temperatures: 423 K (1); 523 K (2); 623 K (3); 723 K (4); 773 K (5); 823 K (6).

As one can see, under radiation wavelengths larger than $\lambda\sim540-550$ nm (the photon energies are less than E_g of the material) the significant increase of the layers transmission coefficient is observed. At $\lambda>800$ nm samples №№1-4 of $Zn_{1-x}Mn_xTe$ films have a transmission coefficient up to 35%. This trend is kept to the values of the radiation wavelengths $\lambda\sim1000$ nm. But samples №№5-6 has a low transmission at all range of wavelengths.

The analysis shows that the maximal value of the optical transmission have the $Zn_{1-x}Mn_xTe$ films obtained at the deposition temperature $T_s = 723$ K. The minimal value corresponds to the $T_s = 823$ K. Probably, the various crystallographic and phase structure of the samples grown under different conditions causes the difference in their transmission coefficient.

Typical reflectance spectra $R(\lambda)$ of $Zn_{1-x}Mn_xTe$ films are shown in the Fig. 2.

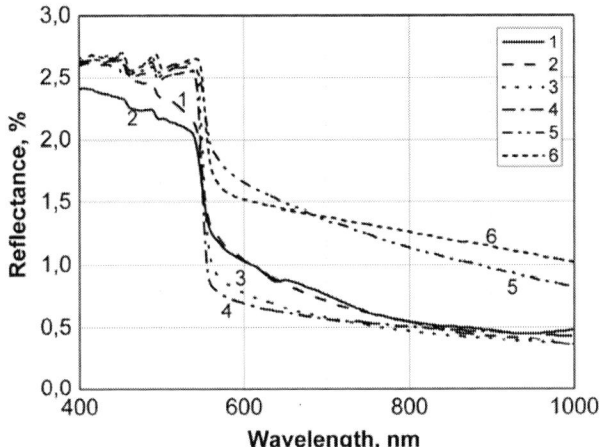

Fig. 2. Reflectance spectra of $Zn_{1-x}Mn_xTe$ films obtained under various substrate temperatures:
423 K (1); 523 K (2); 623 K (3); 723 K (4);
773 K (5); 823 K (6).

The figure analysis shows, all samples of $Zn_{1-x}Mn_xTe$ films are characterized by very low reflection – less to 3 %. The minimal value of the optical reflection have the $Zn_{1-x}Mn_xTe$ films obtained at the deposition temperature T_s = 723 K. The maximal value corresponds to the T_s = 823 K. Low reflectance coefficient which have obtained films, makes them a promising material for the anti-reflection coatings production.

Fig. 3 shows $(\alpha h\nu)^2 - h\nu$ dependences which are used for determine the band gap energy of the $Zn_{1-x}Mn_xS$ films.

As demonstrates Fig. 3, the gap values of the layers obtained at different substrate temperatures is varied. As the substrate temperature increases from 523 K to 823 K the gap of $Zn_{1-x}Mn_xTe$ films changes in the interval from E_g = 2.15 eV to E_g = 2.23 eV. This indicates a change in the manganese content in the $Zn_{1-x}Mn_xTe$ samples, depending on the technological conditions of deposition, especially from substrate temperature.

ACKNOWLEDGEMENT

This work is supported by the Ukraine State Agency for the Science, Innovation and Informatization and Project № 0107U001292 of the Ministry of Science and Education of the Ukraine.

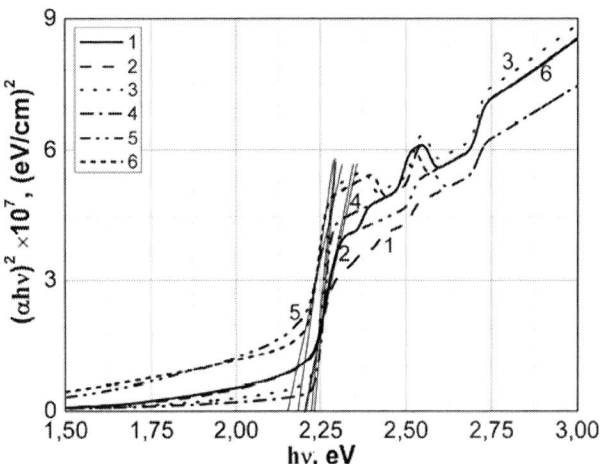

Fig. 3. $(\alpha h\nu)^2 - h\nu$ dependencies of $Zn_{1-x}Mn_xTe$ films obtained under various substrate temperatures:
423 K (1); 523 K (2); 623 K (3); 723 K (4);
773 K (5); 823 K (6).

REFERENCES

[1] R.J. Bandarnayake, J.Y. Lin, H.X. Jiang and C.M. Sonesan, "Synthesis and properties of $Cd_{1-x}Mn_xS$ diluted magnetic semiconductor ultrafine particles", J.Magn.Magn.Mater., vol. 169, pp. 289-302, 1997.

[2] M.C. Tamargo, *II-VI Semiconductor Materials and their Applications*, New York, Taylor & Francis, 2002.

[3] D.I Kurbatov, A.S Opanasyuk, S.M. Duvanov et.al., "Growth kinetics and stoichiometry of ZnS films obtained by close-spaced vacuum sublimation technique", Solid State Sciences, vol. 13, pp. 1068-1071, 2011.

[4] T.S. Moss, M. Balkanski, *Handbook on Semiconductors: Optical Properties of Semiconductors*, Amsterdam, Elsevier, 1994.

[5] D. Kurbatov, A. Opanasyuk, S. Kshnyakina, V. Melnik, V. Nesprava, "Luminescent and optical characteristics of zinc sulfide thin films produced by close-spaced vacuum sublimation", Rom. Journ. Phys., vol. 55 (1-2), pp. 213–219, 2010.

Angular Spectra of Phase Diffraction Gratings Illuminated by Interference Field

V.O. Gnatovskyy[2], S.A. Bugaychuk[1], A.M. Negriyko[1], I.I. Pryadko[1], A.V. Sidorenko[3]

[1] Institute of Physics of National Academy of Sciences of Ukraine, 46, Nauky Ave, Kyiv 03680, Ukraine
[2] Taras Shevchenko National University of Kyiv, 64/13, Volodymyrska Str., Kyiv 01601, Ukraine
[3] Kyivmetroproekt LLC, 16/22, Bohdana Khmelnytskogo str., Kyiv 01030, Ukraine

Abstract - This article proceeds from correlation method for controlling the intensity and angular displacement of the output laser beam. Angular spectra of plane phase diffraction gratings of different relief are investigated under illuminating them by interference field and under its transverse shifting relative to grating. This approach increases the number of control parameters for formation of optical field at the output of the optical arrangement.

I. INTRODUCTION

This article develops the idea of laser beam control through diffraction transformer effects on the interference field. It is aimed at increase in the number of factors of impact and thus at extension of range of problems that can be solved by using the correlation method of forming complex laser beam with desired energy distribution on the target [1].

The case when interference field of two beams (or its diffractive equivalent) illuminates a plane phase diffraction grating shifted parallel to it crosswise to grooves has been considered. The angular spectrum of the formed in this way beam, after the Fourier transform, according to the convolution theorem [2] is defined by convolution of angular spectra of both incident field and a grating. In fact, this is a sort of extension and generalization of well-known investigations in dynamic holography [3, 4], and is realized by means of static phase elements. The difference is that in dynamic holography the phase relief of hologram is unambiguously determined by intensity distribution of the interference field that illuminates some photorefractive material (with possible further displacement of the recorded grating). In our case, the configurations of the incident field and of grating may differ significantly in phase structure of grooves and in their periods.

The complex interference field of two, and in perspective, of several beams, can be replaced by equivalent field of a single beam after its passing through a special computer-designed diffractive element in the form of a synthetic hologram. Thus, within correlation method for formation of beams at the output, the set of control factors can be expanded and diversified. Using diverse diffraction gratings and their positions relative to field can potentially serve to improve laser beam scanning, switching, and so on.

II. THEORETICAL BACKGROUNDS

Let us consider two equal intensity laser beams having a plane wavefront. The electric field of each of them may be described as

$$\vec{E}_{1,2} = \vec{E}\exp\left(i\left(\vec{k}_{1,2}\,\vec{r} - \omega t + \varphi_{1,2}\right)\right),\ \vec{E} = (0, E, 0), \quad (1)$$

where ω is the frequency, $\vec{k}_{1,2}$ are wave vectors of laser beams, and $\varphi_{1,2}$ – their initial phases.

Let the wave vectors of beams have components $\vec{k}_{1,2} = (\pm k\sin\theta, 0, k\cos\theta)$, their interference occurs in the xy-plane, and the intensity of the interference field at an arbitrary time moment, depending on the x-coordinate, will be proportional to

$$\vec{E}_{int}(x) \sim \vec{E}\cos\left(D_{int}x + (\varphi_1 - \varphi_2)/2\right), \quad (2)$$

where the quantity $D_{int} = k\sin\theta$ characterizes the period of interference fringes.

This field, placed in front focal plane of the lens, will produce the angular spectrum in the form of two δ-like peaks at angular distance 2θ of interfering beams convergence, at the back focal plane. Changing phase φ_1 or φ_2 of respective beam results in interference fringes shifting along x-axis.

We define the expression for the complex transmission function of the diffraction grating along x-axis as

$$T(x) = A(x)\exp\left(i\Phi(x)\right), \quad (3)$$

where functions $A(x)$ and $\Phi(x)$ determine, respectively, the amplitude and phase relief of the grating. Angular spectra of plane phase diffraction gratings represent a set of δ-like peaks (diffraction orders) in the back focal plane of the lens. Their location and intensity are determined by the grating period and its phase relief (e.g. [5]).

In this investigation, a plane grating was located in the front focal plane of a lens, on the way of interfering beams. So the field at the back focal plane is

$$\vec{E}_{out} = \hat{F}\left(\vec{E}_{int} \cdot T\right), \tag{4}$$

where $\hat{F}$ denotes Fourier transform. It should be noted that the proposed scheme, compared with the scheme of dynamic holography, can carry a wider range of transformations. In particular, it is possible to use unequal contrasts in interference fringes and in diffraction grating, a variety of their periods, to vary the phase relief in separate grooves of a grating, and so on. These can be used as additional factors of influence on the final result of forming the output radiation.

Among the possible variety of gratings, the following two have been considered. Their angular spectra are shown on Fig. 1.

Their complex transmission functions, respectively, are

$$T_a(x) = \exp\left(i\,\pi/4 \cdot \sin\left(D_{gr}x + \Delta\varphi\right)\right), \tag{5a}$$

$$T_b(x) = \exp\left(i\,\pi/2 \cdot \mathrm{sgn}\left(\sin\left(D_{gr}x + \Delta\varphi\right)\right)\right), \tag{5b}$$

where the quantity D_{gr} characterizes the period of diffraction grating, and $\Delta\varphi$ adjusts the shift of grating grooves relative to interference field fringes. The first grating has sinusoidal phase relief along the x-axis, and phase modulation depth $\pi/2$. Its angular spectrum is shown on Fig. 1a. It is characterized by a single dominant central maximum. The second grating has a rectangular binary phase relief with modulation depth π. Its angular spectrum (Fig. 1b) is largely similar to the spectrum of the incoming interference field.

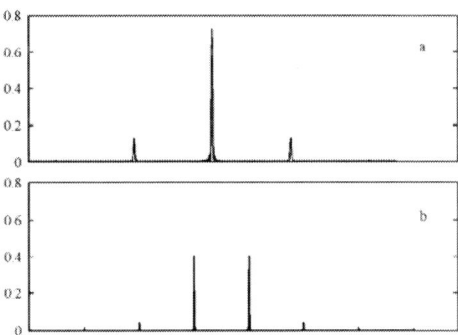

Fig. 1. Angular spectrum of diffraction gratings:
a) sinusoidal   b) binary

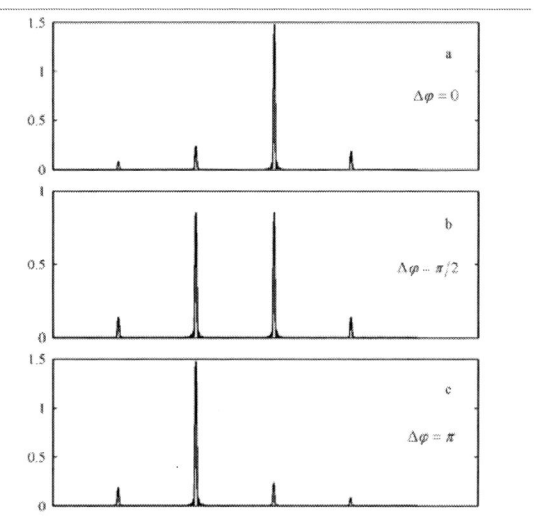

Fig. 2. The angular spectrum of the interference field after passing through the sinusoidal diffraction grating.

Computations for intensity of the field (4) correspond to the case when the gratings with complex transmission functions (5a) and (5b) were placed in the interference field (2), formed with $\varphi_1 - \varphi_2 = 0$. The gratings were shifted then across the field, and this shift was simulated by changing of variable $\Delta\varphi$ in expressions (5). In the case of sinusoidal relief of grating (Fig. 2), there appeared two interesting cases when two identical beams were produced, and when energy pumping into the left (Fig. 2c) or the right (Fig. 2a) beam occurred. This took place under condition $\Delta\varphi = \{\pi/2; \pi; 0\}$ when $D_{gr}/D_{int} = 2$.

The same result has been obtained in the computation of grating diffraction of the field produced not by interference of two beams, but formed by a plane wave of frequency ω passing through the diffraction grating

$$T(x) = \left|\sin\left(D_{int}x\right)\right| \cdot \exp\left(i\pi/2 \cdot \mathrm{sgn}\left(\sin\left(D_{int}x\right)\right)\right). \tag{6}$$

The other case of diffraction of interference field on a binary phase grating with a rectangular relief (Fig. 3) when $D_{gr}/D_{int} = 1$ is interesting because there appears a possibility to unite two beams into a single one (Fig. 3b). It happens when $\Delta\varphi = \pi/2$ due to correction of the wavefront of interference field. The transverse shift according to $\Delta\varphi = 3\pi/16$ in expression (5b) produces even three beams of almost equal intensity (Fig. 3c).

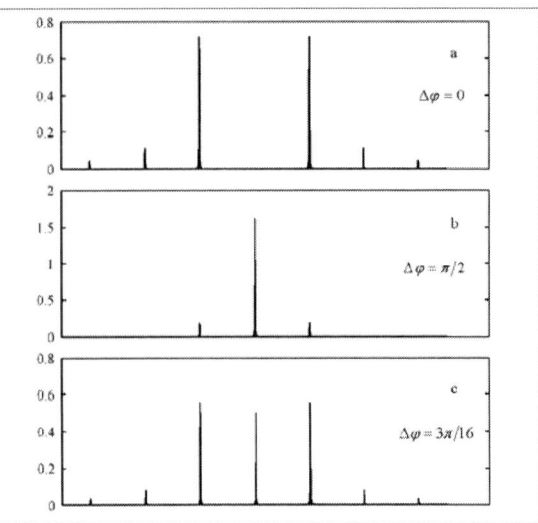

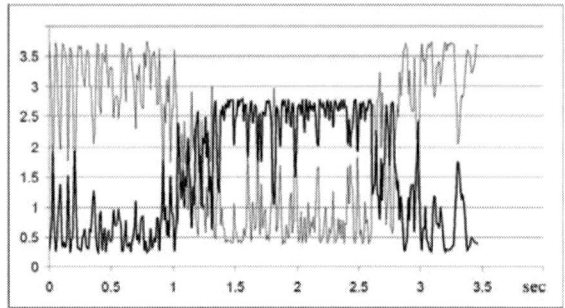

Fig. 4. Energy transfer between two peaks.

Fig. 3. The angular spectrum of the interference field
after passing through the binary diffraction grating.

An example of measurements is shown in Fig. 4. It
illustrates transmutation of different cases of diffraction of
interference field on the type (5a) grating. The time-dynamics
shows, when the grating is shifted, how the right diffraction
maximum (Fig. 2a) decreases, equalizes in intensity to the left
one (Fig. 2b), and then the latter predominates over the former
(Fig. 2c). Under the further reverse regulation, the initial case
(Fig. 2a) is restored.

III. EXPERIMENTAL OBSERVING

In accordance with the above-mentioned calculations, a
series of experimental measurements of intensity of diffraction
maxima in the angular spectra of artificially formed fields were
carried out. A beam of a helium-neon laser was expanded by a
telescopic system, and was filtered by a system of diaphragms.
In such way formed beam with uniform amplitude and the
plane wavefront was directed at beam splitter providing precise
regulation of the angular distance between the beams.

Then interference field of obtained beams encountered a
diffraction grating whose fastening provided the desired
orientation of grating relative to interference field fringes, and
with the help of micro-screws it could be shifted in horizontal
and vertical directions. The desired ratio of the interference
fringes' width to the period of diffraction grating was achieved
by adjusting convergence angle of interfering beams. The
angular spectrum of the resulting distribution of the laser field
was observed in the back focal plane of the lens. The regions of
angular spectrum, after the desired magnification, were
directed by a prismatic system at photoresistors, and their
oscillogram was recorded and processed on a computer.

IV. CONCLUSION

The proposed correlational approach in the case of
illumination of diffraction gratings by periodic light fields
increases the number of control parameters for intensity of
peaks in resulting angular spectrum. It can quite easily be
implemented using digital methods with the help of spatial
light modulators and it appears to be perspective for the tasks
of formation of laser beams with required spatial-angular
characteristics.

REFERENCES

[1] O.V. Gnatovskyy, A.M. Negriyko, V.O. Gnatovskyy, A.V. Sidorenko, "Cross-correlation method for the formation of laser energy fields with complex distribution". *Ukr. J. Phys*, vol. 58/2, pp. 122-125, 2013.

[2] A. Papoulis, *Systems and Transforms with Applications in Optics,* McGraw-Hill, New York, 1968.

[3] S. Odoulov, M. Soskin, and A. Khizhnyak, *Optical oscillators with degenerate four-wave mixing*, Harwood, Chur, Switzerland, 1991.

[4] Yu. Francis and Yin Shizhuo, *Photorefractive optics. Materials, properties and applications*, Academic Press, New York, 2000.

[5] J. W. Goodman, *Introduction to Fourier Optics*, McGraw-Hill, New York, 1968.

Crystal Growth Sector Effect on Dielectrical Properties of Carbamide Doped KDP Crystals

A. N. Levchenko[1], I. M. Pritula[2], V. B. Tyutyunnik[1], A. O. Penkina[1], A. V. Kosinova[2], M. I. Kolybayeva[2]

[1]V. N. Karazin Kharkiv National University, 61022, Svobody Sq. 4, Kharkiv, Ukraine
Phone: (+38057) 707 53 46, e-mail: Alexander.N.Levchenko@univer.kharkov.ua
[2]STC "Institute for Single Crystals" NASU, 61001, Lenin Ave 60, Kharkiv, Ukraine
Phone: (+38057) 341 01 39, e-mail: pritula@isc.kharkov.ua

Abstract-Dependences of the electrical dc conductivity and microwave relative dielectric permittivity of $KH_2PO_4+CH_4N_2O$ crystals on the carbamide concentration are obtained. Features of the influence of the crystal growth sector on the conductivity and microwave dielectric permittivity of the crystals are determined.

I. INTRODUCTION

Potassium dihydrogen phosphate crystals (KH_2PO_4, KDP) are used in quantum electronics as crystal elements of harmonics generators and modulators of the laser radiation.

Some organic additives can improve functional characteristics of KDP crystals. One of such organic additives is carbamide (H_2N-CO-NH_2 or CH_4N_2O) increasing the laser damage threshold of KDP and the intensity of the second harmonic generation of the YAG:Nd^{3+} laser radiation (λ=1.064 µm) by 25 and 20 %, respectively [1].

The characterization of the carbamide-containing KDP crystals was carried out in [1-3], where the influence of carbamide on the dc conductivity, low frequency dielectric constant of crystals and other physical properties has been investigated. However, the influence of the crystal growth sector on the dielectric characteristics of the KDP crystals with carbamide was not given any attention.

The purpose of this work was to determine the growth sector effect on the dc conductivity and microwave dielectric permittivity of KDP+carbamide crystals.

II. EXPERIMENTAL

Pure and doped with carbamide KDP single crystals were grown from water solution onto a point seed by the method of temperature reduction [4]. Molar values of carbamide in the mother liquor were 0.1, 0.5, 1.0 and 5.0 M. The solubility of carbamide at 25° exceeds 20 M/L.

For the conductivity measurements, there were used the samples shaped as a rectangular parallelepiped of $11\times5.5\times5.5$ mm^3. Two opposite surfaces of the samples was coated with good quality graphite served as the electrodes.

For determination of the dc conductivity the resistance R of the crystal samples was measured by means of a teraohmmeter E6-13A. The measurements were performed by a standard method using the two-probe technique to ±8 %. The applied voltage was 100 V. The dependences of the conductivity on the

carbamide concentration were measured at temperature of 25°C.

The conductivity of the crystals was calculated by means of the relation

$$\sigma = \frac{d}{R \cdot S}, \qquad (1)$$

where R is the resistance of the sample, d is the thickness of the sample, S is the area of the crystal sample surface covered by graphite. The thickness d and the area S were 0.55 cm and 0.61 cm^2, respectively.

The conductivity of the crystals was measured at various temperatures ranging from 25 to 140 °C.

The microwave dielectric permittivity measurements were performed in the 8-12 GHz frequency range at 25°C with the help of the waveguide-dielectric resonance method using the technique described in [5].

As a waveguide-dielectric resonator (WDR), there was used the 11×5.5 mm^2 waveguide section with the investigated crystal sample located in the middle of the section.

The estimated accuracy of the dielectric constant determination connected with the WDR resonant frequency and samples thickness measurements (±30 MHz and ±10 µm, respectively) didn't exceed 1%.

For calculation of the dielectric constant values the following equations [1] were used

$$\frac{\gamma_0}{\beta} = +tg\left(\frac{\beta l}{2}\right), \qquad (1)$$

$$\frac{\gamma_0}{\beta} = -ctg\left(\frac{\beta l}{2}\right), \qquad (2)$$

where (1) and (2) are given for the odd and even oscillations, respectively, $\beta=[k_0^2\varepsilon-(m\pi/a)^2]^{1/2}$, where m is the whole numbers; $k_0=2\pi/\lambda_0$, where λ_0 is the resonant wavelength.

III. RESULTS

The dependences of the conductivity of KDP crystals on the carbamide concentration in the solution for the pyramidal and prismatic crystal growth sectors are presented in Fig.1.

$$\sigma = \sigma_0 \times \exp\left[-\frac{E_a}{kT}\right], \qquad (3)$$

where E_a is the activation energy of conductivity, σ_0 is the preexponential factor, k is the Boltzman constant and T is the absolute temperature.

The activation energy values (E_a) calculated from the temperature dependences of the conductivity along the c-direction in the crystal, are presented in Table 1.

The measurements of relative dielectric permittivity (ε_r) of KDP show, that the dielectric constant values (Fig. 4) in the X-band obtained from the equations (1, 2) well enough coincide with the ε_r value known from the literature [8] ($\varepsilon_r = \varepsilon_{33} = 19.7 \pm 0.5$ at 9.4 GHz).

The changes in the ε_r connected with the carbamide incorporation into the crystal don't exceed 5 %. It is less than the changes in the dc conductivity and the change in the dielectric permittivity at frequency of 1 kHz reported in [3].

In Fig. 3 one can see the crystal growth sector effect on the microwave dielectric constant. The concentration dependences of the dielectric constant ε_r for crystal samples made of various growth sectors are different.

It should be mentioned that the minimum of the conductivity value and the maximum of the dielectric constant value for the pyramidal sector are observed at the same value of the carbamide concentration.

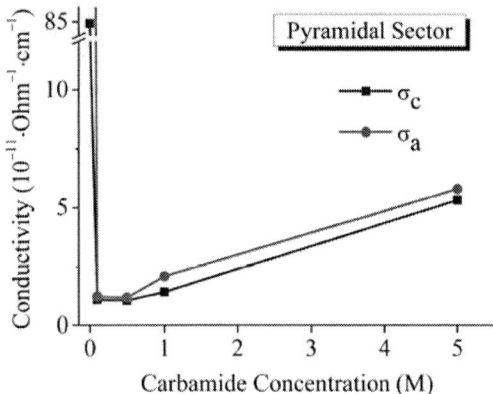

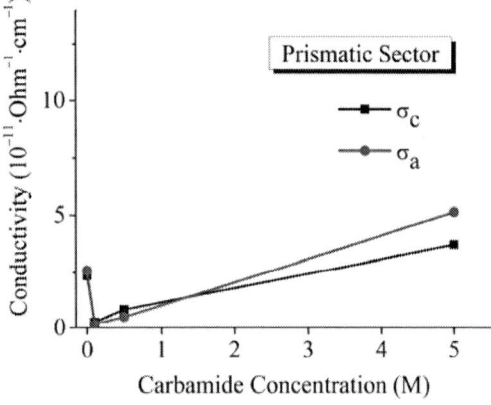

Fig. 1. Concentration dependences of the dc conductivity of the KDP with carbamide additive for the pyramidal and prismatic sectors along the c and a crystal axes (σ_c and σ_a, respectively).

As is seen from the figure, there is a small anisotropy of the conductivity. The conductivity values along the a-direction (σ_a) are more than those along the c-direction (σ_c) at 25°C. This is in correspondence with the results obtained by the previous authors, e.g. [6].

It can be seen, that the concentration dependences for the pyramidal and prismatic growth sectors are different from each other (the crystal growth sector effect).

Incorporation of small quantities of carbamide (0.1-0.5 M in the solution) into the crystal leads to the sharp reduction of the conductivity. It is observed for the samples from both growth sectors.

At the concentration of 5.0 M for samples made of the prismatic sectors, the conductivity becomes higher than the conductivity of pure crystals, while for the samples made of pyramidal sector the conductivity remains considerably lower than the conductivity of the pure crystals.

To obtain the data on charge transport mechanisms, the temperature dependences of the conductivity (Fig. 2) were investigated. It is established that the temperature dependences is described by the expression [6, 7]

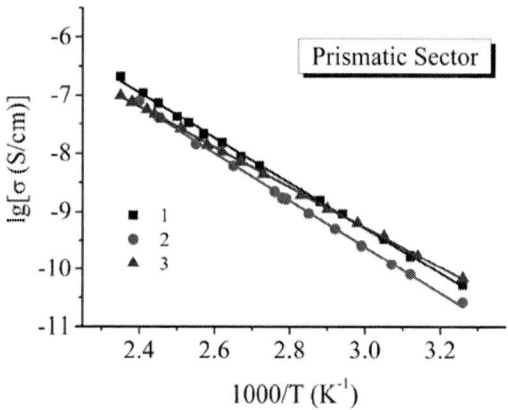

Fig. 2. Temperature dependences of the axial dc conductivity (σ_c) of pure KDP (1), KDP+0.5 M carbamide in the solution (2), KDP+5 M carbamide in the solution (3).

TABLE I
ACTIVATION ENERGY VALUES (E_a) FOR THE ELECTRICAL
DC CONDUCTIVITY OF THE KDP+CARBAMIDE CRYSTALS

Sector	Carbamide (M in Sol)	E_a (eV)
Pyramidal	0 (Pure KDP)	0.617±0.007
	0.5	0.818±0.013
	5.0	0.767±0.007
Prismatic	0 (Pure KDP)	0.771±0.007
	0.5	0.805±0.010
	5.0	0.686±0.003

IV. Discussion

Investigation of the conductivity of KDP [9] shows that the activation energy of the extrinsic conductivity connected with migration of proton vacancies is about $E_a \sim 0.5$ eV, while the activation energy of the intrinsic conductivity is $E_a \geq 0.8$ eV.

In the 0.5-0.7 eV activation energy range both extrinsic and intrinsic conductivity mechanisms take place. In this case the total conductivity is determined by the contribution of its intrinsic and extrinsic component. Low activation energy value means the presence of the extrinsic conductivity component that is probably connected with the migration of proton vacancies. So, the E_a changes (Table 1) from 0. 617 eV for pure crystals up to 0.818 eV for KDP+0.5 M carbamide can be connected with the reduction of number of proton vacancies. It is agreed with the assumption of authors [10] that organic additives in KDP (e.g. L-arginin) occupy interstitial positions and it can create additional hydrogen bonds in KDP.

So, the decrease of the electrical conductivity with the increase of the carbamide concentration could be explained as due to the reduction of proton vacancies caused mainly by the creation of additional hydrogen bonds by the impurity which obstructs the movement of protons [10]. The reduction of number of proton vacancies probably takes place mainly at small carbamide concentrations.

In the same time a carbamide molecule can create more defects and enhance the conductivity of crystal [11]. It probably takes place at large carbamide concentration (> 1 M).

The chemical analysis of the crystals with carbamide shows the difference in concentration of Fe^{3+} and Al^{3+} impurities in various sectors of the crystal [1]. Incorporation of the mentioned impurity ions into the crystal requires a local charge compensation of these ions. The local charge compensation occurs by the formation of proton vacancies during the crystal growth. Therefore, the crystal growth effects apparently are connected with the features of impurity ion incorporation into the KDP+carbamide crystals.

Conclusions

Our investigation shows that the dependences of conductivity on the carbamide concentration are non-monotonic. They considerable depend on the crystal growth sector. The lowest conductivity value observes in the crystals with small (0.1-0.5 M) carbamide concentration. The crystal growth sector effect on the dielectric constant is also observed in the X-band. It is found that the influence of carbamide on the dielectric constant of KDP+carbamide crystals in the X-band is weaker than its influence on the low-frequency (1 kHz) dielectric permittivity and dc conductivity.

References

[1] I. Pritula, A. Kosinova, M. Kolybayeva, V. Puzikov, S. Bondarenko, V. Tkachenko, V. Tsurikov, O. Fesenko, "Optical, structural and microhardness properties of KDP crystals grown from urea-doped solutions," *Mater. Res. Bull.*, vol. 43, pp. 2778–2789, 2008.

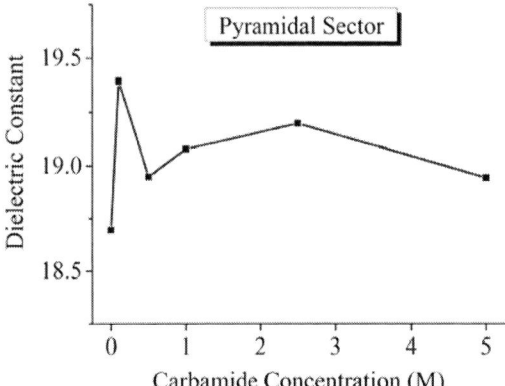

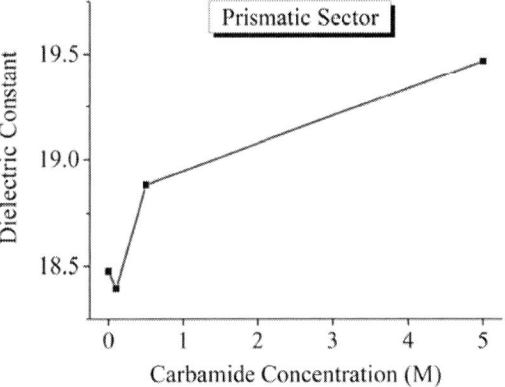

Fig. 3. Dependences of the dielectric constant ε_r on the carbamide concentration in the solution. The electric vector E is directed along the c crystal axis ($E\|c$).

[2] M. Priya, C. M. Padma, T. H. Freeda, C. Mahadevan and C. Balasingh, "Electrical conductivity measurements on gel grown KDP crystals added with urea and thiourea," *Bull. Mater. Sci.*, vol. 24, no. 5, pp. 511–514, 2001.

[3] S. Goma, C. M. Padma, C. K. Mahadevan, "Dielectric parameters of KDP single crystals added with urea," *Mater. Lett.*, vol. 60, pp. 3701-3705, 2006.

[4] N. Zaitseva and L. Carman, "Rapid growth of KDP and DKDP crystals," *Progress in Crystal Growth and Characterization of Materials*, vol. 43, pp. 1-118, 2001.

[5] V. A. Korobkin, N. I. Pyatak, L. I. Babarika, Yu. G. Makeev, Opredeleniye parametrov dielectrikov na sverkhvysokoy chastote s pomoshchyu volnovodno-dielektricheskikh resonansov," *PTE*, no. 3, pp. 169-171, 1976 (in Russian).

[6] L. B. Harris and G. J. Vella, "Conductivity of single crystals of potassium dihydrogen phosphate," *J. Appl. Phys.*, vol. 37, p. 4294, 1966.

[7] M. O'Keefe and C.T. Perrino, "Proton conductivity in pure and doped KH_2PO_4," *J. Phys. Chem. Solids*, vol. 28, pp. 211-218, 1967.

[8] J. T. Milek, M. Neuberger, *Linear Electrooptic Modular Materials*. IFI/Plenum, New York-Washington-London, 1972.

[9] E. D.Yakushkin, E. P. Efremova, A. I. Baranov, "Growth conditions and electric properties of KDP crystals. I. Conductivity measurements," *Kristallografiya*, vol. 46, no. 5, pp. 904-908, 2001.

[10] M. Meena and C. K. Mahadevan, "Growth and electrical characterization of L-arginine added KDP and ADP single crystals," *Cryst. Res. Technol.*, vol. 43, no. 2, pp. 166-172, 2008.

[11] A. A. Assencia and C. Mahadevan, "D.C. electrical conductivity measurements on ADP single crystals added with simple organic compounds," *Bull. Mater. Sci.*, vol. 28, no. 5, pp. 415–418, 2005.

Research of the time of the transverse relaxation using the incoherent NMR-spectrometer

M.O. Pikula, K.V. Shimko

Tavrida National V.I. Vernadsky University, Simferopol, Ukraine

Abstract: Research of the time of the transverse relaxation T_2 using the incoherent NMR-spectrometer and Carr-Purcell sequence. Measurements have shown that this sequence can be used as an estimate of the relaxation time syne the error presents there (~15%).

The system of nuclear spins in magnetically ordered materials; viz. in the ferromagnetic, antiferromagnetic and ferrimagnetic material is in a quite specific condition different from those in the weakly magnetic materials. This fact caused the necessity of special consideration of NMR in magnets. If you attach an external alternating magnetic field to the magnet, then at certain frequencies can be seen a sharp increase in the absorption energy of the body of an external electromagnetic field. Here we are dealing with a resonant or relaxation response of the system. [1]

The impulse responses of NMR and the time of transverse relaxation were researched T_2 in magnetically ordered materials with garnet structure. With a help of a two-pulse Khan Sequence and Carr-Purcell multipulse sequence.

Khan pulse sequence consists of two pulses, 90° and 180°, separated by an interval τ. The sequence Carr-Purcell is an exciting 90-degree pulse, and a number of probing the 180-degree. This sequence allows you to measure the transverse The experiment was carried out on non-coherent pulsed NMR-spectrometer. As a sample of the study the epitaxial iron garnet $Y_3Fe_5O_{12}$ was taken, grown by liquid phase epitaxy on a non-magnetic substrate gadolinium-gallium garnet $Gd_3Ga_5O_{12}$. The orientation of the plane of the membrane is [111]. A geometric size is 5 to 7 mm. The membrane is enriched with the isotope [57]Fe, with a spin equal ½ to 98%. The sample was placed at a constant temperature of 20°C. Several pulses of radio frequency field submitted the sample, and then the response of the system was studied. In the frequency 54.092MHz - 54.660MHz Khan Pulse sequence submitted the sample.

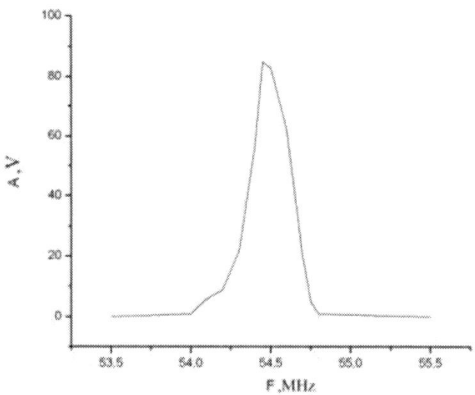

Fig.1. Spectrum [57]Fe.

At the frequency with the maximum amplitude of the echo pulse sequence was submitted, with the changing if the distance between the pulses.

The dependence of the amplitude on the time delay was taken away, and the transverse relaxation time T_2 was measured from the equation:

$$A = A(0)\, \exp(-2\tau/T_2).\ [3]$$

It amounted to 246 microseconds.

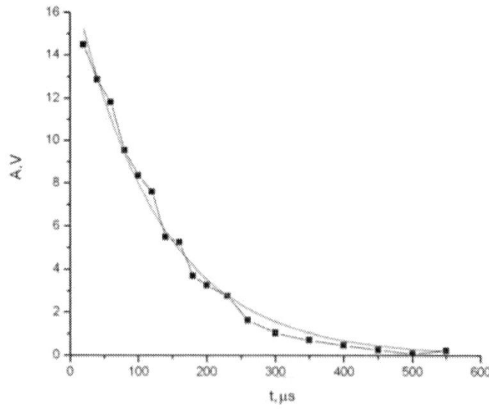

Fig.3. The dependence of the Khan echo amplitude on the time, $T_2 = 246$ µs.

Also at this frequency Carr-Purcell pulse sequence was submitted, with the changing of the pulse duration and the change of the distance between the pulses. However, when using the Carr-Purcell sequence on incoherent installation in the measurements the errors arise.

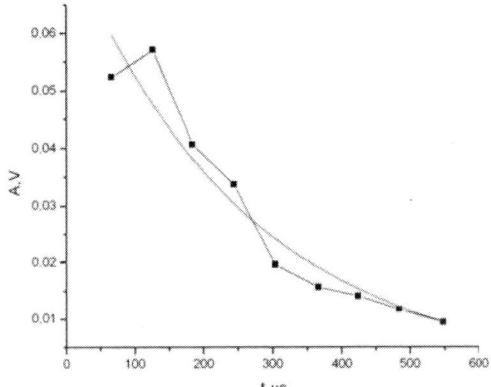

Fig.3. The dependence of the Carr-Purcell echo amplitude on the time, $T_2 = 262$ μs.

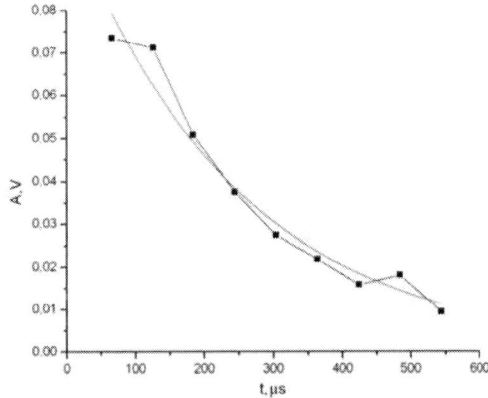

Fig.4. The dependence of the Carr-Purcell echo amplitude on the time, $T_2 = 245$ μs.

The measurement results show that the use of the Carr-Purcell sequence on incoherent set suitable for determining the estimated longitudinal time relaxation.

REFERENCES

[1]   G.S. Krinchik. The physics of magnetic phenomena. Publishing House of Moscow University. M. (1985) (in Russian)
[2]   V.I. Siskin. Nuclear magnetic relaxation. Leningrad University Publishing. L. (1991) (in Russian)
[3]   A.A. Vashman, I.S. Pronin. Nuclear magnetic relaxation spectroscopy. Science. M. (1986) (in Russian)

Ge-Ga-S/Se glasses studied with PALS technique
In application to chalcogenide photonics

A. Ingram[1], H. I. Klym[2], O.I. Shpotyuk[3,4]

[1]Opole University of Technology, Opole, Poland
[2]Lviv Polytechnic National University, Lviv, Ukraine
[3]Lviv Institute of Materials of SRC "Carat", Lviv, Ukraine
[4]Institute of Physics of Jan Dlugosz University, Czestochowa, Poland

Abstract: Free-volume entities in Ge-Ga-S and crystallization behaviour in Ge-Ga-Se chalcogenide glasses caused by thermal annealing at 380°C for 10, 15 and 50 h are studied using positron annihilation lifetime spectroscopy. It is shown that the structural free-volume entities in theses glasses can be adequately described by positron modes determined within two-state trapping model. The observed changes in defect-related component in the fit of experimental lifetime spectra for annealed glasses testifies in a favour of structural fragmentation of larger free volume entities into smaller ones. Because of strong deviation in defect-free bulk positron lifetime τ_b from corresponding additive values proper to boundary constituents, the studied Ge-Ga-S/Se glasses cannot be considered as typical representatives of pseudo-binary cut-section.

Chalcogenide glasses (ChG) play a significant role in modern photonics in view of their wide device applications [1]. For example, $GeSe_2$-Ga_2Se_3 ChGs are mainly used for their wide transparency in the infrared range and also for their high refractive index. Selenium-based glasses have good transparency in the 0.8–16 mm range and can be used in optical systems working in the 3–5 or in the 8–12 mm range [2].

Atomic arrangement in such ChG can be studied with numerous techniques (X-ray diffraction, etc.), while number of probes available to study atomic-deficient distribution is rather limited, especially at a sub-nanometer scale. One of the best techniques capable to probe such fine free volumes is the positron annihilation lifetime spectroscopy (PALS) [3]. In the present paper, we imply the PAL method to study of free-volume structure of Ge-Ga-S/Se ChG within two-states trapping model.

ChGs were prepared from melting mixture of highly pure raw materials (Ge, Ga, and Se/S: 99.999%) in sealed silica ampoule kept under 10^{-6} Pa vacuum. Glass rods were cut into slices of 1 mm thickness and polished for further measurements. The crystallization of the $80GeSe_2$-$20Ga_2Se_3$ was performed with a single step of heat treatment at T_g+10 °C. This temperature has been chosen as an optimal temperature of ceramization as it permits to control the generation by simultaneous nucleation and growth of nanoparticles within the glassy matrix according to the heat treatment time. Thus, glasses were placed in a ventilated furnace, temperature is ± 2 °C for various time varying from 10 to 50 h.

XRD measurements with $CuK_{\alpha1}$ radiation were performed to determine crystalline phases in the studied samples. The measured X-ray beam intensities and reflection angles 2θ were obtained using automatic STOE STADI P diffractometer ("STOE & Cie GmbH, Germany) All measurements were conducted in 2θ-step regime, the profiles of peaks being refined using WinPLOTR software.

Results of XRD measurements for $80GeSe_2$-$20Ga_2Se_3$ annealed at 25 and 50 h and $80GeS_2$-$20Ga_2S_3$ glasses in comparison with theoretical patterns are shown in Fig. 1,a and Fig. 1,b, respectively.

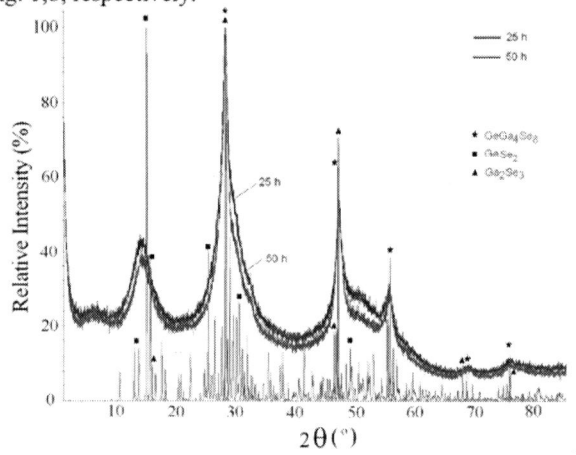

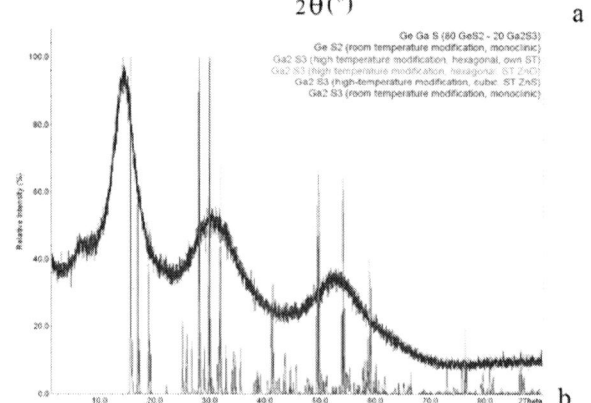

Fig. 1. Comparison of XRD patterns of $80GeSe_2$-$20Ga_2Se_3$ glass annealed at 380 °C for 25 and 50 h (a) and $80GeS_2$-$20Ga_2S_3$ glasses with theoretical patterns $GeGa_4Se_8$, $GeSe_2$ and Ga_2Se_3 (a) and GeS_2 and Ga_2S_3 phases

978-1-4799-0019-0/13 $31.00 © 2013 IEEE

The maximums associated with $GeSe_2$ phase appear on the XRD patterns of thermally annealed $80GeSe_2$-$20Ga_2Se_3$ glass too, but it cannot be well distinguished as separate crystalline peaks even for long annealing durations (50 h). It means that $GeSe_2$ crystals appear only in a small amount, probably as surface inclusions. It can be imagined that $GeSe_2$ crystals grow among glass remainder after Ga_2Se_3 and $GeGa_4Se_8$ crystals extraction as a result of ceramization process.

The PALS spectra were recorded with conventional fast-fast coincidence system (ORTEC) of 230 ps resolution at the temperature $T = 22\ ^oC$ and relative humidity $RH = 35\ \%$. Two identical samples were used to build a sandwich structure needed for PALS measurements. The measured PAL spectra of ChG were processed with standard LT 9.0 computer program, the obtained curve being fitted by two components with τ_1, τ_2 lifetimes and I_1, I_2 intensities ($I_1+I_2=1$). Introducing a third component into the envelope of the fitting curves did not improve decomposition goodness significantly. Therefore, the positron trapping modes in the studied ChG, e.g. average positron lifetimes τ_{av}, positron lifetime in defect-free bulk τ_b, and positron trapping rate in defects κ_d were calculated using a formalism of two-states trapping model [3].

It is shown that with increase of annealing duration from base $80GeSe_2$-$20Ga_2Se_3$ glass to samples annealed for 10 and 25 h, the lifetimes τ_2 decreases and I_2 intensities rather sharply rises, testifying growing behaviour in the number of corresponding free-volume entities. These changes cause the regular extension of the lifetime τ_{av}. But under increase of annealing to 50 h, the I_2 intensity ceases to increase, while lifetime τ_2 appreciably decreases to 0.416 ns. This trend leads to a significant rises of τ_{av}. Other positron trapping parameters (τ_2/τ_b) behave is in harmony with similar ones (Table 1). But ($\tau_2 - \tau_b$) difference, accepted as a size measure for extended free-volume defects where positrons are trapped, naturally decreases with annealing duration.

Table 1. Fitting parameters for PALS spectra of $80GeSe_2$-$20Ga_2Se_3$ and $80GeS_2$-$20Ga_2S_3$ glasses

Sample	Fitting parameters			
	τ_1, ns	I_1, a.u.	τ_2, ns	I_2, a.u.
$80GeSe_2$-$20Ga_2Se_3$/ base glass	0.214	0.69	0.439	0.31
$80GeSe_2$-$20Ga_2Se_3$ / 10 h	0.211	0.67	0.431	0.33
$80GeSe_2$-$20Ga_2Se_3$/25 h	0.216	0.65	0.426	0.35
$80GeSe_2$-$20Ga_2Se_3$/50 h	0.209	0.65	0.416	0.35
$80GeS_2$-$20Ga_2S_3$/ base glass	0.234	0.67	0.481	0.33

Sample	Positron trapping modes				
	τ_{av}, ns	τ_b, ns	κ_d, ns^{-1}	$\tau_2 - \tau_b$, ns	τ_2/τ_b
$80GeSe_2$-$20Ga_2Se_3$/ base	0.28	0.25	0.74	0.19	1.73
$80GeSe_2$-$20Ga_2Se_3$ / 10 h	0.28	0.25	0.80	0.18	1.70
$80GeSe_2$-$20Ga_2Se_3$/25 h	0.29	0.26	0.79	0.17	1.64
$80GeSe_2$-$20Ga_2Se_3$/50 h	0.281	0.252	0.83	0.16	1.65
$80GeS_2$-$20Ga_2S_3$/ base	0.32	0.28	0.73	0.20	1.7

It was shown [4] that coupling of open volumes created by bond free solid angles, belonging to different chalcogen chains

or coordination polyhedrons, can be effective traps for positrons in ChG. During crystallization, the glass structure relaxes towards more thermodynamically favourable state (crystallization shrinkage or densification), eliminating the excess of free volume of neighbouring voids. It means that existing free volume voids either disappear or convert into a greater number of smaller ones. In other words, we can argue that crystallization in $80GeSe_2$-$20Ga_2Se_3$ glass induced by annealing causes fragmentation of larger free-volume entities into smaller ones. Such process is accompanied by essential decrease in τ_2 and corresponding increase in I_2 in full agreement with above scenario.

At values of PALS fitting parameters for $80GeS_2$-$20Ga_2S_3$ system, bulk positron lifetime τ_b is near 0.28 ns and average lifetime τ_{av} is 0.32 ns. It is shown that the size of free-volume trapping centers in the $80GeS_2$-$20Ga_2S_3$ system is higher than in the $80GeS_2$-$20Ga_2Se_3$ glasses ($\tau_2 = 0.439$ ns), weak increases of I_2 intensities testify growing behaviour in the number of corresponding free-volume entities. Other positron trapping parameters for Ge-Ga-S system are similar to Ge-Ga-Se base glasses.

Thus, crystallization behavior of $80GeSe_2$-$20Ga_2Se_3$ glasses during annealing at 380°C for 10 h, 15 h and 50 h indicates the possibility of formation of $GeGa_4Se_8$ and Ga_2Se_3 crystals. Their appearance essentially modifies free-volume structure of ChG leading to specific fragmentation of larger free-volume entities (positron trapping voids) into a greater number of smaller ones. Such effect reveals an increase of the second lifetime component τ_2 and a decrease of its intensities I_2 testifying in a favour of increased number of smaller free volumes. In accordance to PAL spectroscopy, the studied Ge-Ga-S/Se ChG cannot be classified as typical pseudo-binary system based on non-additivity of defect-free bulk lifetimes τ_b of its constituting components ($GeSe_2$ and Ga_2Se_3).

The authors acknowledge support from Science and Technology Centre in Ukraine (project # 5721).

REFERENCES

[1] L. Calvez L., P. Lucas P., et. al. "Influence of gallium and alkali halide addition on the optical and thermo–mechanical properties of $GeSe_2$-Ga_2Se_3 glass", *Appl. Phys. A*, vol. 89, pp. pp. 183-188, 2007.

[2] Ch. Lin, L. Calvez, H. Tao, M. Allix, A. More, X. Zhang, X. Zhao, "Evidence of network demixing in GeS_2–Ga_2S_3 chalcogenide glasses: A phase transformation study" J. Solid State Chemistry, vol. 184, pp. 584-588, 2011.

[3] R. Krause-Rehberg, H.S. Leipner, *Positron annihilation in semiconductor*, Springer-Verlag, Berlin-Heidelberg-New York, 1999.

[4] A. Ingram, R. Golovchak, M. Kostrzewa, S. Wacke, M. Shpotyuk, O. Shpotyuk, "Compositional dependences of average positron lifetime in binary As–S/Se glasses", *Physica B*, vol. 407, pp. 652-655, 2012.

LIDAR hyperspectral system for detecting chemical and biological agents

V. V. Gnidenko[1], V. O. Yatsenko[2]

[1]National Technical University of Ukraine "Kyiv Polytechnic Institute", Kyiv, Ukraine
[2]Space Research Institute of NASU-SSAU, Kyiv, Ukraine

Abstract: The LIDAR hyperspectral system has been developed for remote sensing of chemical and biological agents. It has been tested for detecting and measuring content of chemical and biological agents by using absorption and reflection spectra. The measurements revealed complex scattered signals from different objects on the Earth's surface. The results indicate that danger agents can be detected using optical hyperspectral measurements and laser induced fluorescence.

I. INTRODUCTION

Light detection and ranging (LIDAR) provides active optical remote sensing in backscattering [1-3]. A pulsed laser transmitter is used and light backscattered from molecules and particles is collected by an optical telescope and were detected and range resolved in radar like mode. Chemical and biological content in plant leaves is important characteristic showing plant status. For example, pigment level correlates with power of photosynthetic apparatus and may be used for yield prognosis. Influence of various stresses on photosynthetic apparatus, first of all, induces changes in chlorophyll content. This phenomenon allows to develop effective approaches to fast revealing stresses in phytocenosis of various types using remote estimates of biochemical components content and its dynamics. It is well known that reflectance spectra of plant leaves in optical range is the most informative in respect to chlorophyll content due to the fact that they are formed specific spectrum of the pigments. Moreover, these spectra are the suitable characteristics for remote measurements.

Therefore, an idea to apply these characteristics for realization of agrocenosis monitoring and other types of plant canopies seems to be very attractive. Now there are numerous formulae based on regression relations to calculate the biochemical components concentration in leaves by various spectral coefficients of reflectance or their combinations. Practically, all of them are effective enough when biochemical components concentration is measured under laboratory conditions. When biochemical components content determination is performed for «soil-vegetation» system then estimation results may be distorted. The main cause is a contribution of soil reflectance. There are several approaches to minimize this type of noise by using more complex vegetation indices that include components for correction of soil interference. Their application demands measurements of additional parameters of the system, such as leaf index and spectral coefficients of soil reflectance. This condition makes the procedures less effective and moreover, the problem is not solved completely because of soil reflectance variability, for example, dependences on its moisture. Therefore, it is very important to develop effective methods acceptable for biochemical components measurement in the "soil-vegetation" system.

Application of high resolution spectroscopy for remote sensing of vegetation displayed new characteristics correlating with biochemical content. It appeared that the position of the red edge of spectral curves and the shape of this spectral region depend on the pigment concentration. Various characteristics of red edge region (680-800 nm) have been used. Position of maximum in the 1-st derivative plot of reflectance spectral curve and position of red edge were found to be in regression relation with chlorophyll content. We have revealed a high correlation between the ratio of two maxima in the 1-st derivative plot and chlorophyll concentration. A high resistance of this characteristic to soil contribution has been shown as well. A shortage of this approach is its sensitivity to instrumental noises, which may considerably distort the results, for example, influencing to intensity in maxima and their position. Thus, to develop a device for remote measurements of chlorophyll content in vegetation using the 1st derivative characteristics it is necessary to solve the following problems: 1) to develop a digital filter to noise suppress which permits to distinguish two maxima, 2) to develop an algorithm for automatic determination of maximum positions that depending on chlorophyll concentration as well, 3) to select appropriate characteristics and to develop a computing procedure to calculate the chlorophyll content, 4) to develop a procedure for minimization a soil reflection contribution. This paper is devoted to solving the problems 1-4.

The paper focuses on a methodology for early detection and identification of biochemical agents including various toxins and viruses using network of sensors. The ecological interest is a response to a proliferation of agent development and threats for human health. Additionally, there is a pervasive interest across diverse application areas such as medicine, environmental protection, and vegetation processing to achieve a rapid detection and identification capability of various agents.

II. METHODOLOGY

The following methods, samples and technologies have been used:

- Spectral methods
- Plant samples

- Polymeric films technology for direct microbial community observation
- Advantages and resolutions of polymeric films technology
- Observation of total microbial community architecture
- Visualization of individual components of soil microbial communities
- Reconstruction 3D view of soil microbial community architecture.

Leaves of winter wheat were studied during experiments, and appropriate spectral data had been collected. Variations of chlorophyll content were caused by differences in mineral nutrition and plant age during vegetation. Reflectance spectra were measured by FildSpec spectrometer. The recording system allowed to obtain the spectra in digital form. Digital spectra were recorded with the spectral resolution 1 nm in the range of 400-2500 nm. The pieces of the wheat leaves have been used for recording of reflectance spectra. The spectral data were used for chlorophyll content determination by Arnon's chemical method [4].

A preliminary data processing is based on application of smoothing procedure proposed in [5-8]. To perform data smoothing several moving windows of different size: 5, 7 and 9 point windows have been used. After data smoothing the 1-st derivative was computed using 2-d order polynomial estimated by making use of LS procedure. For this purpose another moving window was applied to smoothed data. Next, the resulting first derivative curve was calculated to estimation of chlorophyll content.

For estimation of biochemical components content in vegetation the following approaches were applied: SVM regression, principal component analysis and neural networks.

III. LIDAR SYSTEM

In the LIDAR approach, a laser pulse is transmitted into vegetation and the backscattered radiation is detected as a function of time by an optical receiver (Fig. 1).

Fig. 1. Onboard optical spectrometer

Our research has been concentrated on technologies such as optical spectroscopy, laser induced fluorescence, pattern recognition, and optimization methods. The optical device is intended to be used for ground measurements, also it can be installed on small airplanes for remote sensing. Spectra training set samples are pre-processed and fed into the SVM algorithm followed by Leave-One-Out Cross-Validation (LOOCV).

The software for detection of chemical and biological agents in spectral mixture has been developed (Fig. 2). The basic conception of LIDAR hyperspectral system is shown on Fig. 3.

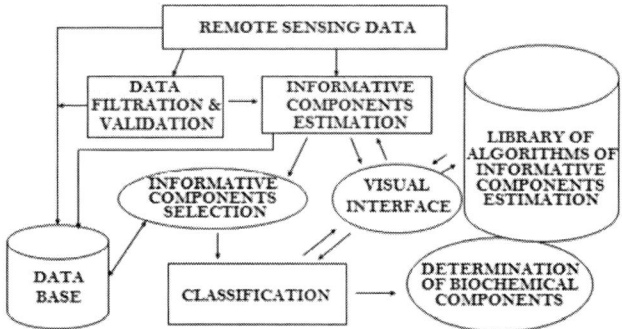

Fig. 2. Software scheme for detection of chemical and biological agents

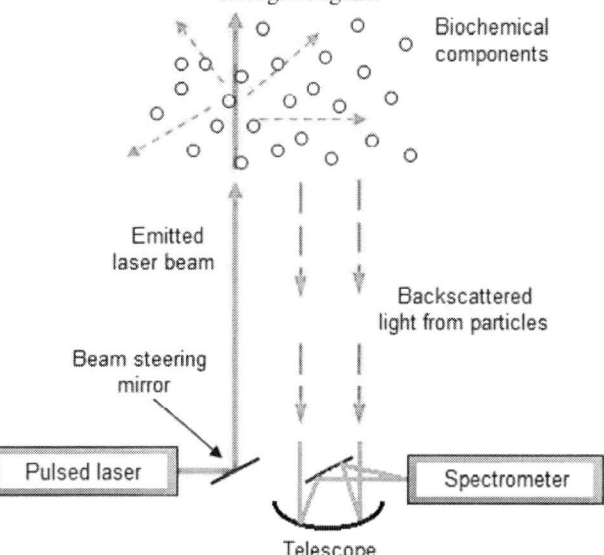

Fig. 3. Basic conception of LIDAR hyperspectral system

It was established during experiments that the noise level inherent to our measuring system was about 0.15% of maximum signal. After computing the 1-st derivative of spectral curves, the level of noise increased depending on the type of calculation procedure applied. Using formulae with 5- and 9-piont moving windows and approximation by 2-nd degree polynomials we have obtained about 4% and 1% noise level at maximum signal level respectively. Preliminary smoothing with 5, 7 and 9-point windows revealed their low effectiveness: the noise levels were 0.7, 0.5 and 0.5 of maximum signal for these procedures, respectively. The 9-point window procedure for computing the 1-st derivative appeared acceptable with respect to distinguishing of maxima (Table 1).

Visual analysis of the plots shown that the ripples are insignificant. There are two maxima in the 1-st derivative plots. Ratio of maximum intensity is appreciably dependent on chlorophyll content in leaves. Its value equals 1 for chlorophyll content 5 mg/dm^2; it becomes greater than 1 for larger pigment concentrations and less than 1 for lower concentrations.

Table 1. Maximum positions in the 1-st derivative plots and chlorophyll values estimated by various methods

Number of experiment	Maximum positions, nm	Chlorophyll content, mg/dm^2		$\Delta_{spectral-chemical}$
		Spectral Method	Chemical method	
1	704	9.35	9.18	0.17
2	705	5.33	5.29	0.04
3	701	3.80	3.75	0.05
4	697	1.54	1.46	0.08

Fig. 4 shows several plots of the 1-st derivative of reflectance spectral curves recorded for winter wheat leaves with various chlorophyll concentrations.

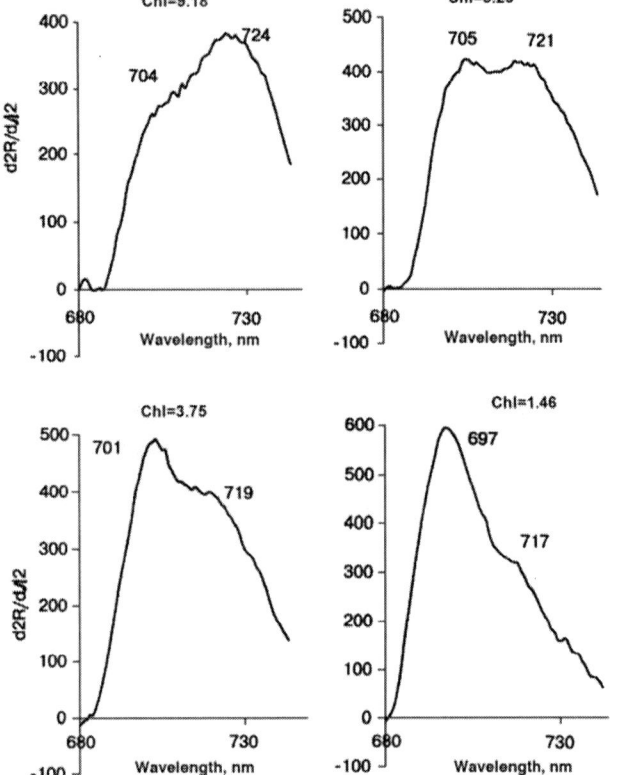

Fig. 4. Plots of the first derivatives of reflectance spectra for winter wheat leaves versus wavelength

IV. CONCLUSIONS

The main purpose of our experimental research is to investigate the influence of small concentrations of chemical and biological agents on reflectance spectra with particular attention paid to the effects of laser induced optical response. Laser experiments have being carried out to give fundamental insight into laser induced fluorescence phenomena. Technologies such as an optical spectroscopy measurement, laser induced fluorescence and microscopy methods have also been used in our experiments.

The algorithms of rapid search and earlier detection of biological/chemical agents using an unmanned aerial vehicle (UAV) have been prepared. Detection-based techniques might be effective in collaboration with classification and clustering techniques for sensor networks (SN).The development of anomaly detection techniques suitable for SN is therefore regarded as an essential research area, which will enable SN to be much more reliable. A few design principles and methods relating to the anomaly detection techniques in SN have been formulated. Informative criteria for anomaly detection from spectral data can be assisted by classification and clustering. The resulting order parameter set is further processed to obtain the minimum number of most informative parameters which represent the most discriminating pattern space for classification.

REFERENCES

[1] R.M. Measures, *Laser Remote Sensing: Fundamentals and Applications*. New York: Wiley, 1984.

[2] E.D. Hinkley, *Laser Monitoring of the Atmosphere*. Berlin: Springer Verlag, 1976.

[3] D.K Killinger, A. Mooradian, *Optical and Laser Remote Sensing*. Berlin: Springer Verlag, 1986.

[4] D.I. Arnon, "Copper enzymes in isolated chloroplasts. Polyphenoloxidase in beta vulgaris", *Plant Physiol.*, vol. 24(1), 1949, pp. 1-15.

[5] O.P. Burlak, J.-P de Vera, V. Yatsenko, N.O. Kozyrovska, "Putative mechanism of bacteria effects on plant photosystem under stress", *Biopolymers and Cell*, vol. 29(1), 2013, pp. 3-10.

[6] M.M. Kharytonov, V.M. Khlopova, S.A. Stankevich, O.V. Titarenko, "Remote and ground-based sensing of air polluted by nitrogen dioxide in the Dnepropetrovsk region (Ukraine)", *Disposal of Dangerous Chemicals in Urban Areas and Mega Cities*, Ed. by I. Barnes, K.J. Rudziński, Dordrecht: Springer, 2013, pp. 291-298.

[7] V. Yatsenko, "Classification, clustering and anomaly detection in sensor networks", International workshop, High School of Economics, Moscow, 2012.

[8] V. Yatsenko, "Active remote sensing of chemical and biological agents: computational models, toxicity testing, and human health risk assessment", 2nd International Conference on Computational Biomedicine, January 24-26, University of Florida, USA, p.7, 2013.

Opto-cryogenic sensitive element with ultrasensitive laser interferometer and microprocessor controller

M. V. Nalyvaychuk [1], V. O. Yatsenko [2]

[1] National Technical University of Ukraine (KPI), Kiev, Ukraine nnv@scs.ntu-kpi.kiev.ua
[2] Space Research Institute of NASU-NSAU Kiev, Ukraine vyatsenko@gmail.com

Abstract: Principles of satellite gravimeter creation to obtain information about gravitational field has been proposed. A mathematical model of the dynamic and stability of controlled superconducting suspension has been developed. Numerical algorithms for gravitational perturbation estimation acting on levitated probe have been proposed. The technology of satellite gravimeter creation based on high-temperature superconducting nanofilms has been proposed.

The solution of many practical problems depends on the widespread use of measurement systems and the principles on which they operate. These problems include the monitoring of natural resources based on analysis of gravity anomalies, the study of global geodynamic processes, the Earth's gravitational field, the movement of the poles of the Earth and others. Details of the gravitational field of the Earth are also needed for many fields of applied science (space exploration, geology, navigation, science of the shape of the Earth). Accurate and rapid detection of geodynamic processes may allow to get new data on the origin and development of critical local and global environmental conditions. Another practical problem is the need for information on undiscovered mineral resources of the Earth.

To solve the problem of detailing Earth's gravitational field using measurements of gravity gradients in three spatial directions from triaxial gravitational gradiometers that can be installed on board of the satellite, and the satellite orbit determination using satellite navigation equipment. The final results can be methodology, algorithms and results of clarify the fine structure of the Earth's gravitational field from a mathematical processing of satellite gradiometric measurements of European satellite GOCE.

Gravimeter is a very delicate instrument [2, 10] to measure the acceleration of gravity g. Presently, the accuracy of the best residential land gravimeters is $10^{-8}\,g$, for sea-based gravimeters $- 10^{-7}\,g$, and for aviation $- 10^{-6}\,g$. Most gravimeters that are produced by industry are based on the properties of stretched springs or elastic properties of springs made of quartz or other special alloys. Their accuracy is not sufficient to solve these problems. Since the gravimeters fault based on traditional principles was essentially exhausted, many developers over the past decades try to use innovative approaches to create super-accurate gravimeter [2, 10]. These efforts can be grouped by the method of hanging proximity sensing weight gravimeters, the use of electric or magnetic forces, the methods used to measure displacement sensing weight gravimeters (optical

registration system, in [9] Goodkind describes one of the first superconducting gravimeters.

As he notes, the basic structure of the superconducting gravimeter remains unchanged for almost 30 years since the first publication [10]. Free state (levitation) of sensing weight of the gravimeter is achieved based on Meissner effect [7, 9].

Figure 1 represents a functional diagram of measuring-computing system: 1 – cryostat with an insert, which is mounted on optical head 2 – optic cable, 3 – focusing lens, 4 – head to the laser diode 5 – power supply laser diode, 6 – unit temperature stabilization, 7 – Interferometer paint – Pen YFP-1, 8 – photodetector 9 – synchronous detector, 10 – a microcontroller with integrated DAC and ADC, 11 – modulator 12 – block keyboard and display. The main components of this system: the superconducting suspension, the optical system for registration of probe position and microprocessor computer unit.

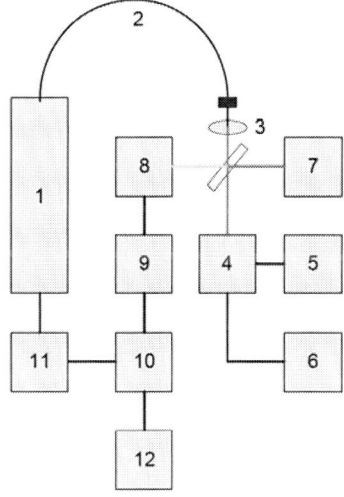

Fig. 1. Functional scheme of the measuring-computing system.

Suspension is coaxial. It contains magnets removed from the coaxial position to the position where their axis is parallel to the axis of suspension. With different options of retaining magnets (two, three, four) was chosen a system of four rare earth permanent magnets with vertical axis. Each magnet in a horizontal plane has a rectangular shape. The vertical magnetic axis of the four sets of magnets was displaced from the axis of suspension so that the resulting space measuring 18 mm in diameter to accommodate the optical sensor. The problem of not suspended vertical position of the free samples that resulted from varying the magnetic properties of sets of permanent magnets, solved with two changes of the structure. One of them - an increase of reciprocation test mass, the other was to thin ferromagnetic ring offset azimuthal magnetic field inhomogeneity suspension. After theoretical and experimental studies of suspension, optical sensors and software measurements were analyzed the feasibility of the chosen design of the project. In part of the suspension the main changes concerned the design of the magnetic system, where instead of placing magnets containing the axis of suspension they had to be shifted from the axis to accommodate her optic laser sensor.

New design working model (Fig. 2) includes four sets of permanent magnets, vertical axis were shifted from the axis of suspension in four radial directions. Test weight has two niobium-titanium ring. The upper plane of the test body was polished as reflecting plane of laser light. The levitation gap, depending on the mass of the test body was from 7 to 15 millimetres. Based on this working model was carried out theoretical and experimental research collaboration suspension system - registration. The influence of the physical state of helium (liquid or gas) to working suspension system was analyzed - registration. The impact of passive filter for accuracy was studied. We analyzed the factors that influence the decrease in stiffness of suspension, in particular, the presence of additional ferromagnetic mass on the free sample

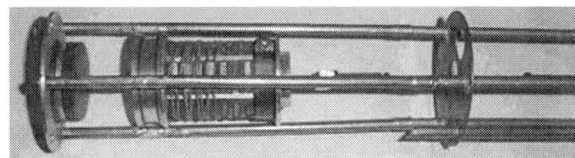

Fig. 2. The superconducting magnetic suspension

Experimental research of collaboration were conducted to identify your gravimeters model properties, dynamic characteristics of magnetic suspension of free trial weight of gravimeters layout (together with a group of the Institute of Physics of NASU) and the revision of the sensor in the magnetic suspension, which was dictated by the experimental work.

REFERENCES

[1] Yatsenko V.A. and Kryuchkov E.I. On the possibility of application of "Magnetic Potential Well" effect for creation of a graviinertial cryogenic device. Journal of Automation and Information Science, V. 34, 2003.-P. 106-119.

[2] W.A. Prothero and J. M. Goodkind. A superconducting gravimeter. – Rev. Sci. Instr, № 2 , Vol. 39, № 9. 1968.– P. 1257 – 1261.

[3] V.Yatsenko and P.Pardallos, Global optimization of cryogenic-optical sensor, in Sensors, Systems, and Next-Generation Satellites, K. W.H. Fujisada, J. Lirie, ed., Proc. SPIE 4550, ãã. 433 - 441, 2001.

[4] V. Yatsenko, M. Nalyvaichuk. Modeling and Optimization of Cryogenic-Optical Gravimeters // International Journal information theories & applications. - 2012. - P. 232-240. Vol. 19. № 3.

[5] V. Yatsenko, M. Nalyvaichuk. Cryogenic-Optical Gravimeter: Principles, Methods and Applications // Kharkov University Vestnik, Ser. Radiophysics and Electronics.-2011.– P. 107-113.

[6] V. Yatsenko, M. Nalyvaichuk, V. Gnidenko, O. Kochkodan. Modeling of Superconducting Gravimeter for Futute Space Missions: Optimization. Control, and Estimation//2nd International Conference on memory of corresponding member of National Academy of Science of Ukraine V. S. Melnik, 4-6 Aprill, 2012. Kyiv, Ukraine. – P.110.

[7] Moon F. C. Superconducting Levitation: Application to bearings and magnetic transportation, John Willey & Sons, NY, 1994, 295p

[8] Kozoriz V. Novel Magnetic Levitation and Propulsion Phenomena, ISBN 966-7108-10-4, Zaporizhya, 1999, 271p

[9] J. M. Goodkind. The superconducting gravimeter. Rev. Sci. Instrum.-Vol. 70, № 11. – P. 4131-4152, 1999.

[10] J. M. Goodkind and R. J. Warburton, IEEE Trans. Magn. MAG – 11, № 2, 1975.

Modern Polarization-Independent Trap Detectors

K. I. Muntean

National Scientific Centre "Institute of Metrology", Kharkov, Ukraine

Abstract: Modern design of advanced optical polarization-independent trap detectors is considered. It is shown that using a trap detector as a transfer standard one must take into account the diffuse scattering of optical radiation by the photosensitive surfaces of a trap detector photodiodes.

At present, the absolute cryogenic radiometers provide the most accurate reproduction of the energy and photometric units of optical radiation, so they are the primary standards in metrology organizations of the leading countries of the world. Operating range of cryogenic radiometers is limited to a few tens of microWatts, while the operating range of instruments that need to verify, extends to the values of hundreds of milliWatts and ones and tens of Watts. Precision transfer standard is necessary to ensure the traceability of reproducible physical units outside the operating range of the cryogenic radiometer.

Photodiode trap detector is the best candidate for the role of transfer standard in the visible and near infrared optical radiation at this time. Trap detector, which was developed in the early 80-ies of the last century, is continuously improved since the area of coherent and non-coherent optical technology is evolving and performance requirements for standards is increasing continuously.

The idea of constructing the trap detector has arisen as a result of the development of silicon photodiodes that have internal quantum efficiency (IQE) is almost equal to unity.

However, the use of these photodiodes in precision metrology was difficult because they had a significant Fresnel reflection of the measured optical radiation from the light-sensitive surface of the photodiode. As a result, the external quantum efficiency (EQE) of single photodiode was unsatisfactory.

E. F. Zalewski and C. R. Duda have implemented a simple idea to improve the external quantum efficiency of the trap detector [1]. They placed the photodiodes in the form of a planar spiral, so that the radiation reflected from one photodiode is directed to the next photodiode. Last photodiode reflects incident light normally, and it passes all the way in opposite direction. All trap detector photodiodes integrated into a parallel circuit, so the trap detector is theoretically equivalent to a photodetector, which has IQE and EQE equal to unity.

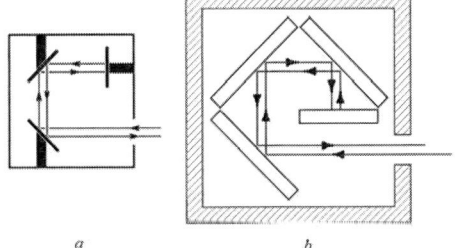

Fig. 1. Trap detectors of the first generation

This design of trap detector has a major drawback, it is very sensitive to the direction of polarization of the incident radia-

tion [2]. The absorption coefficients of the trap detector for p- and s-polarization of the incident optical radiation is significantly different.

Trap detectors of second generation contained three photodiodes, which were placed on a three-dimensional spiral. This arrangement has a much lower sensitivity to the direction of polarization of the incident radiation [3].

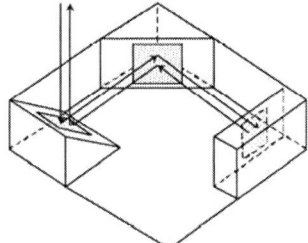

Fig. 2. Trap detector of the second generation

Full equalization of reflection coefficients for p-and s-polarization of the incident radiation beam is obtained only for a plane wave, that propagates precisely along the optical axis of the trap detector. Also trap detector must have a perfect mutual adjustment of light-sensitive surfaces of photodiodes. In reality, the flow of optical radiation is not a plane wave, it has a finite angular divergence. In addition, the flow of optical radiation does not propagate exactly in the optical axis of the trap detector and trap detector has errors of the spatial arrangement of light-sensitive surfaces of the photodiodes. The above factors lead to the fact that part of the incident optical radiation reflected from the trap detector and lost. According to the literature such losses of the measured optical radiation are estimated to be ~ 0.001, which does not meet modern requirements for the reference measurements.

The next step in improving the characteristics of the trap detector was an increase in the number of reflections. In [4] it was proposed to build a spatial spiral consisting of three or four photodiodes, so that the first photodiode of the spiral at the same time would be the last photodiode. This approach allowed us to construct the structure which comprises three (or four) real photodiodes and it is equivalent to the virtual spiral, which includes four (or five) photodiodes. First proposed the concept of "virtual" photodiodes was an important difference between this trap detector design from all previous ones.

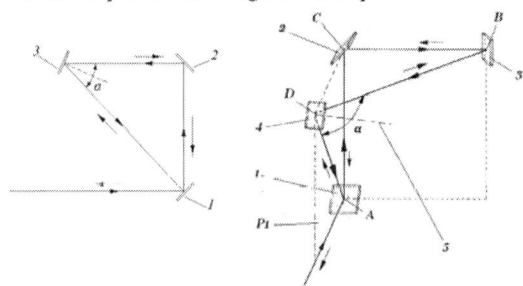

Fig. 3. The design of the trap detectors [4]

978-1-4799-0019-0/13 $31.00 © 2013 IEEE

The paper [4] did not exhaust all reserve increase in the number of reflections. Using the approach proposed in [4], we have developed a design with an even greater number of reflections [5]. This construction comprises four photodiodes which are equivalent to the six photodiode virtual spiral. Spiral is planar, as the proposed design provides a large number of reflections, so that the sensitivity to the direction of polarization is very small. According to the calculations radiation absorption coefficients of s- and p-polarization differ by no more than $3 \cdot 10\text{-}5$.

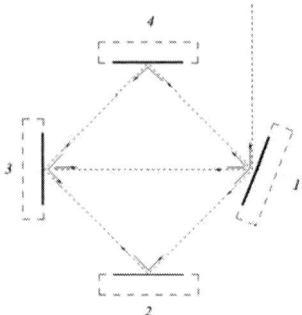

Fig. 4. The design of the trap detector [5]

A comparison of known structures trap detectors shows that they are constructed on the same principle - the entire set of N photodiodes form a single ring configuration (planar or spatial spiral), in which the optical beam is reflected by successively each photodiode. Last photodiode of a configuration is setting normally to the optical beam, causing it to pass again the entire configuration in the opposite direction The optical beam is reflected by each photodiode (except the last photodiode , which returns the beam in the opposite direction) twice, so the total number of reflections is $(2N - 1)$.

A photodiode, which is added in a cyclic configuration increases the number of reflections on the two. The classical scheme of trap detector provides five reflections for three photodiode configurations, then it provides seven reflections for the four photodiode configuration, and so on.

Design of the trap detectors from [4] and [5] provide the use of the same photodiode to create a spiral configuration repeatedly. A special path for the propagation of optical radiation, which is formed in the paper [4], allows to use twice only one photodiode to create the spiral configuration. In [5] two photodiodes are used twice to create a spiral configuration. However, it is impossible to apply this approach to more photodiodes, since the spatial arrangement of all photodiodes of the configurations, except the first photodiode, is strictly interconnected. Only the spatial position of the first photodiode configuration can be changed by changing the angle of incidence of the input optical beam. It is this feature is implemented in [4] and [5].

The number of photodiodes in the classical scheme of the trap detector can not increase indefinitely, as a dark current of the trap detector increases in proportion to the number of photodiodes. The increase of dark current reduces the dynamic range of the trap detector and leads to an increase in the noise that limits the achievable accuracy of the optical radiation.

In designing of the optical trap detectors it is necessary also to take into account such factors as the scattering of the optical radiation by light-sensitive surfaces of the photodiodes.

Our own experimental studies of silicon photodiodes Series S1337-1010 manufactured by HAMAMATSY PHOTONICS KK have shown that the light scattering losses can be more than 0.001, which is unacceptable to the reference measurements.

In the measurements laser ЛГН-302 type, goniometer ГС-5 and laser power meter ИМЛИ-2 were used.

Analyzed photodiode was mounted on the goniometer turn-table/ It was rotated about a vertical axis to set the required angle of incidence of the laser radiation on the photosensitive surface of the photodiode. The angular distribution of the radiation that has been reflected from the photosensitive area of the photodiode, was recorded through the diaphragm collimator by the receiving head of the power meter. The receiving head of the power meter, fitted with a diaphragm collimator was mounted on the housing of the goniometer telescope, which revolved around a turn-table.

Scattered light intensity J is measured in the range 0 ° .... 90 ° every 20 ' at angles of incidence of the primary beam of radiation on photosensitive area of the photodiode near 10 °, 20 °, 30 °, 40 °, 45 °, 50 ° and 55 °. Registered indicatrix $j_i = f(\theta)$ containing an intense central peak and weak wings are almost identical to each other (Fig. 5).

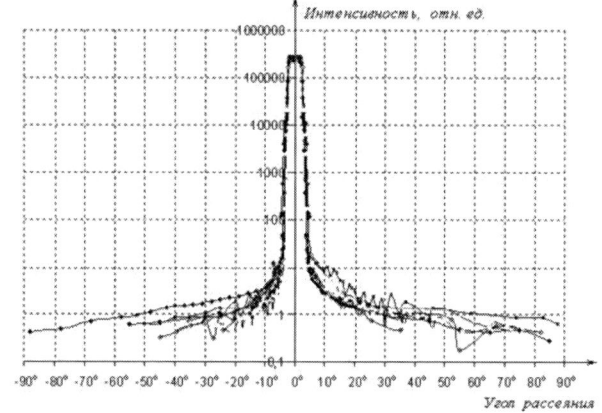

Fig. 5. Imposing the experimental scattering functions

Experimental indicatrix $j_i = f(\theta)$ is the convolution of the actual instrument function h (r) of the meter and the theoretical indicatrix $J_i = f(\theta)$ the scattering of optical radiation by a photodiode photosensitive area.

Analysis of the experimental scattering indicatrix it is conveniently carried out first considering the instrumental function of the meter as δ-function, and then approximate the real two-dimensional instrumental function h (r) of the aperture collimator by axially symmetric rectangular function in half-width ε.

It is known, convolution of the delta function and the Gaussian intensity distribution in half-width θ in semi-logarithmic coordinates is giving the indicatrix of the parabolic form in half-width θ, which vertex is facing up. Rectangular instrumental function in half-width $\varepsilon \gg \theta$ broadens this parabola to the value of ε. Thus, the central peak is observed in the experiment indicatrix corresponds to the specular component of the reflected beam. The presence of the wings of the peak indicates the presence of a diffuse component of the reflected beam.

Characteristics of specular and diffuse components vary considerably are:
- specular component is high-intensity and it is concentrated in a small solid angle;
- diffuse component is low-intensity and it is dispersed over a large solid angle.

The difference between the characteristics of the component greatly facilitates the evaluation of scattering losses.

Given that the convolution integral is linear, and the divergence angle of the incident beam is small and diffuse component has a low-intensity then from width of the central peak of the experimental indicatrix find that the rectangular instrumental function of the aperture collimator in our experiment has the half width $h \approx 2{,}3$ °.

The symmetrized graph of the diffuse component (averaging over seven realizations, smoothing over 11 points, dark current is eliminated) is approximated with the reliability $R^2 = 0{,}96$ by a Gaussian function

$$J(\theta) = 1{,}21 e^{-0{,}000542\ \theta^2}$$

in the half-width $\theta_{0,5} = 36°$. Graphics of the diffuse component and approximating the Gaussian function in logarithmic coordinates are shown in Fig. 6.

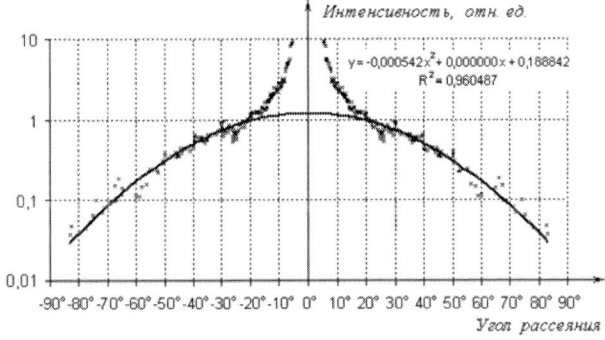

Fig. 6. The diffuse component
and approximating Gaussian function

The half-width of the collimator instrumental function in order of magnitude is smaller half-width of the diffuse component of the experimental indicatrix then instrumental broadening of the true indicatrix is slightly. Representing experimental indicatrix $j = f(\theta)$ in terms of integral moments of the instrumental function $h(r)$ of the collimator and derivatives true indicatrix $J = f(\theta)$ as [6]

$$j(\theta) = m_0\, J(\theta) + \frac{m_2}{2!} J''(\theta) + \ldots \approx 2\,\varepsilon\, A\, e^{-0{,}000542\ \theta^2}\left[1 + \frac{\varepsilon^2}{3}(0{,}001084\theta^2 - 1)\right],$$

find that the instrumental broadening of the true indicatrix does not exceed 0.037%. A similar calculation shows that the broadening due to the smoothing is ($\sim 0.004\%$).

Calculation of the integrated intensity of the axially symmetric scattering function shows that the diffuse component contains about $1{,}7 \cdot 10^{-3}$ of the incident beam energy. Approximately 20% of the scattered radiation is absorbed by photodiodes of the trap detector, but the rest is too large and this contribution to the losses of radiation is dominated.

It is known that by selecting the photodiodes can be reduced radiation diffuse component to 10-4 ... 10-5 [7]. Trap detectors, complete with photodiodes with a small scattering of the incident radiation, are promising as transfer standards for use in the equipment of the national primary standards.

The author is grateful to V.M. Balaban and E.P. Timofeev for helpful discussions.

REFERENCES

1. E. F. Zalewski, C. R. Duda. Silicon photodiode device with 100 % external quantum efficiency. *Appl. Opt.*, 1983, v. 22, No 18, p. 2867-2873.

2. N. P. Fox, Trap Detectors and their Properties, *Metrologia*, 1991, v. 28, 197-202.

3. J. M. Palmer. Alternative Configurations for Trap Detectors. *Metrologia*, 1993, v.30, 327-333.

4. Д. М. Татьянко. Трап-детектор. Патент Украины № 87197, Бюлл. № 12, 2009 г.

5. В. М. Балабан, К. И. Мунтян, Е. П. Тимофеев. Трап-детектор. Патент Украины № 98063, Бюлл. № 7, 2012 г.

6. А. Папулис. Теория систем и преобразований в оптике. "Мир", Москва, 1971 г.

7. M. Sildoja1, F. Manoocheri and E. Ikonen. Reflectance calculations for a predictable quantum efficient detector. *Metrologia*, 2009, № 46, p. 151-154

Thermal Processes in the Bolometric Measurer of Laser Radiation Characteristics

Anastasia O. Pak, *Student Member, IEEE*, Nikolay G. Kokodiy
Quantum Radiophysics Dept., V.N. Karazin Kharkiv National University, Kharkiv, Ukraine
anastasiia.pak@gmail.com

Abstract: In this work there are investigated an influence of nonlinearities of thermophysical parameters of thin metallic bolometers from platinum and nickel in the temperature range from room to the melting temperatures.

The grid measurers are perspective for a wide range of capacities and large apertures of laser beams. Using of the several grids, where bolometers are located under different angles relative to vertical, it is possible to obtain information about several characteristics of the laser radiation - power or pulse energy, intensity distribution in the beam, its diameter, position of the energy center, nature of the radiation polarization, radiation pulse shape and others [1-5].

For correct solve of these problems it is necessary to take into account that the thermal parameters of bolometers' metals depend of their temperature.

At high temperatures, close to the melting point of metals, the nonlinearities occur and influence on results of measurements.

<u>Heating of bolometer by laser radiation (linear mode)</u>. Let us estimate the temperature of laser beam heating of bolometer and find the temperature distribution along the bolometer. Suppose that bolometer is a metal wire with diameter D and length L, located along the axis x. An incident field is the Gaussian beam radiation with intensity distribution:

$$I(r) = I_{\max} \exp\left(r^2 / r_0^2\right),$$

where $I_{\max}$ is intensity at the beam center, r_0 is beam radius, r is radial coordinate.

The radiation power in the beam:

$$P = \pi r_0^2 I_{\max}.$$

Intensity distribution of radiation along the bolometer is described by the formula):

$$I(x) = I_{\max} \exp\left(-\left(x - x_c\right)^2 / r_0^2\right), \qquad (1)$$

where x_c is location of the beam center on bolometer.

Thermal processes in bolometer can be described by the heat equation for the rod, taking into account impact of heat into the environment from the side surface of the wire [6]:

$$\frac{\partial^2 T(x,t)}{\partial x^2} - \gamma^2 T(x,t) - \frac{1}{a}\frac{\partial T(x,t)}{\partial t} = -\frac{4Q_{abs}I(x,t)}{\pi k D}, \qquad (2)$$

where T is temperature, $\gamma^2 = 4\alpha_p / \pi k D^2$, $a = k / c\rho$ – thermal diffusivity coefficient, k – thermal conductivity coefficient, c – specific heat, ρ – material density of bolometer, α_p – linear coefficient of heat exchange with

environment, independent on diameter in case of thin wires [7], Q_{abs} – bolometer absorption efficiency factor. Because of small diameter of bolometer, the radiation intensity is the same across the whole thickness.

The boundary conditions are chosen as follows:

$$\partial T(0,t) / \partial x = \partial T(L,t) / \partial x = 0.$$

In Fig. 1 presented are solutions of equation (2) - the temperature distribution along the bolometer length under the next conditions: bolometer material is platinum (Pt), $L = 60$ mm, $\rho = 21460$ kg/m^3, $k = 72$ W/(m·K), $c = 135$ J/(kg·K), $\alpha_p = 0,03$ W/(m·K), $I_{\max} = 1000$ W/m^2, $x_c = 3$ cm, $r_0 = 1,5$ cm, $Q_{abs} = 1$. The calculations were made for several values of the bolometer diameter – $D_1 = 20$ μm, $D_2 = 40$ μm, $D_3 = 100$ μm. In calculations it was assumed that the values of thermophysical parameters are independent on temperature.

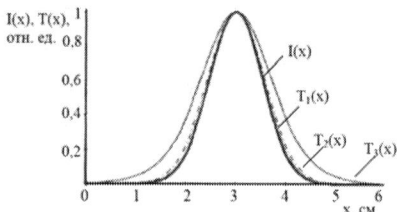

Fig. 1. Intensity and temperature distributions along the bolometer. Functions $T_1(x)$, $T_2(x)$, $T_3(x)$ are for bolometers with diameters $D_1 = 20$ μm, $D_2 = 40$ μm, $D_3 = 100$ μm respectively.

The temperature distribution is shown for the time t_{ycm} when the thermal equilibrium with the environment has been determined. It is known that

$$t_{ycm} \approx 5\tau,$$

where $\tau = mc / \alpha_p L = \pi D^2 c\rho / 4\alpha_p$ – thermal time constant of a radiation receiver [6], where m – the mass of the receiver.

Fig. 1 shows that, when $D_1 = 20$ μm, temperature distribution is closely to the distribution of the radiation intensity.

Thus, for minimizing difference between intensity distribution and temperature distribution, it is necessary to use in receivers bolometers with diameter less than 20 microns.

<u>Nonlinear effects during continuous heating of bolometer.</u> In theoretical research the heat equation (2) is used with taking

into account the temperature dependence of thermal parameters of bolometers) – $c(T)$, $k(T)$, the resistivity $\rho_e(T)$, $Q_{abs}(T)$ and $\alpha_p(T)$.

In Figs. 2-7 presented are temperature distributions along the platinum bolometer ($D = 20\,\mu m$, $L = 60\,mm$) heated by Gaussian radiation beam with the average quadratic radius 15 mm at different times. The maximum intensity of the beam is in the center of bolometer. Radiation is unpolarized in the visible range of wavelengths.

In Figs. 2-3 presented are temperature distributions for the beam power 10 W. At maximum in steady-state mode temperature value is 3° C. For such heating dependence of thermal parameters from the temperature does not appear. Therefore, the linear heat equation and the nonlinear equation give the same results.

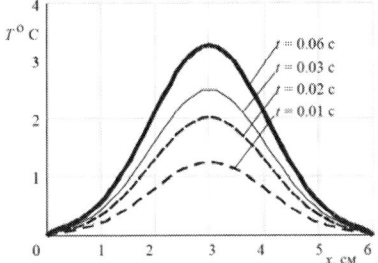

Fig. 2. Temperature distribution along the bolometer (Pt, $D = 20\,\mu m$, $P = 10$ W), linear heat equation.

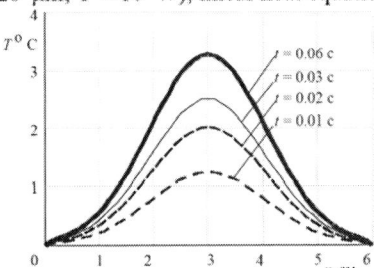

Fig. 3. The same as in Fig. 2, nonlinear heat equation.

In Figs. 4-5 presented are temperature distributions along the bolometer for the case when radiation power in the beam is 10^4 W. The temperature it the maximum is 900° C. The linear equation gives wrong value of temperature heating – 3265° C, in several times greater than actual value.

Thus, taking into account nonlinear properties of the bolometer during heating to high temperatures is necessary. Otherwise, big errors in the results of measurements of energetic characteristics of the laser beams is inevitable.

Fig. 5 shows another one feature of the process of strong heating – reducing of the propagation speed of heat along the bolometer in the heated zone. The distance between curves at the marked time points are smaller than in Fig. 4. This is explained by the «localization of the heat» – effect, discovered by Kurdyumov S.P. *at. al.* in [8].

It consists in the fact that during the heating of material, which thermal conductivity increases with increasing of temperature, the heated zone expands very slowly, and in some cases cannot expand at all during determined period of time.

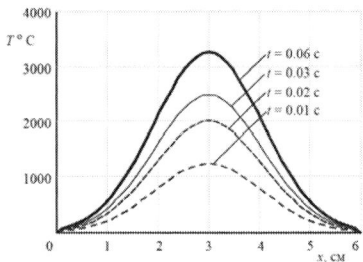

Fig. 4. Temperature distribution along the bolometer (Pt, $D = 20$ μm, $P = 10000$ W), linear heat equation.

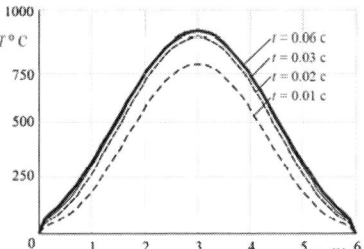

Fig. 5. The same as in Fig. 4, nonlinear heat equation.

For this reason, curve shape of the temperature distribution along the bolometer at high temperature differs from that observed at lower temperatures. It is necessary to take this into account in signal processing algorithms of bolometers.

In Fig. 6 presented are time dependences of temperature in the center of bolometer. The ordinate axis shows its normalized values relative to the stable temperature. The horizontal dashed line corresponds to the level $1 - e^{-1} \approx 0,632$ from the value of stable temperature, where we can evaluate the thermal time constant of heating process of bolometer.

It can be seen, that the presence of temperature dependences of bolometer parameters, leads to strong reducing of the thermal time constant (from 22 ms to 5 ms for the case shown in Fig. 6). The reason of this is growth of the heat exchange coefficient with increasing of temperature.

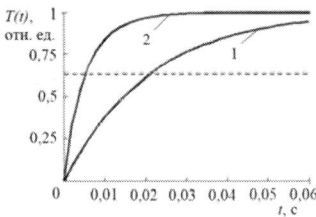

Fig. 6. Time dependence of temperature in center of bolometer (Pt, $D = 20$ μm, $P = 10000$ W): 1 – linear heat equation, 2 – nonlinear heat equation.

Increasing of speed of thermal processes in the bolometer heated zone results in localization of heat in this zone. Thus, the increasing of heat exchange with the temperature has a dual function. From the one hand, the heat transfer to the environment is increased and the temperature of the bolometer is reduced. From the other hand, the thermal time constant decreased, which is promote the spread of heat to the heated zone and the increasing of temperature in this zone.

Nonlinear effects during pulsed heating of bolometer. During a strong impulse heating of the bolometer, when the pulse duration is 1 ms, and pulse energy is 100 J, nonlinear effects can also appear, but weaker than in the case of continuous heating (Fig. 7).

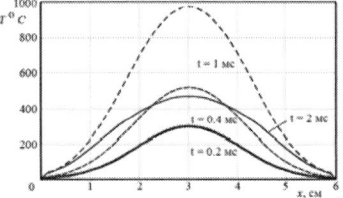

Fig. 7. Temperature distribution along the bolometer (pulse heating) (Pt, $D = 20$ μm, $E = 100$ J), nonlinear heat equation.

The temperature increases during the pulse, but it distribution remains almost the same and differ a little of the distribution of intensity in the beam. After the pulse end temperature decrease and aligned along the bolometer length.

The temperature at the maximum is $1000°$ C – the same as in the case of heating of the bolometer by continuous emission with power of 10000 W. Under pulsed heating the temperature value, which is obtained by solving the linear heat equation, is approximately $1500°$ C, which is differs in 1.5 times than solution of the nonlinear equation, while, during the continuous heating, the difference is 3.5 times.

Consideration of bolometer thermophysical parameters dependence on temperature in signal processing algorithms for grid receiver. The results of research of thermal and optical properties of the bolometer during the heating by laser radiation can take into account the temperature dependence of parameters during the signal processing from the radiation detector. For this case it is necessary to determine the average temperature and power absorbed in each bolometer.

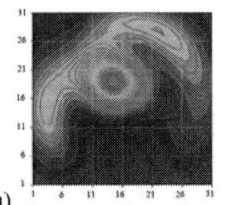 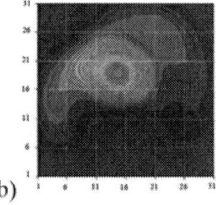

a) b)

Fig. 8. Restored function of intensity distribution of radiation: a) by linear algorithm, b) by nonlinear algorithm.

In Fig. 8 presented are the intensity distributions of projection bulb radiation: Fig. 8a – the results of calculations by linear algorithm, excluding temperature dependence of thermal parameters, Fig. 8b – using the nonlinear algorithm, including temperature dependence of thermal parameters.

Fig. 9a shows the horizontal sections passing through the maximum of graphs, in Fig. 9b – similar vertical sections.

On the edges of the graphs calculated by the linear algorithm (curve 1), there are local maximum and flat areas, which is not presented in the investigated radiation. They are absent in the graphs calculated by the algorithm which takes into account non-linear effects in the bolometers (curve 2).

Thus, taking into account the nonlinear effects in the bolometers in data processing algorithms leads to a more accurate restoration of the picture of the of the intensity distribution in the radiation beam.

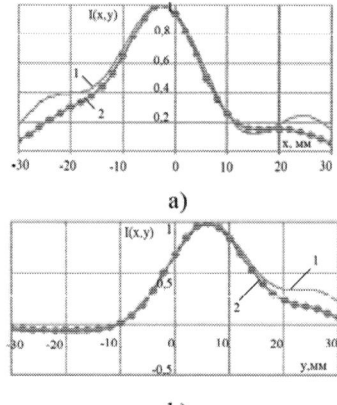

a)

b)

Fig. 9. Restored function of intensity distribution of radiation in a linear (1) and nonlinear (2) algorithms: a) the horizontal section, b) vertical sections.

REFERENCES

[1] N. G. Kokody, D. N. Kokody, «Measurement of the distribution in the beam of thermal radiation», High Temperature, vol. 45, no. 2, pp. 289-295, 2007.

[2] V.M. Kuzmichov, S.V. Pogorelov, V.P. Balkashin, I.A. Priz I.A., P. Kohns, «Measurement of linear or elliptical polarization of the laser radiation by bolometer with three grids», Radiophysics and Radio Astronomy, vol. 14, no. 2, pp. 214-221, 2009.

[3] N. G. Kokody, V. A. Svich, B. A. Timanyuk, S.V. Pogorelov, A. O. Pak, Li Zhenhua, «Direct and inverse problems of heat transfer in laser technology and metrology», Kharkiv National University named after VN Karazina, 240 P., 2012.

[4] A.O. Pak, N. G. Kokody, «Lattice receiver for measuring of the laser radiation characteristics», Vestnik Kharkiv National University named after V.N. Karazina, series «Radiophysics and Electronics», vol. 19, no. 983, pp. 72-76, 2011.

[5] A.B. Katrich, A.V. Khudoshin, «Measurement of spatial-energy characteristics of the laser radiation», Avtometriya, no. 2, pp. 108-110, 1987.

[6] G. Carslaw, V. Jager, «Thermal conductivity of solids», Fizmatgiz, 487 P., 1964.

[7] Perepechai M.P. Research of interaction of intensive infra-red radiation with metal cylinders with the purpose of creation through passage measurer of capacity of laser beams. Candidate's thesises Kharkiv, p.182, 1979.

[8] A.A. Samarskiy, N.V. Zmitrenko, S.P. Kurdyumov, A.P. Mihaylov, «Effect of metastable localization of heat in the environment with nonlinear thermal», vol. 223, no.6, pp. 1344-1347, 1975.

Non-steady-state photoelectromotive force in AlN crystal

M. Bryushinin, V. Kulikov, E. Mokhov, S. Nagalyuk and I. Sokolov

Ioffe Physical Technical Institute, Politekhnicheskaya 26, 194021, St. Petersburg, Russia

Abstract—The experimental investigation of the non-steady-state photoelectromotive force in aluminium nitride crystal is reported. The experiments are performed for two geometries, where arising photocurrent is parallel or perpendicular to the optical axis of the crystal. Dependencies of the signal amplitude versus light intensity, temporal and spatial frequencies are measured. The photoelectric parameters of the material are estimated for the light wavelength $\lambda = 532$ nm.

The photorefractive effect includes the stage of space charge formation, which occurs via migration of charge carriers, i.e. electric current [1]. This current is the basis of the non-steady-state photoelectromotive force (photo-EMF) effect, which reveals itself as an alternating electric signal arising in the sample illuminated by an oscillating interference pattern [2], [3]. Since the current is resulted from the periodic relative shifts of the photoconductivity and space charge gratings, the technique based on this effect allows determination of the number of photoelectric parameters (conductivity, carrier sign, lifetime, diffusion length, and drift mobility) and can be applied for the investigation of both non-centrosymmetric and centrosymmetric media. In this paper we study the non-steady-state photo-EMF effect in AlN bulk crystal and apply this technique for determination the photoelectric properties of the material.

The experimental arrangement used for the investigation of the non-steady-state photo-EMF in AlN crystal is shown in Fig. 1. The second harmonic of Nd:YAG laser ($\lambda = 532$ nm, $P_{out} \simeq 20$ mW) laser was split into two beams forming the interference pattern with spatial frequency K and contrast $m = 0.48$ on the sample's surface. The signal beam was phase modulated with frequency ω and amplitude $\delta = 0.62$ by the electrooptic modulator. The polarization plane of the light was perpendicular to the incidence plane. The beams were expanded in order to illuminate the whole interelectrode spacing. Two sample orientations were studied with the parallel orientation of the grating vector $\mathbf{K}$ and the optical c-axis [0001] in the former case ("longitudinal" $\parallel$-geometry), and with the perpendicular orientation in the latter one ("transversal" $\perp$-geometry). The photocurrent arising in the sample generated the corresponding voltage on the load resistor $R_L = 10$ MΩ. The voltage was amplified and measured using spectrum analyzer or lock-in nanovoltmeter.

The crystal was grown by PVT technique [4] and has pale yellow color. The characteristic dimensions of the sample were $6.0 \times 3.8 \times 3.0$ mm^3. The front and back surfaces (3.8×3.0 mm^2) were polished. The silver paste electrodes were painted on the lateral surfaces (6.0×3.0 mm^2 for $\parallel$-geometry and

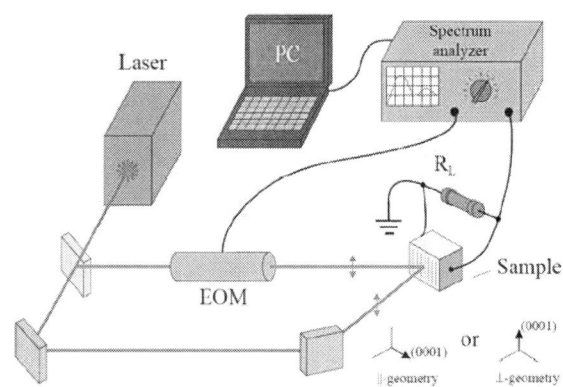

Fig. 1. Experimental setup for investigation of the non-steady-state photo-EMF in semiconductors. EOM is the electrooptic modulator.

6.0×3.8 mm^2 for $\perp$-geometry). The static dielectric constant of the material is of $\epsilon_\parallel = 9.14 - 9.32$ and $\epsilon_\perp = 7.76$.

The presence of the non-steady-state photo-EMF effect in AlN bulk crystal is the first result which should be pointed out. The amplitude of the detected signal is of order of $0.1 - 1$ pA which is noticeably lower than that in the model objects – sillenite crystals Bi$_{12}$Si(Ti,Ge)O$_{20}$, where it reaches 1 nA. [2] Nevertheless this amplitude is quite enough for consistent detection of the signal with signal-to-noise ratio of $1 - 10$. The phase of the detected signal indicates that electrons prevail in the process of photo-EMF excitation at low and moderate spatial frequencies of the interference pattern.

We have measured the frequency transfer functions of the non-steady-state photo-EMF signal (Fig. 2) in the "longitudinal" ($\parallel$) and "transversal" ($\perp$) geometries of the experiments, i.e. for the cases when direction of the space charge field $\mathbf{E_{SC}}$, grating vector $\mathbf{K}$, and photocurrent $\mathbf{J}^\omega$ is parallel or perpendicular to the c-axis. The signal demonstrates typical behavior, namely, there is a linear growth of the amplitude for low frequencies of the phase modulation $\omega < \omega_1$, frequency independent region for higher frequencies $\omega_1 < \omega < \omega_2$, and decay for frequencies $\omega > \omega_2$. These regions are separated by the so-called cut-off frequencies $\omega_1/2\pi = 140 - 700$ Hz and $\omega_2/2\pi = 13 - 30$ kHz. These dependencies are well described by the following expression which is equivalent to that obtained earlier for the simplest model of the semiconductor crystal with one type of partially compensated donor centers

978-1-4799-0019-0/13 $31.00 © 2013 IEEE

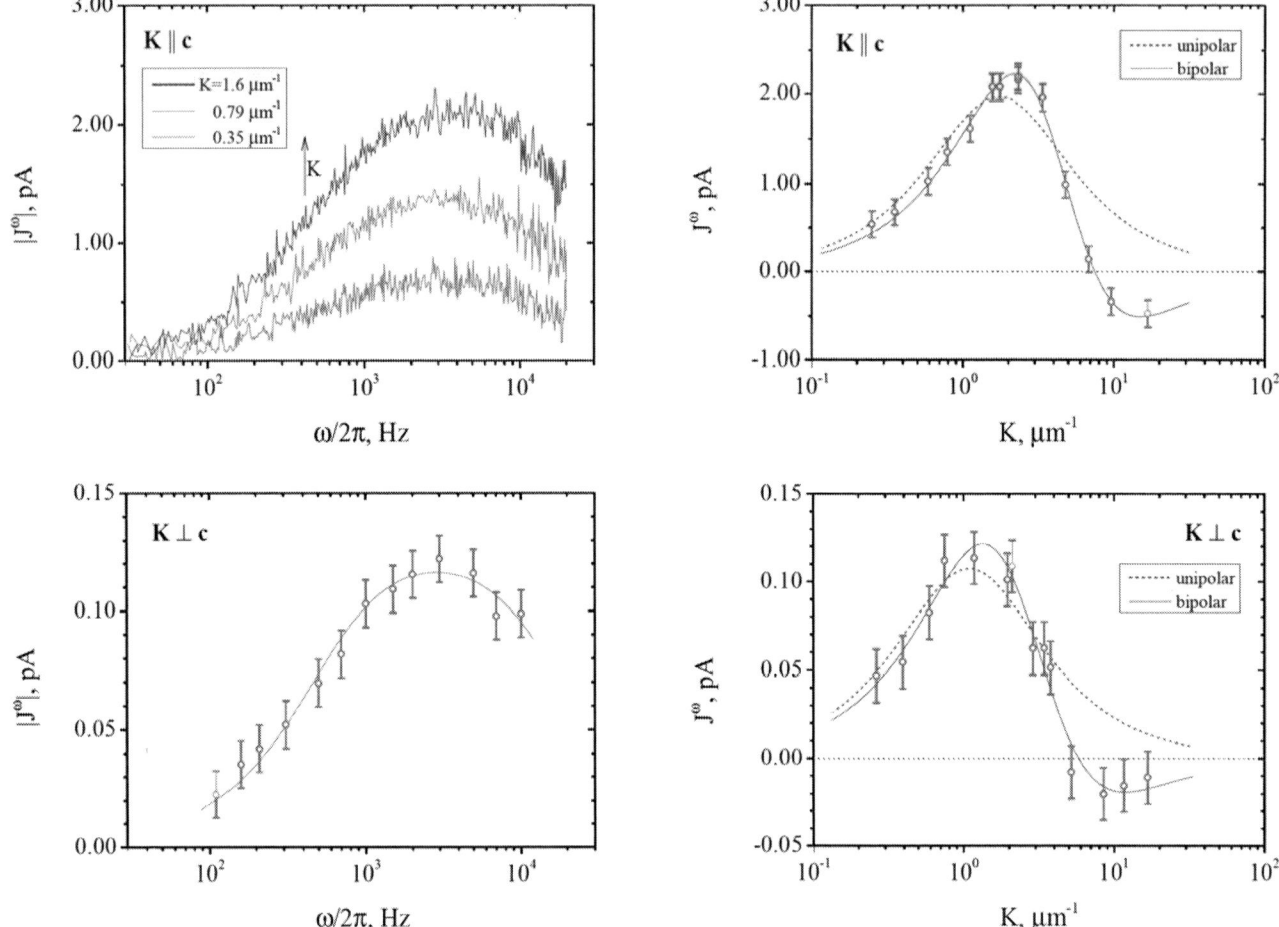

Fig. 2. Frequency transfer functions of the non-steady-state photo-EMF signal in AlN crystal. The dependencies are measured for two geometries: $\mathbf{K} \parallel c$-axis, $K = 0.35\ \mu m^{-1}$, $0.79\ \mu m^{-1}$, $1.6\ \mu m^{-1}$ (top figure), and $\mathbf{K} \perp c$-axis, $K = 2.0\ \mu m^{-1}$ (bottom figure). $I_0 = 110\ mW/cm^2$. The solid line shows approximation by Eq. (1) with $\omega_1/2\pi = 600$ Hz and $\omega_2/2\pi = 14$ kHz.

Fig. 3. Dependence of the non-steady-state photo-EMF amplitude on the spatial frequency of the interference pattern. The dependencies are measured for two geometries: $\mathbf{K} \parallel c$-axis (top figure), and $\mathbf{K} \perp c$-axis (bottom figure). $I_0 = 110\ mW/cm^2$. Dotted and solid lines show approximation by Eqs. (2) and (3).

[3]:

$$J^\omega = J_m^\omega \frac{i\omega(\omega_2 - \omega_1)}{\omega_1\omega_2 - \omega^2 + i\omega(\omega_2 - \omega_1)}, \qquad (1)$$

here J_m^ω is the current amplitude at the maximum of frequency dependence. The linear growth for low modulation frequencies ($\omega/2\pi < 700$ Hz in Fig. 2) is an important manifestation of the adaptive nature of space charge formation in wide-gap semiconductor materials. The signal in this frequency region is small since both the space charge field grating and grating of free carriers (photoconductivity grating) follow the movement of the interference pattern. The spatial shift between the gratings is nearly equal to $\pi/2$ which results in low signal amplitude. For higher modulation frequencies (frequency independent region) the grating with larger relaxation time can be considered as "frozen in", the periodic spatial shifts increase, and the photo-EMF signal reaches its maximum. For very high modulation frequencies ($\omega/2\pi > 13$ kHz in Fig. 2) the oscillation amplitudes of both gratings as well as corresponding relative shifts between them become negligible,

and the resulting current vanishes.

Since the cut-off frequencies are defined by the Maxwell relaxation time τ_M and carrier's lifetime τ [3], the frequency transfer functions can be used for their determination: $\tau_{M\parallel} \simeq 0.23$ ms, $\tau_\parallel = 11\ \mu s$. Using the measured value of Maxwell relaxation time $\tau_M = \epsilon\epsilon_0/\sigma_0$ one can easily estimate the average conductivity σ_0 of the material. For the sample in $\parallel$-geometry it equals: $\sigma_{0\parallel} = 3.6 \times 10^{-9}\ \Omega^{-1}cm^{-1}$.

Characteristic relaxation times $\tau_{M\perp}$ and $\tau_\perp$ for $\perp$-geometry were also estimated: $\tau_{M\perp} \simeq 0.26$ ms, $\tau_\perp \simeq 12\ \mu s$. The corresponding photoconductivity equals $\sigma_{0\perp} \simeq 2.6 \times 10^{-9}\ \Omega^{-1}cm^{-1}$.

The dependence of the signal amplitude versus spatial frequency of the interference pattern, i.e. $J^\omega(K)$, is another important characteristic, that is usually measured in the non-steady-state photo-EMF experiments (Fig. 3). The behavior of the signal can be easily explained: the signal increases for low K due to the growth of space charge field amplitude which is proportional to the so-called diffusion field $E_D = Kk_BT/e$, the signal decrease for high K is resulted from the diffusion blurring of the conductivity grating. The dependencies were fitted by the following simple expression known from the non-steady-state photo-EMF theory: [2]

$$J^{\omega}(K) \propto \frac{\sigma_0 E_D}{1 + (K L_D)^2}. \tag{2}$$

The diffusion length of electrons was estimated from the measured dependence: $L_{D\parallel} = 0.58$ μm and $L_{D\perp} = 0.92$ μm for the $\parallel$- and $\perp$-geometries, respectively. As seen the signal amplitude decays faster than K^{-1} and changes its sign for high spatial frequencies. A similar effect was observed earlier in GaAs crystal at $\lambda = 1.15$ μm [5]. Such behavior is explained by the presence of bipolar photoconductivity: the sign of the signal for low spatial frequencies is determined by the carriers with larger photoconductivity, while the carriers with larger generation rate define the sign of the signal for high K. The expression for photo-EMF amplitude can be written in this case as follows:

$$J^{\omega}(K) \propto \frac{\sigma_0^e E_D}{1 + (K L_D^e)^2} - \frac{\sigma_0^h E_D}{1 + (K L_D^h)^2}, \tag{3}$$

where $\sigma_0^{e,h}$ and $L_D^{e,h}$ is the average photoconductivities and diffusion lengths of electrons and holes, respectively. One can note that this expression fits experimental dependencies much better (Fig. 3), and it describes the change of the signal sign. The best fit was obtained for the following combination of material parameters: $\sigma_{0\parallel}^e / \sigma_{0\parallel}^h = 1.8$, $L_{D\parallel}^e = 0.27$ μm, $L_{D\parallel}^h = 0.18$ μm for $\parallel$-geometry and $\sigma_{0\perp}^e / \sigma_{0\perp}^h = 5.3$, $L_{D\perp}^e = 0.58$ μm, $L_{D\perp}^h = 0.20$ μm for $\perp$-geometry.

Dependence of the non-steady-state photo-EMF versus average light intensity was studied as well [6]. Nonlinear growth of the first cut-off frequency and signal amplitude in the maximum of transfer function was observed: $\omega_1 \propto I_0^{0.67}$ and $J^{\omega} \propto I_0^{0.81}$. These results correlate with the nonlinearity of holographic recording in AlN [7], although the powers of function I^x are different. The second cut-off frequency remains nearly constant, what means the independence of the conductivity relaxation time on light intensity.

To summarize we studied the effect of non-steady-state photo-EMF in new photosensitive material – aluminium nitride bulk crystal. The significant anisotropy of the material was revealed in the experiments with excitation of the photo-EMF along and across [0001] axis. The presence of the bipolar photoconductivity was established from the change of the signal sign for high spatial frequencies. The average photoconductivity of the material was estimated from the measurements of the signal amplitude versus frequency of phase modulation. The dependence of the signal versus spatial frequency provided the estimation of the diffusion lengths of electrons and holes.

The obtained results have demonstrated that the non-steady-state photo-EMF technique can be considered as a powerful tool for investigation of wide band gap semiconductors. A number of photoelectric parameters are estimated using the only arrangement. In contrast to the standard methods of semiconductor characterization the technological problem of ohmic contacts for high-resistive materials is not so significant here, since the effect of the non-steady-state photo-EMF is based on the excitation of an alternating current.

The authors acknowledge financial support from the Ministry of Knowledge Economy of Korea.

REFERENCES

[1] M. Petrov, S. Stepanov, and A. Khomenko, *Photorefractive crystals in coherent optical systems.* Berlin: Springer-Verlag, 1991.
[2] M. Petrov, I. Sokolov, S. Stepanov, and G. Trofimov, "Non-steady-state photo-electro-motive force induced by dynamic gratings in partially compensated photoconductors," *J. Appl. Phys.*, vol. 68, no. 5, pp. 2216–2225, 1990.
[3] I. Sokolov and S. Stepanov, "Non-steady-state photoelectromotive force in crystals with long photocarrier lifetimes," *J. Opt. Soc. Am. B*, vol. 10, no. 8, pp. 1483–1488, 1993.
[4] E. Mokhov, O. Avdeev, I. Barash, T. Chemekova, A. Roenkov, A. Segal, A. Wolfson, Y. Makarov, M. Ramm, and H. Helava, "Sublimation growth of AlN bulk crystals in Ta crucibles," *J. Cryst. Growth*, vol. 281, no. 1, pp. 93–100, 2005.
[5] S. Stepanov and G. Trofimov, "Transient EMF in crystals having ambipolar photoconductivity," *Sov. Phys. Solid State*, vol. 31, no. 1, pp. 49–50, 1989.
[6] M. Bryushinin, V. Kulikov, E. Mokhov, S. Nagalyuk, and I. Sokolov, "Non-steady-state photoelectromotive force in an AlN crystal," *Phys. Rev. B*, vol. 86, no. 8, pp. 085 209–1–5, 2012.
[7] T. Nagai, R. Fujimura, T. Shimura, and K. Kuroda, "Photorefractive effect in undoped aluminum nitride," *Opt. Lett.*, vol. 35, no. 13, pp. 2136–2138, 2010.

Spin controlled optical radiation pressure

Georgiy Tkachenko
Laboratoire Ondes et Matière d'Aquitaine,
Université Bordeaux
F-33400 Talence, France
Email: georgiy.tkachenko@u-bordeaux1.fr

Etienne Brasselet
Laboratoire Ondes et Matière d'Aquitaine,
CNRS, Université Bordeaux
F-33400 Talence, France
Email: e.brasselet@loma.u-bordeaux1.fr

Abstract— **We report on the full control of the optical radiation pressure at fixed photon flux and incident angle by the photon spin. This is done by using transparent chiral liquid crystal droplets that enable a strong coupling between linear and angular degrees of freedom of a light field. From these results, we anticipate optical sorting of particles with different chirality as well as novel optical trapping and micromanipulation strategies.**

I. Introduction

Optical radiation forces are the mechanical manifestation of the transfer of the linear momentum of light to matter, which basically occurs when light is reflected, refracted, scattered or absorbed in the course of its propagation [1]. In particular, the discontinuity of the dielectric permittivity at the interface between two transparent, homogeneous and dielectric media leads to optical radiation pressure that enables the displacement of solids [2] or the deformation of fluid interfaces [3]. Here we report on the full control of the optical radiation pressure at fixed photon flux and incident angle by the photon spin. This is done by using transparent chiral liquid crystal droplets that enable a strong coupling between linear and angular degrees of freedom of light.

Chirality refers to the lack of mirror symmetry. A system whose mirror images cannot be superimposed by rotations and translations in space is said to be chiral. A basic example is our two hands, just like the Greek etymology ('kheir') recalls. Quite naturally, two mirror images of a chiral entity are referred to as right- and left-handed versions of it. This concept holds for light as well, whose polarization handedness is said to be right or left depending on the sense of rotation of the electric field. Under appropriate conditions, the propagation of one of the two circular polarization states through a chiral optically anisotropic material may even be forbidden for a well-defined range of frequencies. An helical arrangement of the optical axis indeed combines dielectric periodic structuration, which leads to a Bragg photonic bandgap, with chirality, which brings circular polarization sensitivity. A famous example is the circular Bragg reflection phenomenon in cholesteric liquid crystals [4], which we use here to experimentally demonstrate how mechanical effects driven by the radiation pressure of light can be fully controlled by the spin of photons. The principle of our experiment is sketched in Fig. 1.

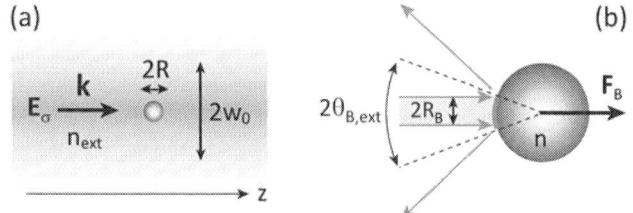

Fig. 1. (a) A radial cholesteric droplet illuminated by on-axis circularly polarized Gaussian beam. (b) Optical radiation force $\mathbf{F_B}$ resulting from the circular Bragg reflection for the appropriate polarization state σ_B, at which the droplet behaves as a perfect mirror over an apex angle $2\theta_{B,ext}$.

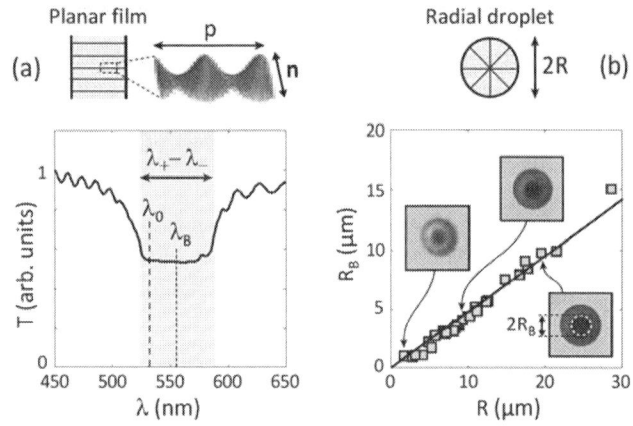

Fig. 2. (a) Unpolarized transmission spectrum of a right-handed cholesteric film with pitch p and uniform alignment of helical axis along the normal to the film. (b) In situ characterization of the reflecting cross-section area of spherical radial cholesteric droplets under uniform illumination at $\lambda_0 = 532$ nm and Bragg incident circular polarization state σ_B.

II. Cholesteric Bragg reflector

The cholesteric mesophase appears when nematic liquid crystals [5], [6], are mixed with certain chiral organic dopants. As a result, the director $\mathbf{n}$ (a unit vector that defines the local averaged molecular orientation) twists in a well-defined direction with a helical pitch p that is the distance over which $\mathbf{n}$ rotates by 2π, see upper sketch in Fig. 2(a). A planar cholesteric film having its helical axis along the normal to the film behaves as a perfect mirror for a normally incident collimated light beam with wavelength inside the polarization photonic bandgap centered on λ_B, $\lambda_- < \lambda_0 < \lambda_+$, see Fig. 2(a), and Bragg circular polarization state referred to as

σ_{B}. In contrast, the orthogonally polarized beam, referred to as anti-Bragg and labeled σ_{AB}, is transmitted through the film up to Fresnel reflection. At fixed incident wavelength λ_0, circular Bragg reflection thus occurs over a range of incidence angle around the normal incidence. This angular range for the external incidence angle is labeled $2\theta_{\mathrm{B,ext}}$, as illustrated in Fig.1(b) in the case of a cholesteric droplet. That is to say, one can consider that total reflection occurs for σ_{B} polarization state when the external incidence angle satisfies $\theta_{\mathrm{ext}} < \theta_{\mathrm{B,ext}} = \arcsin[(n/n_{\mathrm{ext}})\sin\theta_{\mathrm{B}}]$. In the latter expression, $n = (n_{\parallel} + n_{\perp})/2$ is the average refractive indices of the cholesteric with $n_{\parallel,\perp}$ the refractive index parallel and perpendicular to $\mathbf{n}$, n_{ext} ($n_{\mathrm{ext}} < n$ in this work) is the refractive index of the medium in which the cholesteric is immersed. In addition, accounting from $\lambda_{\mathrm{B}} = np$ and $\lambda_+ - \lambda_- = p(n_{\parallel} - n_{\perp})$, one can show that $\theta_{\mathrm{B}} = \arccos(\lambda_0/\lambda_+)$, at least in the limit of small θ_{B}. A spherical cholesteric droplet with radial distribution of the helical axis, as sketched in Fig. 2(b), can therefore be considered as a spherical Bragg mirror with total reflection cross-section area normalized to the geometrical cross-section area being equal to $\sin^2\theta_{\mathrm{B,ext}}$ for σ_{B}-polarized light and to zero for σ_{AB}-polarized light. This allows considering vivid experimental demonstration of spin-dependent optical radiation pressure by using free-floating droplets, as sketched in Fig. 1.

In our experiments, the cholesteric is right-handed with pitch $p = 347$ nm and $n \approx 1.6$ at 589.3 nm wavelength and temperature 20°C. From the transmission spectrum of a uniformly aligned film shown in Fig. 2(a), we measure the intrinsic Bragg angle $\theta_{\mathrm{B}} = 25.1°$ at $\lambda_0 = 532$ nm. This is verified in situ for cholesteric droplets dispersed in glycerol, which ensures parallel anchoring of the director at cholesteric-glycerol interface, hence a radial distribution of the helical axis for $p \ll R$ where R is the droplet radius. For this purpose, we measure the radius R_{B} of the total reflection cross-section area under σ_{B}-polarized light at 532 nm, as a function of R, see Fig. 2(b). Since $R_{\mathrm{B}} = R\sin\theta_{\mathrm{B,ext}}$, we obtain $\theta_{\mathrm{B}} = 25.8°$ from the best linear fit, in good agreement with the spectroscopic characterization.

III. RESULTS AND DISCUSSION

In the first set of experiments we follow the sketch shown in Fig. 1(a) and use spherical radial cholesteric droplets prepared by mechanical stirring at room temperature of a small amount of cholesteric into an isodensity 25.2 wt.% aqueous glycerol solution whose refractive index is 1.365 at 589.3 nm wavelength at 20°C and dynamic viscosity $\eta = 2$ mPa s. This choice indeed eases observation of on-axis light-induced droplet displacement. Typical results are summarized in Fig. 3. Without laser illumination, the droplet is at rest. When the laser is turned on, the droplet almost immediately moves along the beam propagation direction with constant velocity. Partial control of the light-induced motion is obtained by setting the photon spin. The Bragg droplet velocity, v_{B}, is indeed a few times larger than the anti-Bragg one, v_{AB}. Noteworthy, the radial structure of the droplet is unaltered during its

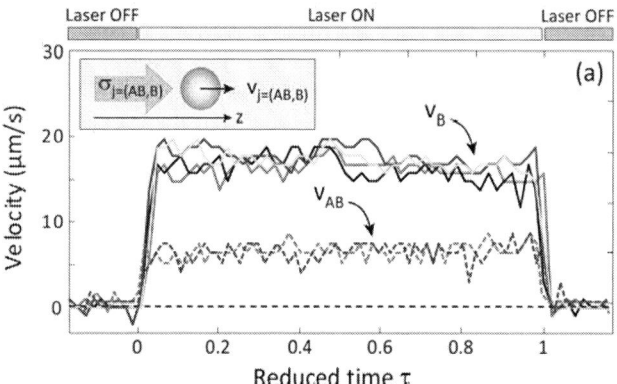

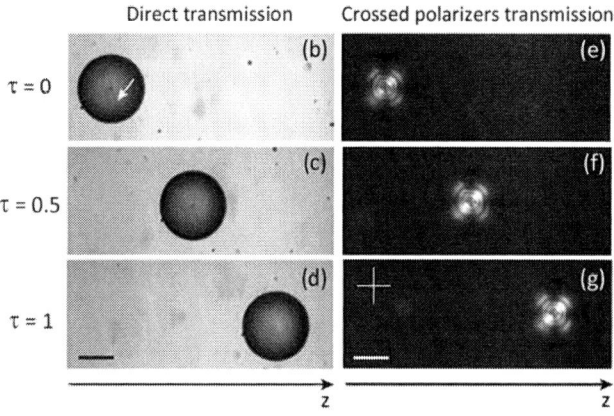

Fig. 3. (a) Droplet velocity dynamics for σ_{B} (v_{B}, solid curves) and σ_{AB} (v_{AB}, dashed curves). Beam power $P = 970$ mW, waist $w_0 = 505$ μm, the droplet radius $R = 28.4$ μm. (b,c,d) Direct transmission snapshots of the droplet for the Bragg case. (e,f,g) Crossed polarizers transmission image, the polarizers direction being indicated on panel (g). Scale bar is 30 μm.

displacement, as demonstrated by the optical transmission image sequences at the early, intermediate and final stages of its laser-induced motion in the Bragg case, see Figs. 3(b,c,d) (direct images) and Figs. 3(e,f,g) (crossed polarized images). A characteristic radial defect is observed in the former case, see white arrow in Fig. 3(b), whereas the unchanged dark cross pattern indicates radially symmetric optical anisotropy in the latter case.

Such partial spin-controlled optomechanical effect in the one-beam experiment is actually explained by considering unavoidable axisymmetric angular redistribution of the optical linear momentum flux away from the propagation direction as the incident beam is refracted by the droplet, thereby leading to a net force directed along the z axis. This interpretation is quantitatively tested by measuring the ratio $v_{\mathrm{AB}}/v_{\mathrm{B}}$ for various R, see square symbols in Fig. 4, and comparing it to the theoretical value predicted by our model described in [7]. According to the model, the velocity ratio equals

$$\frac{v_{\mathrm{AB}}}{v_{\mathrm{B}}} = \frac{\int_0^{\pi/2} f(\theta)\,d\theta}{\int_0^{\theta_{\mathrm{B,ext}}} (1 + \cos 2\theta)\,d\theta + \int_{\theta_{\mathrm{B,ext}}}^{\pi/2} f(\theta)\,d\theta}, \quad (1)$$

with θ - incidence angle, $f(\theta) = 1 + \mathcal{R}(\theta)\cos 2\theta - \mathcal{T}^2(\theta)\frac{\cos\theta_{\mathrm{out}} + \mathcal{R}(\theta)\cos 2\theta}{1 + 2\mathcal{R}(\theta)\cos 2\theta_{\mathrm{int}} + \mathcal{R}^2(\theta)}$, where $\mathcal{R}$ and $\mathcal{T} = 1 - \mathcal{R}$ are the reflectance and transmittance of the droplet interface.

As shown in Fig. 4, we find the experimental value $v_{AB}/v_B = 0.38 \pm 0.075$. From the model, by using $\theta_{B,ext}$ as the only adjustable parameter, we obtain an effective Bragg angle $\theta_B = 15.3°(+3.1°/-2.3°)$. This value is smaller than the ones measured from spectroscopic [Fig. 1(a)] and direct imaging [Fig. 1(b)] measurements. However, this is actually expected recalling that our model assumes perfect Bragg reflection over the effective Bragg angular bandgap. We thus conclude to a fair description of our observations, thereby validating that (i) the physical picture is properly grasped and (ii) the scattering of light by the droplet prevents full control of the optical radiation pressure by the photon spin.

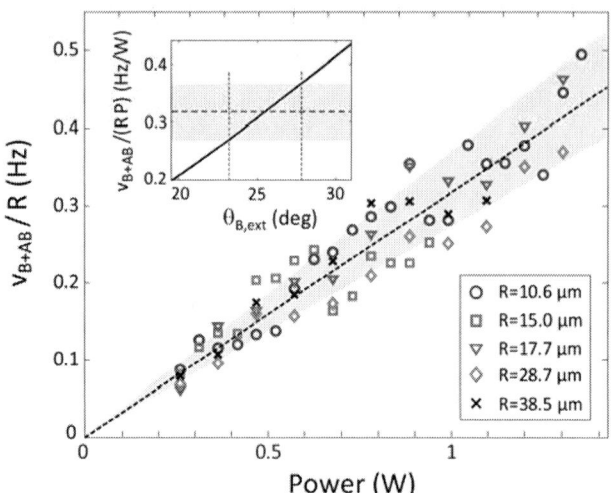

Fig. 5. The ratio v_{B+AB}/R is plotted as a function of total beam power P of one beam for various values of the droplet radius R.

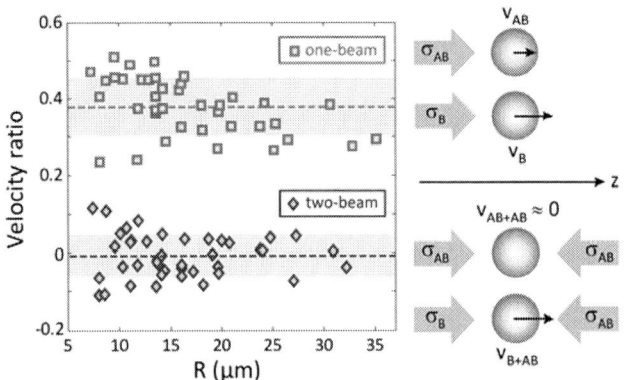

Fig. 4. Experimental droplet velocity ratio v_{AB}/v_B and v_{AB+AB}/v_{B+AB} are plotted as a function of R. Total power of each beam is $P = 780$ mW. Dashed lines refer to mean values, gray areas - standard deviation ranges.

In the second set of experiments we overcome the above-mentioned fundamental limitation by using two coaxial, collimated, circularly polarized and counterpropagating beams with equal powers and waists, one being either σ_B- or σ_{AB}-polarized while the other is σ_{AB}-polarized. When both beams are σ_{AB}-polarized, the individual contributions of the two beams to the total optical scattering force indeed perfectly cancel each other. A nonzero net force is exerted on the droplet otherwise. That is to say, $v_{AB+AB} = 0$ and $v_{B+AB} \neq 0$ are expected, the former case being confirmed experimentally whatever the droplet radius, see diamond symbols in Fig. 4. We measure $v_{AB+AB}/v_{B+AB} = -0.0083 \pm 0.055$, hence achieving full control of the optical radiation pressure exerted on a microscopic object by mere change of the photon spin.

The dependence on P and R of the droplet velocity in the 'Bragg/anti-Bragg' case is shown in Fig. 5. In this figure, the inverse of the characteristic time of the droplet motion, v_{B+AB}/R, is shown as a function of total power P of one beam for various droplets. Our model gives

$$\frac{v_{B+AB}}{R} \approx \frac{n_{ext}}{3\pi\eta w_0^2 c}\left(1-\cos^4\theta_{B,ext}\right)P \qquad (2)$$

with c the speed of light in vacuum. Following Eq. (2), the best linear fit using $\theta_{B,ext}$ as the only adjustable parameter (see dashed line in Fig. 5) gives $\theta_B = 21.5°(+1.9°/-1.9°)$. As emphasized by Eq. (2), a quantitative analysis in the two-beam experiments is more difficult than in the one-beam

case since the fitting procedure involves additional parameters, namely η and w_0. The beam waist is measured from direct imaging whereas the viscosity is taken from tabulated data ($\eta = 2$ mPa s) without measuring it for our actual aqueous glycerol solution. This could explain the difference of $\simeq 6°$ between the values of θ_B obtained from one-beam and two-beam experiments. We conclude that our model gives an overall satisfying description of observations in the two-beam experiments as well.

IV. CONCLUSION

We have presented a quantitative experimental study demonstrating that optical radiation pressure can be fully controlled by the spin angular momentum of light, at fixed photon flux and incidence angle, in contrast to existing optical micromanipulation techniques. A straightforward anticipated application is the development of chiral optical sorting, which would bring enhanced functionality to optical chromatography. Nanoactuation driven by optical radiation pressure is another topic that may benefit from such a polarization control.

REFERENCES

[1] A. Jonas and P. Zemanek, "Light at work: The use of optical forces for particle manipulation, sorting, and analysis," *Electrophoresis*, vol. 29, pp. 4813–4851, 2008.

[2] P. Lebedew, "Untersuchungen über die druckkräfte des lichtes," *Ann. Phys.*, vol. 6, pp. 433–458, 1901.

[3] A. Ashkin and J. M. Dziedzic, "Radiation pressure on a free liquid surface," *Phys. Rev. Lett.*, vol. 30, pp. 139–142, 1973.

[4] D. W. Berreman and T. L. Scheffer, "Bragg reflection of light from single-domain cholesteric liquid crystal films," *Phys. Rev. Lett.*, vol. 25, pp. 577–581, 1970.

[5] G. V. Tkachenko and O. V. Shulika, "Thermal tuning of a thin-film optical filter based on porous silicon and liquid crystal," *Ukrainian Journal of Physical Optics*, vol. 11, no. 4, pp. 260–268, 2010.

[6] G. V. Tkachenko, I. A. Sukhoivanov, O. V. Shulika, and V. Tkachenko, "Tunable optical filter based on nanocomposite (liquid crystal)/(porous silicon)," *Proc. of SPIE*, vol. 8279, pp. 827918–1–827918–9, 2012.

[7] G. Tkachenko and E. Brasselet, "Spin controlled optical radiation pressure," *Phys. Rev. Lett.*, vol. 111, pp. 033605–1–033605–5, 2013.

Biomedical effect the phenomenon of in vivo blood oxyhemoglobin photodissociation

M.M. Asimov[*], R.M. Asimov[**], D.B. Vladimirov[*], A.N. Rubinov[*]

(Invited Paper)

[*]Institute of Physics National Academy of Science of Belarus,
68 Nezavisimosti pros., 220072, Minsk, Belarus
E-mail: m.asimov@dragon.bas-net.by
[**]"Sensotronic Ltd.", Belarus High Technologies Park,
11 Kulman Str, 220100 Minsk, Belarus

Abstract: The results in vivo investigation the phenomenon of laser-induced photodissociation of blood oxyhemoglobin and its biomedical applications are presented. New method of control in individual response to the effect of laser radiation is proposed. Novel optical method of determination the therapeutic "dose" based on the response of changes in tissue oxygen concentration in dependence on wavelength and intensity of laser radiation has been developed. It is shown that in order to make the methods of phototherapy as well as laser therapy really efficient one has to control the oxygen concentration in tissue keeping it at the necessary level.

INTRODUCTION

Therapeutic effect of low intensity laser radiation now is a well-established fact and in spite of that the mechanism of this phenomenon still remains topic of discussion this phenomenon is widely used in clinical practice. Considering the biological effect of "Laser-Tissue" interaction the phenomenon of in vivo blood oxyhemoglobin (HbO_2) photodissociation should be taking into account.

The role of blood HbO_2 photodissociation in the biological response and the mechanism of therapeutic effect of low intensity laser radiation were proposed in [1-3]. It was shown that the phenomenon of laser-induced photodissociation of blood HbO_2 gives unique possibility controlling local concentration of molecular oxygen in tissue.

Molecular oxygen plays a key role in aerobic cell metabolism and energy production in biological tissue. Controlling this mechanism provides unique possibility of stimulation biological response on laser radiation and reaching desired therapeutic effect.

The problem of controlling the local tissue oxygen concentration and keeping it at the necessary for normal cell metabolism level is an actual problem in modern medicine [4].

The following criteria of tissue oxygen tension ($TcPO_2$) in clinical practice now are accepted:

- $TcPO_2 \sim 40 - 100$ mmHg - normal cell metabolism;

- $20 > TcPO_2 < 40$ mmHg – tissue hypoxia;

- $TcPO2 < 20$ mmHg –deep hypoxia and tissue necroses.

Different methods for elimination of tissue hypoxia in clinical practice are used. Hyperventilation of lung with pure oxygen at normal atmospheric pressure is the oldest method of oxygenation, but its low efficiency limits wide application.

More effective method is hyperbaric oxygenation (HBO) that is based on blood saturation with oxygen under high pressure. The method of HBO may cause oxygen toxemia that limits it routine application in clinical practice [5].

In this paper a new optical technology in local tissue oxygenation is presented. The results of the effect of laser radiation on blood HbO_2 and it photodissociation are discussed.

New method of determination of therapeutic "dose" during laser irradiation that based on the use of changes in oxygen concentration as a feedback signal is developed. It is shown that photodissociation of blood HbO_2 provides a unique possibility in additional oxygen extraction and restoring normal cell metabolism.

LASER-INDUCED TISSUE OXYGENATION

As it well known amount of oxygen delivered by microcirculation for cell metabolism is the function of:

$$\Sigma\, O_2\,(TcPO_2) = f(F(HbO_2)*[O_2])$$

Here HbO_2 is the value of arterial blood oxyhemoglobin and $[O_2]$ - is the concentration of oxygen in plasma. Deterioration of the blood microcirculation required extra oxygen supply to provide the demands of normal cell metabolism. This could be reached by in vivo laser-induced photodissosiation of HbO_2 directly at the zone were it is necessary to increase the local concentration of free molecular oxygen.

As a result we obtain total concentration of oxygen delivering in conventional way and due to laser-induced photodissociation of blood HbO_2:

$$\Sigma[O_2] = [O_2] + [O_2^{hv}]$$

This process clearly demonstrates that in vivo photodissociation of blood HbO_2 may be used for extraction of additional amount of molecular oxygen in tissue [6-8]. This process is illustrated in fig.1.

978-1-4799-0019-0/13 $31.00 © 2013 IEEE

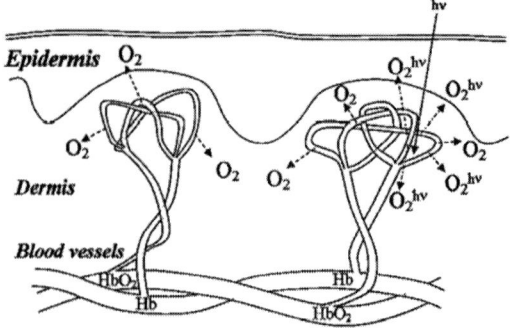

Fig.1 - Optical method of additional oxygen extraction from HbO$_2$ and laser-induced tissue oxygenation.

Additional oxygen extraction allows develop novel optical technology in elimination of tissue hypoxia and restoring normal cell metabolism.

EXPARIMENTAL

Experimental investigation the phenomenon of laser-induced tissue oxygenation has been carried out using transcutaneous oxygen monitor (TCOM) - "Radiometer" TCM-4 (Fig.2). Direct in vivo measurements of tissue oxygen tension TcPO$_2$ under irradiation by He-Ne laser at the power of 1mW has been taken [9]. For this purpose usually uses Clark-type polarographic sensor ("TcPO$_2$ electrode", see fig. 2) that consist of a silver anode, electrolyte, and an oxygen permeable membrane; heating section and electronic system for measuring and controlling the sensor temperature.

First a background tissue oxygen tension was measured. Then diode laser radiation at the power of 15 mW was applied. The kinetics of tissue oxygenation in dependence of the time irradiation tension was experimentally investigated.

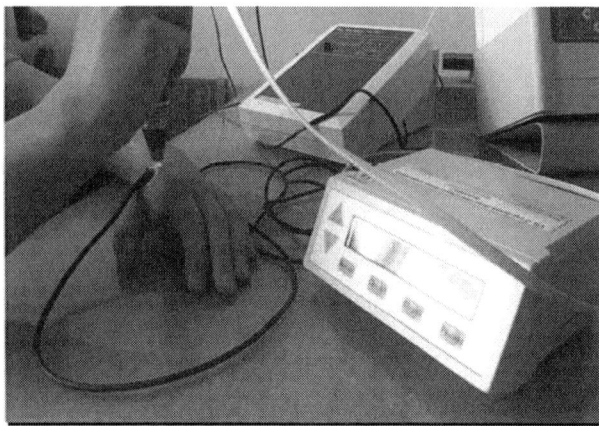

Fig. 2 – Experimental setup for measuring local tissue oxygen concentration directly at the zone of laser irradiation

Obtained experimental results demonstrate that the process of oxygen extraction take place immediately at laser light absorption and remains constant during the irradiation [10].

RESULTS AND DISCUSSION

The results obtained were normalized to the initial oxygen tension value (Fig. 3).

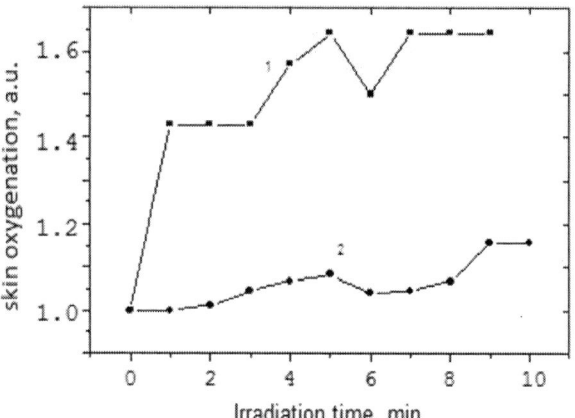

Fig. 3 - The kinetics of laser-induced tissue oxygenation during laser irradiation in norm blood microcirculation -1, and in artificially induced ischemia – 2

As it seen from fig.3 during laser irradiation the value of tissue oxygenation increases and exceeds its initial level by 1.6 times (curve 1) after ten minutes of illumination. In case of induced ischemia additional extraction of oxygen also is observed. This result clearly demonstrates that laser-induced tissue oxygenation could be applied in clinical practice for restoration of normal cell metabolism in tissue with damaged microcirculation.

The results of calculations demonstrate that in order to reach the experimentally observed rise of TcPO$_2$ by ~1, 6 times on the tissue surface, the calculation indicating the increase of oxygen release rate from arterial HbO$_2$ into blood plasma should increase by approximately 4,3 times.

Photodissociation of HbO$_2$ induced by laser radiation and release rate of free molecular oxygen into blood plasma has been measured experimentally in vivo using highly sensitive pulse oximeter [11]. The oxygen released from HbO$_2$ primarily increases the PO$_2$ of blood plasma and then O$_2$ diffuses into the tissue.

It is exciting that the value of PO$_2$ in blood plasma reached by laser-induced photodissociation of HbO$_2$ is comparable to that one typically reached by the method of HBO [12].

The distribution of TcPO$_2$ in the volume at the irradiation zone depends on the time of exposure and the tissue properties.

The comparison of calculated results with experimental data demonstrates that kinetics of TcPO$_2$ considered in relation to time of elimination by laser radiation gives possibility to determine O$_2$ diffusion coefficient in tissue.

This means that one could calculate and determine how to reach desirable level of TcPO$_2$ in zones with disturbed blood microcirculation such as solid tumour, burn or wounds. So it's possible to determine optimal parameters of irradiation taking into account the volume that has to be oxygenated and the time of elimination.

Thus suggested novel method can eliminate the deficit of oxygen until the new vascular net in tissue is restored. This result could be applied for those pathologies where elimination of tissue hypoxia is critical.

Extra oxygen increases the rate of collagen deposition, epithelization and improve healing of split thickness grafts. Increased subcutaneous $TcPO_2$ has also improved bacterial defenses.

Thus a unique possibility in selective and local increase of the concentration of free molecular oxygen in tissue which enhances metabolism of cells has been developed.

Laser-induced enrichment the tissue oxygenation stimulates cell metabolism and allows developing new effective methods of therapy as well as laser therapy of pathologies where elimination of local tissue hypoxia is critical.

Laser-induced photodissociation of HbO_2 may serve as a unique method in laser therapy for optical increase of free molecular oxygen local concentration in tissue which significantly enhances cell metabolism.

It is valuable that even in the case of ischemia it is possible to extract additional oxygen from arterial blood and restore normal cell metabolism. Thus laser-induced tissue oxygenation allows using optical method eliminate the deficit of oxygen until the new vascular net in tissue is restored.

The obtained results provide experimental argumentation for considering the primary mechanism of biostimulation and therapeutic effect of low energy laser radiation that could be based on increasing tissue local oxygen concentration directly in the zone of irradiation.

This phenomenon allows developing an objective method of controlling the efficiency of treatment by laser phototherapy. Now in clinical application the parameters of laser radiation can be tuned to optimal wavelength, power and exposition time taking into account optical characteristics of the patient skin.

The obtained results also show the way of increasing the efficiency of biostimulation and the therapeutic effect of low energy laser radiation based on its combination with the method of oxygen hyperventilation therapy.

An important conclusion can also be drawn from the obtained results. In interpretation of the biostimulation and therapeutic effects of laser radiation the phenomenon of induced photodissociation of blood HbO_2 should be taken into account. Control the value of tissue oxygenation becomes important in reaching maximal therapeutic effect.

CONCLUSION

The new optical method of local tissue hypoxia elimination has been developed. The value of tissue O_2 concentration increases significantly during the laser irradiation.

It is shown that establishing the therapeutic "dose" of laser radiation could be based on controlling the local concentration of free oxygen in tissue.

To make the phototherapy as well as laser therapy methods really efficient one has to control the O_2 concentration in tissue keeping it at the necessary level. This goal could be reached by the use of laser-induced photodissociation of HbO_2 in tissue blood vessels.

The efficiency of developing method is comparable to the method of HBO gaining advantages in local action.

Novel method of determination of optical "dose" based on using the changes in tissue oxygen concentration as a feedback signal for the optimization of low intensity laser radiation therapeutic effect has been developed.

Photodissociation of oxyhemoglobin, the main biological function of which is molecular oxygen transport gives a unique possibility of additional oxygen supply and allows develop laser-optical method of tissue hypoxia elimination that restores normal cell metabolism.

REFERENCES

[1] S.Takas, S. Stojanovich, *Diagnostic and biostimulating lasers.* Med. Pregl., vol. 51, № 5-6, pp. 245-249,1998.

[2] G.D. Baxter, *Therapeutic lasers: Theory and Practice*, Edinburgh; New-York, 1994

[3] J. Tuner and L. Hode *Laser Therapy: Clinical practice and scientific background,* Prima Books AB, 2002.

[4] M.M. Asimov, R.M. Asimov, A.N. Rubinov, "*Action spectra of laser radiation on hemoglobin of skin blood vessels*", Journal of Applied Spectroscopy. v. 65. № 6,877-8880, (1998)

[5] M.M. Asimov, A.N. Korolevich, E.E. Konstantinova, "*Investigation of the kinetics of tissue oxygenation under the effect of low intensity laser radiation,*" J of Appl. Spectr., vol.74, pp.120-125, 2007.

[6] McGrath, et al., "Compositions and method of tissue superoxygenation", U.S. Pat. No. 6,649,145, 2003.

[7] M.M. Asimov, R.M. Asimov, A.N. Rubinov, "*Investigation of the efficiency of laser action on hemoglobin and oxyhemoglobin in the skin blood vessels,*" Proc. SPIE., vol. 3254, pp. 407 – 412, 1998.

[8] R.A. Abergel, R.F. Lyons, J.C Castel, et all. *Biostimulation of wound healing by lasers: Experimental approaches in animal models and in fibroplast cultures,* J. Dermatol. Surg. Oncol., vol.3, pp. 127-133, 1987.

[9] V.I. Kravchenko, S.A. Mamilov, O.N. Minov, Yu. S. Plaksiy. "*Investigations of correlation of optical signals during measuring an arterial blood oxygen saturation,*" The Ukrainian Journal of Medical Technique and Technologies, vol.1, pp. 59-61, 1999.

[10] M.M. Asimov, R.M. Asimov, A.N. Rubinov, "*Action spectra of laser radiation on hemoglobin of skin blood vessels,*" Journal of Applied Spectroscopy, vol. 65 (6), pp. 877-880, 1998.

[11] M.M. Asimov, "*Laser-induced Photodissociation of Hemoglobin Complexes with Gas Ligands and its Biomedical Applications,*" Proceedings of "LTL Plovdiv 2005", IV International Symposium Laser Technologies and Lasers, Plovdiv, Bulgaria, pp. 3-11, 2005.

[12] P.S. Grim, "*Hyperbaric Oxygen Therapy*", JAMA. vol.263, pp. 2216-2220, 1990.

Method of generating a pulsed X-ray

M. I. Samoylovich[1], N. V. Tcherniega[2], A. D. Kudryavtseva[2], A. F. Belyanin[1]

[1] Central Research Technological Institute «TECHNOMASH», Moscow, Russia
[2] P.N. Lebedev Physical Institute of the Russian Academy of Sciences, Moscow, Russia

Abstract: Nonlinear optical effects in lattice SiO_2 nanosphere packings under pulse laser excitation are experimentally studied. Pulse generation of narrow-beam X-ray radiation is determined in this samples. The possibility of using this effect in medicine are considered.

The studies relates to the field of Radiological, and can be used in medicine, medical imaging, inspection of materials, X-ray microscopy. In the medical literature described and widely discussed such issues with endoscopic studies related to early-stage cancer (or other) diseases, by using different types of external exposure to radiation, including laser. It is known that the most effective method of radiation exposure (at minimum injury to healthy cells) is the x-rays aimed directly (locally) to the affected area with not very high energies (and therefore, with comparatively small depth of penetration) and precisely controlled local dose.

Until now, such methods had not been developed because of the lack of miniature x-ray sources. Listed in our articles data used to generate a focused pulsed X-rays (photons with energies of the order of 80–100 keV) metamaterials based on opal matrices show the reality of the creation of endoscopic devices using fiber-optic cables to the size input devices themselves to generate not more than 1–1.5 mm both in length and in diameter.

Recently in Central Scientific Technology Institute "Technomash" a method has been developed (and patented) for generating pulsed X-ray radiation with a small angular divergence, which is based on a pulsed ruby laser with power density 0.25–2.0 GW/cm^2 interaction with opal matrix – an ordered structure of X-ray amorphous silica (SiO_2) nanospheres with diameters in the range of 200–350 nm [1–4]. Variations in the diameter of a single sample were not more than ± 3%). Voids between spheres were filled with the material with dielectric constant value (ε) more than 2.5. Filling factor was in the range of 30–85%. Due to the opal matrix structure as a densely packed lattice of SiO_2 nanospheres (Fig. 1), the samples are both photonic structures with a system of photonic band gaps. If the voids between spheres are filled with dielectric materials with large ε' or electrically (magnetically) active materials, such substances may be considered as metamaterials.

Ruby laser pulse (λ=694.3 nm, τ=20 ns, E_{max}=0.3 J) has been used as a source of excitation. Exciting light has been focused into the material by lenses with different focal lengths (50, 90, and 150 mm). The sample distance from focusing system and exciting light energy were also changed. It allowed us to vary the power density at the entrance of the sample and field distribution in microvoids inside the sample (Fig. 2).

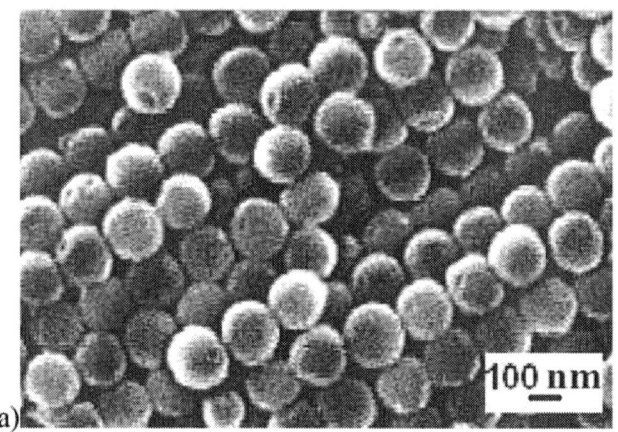

a)

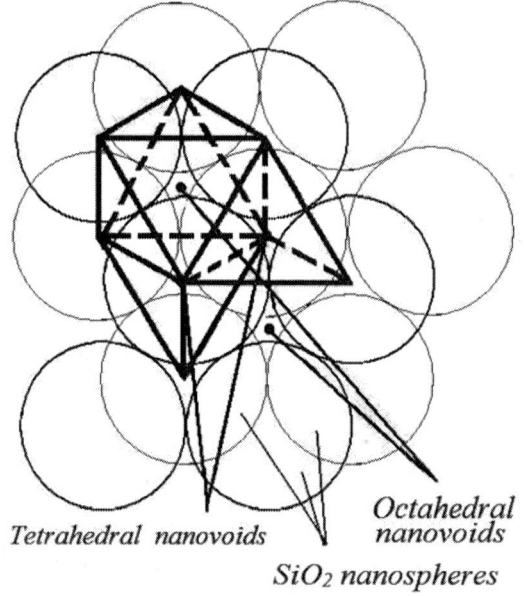

b)

Tetrahedral nanovoids

Octahedral nanovoids

SiO₂ nanospheres

Fig. 1. a) Structure (SEM) of the volume opal matrix sample surface (ordered package of SiO_2 nanospheres).
b) Tetrahedral and octahedral nanovoids (in accordance with a number of spheres, forming void), produced by SiO_2 nanospheres.

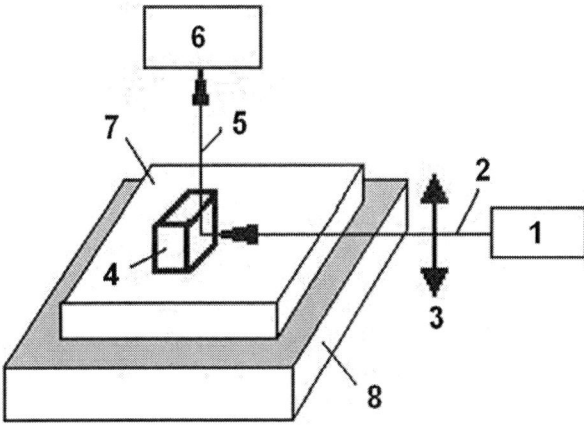

Fig. 2. Experimental setup: 1 – laser, 2 – laser beam, 3 – optical system, focusing laser light to the sample, 4 – opal matrix sample (ordered structure consisting of silica nanospheres with diameter 200–400 nm), 5 – X-ray emission, 6 – system for X-ray emission registration, 7 – metal plate, 8 – cell with liquid nitrogen.

Opal matrix is produced as a sample with plane parallel surfaces, contacting with piezoelectric plates, made of single crystals with large (more than 0.2 %) electromechanical coupling coefficient, which allow to create selected direction for acoustic waves due to photon-phonon interaction. As a result we can obtain reduction of the X-ray emission pulses angles distribution.

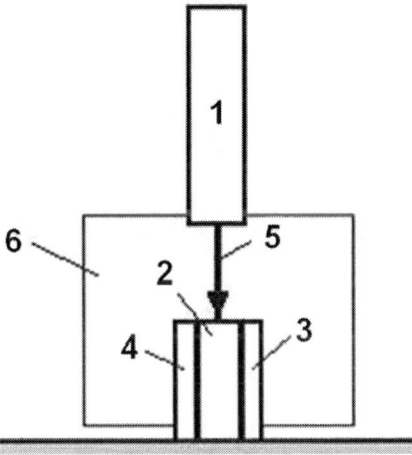

Fig. 3. Scheme of the device for directed pulsed X-ray emission generation: 1 – laser, 2 – opal matrix sample (ordered structure of silica nanospheres with diameter 200–400 nm) with plane parallel surfaces, hold together with plates, made of piezoelectric material (3, 4), 5 – pump beam, 6 – X-ray cassette with film.

Suggested device for directed pulsed X-ray generation includes external power source – ruby laser; target – opal matrix sample, produced with plane parallel surfaces, which, in

turn, contact with plates of piezoelectric materials, made of crystals with large (more than 0.2 %) electromechanical coupling coefficient. Opal matrix sample and piezoelectric plates are hold together with the help of mechanical clip (clamp) or special glue.

Basically, technical realization can be demonstrated by the Fig. 3, in which scheme of the device for directed pulsed X-ray emission generation is shown.

It is in such design the opal matrix sample, consisting of ordered SiO_2 nanospheres with under power laser light action is possible to generate directed X-ray emission (Fig. 4, 5) in the determined direction for excited emission at the condition of simultaneous exciting in the sample strong coherent acoustic vibrations of nanoparticles ensemble. These vibrations frequencies are defined by the nanospheres dimensions (200–350 nm) are in the range of 6–20 GHz.

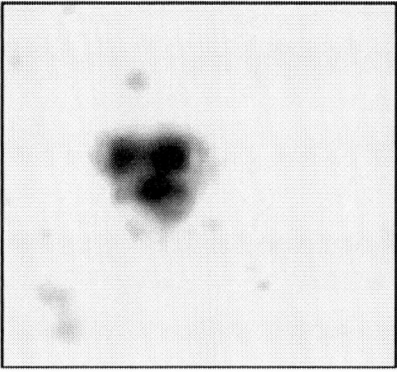

Fig. 4. X-ray picture, obtained from the opal matrix sample, consisting of ordered package of SiO_2 nanospheres with diameter 250 nm for pumping light intensity 0.2–0.4 GW/cm^2. Three single pulses are shown (background is filtered). Distance from the source to the X-ray film is 5 cm.

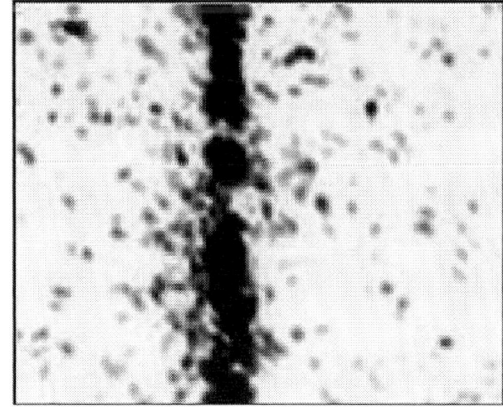

Fig. 5. Picture of X-ray film blackening at 3^x magnification for long-term exposure by pulsed X-ray emission with divergence less than 10^{-3} rad. Distance from the source to the film is 15 cm.

Lithium niobate (LiNbO$_3$) or barium titanate (BaTiO$_3$) have been used as piezoelectric single crystals. Opal matrix sample is held together with piezoelectric plates by different methods. Suggested methods provide resistance of the device to the destruction from mechanical stress, arising as a result of laser light action to the composite structure.

X-ray emission, registered at the X-ray film, is a result of pulsed laser light of nanosecond range interaction with 3-D photon-phonon material – lattice package of SiO$_2$ nanospheres. This method gives possibility to form pulsed X-ray emission with divergence less than 10^{-3} rad while intensity in single pulse is 0.3–0.6 W for duration of fractions of a millisecond. Thus, power is near to that of generated by micrifocus X-ray tube.

If necessary, input of the laser radiation into the optical fiber with core diameter 100–500 μm can be realized with the help of standard schemes. This method is known to be used for many practical tasks (for instance, welding of plastic and stainless steel, soldering in semiconductor, electronic and auto industry, light transmission to the subject in medicine and so on). For this purpose objectives of high quality and short focus coated lenses are used. It allows to realize light input into fiber with minimum losses (less than 15–20 %). In the experiments, performed for conditions optimization (maximum efficiency of ruby laser light conversion into high energy emission) quartz lenses with focuses from 5 to 40 mm and optical fibers with core diameter from 100 to 500 μm have been used.

Till now none of these methods have been developed for producing the miniature sources of X-ray emission. Data from the first part and description of the X-ray emission (with quant energy 80–100 keV) using metamaterials on the base of opal matrices show the real possibility of the endoscopic device creation using optical fiber cables with dimensions of being inserted part of the device not more than 1–1.5 mm both for length and in diameter. Possible setup is shown in Fig. 6. Using of fiber optics system of such device with some peculiarities of construction will allow not only observe affected places, but register Raman spectra for control of the efficiency and peculiarities of local X-ray action.

In medical literature there are detailed descriptions and discussions of the problem, arising in the endoscopic examinations, connected with identification of small areas of inner organs related to the early stages of cancer or another illnesses and different types of outer action with the help of different kinds of radiation including laser. First problem, as it can be seen from numerous publications, can be solved with the help of Raman spectroscopy, because Raman spectra of healthy and sick cells differ each other and sensitivity of this method is much higher than in usual medical practice. As for the second problem, it is known that the most effective method of radiation action (with minimum damage of healthy cells) is X-ray radiation, directed locally to the affected area with moderate energies (consequently, not very large deepness of penetration) and exactly controllable doze.

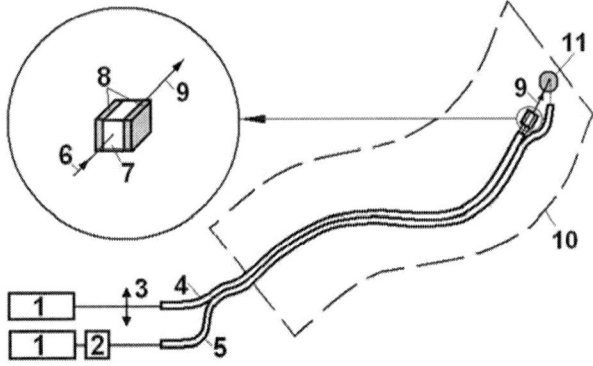

Fig. 6. Scheme of two-fiber endoscopic device for generating directional pulsed X-ray radiation: 1 – laser, 2 – imaging system of the object or the Raman spectra were recorded using a Raman spectrometer, equipped with fiber-optic detection system, 3 – optical system focusing the laser light, 4, 5 – fiber-optic cables, 6 – supply of laser pumping through the fiber, 7 – sample opal matrix (ordered structure of silica nanospheres with a diameter of 200–300 nm), 8 – thin sheets of piezoelectric material (lithium niobate, barium titanate, SBN and other), 9 – directed pulsed X-rays, 10 – investigated the cavity, 11 – piece cavity treated by X-rays in combination with acoustic.

REFERENCES

[1] N. V. Tcherniega, M. I. Samoylovich, A. D. Kudryavtseva, A. F. Belyanin, P. V. Pashchenko, N. N. Dzbanovski. "Stimulated scattering caused by the interaction of light with morphology-dependent acoustic resonance," *Optics Letters.* 2010, vol. 35, issue 3, pp. 300–302.

[2] N. V. Tcherniega, M. I. Samoylovich, A. F. Belyanin, A. D. Kudryavtseva, and S. M. Klescheva, "Generation of electromagnetic and acoustic emissions in nanostructured systems," *Journal of "Nano and microsystem technique".* 2011, № 4, pp. 21–31. (In Russian).

[3] N. V. Tcherniega, A. D. Kudryavtseva, M. I. Samoylovich, A. S. Shevchuk, S. M. Kleshcheva. "Stimulated Low-Frequency Raman Scattering in Nanostructures," *Optoelectronics, Instrumentation and date processing.* 2012, vol. 48, № 3, pp. 39–45.

[4] A. D. Kudryavtseva, N. V. Tcherniega, M. I. Samoylovich, A. S. Shevchuk. "Photon-Phonon Interaction in Nanostructured Systems," *International Journal of Thermophysics.* 2012, vol. 33, pp. 2194–2202.

Optical sensor network for detecting chemical and biological agents

V. A. Yatsenko

Space Research Institute of NASU-SSAU, Kyiv, Ukraine

Abstract: This paper describes the concept of optical sensor networks which has been made viable by the convergence of optical technology, wireless communications, polymeric films technology for direct microbial community observation, classification by Support Vector Machines, and digital electronics. It focuses on a new methodology of optical sensor network for detection and identification of biological and chemical agents including various toxins and viruses.

I. INTRODUCTION

The ecological interest is in response to a proliferation of agents development and threats for human health. Additionally, there is a pervasive interest across diverse application areas such as medicine, environmental protection, and vegetation processing to achieve a rapid detection and identification capability of various agents.

II. PROBLEMS

The following problems have been considered:

- Theoretical investigations: methods, models, and algorithms for remote sensing of chemical and biological agents (early warning chemical sensing).
- Development of spectral data processing methods.
- Analysis of laser emission impact on absorption and reflectance spectra.
- Development of new approaches for earlier detection of biochemical agents.
- Development of models for on-board detection and identification of biological and chemical agents.
- Development of optical device for detection and identification of biological and chemical agents including various toxins and viruses

III. OPTICAL DEVICE

This device is intended to be used for ground measurements or can be installed on small airplanes for remote sensing. Technologies such as optical spectroscopy measurements, laser induced fluorescence, pattern recognition and optimization methods are used in our device.

Technologies such as optical spectroscopy measurements, laser induced fluorescence, pattern recognition and optimization methods are used in our device (Fig. 1).

785nm laser is used in order to reduce the amount of background fluorescence encountered in biological samples as this wavelength is said to be in the 'Biological NIR-window' (approx. $\lambda = 750$-1100nm).

Fig.1

III. SENSOR NETWORK

Sensor networks may consist of many different types of sensors such as optical, Raman spectroscopy, visual, infrared, and radar, which are able to monitor a wide variety of ambient conditions.

A new methodology and algorithms are feasible for cooperative execution of complex tasks by multiple autonomous unmanned vehicles (both aerial and ground during intricate missions. The UAV provide a platform for developing new sensors and techniques for earlier detecting biological and chemical agents.

Nuclear, biological and chemical attack detection and reconnaissance: In chemical and biological warfare, being close to ground zero is important for timely and accurate detection of the agents. Optical sensor networks deployed in the friendly region and used as a chemical or biological warning system can provide the friendly forces with critical reaction time, which drops casualties drastically.

IV. CONCLUSIONS

The conception of the hypespectrometer has been proposed. It is based on the novel principles of information processing.

The fluorescence spectra exibit specific signature that can be used for the identification of chemical and biological agents.

We shown that vegetation in the path of filaments induced by intense femtosecond laser pulse propagating in air could be fragmented and result in the emission of characteristic fluorescence spectra from the excited fragments.

REFERENCES

[1] V. Yatsenko, N. Kozyrovska et. al. Putative mechanism of bacteria effects on plant photosystem under stress // Biopolymers and Cell. - 2012. - Vol. 28. № 6.

Diagnostics of biomedical agents by whispering gallery mode optical resonance based sensor

Vladimir A. Saetchnikov[1], *Member, OSA,* Elina A. Tcherniavskaia[1], Anton V. Saetchnikov[1],
Gustav Schweiger[2], *Member, IEEE,* Andreas Ostendorf[2], *Member, IEEE*
[1]Belarusian State University, Minsk, Belarus
[2]Ruhr-Universitaet, Bochum, Germany

Abstract: Experimental data on optical resonance spectra of whispering gallery modes of dielectric microspheres in antibiotic solutions under varied in wide range concentration are represented. Optical resonance was demonstrated could be detected at a laser power of less than 1 microwatt. Several antibiotics of different generations: Amoxicillin, Azithromycin, Cephazolin, Chloramphenicol, Levofloxacin, Lincomicin Benzylpenicillin, Riphampicon both in de-ionized water and physiological solution had been used for measurements. Both spectral shift and the structure of resonance spectra were of specific interest in this investigation. Drag identification has been performed by developed multilayer perceptron network. The network topology was designed included: a number of the hidden layers of multilayered perceptron, a number of neurons in each of layers, a method of training of a neural network, activation functions of layers, type and size of a deviation of the received values from required values. For a network training the method of the back propagation error in various modifications has been used. Input vectors correspond to 6 classes of biological substances under investigation. The result of classification was considered as positive when each of the region, representing a certain substance in a space: relative spectral shift of an optical resonance maxima - relative efficiency of excitation of WGM, was singly connected.

Recently a number of evanescent wave optical sensors have been developed and used for label-free biomolecule detection in sensing systems with extremely high sensitivity [1–8]. A novel emerging method for the label-free analysis of nanoparticles and biomolecules in liquid fluids using optical micro cavity resonance of whispering-gallery-type modes is being developed [9–12]. Several schemes of experimental realization of the method have been tested [13–16]. The most promising looks a scheme based on polymer microspheres fixed by adhesive on the evanescence wave coupling element [12]. The sensitivity of the developed scheme has been tested by monitoring the whispering gallery mode spectral shift. Water solutions of ethanol, HCl, glucose, vitamin C and biotin have been used [13–16]. Particular efforts were made for an optimal geometry for micro resonance observation under extremely low power of tunable laser exciting resonance. It was demonstrated that optical resonance under optimal geometry could be detected under the laser power of less then 1 microwatt. Material of microsphere the most appropriate for microbial application was also under investigation. Resonance shifts of C reactive protein water solutions as well as albumin solutions in pure water and with HCl modelling blood have

been investigated in developed experimental geometry [13–16]. Introducing controlled amount of nano particles (50 nm in diameter glass gel solution) into microsphere surrounding was accompanied also by correlative resonance shift. The most attention was concentrated on development of a technique for recognition and identification of nanoparticles and biomolecules in liquid fluids [8–12]. We demonstrated that the only spectral shift is not sufficient for identification of biological agents by developed approach. So classifier based on probabilistic neural network for biological agents and micro/nano particles classification has been developed [8–12]. It was tested on solutions of different biological agents. Here we represent an improved version of classifier based on multilayer perceptron and results of data processing for antibiotics of different generation in various solutions.

The developed technique to determine parameters of solutions of the biological agents, based on whispering-gallery modes optical resonance are the following. While tuning the laser wavelength images were recorded as avi-file. All sequences were broken into single frames and the location of the resonance was allocated in each frame. The image was filtered for noise reduction and integrated over two coordinates for evaluation of integrated energy of a measured signal. As input data the following parameters were used: normalized by free spectral range resonance shift of whispering-gallery modes and a relative efficiency of whispering-gallery modes excitation. The last value can be interpreted as the intensity of resonant spectrum integrated over the free spectral range and normalized by the intensity maximum within this free spectral range.

We have used several schemes of experimental realization of the method [13–16]. Here the newer version of the scheme has been developed [12]. It uses thin adhesive layer on the surface being in the field of evanescence wave. Microspheres from PMMA were fixed just by this layer. The layer was putting by spin-coating technology. Compact spin-coater system with digital dosage was used to put and dry previously a thin film of adhesive on the surface of substrate or directly on the coupling element. After that, microspheres were superimposed on the surface of adhesive layer and final drying procedure by free solvent evaporation during 12 hours followed. Under optimized experimentally parameters of the process microspheres were reliably fixed as it was tested. The most part of their surface appeared in contact with tested

978-1-4799-0019-0/13 $31.00 © 2013 IEEE

solution and so can react to variation of the solute concentration.

The spheres used in these experiments were 50 ÷ 120 micron in diameter. The light from a tuneable diode laser (New Focus, 680 nm) is coupled into the microsphere through a prism. Standard right angle turning glass prisms with a refractive index of 1.51 were used. The microsphere is submerged into a fluidic cell and brought into contact with the prism. The cell contains initially de-ionized (DI) water or physiological solution. To vary the refractive index, a solution of ethanol and water is incrementally added to the fluidic cell with a digital syringe. Following each injection, the WGM modes are monitored until equilibrium is reached, and then, the subsequent injection is made. To observe the WGM, the laser repeatedly scans across a spectral range of approximately 150 pm at a frequency of about 0.1 Hz. Light scattered by the micro sphere is collected through a microscope by a CMOS camera or photodiode and monitored with a data acquisition card and computer.

Laser beam was sharply focussed on the single microsphere to increase the contrast and intensity of the resonance scattering signal. When the wavelength of the tuneable laser corresponds to a resonance of the sphere, the power of the light scattered by the micro sphere increases, and a spectral maximum indicating the WGM spectral position is recorded. The width of such a resonance after filtering is used to estimate the resonance quality. The sensitivity of the scheme was tested to determine refractive index variation by monitoring the magnitude of the whispering gallery modes (WGM) spectral shift as in [13–16].

A constructing a qualifier on the basis of a neural network has been passed following stages: data preprocessing, network designing and training, diagnostics of a network performance. We will explain below in more details listed stages, with reference to the decision of our problem.

The data were obtained in the form of the video file form CMOS camera in a format *.avi. All sequence was broken into frames where the area of a resonance was allocated. The image was filtered for noise reduction and integrated on two coordinates for evaluation of integrated energy of a measured signal. As the entrance data following signal parameters were used: relative (to a free spectral range) spectral shift of frequency of WGM optical resonance in microsphere and relative efficiency of WGM excitation obtained within a free spectral range which depended on both type concentration of investigated agents. Last parameter was defined as normalized to an integrated resonant spectrum within a free spectral range. Then we broke the data set into two subsets – training and tested (randomly). The data before submitting on a network input ware preprocessed (normalized and standardized).

The network topology was designed: a number of the hidden layers of multilayered perceptron, a number of neurons in each of layers, a method of training of a neural network, activation functions of layers, type and size of a deviation of the received values from required values. For a network training the method of the back propagation error in various modifications has been used. Input vectors correspond to 6

classes of biological substances under investigation. The result of classification was considered as positive when each of the region, representing a certain substance in a space: relative spectral shift of an optical resonance maxima - relative efficiency of excitation of WGM, was singly connected.

To obtain three-dimensional classification diagram with maximum accuracy all experimental data were used. It is possible to increase the accuracy of an assessment by increasing experimental sampling with a smaller step of a concentration variation.

Several antibiotics of different generations: Benzylpenicillin, Cefazolin, Amoxicillin, Azithromycin, Chloramphenicol, Levofloxacin both in de-ionized water and physiological solution have been used for measurements. They were calculated as:

$$(\nu - \nu_0)/\Delta\nu_{FSR}$$
$$\int_{FSR} I(\nu)/I_{max}d\nu \qquad (1)$$

where $I(\nu)$ - - intensity of the radiation disseminated by microsphere on the frequency ν, $\Delta\nu_{FSR}$ - free spectral range, I_{max} - the maximum intensity of the radiation disseminated by microsphere within a free spectral interval. Spectrum structure were fairly complicated due to relatively high Q factor ($\sim 10^5$), but identification of resonance maxima and their spectral shifts have been surely evaluated. Multiple store of spectrum with statistical preprocessing have been used. All the experiments were performed using the same microsphere as a sensitive sell.

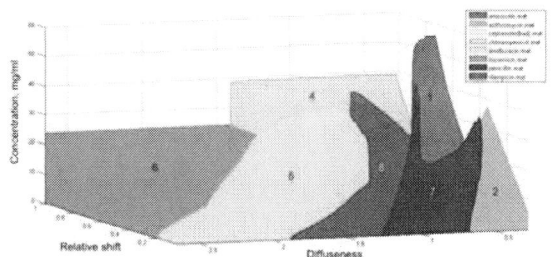

Fig. 1. Results of classification for water solution of Amoxicillin (1), Azithromycin (2), Cefazolin (3), Chloramphenicol (4), Levofloxacin (5), Lincomicin (6), Benzylpenicillin (7), Riphampicon (8).

Spectral data have been processed by procedures explained above and the results of classification by neural network have been represented as three dimensional diagram of a certain drug concentration, identified by neural network and depend on two WGM optical resonance parameters. Estimated sensitivity of the method was about 6 mkg/l. Optimized parameters for classification of agents under investigation were the following: the maximum number of training cycles – 50000; training criterion of 0.00001, activation function for all layers – nonlinear sigmoidal logistic type (logsig). For training of a network the method of gradient descent with adaptation of speed of training parameter (GDA), number of the hidden layers – 3 with 17 neurons in each was used. In input and output layers the number of neurons corresponded to dimension of input and output signals (2 in input and 1 in output). 11212 epochs of training were required for a training of a network. As a result the neural network constructed by developed algorithm, classified antibiotics of different classes with accuracy till approximately 99 %.

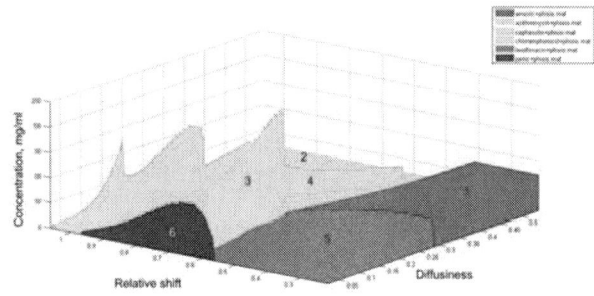

Fig. 2. Results of classification for physiological solution of Amoxicillin (1), Azithromycin (2), Cefazolin (3), Chloramphenicol (4), Levofloxacin (5), Benzylpenicillin (6).

Thus, a technique of detection and identification of drugs on the basis of spectroscopy of an optical resonance of WGM has been developed. Reasonability of the use of multilayered perceptron for processing and interpretation of the data obtained by a biosensor based on WGM optical resonance has been confirmed.

REFERENCES

[1] A. B. Matsko and V. S. Ilchenko, "Optical resonators with whispering-gallery modes—Part I: Basics," IIEEE J. Sel. Top. Quantum Electron., vol. 12, no 1, pp. 3–14, 2006.

[2] V. S. Ilchenko and A. B. Matsko, "Optical resonators with whispering-gallery modes—Part II: Applications,"IEEE J. Sel. Top. Quantum Electron., vol. 12, no 1, pp. 15–32, 2006.

[3] F. Vollmer and S. Arnold, "Whispering-gallery-mode biosensing: label-free detection down to single molecules," Nat. Meth., vol.5, no 7, pp. 591–596 2008.

[4] X. D. Fan, I. M. White, S. I. Shopoua, H. Y. Zhu, J. D. Suter, and Y. Z. Sun, "Sensitive optical biosensors for unlabeled targets: A review," Anal. Chim. Acta, vol. 620, no 7, pp. 8–26, 2008.

[5] M. A. Santiago-Cordoba, M. Cetinkaya, S. V. Boriskina, F. Vollmer, M. C. Demirel, "Ultrasensitive detection of a protein by optical trapping in a photonic-plasmonic microcavity," J. Biophotonics, vol. 5, no 8–9, pp. 629–638, 2012.

[6] S. Arnold, V. R. Dantham, C. Barbre, B. A. Garetz, X. Fan, "Periodic plasmonic enhancing epitopes on a whispering gallery mode biosensor" Optics express, vol. 20, no 24, pp. 26147-26160. 2012.

[7] V. R. Dantham, S. Holler, V. Kolchenko, Z. Wan and S. Arnold, "Taking whispering gallery-mode single virus detection and sizing to the limit," Appl. Phys. Lett., vol. 101, pp. 043704-1–043704-4, 2012.

[8] M. Baaske and F. Vollmer, "Optical Resonator Biosensors: Molecular Diagnostic and Nanoparticle Detection on an Integrated Platform," Chem. Phys. Chem., vol. 13, pp. 427 – 436, 2012.

[9] V.A. Saetchnikov, E.A. Tcherniavskaia, G. Schweiger and A. Ostendorf," Classification of the micro and nanoparticles and biological agents by neural network analysis of the parameters of optical resonance of whispering gallery mode in dielectric microspheres,", Proceeding of the SPIE, vol. 8090, - pp. 80900R1- 80900R11, 2011.

[10] E.A. Tcherniavskaia, V.A. Saetchnikov,"Application of neural networks for classification of biological compounds from the characteristics of whispering-gallery-mode optical resonance," Journal of Applied Spectroscopy, vol. 78, no 3, pp. 457-460, 2011.

[11] V.A. Saetchnikov, E.A. Tcherniavskaia, G. Schweiger, A. Ostendorf and A.V. Saetchnikov, "Neural Network analysis of the resonance whispering gallery mode characteristics of biological agents," Nonlinear Phenomena in Complex Systems, vol. 14, no 3, pp. 253–263, 2011.

[12] V. A. Saetchnikov, E. A. Tcherniavskaia, G. Schweiger and A. Ostendorf, „Classification of antibiotics by neural network analysis of optical resonance data of whispering gallery modes in dielectric microsphere", Nanophotonics IV, Proceeding of the SPIE, vol. 8424, pp. 345-356, 2012.

[13] E.A. Tcherniavskaia, V.A. Saetchnikov, "Detection and identification of microparticles/nanoparticles and blood components using optical resonance of whispering-gallery modes in microspheres," Journal of Applied Spectroscopy, vol. 77, no 5, pp. 692-699, 2010.

[14] V.A. Saetchnikov, E.A. Tcherniavskaia, "Using optical resonance of whispering gallery modes in microspheres for real-time detection and identification of biological compounds," Journal of Applied Spectroscopy, vol. 77, no 5, pp. 714-721, 2010.

[15] V.A. Saetchnikov, E.A. Tcherniavskaia, "Analysis of the biochemical composition of biological tissue in vivo by the diffuse light scattering method," Journal of Applied Spectroscopy, vol. 77, no 6, pp. 878-886, 2011.

[16] V.A. Saetchnikov, E.A. Tcherniavskaia, G. Schweiger and A. Ostendor, "Optical micro resonance based sensor schemes for detection and identification of nano particles and biological agents in situ," Nanophotonics III, Proceeding of the SPIE, vol. pp. 7712, 771221-771232, 2010.

Photo-Induced Birefringence Investigation of Azo Polymer with Cyano Substituted group

E. Bagherzadeh Khajeh Marjan[1], S. Ahmadi Kandjani[1], M. S. Zakerhamidi[1]

[1]Research Institute for Applied Physics and Astronomy, University of Tabriz, Tabriz,Iran

Abstract: Light-induced behavior of polymethacrylate with side chain mesogenic groups of cyano azobenzene was investigated. The intensity of absorption peaks increase when the polymer film was illuminated by pump beam. The birefringence induced by high powers of linearly polarized light shows extraordinary behavior. When the polymer film was kept in the dark, a long-term stability was revealed. The birefringence can be induced by illumination of circularly polarized light. OPM photographs show liquid crystalline behavior of polymer.

Azo dyes are well-known class of colorants and have many applications in paper printing, optical data storage and dye-sensitized solar cells [1,2]. Optical properties of azobenzenes due to their flat aromatic structures and their ability to undergo photoisomerization, make them active compounds for a variety of applications[3,4]. When azobenzene chromophores are illuminated with linearly polarized light, they orientate in the direction perpendicular to the electric field of the incoming light. This process results in remarkable changes in their molecular shape, dipole moments and color[5,6].

Cyanoazobenzene chromophores are push-pull derivatives of the pseudo-stilbene type which show a strong overlapping in the absorption spectra of cis and trans isomers[7]. The addition of the cyano group will result in a larger cis fraction[8].

The investigated polymer was methacrylate {[{4 - [(E) - (4-cyanophenyl) diazenylo] phenyl} (methyl) amino] ethoxy} ethyl(Mn:3190g/mol , Mw:7996g/mol, Tg:72 ℃). It was synthesized by Stanisław Kucharski and co-workers (Institute of Organic and Polymer Technology, Wrocław Technical University, Poland) [7]. The chemical structure of the polymer is shown in Figure 1. It contains cyanoazobenzene chromophore as a mesogenic side group and methacrylate main chain as a linear backbone, abbreviated as pCNOhomo. Thin films (0.5μm) on glass substrates were prepared by spin-coating of the polymer from THF solution with a concentration of 50 mg/ml.

Double beam Shimadzu UV-2450 Scan UV–Visible spectrophotometer was used to record the absorption spectra over wavelength range of 350–700 nm.

The photo-induced orientation behavior of the pCNOhomo polymer film was studied through the birefringence measurement. The experimental setup for photo-induced birefringence of the film can be seen in figure 2. A linearly polarized laser beam obtained by using a Glan prism from a diode laser with the intensity of 150 mW/cm^2 was used as a pump beam (532nm). For birefringence measurements, a low power He–Ne laser at 633nm was used as a probe beam. The sample was placed between a pair of crossed polarizers, which were set to be 45° with respect to the polarization angle of the

incident pump beam. The angle between incident laser beam and probe beam was fixed to be 45°. The photo-induced birefringence of the polymer film was monitored by transmittance of probe beam. To investigate the recording speed of optical information on the film, the changes of the birefringence was measured as a function of the power of pump beam. The photo-induced birefringence was also studied for circularly polarized light.

Thermal properties were examined using a Zeiss optical polarization microscope (OPM) equipped with a Linkam THMSG 600 heating and cooling stage and Linkam THMS 93 programmable temperature-controller.

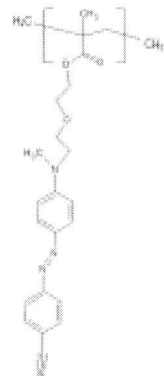

Fig. 1. Chemical structure of pCNOhomo

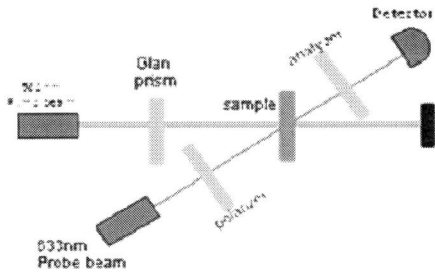

Fig. 2. Experimental set-up to measure photo-induced birefringence.

Absorption spectra of polymer film before and after illumination with pump beam at 532 nm is shown in figure 3. The peaks are located around 430 nm, corresponding to overlapping of π- π* and n- π* transitions of azobenzene groups. It is seen that the intensity of absorption peak after illumination, is increased. In this polymer the fraction of cis isomers are higher than trans isomers and after illumination of pump beam, photoisomerization process and thermal relaxation of cis to trans isomers cause increase in intensity of absorption peaks. The increase in trans fraction could guide the polymer system to liquid crystalline order.

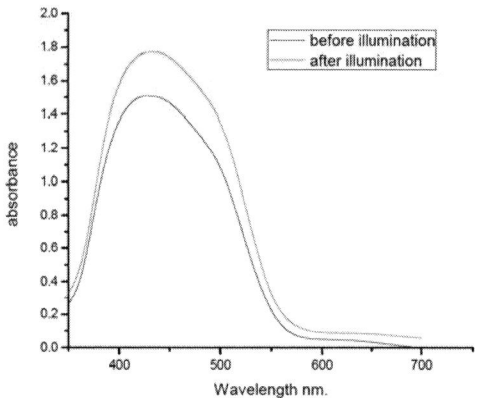

Fig. 3. Absorption spectra of polymer film under illumination of laser diode

Figure 4 shows the photo-induced birefringence of the pCNOhomo polymer film measured by pump beam at 532nm. Before switching on the pump beam, the signal of probe beam was above zero as shown in figure 4. It means that the polymer film shows anisotropy before illumination of pump beam. After irradiation, the polymer showed high photo-induced birefringence. A sudden increase and relaxation of transmitted beam under illumination was due to reorientation of azo molecules and might be because of phase transition to more order. The pCNOhomo polymer film shows a typical feature of side chain liquid crystalline polymers[7]. When the azo groups are subjected to a polarized light they orientate in the direction forced by the polarized light and reach a certain directional order. On the other hand, it is known that in liquid crystalline materials there is an intrinsic thermodynamic tendency to self-organize into ordered phases [9]. When the linearly polarized pump beam was stopped, the photo induced birefringence remained stable and liquid crystalline domains did not relax and also no thermal relaxation occurred, which can be a verification of photo induced phase transition to liquid crystallinity order.

This birefringence behavior also seems to be attributed to the structure of the donor and acceptor of the photosensitive azobenzene molecules and the mesogenic behavior of cyano azo group as side chain. The good thermal stability behavior could also be due to high T_g of polymer, too[10].

It is typical that when the sample is irradiated with circularly polarized light, the birefringence is erased, but in pCNOhomo polymer film, circularly polarized pump beam induce birefringence. As shown in figure4, switching off the pump beam leads to a considerable decrease on the transmittion beam. It could mean that under illumination of circularly polarized light, the domains could not reach order completely. Nevertheless, when the sample was kept in the dark for a long time, no change in transmitted beam was seen. This behavior is also a verification of photo-induced phase transition under illumination of circularly polarized pump beam.

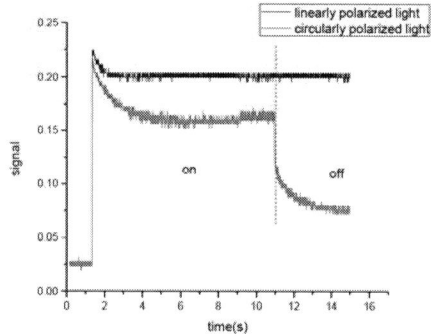

Fig. 4. Photo-induced birefringence for the polymer film with a linearly polarized (150mW) and circularly polarized laser beam(80mW).

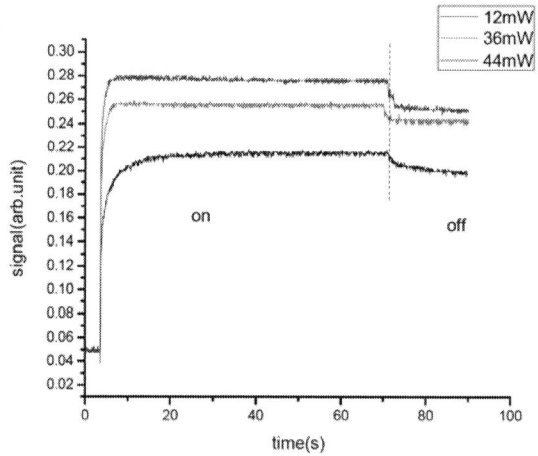

Fig. 5. Photo-induced birefringence of the polymer film at various powers of linearly polarized pump beam.

The birefringence was also induced in polymer film at lower pump beam intensities of linearly polarized light. However, once the irradiation was switched off, a slight decrease on the transmittion beam took place. As seen in figure5, the birefringence was induced with high rate at high light powers.

The phase behaviour of pCNOhomo polymer was determined by Optical Polarized Microscopy (OPM). The polymer was heated up to 200 °C and then cooled down to room temperature at the rate of 5 °C/min. When the polymer was cooled from it's isotropic state, optical anisotropy emerged from the dark background of the isotropic liquid. The fine texture grow to large monodomain and clearly needle-shaped textures of smectic phase was observed(figure 6). The texture was remained stable upon cooling to room temperature.

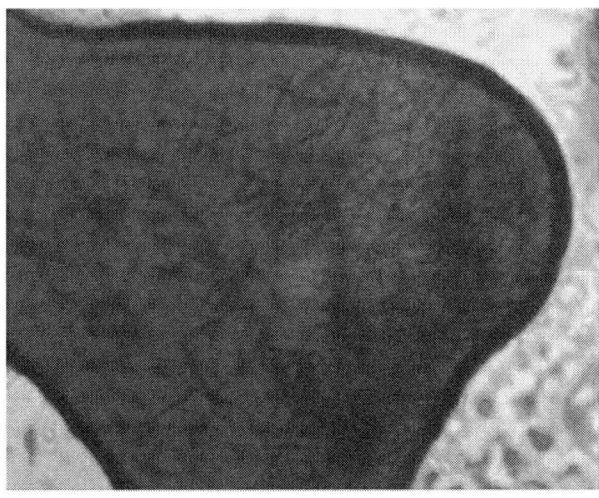

Fig. 6. OPM photographs of polymer during cooling at 140°C

Optical properties of polymethacrylate bearing cyano azobenzene as side chain was studied. The intensity of absorption peaks increased after illumination of pump beam due to photoisomerization process and probable liquid crystalline order. The birefringence was induced during illumination of linearly polarized pump beam. The atypical behavior of photo-induced birefringence was due to reorientation of azo molecules and probable phase transition. The birefringence was not erased by circularly polarized light. It shows that circularly polarized light can also orientate the azo molecules and probably induce a phase transition in polymer film. Photo-induced birefringence decreased with lowering light power. Liquid crystalline behavior of pCNOhomo polymer was determined by OPM.

REFERENCES

[1] Chen B.-Y., "Understanding decolorization characteristics of reactive azo dyes by pseudomonas luteola: toxicity and kinetics," *Process Biochem.*, 2002. Vol. 38. No. 3. P. 437-446.

[2] Dinçalp H., Yavuz S., Haklı Ö., Zafer C., Özsoyc C., Durucasu İ., İçli S., "Optical and photovoltaic properties of salicylaldimine-based azo ligands", *Photochem. Photobiol., A.*, 2010. Vol. 210. No.1. P. 8-16.

[3] Yager K.G., Barrett C. J., "Novel photo-switching using azobenzene functional materials", *Photochem. Photobiol., A.*, 2006. Vol. 182. P. 250–261.

[4] Kaczmarek H., Borowska M.Z., Chylinska M., Kowalonek J., Wolnicka M., "Effect of azobenzene derivatives on the photochemical stability of poly (methyl methacrylate) films", *Polym. Degrad. Stab.*, 2012. Vol.97. No.8. P. 1305-1313.

[5] Sin S. L., Gan L. H., Hu X., Tam K. C., Gan Y. Y., "Photochemical and thermal isomerizations of azobenzene-containing amphiphilic diblock copolymers in aqueous micellar aggregates and in film" , *Macromolecules,* 2005. Vol. 38. No. 9. P. 3943-3948.

[6] Wang Y., He Y., Wang X., "Photo-induced orientation and cooperative motion of three epoxy-based azo polymers" , *Polym. Bull.*, 2012. Vol. 68. P. 1731–1746.

[7] Serwadczak M., Wu¨bbenhorst M., Kucharski S., "Photochromism and diffraction grating in cyanoazobenzene polymer films", *Proc.of SPIE.*, 2006.Vol. 6335.

[8] Yan J., Liu L., Ji L., Ye M., Xu L., Wang W., "Photo-isomerization induced rapid photo-degradation of optical nonlinearity in cyano substituted stilbene derivative doped poled polymer, " *Physics D: Applied Physics*, 2004.Vol. 37. No.12. P. 1597–1602.

[9] García T., López L., Rodríguez-González R., Martínez-Ponce G., Solano C., Navarro-Rodríguez D., "Liquid-crystalline polymers bearing phenylene(azobenzene) moieties substituted with an electron-donor or electron-acceptor lateral group. Synthesis, mesomorphic behavior and photo-induced isomerization", *polymer*, 2012. Vol. 53. No.10. P. 2049-2061.

[10] Chen X., Zhang Y., Liu B., Zhang J., Wang H., Zhang W., Chen Q., Pei S., Jiang Z., "Novel photoactive hyperbranched poly(aryl ether)s containing azobenzene chromophores for optical storage", *Mater. Chem.,* 2008. Vol. 18. P. 5019-5026.

978-1-4799-0019-0/13 $31.00 © 2013 IEEE

The change response of holographic glucose sensors when changing pH of the solution

V. A. Postnikov[1], A. V. Kraiski[2], M. A. Shevchenko[2]
[1] Institute of Physico-chemical medicine, Moscow, Russia
[2] P.N.Lebedev Physical Institute RAS, Moscow, Russia
[*] Corresponding author: vladpostnikov@mail.ru

Abstract: We have examined changes in the responses of holographic sensors with phenylboronic acid while pH of solution were increasing in water HCl solution and in solutions of NaCl (0,1-0,4 M). Effect ionic strength results in a significant change in sensors response, that is the result of compression of the matrix sensors.

Denisyuk hologram, recorded in the grafted onto the glass plate hydrogel matrices, with an appropriate choice of the matrix components are a holographic sensor of solutions. After the contact with an analyte, the matrix changes its configuration, and the distance between layers of silver in the hologram are changing. As a result, the wavelength of a reflected light also changes.[1,4]

The holographic glucose sensors were synthesized by recording a hologram of a plane mirror in a hydrogel matrix consisting of three-dimensional copolymer of acrylamide (AA), N(ε)-methacryloyl-lysine, N-acryloyl-D-2-glucosamine, N-acryloyl-3-aminophenylboronic acid, and N,N'-methylene-bis-acrylamide as crosslinking agent (AA-m1,m2,m3,m4; mol.%) by the irradiation LED laser (532 nm, 1500 mW).[5]

The sensitivity of holographic glucose sensors based aminophenylboronic acid depends on the composition of the matrix sensor, and components of the solution, wherein the glucose is determined. Swelling (or contraction) of hydrogel matrices is determined by changes in the total degree of ionization of the polymer molecules. According to Fig. 1, the wavelength of the reflected light from sensors of different compositions varies with increasing solution pH, reflecting changes swelling matrix sensors. Differences in the response of these sensors at titration of HCl solution correspond to pK_a of components in polymeric matrices. The response of sensors without lysine derivative (0061, 0660,5) practically does not change until the pH = pK_a of phenylboronic acid, when the charge changes and matrices swelling takes place. Sensor response (swelling) containing lysine (6061, 6661) at a pH of less than 4 is determined by the positively charged amino groups. At pH nearly 7.4 the share of positively charged amino groups is decreased, but the share of negatively charged carboxyl groups and negatively charged ions phenylboronic acid is increased. The total charge of the matrix reaches 0 and the matrix is compressed. At pH of more than 7 the charge of the polymer is determined by ionized carboxyl and phenylboronic acid and sensor response corresponds to swelling.

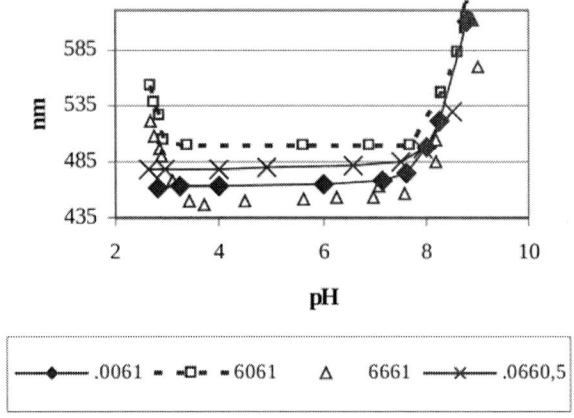

Fig. 1. Changing the responses sensors in aqueous solution while the titration is carried out. (solution HCl 0,01M, NaOH 0,1M; The composition of sensors: m1,m2,m3,m4; mol.%, when m1- N(ε)-methacryloyl-lysine, m2 - N-acryloyl-2-glucosamine, m3 - N-acryloyl-3-aminophenylboronic acid, m4 - N, N'-methylene-bis-acrylamide)

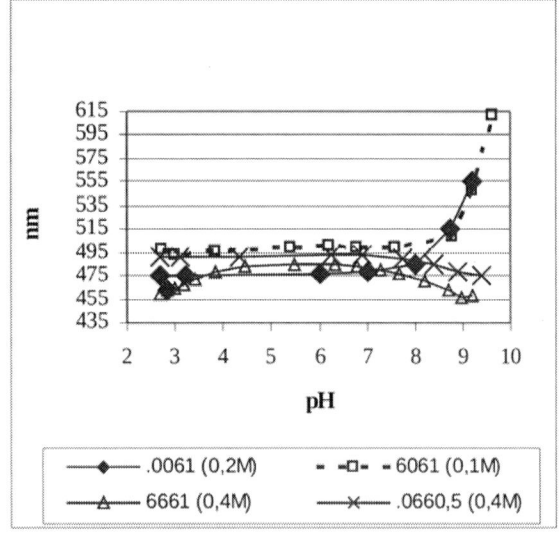

Fig. 2. The response of the sensors at changing of pH in NaCl solutions . (The same composition of sensors) .

[5] Kraiskii AV, Postnikov VA, Sultanov TT, Mironova TV, Kraiskii AA. On optical properties of holographic sensors based on silver emulsions. In Conference Proceedings - 11th CAOL, Kharkov, Ukraine, September 5 – 8, 2011 , art. no. 5634214, 191-192

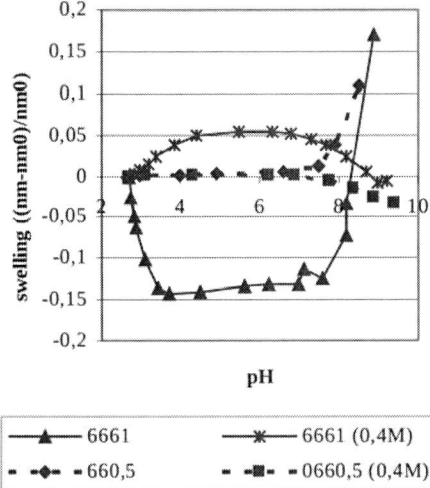

| ▲ 6661 | ✱ 6661 (0,4M) |
| ◆ 660,5 | ■ 0660,5 (0,4M) |

Fig. 3. Swelling of the sensors at changing of pH in NaCl solutions . (The same composition of sensors) .

The responses of sensors have demonstrated a clear dependence of the gel swelling degree on the salt concentration in aqueous solutions. Fig. 2 shows the value of the sensor signal at different sensor composition and concentrations of NaCl. The sensor characteristics demonstrate a "salting out" effect of NaCl salt over the range where matrices have high negative degree of ionization. According to Fig.3, the response of the sensor (6661) in the presence of salt depends on the pH of the solution reaching its maximum at around pH 7, and then starts to decrease.

ACKNOWLEDGMENTS

This work was supported by grant of the Program of Fundamental Studies "Fundamental Sciences for Medicine" of the Presidium of RAS.

REFERENCES

[1] Millington R, Mayes A, Blyth J, and Lowe C. A Holographic sensor for Proteases. Anal. Chem. 1995; 67,4229-4233.

[2] Marshall A, Young D, Kabilan S, Hussain A, Blyth J. and Lowe C. Holographic sensors for the determination of ionic strength. Analytica Chimica Acta.2004; 527(1), 13-20.

[3] Lee M-C, Kabilan S, Hussain A, Yang X, Blyth J, and Lowe CR. Glucose-Sensitive Holographic sensors for Monitoring Bacterial Growth. Anal. Chem. 2004; 76 (19), 5748-5755.

[4] Kraiskii AV, Postnikov VA, Sultanov TT, Khamidulin AV. Holographic sensors for diagnostics of solution components. Quantum Electronics. 2010; 40 (2), 178 -182.

Application of the digital holographic interference and electron microscopy methods for study of rats' blood erythrocytes 3D morphology in the condition of hard metal salts effect

T V. Tishko[1], D.N. Tishko[1], V.P. Titar[1], *Member, IEEE*, O.O. Prikhodko[2], V.I. Bumeister[2]

[1]V.N. Karazin Kharkov National University, Kharkov, Ukraine
[2]Sumi Medical University, Sumi, Ukraine

Abstract: We use the combination of the digital holographic interference and electron microscopy methods for study of 3D morphology of blood erythrocytes of rats that took water with high concentrations of heavy metal salts for a month. The obtained results show essential morphological changes of blood erythrocytes in experimental group of animals, increase of the quantity of transformed and degenerative erythrocytes, decrease of the average diameter of normal erythrocytes, transformations of shapes of normal erythrocytes in the direction of sphericity coefficients increase.

A cell is a unit of a living organism that reflects and determines the status and functioning of a biological system as a whole. The blood cells are of special interest. Blood unites the operation and functioning of all organs of a living organism. Study of morphological features of blood erythrocytes gives important information on their functional possibilities, decreasing of erythrocytes functionality in different pathologies and under influence of different external damaging factors. Erythrocyte is the cell of a living organism whose functionality is determined by its 3D shape. Solution of the problem of erythrocyte 3D morphology study is directly connected with the development of the methods of 3D imaging transparent biological microscopic objects. Using optical holographic interferometry methods in microscopy with the digital processing of images makes it possible to obtain 3D images of native cells and to measure their morphological parameters [1, 3]. The results of the investigations lately carried out by the electron microscopy method [2] and by the method of digital holographic interference microscopy [1, 3, 4] have shown that, in addition to hematological diseases, the diseases of various genesis and external damaging factors can be the reason for modifications of erythrocyte shapes.

In this work we use the combination of the digital holographic interference and electron microscopy methods for study of 3D morphology of rats' blood erythrocytes in the condition of heavy metal salts effect.

Present level of development of industry, transport is the potential reason of serious ecological problems which form serious danger for health of population.

Salts of heavy metals are the most ecologically dangerous. Environment pollution results in accumulation of heavy metals in water, soil, air, and, as a result, in organisms of people and animals. Such heavy metals can damage biostructures of different organs and system, including blood erythrocytes.

Study of functionality of rats' erythrocytes in the condition of heavy metal salts effect is a model for study of the pathological process in a human organism because their erythrocytes by the morphological and functional properties correspond to human erythrocytes.

White adult male rats were used in our experiment. The experiment with the rats was carried out on the base of the Medical University of the Sumy State University (Ukraine) in accordance to the requirements of the European Convention for Protection of Vertebrate Animals used for Experimental and Other Scientific Purposes (Strasbourg, 1986).

The rats (8 months old) were divided in two groups: experimental and control. The rats in the experimental group (10 rats) took drinking water with high concentrations (that in 10 times exceed the concentrations in ordinary water) of the salts of zinc, copper, iron, manganese, lead: $ZnCl_2 - 5$ mg/l, $CuSO_4 \times 5H_2O - 5$ mg/l, $FeSO_4 - 10$ mg/lл, $MnCl_2 \times 4H_2O - 1$ mg/l, $Pb(NO_3)_2 - 3$ mg/l. The rats in the control group took drinking water of ordinary quality.

The experiment lasted one month. Blood erythrocytes were the material for study. Erythrocytes were taken from a tail vein after the end of the experiment.

Study of the erythrocyte morphology was carried out using a scanning electron microscope REM 106" with a low vacuum chamber, photos were made with the magnification from ×2000 to ×40000. The samples for the study were prepared by the special procedure. The obtained electron photos of erythrocytes were analyzed using a computer analyzing system. Percentage of different morphological types of erythrocytes was calculated from a thousand of the cells. Calculations were performed for the fragment of the sample where erythrocytes were situated in one layer.

The results obtained using the electron microscopy method show, that 93 % of rats' erythrocytes in control group are

978-1-4799-0019-0/13 $31.00 © 2013 IEEE

biconcave disk full-functional erythrocytes which correspond to the norm. The remaining 7% of the erythrocytes are different types of transformed and degenerative erythrocytes. In Fig.1 one can see the fragment of a sample with erythrocytes of a rat from the control group. The average diameter of the erythrocytes in the control group is about 5 μm.

Essential changes of erythrocyte morphology are observed in the experimental group of rats that took heavy metal salts for a month. Biconcave disk erythrocytes dominate in the samples but their percentage reduced to 71 %. The quantity of transformed and degenerative erythrocytes increased to 29%. In Fig. 2 one can see the fragment of a sample with erythrocytes of a rat from the experimental group with different types of transformed and degenerative erythrocytes. Results of erythrocyte morphometry in the experimental group show decrease of the average diameter of normal biconcave disk erythrocytes on 3.8 % in comparison with the control group.

So, the method of scanning electron microscopy has found that effect of heavy metal salts resulted in increase of the

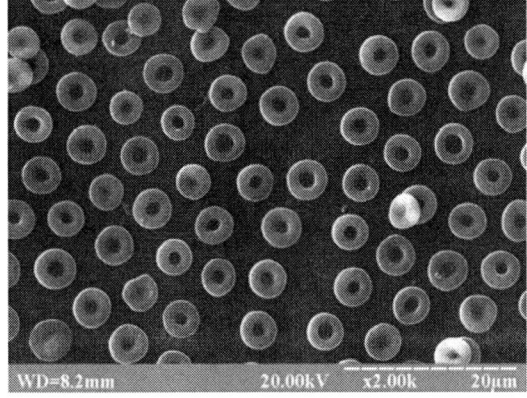

Fig. 1. Erythrocytes of a rat from the control group. It is seen the domination of normal biconcave disk erythrocytes.

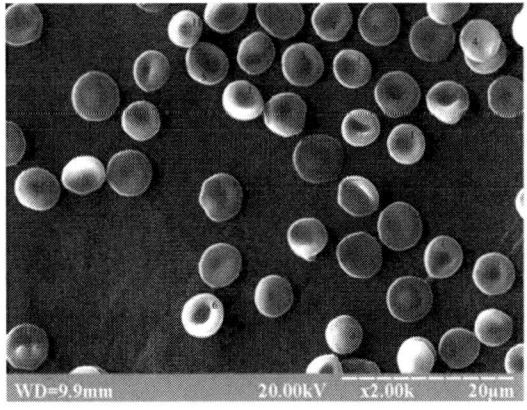

Fig. 2. Erythrocytes of a rat from the experimental group. One can see the presence of different shapes of transformed erythrocytes.

percentage of transformed and degenerative erythrocytes which

are less functional, and 3,8 % decrease of average diameter of the normal erythrocytes.

The erythrocyte morphology study was also carried out with the use of the digital holographic interference microscope (DHIM) in the Laboratory of Holography, V.N. Karazin Kharkov National University. The DHIM resolution is considerably smaller than the resolution of the electron microscope though the DHIM method is a nondestructive quantitative method allowing 3D imaging of native cells [3]. The DHIM consists of three main units: holographic microinterferometer, digital video camera and computer. The real –time interferograms of the samples under study are recorded by the digital camera. The digital interferograms are computer processed using the mathematical algorithms that make it possible to reconstruct the 3D images of erythrocytes and to measure their morphological parameters. He-Ne laser with the radiation wavelength equal 0.63 μm is used as a source of light in the DHIM.

Dry untreated blood smears on glass substrates were used in erythrocyte morphology study using the DHIM. Interferograms of erythrocytes were registered using the objective 40 x 0, 65.

Blood erythrocyte is a transparent microscopic object with homogeneous distribution of refractive index that inserts only phase increments into the light wave passing through it. These phase increments are observed as curvatures of straight interference fringes on their interferograms. In Fig.3 one can see a typical interferogram of a native blood erythrocyte. The thicknesses of erythrocyte $t_i(x,y)$ are proportional to shifts of interference fringes in the corresponding points $h_i(x',y')$:

$$t_i(x,y) = h_i(x',y') \cdot \lambda / T \cdot \Delta n,$$

where $h_i(x',y')$ – shift of an interference fringe; λ – light wavelength; T – period of the system of interference fringes; Δn – difference of refractive indices of a microscopic object and its environment.

Application of the DHIM method makes it possible to detect and quantitatively estimate small pathological transformations of main mass of normal erythrocytes which result in decrease of their surface area and shape changes. These small morphological changes are of great importance for the system of blood consisting of colossal number of erythrocytes. 3D shape of erythrocyte and its morphological functionality is characterized by the sphericity coefficient k, which is introduced as a ratio of the erythrocyte thickness at the center t_c to the thickness t_r at half of its radius r (Fig. 4).

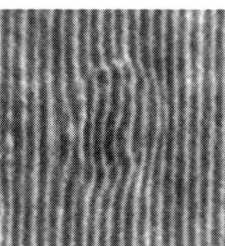

Fig. 3. Typical intereferogram of an erythrocyte.

There are three main morphological types of normal

erythrocytes: biconcave disk, flat disk, and spherocytes. Fig. 5 shows 3D images of these main morphological types and their sphericity coefficients. For the biconcave erythrocytes, flat disks, and spherocytes, the sphericity coefficient is less than unity, about unity, and greater than unity, respectively. The increase of the sphericity coefficient corresponds to the decrease of the surface area of the erythrocytes under its given volume, and to decrease of their functionality [4]. Changing erythrocyte shape from the shape of a biconcave disk to the spherical shape under a given volume results in some decrease of erythrocyte diameter.

The spericity coefficient of erythrocytes is measured under computer processing the interferogram. The spericity coefficient is calculated for every erythrocyte, and the average one for a blood smear.

It was detected that the average sphericity coefficients of erythrocytes in the control group of rats was equal to 0.41. This means that in blood of rats in control group normal biconcave disk erythrocytes dominate. The main mass of erythrocytes of rats in the experimental group undergone essential morphological transformations. The average sphericity coefficient of erythrocytes in this group is equal to 0.81. In Fig.6 one can see images of fragments of native blood smears of rats from the control (up) and the experimental (down) groups obtained using the DHIM.

Combination of the digital holographic interference and electron microscopy methods for study of 3D morphology of rats' blood erythrocytes has allowed detecting essential morphological changes of blood erythrocytes of rats in the condition of heavy metal salts effect These morphological changes are non-specific, and appear as increase of percentage of transformed and degenerative erythrocytes, and transformations of shapes of the main mass of normal

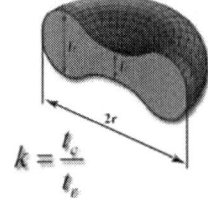

Fig. 4. The sphericity coefficient k.

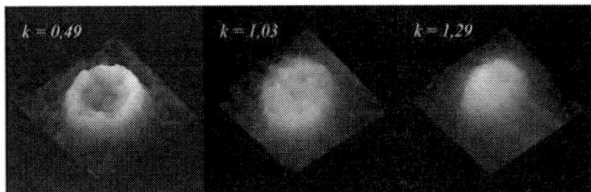

Fig. 5. Three main morphological types of blood erythrocytes and their sphericity coefficients. From left to right: biconcave disk; flat disk; spherocyte.

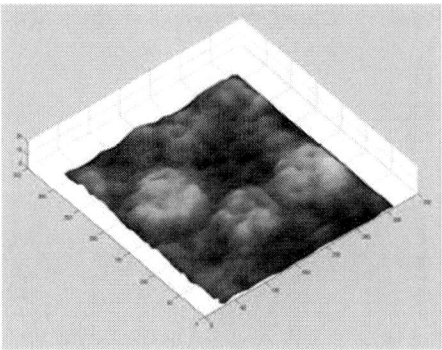

Fig. 6. Fragments of native blood smears of rats from the control (up) and the experimental (down) groups obtained using the DHIM.

erythrocytes in the direction of increase of their sphericity coefficients. Such morphological changes decrease blood erythrocyte functionality, and can be the reason of different functional disorder in an organism.

REFERENCES

[1] T.V. Tishko, V.P. Titar, D.N. Tishko, "Holographic methods of three-dimensional visualization of microscopic phase objects," *J. Opt. Technol.*, vol. 72, no. 2, pp. 203-209, 2005.

[2] Novitsky V.V., Ryazantseva N.V., Stepovaya E.A., Bystritsky L.D. Tkachenko S.B., *Atlas. Clinical Erythrocyte Pathomorphosis* Tomsk : State University Publishers, Ru; 2003.

[3] T.V. Tishko, D.N. Tishko, V.P. Titar, *Holographic microscopy of phase microscopic objects. Theory and practice,* London- Singapore; World Scientific, 2011.

[4] T.V. Tishko, D.N. Tishko, V.P. Titar, K.V.Nosov, "Digital holographic interference microscopy in the study of 3D morphology of human blood erythrocytes", *Laser Physics,* vol. 18, no. 4, pp. 1-5, 2008.

Nonlinear holographic model of physiological optics

V.P. Titar, *Member, IEEE*

V.N. Karazin Kharkov National University, Kharkov, Ukraine

Abstract: A nonlinear dynamic holographic model of visual perception is proposed. The model is based on the effects and properties of laser radiation interacting with nonlinear quasi crystal structures of the retina.

Lasers are widely used in ophthalmology for diagnostics, correction, and eye surgery now. Though there is no clear understanding the mechanisms of interaction of laser radiation with the retina of human eye. The model of interaction of laser radiation with eye media is proposed.

Spatial information inherent in an observable pattern undergoes three processing stages: optical, neurologic and psychophysical [1]. On the first stage, optical processing is performed by eye optics, which forms an image on the retina. As in any other optical imaging system, the image quality is defined by several factors, namely: a finite aperture of the system and its pupil; aberrations, due to deviation from the ideal curvature of the refracting surface; defocusing errors, and other effects such as intraocular scattering. All these effects can be completely described in terms of physical optics.

The second stage is related to the neural activity. It starts on the retina where the image is detected by the photoreceptor mosaic. Then the signals are transmitted over the neural network of the ganglion cells to the nerve fibers, and finally to the neurons of the visual cortex area. By now, the nature of this processing stage has been understood only partially. The distortions arising at this stage may be caused by a finite number of elements in the detector array, by the shot noise due to the quantum nature of light, and by the structure of the neural network, which includes the effects of the interaction between the internal signals generated by adjacent groups of receptors in the form of a spatial addition or suppression.

The third stage is a psycho-physical in nature and, therefore, could not be quantitative analyzed.

Of course, the complete system is nonlinear. The difficulties arising from the nonlinearity of the visual system may be solved by using a perturbation procedure in which the observed pattern has low contrast, that is, contains a medium level of light and small fluctuations around it. Then the system becomes approximately linear and can be fully characterized by its impulse response. When the system is linear and spatially invariant, it can be fully characterized by its response to a sinusoidal signal. The response to this input signal is also sinusoidal, having the same frequency but with a reduced amplitude and phase shifted.

The layers of the retina have a pronounced anisotropy, thereby nerve fiber layer behaves as a positive uniaxial crystal, and external segments of the rods and cones behave as a negative uniaxial crystal. The optical axes of the crystals are mutually perpendicular. If such a system is illuminated by a plane monochromatic wave of arbitrary polarization four linearly polarized waves: two ordinary and two extraordinary will be produced in the outer segments of photoreceptors. We will show below that their interaction with photosensitive medium (rhodopsin in rods and in iodopsin cones) leads to formation of dynamic holograms in the receptor layer of the retina [2].

The molecules of visual pigments packed in disks might be thought of as a dynamic recording medium, since their illumination leads to a substantially instantaneous realignment of the chromoprotein globule hromoproteidov and, as a consequence, to a rapid change of refractive index of the discs. The change of the refractive index caused by any part of the photon flow passing through a dynamic environment affects the conditions for the passage of other photons. In other words, the processes of image recording and reading inherent in the dynamic holography take place in the outer segments of the photoreceptor cells, which are interconnected in time. Any portion of the light beam incident on the retina in an arbitrarily time interval varies in accord with the refractive index change caused by the previous part of the beam, i.e. reads a recently recorded image and at the same time writes a new image corresponding to the intensity distribution in the light beam arising during recording, and so on.

Three-dimensional phase hologram recorded in the anisotropic retina medium is defined by variations in the permittivity tensor, i.e. its form is essentially anisotropic. The properties of the holograms recorded in anisotropic crystals have been most extensively studied in [2].

In general, hologram recording requires the presence of at least two coherent waves: the reference and object waves. In an anisotropic birefringent medium a diffraction grating can be created by full interaction of the ordinary and extraordinary waves. [3]

As is well known, for an ordinary wave, which polarization vector is perpendicular to an optical axis of an anisotropic medium, the refractive index does not dependent upon direction of wave propagation. For extraordinary wave which polarization vector is directed along the optical axis, the refractive index n_e depends on the propagation direction. Thus, the ordinary and extraordinary waves can not interfere, since the polarization vectors are perpendicular to one another. However, in the outer segments of photoreceptor cells, there exist four waves created by incident wave of arbitrary polarization: two ordinary and two extraordinary, so that interference between the respective pairs of waves with the same polarization may always occur. For example, two extraordinary waves will create interference field and write down the corresponding diffraction grating with the wave vector $k = k_{1o} - k_{2e}$, where k_{1o} and k_{2e} are wave vectors of the first and the second incident extraordinary waves. Since the refractive index n_o depends upon the propagation direction, diffraction of initial recording waves on this grid will be of the extraordinary type.

The energy conservation is fulfilled automatically, since the frequencies of the four photons are the same. A specific feature of the anisotropic self-diffraction is the possibility of wave front rotation in the scheme with the two initial beams [4]. The second feature of the anisotropic self-diffraction is the significant difference between propagation angle of diffracted wave and the convergence angle of initial waves. The scattering directions for the anisotropic and isotropic diffraction are opposite. Thus, in an anisotropic medium consisting of photosensitive outer segments of photoreceptors, each pair of waves $E_{o(ref)}$, which is converted from $E_{e(ins)}$, and

$E_{o(ins)}$ creates an interference pattern of the dynamic hologram.

According to the mode theory of related holograms [5], a nonlinear selection of non-monochromatic radiation after reflection from anisotropic periodic structures similar to the outer segment of rods and cones may be possible. This selection is may be explained by the fact that gradients of the anisotropic structure reflects with transformation of the wave front only the component of stimulation light, which coincides with spatial periods of these structures.

Assume that horizontal, bipolar or ganglion cells are allocated in nodes of holographic gratings formed between the membranes in the retina. Since in the foveal region each cone corresponds to a ganglion cell, the information concerning the signal of a definite wavelength is transmitted over the ganglion cells. Thus, the factor stipulating effectiveness of light signal perception has a resonant nature. Only beat signals in the interference pattern of the incident and reflected radiation are resonant when interacting with respective horizontal, bipolar and ganglion cells through which signals in the frequency band corresponding to one of the four holographic gratings are passing.

One more feature of anisotropic hologram is that due to the difference of the refractive indices of the ordinary and extraordinary waves the Bragg conditions are for different angles of light incidence on the hologram. Based on the property, a wave with an arbitrary polarization state can be written at one hologram in an anisotropic medium due to the writing of separate and independent holograms corresponding to the ordinary and extraordinary waves.

Thus, geometry of hologram reading process for anisotropic diffraction differs from that for the isotropic case. Such differences create favorable conditions for hologram reading by radiation whose wavelength is different from that used during recording. In this way a non-destructive reading, when the reading wave does not erase previously recorded information, can be carried out.

By analogy with biorhodopsin [6-7] crystal structure of the retina is capable of stimulated Mandelstam - Brillouin scattering.

If such a structure is illuminated by a wave with an inhomogeneous wave front, the wave will be reflected from the layer of rods and cones, as conjugate to the incident wave. A system with these properties is known as Brillouin mirror.

The refractive index of the photosensitive medium used for the conventional holographic recording is proportional to the intensity of incident radiation. Therefore, light blobs, formed by the interference of radiation incident on the hologram, are recorded as uniformly filling the medium. Combs of sound waves following each other at some distance, reflect light incident on them, similar to a system of metal mirrors at 3D holograms.

The change of the photosensitive medium response, of course, alters interaction between reconstructing radiation and a hologram structure. While on uniformly filled clusters of the refractive index, the radiation can only be refracted and, therefore, reconstructed wave propagates in the same direction as the incident, for the Brillouin mirror the same clusters are cross-modulated by sound waves and reflect light strongly in the opposite direction. If direction of waves propagation is reversed, while the interference pattern does not vary, the situation corresponds to the transition to the conjugate wave.

The effect of photon echo is observed in resonant media, that is, media in which the wavelength of absorption-emission line corresponding to lower level transitions coincides with the wavelength of light used for the hologram exposure. The echo generation can be interpreted as the ability to store and reproduce time parameters of the electromagnetic field, i.e. the time interval between successive pulses. Furthermore, the medium remembers the spatial phase distribution for the first pulse.

Shtyrkov and Samartsev [8] using the concept of the photon echo had proposed to write resonant dynamic holograms by object and reference radiation pulses, not overlapping in time. In this method, an object pulse is directed to a resonant media at time $t = 0$, which transfers some of the atoms from the ground state to the excited state. The theory of photon echo predicts that in this state, the phase of atom oscillation is correlated with the phase of the object wave for the transverse relaxation time.

Holography based on recording in resonant media, in which the spatial memory hologram is combined with the temporary memory of the photon echo, opens a new opportunity to store and reconstruct the processes associated with changing conditions in time and space.

It was shown in [8] that under the condition of the nonlinear coherent interaction of object and reference light pulses with resonant medium, the pulses may be separated from each other by a time interval $\tau < T_1, T_2$ but the hologram can be formed. T_1 is the longitudinal relaxation time, which characterizes the rate of energy transfer from the ensemble of active particles to the thermostat and vice versa, T_2 is a transverse relaxation time, which is called the phase relaxation. The dynamic holograms formed under these conditions, are known as echo-holograms [8]. Such holograms contain information on the spatial and temporal characteristics of the object wave. Reading of such holograms is performed automatically by adjustment of time period after recording. Such a hologram may be considered as 4D hologram.

As a result, the field of the primary light echo is the complex conjugate to the field of first exciting pulse (which is the object wave), and their wave front carries information concerning the pseudoscopic image of the object. The field of stimulated photon echo is proportional to the field of the second (object) of a pulse, and its wave front carries the information of the real image of the object.

In the vision process, the human eye makes the so-called "saccadic" (jump) movements, moving the fixation point over the plane of the observed object. This corresponds to the operation of "scanning phase center" known in the location and in the eye optical system two signals are arriving:

- the field reflected from the point of fixation (the so-called brilliant point), which contains the phase information of the distance to the object of this section;

- blurred field from the area surrounding this the point.

The first field may be used as a reference wave (in models with a reference wave formed by a part of the object) during the formation of the hologram on the retina, and the second field is an object wave, since it contains the amplitude and phase information is the amplitude and phase information of the object.

When saccadic eye movements relative to the observed object is excluded, hologram is not formed, since the reference signal is absent. From the experiments it is known that this breaks the visible picture of the world.

In a series of papers [9-11] it was shown a constant visual stimulation of the retina generally breaks the process of visual perception completely.

In [11] it was noted that any light intensity acting on the retina more than 1-2 seconds in a stabilized presentation, does not cause visual aftervision. For example, in experiments carried out under artificial image stabilization on the retina [9] the subject could not see the test object after 1-3 seconds after the start of the experiment. In addition, the object was invisible during entire experiment. Thus in this a blank field is observed without any details.

In the experiments carried out by Gibson [12], a man by will effort fixed his gaze on an arbitrary point, and in a few seconds the image fades, the contours of objects blur, depth of the image disappears. According to Gibson, the visible world is converted to a visible field.

REFERENCES

[1]   R Cornsweet T. N. Visual Perception.-New York : Academic Press, 1970. - 338 p.

[2]   Titar V. P., Spatchenko O.V. Modeling macular polarization effects in the frame of the holographic model of physiological optics// Applied radioelectronics. – 2009. – V. 8, № 1. –P. 53-61.

[3]   Stepanov S. I., Kamlishin A. A., Petrov M.P. // FTT. – 1977. – V. 19, № 3. – P. 721-726.

[4]   Lesthev A. A., Sidorovich V. G. To the theory of light wave transformation by reflective volume holograms // Optics and spectroscopy. –1978. – T. 48, № 2. – P. 302-307.

[5]   Sidorovich V. G. To the theory of Brillouin mirror // J. Tech. Phys, – 1976. – V. 46, № 10. – P. 2168-2171.

[6]   Vsevolodov N. N. Biopigments – photo detectors. BR photometry – M. : Nauka, 1988. – 222 c.

[7]   Vsevolodov N. N., Djukova T. V., Korchemskaya V. B., Taranenko V. B. Non-linear Veigert-effect in the "Biochrom" films on the basis of BR // Ukr. Phys. J. – V. 29, № 7. – P. 1107-1109.

[8]   Shtirkov E. I., Samartsev V. V. Resonance dynamic holography and optical super radiation.// Electromagnetic super radiation. – Kazan, 1975. – C. 398-417.

[9]   Andreeva E. A., Vergilis N.Yu. Mechanisms of eyes motion in the vision process // Моторные components of vision. – M., 1975. – C. 3-55.

[10] Vergelis N. Yu., Andreeva E. A. Study of eye-moving system transformation in the vision process // Psychology of perception. – M.: Nauka, 1989. – C. 70-84.

[11] Yarbus A. L. The role of eyes motion in the vision process. - M.: Nauka, 1965. – c.

[12] Gibson E. Ecological approach to visual reception. – M.: Progress, 1988. – 462 c.

Picosecond 175 ~ 210 nm tunable deep-ultraviolet laser

(Invited Paper)

Shen-Jin Zhang[1], Yong Bo[1], Feng-Feng Zhang[1], Feng Yang[1], Zhi Xu[1,3], Feng-Liang Xu[1], Zhi-Min Wang[1], Qin-Jun Peng[1], Jing-Yuan Zhang[1], Xiao-Yang Wang[2], Chuang-Tian Chen[2], Wen-Qiang Lei[1], and Zu-Yan Xu[1]

[1]Research Center for Laser Physics and Technology, Key Lab of Functional Crystal and Laser Technology, Technical Institute of Physics and Chemistry, Chinese Academy of Sciences, Beijing, 100190, China
[2]Beijing Center for Crystal R&D, Key Lab of Functional Crystal and Laser Technology, Technical Institute of Physics and Chemistry, Chinese Academy of Sciences, Beijing, 100190, China
[3]University of Chinese Academy of Sciences, Beijing, 100190, China

Abstract: A picosecond tunable deep-ultraviolet laser by two stages of second harmonic generation (SHG) was developed with a fundamental tunable Ti:sapphire laser from 700 to 840 nm. In the first SHG stage, two β-BaB$_2$O$_4$ crystals was adopted to compensate the walking of the beam direction due to frequency-tuning, which improved beam quality and increased output power. The tunable wavelength of first SHG stage was 350 ~ 420 nm, and the maximum output power was 1.1 W at 386 nm. In the second SHG stage with a KBe$_2$BO$_3$F$_2$ as nonlinear optical crystal, the tunable range was 175 ~ 210 nm, and the highest power was 3.72 mW at 193 nm.

Deep-ultraviolet (DUV) lasers have very useful for angle-resolved photoemission spectroscopy (ARPES), and photoemission electron microscopy (PEEM), etc.[1,2,3] DUV lasers generated by second harmonic generation (SHG) in KBe$_2$BO$_3$F$_2$ (KBBF) crystal possess the practical and precise function, which greatly attractes the attention of scientists.

Compared with sum frequency method, frequency doubling in KBBF crystal to generate tunable deep-ultraviolet (DUV) lasers is a simple and convenient method which is important for practical applications. So far, some researches on tunable DUV lasers have been reported. [4,5,6] These reports, however, focus on the nanosecond (ns) and femtosecond (fs) light sources, which can not meet some specific applications, such as the application in ARPES. On the other hand, the previous DUV lasers applied in ARPES [2, 7] operated at fixed wavelength, which limits the wider application of DUV lasers.

In this letter, by two stages of SHG, we have developed a picosecond (ps) DUV laser with the tuning range from 175 to 185 nm. The maximun output power of the tunable DUV laser was 4.83 mW at 193 nm, and the 193 nm has a stable 2.88 mW output. To our knowledge, it is the first time that a tunable ps DUV laser was used as a light source for ARPES.

Fig.2 shows the scheme of the ps 175~210 nm tunable laser. A picosecond Ti:sapphire laser (Tsunami, Spectra-Physics) with tunable range from 700~840 nm was used as the pump source with a pulse width of about 2 ps (FWHM) and a repetition rate of 80 MHz.

In order to obtain a optimal output of second harmonic and compensate the walking of the beam direction due to frequency-tuning, a walk off compensation (WOC)

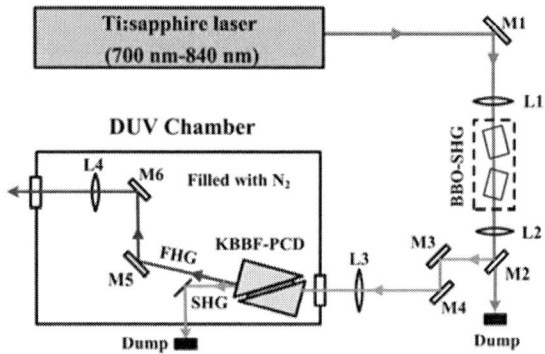

Fig. 1. Schematic layout of picosecond 175 ~ 210 nm tunable DUV laser. M1: high reflection at 690-850 nm; M2, M3, M4: dichroic splitting mirrors, 45° high reflection at 340-430 nm and antireflection at 690-850 nm; M5, M6: 45° high reflection at 175-210 nm.

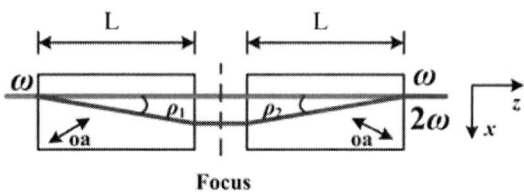

Fig. 2. Schematic layout of walk off compensation configuration.

configuration composed of two type-I phase-matched BBO crystals was adopted, as shown in Fig.2. The thickness of both BBO crystals is 6 mm, and the cutting angle is 30.8°, approximately being middle of the calculated phase-matching angles of 33.7° and 27.9° for frequency doubling at 700 and 840 nm. During the SHG process, the focus of the fundamental light is located in the middle of the two BBO crystals. Fig.3 shows the average output power in the first SHG stage from 350~420 nm versus the fundamental pump light. A maximum output power of 1.1 W was obtained at 386 nm, corresponding to a conversion efficiency of 40%, and the beam quality factors was 1.8 at the horizontal axis and 1.1 at the vertical axis, respectively.

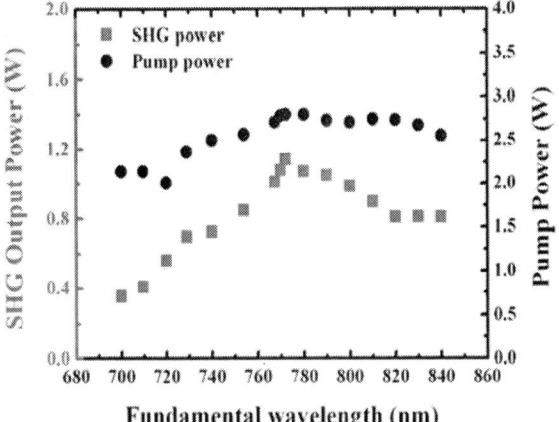

Fig. 3. Output power in the first SHG stage from 350-420 nm as a function of fundamental wavelength.

In order to generate a DUV laser with tuning range from 175 ~ 210 nm, a KBBF prism-coupled device (KBBF-PCD) was adopted in the second SHG stage. For the KBBF-PCD, the apex angle of both SiO_2 prism and CaF_2 prism are designed to be 60°. In the experiment, two KBBF-PCDs were adopted with the KBBF thickness of 0.87 mm and 1.45 mm, respectively. Due to the large absorption of DUV light (especially below 185 nm) in air, the generation and propagation of DUV are enclosed in a hermetical chamber filled with pure N_2 (> 99.999%).

The DUV output powers under different thickness of KBBF are shown in Fig.4. We can see that a thicker crystal is beneficial to achieving higher output power. The maximum output power was 4.83 mW at 193 nm by adopting a 1.45 mm thickness KBBF crystal, and the output powers in the whole tuning range were larger than 0.2 mW. The lower DUV output power in the range at longer and shorter wavelengths respect to 193 nm is mainly attributed to the lower fundamental power as well as SH light. Meanwhile, at shorter wavelength than 193 nm, the higher absorption loss in KBBF crystal and CaF_2 also result in the drop of the DUV output power. Considering the practical applications, two mirrors with 45° high reflection at 175-210 nm were used to align the propagation direction of the generated DUV light, and a CaF_2 lens was adopted to collimate the diverging DUV light. Sometimes, in order to obtain an appropriate beam diameter, more CaF_2 lenses will be used.

Finally, by adopting a 1.45 mm thickness KBBF crystal, we test the DUV power stability in two hours. Fig.5 shows the stability measurements of 175 nm, 193 nm and 210 nm, and the fluctuations are less than 3.5%, 2% and 3% respectively.

Fig.6 is a photograph of the ps 175 ~ 210 nm tunable laser, and it has been used as a light source for an ARPES for the first time. Based on this equipment, it is possible to clarify the final state effect of the photoelectron [8, 9], the overall band structure and the three dimensional dispersion curves in the momentum space [10].

In conclusion, by adoption a BBO walk off compensation configuration in the first SHG stage and a KBF-PCD in the second SHG, we demonstrate a practical and precise picosecond 175 ~ 210 nm tunable laser. With the picosecond tunable DUV laser, a photon energy tunable ARPES has been developed, and it will be a more powerful tool for probing the electronic structures of superconductors, topological insulators and other materials with low work function.

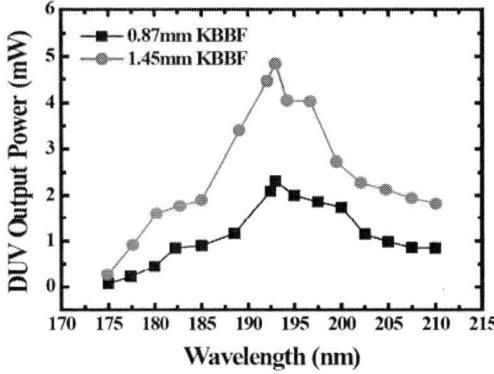

Fig. 4. Output power as a function of wavelength at two KBBF crystals with different thickness.

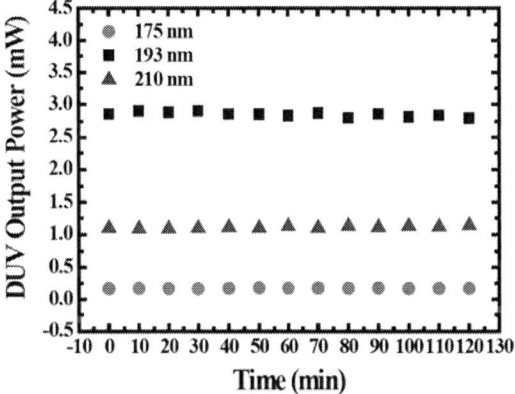

Fig. 5. Stability measurement at 175 nm, 193 nm and 210 nm, respectively.

Fig. 6. A photograph of ps 175 ~ 210 nm tunable DUV laser.

978-1-4799-0019-0/13 $31.00 © 2013 IEEE   427

REFERENCES

[1] G.D. Liu, G.L. Wang, Y. Zhu, H.B. Zhang, G.C. Zhang, X.Y Wang, Y. Zhou, W.T. Zhang, H.Y. Liu, L. Zhao, J.Q. Meng, X.L. Dong, C.T. Chen, Z.Y. Xu, and X.J. Zhou, "Development of a vacuum ultraviolet laser-based angle-resolved photoemission system with a super-high energy resolution better than 1 meV," *Rev. Sci. Instrum.*, vol. 79, no.2, pp. 023105, 2008

[2] L. Jin, Q. Fu, R. Mu, D. Tan, and X. Bao, "Pb interaction underneath a graphene layer on Ru(0001) and its effect on graphene oxidation," *Phys. Chem. Chem. Phys.*, vol 13, pp. 16655-16660, 2011.

[3] Q.J. Peng, Z.M. Wang, S.J. Zhang, F.F. Zhang, F. Yang, D.F. Cui, and Z.Y. Xu, "Recent progress on deep-ultraviolet coherent light sources," *International Conference on Advanced Optoelectronics and Lasers (CAOL).* I.A. Sukhoivanov and O.V. Shulika (Ukraine: IEEE 2010), pp. 28-32.

[4] H. Zhang, G. Wang, L. Guo, A. Geng, Y. Bo, D. Cui, Z. Xu, R. Li, Y. Zhu, X. Wang, and C. Chen, "175 to 210 nm widely tunable deep-ultraviolet light generation based on KBBF crystal", *Appl. Phys.* B, vol. 93, pp. 323-326, 2008.

[5] X. Zhang, Z.M. Wang, G.L. Wang, Y. Zhu, Z.Y. Xu, and C.T. Chen, "Widely tunable and high-average-power fourth-harmonic generation of a Ti:sapphire laser with a $KBe_2BO_3F_2$ prism-couple device", *Opt. Lett.*, vol. 34, no.9, pp.1342-1344, 2009.

[6] T. Kanai, X.Y. Wang, S. Adachi, S. Watanabe, and C.T. Chen, "Watt-level tunable deep ultraviolet light source by a KBBF prism-coupled device", *Opt. Express*, vol.17, no.10, pp. 8696-8703, 2009.

[7] T. Kiss, T. Shimojima, K. Ishizaka, A. Chainani, T. Togashi, T. Kanai, X.Y. Wang, C.T. Chen, S. Watanabe, and S. Shin, "A versatile system for ultrahigh resolution, low temperature, and polarization dependent Laser-angle-resolved photoemission spectroscopy", *Rev. Sci. Instrum.*, vol. 79, no.2, pp. 023106, 2008.

[8] T. Yamasaki, K. Yamazaki, A. Ino, M. Arita, H. Namatame, M. Taniguchi, A. Fujimori, Z.X. Shen, M. Ishikado, and S. Uchida, "Unmasking the nodal quasiparticle dynamics in cuprate superconductors using low-energy photoemission", *Phys. Rev.* B, vol. 75, no. 14, pp. 140513(R), 2007.

[9] J.W. Harter, P.D.C. King, E.J. Monkman, D.E. Shai, Y.F. Nie, M. Uchida, B. Burganov, S. Chatterjee, and K.M. Shen, "A tunable low-energy photon source for high-resolution angle-resolved photoemission spectroscopy", *Rev. Sci. Instrum.*, vol. 83, no. 11, pp.113103, 2012.

[10] S. Kimura, T. Ito, M. Sakai, E. Nakamura, N. Konda, T. Horigome, K. Hayashi, M. Hosaka, M. Katoh, T. Goto, T. Ejima, and K. Soda, "SAMREI: A nocal variably polarized angle-resolved photoemission beamline in the VUV region at UVSOR- Ⅱ", *Rev. Sci. Instrum.*, vol. 81, no. 5, pp. 053104, 2010.